The Collected Papers of
Raymond D. Mindlin
Volume II

H. Deresiewicz M.P. Bieniek F.L. DiMaggio
Editors

The Collected Papers of Raymond D. Mindlin Volume II

The Late James Kip Finch Professor Emeritus
of Applied Science, Columbia University

Springer-Verlag
New York Berlin Heidelberg
London Paris Tokyo

H. Deresiewicz
Department of Mechanical Engineering
M.P. Bieniek
F.L. DiMaggio
Department of Civil Engineering
and Engineering Mechanics
Columbia University
New York, NY 10027
USA

Library of Congress Cataloging-in-Publication Data
Mindlin, Raymond D. (Raymond David), 1906–1987.
[Works. 1989]
The collected papers of Raymond D. Mindlin.
p. cm.
Bibliography: p.
1. Mechanics, Applied—Bibliography. 2. Mindlin, Raymond D.
(Raymond David), 1906–1987—Bibliography. I. Title.
Z7144.M4M56 1989
[TA350]
016.6201—dc19 88-38194

Printed on acid-free paper

Printed and bound by Edwards Brothers, Inc., Ann Arbor, Michigan.
Printed in the United States of America.

9 8 7 6 5 4 3 2 1

ISBN 0-387-96933-0 Springer-Verlag New York Berlin Heidelberg
ISBN 3-540-96933-0 Springer-Verlag Berlin Heidelberg New York

Preface

The present volume makes available in one convenient place all the research papers published by Raymond D. Mindlin in the course of a half-century of professional activity. Included, in addition, is the full text of three of his extended review articles, because these remain timely as well as historically important.

Also to be found here is a brief sketch, written by one of us, his student and co-author, colleague and friend. It contains Mindlin's basic biographical data, although it dwells primarily on an analysis of his work and a chronicling of his accomplishments.

The papers reprinted herein are reproduced photographically as they appeared originally. We have excluded several short review articles; moreover, to our regret, because of constraints of space we have been unable to print three extended reviews. These are: "The photoelastic method of stress analysis," published in volume 10 of the *Journal of Applied Physics*; a short monograph, "Dynamics of package cushioning," which may be found in volume 24 of the *Bell System Technical Journal*; and the chapter entitled "Analogies," written for the *Handbook of Experimental Stress Analysis.*

In compiling the list of Publications (pp. xxi–xxviii) we have omitted abstracts and brief discussions; we have, however, included several early comprehensive discussions, deeming them, in substance and in length, each the equal of an independent paper. We offer this list in the firm belief that it represents the definitive bibliography of Mindlin's oeuvre, accurate and complete. The entries are arranged in the order of their appearance in print, rather than the order of submission (in many instances unavailable) or of composition (known only with unacceptable uncertainty). Happily, the sequences are known to differ in but a few particulars.

The reader seeking a thematic approach to the wealth of material presented here will find it useful to consult the topical arrangement given below. (The numbers in brackets refer to the corresponding items in the list of Publications.)

I. Photoelasticity and experimental mechanics
   [1], [8], [10], [11], [16], [17], [18], [19], [20], [24], [26], [27], [31], [36], [113]

II. Classical elastostatics
   Two-dimensional: [9], [12], [13], [14], [22], [24], [29]
   Three-dimensional: [3], [4], [5], [32], [33], [42], [110], [112]

III. Vibrations and waves
[21], [25], [30], [49], [54], [94], [115], [128]

IV. Frictional contact and granular media
[28], [38], [41], [48], [56]

V. Rods, bars, beams, cylinders
[37], [60], [61], [62], [72], [114], [117], [118]

VI. Isotropic plates
[34], [46], [47], [51], [52], [55], [58], [59], [63], [64], [76], [123], [129]

VII. Crystal plates
[35], [43], [45], [50], [53], [54], [57], [65], [66], [69], [70], [71], [73], [77], [81], [84], [85], [86], [89], [90], [98], [119], [121], [122], [124], [125], [127]

VIII. Piezoelectric plates
[39], [44], [67], [68], [100], [103], [105], [107], [108], [109], [126]

IX. Generalized continua
[74], [75], [78], [79], [80], [82], [83], [87], [106]

X. Elastic dielectrics
[88], [95], [97], [99], [101], [102], [106], [111], [120]

XI. Lattice theory
[91], [92], [93], [95], [96], [105]

XII. Structural and geotechnical mechanics
[2], [6], [7], [15]

XIII. Other
[40]

XIV. Review articles
[16], [20], [21], [23], [31], [36], [48], [64], [97], [105], [116], [121]

Ray Mindlin's place in the annals of mechanics is made secure by his scientific output. But it seems to us that to omit mention of his influence as teacher, fount of inspiration, and example to emulate would leave a grossly incomplete picture of the man and the effect he has wrought on the field in which he labored with so much dedication and such signal distinction. He nurtured some thirty students to the doctoral degree, giving unstintingly of his time and energies, devoting himself to the task of showing the way. Many of those he trained occupy prestigious positions in universities and in industrial laboratories; not a few have gone on to direct their own research students and these, in turn, have produced others, in an ever more ramified "genealogy."

We are confident we speak for all his students when we say that they have always viewed him with warmth and genuine affection and, above all, taken great pride in acknowledging themselves his disciples.

H. Deresiewicz
M. P. Bieniek
F. L. DiMaggio

Volume II

Contents*

Papers numbered [78] - [129] appear in this volume; papers [1] - [78] will be found in Volume I.

* Numbers in brackets refer to the Publications of Raymond D. Mindlin (see pages xxi–xxviii).

60th Birthday Celebration (Minneapolis, Minnesota, 1966)

Seated, l. to r.: L. E. Goodman, Mrs. Goodman, D. C. Drucker, Mrs. Mindlin, Mindlin, M. Hetényi, G. Herrmann, Mrs. Lubkin. *Standing, in the same order*: H. Deresiewicz, A. G. Lubowe, D. H. Cheng, J. J. McCoy, H. D. McNiven, M. A. Medick, H. F. Tiersten, J. Tasi, D. C. Gazis, Y.-H. Pao, L. Rongved, E. Sternberg, T. R. Kane, E. A. Fox, P. C. Y. Lee, J. Duffy, and J. L. Lubkin.

A Bio/Bibliographical Sketch

Raymond David Mindlin was born on September 17, 1906, in New York City, the second of three sons of Henry, a prosperous businessman, and Beatrice (née Levy). There are clear indications that the family, which subsequently came also to include a half-sister, was supportive and closely knit, and that Ray, as he came to be known to his friends and colleagues, enjoyed a tranquil childhood.

For his secondary education the young man was sent to the highly regarded Ethical Culture School, of which he had fond recollections in later years. Upon graduating in 1924 he enrolled at Columbia, beginning an association with the University that was to last for more than half a century. Four years later (during which time he distinguished himself as a sprinter on the varsity track team) he received the first of his four earned degrees, a B. A., followed by a B. S. in 1931, and in 1932 by a C. E. and the Illig medal for "proficiency in scholarship."

The economic depression then gripping the land was a major reason why he remained for graduate work, funded by the modest stipend that accompanied the research assistantship to which he was appointed. Love's *Treatise on the Mathematical Theory on Elasticity*, the fourth and last edition of which had appeared not long before, came to his hand, and from then on the direction of his professional work was firmly set.

Applied mechanics was not strongly represented on Columbia's faculty; in fact, there was little substantive research in it in the United States, and essentially no graduate instruction, until the advent on the scene of Stephen Timoshenko, exiled by the Bolshevik revolution in Russia. Some time after his appointment as professor at the University of Michigan in 1927, he organized a series of summer courses, in which each year one or another of the foremost names in this field, including L. Prandtl, R. V. Southwell, and H. M. Westergaard, participated. Mindlin attended in 1933, '34, and '35, and there is no doubt that the experience at Ann Arbor served to confirm him in his choice of his life's work.

His initial publication [1]* dates from the middle of this period, a solo, full-length paper describing a new type of polariscope for photoelastic analysis. April of 1935 saw the appearance of a lengthy discussion, written

* Numbers in brackets refer to corresponding entries in the list of publications (pages xxi–xxviii).

with Westergaard, of a paper on torsion of structural beams. During the year, Mindlin spent all available time in his basement office struggling to master Love's grand opus, a massive, difficult, demanding book, from time to time having to contend with complaints, from a faculty lacking in theoretical training and outlook, that the young man was wasting time. It was James Kip Finch, his Department chairman (and in later years a noted historian of engineering as well as Dean of the Engineering School), who served both as his protector and source of encouragement by issuing an edict to his colleagues, "Leave him alone, maybe something will come of it." Mindlin, in his turn, never forgot Finch's kindness and support, and it was at his persistent urging that the University, 35 years later, created the Finch chair (with Mindlin its first occupant).

For his doctoral research Mindlin set himself a fundamental problem in theoretical elasticity: determining the stresses in an elastic half-space subjected to a sub-surface point load. Working without any guidance at Columbia, he succeeded in finding the solution by employing the method of nuclei of strain. The results, nowadays referred to as "Mindlin's problem," represent a generalization of the two classical 19th century solutions respectively associated with the names of Kelvin and Boussinesq, and have become the basis for analytical formulations widely employed in geotechnical engineering. At Timoshenko's urging, in order to establish the author's priority of discovery, a summary of the results was dispatched to the *Comptes Rendus* of the *Académie des Sciences* in Paris, noted for almost instantaneous publication, where it appeared [3] in September of 1935. The full paper [4] was published in *Physics* (now the *Journal of Applied Physics*) in 1936, the year Mindlin received the Ph.D. degree. It was also the year in which he published his one and only paper to appear in a periodical devoted to pure mathematics, in essence a completeness proof of the Papkovich representation in classical elasticity [5]. It is interesting to note that nearly twenty years later he returned to the problem bearing his name and offered a solution by means of a systematic procedure grounded in potential theory [42].

Academic advancement, it appears, proceeded at a rather leisurely tempo in those days. Mindlin remained an assistant for another two years, at which point he was elevated to instructor in civil engineering, and only in 1940 did he receive promotion to assistant professor. His research in the six years following receipt of the doctorate was a mix of analytical and experimental endeavors, with emphasis on the photoelastic method of stress analysis. It included a work on three-dimensional photoelasticity [17], published jointly with D. C. Drucker, his first doctoral student, in which the method of oblique incidence was first suggested. An important part of current laboratory technique is based on this and a companion and two subsequent papers [18], [26], [27].

In 1942 Mindlin was co-opted by the Applied Physics Laboratory in Silver Spring (a Maryland suburb of the nation's capital), an institution engaged in naval ordnance work. There he played a significant role in the development of the proximity fuse, one of the major achievements in the scientific war effort. For his part in its success he was presented, shortly after the end of the war, with the Presidential Medal for Merit, the highest decoration awarded to civilians.

The swiftness of his climb up the academic ladder now made up for the earlier glacial pace: he came back to Columbia in 1945 as an associate professor, and two years later attained the rank of professor.

The year of his return to academia also marked the publication, in the *Bell System Technical Journal*, of a small monograph on the dynamics of package cushioning [21], which quickly became a vade mecum for designers of containers for protection of electronic equipment from mechanical shock and vibration. It was the first of several papers stimulated by his eight-year association as consultant to the Bell Telephone Laboratories.

A most important paper arising from his symbiosis with the Bell Laboratories stemmed from a perennial effort toward rational design of the carbon microphone. Published in 1949, it gave new direction to research on frictional phenomena at surfaces in contact [28]. The work, its basic consequences first tested and verified experimentally [38] and then generalized [41], led to an understanding of the process of wear due to local sliding and gave a mighty impetus to world-wide researches, analytical and experimental, in the industrially vital areas of sliding and rolling contacts, ball bearings, railway wheels, and the mechanics of granular media [56].

Shortly thereafter Mindlin began his trail-blazing studies, which he was to continue intermittently for more than two decades, on waves and vibrations in isotropic elastic plates [34] and, concommitantly, on high-frequency vibrations of crystal plates [35], soon extending his studies to account for the piezoelectric effect [39]. The method he had employed of reducing the exact, three-dimensional equations to describe the dynamics of plates, by expanding in series of the thickness coordinate (or in terms of selected functions thereof) and truncating the series, he also applied, suitably altered, to dynamical problems of bars [37], [60], [62], stimulating, as in the case of plates, an ever growing literature of the subject.

The work on isotropic bars and plates led to results in use world-wide in development and design of electromechanical filters and solid delay lines. But it was his studies on crystal plates that caused the greatest stir, for they were pioneering papers in a vexingly difficult mathematical field. In them, he elucidated a very complicated phenomenon of great technical importance and thereby pointed the way to major improvements in the design and performance of quartz crystals for filter circuits. The U.S. Army Signal Corps, long-time sponsor of the research on crystal plate vibrations, persuaded him to prepare a monograph on the subject by granting him a sole-supplier contract for the purpose. In 1955 the task was completed to the acclaim of the sponsor—it was in the form of a ca. 170-page-long report entitled "An Introduction to the Mathematical Theory of Vibrations of Elastic Plates." Yet even as its author was composing the text, better ways of doing the old were occurring to him and new ideas for what should, and could, be done were springing to his mind. Indeed, despite much subsequent work in other areas of mechanics, he kept returning to this subject to the very end of his life. The monograph seemed to haunt him, for he intermittently worked on a revision–more accurately, a complete reworking–of it. A few months before his death he spoke of plans to add yet another major part to what he had already written before he would consider the job done; ever the realist, he added, "But don't hold your breath!"*

It is strikingly clear in retrospect that the period of the late '40s and early '50s was a particularly fruitful one for him. In addition to the basic papers on elastic contact and on high-frequency vibrations of plates and bars, a pioneering article on the interpretation of optical birefringence in viscoelastic materials [27] and a seminal paper on the interaction of an elastic shell with a surrounding fluid [40], the second of these written in collaboration with his colleague H. H. Bleich, date to that time. Of evident importance to the Navy, the latter work triggered a great deal of subsequent research on the response of submarine hulls to shock loading. Also sandwiched into that period is the lengthy article on analogies [31], with another of his colleagues, M. G. Salvadori, as co-author, which appeared as a chapter in the *Handbook of Experimental Stress Analysis.*

Beginning in 1962, in response to what purported to be, but were seen by Mindlin as erroneously based, expanded theories of anisotropic materials, he was led to construct, first, a theory that took account of couple-stresses [74], characteristically applying it immediately thereafter to a two-dimensional case that seemed simple enough to be susceptible of experimental validation [75]. Two years later, in a paper on micro-structure [78], he treated a continuum endowed at each point with an internal displacement field, and showed that the lowest order of this formulation results in a theory that includes the optical lattice modes at the long-wavelength limit. In a paper published the following year [83], he formulated a linear theory of equilibrium of an elastic solid in which the stored energy depends on the strain and its first and second gradients. This proved to be the first continuum description capable of accounting for surface tension, without resorting to the artificial construct of a vanishingly thin surface membrane. The papers contributed to the start of lively activity in this field of generalized elastic continua, and stimulated applications to areas as diverse as the mechanics of laminated and of fiber-reinforced materials, and of framed structures.

In 1967 he was appointed James Kip Finch Professor of Applied Science, the first holder of a chair named in honor of the patron and protector of his days as a graduated student. He was to hold this post until his retirement in 1975.

A series of papers initiated with [88] introduced the contribution of the

* Shortly after making this observation, aware of the gravity of what proved to be his last illness, he entrusted the manuscript for completion (and, it is fervently hoped, timely publication) to P. C. Y. Lee, one of his last doctoral students.

polarization gradient (in addition to the strain and the polarization itself) to the stored energy of a solid. Mindlin was able to show that this augmented formulation of the elastic dielectric continuum accomodates the surface energy of deformation and polarization, predicts the experimentally observed anomalous capacitance of thin dielectric films [95], and accounts for acoustical and, when the magnetic field is included, optical activity [99].

More or less concurrently came one other group of highly innovative researches, in which his aim was to bridge the gap between continuum and lattice theories. Thus, he was able to show [91] that a model of a continuum, formed by two or more interpenetrating media, in which the stored energy depends not only on the strains of the individual constituents but also on their relative displacements, and simulating a (non-Bravais) lattice with a basis, contains optical as well as acoustical branches in the vibrational spectrum. In [95], a paper mentioned above, he demonstrated that his augmented theory of the elastic dielectric continuum yields the correct long-wave limit of a monatomic lattice theory with a shell model, and in [101], by endowing the model of [91] with individual polarizations, that the resulting continuum can yield the long-wave limit of a diatomic lattice with shell model.

The steady stream of his research was interrupted early in 1969 by illness, and he was obliged to submit to cardiac surgery. (This was before the coronary bypass operation became part of clinical procedure.) As he made amply clear at the time, he resented the lengthy convalescence because it was inhibiting his work. But by the end of that year he was back at his desk, and a glance at his list of publications after that date shows that he was able to resume his enviable scientific productivity.

His last hurrah, just a year before his death, came in the form of two papers, one dealing with free vibrations of a rectangular parallelepiped [128] and the other, with vibrations of a rectangular plate with all four edges free [129], classic problems against which he had butted his head, in vain, decades earlier. At long last he managed to construct solutions, apparently by inspired guesswork, reminiscent of the *tour de force* of his doctoral dissertation 50 years earlier. He was manifestly elated, particularly by what turned out to be his final publication, characterizing the accomplishment in the following words: "It's a pity there aren't many around, anymore, who would appreciate what an extraordinary feat that solution is." His letter, accompanying a photocopy of the work in holograph and dated January 9, 1986, continues, in evident allusion to the 19th century origin of the problem, "I'm reminded of Tony Biot [Maurice Anthony Biot] who was frequently a generation ahead of his time. I, on the other hand, seem to be some generations behind."

The preceeding narrative, though sketchy, has, I hope, given some indication of the breadth and depth of Mindlin's scholarship. Those who worked with him, and those others who knew him and his work, quickly came to appreciate his profound insights into the physics of any given situation, his surefootedness in discerning what is central and what is peripheral, and his multifaceted analytical and experimental skills. Nevertheless, a disservice would be done the subject of this account (as well as its reader) if nothing were said about him as colleague, mentor, and human being.

His professional colleagues treated him with deference at times bordering on awe. He was secure in the knowledge of his own worth, but wore the mantle of his eminence with genuine modesty. He was generous in giving or sharing credit, and unfailingly courteous to peers seeking his opinion or advice. Basically a shy, reserved person, he was, invariably and unexceptionally, the consummate gentleman.

His many disciples revered him and, one suspects, in his quiet way he enjoyed being the father figure. To celebrate his 60th birthday, his former doctoral students tendered him a dinner in Minneapolis, where most had journeyed ostensibly to attend a professional meeting. In 1974, in anticipation of his imminent retirement, he was presented with a published book, entitled *R. D. Mindlin and Applied Mechanics.* In eight substantial chapters, by fifteen of his erstwhile students, this novel kind of Festschrift contains summaries of those areas on which Mindlin's own endeavors have exercised a profound and lasting effect.

But it was by no means a one-way relationship. Whenever a former student of his was being feted, as such things tended to occur with increasing frequency as the years passed, Ray Mindlin would be inconspicuously present among the well-wishers. Perhaps the loyalty and affection made up to some extent for not having any children of his own. He married twice, each time losing a beloved mate by untimely death due to illness.

He served with devotion the profession which he made his life's work, through his research, his teaching, his advisory capacity to numerous government agencies, and his activities in various scientific and technical societies. Among the latter, mention is warranted of the following positions he held at various times: In the American Society of Mechanical Engineers (ASME), chairman, Applied Mechanics Division; member, Publications Committee, Engineering Societies Monographs Committee, Advisory Board of *Applied Mechanics Reviews.* In the American Society of Civil Engineers (ASCE), chairman, Committee on Applied Mechanics of the Structural Engineering Division (precursor of the Engineering Mechanics Division). In the Society for Experimental Stress Analysis (SESA), co-founder and president; member, executive committee. In the American Institute of Physics, associate editor, *Journal of Mathematical Physics.* Also, he was member of: the U.S. National Committee for Theoretical and Applied Mechanics; the General Assembly of the International Union of Theoretical and Applied Mechanics; the American Physical Society.

To be sure, his remarkable intellectual fecundity and gentle personality did not go unappreciated by his peers, and recognition came in the form of copious honors and awards. He was elected Fellow of the American Academy of Arts and Sciences (1958), of the ASME (1962), and of the Accoustical Society of America (1963); member of the National Academy of Engineering (1966) and of the National Academy of Sciences (1973); and Honorary Member of the ASME (1969).

He received the Research Prize (1958) and the von Karman Medal (1961) of the ASCE; the Timoshenko Medal (1964) and the ASME Medal (1976) from the ASME; the Trent-Crede Award of the Acoustical Society of America (1971); the Frocht Award of the SESA (1974); the Great Teacher Award (1960) and the Egleston Medal (1971) from Columbia University; and an Honorary D.Sc. degree from Northwestern University (1975); also, a Naval Ordnance Development Award (1945) and the C. B. Sawyer Award of the Army Electronics Command (1967). Remarkably, prestigious as these are, all but one were sandwiched between his most noteworthy honors, both bestowed by the U.S. Government: the Presidential Medal for Merit (1946), as noted earlier the highest civilian decoration of the second World War, and his ultimate accolade, the National Medal of Science, which he received in 1979.

Mindlin died on November 22, 1987, at age 81, in consequence of a debilitating illness that lingered for about a year. And yet, to the last, his mind remained acute and active, almost as if divorced from his frail body. His passing was mourned by all who knew him, but his name will live through his work.

Chaucer's lines, written about another scholar, of days long gone, might just as aptly have been indited for Raymond Mindlin:

Of study took he utmost care and heed
Not one word spoke he more than was his need
And that was said in fullest reverence
And short and quick and full of high good sense
Pregnant of moral virtue was his speech
And gladly would he learn and gladly teach.

H. Deresiewicz

Surrounded by well-wishers observing his 68th birthday and marking his forthcoming mandatory retirement from Columbia (Boulder, Colorado, 1974)

Seated, l. to r., front row: K. Robinson, F. L. DiMaggio, N. Perrone, R. Skalak, P. C. Y. Lee, H. F. Tiersten; *second row*: G. Herrmann, Mrs. Skalak, Mrs. Pao, Mindlin, Mrs. Mindlin, M. G. Salvadori, Mrs. Albrecht, J. H. Weiner.

Standing, in the same order: R. L. Schiffman, J. M. Garrelts, L. E. Goodman, J. J. McCoy, H. D. McNiven, M. L. Baron, E. H. Lee, B. A. Boley, J. Achenbach, B. Albrecht, J. Miklowitz (in rear), M. A. Medick, H. Liebowitz, D. C. Drucker, A. G. Lubowe, D. C. Gazis, J. Tasi, P. M. Naghdi, unidentified (in rear), N. Basdekas, J. Duffy, E. Sternberg, H. H. Bleich, and Y.-H. Pao.

Publications of Raymond D. Mindlin

[1] A reflection polariscope for photoelastic analysis, *Review of Scientific Instruments*, **5**, 224–228 (1934).

[2] Structural beams in torsion (with H. M. Westergaard), discussion of paper by I. Lyse and B. G. Johnston, *Proceedings of the American Society of Civil Engineers*, **61**, 509–512 (1935).

[3] Contribution au problème d'équilibre de l'élasticité d'un solide indéfinie limité par un plan, *Comptes Rendus de l'Académie des Sciences*, Paris, **201**, 536–537 (1935).

[4] Force at a point in the interior of a semi-infinite solid, *Physics*, **7**, 195–202 (1936).

[5] Note on the Galerkin and Papkovitch stress functions, *Bulletin of the American Mathematical Society*, **42**, 373–376 (1936).

[6] Stresses around circular holes in dams and buttresses, discussion of paper by I. K. Silverman, *Proceedings of the American Society of Civil Engineers*, **63**, 399–403 (1937).

[7] Pressure distributions on retaining walls, discussion of paper by M. G. Spangler, *Proceedings of the International Conference on Soil Mechanics and Foundation Engineering*, Cambridge, Mass.-1936, **III**, 155–156 (1937).

[8] Analysis of doubly refracting materials with circularly and elliptically polarized light, *Journal of the Optical Society of America*, **27**, 288–291 (1937).

[9] Stress systems in a circular disk under radial forces, *Journal of Applied Mechanics*, **4**, A-115–A-118 (1937).

[10] Distortion of the photoelastic fringe pattern in an optically unbalanced polariscope, *Journal of Applied Mechanics*, **4**, A-170–A-172 (1937).

[11] The equiangular strain-rosette, *Civil Engineering*, **8**, 546–547 (1938).

[12] Stresses in a heavy disk suspended from an eccentric peg, *Journal of Applied Physics*, **9**, 714–717 (1938).

[13] Stresses in an eccentrically rotating disk, *Philosophical Magazine*, Ser. 7, **xxvi**, 713–719 (1938).

[14] Gravitational stresses in bipolar coordinates, *Proceedings of the Fifth International Congress for Applied Mechanics*, Cambridge, Mass.-1938 112–116 (1939).

[15] Stress distribution around a tunnel, *Proceedings of the American Society of Civil Engineers*, **65**, 619–642 (1939); reprinted in *Transactions American Society of Civil Engineers*, **105**, 1117–1140 (1940).

*[16] A review of the photoelastic method of stress analysis, *Journal of Applied Physics*, **10**, 222–241, 273–294 (1939).

[17] Stress analysis by three-dimensional photoelastic methods (with D. C. Drucker), *Journal of Applied Physics*, **11**, 724–732 (1940).

[18] An extension of the photoelastic method of stress measurement to plates in transverse bending, discussion of paper by J. N. Goodier and G. H. Lee, *Journal of Applied Mechanics*, **8**, A-187–A-189 (1941).

[19] Optical aspects of three-dimensional photoelasticity, *Journal of the Franklin Institute*, **233**, 349–363 (1942).

*[20] The present status of three-dimensional photoelasticity, *Civil Engineering*, **12**, 255–258 (1942).

*[21] Dynamics of package cushioning, *Bell System Technical Journal*, **24**, 353–461 (1945).

[22] The analogy between multiply-connected slices and slabs, *Quarterly of Applied Mathematics*, **IV**, 279–290 (1946).

*[23] Activities of the Society for Experimental Stress Analysis, *Proceedings of the Society for Experimental Stress Analysis*, **IV**, no. 2, ix–xi (1947).

[24] Stress distribution around a hole near the edge of a plate under tension, *Proceedings of the Society for Experimental Stress Analysis*, **V**, no. 2, 56–68 (1948).

[25] Response of damped elastic systems to transient disturbances (with F. W. Stubner and H. L. Cooper), *Proceedings of the Society for Experimental Stress Analysis*, **V**, no. 2, 69–87 (1948).

[26] The optical equations of three-dimensional photoelasticity (with L. E. Goodman), *Journal of Applied Physics*, **20**, 89–95 (1949).

[27] A mathematical theory of photo-viscoelasticity, *Journal of Applied Physics*, **20**, 206–216 (1949).

[28] Compliance of elastic bodies in contact, *Journal of Applied Mechanics*, **16**, 259–268 (1949).

[29] Thermoelastic stress around a cylindrical inclusion of elliptic cross section (with H. L. Cooper), *Journal of Applied Mechanics*, **17**, 265–268 (1950).

[30] Beam vibrations with time-dependent boundary conditions (with L. E. Goodman), *Journal of Applied Mechanics*, **17**, 377–380 (1950).

*[31] Analogies (with M. G. Salvadori); in M. Hetényi(ed), *Handbook of Experimental Stress Analysis*, John Wiley & Sons, New York, 700–827 (1950).

[32] Nuclei of strain in the semi-infinite solid (with D. H. Cheng), *Journal of Applied Physics*, **21**, 926–930 (1950).

* Not part of this collection.

[33] Thermoelastic stress in the semi-infinite solid (with D. H. Cheng), *Journal of Applied Physics*, **21**, 931–933 (1950).

[34] Influence of rotatory inertia and shear on flexural motions of isotropic, elastic plates, *Journal of Applied Mechanics*, **18**, 31–38 (1951).

[35] Thickness-shear and flexural vibrations of crystal plates, *Journal of Applied Physics*, **22**, 316–323 (1951).

*[36] Development of the mathematical theory of three-dimensional photoelasticity, *Applied Mechanics Reviews*, **4**, 537–539 (1951).

[37] A one-dimensional theory of compressional waves in an elastic rod (with G. Herrmann), *Proceedings of the First U.S. National Congress of Applied Mechanics-1951*, 187–191 (1952).

[38] Effects of an oscillating tangential force on the contact surfaces of elastic spheres (with W. P. Mason, T. F. Osmer, and H. Deresiewicz), *Proceedings of the First U.S. National Congress of Applied Mechanics-*1951, 203–208 (1952).

[39] Forced thickness-shear and flexural vibrations of piezoelectric crystal plates, *Journal of Applied Physics*, **23**, 83–88 (1952).

[40] Response of an elastic cylindrical shell to a transverse, step shock wave (with H. H. Bleich), *Journal of Applied Mechanics*, **20**, 189–195 (1953).

[41] Elastic spheres in contact under varying oblique forces (with H. Deresiewicz), *Journal of Applied Mechanics*, **20**, 327–344 (1953).

[42] Force at a point in the interior of a semi-infinite solid, *Proceedings of the First Midwestern Conference on Solid Mechanics*, 56–59 (1953).

[43] Thickness-shear and flexural vibrations of contoured crystal plates (with M. Forray), *Journal of Applied Physics*, **25**, 12–20 (1954).

[44] Thickness-shear vibrations of piezoelectric crystal plates with incomplete electrodes (with H. Deresiewicz), *Journal of Applied Physics*, **25**, 21–24 (1954).

[45] Suppression of overtones of thickness-shear and flexural vibrations of crystal plates (with H. Deresiewicz), *Journal of Applied Physics*, **25**, 25–27 (1954).

[46] Thickness-shear and flexural vibrations of a circular disk (with H. Deresiewicz), *Journal of Applied Physics*, **25**, 1329–1332 (1954).

[47] Axially symmetric flexural vibrations of a circular disk (with H. Deresiewicz), *Journal of Applied Mechanics*, **22**, 86–88 (1955).

*[48] Mechanics of granular media, *Proceedings of the Second U.S. National Congress of Applied Mechanics*-1954, 13–20 (1955).

[49] Timoshenko's shear coefficient for flexural vibrations of beams (with H. Deresiewicz), *Proceedings of the Second U.S. National Congress of Applied Mechanics*-1954, 175–178 (1955).

[50] Thickness-shear and flexural vibrations of rectangular crystal plates (with H. Deresiewicz), *Journal of Applied Physics*, **26**, 1435–1442 (1955).

* Not part of this collection.

[51] High-frequency extensional vibrations of plates (with T. R. Kane), *Journal of Applied Mechanics*, **23**, 277–283 (1956).

[52] Flexural vibrations of rectangular plates (with A. Schacknow and H. Deresiewicz), *Journal of Applied Mechanics*, **23**, 430–436 (1956).

[53] Simple modes of vibration of crystals, *Journal of Applied Physics*, **27**, 1462–1466 (1956).

[54] Waves on the surface of a crystal (with H. Deresiewicz), *Journal of Applied Physics*, **28**, 669–671 (1957).

[55] Influence of width on velocities of long waves in plates (with D. C. Gazis), *Journal of Applied Mechanics*, **24**, 541–546 (1957).

[56] Stress-strain relations and vibrations of a granular medium (with J. Duffy), *Journal of Applied Mechanics*, **24**, 585–593 (1957).

[57] Vibrations of a monoclinic crystal plate (with E. G. Newman), *Journal of the Acoustical Society of America*, **29**, 1206–1218 (1957).

[58] Vibrations of an infinite, elastic plate at its cut-off frequencies, *Proceedings of the Third U.S. National Congress of Applied Mechanics*, 225–226 (1958).

[59] Extensional vibrations of elastic plates (with M. A. Medick), *Journal of Applied Mechanics*, **26**, 561–569 (1959).

[60] Axially symmetric waves in elastic rods (with H. D. McNiven), *Journal of Applied Mechanics*, **27**, 145–151 (1960).

[61] Vibrations and waves in elastic bars of rectangular cross section (with E. A. Fox), *Journal of Applied Mechanics*, **27**, 152–158 (1960).

[62] Dispersion of flexural waves in an elastic, circular cylinder (with Y.-H. Pao), *Journal of Applied Mechanics*, **27**, 513–520 (1960).

[63] Extensional vibrations and waves in a circular disk and a semi-infinite plate (with D. C. Gazis), *Journal of Applied Mechanics*, **27**, 541–547 (1960).

[64] Waves and vibrations in isotropic, elastic plates; in J. N. Goodier and N. J. Hoff (eds), *Structural Mechanics*, Pergamon Press, Oxford, 199–232 (1960).

[65] High frequency vibrations of crystal plates, *Quarterly of Applied Mathematics*, **XIX**, 51–61 (1961).

[66] On the equations of extensional motion of crystal plates (with H. L. Cooper), *Quarterly of Applied Mathematics*, **XIX**, 111–118 (1961).

[67] On the equations of motion of piezoelectric crystals; in *Problems of Continuum Mechanics* (N. I. Muskhelishvili 70th Birthday Volume), Society for Industrial and Applied Mathematics, Philadelphia, 282–290 (1961).

[68] Forced vibrations of piezoelectric crystal plates (with H. F. Tiersten), *Quarterly of Applied Mathematics*, **XX**, 107–119 (1962).

[69] Vibrations of an infinite, monoclinic crystal plate at high frequencies and long wavelengths (with R. K. Kaul), *Journal of the Acoustical Society of America*, **34**, 1895–1901 (1962).

[70] Frequency spectrum of a monoclinic crystal plate (with R. K. Kaul), *Journal of the Acoustical Society of America*, **34**, 1902–1910 (1962).

[71] Extensional vibrations of thin quartz disks (with A. G. Lubowe), *Journal of the Acoustical Society of America*, **34**, 1911–1918 (1962).

[72] Dispersion of axially symmetric waves in elastic rods (with M. Onoe and H. D. McNiven), *Journal of Applied Mechanics*, **29**, 729–734 (1962).

[73] Strong resonances of rectangular AT-cut quartz plates (with D. C. Gazis), *Proceedings of the Fourth U.S. National Congress of Applied Mechanics*, 305–310 (1962).

[74] Effects of couple-stresses in linear elasticity (with H. F. Tiersten), *Archive for Rational Mechanics and Analysis*, **11**, 415–448 (1962).

[75] Influence of couple-stresses on stress concentrations, *Experimental Mechanics*, **3**, 1–7 (1963).

[76] Extensional waves along the edge of an elastic plate (with J. J. McCoy), *Journal of Applied Mechanics*, **30**, 75–78 (1963).

[77] High frequency vibrations of plated, crystal plates; in *Progress in Applied Mechanics* (The Prager Anniversary Volume), Macmillan, New York, 73–84 (1963).

[78] Micro-structure in linear elasticity, *Archive for Rational Mechanics and Analysis*, **16**, 51–78 (1964).

[79] Complex representation of displacements and stresses in plane strain with couple-stresses; in *Applications of the Theory of Functions in Continuum Mechanics* (Proceedings of the International Symposium, Tbilisi, U.S.S.R., 1963), Nauka, Moscow, 256–259 (1965).

[80] On the equations of elastic materials with micro-structure, *International Journal of Solids and Structures*, **1**, 73–78 (1965).

[81] Thickness-twist vibrations of an infinite, monoclinic, crystal plate, *International Journal of Solids and Structures*, **1**, 141–145 (1965).

[82] Stress functions for a Cosserat continuum, *International Journal of Solids and Structures*, **1**, 265–271 (1965).

[83] Second gradient of strain and surface-tension in linear elasticity, *International Journal of Solids and Structures*, **1**, 417–438 (1965).

[84] Thickness-shear and flexural vibrations of partially plated, crystal plates (with P. C. Y. Lee), *International Journal of Solids and Structures*, **2**, 125–139 (1966).

[85] Bechmann's number for harmonic overtones of thickness/twist vibrations of rotated-Y-cut quartz plates, *Journal of the Acoustical Society of America*, **41**, 969–973 (1967).

[86] Anharmonic, thickness-twist overtones of thickness-shear and flexural vibrations of rectangular, *AT*-cut quartz plates (with W. J. Spencer), *Journal of the Acoustical Society of America*, **42**, 1268–1277 (1967).

[87] On first strain-gradient theories in linear elasticity (with N. N. Eshel), *International Journal of Solids and Structures*, **4**, 109–124 (1968).

[88] Polarization gradient in elastic dielectrics, *International Journal of Solids and Structures*, **4**, 637–642 (1968).

[89] Optimum sizes and shapes of electrodes for quartz resonators, *Journal of the Acoustical Society of America*, **43**, 1329–1331 (1968).

[90] Face-shear waves in rotated Y-cut quartz plates; in F. P. J. Rimrott and J. Schwaighofer (eds), *Mechanics of the Solid State* (Joseph Marin Memorial Volume), University of Toronto Press, Toronto, 143–145 (1968).

[91] Theories of elastic continua and crystal lattice theories; in E. Kröner (ed), *Mechanics of Generalized Continua*, (IUTAM Symposium, Freudenstadt-Stuttgart, 1967), Springer-Verlag, Berlin, 312–320 (1968).

[92] Lattice and continuum theories of simple modes of vibration in cubic crystal plates and bars; in *Dynamics of Structured Solids*, American Society of Mechanical Engineers, New York, 2–9 (1968).

[93] Crystal lattice theory of torsion of a rectangular bar of simple cubic structure, *Indian Journal of Mechanics and Mathematics*, special issue (B. Sen 70th Birthday Volume), part I, 35–40 (1968).

[94] Note on vibrations of prisms and polygonal plates; in *Problems of Hydrodynamics and Continuum Mechanics* (L. I. Sedov 60th Birthday Volume), *Society for Industrial and Applied Mathematics*, Philadelphia, 485–488 (1969).

[95] Continuum and lattice theories of influence of electromechanical coupling on capacitance of thin dielectric films, *International Journal of Solids and Structures*, **5**, 1197–1208 (1969).

[96] Lattice theory of shear modes of vibration and torsional equilibrium of simple-cubic crystal plates and bars, *International Journal of Solids and Structures*, **6**, 725–738 (1970).

[97] Polarization gradient in elastic dielectrics, *Courses and Lectures* No. 24, International Centre for Mechanical Sciences, Udine, Italy, 3–55 (1970).

[98] Thickness-twist vibrations of a quartz strip, *International Journal of Solids and Structures*, **7**, 1–4 (1971).

[99] Acoustical and optical activity in alpha quartz (with R. A. Toupin), *International Journal of Solids and Structures*, **7**, 1219–1227 (1971).

[100] Electromechanical vibrations of centrosymmetric cubic crystal plates, *PMM–Journal of Mechanics and Applied Mathematics*, **35**, 404–408 (1971).

[101] A continuum theory of a diatomic, elastic dielectric, *International Journal of Solids and Structures*, **8**, 369–383 (1972).

[102] Coupled elastic and electromagnetic fields in a diatomic, dielectric continuum, *International Journal of Solids and Structures*, **8**, 401–408 (1972).

[103] High frequency vibrations of piezoelectric crystal plates, *International Journal of Solids and Structures*, **8**, 895–906 (1972).

[104] Galerkin stress functions for non-local theories of elasticity, *International Journal of Solids and Structures*, **8**, 1407–1411 (1972).
[105] Elasticity, piezoelectricity and crystal lattice dynamics, *Journal of Elasticity*, **2**, 217–282 (1972).
[106] On the electrostatic potential of a point charge in a dielectric solid, *International Journal of Solids and Structures*, **9**, 233–235 (1973).
[107] Electromagnetic radiation from a vibrating quartz plate, *International Journal of Solids and Structures*, **9**, 697–702 (1973).
[108] Coupled piezoelectric vibrations of quartz plates, *International Journal of Solids and Structures*, **10**, 453–459 (1974).
[109] Equations of high frequency vibrations of thermopiezoelectric crystal plates, *International Journal of Solids and Structures*, **10**, 625–637 (1974).
[110] Note on torsional rigidity of elastic bars, *Mechanics Research Communications*, **1**, 57–59 (1974).
[111] Electromagnetic radiation from a vibrating, elastic sphere, *International Journal of Solids and Structures*, **10**, 1307–1314 (1974).
[112] Solution of St. Venant's torsion problem by power series, *International Journal of Solids and Structures*, **11**, 321–328 (1975).
[113] Rotation-dependence of the photoelastic effect; in J. Kestens (ed), *The Photoelastic Effect and its Applications*, Springer-Verlag, Berlin, 376–388 (1975).
[114] Low frequency vibrations of elastic bars, *International Journal of Solids and Structures*, **12**, 27–49 (1976).
[115] Note on axial-shear and contour vibrations of prisms, *International Journal of Solids and Structures*, **12**, 99–105 (1976).
*[116] Education and research in mechanics in the U.S.A., *Técnica* (Instituto Superior Técnico, Lisbon), **38**, 1–3 (1976).
[117] Torsion and flexure of piezoelectric crystal bars; in J. D. Achenbach, Y.-H. Pao and H. F. Tiersten (eds), *Application of Elastic Waves in Electrical Devices, Non-Destructive Testing and Seismology*, Northwestern University, Evanston, Illinois, 83–96 (1976).
[118] Frequencies of vibration of beams by Schelkunoff iteration with the HP-65 calculator, *Computers & Structures*, **7**, 639–650 (1977).
[119] On thickness-twist vibrations of a quartz strip; in *Some Aspects of Mechanics of Continua* (B. Sen Memorial Volume), Department of Mathematics, Jadavpur University, Calcutta, Part I, 29–31 (1977).
[120] A variational principle for the equations of piezoelectromagnetism in a compound continuum; in *Kompleksnii Analiz i ego Prilozhenia* (I. N. Vekua 70th Birthday Volume), Academy of Sciences U.S.S.R.; Nauka, Moscow, 397–400 (1978).
*[121] The sesquicentennial of the first crystal plate equations, *Proceedings*

* Not part of this collection.

of the 33rd Annual Symposium on Frequency Control, Electronic Industries Association, Washington, 1–3 (1979).

[122] Vibrations of quartz plates with the HP-67 pocket calculator, *Computers & Structures*, **10**, 751–759 (1979).

[123] On Reissner's equations for sandwich plates; in S. Nemat-Nasser (ed), *Mechanics Today*, **5** (E. Reissner 65th Birthday Volume), Pergamon Press, Oxford, 315–328 (1980).

[124] High frequency vibrations of quartz plates by expansion in series of Ekstein functions, *International Journal of Solids and Structures*, **16**, 785–791 (1980).

[125] Third overtone quartz resonator, *International Journal of Solids and Structures*, **18**, 809–817 (1982).

[126] Frequencies of piezoelectrically forced vibrations of electroded, doubly rotated, quartz plates, *International Journal of Solids and Structures*, **20**, 141–157 (1984).

[127] Vibrations of doubly-rotated-cut quartz plates with monoclinic symmetry, *International Journal of Solids and Structures*, **21**, 597–607 (1985).

[128] On vibrations of rectangular parallelepipeds, *International Journal of Solids and Structures*, **22**, 1423–1430 (1986).

[129] Flexural vibrations of rectangular plates with free edges, *Mechanics Research Communications*, **13**, 349–357 (1986).

Offprint from "Archive for Rational Mechanics and Analysis", Volume 16, Number 1, 1964, P. 51—78

Springer-Verlag, Berlin · Göttingen · Heidelberg

Micro-structure in Linear Elasticity

R. D. MINDLIN

Communicated by R. A. TOUPIN

Contents

Introduction

In this paper, there is formulated a linear theory of a three-dimensional, elastic continuum which has some of the properties of a crystal lattice as a result of the inclusion, in the theory, of the idea of the unit cell. The equations yield wave-dispersion relations with acoustic and optical branches of the same character as those found at long wave-lengths in crystal lattice theories and observed in neutron scattering experiments. Although specific solutions are not exhibited in detail, it is apparent from the form of the equations that there will be interesting surface effects under conditions of both motion and equilibrium.

The unit cell may also be interpreted as a molecule of a polymer, a crystallite of a polycrystal or a grain of a granular material. The mathematical model of the cell is a linear version of ERICKSEN & TRUESDELL's deformable directors [*1*]. If the cell is made rigid, the equations reduce to those of a linear COSSERAT continuum [*2*].

The method of derivation of the equations is analogous to one used in deducing two-dimensional equations of high frequency vibrations of plates from the three-dimensional equations of classical linear elasticity. By the same technique as that employed in passing from high order theories of plates to classical theories of plates, the equations are shown to reduce, at low frequencies and very long wave-lengths in isotropic materials, to those of an elastic continuum with potential energy-density dependent on strain and strain gradient and kinetic energy-density dependent on velocity and velocity gradient.

A linear form of TOUPIN's generalization [*8*, Section *7*] of couple-stress theory [*5*—*10*] is obtained by eliminating the difference between the deformations of the unit cell and the surrounding medium; and linear couple-stress theory itself is obtained by eliminating, further, the symmetric part of the strain gradient. Both of these special cases are also limited to low frequencies and very long wave-lengths.

1. Kinematics

Consider a material volume V, bounded by a surface S, with X_i, $i=1, 2, 3$, the rectangular components of the material position vector, measured from a fixed origin, and x_i the components, in the same rectangular frame, of the spatial position vector. The components of displacement of a material particle are defined as

$$u_i \equiv x_i - X_i. \tag{1.1}$$

Embedded in each material particle there is assumed to be a *micro-volume* V' in which X'_i and x'_i are the components of the material and spatial position vectors, respectively, referred to axes parallel to those of the x_i, with origin fixed in the particle: so that the origin of the coordinates x'_i moves with the displacement $\boldsymbol{u}$. A *micro-displacement* $\boldsymbol{u}'$ is defined; with components

$$u'_i \equiv x'_i - X'_i. \tag{1.2}$$

The absolute values of the displacement-gradients are assumed to be small in comparison with unity:

$$\left|\frac{\partial u_j}{\partial X_i}\right| \ll 1, \qquad \left|\frac{\partial u'_j}{\partial X'_i}\right| \ll 1, \tag{1.3}$$

so that we may write

$$\frac{\partial u_j}{\partial X_i} \approx \frac{\partial u_j}{\partial x_i} \equiv \partial_i u_j, \qquad u_j = u_j(x_i, t), \tag{1.4}$$

$$\frac{\partial u'_j}{\partial X'_i} \approx \frac{\partial u'_j}{\partial x'_i} \equiv \partial'_i u'_j, \qquad u'_j = u'_j(x_i, x'_i, t), \tag{1.5}$$

where t is the time.

Assume that the micro-displacement can be expressed as a sum of products of specified functions of the x'_i and arbitrary functions of the x_i and t. As an approximation, retain only a single, linear term of the series:

$$u'_j = x'_k \psi_{kj}, \tag{1.6}$$

where ψ_{kj} is a function of the x_i and t only. Then the displacement-gradient in the micro-medium is

$$\partial'_i u'_j = \psi_{ij}, \tag{1.7}$$

i.e., the *micro-deformation* ψ_{ij} is taken to be homogeneous in the *micro-medium* V' and non-homogeneous in the *macro-medium* V. In view of $(1.3)_2$, $|\psi_{ij}| \ll 1$. The symmetric part of ψ_{ij} is the *micro-strain:*

$$\psi_{(ij)} \equiv \tfrac{1}{2}(\psi_{ij} + \psi_{ji}) \tag{1.8}$$

and the antisymmetric part is the *micro-rotation:*

$$\psi_{[ij]} \equiv \tfrac{1}{2}(\psi_{ij} - \psi_{ji}). \tag{1.9}$$

An alternative interpretation of the micro-deformation is that the quantities ψ_{ij} are proportional to the components of the displacements of the tips of deformable *directors*, as described by ERICKSEN & TRUESDELL [*1*]. The $\psi_{[ij]}$ then are the components of rotation of the COSSERAT *trièdre* [*2*, p. 122].

We define the usual strain (now the *macro-strain*)

$$\varepsilon_{ij} \equiv \tfrac{1}{2}(\partial_i u_j + \partial_j u_i); \tag{1.10}$$

and also a *relative deformation* (the difference between the macro-displacement-gradient and the micro-deformation)

$$\gamma_{ij} \equiv \partial_i u_j - \psi_{ij}, \tag{1.11}$$

and a *micro-deformation gradient* (the macro-gradient of the micro-deformation):

$$\varkappa_{ijk} \equiv \partial_i \psi_{jk}. \tag{1.12}$$

All three of the tensors ε_{ij}, γ_{ij} and $\varkappa_{ijk}$ are independent of the micro-coordinates x'_i. Typical components of γ_{ij} and $\varkappa_{ijk}$ are illustrated in Figs. 1 and 2.

The u_i and ψ_{ij} are assumed to be single-valued functions of the x_i, leading to the compatibility equations

$$e_{mik} e_{nlj} \partial_i \partial_j \varepsilon_{kl} = 0, \tag{1.13}$$

$$e_{mij} \partial_i \varkappa_{jkl} = 0, \tag{1.14}$$

$$\partial_i(\varepsilon_{jk} + \omega_{jk} - \gamma_{jk}) = \varkappa_{ijk}, \tag{1.15}$$

where ω_{ij} is the *macro-rotation:*

$$\omega_{ij} \equiv \tfrac{1}{2}(\partial_i u_j - \partial_j u_i) \tag{1.16}$$

and e_{ijk} is the alternating tensor.

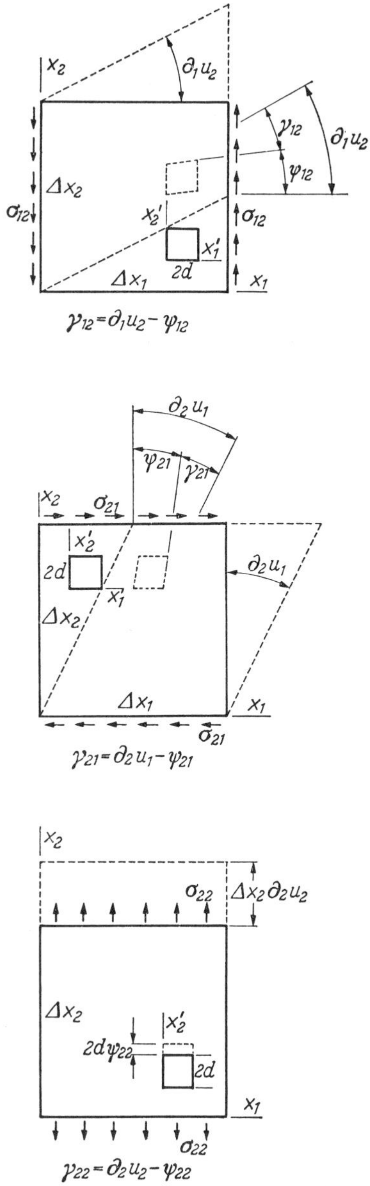

Fig. 1. Typical components of relative stress σ_{ij}, displacement-gradient $\partial_i u_j$, micro-deformation ψ_{ij} and relative deformation γ_{ij}

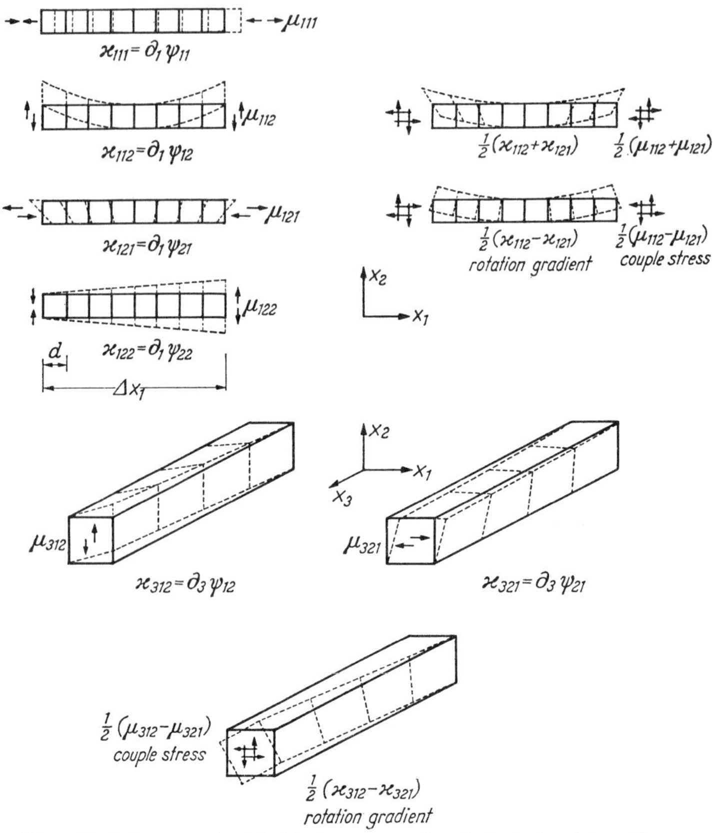

Fig. 2. Typical components of double stress μ_{ijk} and gradient of micro-deformation $\varkappa_{ijk}$

2. Kinetic and potential energies

Let the micro-medium be a parallelepiped with volume V' and edges of lengths $2d_i$ and direction cosines l_{ij} with respect to the axes x'_i. Let x''_i be oblique Cartesian coordinates parallel to the edges d_i, respectively. Then [3, p. 153]

$$x'_i = l_{ij}x''_j, \tag{2.1$_1$}$$

$$V' = 8\,\|l_{ij}l_{ik}\|^{\frac{1}{2}}\,d_1 d_2 d_3, \tag{2.1$_2$}$$

$$dV' = \|l_{ij}l_{ik}\|^{\frac{1}{2}}\,dx''_1\,dx''_2\,dx''_3. \tag{2.1$_3$}$$

Let ϱ_M be the mass of macro-material per unit macro-volume and let ϱ' be the mass of micro-material per unit macro-volume. We define a kinetic energy-density (kinetic energy per unit macro-volume):

$$T \equiv \frac{1}{2}\varrho_M \dot{u}_j\dot{u}_j + \frac{1}{V'}\int\limits_{V'} \frac{1}{2}\varrho'(\dot{u}_j+\dot{u}'_j)(\dot{u}_j+\dot{u}'_j)\,dV', \tag{2.2}$$

where the dot designates differentiation with respect to time. Upon substituting (1.6) and (2.1) in (2.2) and performing the integration, we find

$$T = \tfrac{1}{2}\varrho\dot{u}_j\dot{u}_j + \tfrac{1}{6}\varrho' d^2_{kl}\dot{\psi}_{kj}\dot{\psi}_{lj}, \tag{2.3$_1$}$$

where

$$\varrho \equiv \varrho_M + \varrho', \tag{2.3$_2$}$$

$$d_{kl}^2 \equiv d_p d_q (\delta_{p1}\delta_{q1} l_{k1} l_{l1} + \delta_{p2}\delta_{q2} l_{k2} l_{l2} + \delta_{p3}\delta_{q3} l_{k3} l_{l3}) = d_{lk}^2 \tag{2.3$_3$}$$

and δ_{ij} is the Kronecker symbol. In the case of a cube with edges of length $2d$ parallel to the axes of x_i',

$$l_{ij} = \delta_{ij}, \qquad d_1 = d_2 = d_3 \equiv d. \tag{2.4}$$

Then the second term in $(2.3)_1$ reduces to $\frac{1}{6}\varrho' d^2 \dot{\psi}_{ij}\dot{\psi}_{ij}$. If the material is composed wholly of unit cells, $\varrho_M = 0$. Then $\varrho' = \varrho$.

For the potential energy-density (potential energy per unit macro-volume), we assume a function, W, of the forty-two variables $\varepsilon_{ij}, \gamma_{ij}, \varkappa_{ijk}$:

$$W = W(\varepsilon_{ij}, \gamma_{ij}, \varkappa_{ijk}). \tag{2.5}$$

A small, rigid rotation of the deformed body is described by a rotation, ω_{ij} = constant, of the macro-material and an equal rotation $\psi_{[ij]}$ of the micro-material. The associated displacements are

$$u_j = x_i \omega_{ij}, \qquad u_j' = x_j' \psi_{[ij]}. \tag{2.6}$$

The addition of such a displacement leaves W unchanged since the added $\varepsilon_{ij}, \gamma_{ij}$ and $\varkappa_{ijk}$ are zero.

The assumptions (1.6), (2.2) and (2.5) are the minimum that will lead to equations which yield the desired dispersion relations for plane waves: including longitudinal and transverse acoustic and optical branches. More or less than (1.6), (2.2) and (2.5) would be more or less than what is required.

The unit cell is taken to be a parallelepiped in order to represent the unit cell of a crystal lattice. However, another shape would only change the tensor d_{kl}^2. Also, the cell can be interpreted as a molecule of a polymer, a crystallite of a polycrystal or a grain of a granular material.

3. Variational equation of motion

We write Hamilton's principle for independent variations δu_i and $\delta\psi_{ij}$ between fixed limits of u_i and ψ_{ij} at times t_0 and t_1:

$$\delta \int_{t_0}^{t_1} (\mathcal{T} - \mathcal{W})\, dt + \int_{t_0}^{t_1} \delta\mathcal{W}_1\, dt = 0, \tag{3.1}$$

where $\mathcal{T}$ and $\mathcal{W}$ are the total kinetic and potential energies:

$$\mathcal{T} \equiv \int_V T\, dV, \qquad \mathcal{W} \equiv \int_V W\, dV \tag{3.2}$$

and $\delta\mathcal{W}_1$ is the variation of the work done by external forces.

In the usual way [4, p. 166], we find, from $(2.3)_1$ and $(3.2)_1$,

$$\delta \int_{t_0}^{t_1} \mathcal{T}\, dt = -\int_{t_0}^{t_1} dt \int_V (\varrho \ddot{u}_j \delta u_j + \tfrac{1}{3}\varrho' d_{jl}^2 \ddot{\psi}_{lk} \delta\psi_{jk})\, dV. \tag{3.3}$$

As for the variation of potential energy, we first define

$$\tau_{ij} \equiv \frac{\partial W}{\partial \varepsilon_{ij}} = \tau_{ji}, \tag{3.4$_1$}$$

$$\sigma_{ij} \equiv \frac{\partial W}{\partial \gamma_{ij}}, \tag{3.4$_2$}$$

$$\mu_{ijk} \equiv \frac{\partial W}{\partial \varkappa_{ijk}}. \tag{3.4$_3$}$$

Then

$$\begin{aligned}\delta W &= \tau_{ij}\delta\varepsilon_{ij} + \sigma_{ij}\delta\gamma_{ij} + \mu_{ijk}\delta\varkappa_{ijk},\\ &= \tau_{ij}\partial_i\delta u_j + \sigma_{ij}(\partial_i\delta u_j - \delta\psi_{ij}) + \mu_{ijk}\partial_i\delta\psi_{jk},\\ &= \partial_i[(\tau_{ij}+\sigma_{ij})\,\delta u_j] - \partial_i(\tau_{ij}+\sigma_{ij})\,\delta u_j - \sigma_{ij}\delta\psi_{ij} + \partial_i(\mu_{ijk}\delta\psi_{jk}) -\\ &\quad - \partial_i\mu_{ijk}\delta\psi_{jk}.\end{aligned}$$

Applying the divergence theorem, we find

$$\begin{aligned}\delta\mathscr{W} &= \int_V \delta W\, dV\\ &= -\int_V \partial_i(\tau_{ij}+\sigma_{ij})\,\delta u_j dV - \int_V (\partial_i\mu_{ijk}+\sigma_{jk})\,\delta\psi_{jk} dV +\\ &\quad + \int_S n_i(\tau_{ij}+\sigma_{ij})\,\delta u_j dS + \int_S n_i\mu_{ijk}\delta\psi_{jk} dS.\end{aligned} \tag{3.5}$$

The form of (3.5) is the motivation for the adoption of the following form for the variation of work done by external forces:

$$\delta\mathscr{W}_1 = \int_V f_j\,\delta u_j dV + \int_V \Phi_{jk}\delta\psi_{jk} dV + \int_S t_j\delta u_j dS + \int_S T_{jk}\delta\psi_{jk} dS. \tag{3.6}$$

The definitions of u_j and ψ_{jk}, and the fact that the integrands of the volume and surface integrals represent variations of work per unit volume and area, yield the physical significances of the coefficients of δu_j and $\delta\psi_{jk}$. Thus, f_j is the body force per unit volume and t_j is the surface force per unit area (stress-vector or traction); Φ_{jk} is to be interpreted as a double force per unit volume [*4*, p. 187] and T_{jk} as a double force per unit area. The diagonal terms of Φ_{jk} and T_{jk} are double forces without moment and the off-diagonal terms are double forces with moment. The antisymmetric part $\Phi_{[jk]}$ of the *body double force* Φ_{jk} is the body couple. The antisymmetric part $T_{[jk]}$ of the *double traction* T_{jk} is the Cosserat couple-stress vector. In both Φ_{jk} and T_{jk}, the first subscript gives the orientation of the lever arm between the forces and the second subscript gives the orientation of the forces. Across a surface with its outward normal in the positive direction, the force at the positive end of the lever arm acts in the positive direction. ("Positive" refers to the positive sense of the coordinate axis parallel to the lever arm or force). Across a surface with its outward normal in the negative direction, the directions of the forces are reversed.

Substituting (3.3), (3.5) and (3.6) in (3.1), and dropping the integration with respect to time, we obtain the variational equation of motion:

$$\begin{aligned}&\int_V (\partial_i\tau_{ij} + \partial_i\sigma_{ij} + f_j - \varrho\ddot{u}_j)\,\delta u_j dV +\\ &\quad + \int_V (\partial_i\mu_{ijk} + \sigma_{jk} + \Phi_{jk} - \tfrac{1}{3}\varrho' d_{jl}^2\ddot{\psi}_{lk})\,\delta\psi_{jk} dV +\\ &\quad + \int_S [t_j - n_i(\tau_{ij}+\sigma_{ij})]\,\delta u_j\, dS + \int_S (T_{jk} - n_i\mu_{ijk})\,\delta\psi_{jk} dS = 0.\end{aligned} \tag{3.7}$$

4. Stress-equations of motion and boundary conditions

From the variational equation of motion, there follow immediately the twelve stress-equations of motion:

$$\partial_i(\tau_{ij}+\sigma_{ij})+f_j=\varrho\ddot{u}_j, \qquad (4.1)_1$$

$$\partial_i\mu_{ijk}+\sigma_{jk}+\Phi_{jk}=\tfrac{1}{3}\varrho' d^2_{lj}\ddot{\psi}_{lk}, \qquad (4.1)_2$$

and the twelve traction boundary conditions:

$$t_j=n_i(\tau_{ij}+\sigma_{ij}), \qquad (4.2)_1$$

$$T_{jk}=n_i\mu_{ijk}. \qquad (4.2)_2$$

In view of (3.4), (4.2) and the significance of $\varepsilon_{ij}, \gamma_{ij}, \varkappa_{ijk}$, appropriate terminology appears to be *Cauchy stress* for τ_{ij}, *relative stress* for σ_{ij} and *double stress* for μ_{ijk}. The twenty-seven components of μ_{ijk} are interpreted as double forces per unit area. The first subscript of a μ_{ijk} designates the normal to the surface across which the component acts; the second and third subscripts have the same significance as the two subscripts of T_{jk}. Typical components of μ_{ijk} are illustrated in Fig. 2.

The linear equations of a Cosserat continuum [*2*] are obtained by setting $\psi_{(ij)}=0$. Then $\sigma_{(ij)}=\tau_{ij}$ and $\mu_{i(jk)}=0$; and there remain $\mu_{i[jk]}$ (the Cosserat couple-stress) and $\sigma_{[ij]}$: which has been regarded as the antisymmetric part of an asymmetric stress τ_{ij}. However, in the present theory, the Cauchy stress, τ_{ij}, is symmetric and $\sigma_{[ij]}$ is the antisymmetric part of an asymmetric relative stress σ_{ij}.

Besides containing the linear equations of a Cosserat continuum as a special case, Eqs. (2.5), (3.4) and (4.1) also include, as low frequency, very long wave length approximations, linear versions of the equations of couple-stress theory [*5*—*10*] and Toupin's generalization of couple-stress theory [*8*, Section *7*]. These are considered in Sections 9—12.

If additional terms were retained in the series expansion (1.6) of the micro-displacement u'_j, higher order stresses would appear. In addition to stresses corresponding to double forces per unit area, there would be stresses corresponding to n-tuple forces per unit area. All of the latter would be self equilibrating; whereas, of the μ_{ijk}, only the $\mu_{i(jk)}$ are self equilibrating.

5. Constitutive equations

For the potential energy-density we take a homogeneous, quadratic function of the forty-two variables $\varepsilon_{ij}, \gamma_{ij}, \varkappa_{ijk}$:

$$\begin{aligned} W=&\tfrac{1}{2}c_{ijkl}\varepsilon_{ij}\varepsilon_{kl}+\tfrac{1}{2}b_{ijkl}\gamma_{ij}\gamma_{kl}+\tfrac{1}{2}a_{ijklmn}\varkappa_{ijk}\varkappa_{lmn}+\\ &+d_{ijklm}\gamma_{ij}\varkappa_{klm}+f_{ijklm}\varkappa_{ijk}\varepsilon_{lm}+g_{ijkl}\gamma_{ij}\varepsilon_{kl}. \end{aligned} \qquad (5.1)$$

Only $\frac{1}{2}\times 42\times 43=903$ of the $42\times 42=1764$ coefficients in (5.1) are independent. The number of coefficients, the relations among them and the number of independent ones are given in the following table:

$$
\begin{aligned}
c_{ijkl}=c_{klij}=c_{jikl}&: \quad 9\times 9-60= \ \ 21\\
b_{ijkl}=b_{klij}&: \quad 9\times 9-36= \ \ 45\\
a_{ijklmn}=a_{lmnijk}&: 27\times 27-351=378\\
d_{ijklm}&: \quad\quad 9\times 27=243\\
f_{ijklm}=f_{ijkml}&: \quad 9\times 27-81=162\\
g_{ijkl}=g_{ijlk}&: \quad 9\times 9-27= \ \ 54\\
&\qquad\qquad\qquad\ \ \overline{903}
\end{aligned}
\tag{5.2}
$$

From (5.1) and (3.4):

$$\tau_{pq}=c_{pqij}\varepsilon_{ij}+g_{ijpq}\gamma_{ij}+f_{ijkpq}\varkappa_{ijk}, \tag{5.3$_1$}$$

$$\sigma_{pq}=g_{pqij}\varepsilon_{ij}+b_{ijpq}\gamma_{ij}+d_{pqijk}\varkappa_{ijk}, \tag{5.3$_2$}$$

$$\mu_{pqr}=f_{pqrij}\varepsilon_{ij}+d_{ijpqr}\gamma_{ij}+a_{pqrijk}\varkappa_{ijk}. \tag{5.3$_3$}$$

In the case of a centrosymmetric, isotropic material (referred to as isotropic in the sequel) the number of independent coefficients is greatly reduced. As there are no isotropic tensors of odd rank, d_{ijklm} and f_{ijklm} must vanish. The remaining coefficients must be homogeneous, linear functions of products of Kronecker deltas. There are three independent products of two Kronecker deltas and fifteen independent products of three Kronecker deltas. Hence

$$c_{ijkl}=\lambda\delta_{ij}\delta_{kl}+\mu_1\delta_{ik}\delta_{jl}+\mu_2\delta_{il}\delta_{jk}, \tag{5.4$_1$}$$

$$b_{ijkl}=b_1\delta_{ij}\delta_{kl}+b_2\delta_{ik}\delta_{jl}+b_3\delta_{il}\delta_{jk}, \tag{5.4$_2$}$$

$$g_{ijkl}=g_1\delta_{ij}\delta_{kl}+g_2\delta_{ik}\delta_{jl}+g_3\delta_{il}\delta_{jk}, \tag{5.4$_3$}$$

$$
\begin{aligned}
a_{ijklmn}=\ &a_1\delta_{ij}\delta_{kl}\delta_{mn}+a_2\delta_{ij}\delta_{km}\delta_{nl}+a_3\delta_{ij}\delta_{kn}\delta_{lm}+\\
&+a_4\delta_{jk}\delta_{il}\delta_{mn}+a_5\delta_{jk}\delta_{im}\delta_{nl}+a_6\delta_{jk}\delta_{in}\delta_{lm}+\\
&+a_7\delta_{ki}\delta_{jl}\delta_{mn}+a_8\delta_{ki}\delta_{jm}\delta_{nl}+a_9\delta_{ki}\delta_{jn}\delta_{lm}+\\
&+a_{10}\delta_{il}\delta_{jm}\delta_{kn}+a_{11}\delta_{jl}\delta_{km}\delta_{in}+a_{12}\delta_{kl}\delta_{im}\delta_{jn}+\\
&+a_{13}\delta_{il}\delta_{jn}\delta_{km}+a_{14}\delta_{jl}\delta_{kn}\delta_{im}+a_{15}\delta_{kl}\delta_{in}\delta_{jm}.
\end{aligned}
\tag{5.4$_4$}
$$

The conditions (5.2) require the six relations

$$\mu_1=\mu_2\equiv\mu, \qquad g_2=g_3,$$

$$a_1=a_6, \qquad a_2=a_9, \qquad a_5=a_7, \qquad a_{11}=a_{12},$$

leaving eighteen independent coefficients. Thus, the potential energy density reduces to

$$
\begin{aligned}
W=\ &\tfrac{1}{2}\lambda\varepsilon_{ii}\varepsilon_{jj}+\mu\varepsilon_{ij}\varepsilon_{ij}+\tfrac{1}{2}b_1\gamma_{ii}\gamma_{jj}+\tfrac{1}{2}b_2\gamma_{ij}\gamma_{ij}+\\
&+\tfrac{1}{2}b_3\gamma_{ij}\gamma_{ji}+g_1\gamma_{ii}\varepsilon_{jj}+g_2(\gamma_{ij}+\gamma_{ji})\varepsilon_{ij}+\\
&+a_1\varkappa_{iik}\varkappa_{kjj}+a_2\varkappa_{iik}\varkappa_{jkj}+\tfrac{1}{2}a_3\varkappa_{iik}\varkappa_{jjk}+\tfrac{1}{2}a_4\varkappa_{ijj}\varkappa_{ikk}+\\
&+a_5\varkappa_{ijj}\varkappa_{kik}+\tfrac{1}{2}a_8\varkappa_{iji}\varkappa_{kjk}+\tfrac{1}{2}a_{10}\varkappa_{ijk}\varkappa_{ijk}+a_{11}\varkappa_{ijk}\varkappa_{jki}+\\
&+\tfrac{1}{2}a_{13}\varkappa_{ijk}\varkappa_{ikj}+\tfrac{1}{2}a_{14}\varkappa_{ijk}\varkappa_{jik}+\tfrac{1}{2}a_{15}\varkappa_{ijk}\varkappa_{kji},
\end{aligned}
\tag{5.5}
$$

and the constitutive equations become

$$\tau_{pq} = \lambda\,\delta_{pq}\,\varepsilon_{ii} + 2\mu\,\varepsilon_{pq} + g_1\,\delta_{pq}\gamma_{ii} + g_2(\gamma_{pq} + \gamma_{qp}), \tag{5.6$_1$}$$

$$\sigma_{pq} = g_1\,\delta_{pq}\,\varepsilon_{ii} + 2g_2\,\varepsilon_{pq} + b_1\,\delta_{pq}\gamma_{ii} + b_2\gamma_{pq} + b_3\gamma_{qp}, \tag{5.6$_2$}$$

$$\begin{aligned}\mu_{pqr} = {} & a_1(\varkappa_{iip}\,\delta_{qr} + \varkappa_{rii}\,\delta_{pq}) + a_2(\varkappa_{iiq}\,\delta_{pr} + \varkappa_{iri}\,\delta_{pq}) + a_3\varkappa_{iir}\,\delta_{pq} + \\ & + a_4\varkappa_{pii}\,\delta_{qr} + a_5(\varkappa_{qii}\,\delta_{pr} + \varkappa_{ipi}\,\delta_{qr}) + a_8\varkappa_{iqi}\,\delta_{pr} + a_{10}\varkappa_{pqr} + \\ & + a_{11}(\varkappa_{rpq} + \varkappa_{qrp}) + a_{13}\varkappa_{prq} + a_{14}\varkappa_{qpr} + a_{15}\varkappa_{rqp}.\end{aligned} \tag{5.6$_3$}$$

6. Displacement-equations of motion

We may obtain twelve equations of motion on the twelve variables u_i and ψ_{ij} by first inserting (1.10), (1.11) and (1.12) into the constitutive equations and then the latter into the stress-equations of motion (4.1). There is no necessity to assume spatially homogeneous material properties. In fact, the coefficients of elasticity and the densities may be taken as periodic functions, of oblique coordinates parallel to the edges of the unit cell, of periods $2d_i$. This would represent the periodic structure of a crystal lattice. However, the equations would then be highly intractable; whereas some of their main features are exhibited with *macro-homogeneous* material properties, at least for wave-lengths greater than the dimensions of the unit cell. The isotropic case is especially simple; but nevertheless it still contains many of the novel properties of the macro-homogeneous material.

In the case of isotropy, the constitutive Eqs. (5.6) apply and also (2.4). Then, for the macro-homogeneous, isotropic material the equations on u_i and ψ_{ij} are

$$\begin{aligned}&(\mu + 2g_2 + b_2)\,\partial_j\partial_j u_i + (\lambda + \mu + 2g_1 + 2g_2 + b_1 + b_3)\,\partial_i\partial_j u_j - \\ &\quad - (g_1 + b_1)\,\partial_i\psi_{jj} - (g_2 + b_2)\,\partial_j\psi_{ji} - (g_2 + b_3)\,\partial_j\psi_{ij} + f_i = \varrho\ddot{u}_i,\end{aligned} \tag{6.1}$$

$$\begin{aligned}&(a_1 + a_5)\,(\partial_k\partial_l\psi_{kl}\delta_{ij} + \partial_i\partial_j\psi_{kk}) + (a_2 + a_{11})\,(\partial_j\partial_k\psi_{ki} + \partial_i\partial_k\psi_{jk}) + \\ &\quad + (a_3 + a_{14})\,\partial_i\partial_k\psi_{kj} + a_4\,\partial_k\partial_k\psi_{ll}\delta_{ij} + (a_8 + a_{15})\,\partial_j\partial_k\psi_{ik} + \\ &\quad + a_{10}\,\partial_k\partial_k\psi_{ij} + a_{13}\,\partial_k\partial_k\psi_{ji} + g_1\,\partial_k u_k\,\delta_{ij} + g_2(\partial_i u_j + \partial_j u_i) + \\ &\quad + b_1(\partial_k u_k - \psi_{kk})\,\delta_{ij} + b_2(\partial_i u_j - \psi_{ij}) + b_3(\partial_j u_i - \psi_{ji}) + \Phi_{ij} = \tfrac{1}{3}\varrho' d^2\ddot{\psi}_{ij}.\end{aligned} \tag{6.2}$$

7. Micro-vibrations

Consider solutions of (6.1) and (6.2) of the form

$$u_i = f_i = \Phi_{ij} = 0, \qquad \psi_{ij} = A_{ij}\,e^{i\omega t}, \tag{7.1}$$

where the A_{ij} are constants. Then (6.1) are satisfied identically and (6.2) become

$$b_1\delta_{ij}A_{kk} + b_2A_{ij} + b_3A_{ji} = \tfrac{1}{3}\varrho' d^2\omega^2 A_{ij}, \tag{7.2}$$

which admit the following solutions:

dilatational mode:

$$A_{11} = A_{22} = A_{33}; \qquad A_{ij} = 0, \qquad i \neq j; \tag{7.3$_1$}$$

$$\omega_d^2 = 3\,(3\,b_1 + b_2 + b_3)/\varrho' d^2; \tag{7.3$_2$}$$

shear modes:

$$A_{ij}=A_{ji}, \qquad i\neq j; \qquad A_{ij}=0, \qquad i=j; \tag{7.4$_1$}$$

$$\omega_s^2=3(b_2+b_3)/\varrho' d^2; \tag{7.4$_2$}$$

equivoluminal extensional modes:

$$A_{ii}=0; \qquad A_{ij}=0, \qquad i\neq j; \tag{7.5$_1$}$$

$$\omega_s^2=3(b_2+b_3)/\varrho' d^2; \tag{7.5$_2$}$$

rotational modes:

$$A_{ij}=-A_{ji}, \tag{7.6$_1$}$$

$$\omega_r^2=3(b_2-b_3)/\varrho' d^2. \tag{7.6$_2$}$$

The restriction of the potential energy-density to be positive definite requires

$$3b_1+b_2+b_3>0, \qquad b_2+b_3>0, \qquad b_2-b_3>0.$$

Hence ω_d, ω_s and ω_r are real frequencies. The corresponding modes are analogous to the simple thickness-modes of vibration of a plate. Just as the latter are independent of the coordinates in the plane of the plate, so are the micro-modes independent of the coordinates x_i of the three-dimensional continuum with micro-structure. Extensional and flexural waves in a plate couple with thickness-modes, at high frequencies, to form the higher branches of the dispersion relations for a plate. Analogously, we may expect longitudinal and transverse acoustic waves, in the three-dimensional continuum with micro-structure, to couple with the micro-modes to form optical branches.

8. Plane waves, long wave-length

Consider solutions of (6.1) and (6.2) with f_i and Φ_{ij} zero and

$$u_i=u_i(x_1,t), \qquad \psi_{ij}=\psi_{ij}(x_1,t). \tag{8.1}$$

By means of linear combinations, the twelve equations may be composed into three independent equations and three independent systems of three equations each:

shear optical I:

$$(a_{10}+a_{13})\,\partial_1\partial_1\psi_{(23)}-(b_2+b_3)\,\psi_{(23)}=\tfrac{1}{3}\varrho' d^2\ddot{\psi}_{(23)}; \tag{8.2}$$

shear optical II: the same as (8.2) except that $\psi_{(23)}$ is replaced with $\psi_{22}-\psi_{33}$;

rotational optical:

$$(a_{10}-a_{13})\,\partial_1\partial_1\psi_{[23]}-(b_2-b_3)\,\psi_{[23]}=\tfrac{1}{3}\varrho' d^2\ddot{\psi}_{[23]}; \tag{8.3}$$

longitudinal system:

$$\begin{gathered} k_{11}\partial_1\partial_1 u_1-k_{12}\partial_1\psi_{11}^D-k_{13}\partial_1\psi=\varrho\ddot{u}_1,\\ k_{21}\partial_1 u_1+k_{22}\partial_1\partial_1\psi_{11}^D-k'_{22}\psi_{11}^D+k_{23}\partial_1\partial_1\psi=\tfrac{1}{2}\varrho' d^2\ddot{\psi}_{11}^D,\\ k_{31}\partial_1 u_1+k_{32}\partial_1\partial_1\psi_{11}^D+k_{33}\partial_1\partial_1\psi-k'_{33}\psi=\varrho' d^2\ddot{\psi}, \end{gathered} \tag{8.4}$$

where $\psi = \frac{1}{3}\psi_{ii}$, $\psi_{11}^D = \psi_{11} - \psi$ and

$$k_{11} = \lambda + 2\mu + 2g_1 + 4g_2 + b_1 + b_2 + b_3,$$
$$k_{22} = 2a_2 + a_3 + a_8 + \tfrac{3}{2}a_{10} + 2a_{11} + \tfrac{3}{2}a_{13} + a_{14} + a_{15},$$
$$k_{33} = 6a_1 + 2a_2 + a_3 + 9a_4 + 6a_5 + a_8 + 3a_{10} + 2a_{11} + 3a_{13} + a_{14} + a_{15},$$
$$k_{23} = k_{32} = 3a_1 + 2a_2 + a_3 + 3a_5 + a_8 + 2a_{11} + a_{14} + a_{15},$$
$$k_{31} = k_{13} = 3g_1 + 2g_2 + 3b_1 + b_2 + b_3,$$
$$k_{12} = k_{21} = 2g_2 + b_2 + b_3,$$
$$k'_{22} = \tfrac{3}{2}(b_2 + b_3),$$
$$k'_{33} = 3(3b_1 + b_2 + b_3);$$

transverse system I:

$$\bar{k}_{11}\partial_1\partial_1 u_2 - \bar{k}_{12}\partial_1\psi_{(12)} - \bar{k}_{13}\partial_1\psi_{[12]} = \varrho\ddot{u}_2,$$
$$\bar{k}_{21}\partial_1 u_2 + \bar{k}_{22}\partial_1\partial_1\psi_{(12)} - \bar{k}'_{22}\psi_{(12)} + \bar{k}_{23}\partial_1\partial_1\psi_{[12]} = \tfrac{2}{3}\varrho' d^2\ddot{\psi}_{(12)}, \tag{8.5}$$
$$\bar{k}_{31}\partial_1 u_2 + \bar{k}_{32}\partial_1\partial_1\psi_{(12)} + \bar{k}_{33}\partial_1\partial_1\psi_{[12]} - \bar{k}'_{33}\psi_{[12]} = \tfrac{2}{3}\varrho' d^2\ddot{\psi}_{[12]},$$

where

$$\bar{k}_{11} = \mu + 2g_2 + b_2,$$
$$\bar{k}_{22} = 2a_2 + a_3 + a_8 + 2a_{10} + 2a_{11} + 2a_{13} + a_{14} + a_{15},$$
$$\bar{k}_{33} = -2a_2 + a_3 + a_8 + 2a_{10} - 2a_{11} - 2a_{13} + a_{14} + a_{15},$$
$$\bar{k}_{23} = \bar{k}_{32} = a_3 - a_8 + a_{14} - a_{15},$$
$$\bar{k}_{31} = \bar{k}_{13} = b_2 - b_3,$$
$$\bar{k}_{12} = \bar{k}_{21} = 2g_2 + b_2 + b_3,$$
$$\bar{k}'_{22} = 2(b_2 + b_3),$$
$$\bar{k}'_{33} = 2(b_2 - b_3),$$

transverse system II: the same as transverse system I except that u_2, $\psi_{(12)}$ and $\psi_{[12]}$ are replaced with u_3, $\psi_{(13)}$ and $\psi_{[13]}$, respectively.

If, now, (8.1) are specialized to the plane waves

$$u_i = A_i i \exp[i(\xi x_1 - \omega t)], \qquad \psi_{ij} = B_{ij}\exp[i(\xi x_1 - \omega t)], \tag{8.6}$$

four dispersion relations (ω *vs.* ξ) result:

shear optical waves (*SO*) (twice):

$$\tfrac{1}{3}\varrho' d^2\omega^2 = b_2 + b_3 + (a_{10} + a_{13})\xi^2; \tag{8.7}$$

rotational optical waves (*RO*):

$$\tfrac{1}{3}\varrho' d^2\omega^2 = b_2 - b_3 + (a_{10} - a_{13})\xi^2; \tag{8.8}$$

longitudinal waves (*LA*, *LO*, *LDO*):

$$\begin{Vmatrix} k_{11}\xi^2-\varrho\omega^2 & k_{12}\xi & k_{13}\xi \\ k_{21}\xi & k_{22}\xi^2+k'_{22}-\frac{1}{2}\varrho' d^2\omega^2 & k_{23}\xi^2 \\ k_{31}\xi & k_{32}\xi^2 & k_{33}\xi^2+k'_{33}-\varrho' d^2\omega^2 \end{Vmatrix}=0; \quad (8.9)$$

transverse waves (*TA*, *TO*, *TRO*) (twice):

$$\begin{Vmatrix} \bar{k}_{11}\xi^2-\varrho\omega^2 & \bar{k}_{12}\xi & \bar{k}_{13}\xi \\ \bar{k}_{21}\xi & \bar{k}_{22}\xi^2+\bar{k}'_{22}-\frac{2}{3}\varrho' d^2\omega^2 & \bar{k}_{23}\xi^2 \\ \bar{k}_{31}\xi & \bar{k}_{32}\xi^2 & \bar{k}_{33}\xi^2+\bar{k}'_{33}-\frac{2}{3}\varrho' d^2\omega^2 \end{Vmatrix}=0. \quad (8.10)$$

These dispersion relations are similar to those encountered in a second order theory of extensional waves in plates [*11*]. The relations (8.7) and (8.8) are like that for the second face-shear mode in the plate. The relations (8.9) and (8.10) are like that for the coupled extensional, thickness-stretch and symmetric thickness-shear modes in the plate; *i.e.*, one acoustic and two optical branches in each of (8.9) and (8.10).

In the dispersion relations (8.7) and (8.8) for the non-coupled modes, there are cut-off frequencies ω_s and ω_r, respectively, at which the group velocity $(d\omega/d\xi)$ is zero. Positive definiteness of W requires

$$a_{10}+a_{13}>0, \qquad a_{10}-a_{13}>0. \quad (8.11)$$

Hence, the frequencies increase, from cut-off, with increasing real wave numbers. Below the cut-off frequencies, the wave numbers are pure imaginary with cut-off values

$$\xi=\pm i\left(\frac{b_2+b_3}{a_{10}+a_{13}}\right)^{\frac{1}{2}}, \qquad \xi=\pm i\left(\frac{b_2-b_3}{a_{10}-a_{13}}\right)^{\frac{1}{2}}, \quad (8.12)$$

respectively, at zero frequency.

The behavior of the acoustic branches in (8.9) and (8.10), at low frequencies, is described by ω'_i, ω''_i, ω'''_i ($i=1$ for longitudinal and $i=2$ for transverse): the values, at $\omega=0$ and $\xi=0$, of the first, second and third derivatives of ω with respect to ξ. We find

$$\omega'_i=\tilde{v}_i, \qquad \omega''_i=0, \qquad \omega'''_i=3\tilde{v}_i(\tilde{l}_i^2-h_i^2), \quad (8.13)$$

where

$$\tilde{v}_1^2=(\tilde{\lambda}+2\tilde{\mu})/\varrho, \qquad \tilde{v}_2^2=\tilde{\mu}/\varrho, \quad (8.14)$$

$$\tilde{\lambda}+2\tilde{\mu}=\lambda+2\mu-\frac{8g_2^2}{3(b_2+b_3)}-\frac{(3g_1+2g_2)^2}{3(3b_1+b_2+b_3)}, \quad (8.15)_1$$

$$\tilde{\mu}=\mu-\frac{2g_2^2}{b_2+b_3}, \quad (8.15)_2$$

$$\tilde{l}_1^2=2(\tilde{a}_1+\tilde{a}_2+\tilde{a}_3+\tilde{a}_4+\tilde{a}_5)/(\tilde{\lambda}+2\tilde{\mu}), \qquad \tilde{l}_2^2=2(\tilde{a}_3+\tilde{a}_4)/\tilde{\mu}, \quad (8.16)$$

$$h_1^2=\varrho' d^2[2\alpha^2+(\alpha+\beta)^2]/3\varrho, \qquad h_2^2=\varrho' d^2(1+\beta^2)/6\varrho, \quad (8.17)$$

$$\alpha=\frac{1}{b_2+b_3}\left(g_1-\frac{b_1(3g_1+2g_2)}{3b_1+b_2+b_3}\right), \qquad \beta=1+\frac{2g_2}{b_2+b_3}, \quad (8.18)$$

$$
\begin{aligned}
\tilde{a}_1 = \tfrac{1}{2}[&(1+\beta)(3\alpha+\beta)a_1 + (1+2\alpha\beta+\beta^2)a_2 - \tfrac{1}{2}(1+\beta)(1-2\alpha-\beta)a_3 - \\
&-(1-\beta)(3\alpha+\beta)a_5 - \tfrac{1}{2}(1-\beta)(1+2\alpha+\beta)a_8 + 2\alpha\beta a_{11} - \alpha(1-\beta)a_{14} + \\
&+\alpha(1+\beta)a_{15}], \\
\tilde{a}_2 = \tfrac{1}{2}\{&-(1-2\alpha-\beta)(3\alpha+\beta)a_1 - \tfrac{1}{2}[1-(2\alpha+\beta)^2]a_2 + \tfrac{1}{4}(1-2\alpha-\beta)^2 a_3 + \\
&+(3\alpha+\beta)^2 a_4 + (3\alpha+\beta)(1+2\alpha+\beta)a_5 + \tfrac{1}{4}(1+2\alpha+\beta)^2 a_8 + \\
&+\alpha(3\alpha+2\beta)a_{10} + 2\alpha(\alpha+\beta)a_{11} + \alpha(3\alpha+2\beta)a_{13} + \alpha(1+\alpha+\beta)a_{14} - \\
&-\alpha(1-\alpha-\beta)a_{15}\}, \\
\tilde{a}_3 = \tfrac{1}{4}[&-(1-\beta^2)a_2 + \tfrac{1}{2}(1+\beta)^2 a_3 + \tfrac{1}{2}(1-\beta)^2 a_8], \\
\tilde{a}_4 = \tfrac{1}{4}[&(1+\beta^2)a_{10} - (1-\beta^2)(a_{11}+a_{13}) + \tfrac{1}{2}(1+\beta)^2 a_{14} + \tfrac{1}{2}(1-\beta)^2 a_{15}], \\
\tilde{a}_5 = \tfrac{1}{4}[&-(1-\beta^2)a_{10} + (1+3\beta^2)a_{11} + (1+\beta^2)a_{13} - \tfrac{1}{2}(1+2\beta-3\beta^2)a_{14} - \\
&-\tfrac{1}{2}(1-2\beta-3\beta^2)a_{15}].
\end{aligned}
\tag{8.19}
$$

Positive definiteness of W requires $\tilde{\mu}$, $\tilde{\lambda}+2\tilde{\mu}$, $\tilde{l}_i^2>0$, while $h_i^2>0$ by inspection.

It may be seen that the limiting group velocities $\tilde{v}_i$ are less than those that would be calculated from the strain-stiffnesses $\lambda+2\mu$ and μ. This phenomenon is due to the compliance of the unit cell and has been found in a theory of crystal lattices by GAZIS & WALLIS [*12*]. Inasmuch as $\omega_i''=0$, the group velocities at zero frequency are maxima or minima. Which one occurs depends on whether ω_i''' is greater or less than zero; and this, in turn, depends on whether $\tilde{l}^2$ is greater or less than h^2. Now $\tilde{l}_i^2$ and h_i^2 are positive quantities that are length-properties of the material — depending on stiffness ratios, density ratio and the size of the unit cell. Although d is probably smaller than the $\tilde{l}_i$, the density ratio ϱ'/ϱ and the stiffness ratios α and β can make either $\tilde{l}$ or h the greater. Hence, as the frequency increases from zero, both group velocities can increase or both can decrease or one can increase and the other decrease — depending on the properties of the material. There is no analogue in the theory of homogeneous plates because they do not have multiple stiffnesses and densities. The phenomenon does occur, however, in sandwich plates [*13*] and it has also been found in a theory of crystal lattices, with complex interatomic interactions, by GAZIS & WALLIS [*14*].

At the short wave-length limit ($\xi\to\infty$), the asymptotic values of the group velocities of the acoustic branches are

$$
\frac{1}{2d}\left\{\frac{k_{22}+k_{33}-[(k_{22}-k_{33})^2+4k_{23}^2]^{\frac{1}{2}}}{\frac{1}{3}\varrho'}\right\}^{\frac{1}{2}}
\tag{8.20}
$$

from (8.9) and the same expression, with k replaced by $\bar{k}$, from (8.10). These can be much smaller than $(8.13)_1$ if

$$
4(k_{22}k_{33}-k_{23}^2)\ll(k_{22}+k_{33})^2,
\tag{8.21}
$$

$$
4(\bar{k}_{22}\bar{k}_{33}-\bar{k}_{23}^2)\ll(\bar{k}_{22}+\bar{k}_{33})^2,
\tag{8.22}
$$

which appear to be possible.

Regarding the optical branches in the longitudinal and transverse systems, the former have long wave-length cut-off frequencies ω_s and ω_d while the corresponding quantities for the latter are ω_s and ω_r. Thus, the two systems have one cut-off frequency in common. One of the two modes is shear and the other is equivoluminal extension. As in the case of plates, the group velocities of all four optical branches are zero at the long wave-length cut-off frequencies except in the unlikely circumstance of coincidence of cut-off frequencies within a system ($b_1=0$ for longitudinal modes or $b_3=0$ for transverse modes). Another parallel to the situation in plates is that, as ξ increases from zero, the behavior of the optical branches is very sensitive to small changes in the ratios of material properties. One possibility is that both lower optical branches have phase and group velocities of opposite sign; *i.e.*, diminishing ω with increasing ξ. With further increase of ξ, the absolute values of the group velocities would pass through maxima, then drop to zero and then increase; *i.e.*, the dispersion curves first would have a point of inflexion and then a minimum.

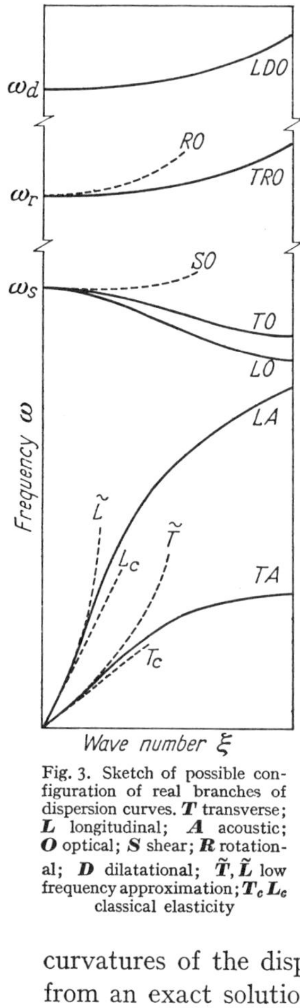

Fig. 3. Sketch of possible configuration of real branches of dispersion curves. ***T*** transverse; ***L*** longitudinal; ***A*** acoustic; ***O*** optical; ***S*** shear; ***R*** rotational; ***D*** dilatational; $\tilde{T}, \tilde{L}$ low frequency approximation; T_c L_c classical elasticity

A sketch of a possible configuration of the real segments of the dispersion curves is shown in Fig. 3. The four lowest branches (*TA*, *LA*, *LO*, *TO*) are remarkably similar to those obtained by BROCKHOUSE & IYENGAR [*15*, Fig. 5] from measurements of neutron scattering in germanium.

At the corresponding stage in the development of equations of high frequency vibrations of plates, it is expedient to introduce correction factors to compensate, as well as possible within the framework of the theory, for errors introduced by the restrictive assumption regarding the variation of displacement through the thickness of the plate. The analogous restriction, here, is the assumed homogeneous deformation of the unit cell. The values of the correction factors are obtained, in the theory of plates, by matching appropriate points, slopes and curvatures of the dispersion curves with the corresponding quantities obtained from an exact solution of the three-dimensional equations. Since the analogue of the latter does not exist, in the present case, such an adjustment cannot be made. An alternative is to use experimental data.

There is another aspect of the theory of plates that should be mentioned. It is possible for a thickness-mode with $n+m$ nodal planes to have a frequency lower than one with n nodal planes. For example, in an isotropic plate, the thickness-shear mode with two nodal planes has a frequency lower than that of the thickness-stretch mode with one nodal plane if Poisson's ratio is greater than one-third. Thus, a better approximation is obtained if a sufficient number of terms is retained, in the series expansion of the displacement, to accomodate this contingency [*11*]. The analogue, in the present case, is the possibility of the appearance of a micro-mode with frequency lower than that of the dilatational micro-mode (7.3) if additional terms are retained in the series expansion (1.6). However, because of the complications that would ensue, such a step does not appear to be warranted at this time.

9. Low frequency, very long wave-length approximation: Form I

This section and the following three are devoted to discussions and derivations of equations of motion simpler than (6.1) and (6.2) but limited, in application, to much lower frequencies and much longer wave-lengths.

As noted previously, when thickness-shear and thickness-stretch deformations and the associated inertias are taken into account in the theory of plates [*16*, *17*, *18*], thickness-modes of vibration, analogous to the micro-modes, are obtained as well as flexural and extensional modes analogous to the transverse and longitudinal acoustic modes. At low frequencies, in comparison with the frequencies of the thickness-modes, and at long wave-lengths, in comparison with the thickness of the plate, the coupling of the flexural and extensional modes with the thickness-modes is negligible. As the frequencies of the flexural and extensional modes approach zero, the thickness-shear deformation approaches zero but the thickness-stretch deformation does not; rather, it is the stress associated with thickness-stretch that approaches zero. Thus, the antisymmetric and symmetric parts of deformation and stress have to be treated differently in passing from high frequency equations to the classical, low frequency equations. To obtain equations valid at low frequencies in the case of flexure, the thickness-shear deformation is made to approach zero by passing to a limit as the associated modulus of elasticity approaches infinity. The product of the two is indeterminate and this leaves the thickness-shear stress indeterminate in the constitutive equations. In the case of extension, the thickness-stress is set equal to zero and the resulting constitutive equation is used to eliminate the thickness-strain from the remaining equations.

In both flexure and extension of homogeneous plates, the thickness velocities are set equal to zero in the kinetic energy, for the low frequency approximation, because their contributions are negligibly small at the low frequencies to which the resulting equations are restricted owing to the suppression of the thickness-shear deformation and the omission of the thickness-stretch stress [*18*, *19*]. The same is not true of non-homogeneous plates. For example, in a sandwich plate the rotatory inertia of the facings, about the middle plane of the plate, can be of paramount importance, even at low frequencies, for certain combinations of stiffness ratios, density ratio and distance between facings [*13*].

Now, the thickness velocities are analogous to the micro-velocities $\dot{\psi}_{ij}$; the thickness of the plate is analogous to dimensions $2d_i$ of the unit cell; the thickness-shear deformation is analogous to the antisymmetric part of the relative deformation $(\gamma_{[ij]})$; the thickness-shear moduli are analogous to the $b_{ij[kl]}$; and the stress associated with thickness-stretch is analogous to the symmetric part of the relative stress $(\sigma_{(ij)})$. The known process of descending from high frequency equations of plates to low frequency, long wave-length approximations can serve as a guide to the treatment of analogous terms in the equations of the elastic continuum with micro-structure. However, regardless of the process or of the theory of plates, the test of the validity of the resulting equations of motion is that they yield the same dispersion relations that are found in the limit as $\omega \to 0$, $\xi \to 0$ for the acoustic branches of the general equations. Thus, in the isotropic case, the values of ω_i', ω_i'' and ω_i''', in (8.13), must be reproduced exactly. Attention will be confined, here, to this case because it is much simpler than the anisotropic form and because (8.13) are available for the final test.

Inspection of $(8.13)_3$ shows that, as in the case of sandwich plates, it is not permissible to discard the micro-velocities $\dot{\psi}_{ij}$. Their effect is contained in the h_i^2 and, as remarked in the discussion following (8.19), h^2 can be less than or greater than $\tilde{l}^2$ depending on stiffness ratios, the density ratio and the dimensions of the unit cell. Hence, omission of the $\dot{\psi}_{ij}$ would preclude the reproduction of the low frequency behavior. The remainder of the process, however, can follow the analogy with homogeneous plates.

Thus, we let

$$\sigma_{(ij)} = 0, \tag{9.1}$$

$$b_2 - b_3 \to \infty, \qquad \gamma_{[ij]} \to 0, \tag{9.2}$$

and proceed to find the effect of these assumptions on the remaining terms.

The isotropic, constitutive equations for τ_{pq} and σ_{pq}, separated into symmetric and antisymmetric parts, are

$$\tau_{pq} = \lambda \delta_{pq} \varepsilon_{ii} + 2\mu \varepsilon_{pq} + g_1 \delta_{pq} \gamma_{ii} + 2 g_2 \gamma_{(pq)}, \tag{9.3}$$

$$\sigma_{(pq)} = g_1 \delta_{pq} \varepsilon_{ii} + 2 g_2 \varepsilon_{pq} + b_1 \delta_{pq} \gamma_{ii} + (b_2 + b_3) \gamma_{(pq)}, \tag{9.4}$$

$$\sigma_{[pq]} = (b_2 - b_3) \gamma_{[pq]}. \tag{9.5}$$

Then, with (9.2), $\sigma_{[pq]}$ is indeterminate in (9.5) and, with (9.1), (9.4) may be solved for $\gamma_{(pq)}$ in terms of ε_{pq}:

$$\gamma_{(pq)} = -\alpha \delta_{pq} \varepsilon_{ii} + (1-\beta) \varepsilon_{pq}, \tag{9.6}$$

where α and β are given in (8.18).

With regard to $\varkappa_{ijk}$, we note first that, since

$$\gamma_{pq} = \partial_p u_q - \psi_{pq}$$

and $\gamma_{[pq]}$ is now zero, we are left with

$$\psi_{[pq]} = \omega_{pq}, \qquad \psi_{(pq)} = \varepsilon_{pq} - \gamma_{(pq)}; \tag{9.7}$$

or, using the expression (9.6) for $\gamma_{(pq)}$, we have

$$\psi_{(pq)} = \alpha \delta_{pq} \varepsilon_{ii} + \beta \varepsilon_{pq}. \tag{9.8}$$

Accordingly, $\varkappa_{ijk} \equiv \partial_i \psi_{jk} = \partial_i \psi_{(jk)} + \partial_i \psi_{[jk]}$ reduces to

$$\varkappa_{ijk} \to \alpha \tilde{\varkappa}_{ill} \delta_{jk} + \tfrac{1}{2}(1+\beta) \tilde{\varkappa}_{ijk} - \tfrac{1}{2}(1-\beta) \tilde{\varkappa}_{ikj}, \tag{9.9}$$

where

$$\tilde{\varkappa}_{ijk} \equiv \partial_i \partial_j u_k = \tilde{\varkappa}_{jik}. \tag{9.10}$$

Thus, the part of the potential energy-density that is a function of $\varkappa_{ijk}$ becomes a function of $\tilde{\varkappa}_{ijk}$: the second gradient of the displacement. The eighteen components of $\tilde{\varkappa}_{ijk}$ may be resolved, in more than one way, into tensors whose components are independent linear combinations of the $\partial_i \partial_j u_k$; so that other forms of the energy-density, for the low frequency approximation, are possible. These are treated in Sections 11 and 12.

Upon inserting (9.9), (9.6) and (9.2) in (5.5), we find Form I of the low frequency approximation for the potential energy-density:

$$W \to \tilde{W} = \tfrac{1}{2}\tilde{\lambda}\,\varepsilon_{ii}\varepsilon_{jj} + \tilde{\mu}\,\varepsilon_{ij}\varepsilon_{ij} + \tilde{a}_1\tilde{\varkappa}_{iik}\tilde{\varkappa}_{kjj} + \tilde{a}_2\tilde{\varkappa}_{ijj}\tilde{\varkappa}_{ikk} + \\ + \tilde{a}_3\tilde{\varkappa}_{iik}\tilde{\varkappa}_{jjk} + \tilde{a}_4\tilde{\varkappa}_{ijk}\tilde{\varkappa}_{ijk} + \tilde{a}_5\tilde{\varkappa}_{ijk}\tilde{\varkappa}_{kji}, \tag{9.11}$$

where $\tilde{\lambda}$ and $\tilde{\mu}$ are given in (8.15) and $\tilde{a}_1 \dots \tilde{a}_5$ in (8.19). The appearance of these coefficients is preliminary evidence of the validity of the process.

We define new stresses:

$$\tilde{\tau}_{ij} \equiv \frac{\partial \tilde{W}}{\partial \varepsilon_{ij}} = \tilde{\tau}_{ji}, \tag{9.12$_1$}$$

$$\tilde{\mu}_{ijk} \equiv \frac{\partial \tilde{W}}{\partial \tilde{\varkappa}_{ijk}} = \tilde{\mu}_{jik}. \tag{9.12$_2$}$$

Then

$$\tilde{\tau}_{pq} = \tilde{\lambda}\,\delta_{pq}\varepsilon_{ii} + 2\tilde{\mu}\,\varepsilon_{pq}, \tag{9.13$_1$}$$

$$\tilde{\mu}_{pqr} = \tfrac{1}{2}\tilde{a}_1(\tilde{\varkappa}_{iip}\delta_{qr} + 2\tilde{\varkappa}_{rii}\delta_{pq} + \tilde{\varkappa}_{iiq}\delta_{pr}) + \tilde{a}_2(\tilde{\varkappa}_{pii}\delta_{qr} + \tilde{\varkappa}_{qii}\delta_{pr}) + \\ + 2\tilde{a}_3\tilde{\varkappa}_{iir}\delta_{pq} + 2\tilde{a}_4\tilde{\varkappa}_{pqr} + \tilde{a}_5(\tilde{\varkappa}_{rqp} + \tilde{\varkappa}_{rpq}). \tag{9.13$_2$}$$

The variational equation of motion is now obtained from Hamilton's principle with independent variations δu_i alone since, by $(9.7)_1$ and (9.8), the ψ_{ij} are no longer independent of the u_i.

The variation of the potential energy-density is

$$\begin{aligned} \delta W &= \tilde{\tau}_{ij}\,\delta\varepsilon_{ij} + \tilde{\mu}_{ijk}\,\delta\tilde{\varkappa}_{ijk}, \\ &= \tilde{\tau}_{ij}\,\partial_i\delta u_j + \tilde{\mu}_{ijk}\,\partial_i\partial_j\delta u_k, \\ &= \partial_j[(\tilde{\tau}_{jk} - \partial_i\tilde{\mu}_{ijk})\,\delta u_k] - \partial_j(\tilde{\tau}_{jk} - \partial_i\tilde{\mu}_{ijk})\,\delta u_k + \partial_i(\tilde{\mu}_{ijk}\,\partial_j\delta u_k). \end{aligned} \tag{9.14}$$

Hence

$$\int_V \delta\tilde{W}\,dV = \int_S n_j(\tilde{\tau}_{jk} - \partial_i\tilde{\mu}_{ijk})\,\delta u_k\,dS - \\ - \int_V \partial_j(\tilde{\tau}_{jk} - \partial_i\tilde{\mu}_{ijk})\,\partial u_k\,dV + \int_S n_i\tilde{\mu}_{ijk}\,\partial_j\delta u_k\,dS. \tag{9.15}$$

Now, in the last integral of (9.15), the variation $\partial_j\delta u_k$ is not independent of δu_j on S: only its normal component $n_j\partial_j\delta u_k$ is independent. We separate the latter:

$$n_i\tilde{\mu}_{ijk}\,\partial_j\delta u_k = n_i\tilde{\mu}_{ijk}D_j\,\delta u_k + n_i\tilde{\mu}_{ijk}n_jD\,\delta u_k, \tag{9.16}$$

where

$$D_j \equiv (\delta_{jl} - n_jn_l)\,\partial_l, \qquad D \equiv n_l\partial_l. \tag{9.17}$$

The terms in (9.16) may be resolved, further, in more than one way. In this section we follow Toupin [*8*] and reserve an alternative resolution [*9*] for Section 12. Thus, for the first term on the right hand side of (9.16), which contains the non-independent variation $D_j\delta u_k$, we write

$$n_i\tilde{\mu}_{ijk}D_j\,\delta u_k = D_j(n_i\tilde{\mu}_{ijk}\,\delta u_k) - n_iD_j\tilde{\mu}_{ijk}\,\delta u_k - (D_jn_i)\,\tilde{\mu}_{ijk}\,\delta u_k. \tag{9.18}$$

The last two terms in (9.18) now contain the independent variation δu_k. For the preceding term, we note that, on the surface S,

$$D_j(n_i\tilde{\mu}_{ijk}\,\delta u_k) = (D_ln_l)\,n_jn_i\tilde{\mu}_{ijk}\,\delta u_k + n_qe_{qpm}\,\partial_p(e_{mlj}n_ln_i\tilde{\mu}_{ijk}\,\delta u_k). \tag{9.19}$$

By Stokes's theorem, the integral, over a smooth surface, of the last term in (9.19) vanishes. If the surface has an edge C, formed by the intersection of two portions, S_1 and S_2, of S, Stokes's theorem gives

$$\int_S n_q e_{qpm} \partial_p (e_{mlj} n_l n_i \tilde{\mu}_{ijk} \delta u_k) \, dS = \oint_C [n_i m_j \tilde{\mu}_{ijk}] \, \delta u_k \, ds, \qquad (9.20)$$

where $m_j = e_{mlj} s_m n_l$ and the s_m are the components of the unit vector tangent to C. The bold face brackets **[ ]** in (9.20) indicate that the enclosed quantity is the difference between the values on S_1 and S_2.

Finally, we note that, in the first surface integral in (9.15), we may write

$$n_j \partial_i \tilde{\mu}_{ijk} = n_j D_i \tilde{\mu}_{ijk} + n_i n_j D \tilde{\mu}_{ijk}. \qquad (9.21)$$

Then, assembling the results in (9.15)–(9.21), we find

$$\begin{aligned} \int_V \tilde{W} \, dV = & -\int_V \partial_j (\tilde{\tau}_{jk} - \partial_i \tilde{\mu}_{ijk}) \, \delta u_k \, dV + \\ & + \int_S [n_j \tilde{\tau}_{jk} - n_i n_j D \tilde{\mu}_{ijk} - 2 n_j D_i \tilde{\mu}_{ijk} + (n_i n_j D_l n_l - D_j n_i) \tilde{\mu}_{ijk}] \, \delta u_k \, dS + \\ & + \int_S n_i n_j \tilde{\mu}_{ijk} D \, \delta u_k \, dS + \oint_C [n_i m_j \tilde{\mu}_{ijk}] \, \delta u_k \, ds. \end{aligned} \qquad (9.22)$$

This form suggests, for the variation of work done by external forces,

$$\delta \mathscr{W}_1 = \int_V F_k \, \delta u_k \, dV + \int_S \tilde{P}_k \, \delta u_k \, dS + \int_S \tilde{R}_k D \, \delta u_k \, dS + \oint_C \tilde{E}_k \, \delta u_k \, ds. \qquad (9.23)$$

As for the kinetic energy, the micro-velocity $\dot{\psi}_{ij}$, in $(2.3)_1$, must be replaced by a linear function of macro-velocity gradients:

$$\dot{\psi}_{ij} \to h_{ijkl} \, \partial_k \dot{u}_l, \qquad (9.24)$$

where

$$h_{ijkl} \equiv \tfrac{1}{2} (\delta_{ik} \delta_{jl} - \delta_{il} \delta_{jk}) + \alpha \, \delta_{ij} \delta_{kl} + \tfrac{1}{2} \beta (\delta_{ik} \delta_{jl} + \delta_{il} \delta_{jk}), \qquad (9.25)$$

so as to satisfy $(9.7)_1$ and (9.8). Then the kinetic energy-density $(2.3)_1$ becomes

$$\begin{aligned} T & = \tfrac{1}{2} \varrho \dot{u}_j \dot{u}_j + \tfrac{1}{6} \varrho' \tilde{d}^2_{pkmn} \partial_m \dot{u}_n \partial_p \dot{u}_k, \\ & = \tfrac{1}{2} \varrho \dot{u}_j \dot{u}_j + \tfrac{1}{6} \partial_p [\varrho' \tilde{d}^2_{pkmn} (\partial_m \dot{u}_n) \dot{u}_k] - \tfrac{1}{6} \partial_p (\varrho' \tilde{d}^2_{pkmn} \partial_m \dot{u}_n) \dot{u}_k, \end{aligned} \qquad (9.26)$$

where

$$\begin{aligned} \tilde{d}^2_{pkmn} & \equiv d^2_{jl} h_{lqpk} h_{jqmn} = \tilde{d}^2_{mnpk} \\ & = \tfrac{1}{2} d^2 [\delta_{pm} \delta_{kn} - \delta_{pn} \delta_{km} + 2\alpha (3\alpha + 2\beta) \delta_{pk} \delta_{mn} + \beta^2 (\delta_{pm} \delta_{kn} + \delta_{pn} \delta_{km})]. \end{aligned} \qquad (9.27)$$

The total kinetic energy is

$$\begin{aligned} \mathscr{T} = \int_V T \, dV = & \int_V [\tfrac{1}{2} \varrho \dot{u}_j \dot{u}_j - \tfrac{1}{6} \partial_p (\varrho' \tilde{d}^2_{pkmn} \partial_m \dot{u}_n) \dot{u}_k] \, dV + \\ & + \int_S \tfrac{1}{6} \varrho' n_p \tilde{d}^2_{pkmn} (\partial_m \dot{u}_n) \dot{u}_k \, dS; \end{aligned} \qquad (9.28)$$

from which

$$\begin{aligned} \delta \int_{t_0}^{t_1} \mathscr{T} \, dt = & -\int_{t_0}^{t_1} dt \int_V [\varrho \ddot{u}_k - \tfrac{1}{3} \partial_p (\varrho' d^2_{pkmn} \partial_m \ddot{u}_n)] \, \delta u_k \, dV - \\ & - \int_{t_0}^{t_1} dt \int_S \tfrac{1}{3} \varrho' n_p d^2_{pkmn} (D_m \ddot{u}_n + n_m D \ddot{u}_n) \, \delta u_k \, dS. \end{aligned} \qquad (9.29)$$

The variational equation of motion is formed from (9.22), (9.23) and (9.29), from which follow the stress-equations of motion and boundary conditions:

$$\partial_j(\tilde{\tau}_{jk} - \partial_i \tilde{\mu}_{ijk}) + F_k = \varrho \ddot{u}_k - \tfrac{1}{3}\partial_p(\varrho' \tilde{d}^2_{pkmn}\partial_m \ddot{u}_n), \qquad (9.30)_1$$

$$\begin{aligned} n_j\tilde{\tau}_{jk} - n_i n_j D\tilde{\mu}_{ijk} - 2n_j D_i\tilde{\mu}_{ijk} + (n_i n_j D_l n_l - D_j n_i)\tilde{\mu}_{ijk} + \\ + \tfrac{1}{3}\varrho' n_p \tilde{d}^2_{pkmn}(D_m\ddot{u}_n + n_m D\ddot{u}_n) = \tilde{P}_k, \end{aligned} \qquad (9.30)_2$$

$$n_i n_j \tilde{\mu}_{ijk} = \tilde{R}_k, \qquad (9.30)_3$$

$$[n_i m_j \tilde{\mu}_{ijk}] = \tilde{E}_k. \qquad (9.30)_4$$

The displacement-equations of motion are obtained by first replacing $2\varepsilon_{ij}$ with $\partial_i u_j + \partial_j u_i$ and $\tilde{\varkappa}_{ijk}$ with $\partial_i\partial_j u_k$, in (9.13), and then substituting the latter in $(9.30)_1$. The result is

$$\begin{aligned} (\tilde{\lambda} + 2\tilde{\mu})(1 - \tilde{l}_1^2 \nabla^2)\boldsymbol{\nabla\nabla}\cdot\boldsymbol{u} - \tilde{\mu}(1 - \tilde{l}_2^2\nabla^2)\boldsymbol{\nabla}\times\boldsymbol{\nabla}\times\boldsymbol{u} + \boldsymbol{F} \\ = \varrho(\ddot{\boldsymbol{u}} - h_1^2\boldsymbol{\nabla\nabla}\cdot\ddot{\boldsymbol{u}} + h_2^2\boldsymbol{\nabla}\times\boldsymbol{\nabla}\times\ddot{\boldsymbol{u}}), \end{aligned} \qquad (9.31)$$

where the $\tilde{l}_i^2$ and h_i^2 are defined in (8.16) and (8.17).

Omitting the body force and taking the divergence and curl of (9.31), we find the equations governing the propagation of dilatation and rotation:

$$\tilde{v}_1^2(1 - \tilde{l}_1^2\nabla^2)\nabla^2\boldsymbol{\nabla}\cdot\boldsymbol{u} = (1 - h_1^2\nabla^2)\boldsymbol{\nabla}\cdot\ddot{\boldsymbol{u}}, \qquad (9.32)_1$$

$$\tilde{v}_2^2(1 - \tilde{l}_2^2\nabla^2)\nabla^2\boldsymbol{\nabla}\times\boldsymbol{u} = (1 - h_2^2\nabla^2)\boldsymbol{\nabla}\times\ddot{\boldsymbol{u}}, \qquad (9.32)_2$$

where the $\tilde{v}_i^2$ are defined in (8.14). For the plane waves

$$(\boldsymbol{\nabla}\cdot\boldsymbol{u}, \nabla\times\boldsymbol{u}) = (A, \boldsymbol{A})\exp[i(\xi\,\boldsymbol{n}\cdot\boldsymbol{r} - \omega t)], \qquad (9.33)$$

the dispersion relations are

$$\omega_i^2 = \tilde{v}_i^2\xi^2(1 + \tilde{l}_i^2\xi^2)/(1 + h_i^2\xi^2), \qquad (9.34)$$

from which follow exactly the properties (8.13). Thus the validity of the approximate equations for low frequencies and very long wave lengths is established. The dispersion relations (9.34) are illustrated in Fig. 3 by the dashed curves labelled $\tilde{L}$ and $\tilde{T}$.

10. Relation to Toupin's generalization of couple-stress theory

The theory of elasticity with couple-stresses, which is considered in [*5*—*10*], is based on the same kinematics as is classical elasticity; but the potential energy-density is assumed to be a function of the strain and the curl of the strain instead of the strain alone. In the linear theory, the components of the curl of the strain are the same as the components of the gradient of the rotation: eight independent linear combinations of the eighteen components of the second gradient of the displacement. For the equilibrium case, Toupin [*8*, Section *7*] has generalized the theory to include all eighteen components. If the inertia terms are omitted, (9.30) are identical, in form, with Toupin's Eqs. (7.8)—(7.11)*.

* Note that, by definition, Toupin's β^{pqr} is symmetric in the second and third indices, whereas $\tilde{\mu}_{ijk}$ is symmetric in the first two indices. Note, also, that b should be replaced by $-b$ in Toupin's Eqs. *A*, *B*, (7.9) and (7.19).

However, although the form is the same, there are some significant differences. Equations (9.30) pertain to a low frequency, very long wave-length approximation to the equations of a material with micro-structure and the effect of the micro-structure survives, in both the potential and kinetic energy-densities, through the contribution of the symmetric part of the relative deformation. This part, $\gamma_{(ij)}$, can, in fact be traced to $\tilde{\tau}_{ij}$ and $\tilde{\mu}_{ijk}$, in (9.30), through the coefficients α and β in $\widetilde{W}$. Similarly the contribution of $\gamma_{(ij)}$ to the acceleration terms in (9.30) can be traced through the coefficients α and β in h_{ijkl} and then in $\tilde{d}^2_{ijkl}$. On the other hand, TOUPIN's equations do not stem from considerations of microstructure.

The equations of the material with micro-structure can be reduced to those of a material without micro-structure (*i.e.*, to a *micro-homogeneous* material) by causing the micro-medium to merge with the macro-medium. This may be accomplished (just as readily, in this case, for the anisotropic as for the isotropic medium) by passing to the limit as

$$b_{ijkl} \to \infty, \qquad \gamma_{ij} \to 0, \tag{10.1}$$

instead of (9.1) and (9.2); and, at the same time, $\varrho \to \varrho'$, so as to remove the distinction between micro- and macro-densities $(i.e., \varrho_M \to 0$ in $(2.3)_2)$. Then

$$\psi_{ij} \to \partial_i u_j, \qquad \varkappa_{ijk} \to \partial_i \partial_j u_k \equiv \tilde{\varkappa}_{ijk}, \tag{10.2}$$

instead of $(9.7)_1$, (9.8) and (9.9). Accordingly, from (5.1),

$$W \to \widetilde{W}^0 = \tfrac{1}{2} \mathring{c}_{ijkl}\, \varepsilon_{ij}\, \varepsilon_{kl} + \tfrac{1}{2} \mathring{a}_{ijklmn} \tilde{\varkappa}_{ijk} \tilde{\varkappa}_{lmn} + \mathring{f}_{ijklm} \tilde{\varkappa}_{ijk}\, \varepsilon_{lm} \tag{10.3}$$

and we define new stresses

$$\tilde{\tau}^0_{ij} \equiv \frac{\partial \widetilde{W}^0}{\partial \varepsilon_{ij}} = \tilde{\tau}^0_{ji}, \tag{10.4$_1$}$$

$$\tilde{\mu}^0_{ijk} \equiv \frac{\partial \widetilde{W}^0}{\partial \tilde{\varkappa}_{ijk}} = \tilde{\mu}^0_{jik}. \tag{10.4$_2$}$$

Also, in the kinetic energy-density (9.26),

$$\tilde{d}^2_{pkmn} \to \tilde{d}^2_{pm}\, \delta_{kn}, \tag{10.5}$$

where the d^2_{pm} are again given by $(2.3)_3$.

In the isotropic case $d^2_{pm} = d^2 \delta_{pm}$ and $\widetilde{W}^0, \tilde{\tau}^0_{ij}, \tilde{\mu}^0_{ijk}$ have the same form as $\widetilde{W}, \tilde{\tau}_{ij}, \tilde{\mu}_{ijk}$ but the coefficients are $\lambda, \mu, \mathring{a}_1 \ldots \mathring{a}_5$ instead of $\tilde{\lambda}, \tilde{\mu}, \tilde{a}_1 \ldots \tilde{a}_5$.

The formulation of the variational equation of motion proceeds as before and we arrive at the stress-equations of motion and boundary conditions:

$$\partial_j (\tilde{\tau}^0_{jk} - \partial_i \tilde{\mu}^0_{ijk}) + F_k = \varrho' \ddot{u}_k - \tfrac{1}{3} \partial_p (\varrho' d^2_{pm}\, \partial_m \ddot{u}_k), \tag{10.6$_1$}$$

$$\begin{aligned} n_j \tilde{\tau}^0_{jk} - n_i n_j D \tilde{\mu}^0_{ijk} - 2 n_j D_i \tilde{\mu}^0_{ijk} + (n_i n_j D_l n_l - D_j n_i) \tilde{\mu}^0_{ijk} + \\ + \tfrac{1}{3} \varrho' n_p d^2_{pm} (D_m \ddot{u}_k + n_m D \ddot{u}_k) = \widetilde{P}^0_k, \end{aligned} \tag{10.6$_2$}$$

$$n_i n_j \tilde{\mu}^0_{ijk} = \widetilde{R}^0_k, \tag{10.6$_3$}$$

$$[n_i m_j \tilde{\mu}^0_{ijk}] = \widetilde{E}^0_k. \tag{10.6$_4$}$$

Without the acceleration terms, (10.6) are now precisely the linear form of TOUPIN's equations; and (9.30), without the acceleration terms, differ only in that the coefficients in $\widetilde{W}$ and $\widetilde{W}^0$ (and hence in the stresses) have different meanings. However, with the acceleration terms included, the *forms* are actually different: the fourth rank tensor $\tilde{d}^2_{pkmn}$ in (9.30) is replaced by the second rank tensor d^2_{pm} in (10.6) so that there are fewer coefficients in the latter.

In the isotropic case, it is only necessary to let

$$\tilde{\lambda}\to\lambda, \quad \tilde{\mu}\to\mu, \quad \alpha\to 0, \quad \beta\to 1 \quad (\text{or } g_1\to 0,\ g_2\to 0) \tag{10.7}$$

in the equations of Section 9, to reach the equations of the micro-homogeneous medium. Thus, (9.31) reduces to

$$(\lambda+2\mu)(1-\mathring{l}_1^2\nabla^2)\nabla\nabla\cdot\boldsymbol{u}-\mu(1-\mathring{l}_2^2\nabla^2)\nabla\times\nabla\times\boldsymbol{u}+\boldsymbol{F}=\varrho'(1-\tfrac{1}{3}d^2\nabla^2)\ddot{\boldsymbol{u}}, \tag{10.8}$$

where $\mathring{l}_1^2$ and $\mathring{l}_2^2$ are obtained from $\tilde{l}_1^2$ and $\tilde{l}_2^2$, in (8.16), by employing (10.7). It will be observed that the left hand sides of (9.31) and (10.8) have the same form, but the right hand sides have different forms because, with (10.7), $h_1^2=h_2^2=\frac{1}{3}d^2$. The dispersion relations for plane waves, from (10.8) are

$$\omega_i^2=v_i^2\xi^2(1+\mathring{l}_i^2\xi^2)/(1+\tfrac{1}{3}d^2\xi^2), \tag{10.9}$$

where $v_1^2=(\lambda+2\mu)/\varrho'$, $v_2^2=\mu/\varrho'$; and the properties at $\omega=0$, $\xi=0$ are

$$\omega_i'=v_i, \quad \omega_i''=0, \quad \omega_i'''=3v_i(\mathring{l}_i^2-\tfrac{1}{3}d^2) \tag{10.10}$$

instead of (8.13).

The difference between the equations of the micro-homogeneous medium and the equations of the low frequency, very long wave-length approximation is similar to the difference between plane strain and plane stress; or, more appropriately in the present context, to the difference between equations of low frequency extensional vibrations of plates with the thickness of the plate constrained and not constrained to remain constant. In the case of equilibrium, the difference is solely in the physical interpretation of the elastic stiffnesses. With stiffnesses determined by experiments falling within the restrictions of the equations, the two equilibrium theories would be indistinguishable. For example, the numerical quantity that would be assigned to the stiffness $\tilde{\mu}$, in one theory, would be assigned to μ in the other. In the case of motion, however, there is an essential difference as the number of coefficients is not the same in the two theories unless $g_1=g_2=0$, which is analogous to zero Poisson's ratio.

11. Low frequency, very long wave-length approximation: Form II

As noted before, the eighteen components $\partial_i\partial_j u_k$ may be arranged in independent linear combinations which form tensors. One such, indicated by Toupin [*8*, p. 404], is the gradient of the strain:

$$\hat{\varkappa}_{ijk}\equiv\partial_i\varepsilon_{jk}=\tfrac{1}{2}(\partial_i\partial_j u_k+\partial_i\partial_k u_j)=\hat{\varkappa}_{ikj}. \tag{11.1}$$

The potential energy-density, for the low frequency, very long wave-length approximation, may be expressed as a function of ε_{ij} and $\hat{\varkappa}_{ijk}$ by setting

$$\tilde{\varkappa}_{ijk}=\hat{\varkappa}_{ijk}+\hat{\varkappa}_{jki}-\hat{\varkappa}_{kij} \tag{11.2}$$

in (9.11), with the result:

$$\begin{aligned}W\to\hat{W}=\tfrac{1}{2}\tilde{\lambda}\,\varepsilon_{ii}\varepsilon_{jj}+\tilde{\mu}\,\varepsilon_{ij}\varepsilon_{ij}+\hat{a}_1\hat{\varkappa}_{iik}\hat{\varkappa}_{kjj}+\hat{a}_2\hat{\varkappa}_{ijj}\hat{\varkappa}_{ikk}+\hat{a}_3\hat{\varkappa}_{iik}\hat{\varkappa}_{jjk}+\\+\hat{a}_4\hat{\varkappa}_{ijk}\hat{\varkappa}_{ijk}+\hat{a}_5\hat{\varkappa}_{ijk}\hat{\varkappa}_{kji},\end{aligned} \tag{11.3}$$

where

$$\hat{a}_1 = 2\tilde{a}_1 - 4\tilde{a}_3, \qquad \hat{a}_2 = -\tilde{a}_1 + \tilde{a}_2 + \tilde{a}_3,$$
$$\hat{a}_3 = 4\tilde{a}_3, \qquad \hat{a}_4 = 3\tilde{a}_4 - \tilde{a}_5, \qquad \hat{a}_5 = -2\tilde{a}_4 + 2\tilde{a}_5. \tag{11.4}$$

New stresses are defined by

$$\hat{\tau}_{ij} = \frac{\partial \hat{W}}{\partial \varepsilon_{ij}} = \hat{\tau}_{ji}, \tag{11.5$_1$}$$

$$\hat{\mu}_{ijk} = \frac{\partial \hat{W}}{\partial \hat{\varkappa}_{ijk}} = \hat{\mu}_{ikj}, \tag{11.5$_2$}$$

whence

$$\hat{\tau}_{pq} = \tilde{\lambda}\,\delta_{pq}\,\varepsilon_{ii} + 2\tilde{\mu}\,\varepsilon_{pq}, \tag{11.6$_1$}$$

$$\hat{\mu}_{pqr} = \tfrac{1}{2}\hat{a}_1(\delta_{pq}\hat{\varkappa}_{rii} + 2\delta_{qr}\hat{\varkappa}_{iip} + \delta_{rp}\hat{\varkappa}_{qii}) + 2\hat{a}_2\,\delta_{qr}\hat{\varkappa}_{pii} +$$
$$+ \hat{a}_3(\delta_{pq}\hat{\varkappa}_{iir} + \delta_{pr}\hat{\varkappa}_{iiq}) + 2\hat{a}_4\hat{\varkappa}_{pqr} + \hat{a}_5(\hat{\varkappa}_{rpq} + \hat{\varkappa}_{qrp}). \tag{11.6$_2$}$$

The variation of the potential energy-density is

$$\delta\hat{W} = \hat{\tau}_{ij}\,\delta\varepsilon_{ij} + \hat{\mu}_{ijk}\,\delta\hat{\varkappa}_{ijk} = \hat{\tau}_{ij}\,\partial_i\,\delta u_j + \hat{\mu}_{ijk}\,\partial_i\,\partial_j\,\delta u_k$$
$$= \partial_j[(\hat{\tau}_{jk} - \partial_i\hat{\mu}_{ijk})\,\delta u_k] - \partial_j(\hat{\tau}_{jk} - \partial_i\hat{\mu}_{ijk})\,\delta u_k + \partial_i(\hat{\mu}_{ijk}\,\partial_j\,\delta u_k). \tag{11.7}$$

Now, $(11.7)_3$ has the same form as $(9.14)_3$ and the kinetic energy density (9.26) is unchanged. Hence, the variational equation of motion has the same form as in Section 9 and leads to boundary conditions like those in (9.30), but with $2n_j D_i$ replaced by $(n_j D_i + n_i D_j)$, and stress-equations of motion:

$$\partial_j(\hat{\tau}_{jk} - \partial_i\,\hat{\mu}_{ijk}) + F_k = \varrho\,\ddot{u}_k - \tfrac{1}{3}\partial_p(\varrho'\,\tilde{d}^2_{pkmn}\,\partial_m\ddot{u}_n). \tag{11.8}$$

Recalling that $\tilde{\mu}_{ijk} = \tilde{\mu}_{jik}$ whereas $\hat{\mu}_{ijk} = \hat{\mu}_{ikj}$, the quantity in parentheses on the left hand side of $(9.30)_1$ is not symmetric but the corresponding quantity in (11.8) is symmetric. The latter is a more convenient form for the introduction of a stress function of the Airy type.

To get the displacement-equation of motion, substitute $(11.1)_2$ and (1.10) in (11.6) and the latter in (11.8). The result is

$$(\tilde{\lambda} + 2\tilde{\mu})(1 - \hat{l}_1^2\nabla^2)\,\boldsymbol{\nabla}\boldsymbol{\nabla}\cdot\boldsymbol{u} - \tilde{\mu}(1 - \hat{l}_2^2\nabla^2)\,\boldsymbol{\nabla}\times\boldsymbol{\nabla}\times\boldsymbol{u} + \boldsymbol{F}$$
$$= \varrho\,(\ddot{\boldsymbol{u}} - h_1^2\boldsymbol{\nabla}\boldsymbol{\nabla}\cdot\ddot{\boldsymbol{u}} + h_2^2\boldsymbol{\nabla}\times\boldsymbol{\nabla}\times\ddot{\boldsymbol{u}}), \tag{11.9}$$

where

$$\hat{l}_1^2 = 2(\hat{a}_1 + \hat{a}_2 + \hat{a}_3 + \hat{a}_4 + \hat{a}_5)/(\tilde{\lambda} + 2\tilde{\mu}), \qquad \hat{l}_2^2 = (\hat{a}_3 + 2\hat{a}_4 + \hat{a}_5)/2\tilde{\mu}. \tag{11.10}$$

In view of (11.4), $\hat{l}_i^2 = \tilde{l}_i^2$, so that the displacement-equations of motion (11.9) and (9.31) are identical.

12. Low frequency, very long wave-length approximation: Form III

For some purposes it is advantageous to separate the curl of the strain (or the gradient of the rotation):

$$\bar{\varkappa}_{ij} = e_{jlm}\,\partial_l\,\varepsilon_{mi} = \tfrac{1}{2}e_{jlm}\,\partial_i\,\partial_l\,u_m, \tag{12.1}$$

from $\partial_i\partial_j u_k$ as $\bar{\varkappa}_{ij}$ is the part of $\partial_i\partial_j u_k$ that gives rise to couple-stresses. The double stress is separated, thereby, into a non-self-equilibrating part and a self-equilibrating part. Now, $\bar{\varkappa}_{ii} = 0$. Hence $\bar{\varkappa}_{ij}$ has only eight independent components. (They are the components of the dyadic $\boldsymbol{\varkappa}$ in Reference [*9*].) The remaining ten linear combinations of the $\partial_i \partial_j u_k$ were considered separately by JARAMILLO [*20*]. They can be expressed as

$$\bar{\bar{\varkappa}}_{ijk} = \hat{\varkappa}_{ijk} + \tfrac{1}{3} e_{ilj} \bar{\varkappa}_{kl} + \tfrac{1}{3} e_{ilk} \bar{\varkappa}_{jl} = \tfrac{1}{3} (\partial_i \partial_j u_k + \partial_k \partial_i u_j + \partial_j \partial_k u_i). \tag{12.2}$$

Thus, $\bar{\bar{\varkappa}}_{ijk} = \bar{\bar{\varkappa}}_{kij} = \bar{\bar{\varkappa}}_{jki} = \bar{\bar{\varkappa}}_{jik}$; i.e., $\bar{\bar{\varkappa}}_{ijk}$ is fully symmetric.

The potential energy-density, for the low frequency, very long wave-length approximation, may be expressed as a function of ε_{ij}, $\bar{\varkappa}_{ij}$ and $\bar{\bar{\varkappa}}_{ijk}$ by setting

$$\hat{\varkappa}_{ijk} = \bar{\bar{\varkappa}}_{ijk} - \tfrac{1}{3} e_{ilj} \bar{\varkappa}_{kl} - \tfrac{1}{3} e_{ilk} \bar{\varkappa}_{jl} \tag{12.3}$$

in (11.3). The result is

$$\begin{aligned} W \to \bar{W} = {} & \tfrac{1}{2} \tilde{\lambda} \varepsilon_{ii} \varepsilon_{jj} + \tilde{\mu}\, \varepsilon_{ij} \varepsilon_{ij} + 2 \bar{d}_1 \bar{\varkappa}_{ij} \bar{\varkappa}_{ij} + 2 \bar{d}_2 \bar{\varkappa}_{ij} \bar{\varkappa}_{ji} + \\ & + \tfrac{3}{2} \bar{a}_1 \bar{\bar{\varkappa}}_{iij} \bar{\bar{\varkappa}}_{kkj} + \bar{a}_2 \bar{\bar{\varkappa}}_{ijk} \bar{\bar{\varkappa}}_{ijk} + \bar{f} e_{ijk} \bar{\varkappa}_{ij} \bar{\bar{\varkappa}}_{kll}, \end{aligned} \tag{12.4}$$

where

$$\begin{aligned} 18 \bar{d}_1 &= -2 \hat{a}_1 + 4 \hat{a}_2 + \hat{a}_3 + 6 \hat{a}_4 - 3 \hat{a}_5, \\ 18 \bar{d}_2 &= 2 \hat{a}_1 - 4 \hat{a}_2 - \hat{a}_3, \qquad 3 \bar{a}_1 = 2 (\hat{a}_1 + \hat{a}_2 + \hat{a}_3), \\ \bar{a}_2 &= \hat{a}_4 + \hat{a}_5, \qquad 3 \bar{f} = \hat{a}_1 + 4 \hat{a}_2 - 2 \hat{a}_3. \end{aligned} \tag{12.5}$$

The definitions

$$\bar{\tau}_{ij} \equiv \frac{\partial \bar{W}}{\partial \varepsilon_{ij}} = \bar{\tau}_{ji}, \tag{12.6$_1$}$$

$$\bar{\mu}_{ij} \equiv \frac{\partial \bar{W}}{\partial \bar{\varkappa}_{ij}}, \qquad \bar{\mu}_{ii} = 0, \tag{12.6$_2$}$$

$$\bar{\bar{\mu}}_{ijk} \equiv \frac{\partial \bar{W}}{\partial \bar{\bar{\varkappa}}_{ijk}} = \bar{\bar{\mu}}_{kij} = \bar{\bar{\mu}}_{jki} = \bar{\bar{\mu}}_{jik}, \tag{12.6$_3$}$$

where $\bar{\mu}_{ij}$ is the couple-stress deviator, lead to

$$\bar{\tau}_{pq} = \tilde{\lambda} \delta_{pq} \varepsilon_{ii} + 2 \tilde{\mu} \varepsilon_{pq}, \tag{12.7$_1$}$$

$$\bar{\mu}_{pq} = 4 \bar{d}_1 \bar{\varkappa}_{pq} + 4 \bar{d}_2 \bar{\varkappa}_{qp} + \bar{f} e_{pqi} \bar{\bar{\varkappa}}_{ijj}, \tag{12.7$_2$}$$

$$\begin{aligned} \bar{\bar{\mu}}_{pqr} = {} & \bar{a}_1 (\bar{\bar{\varkappa}}_{iir} \delta_{pq} + \bar{\bar{\varkappa}}_{iip} \delta_{qr} + \bar{\bar{\varkappa}}_{iiq} \delta_{rp}) + 2 \bar{a}_2 \bar{\bar{\varkappa}}_{pqr} + \\ & + \tfrac{1}{3} \bar{f} \bar{\varkappa}_{ij} (\delta_{pq} e_{ijr} + \delta_{qr} e_{ijp} + \delta_{rp} e_{ijq}). \end{aligned} \tag{12.7$_3$}$$

The variation of the potential energy-density now takes the form

$$\begin{aligned} \delta \bar{W} &= \bar{\tau}_{ij} \delta \varepsilon_{ij} + \bar{\mu}_{ij} \delta \bar{\varkappa}_{ij} + \bar{\bar{\mu}}_{ijk} \delta \bar{\bar{\varkappa}}_{ijk}, \\ &= \partial_j [(\bar{\tau}_{jk} - \partial_i \bar{\mu}^*_{ijk}) \delta u_k] - \partial_j (\bar{\tau}_{ij} - \partial_i \bar{\mu}^*_{ijk}) \delta u_k + \partial_i (\bar{\mu}^*_{ijk} \partial u_k), \end{aligned} \tag{12.8}$$

where

$$\bar{\mu}^*_{ijk} \equiv \tfrac{1}{2} e_{jkl} \bar{\mu}_{il} + \bar{\bar{\mu}}_{ijk}. \tag{12.9}$$

Again, (12.8) has the same form as (9.14) and so we can find stress-equations of motion and boundary conditions of the same form as (9.30), but with $\tilde{\mu}_{ijk}$ replaced by $\bar{\mu}^*_{ijk}$. Such a form of the boundary conditions cannot be compared directly with the results of Reference [*9*] because, there, one of the independent variations was taken to be the tangential component of rotation — which here is embedded in the normal derivative $D \delta u_k$. To get the alternative form of the boundary conditions, we can return to (9.16) and further resolve:

$$D\,\delta u_k = 2\,\delta w_i n_j e_{ijk} + D_k(n_i\,\delta u_i) - (D_k n_i)\,\delta u_i + n_k\,\delta\varepsilon_{nn}, \qquad (12.10)$$

where $w_i (\equiv \frac{1}{2} e_{ilm} \partial_l u_m)$ is the rotation and ε_{nn} (not summed) is the normal component of strain $n_i n_j \varepsilon_{ij}$. Then integration by parts and application of the divergence theorem and Stokes's theorem leads, in the notation of Reference [9], to

$$\begin{aligned}
\delta\overline{W} = &-\int_V (\boldsymbol{\nabla}\cdot\overline{\boldsymbol{\tau}} + \tfrac{1}{2}\boldsymbol{\nabla}\times\boldsymbol{\nabla}\cdot\overline{\boldsymbol{\mu}} - \boldsymbol{\nabla}\cdot\overline{\overline{\boldsymbol{\mu}}}\cdot\boldsymbol{\nabla})\cdot\delta\boldsymbol{u}\,dV + \\
&+\int_S \{\boldsymbol{n}\cdot\overline{\boldsymbol{\tau}} + \tfrac{1}{2}\boldsymbol{n}\times(\boldsymbol{\nabla}\cdot\overline{\boldsymbol{\mu}} - \boldsymbol{\nabla}\overline{\mu}_{nn}) - (\boldsymbol{\nabla}\cdot\overline{\overline{\boldsymbol{\mu}}})\cdot\boldsymbol{n} - \\
&-\boldsymbol{n}\cdot\boldsymbol{\nabla}\times[\boldsymbol{n}\times(\boldsymbol{n}\cdot\overline{\overline{\boldsymbol{\mu}}} + \boldsymbol{n}\cdot\overline{\overline{\boldsymbol{\mu}}}\cdot\boldsymbol{n}\boldsymbol{n})]\}\cdot\delta\boldsymbol{u}\,dS + \qquad (12.11)\\
&+\int_S [\boldsymbol{n}\cdot\overline{\boldsymbol{\mu}}\times\boldsymbol{n} + 2\boldsymbol{n}\times(\boldsymbol{n}\cdot\overline{\overline{\boldsymbol{\mu}}}\cdot\boldsymbol{n})\times\boldsymbol{n}]\cdot(\delta\boldsymbol{w}\times\boldsymbol{n})\,dS + \int_S \boldsymbol{n}\boldsymbol{n}:\overline{\overline{\boldsymbol{\mu}}}\cdot\boldsymbol{n}\,\delta\varepsilon_{nn}\,dS + \\
&+\oint_C [\tfrac{1}{2}\overline{\boldsymbol{\mu}}_{nn}\boldsymbol{s} + (\boldsymbol{s}\times\boldsymbol{n})\cdot(\boldsymbol{n}\cdot\overline{\overline{\boldsymbol{\mu}}} + \boldsymbol{n}\cdot\overline{\overline{\boldsymbol{\mu}}}\cdot\boldsymbol{n}\boldsymbol{n})]\cdot\delta\boldsymbol{u}\,ds,
\end{aligned}$$

where $\boldsymbol{s}$ is a unit vector tangent to the edge C.

The variation of work done by external forces is now taken to be

$$\delta\mathscr{W}_1 = \int_V \boldsymbol{F}\cdot\delta\boldsymbol{u}\,dV + \int_S (\boldsymbol{P}\cdot\delta\boldsymbol{u} + \boldsymbol{Q}\cdot\delta\boldsymbol{w}\times\boldsymbol{n} + R\,\delta\varepsilon_{nn})\,dS + \oint_C \boldsymbol{E}\cdot\delta\boldsymbol{u}\,ds. \qquad (12.12)$$

Here $\boldsymbol{Q}$ is the tangential component of the couple-stress vector and R is a double force per unit area, without moment, normal to S.

The variational equation of motion then yields the stress-equation of motion

$$\boldsymbol{\nabla}\cdot\overline{\boldsymbol{\tau}} + \tfrac{1}{2}\boldsymbol{\nabla}\times\boldsymbol{\nabla}\cdot\overline{\boldsymbol{\mu}} - \boldsymbol{\nabla}\cdot\overline{\overline{\boldsymbol{\mu}}}\cdot\boldsymbol{\nabla} + \boldsymbol{F} = \varrho\,\ddot{\boldsymbol{u}} - \tfrac{1}{3}\boldsymbol{\nabla}\cdot(\varrho'\,\tilde{\boldsymbol{d}}^2:\boldsymbol{\nabla}\ddot{\boldsymbol{u}}) \qquad (12.13)$$

and the boundary conditions

$$\begin{aligned}
&\boldsymbol{n}\cdot\overline{\boldsymbol{\tau}} + \tfrac{1}{2}\boldsymbol{n}\times(\boldsymbol{\nabla}\cdot\overline{\boldsymbol{\mu}} - \boldsymbol{\nabla}\overline{\mu}_{nn}) - \\
&\quad - (\boldsymbol{\nabla}\cdot\overline{\overline{\boldsymbol{\mu}}})\cdot\boldsymbol{n} - \boldsymbol{n}\cdot\boldsymbol{\nabla}\times[\boldsymbol{n}\times(\boldsymbol{n}\cdot\overline{\overline{\boldsymbol{\mu}}} + \boldsymbol{n}\cdot\overline{\overline{\boldsymbol{\mu}}}\cdot\boldsymbol{n}\boldsymbol{n})] + \tfrac{1}{3}\varrho'\,\boldsymbol{n}\cdot\tilde{\boldsymbol{d}}^2:\boldsymbol{\nabla}\ddot{\boldsymbol{u}} = \boldsymbol{P}, \qquad (12.14)_1\\
&\boldsymbol{n}\cdot\overline{\boldsymbol{\mu}}\times\boldsymbol{n} + 2\boldsymbol{n}\times(\boldsymbol{n}\cdot\overline{\overline{\boldsymbol{\mu}}}\cdot\boldsymbol{n})\times\boldsymbol{n} = \boldsymbol{Q}, \qquad (12.14)_2\\
&\boldsymbol{n}\boldsymbol{n}:\overline{\overline{\boldsymbol{\mu}}}\cdot\boldsymbol{n} = R, \qquad (12.14)_3\\
&[\tfrac{1}{2}\overline{\mu}_{nn}\boldsymbol{s} + (\boldsymbol{s}\times\boldsymbol{n})\cdot(\boldsymbol{n}\cdot\overline{\overline{\boldsymbol{\mu}}} + \boldsymbol{n}\cdot\overline{\overline{\boldsymbol{\mu}}}\cdot\boldsymbol{n}\boldsymbol{n})] = \boldsymbol{E}. \qquad (12.14)_4
\end{aligned}$$

The displacement equation of motion is obtained by substituting (1.10), $(12.1)_2$ and $(12.2)_2$ in (12.7) and the latter in (12.13). The result is

$$\begin{aligned}
(\tilde{\lambda} + 2\tilde{\mu})\,(1 - \bar{l}_1^2\,\nabla^2)\,\boldsymbol{\nabla}\boldsymbol{\nabla}\cdot\boldsymbol{u} - \tilde{\mu}\,(1 - \bar{l}_2^2\,\nabla^2)\,\boldsymbol{\nabla}\times\boldsymbol{\nabla}\times\boldsymbol{u} + \boldsymbol{F}&\\
= \varrho\,(\ddot{\boldsymbol{u}} - h_1^2\boldsymbol{\nabla}\boldsymbol{\nabla}\cdot\ddot{\boldsymbol{u}} + h_2^2\boldsymbol{\nabla}\times\boldsymbol{\nabla}\times\ddot{\boldsymbol{u}}),& \qquad (12.15)
\end{aligned}$$

where

$$\bar{l}_1^2 = (3\bar{a}_1 + 2\bar{a}_2)/(\tilde{\lambda} + 2\tilde{\mu}), \qquad \bar{l}_2^2 = (3\,\bar{d}_1 + \bar{a}_1 + 2\bar{a}_2 - \bar{f})/3\tilde{\mu}. \qquad (12.16)$$

In view of (12.5) and (11.4),

$$\bar{l}_i^2 = \hat{l}_i^2 = \tilde{l}_i^2 = l_i^2, \quad \text{say}; \qquad (12.17)$$

so that all three forms of the low frequency, very long wave-length approximation yield the same displacement-equation of motion. Necessary and sufficient conditions for positive definiteness of $\bar{W}(=\hat{W}=\tilde{W})$ are

$$\begin{aligned} &\tilde{\mu}>0, \quad 3\tilde{\lambda}+2\tilde{\mu}>0, \\ &\bar{d}_1>0, \quad -\bar{d}_1<\bar{d}_2<\bar{d}_1, \\ &\bar{a}_2>0, \quad 3\bar{a}_1+2\bar{a}_2>0, \quad \bar{f}<0. \end{aligned} \tag{12.18}$$

Hence $l_i^2>0$ and we have already seen that $h_i^2>0$.

The equations that were considered in Reference [9] are obtained from those of this section by setting $\bar{\bar{\varkappa}}_{ijk}=0$ in the potential energy-density and $\varrho'=0$ in the kinetic energy density. Accordingly, the limitation to low frequencies, long wave-lengths and large dimensions is more severe than was apparent previously.

13. Solution of the approximate equations of equilibrium

In this section it is proved that any solution, $\boldsymbol{u}$, of the equation

$$(\tilde{\lambda}+2\tilde{\mu})(1-l_1^2\nabla^2)\boldsymbol{\nabla}\boldsymbol{\nabla}\cdot\boldsymbol{u}-\tilde{\mu}(1-l_2^2\nabla^2)\boldsymbol{\nabla}\times\boldsymbol{\nabla}\times\boldsymbol{u}+\boldsymbol{F}=0, \tag{13.1}$$

in a region V bounded by a surface S, can be expressed as

$$\boldsymbol{u}=\boldsymbol{B}-l_2^2\boldsymbol{\nabla}\boldsymbol{\nabla}\cdot\boldsymbol{B}-\tfrac{1}{2}(k_1-l_1^2\nabla^2)\boldsymbol{\nabla}[\boldsymbol{r}\cdot(1-l_2^2\nabla^2)\boldsymbol{B}+B_0], \tag{13.2$_1$}$$

where

$$\tilde{\mu}(1-l_2^2\nabla^2)\nabla^2\boldsymbol{B}=-\boldsymbol{F}, \tag{13.2$_2$}$$

$$\tilde{\mu}(1-l_1^2\nabla^2)\nabla^2 B_0=\boldsymbol{r}\cdot(1-l_1^2\nabla^2)\boldsymbol{F}-4l_1^2\boldsymbol{\nabla}\cdot\boldsymbol{F}, \tag{13.2$_3$}$$

$$k_1=(\tilde{\lambda}+\tilde{\mu})/(\tilde{\lambda}+2\tilde{\mu})$$

and $\boldsymbol{r}$ is the position vector.

Consider a field point $P(x, y, z)$ and a source point $Q(\xi, \eta, \zeta)$ and define

$$4\pi\boldsymbol{U}_P\equiv-\int_V r_1^{-1}\boldsymbol{u}_Q\,dV_Q, \tag{13.3}$$

where

$$r_1^2=(x-\xi)^2+(y-\eta)^2+(z-\zeta)^2, \qquad dV_Q=d\xi\,d\eta\,d\zeta.$$

Then $\nabla^2\boldsymbol{U}=\boldsymbol{u}$ [3, p. 210], or

$$\boldsymbol{\nabla}\boldsymbol{\nabla}\cdot\boldsymbol{U}-\boldsymbol{\nabla}\times\boldsymbol{\nabla}\times\boldsymbol{U}=\boldsymbol{u}. \tag{13.4}$$

Define

$$\psi\equiv\boldsymbol{\nabla}\cdot\boldsymbol{U}, \qquad \boldsymbol{H}=-\boldsymbol{\nabla}\times\boldsymbol{U}, \qquad (\boldsymbol{\nabla}\cdot\boldsymbol{H}=0). \tag{13.5}$$

Then, from (13.4),

$$\boldsymbol{u}=\nabla\psi+\boldsymbol{\nabla}\times\boldsymbol{H}, \qquad \boldsymbol{\nabla}\cdot\boldsymbol{H}=0, \tag{13.6}$$

which is Helmholtz's resolution. Substituting (13.6) in $(13.2)_1$, we have

$$\tilde{\mu}\nabla^2[k(1-l_1^2\nabla^2)\boldsymbol{\nabla}\psi+(1-l_2^2\nabla^2)\boldsymbol{\nabla}\times\boldsymbol{H}]+\boldsymbol{F}=0, \tag{13.7}$$

where $k=(\tilde{\lambda}+2\tilde{\mu})/\tilde{\mu}$.

Define

$$4\pi l_2^2\boldsymbol{B}'_P\equiv\int_V r_1^{-1}e^{-r_1/l_2}[k(1-l_1^2\nabla^2)\boldsymbol{\nabla}\psi+(1-l_2^2\nabla^2)\boldsymbol{\nabla}\times\boldsymbol{H}]_Q\,dV_Q. \tag{13.8}$$

Then [3, p. 210]

$$(1 - l_2^2 \nabla^2) \boldsymbol{B}' = k(1 - l_1^2 \nabla^2) \boldsymbol{\nabla} \psi + (1 - l_2^2 \nabla^2) \boldsymbol{\nabla} \times \boldsymbol{H} \tag{13.9}$$

and, from (13.9) and (13.7),

$$\tilde{\mu}(1 - l_2^2 \nabla^2) \nabla^2 \boldsymbol{B}' = -\boldsymbol{F}. \tag{13.10}$$

Also, the divergence of (13.9) yields

$$(1 - l_2^2 \nabla^2) \boldsymbol{\nabla} \cdot \boldsymbol{B}' = k(1 - l_1^2 \nabla^2) \nabla^2 \psi. \tag{13.11}$$

Define

$$2k\psi^* \equiv \boldsymbol{r} \cdot (1 - l_2^2 \nabla^2) \boldsymbol{B}'. \tag{13.12}$$

Then, using (13.10), we have

$$2k\tilde{\mu}(1 - l_1^2 \nabla^2) \nabla^2 \psi^* = 4 l_1^2 \boldsymbol{\nabla} \cdot \boldsymbol{F} - \boldsymbol{r} \cdot (1 - l_1^2 \nabla^2) \boldsymbol{F} + 2\tilde{\mu}(1 - l_2^2 \nabla^2) \boldsymbol{\nabla} \cdot \boldsymbol{B}'. \tag{13.13}$$

Define

$$B_0 = 2k(\psi - \psi^*) \tag{13.14}$$

and find

$$\tilde{\mu}(1 - l_1^2 \nabla^2) \nabla^2 B_0 = \boldsymbol{r} \cdot (1 - l_1^2 \nabla^2) \boldsymbol{F} - 4 l_1^2 \boldsymbol{\nabla} \cdot \boldsymbol{F} \tag{13.15}$$

by using (13.11) and (13.13). Also, from (13.14) and (13.12),

$$2k\psi = \boldsymbol{r} \cdot (1 - l_2^2 \nabla^2) \boldsymbol{B}' + B_0. \tag{13.16}$$

Now, define

$$\boldsymbol{D} \equiv \boldsymbol{B}' - l_2^2 \boldsymbol{\nabla} \boldsymbol{\nabla} \cdot B' - k(1 - l_1^2 \nabla^2) \boldsymbol{\nabla} \psi. \tag{13.17}$$

By (13.11), $\boldsymbol{\nabla} \cdot \boldsymbol{D} = 0$; which is a necessary and sufficient condition for the existence of a function $\boldsymbol{H}^*$ such that $\boldsymbol{\nabla} \times \boldsymbol{H}^* = \boldsymbol{D}$, *i.e.*,

$$\boldsymbol{\nabla} \times \boldsymbol{H}^* = \boldsymbol{B}' - l_2^2 \boldsymbol{\nabla} \boldsymbol{\nabla} \cdot \boldsymbol{B}' - k(1 - l_1^2 \nabla^2) \boldsymbol{\nabla} \psi. \tag{13.18}$$

Making use of (13.9) and (13.11) we find

$$(1 - l_2^2 \nabla^2) \boldsymbol{\nabla} \times \boldsymbol{H}^* = (1 - l_2^2 \nabla^2) \boldsymbol{\nabla} \times \boldsymbol{H}. \tag{13.19}$$

Next, define

$$\boldsymbol{B}'' \equiv \boldsymbol{\nabla} \times \boldsymbol{H} - \boldsymbol{\nabla} \times \boldsymbol{H}^* \tag{13.20}$$

and note that, by (13.19) and (13.20),

$$(1 - l_2^2 \nabla^2) \boldsymbol{B}'' = 0, \qquad \boldsymbol{\nabla} \cdot \boldsymbol{B}'' = 0. \tag{13.21}$$

From (13.20), using (13.18) and then (13.16),

$$\nabla \times \boldsymbol{H} = \boldsymbol{B}'' + \boldsymbol{B}' - l_2^2 \boldsymbol{\nabla} \boldsymbol{\nabla} \cdot \boldsymbol{B}' - \tfrac{1}{2}(1 - l_1^2 \nabla^2) \boldsymbol{\nabla} [\boldsymbol{r} \cdot (1 - l_2^2 \nabla^2 \boldsymbol{B}') + B_0]. \tag{13.22}$$

Then substitute (13.22) and (13.16) in $(13.6)_1$ to get

$$\boldsymbol{u} = \boldsymbol{B}'' + \boldsymbol{B}' - l_2^2 \boldsymbol{\nabla} \boldsymbol{\nabla} \cdot (\boldsymbol{B}'' + \boldsymbol{B}') - \tfrac{1}{2}(k_1 - l_1^2 \nabla^2) \boldsymbol{\nabla} [\boldsymbol{r} \cdot (1 - l_2^2 \nabla^2) \boldsymbol{B}' + B_0]. \tag{13.23}$$

Finally, define

$$\boldsymbol{B} \equiv \boldsymbol{B}' + \boldsymbol{B}''. \tag{13.24}$$

In view of (13.21), we may write (13.23) and (13.10) in the form of $(13.2)_1$ and $(13.2)_2$; and (13.15) is already in the form $(13.2)_3$. Thus Eqs. (13.2) are a complete solution of (13.1). If $l_1^2 = 0$, (13.1) reduces to the equilibrium equation of couple-stress theory and (13.2) is the solution found in [9]. If both l_1^2 and l_2^2 are zero,

(13.1) is the classical equation of equilibrium with body force and (13.2) is the solution found in [21]. If, in addition, the body force is zero, (13.2) is PAPKOVITCH's solution [22]. The proof follows, generally, that in [23] but with an improvement as a result of an illuminating criticism by E. STERNBERG.

14. Concentrated force according to the approximate equations

In an infinite region V, let the body force be zero outside a finite region V_0 which contains the origin and a non-vanishing field of parallel forces $\boldsymbol{F}$. A concentrated force is defined by

$$\boldsymbol{P} = \lim_{V_0 \to 0} \int_V \boldsymbol{F}_Q \, dV_Q. \tag{14.1}$$

In [9] it was shown that, in an infinite region, solutions of equations of the type $(13.2)_2$ and $(13.2)_3$ are

$$4\pi \tilde{\mu} \boldsymbol{B} = \int_V r_1^{-1} (1 - e^{-r_1/l_2}) \boldsymbol{F}_Q \, dV_Q, \tag{14.2}$$

$$4\pi \tilde{\mu} B_0 = -\int_V r_1^{-1} (1 - e^{-r_1/l_1}) \left[\boldsymbol{r}' \cdot (1 - l_1^2 \nabla_Q^2) \boldsymbol{F}_Q - 4 l_1^2 \boldsymbol{\nabla}_Q \cdot \boldsymbol{F}_Q\right] dV_Q, \tag{14.3}$$

where $r' = \sqrt{\xi^2 + \eta^2 + \zeta^2}$. Now

$$\lim_{V_0 \to 0} r_1 = r, \qquad \lim_{V_0 \to 0} r' = 0. \tag{14.4}$$

Hence, for the concentrated force, (14.2) reduces to

$$4\pi \tilde{\mu} \boldsymbol{B} = r^{-1} (1 - e^{-r/l_2}) \boldsymbol{P}. \tag{14.5}$$

In (14.3) the term in the integrand of the form $\psi(r_1) \boldsymbol{\nabla}_Q \cdot \boldsymbol{F}_Q$ is transformed according to

$$\int_V \psi \boldsymbol{\nabla}_Q \cdot \boldsymbol{F}_Q \, dV_Q = \int_V \left[\boldsymbol{\nabla}_Q \cdot (\psi \boldsymbol{F}_Q) - \boldsymbol{F}_Q \cdot \boldsymbol{\nabla}_Q \psi\right] dV_Q = \int_S \psi \boldsymbol{n} \cdot \boldsymbol{F}_Q \, dS - \int_V \boldsymbol{F}_Q \cdot \boldsymbol{\nabla}_Q \psi \, dV_Q. \tag{14.6}$$

The surface integral in (14.6) vanishes because $\boldsymbol{F} = 0$ outside V_0. Also

$$\lim_{V_0 \to 0} \int_V \boldsymbol{F}_Q \cdot \boldsymbol{\nabla}_Q \psi(r_1) \, dV_Q = -\boldsymbol{P} \cdot \boldsymbol{\nabla} \psi(r). \tag{14.7}$$

Hence, for the concentrated force,

$$\pi \tilde{\mu} B_0 = l_1^2 \boldsymbol{P} \cdot \boldsymbol{\nabla} \left[r^{-1} (1 - e^{-r/l_1})\right]. \tag{14.8}$$

Equations (14.5) and (14.8) constitute the solution of (13.1) for the concentrated force. If $l_1^2 = 0$ then $B_0 = 0$ and the solution reduces to that found in [9]. If, in addition, $l_2^2 = 0$, the solution reduces to KELVIN's [4, p. 183].

Acknowledgement. I wish to thank Dr. R. A. TOUPIN for many valuable discussions.

References

[1] ERICKSEN, J. L., & C. TRUESDELL, Exact theory of stress and strain in rods and shells. Arch. Rational Mech. Anal. **1**, 295–323 (1958).

[2] COSSERAT, E. et F., Théorie des Corps Déformables. Paris: A. Hermann & Fils 1909. For the linear, two-dimensional theory, with applications, see H. SCHAEFER, Versuch einer Elastizitätstheorie des Zweidimensionalen ebenen Cosserat-Kontinuums. Miszellaneen der Angewandten Mechanik, pp. 277–292. Berlin: Akademie-Verlag 1962.

[3] JEFFREYS, H., & B. S, Methods of Mathematical Physics, Second Edition. Cambridge: University Press 1950.
[4] LOVE, A. E. H., A Treatise on the Mathematical Theory of Elasticity, Fourth Edition. Cambridge: University Press 1927.
[5] GRIOLI, G., Elasticità asimmetrica. Ann. di Mat. pura ed appl., Ser. IV **50**, 389—417 (1960).
[6] AERO, E. L., & E. V. KUVSHINSKII, Fundamental equations of the theory of elastic media with rotationally interacting particles. Fizika Tverdogo Tela **2**, 1399—1409 (1960); Translation: Soviet Physics Solid State **2**, 1272—1281 (1961).
[7] RAJAGOPAL, E. S., The existence of interfacial couples in infinitesimal elasticity. Ann. der Physik **6**, 192—201 (1960).
[8] TOUPIN, R. A., Elastic materials with couple-stresses. Arch. Rational Mech. Anal. **11**, 385—414 (1962).
[9] MINDLIN, R. D., & H. F. TIERSTEN, Effects of couple-stresses in linear elasticity. Arch. Rational Mech. Anal. **11**, 415—448 (1962).
[10] MINDLIN, R. D., Influence of couple-stresses on stress-concentrations. Experimental Mechanics **3**, 1—7 (1963).
[11] MINDLIN, R. D., & M. A. MEDICK, Extensional vibrations of elastic plates. J. Appl. Mech. **26**, 561—569 (1959).
[12] GAZIS, D. C., & R. F. WALLIS, Extensional waves in cubic crystal plates. Proc. 4th U.S.Nat. Cong. Appl. Mech., 161—168 (1962).
[13] MINDLIN, R. D., High frequency vibrations of plated, crystal plates. Progress in Applied Mechanics, pp. 73—84. New York: Macmillan Co. 1963.
[14] GAZIS, D. C., & R. F. WALLIS, Surface vibrational modes in crystal lattices with complex interatomic interactions. J. Math. Phys. **3**, 190—199 (1962).
[15] BROCKHOUSE, B. N., & P. K. IYENGAR, Normal modes of germanium by neutron spectroscopy. Phys. Rev. **111**, 747—754 (1958).
[16] MINDLIN, R. D., Thickness-shear and flexural vibrations of crystal plates. J. Appl. Phys. **22**, 316—323 (1951).
[17] MINDLIN, R. D., High frequency vibrations of crystal plates. Q. Appl. Math. **19**, 51—61 (1961).
[18] KANE, T. R., & R. D. MINDLIN, High frequency extensional vibrations of plates. J. Appl. Mech. **23**, 277—283 (1956).
[19] MINDLIN, R. D., Influence of rotatory inertia and shear on flexural vibrations of isotropic, elastic plates. J. Appl. Mech. **18**, 31—38 (1951).
[20] JARAMILLO, T. J., A generalization of the energy function of elasticity theory. Dissertation, Department of Mathematics, University of Chicago, 1929.
[21] MINDLIN, R. D., Force at a point in the interior of a semi-infinite solid. Proc. First Midwestern Conf. on Solid Mech., Urbana, Illinois, 56—59 (1953).
[22] PAPKOVITCH, P. F., The representation of the general integral of the fundamental equations of elasticity theory in terms of harmonic functions. Izv. Akad. Nauk SSSR, Phys.-Math. Ser. **10**, 1425 (1932).
[23] MINDLIN, R. D., Note on the Galerkin and Papkovitch stress functions. Bull. Amer. Math. Soc. **42**, 373—376 (1936).

Department of Civil Engineering
Columbia University
New York 27, New York

(Received November 6, 1963)

COMPLEX REPRESENTATION OF DISPLACEMENTS AND STRESSES IN PLANE STRAIN WITH COUPLE-STRESSES

R. D. MINDLIN

Columbia University, New York, N. Y., U.S.A.

In this paper, the method of Muskhelishvili [1] is employed to express the displacements and stresses, in plane elasticity with couple-stresses, in terms of functions of a complex variable and a Helmholtz function.

When couple-stresses are taken into account, the two-dimensional stress equations of equilibrium are [2]

$$
\begin{aligned}
&\frac{\partial X_x}{\partial x} + \frac{\partial X_y}{\partial y} = 0,\\
(1)\qquad &\frac{\partial Y_x}{\partial x} + \frac{\partial Y_y}{\partial y} = 0,\\
&\frac{\partial M_x}{\partial x} + \frac{\partial M_y}{\partial y} + Y_x - X_y = 0,
\end{aligned}
$$

where X_x, Y_y, Y_x, X_y are the components of the asymmetric Cosserat stress tensor and M_x, M_y are the components of the Cosserat couple-stress tensor in the plane of x and y.

The stress-displacement relations for a centro-symmetric, isotropic, elastic material, in a state of plane strain, are [3]:

$$
\begin{aligned}
&X_x = \lambda\left(\frac{\partial u}{\partial x} + \frac{\partial v}{\partial y}\right) + 2\mu\frac{\partial u}{\partial x},\\
&Y_y = \lambda\left(\frac{\partial u}{\partial x} + \frac{\partial v}{\partial y}\right) + 2\mu\frac{\partial v}{\partial y},\\
(2)\qquad &Y_x + X_y = 2\mu\left(\frac{\partial v}{\partial x} + \frac{\partial u}{\partial y}\right),\\
&M_x = 2\eta\frac{\partial}{\partial x}\left(\frac{\partial v}{\partial x} - \frac{\partial u}{\partial y}\right),\\
&M_y = 2\eta\frac{\partial}{\partial y}\left(\frac{\partial v}{\partial x} - \frac{\partial u}{\partial y}\right),
\end{aligned}
$$

where λ and μ are Lamé's constants and 4η is the modulus of flexure: the ratio of bending moment to curvature of an element.

It has been shown [4] that a general solution of Eqs. (1) and (2) can be expressed in terms of the Airy stress function U and another function V, proportional to the rotation, as follows:

$$X_x = \frac{\partial^2 U}{\partial y^2} - \frac{\partial^2 V}{\partial x \partial y}, \qquad Y_y = \frac{\partial^2 U}{\partial x^2} + \frac{\partial^2 V}{\partial x \partial y},$$

$$(3) \qquad Y_x = -\frac{\partial^2 U}{\partial x \partial y} - \frac{\partial^2 V}{\partial y^2}, \qquad X_y = -\frac{\partial^2 U}{\partial x \partial y} + \frac{\partial^2 V}{\partial x^2},$$

$$M_x = \frac{\partial V}{\partial x}, \qquad M_y = \frac{\partial V}{\partial y}.$$

The functions U and V must satisfy the equations

$$\nabla^4 U = 0, \qquad (1 - \mathfrak{l}^2\nabla^2)\nabla^2 V = 0,$$

$$(4) \qquad \frac{\partial}{\partial x}(V - \mathfrak{l}^2\nabla^2 V) = -2(1-\sigma)\mathfrak{l}^2 \frac{\partial}{\partial y}\nabla^2 U,$$

$$\frac{\partial}{\partial y}(V - \mathfrak{l}^2\nabla^2 V) = 2\ (1-\sigma)\mathfrak{l}^2 \frac{\partial}{\partial x}\nabla^2 U,$$

where ∇^2 is Laplace's operator, σ is Poisson's ratio and $\mathfrak{l}^2 = \eta/\mu$. Thus, $\mathfrak{l}$ is a material constant with the dimension of length.

As these equations differ from those of classical plane strain only through the presence of the function V, it is a straightforward matter to follow Muskhelishvili's procedure for the classical case to find the complex form of the displacements and stresses.

Define

$$(5) \qquad P = \nabla^2 U, \qquad Q = \frac{V - \mathfrak{l}^2\nabla^2 V}{2(1-\sigma)\mathfrak{l}^2}.$$

Then, by Eqs. (4), P and Q satisfy the Cauchy-Riemann equations. We may define a function φ of the complex variable $z = x + iy$,

$$(6) \qquad \varphi(z) = p + iq = \frac{1}{4}\int (P + iQ)\, dz,$$

whence

$$(7) \qquad P = 4\Re\mathfrak{e}\, \varphi'(z) = 2[\varphi'(z) + \overline{\varphi'(z)}],$$

$$(8) \qquad Q = 4\Im\mathfrak{m}\, \varphi'(z) = 2[\ \ '(z) - \overline{\varphi'(z)}].$$

From Eq. (7) and the firs of Eqs. (5), as in the case of classical **elasticity** [1: *Eq. (31.2)*],

$$(9) \qquad 2U = \bar{z}\varphi(z) + z\overline{\varphi(z)} + \chi(z) + \overline{\chi(z)}.$$

From Eq. (8) and the second of Eqs. (5),

$$(10) \qquad V - \mathfrak{l}^2\nabla^2 V = 4(1-\sigma)\mathfrak{l}^2[\varphi'(z) - \overline{\varphi'(z)}],$$

a complete solution of which is

$$(11) \qquad V = 4(1-\sigma)\mathfrak{l}^2[\varphi'(z) - \overline{\varphi'(z)}] + H(z, \bar{z}),$$

where $H(z, \bar{z})$ is a Helmholtz function satisfying the equation

$$(12) \qquad 4\mathfrak{l}^2 \frac{\partial^2 H}{\partial z \partial \bar{z}} - H = 0.$$

The first two of Eqs. (2) and (3) may be solved for the displacements by the same method as Muskhelishvili's, with the result

$$
(13)\qquad
\begin{aligned}
2\mu u &= -\frac{\partial U}{\partial x} + \frac{2(\lambda+2\mu)}{\lambda+\mu}\,p - \frac{\partial V}{\partial y},\\
2\mu v &= -\frac{\partial U}{\partial y} + \frac{2(\lambda+2\mu)}{\lambda+\mu}\,q + \frac{\partial V}{\partial x},
\end{aligned}
$$

which differs from Muskhelishvili's Eq. (30.8) [1] only in the presence of the function V.

The companion to

$$(14)\qquad \frac{\partial U}{\partial x} + i\frac{\partial U}{\partial y} = \varphi(z) + z\overline{\varphi'(z)} + \overline{\psi(z)},$$

where $\psi(z) = d\chi/dz$, is, from Eq. (11),

$$(15)\qquad \frac{\partial V}{\partial x} + i\frac{\partial V}{\partial y} = -8(1-\sigma)\,l^2\,\overline{\varphi''(z)} + 2\frac{\partial H}{\partial \bar z}.$$

Hence,

$$(16)\qquad 2\mu(u+iv) = (3-4\sigma)\varphi(z) - z\overline{\varphi'(z)} - 8i(1-\sigma)\,l^2\,\overline{\varphi''(z)} + 2i\frac{\partial H}{\partial \bar z},$$

which is the desired result for the displacements.

As for the stresses, we have, from Eqs. (3),

$$
\begin{aligned}
X_x + Y_y &= \nabla^2 U = 4\frac{\partial^2 U}{\partial z\,\partial \bar z},\\
Y_y - X_x + i(Y_x + X_y) &= \left(\frac{\partial^2}{\partial x^2} - \frac{\partial^2}{\partial y^2} - 2i\frac{\partial^2}{\partial x\,\partial y}\right)(U+iV) =\\
&= 4\frac{\partial^2}{\partial z^2}(U+iV),\\
Y_x - X_y &= -\nabla^2 V = -4\frac{\partial^2 V}{\partial z\,\partial \bar z},\\
M_x + iM_y &= \frac{\partial V}{\partial x} + i\frac{\partial V}{\partial y} = 2\frac{\partial V}{\partial \bar z}.
\end{aligned}
$$

Then, from Eqs. (9), (11) and (12), these become

$$
\begin{aligned}
X_x + Y_y &= 2[\varphi'(z) + \overline{\varphi'(z)}],\\
Y_y - X_x + i(Y_x + X_y) &= 2[\bar z\varphi''(z) + \psi'(z)] +\\
&\quad + 16i(1-\sigma)\,l^2\varphi'''(z) + 4i\frac{\partial^2 H}{\partial z^2},\\
Y_x - X_y &= -\frac{H}{l^2},\\
M_x + iM_y &= -8(1-\sigma)\,l^2\,\overline{\varphi''(z)} + 2\frac{\partial H}{\partial \bar z},
\end{aligned}
$$

which are the required results for the stresses.

REFERENCES

1. Н. И. Мусхелишвили. Некоторые основные задачи математической теории упругости. Изд. 4. Москва, Изд-во Акад. наук СССР, 1954 [Translation: N. I. Muskhelishvili. Some Basic Problems of the Mathematical Theory of Elasticity. Groningen, P. Noordhoff, 1963].
2. E. et F. Cosserat. Théorie des corps déformables. Paris, A. Hermann et Fils, 1909.
3. Э. Л. Аэро, Е. В. Кувшинский. Основные уравнения теории упругости сред с вращательным взаимодействием частиц.— Физика твердого тела, 1960, 2, № 7, 1399—1409.
4. R. D. Mindlin. Influence of Couple-stresses on Stress Concentration.— Exptl Mech., 1963, 3, № 1, 1—7.

Reprinted from
Applications of the Theory of Functions in Continuum Mechanics
Proceedings of the International Symposium, Tbilisi, U.S.S.R.-1963
Nauka, Moscow (1965) 256-259

Int. J. Solids Structures, 1965, Vol. 1, pp. 73 to 78. Pergamon Press Ltd. Printed in Great Britain

ON THE EQUATIONS OF ELASTIC MATERIALS WITH MICRO-STRUCTURE

R. D. MINDLIN

Department of Civil Engineering, Columbia University, New York, N.Y.

Abstract—In the classical theory of elasticity it is possible to reduce the displacement equations of motion to Lamé's form by means of the Helmholtz resolution. In this paper the analogous reduction is effected for equations, derived recently [1], of an isotropic, elastic material with micro-structure.

INTRODUCTION

IN THE classical theory of elasticity, the displacement-equation of motion,

$$(\lambda+2\mu)\nabla\nabla\cdot\mathbf{u}-\mu\nabla\times\nabla\times\mathbf{u}+\mathbf{f} = \rho\ddot{\mathbf{u}}, \tag{1}$$

is reduced, by means of the Helmholtz resolutions

$$\mathbf{u} = \nabla\varphi_1+\nabla\times\mathbf{H}_1, \qquad \nabla\cdot\mathbf{H}_1 = 0, \tag{2a}$$

$$\mathbf{f} = \nabla\varphi_1^*+\nabla\times\mathbf{H}_1^*, \qquad \nabla\cdot\mathbf{H}_1^* = 0, \tag{2b}$$

to Lamé's form:

$$(\lambda+2\mu)\nabla^2\varphi_1+\varphi_1^* = \rho\ddot{\varphi}_1, \tag{3a}$$

$$\mu\nabla^2\mathbf{H}_1+\mathbf{H}_1^* = \rho\ddot{\mathbf{H}}_1. \tag{3b}$$

In this paper, the analogous reduction is effected for equations, derived recently [1], of an isotropic, elastic material with micro-structure. The new equations have, in addition to the displacement vector $\mathbf{u}$, a second dependent variable: the micro-deformation dyadic $\boldsymbol{\psi}$. The variables $\mathbf{u}$ and $\boldsymbol{\psi}$ are governed by the equations†

$$\begin{aligned}(\mu+2g_2+b_2)\nabla^2\mathbf{u}&+(\lambda+\mu+2g_1+2g_2+b_1+b_3)\nabla\nabla\cdot\mathbf{u}\\ &-(g_1+b_1)\nabla(\mathbf{I}:\boldsymbol{\psi})-(g_2+b_2)\nabla\cdot\boldsymbol{\psi}-(g_2+b_3)\boldsymbol{\psi}\cdot\nabla+\mathbf{f} = \rho\ddot{\mathbf{u}},\end{aligned} \tag{4a}$$

$$\begin{aligned}(a_1+a_5)[\mathbf{I}\nabla\cdot\boldsymbol{\psi}\cdot\nabla+\nabla\nabla(\mathbf{I}:\boldsymbol{\psi})]&+(a_2+a_{11})(\nabla\cdot\boldsymbol{\psi}\nabla+\nabla\boldsymbol{\psi}\cdot\nabla)\\ &+(a_3+a_{14})\nabla\nabla\cdot\boldsymbol{\psi}+a_4\mathbf{I}\nabla^2(\mathbf{I}:\boldsymbol{\psi})+(a_8+a_{15})\boldsymbol{\psi}\cdot\nabla\nabla+a_{10}\nabla^2\boldsymbol{\psi}\\ &+a_{13}\nabla^2\boldsymbol{\psi}_C+g_1\mathbf{I}\nabla\cdot\mathbf{u}+g_2(\nabla\mathbf{u}+\mathbf{u}\nabla)+b_1\mathbf{I}(\nabla\cdot\mathbf{u}+\mathbf{I}:\boldsymbol{\psi})\\ &+b_2(\nabla\mathbf{u}-\boldsymbol{\psi})+b_3(\mathbf{u}\nabla-\boldsymbol{\psi}_C)+\boldsymbol{\Phi} = \tfrac{1}{3}\rho' d^2\ddot{\boldsymbol{\psi}},\end{aligned} \tag{4b}$$

where $\boldsymbol{\psi}_C$ is the conjugate of $\boldsymbol{\psi}$, $\mathbf{I}$ is the idemfactor, $\boldsymbol{\Phi}$ is the body double force dyadic, ρ' is the mass density of the micro-material, d is the half-length of the unit micro-cell and a_i, b_i, g_i are material constants.

Whereas (1) is equivalent to three scalar equations, (4) are equivalent to twelve scalar equations.

† See [1], equations (6.1) and (6.2).

ALTERNATIVE FORM OF EQUATIONS

A complete dyadic, $\mathbf{\Delta}$, may be resolved into the sum of its symmetric part $\mathbf{\Delta}^S$ and antisymmetric part $\mathbf{\Delta}^A$:

$$\mathbf{\Delta} = \mathbf{\Delta}^S + \mathbf{\Delta}^A, \qquad \mathbf{\Delta}^S = \tfrac{1}{2}(\mathbf{\Delta} + \mathbf{\Delta}_C), \qquad \mathbf{\Delta}^A = \tfrac{1}{2}(\mathbf{\Delta} - \mathbf{\Delta}_C).$$

The symmetric part may be resolved, further, into the sum of its deviator $\mathbf{\Delta}^{SD}$ and its spherical part $\mathbf{I}\Delta_S$:

$$\mathbf{\Delta}^S = \mathbf{\Delta}^{SD} + \mathbf{I}\Delta_S, \qquad \mathbf{\Delta}^{SD} = \mathbf{\Delta}^S - \tfrac{1}{3}\mathbf{II} : \mathbf{\Delta}, \qquad \Delta_S = \tfrac{1}{3}\mathbf{I} : \mathbf{\Delta}.$$

Thus,

$$\mathbf{\Delta} = \mathbf{\Delta}^{SD} + \mathbf{I}\Delta_S + \mathbf{\Delta}^A. \tag{5}$$

Necessary and sufficient conditions for $\mathbf{\Delta} = 0$ are

$$\mathbf{\Delta}^{SD} = 0, \qquad \Delta_S = 0, \qquad \mathbf{\Delta}^A = 0. \tag{6}$$

Upon resolving the dyadics $\boldsymbol{\psi}$ and $\mathbf{\Phi}$, in (4), in accordance with (5), and applying (6) to the resulting equations, the following equations on $\mathbf{u}$, $\boldsymbol{\psi}^{SD}$, ψ_S and $\boldsymbol{\psi}^A$ are obtained:

$$k_{11}\nabla\nabla\cdot\mathbf{u} - k_{12}\nabla\cdot\boldsymbol{\psi}^{SD} - k_{13}\nabla\psi_S - \bar{k}_{11}\nabla\times\nabla\times\mathbf{u} - \bar{k}_{13}\nabla\cdot\boldsymbol{\psi}_C^A + \mathbf{f} = \rho\ddot{\mathbf{u}}, \tag{7a}$$

$$k_{21}(\nabla\mathbf{u})^{SD} + k_{22}(\nabla\nabla\cdot\boldsymbol{\psi}^{SD})^{SD} - \tfrac{2}{3}k'_{22}\boldsymbol{\psi}^{SD} + k_{23}(\nabla\nabla\psi_S)^D$$
$$+\bar{k}_{23}(\nabla\nabla\cdot\boldsymbol{\psi}^A)^S + (a_{10}+a_{13})[\nabla^2\boldsymbol{\psi}^{SD} - \tfrac{3}{2}(\nabla\nabla\cdot\boldsymbol{\psi}^{SD})^{SD}] + \mathbf{\Phi}^{SD} = \tfrac{1}{3}\rho' d^2\ddot{\boldsymbol{\psi}}^{SD}, \tag{7b}$$

$$k_{31}\nabla\cdot\mathbf{u} + k_{32}\nabla\cdot\boldsymbol{\psi}^D\cdot\nabla + k_{33}\nabla^2\psi_S - k'_{33}\psi_S + 3\Phi_S = \rho' d^2\ddot{\psi}_S, \tag{7c}$$

$$\bar{k}_{31}(\nabla\mathbf{u})^A + \bar{k}_{32}(\nabla\nabla\cdot\boldsymbol{\psi}^{SD})^A + \bar{k}_{33}(\nabla\nabla\cdot\boldsymbol{\psi}^A)^A - \tfrac{1}{2}\bar{k}'_{33}\boldsymbol{\psi}^A$$
$$+(a_{10}-a_{13})[\nabla^2\boldsymbol{\psi}^A - 2(\nabla\nabla\cdot\boldsymbol{\psi}^A)^A] + \mathbf{\Phi}^A = \tfrac{1}{3}\rho' d^2\ddot{\boldsymbol{\psi}}^A, \tag{7d}$$

where

$$k_{11} = \lambda + 2\mu + 2g_1 + 4g_2 + b_1 + b_2 + b_3,$$
$$k_{22} = 2a_2 + a_3 + a_8 + \tfrac{3}{2}a_{10} + 2a_{11} + \tfrac{3}{2}a_{13} + a_{14} + a_{15},$$
$$k_{33} = 6a_1 + 2a_2 + a_3 + 9a_4 + 6a_5 + a_8 + 3a_{10} + 2a_{11} + 3a_{13} + a_{14} + a_{15},$$
$$k_{23} = k_{32} = 3a_1 + 2a_2 + a_3 + 3a_5 + a_8 + 2a_{11} + a_{14} + a_{15},$$
$$k_{31} = k_{13} = 3g_1 + 2g_2 + 3b_1 + b_2 + b_3,$$
$$k_{12} = k_{21} = 2g_2 + b_2 + b_3,$$
$$k'_{22} = \tfrac{3}{2}(b_2 + b_3),$$
$$k'_{33} = 3(3b_1 + b_2 + b_3),$$
$$\bar{k}_{11} = \mu + 2g_2 + b_2,$$
$$\bar{k}_{22} = 2a_2 + a_3 + a_8 + 2a_{10} + 2a_{11} + 2a_{13} + a_{14} + a_{15},$$
$$\bar{k}_{33} = -2a_2 + a_3 + a_8 + 2a_{10} - 2a_{11} - 2a_{13} + a_{14} + a_{15},$$
$$\bar{k}_{23} = \bar{k}_{32} = a_3 - a_8 + a_{14} - a_{15},$$

$$\bar{k}_{31} = \bar{k}_{13} = b_2 - b_3,$$
$$\bar{k}_{12} = \bar{k}_{21} = 2g_2 + b_2 + b_3,$$
$$\bar{k}'_{22} = 2(b_2 + b_3),$$
$$\bar{k}'_{33} = 2(b_2 - b_3).$$

ANALOGUES OF HELMHOLTZ RESOLUTION

To find a resolution of the dyadic $\boldsymbol{\psi}$ (or $\boldsymbol{\Phi}$) analogous to that of the vector $\mathbf{u}$ (or $\mathbf{f}$) in (2), first define

$$4\pi\boldsymbol{\psi}' \equiv -\int_V r^{-1}\boldsymbol{\psi}_Q \, dV_Q, \tag{8}$$

where $r\ (= \sqrt{[(x-\xi)^2+(y-\eta)^2+(z-\zeta)^2]})$ is the distance between a field point $P(x, y, z)$ and a source point $Q(\xi, \eta, \zeta)$, $\boldsymbol{\psi}_Q = \boldsymbol{\psi}(\xi, \eta, \zeta)$ and $dV_Q = d\xi\, d\eta\, d\zeta$. Then

$$\boldsymbol{\psi} = \nabla^2\boldsymbol{\psi}' = \nabla\nabla\cdot\boldsymbol{\psi}' - \nabla\times\nabla\times\boldsymbol{\psi}'.$$

Define

$$\mathbf{G}' \equiv \nabla\cdot\boldsymbol{\psi}', \qquad \boldsymbol{\Lambda}' \equiv -\nabla\times\boldsymbol{\psi}'.$$

Then a first resolution is

$$\boldsymbol{\psi} = \nabla\mathbf{G}' + \nabla\times\boldsymbol{\Lambda}', \qquad \nabla\cdot\boldsymbol{\Lambda}' = 0. \tag{9}$$

Now, define

$$4\pi\boldsymbol{\Lambda}'' \equiv -\int_V r^{-1}\boldsymbol{\Lambda}'_Q \, dV_Q,$$

so that

$$\boldsymbol{\Lambda}' = \nabla^2\boldsymbol{\Lambda}'' = (\boldsymbol{\Lambda}''\cdot\nabla)\nabla - \boldsymbol{\Lambda}''\times\nabla\times\nabla.$$

Define

$$\mathbf{G}'' \equiv \boldsymbol{\Lambda}''\cdot\nabla, \qquad \boldsymbol{\Lambda}''' \equiv -\boldsymbol{\Lambda}''\times\nabla.$$

Then

$$\boldsymbol{\Lambda}' = \mathbf{G}''\nabla + \boldsymbol{\Lambda}'''\times\nabla, \qquad \boldsymbol{\Lambda}'''\cdot\nabla = 0, \qquad \nabla^2\nabla\cdot\boldsymbol{\Lambda}''' = 0. \tag{10}$$

Substituting (10) in (9), we find a second form:

$$\boldsymbol{\psi} = \nabla\mathbf{G}' + \nabla\times(\mathbf{G}''\nabla) + \nabla\times\boldsymbol{\Lambda}'''\times\nabla, \qquad \boldsymbol{\Lambda}'''\cdot\nabla = 0, \qquad \nabla^2\nabla\cdot\boldsymbol{\Lambda}''' = 0. \tag{11}$$

Consider, now, the symmetric part of $\boldsymbol{\psi}$:

$$\boldsymbol{\psi}^S = \nabla\mathbf{G} + \mathbf{G}\nabla + \nabla\times\boldsymbol{\Lambda}\times\nabla, \tag{12}$$

where

$$\mathbf{G} \equiv \tfrac{1}{2}(\mathbf{G}' + \nabla\times\mathbf{G}''), \qquad \boldsymbol{\Lambda} \equiv \tfrac{1}{2}(\boldsymbol{\Lambda}''' + \boldsymbol{\Lambda}'''_C) = \boldsymbol{\Lambda}_C \tag{13}$$

and we note that, in view of the second and third parts of (11),

$$\nabla\cdot\boldsymbol{\Lambda}\cdot\nabla = 0, \qquad \nabla^2\nabla\cdot\boldsymbol{\Lambda} = 0. \tag{14}$$

Further, define

$$4\pi \mathbf{G}''' \equiv -\int_V r^{-1}\mathbf{G}_Q \, \mathrm{d}V_Q.$$

Then

$$\mathbf{G} = \nabla^2\mathbf{G}''' = \nabla\nabla\cdot\mathbf{G}''' - \nabla\times\nabla\times\mathbf{G}'''.$$

Define

$$\chi \equiv 2\nabla\cdot\mathbf{G}''', \qquad \mathbf{H}_2 \equiv -\nabla\times\mathbf{G}'''.$$

Then

$$\mathbf{G} = \tfrac{1}{2}\nabla\chi + \nabla\times\mathbf{H}_2, \qquad \nabla\cdot\mathbf{H}_2 = 0. \tag{15}$$

Upon substituting (15) in (12), we find

$$\boldsymbol{\psi}^S = \nabla\nabla\chi + 2(\nabla\nabla\times)^S\mathbf{H}_2 + \nabla\times\boldsymbol{\Lambda}\times\nabla, \tag{16}$$

$$\nabla\cdot\mathbf{H}_2 = 0, \qquad \boldsymbol{\Lambda} = \boldsymbol{\Lambda}_C, \qquad \nabla\cdot\boldsymbol{\Lambda}\cdot\nabla = 0, \qquad \nabla^2\nabla\cdot\boldsymbol{\Lambda} = 0,$$

where the symbol $(\nabla\nabla\times)^S$ is defined according to

$$(\nabla\nabla\times)^S\mathbf{H} \equiv \tfrac{1}{2}[\nabla(\nabla\times\mathbf{H}) + (\nabla\times\mathbf{H})\nabla].$$

The spherical part of $\boldsymbol{\psi}^S$ is

$$\psi_S \equiv \tfrac{1}{3}\mathbf{I}:\boldsymbol{\psi}^S = \tfrac{1}{3}\nabla^2(\chi - \mathbf{I}:\boldsymbol{\Lambda})$$

and the deviator of $\boldsymbol{\psi}^S$ is

$$\begin{aligned}\boldsymbol{\psi}^{SD} &\equiv \boldsymbol{\psi}^S - \mathbf{I}\psi_S,\\ &= \nabla\nabla\chi + 2(\nabla\nabla\times)^S\mathbf{H}_2 + \nabla\times\boldsymbol{\Lambda}\times\nabla - \tfrac{1}{3}\mathbf{I}\nabla^2(\chi - \mathbf{I}:\boldsymbol{\Lambda}),\\ &= (\nabla\nabla)^D\chi + 2(\nabla\nabla\times)^S\mathbf{H}_2 + (\nabla\times\boldsymbol{\Lambda}\times\nabla)^D,\end{aligned}$$

where $(\ )^D$ designates the deviatoric part.

It is convenient to define

$$\varphi_2 \equiv \tfrac{1}{3}(2\chi + \mathbf{I}:\boldsymbol{\Lambda}), \qquad \varphi_3 \equiv \tfrac{1}{3}(\chi - \mathbf{I}:\boldsymbol{\Lambda}) \tag{17}$$

and the symmetric deviator

$$\boldsymbol{\Gamma} \equiv (\nabla\times\boldsymbol{\Lambda}\times\nabla)^D - \tfrac{1}{2}(\nabla\nabla)^D\varphi_2 + (\nabla\nabla)^D\varphi_3. \tag{18}$$

Then

$$\boldsymbol{\psi}^{SD} = \boldsymbol{\Gamma} + \tfrac{3}{2}(\nabla\nabla)^D\varphi_2 + 2(\nabla\nabla\times)^S\mathbf{H}_2, \qquad \nabla\cdot\mathbf{H}_2 = 0, \tag{19}$$

$$\psi_S = \nabla^2\varphi_3. \tag{20}$$

As for the antisymmetric part of $\boldsymbol{\psi}$, we have, from (11),

$$\boldsymbol{\psi}^A = \nabla\mathbf{F} - \mathbf{F}\nabla + \tfrac{1}{2}\nabla\times(\boldsymbol{\Lambda}''' - \boldsymbol{\Lambda}'''_C)\times\nabla, \tag{21}$$

where

$$\mathbf{F} \equiv \tfrac{1}{2}(\mathbf{G}' - \nabla\times\mathbf{G}'').$$

Define

$$4\pi \mathbf{F}' \equiv -\int_V r^{-1} \mathbf{F}_Q \, dV_Q.$$

Then

$$\mathbf{F} = \nabla^2 \mathbf{F}' = \nabla\nabla \cdot \mathbf{F}' - \nabla \times \nabla \times \mathbf{F}'.$$

Define

$$\chi' \equiv \nabla \cdot \mathbf{F}', \qquad \mathbf{H}_3 \equiv -\nabla \times \mathbf{F}'.$$

Then

$$\mathbf{F} = \nabla\chi' + \nabla \times \mathbf{H}_3, \qquad \nabla \cdot \mathbf{H}_3 = 0 \tag{22}$$

and, from (22) and (21),

$$\boldsymbol{\psi}^A = \nabla(\nabla \times \mathbf{H}_3) - (\nabla \times \mathbf{H}_3)\nabla + \tfrac{1}{2}\nabla \times (\boldsymbol{\Lambda}''' - \boldsymbol{\Lambda}'''_C) \times \nabla, \tag{23}$$

$$\nabla \cdot \mathbf{H}_3 = 0, \qquad \nabla^2 \nabla \cdot \boldsymbol{\Lambda}''' = 0, \qquad \boldsymbol{\Lambda}''' \cdot \nabla = 0. \tag{24}$$

Now, in view of (24), we have, from (23),

$$\mathbf{I} \times \mathbf{I} : \boldsymbol{\psi}^A = 2\nabla^2 \mathbf{H}_3 + \nabla \Upsilon,$$

where

$$\Upsilon \equiv -\mathbf{I} \times \mathbf{I} \,\vdots\, \nabla\boldsymbol{\Lambda}'''.$$

Then

$$\boldsymbol{\psi}^A = \mathbf{I} \times (\tfrac{1}{2}\nabla\Upsilon + \nabla^2 \mathbf{H}_3), \qquad \nabla \cdot \mathbf{H}_3 = 0. \tag{25}$$

Assembling the results in (19), (20), (25), we have a special form of resolution of $\boldsymbol{\psi}$ suitable for the present purpose:

$$\begin{aligned}
\boldsymbol{\psi} &= \boldsymbol{\psi}^{SD} + \psi_S \mathbf{I} + \boldsymbol{\psi}^A, \\
&= \boldsymbol{\Gamma} + \tfrac{3}{2}(\nabla\nabla)^D \varphi_2 + 2(\nabla\nabla \times)^S \mathbf{H}_2 + \mathbf{I}\nabla^2\varphi_3 + \mathbf{I} \times (\tfrac{1}{2}\nabla\Upsilon + \nabla^2 \mathbf{H}_3), \qquad (26a) \\
&\nabla \cdot \mathbf{H}_2 = 0, \qquad \nabla \cdot \mathbf{H}_3 = 0, \qquad \boldsymbol{\Gamma} = \boldsymbol{\Gamma}_C, \qquad \mathbf{I} : \boldsymbol{\Gamma} = 0.
\end{aligned}$$

Similarly, we may express the body double force dyadic $\boldsymbol{\Phi}$ as

$$\boldsymbol{\Phi} = \boldsymbol{\Phi}^{SD} + \mathbf{I}\Phi_S + \boldsymbol{\Phi}^A, \tag{26b}$$

where

$$\begin{aligned}
\boldsymbol{\Phi}^{SD} &= \boldsymbol{\Gamma}^* + \tfrac{3}{2}(\nabla\nabla)^D \varphi_2^* + 2(\nabla\nabla \times)^S \mathbf{H}_2^*, \qquad \nabla \cdot \mathbf{H}_2^* = 0, \\
\Phi_S &= \nabla^2 \varphi_3^*, \\
\boldsymbol{\Phi}^A &= \mathbf{I} \times (\tfrac{1}{2}\nabla\Upsilon^* + \nabla^2 \mathbf{H}_3^*), \qquad \nabla \cdot \mathbf{H}_3^* = 0,
\end{aligned}$$

and φ_2^*, φ_3^*, $\mathbf{H}_2^*$, $\mathbf{H}_3^*$, $\boldsymbol{\Gamma}^*$, Υ^* are defined in the same way as their counterparts in $\boldsymbol{\psi}$.

EQUATIONS ON φ_i, $\mathbf{H}_i$, Y AND Γ

Upon substituting (2) and (26) in (7), we find

$$\nabla[\nabla^2(k_{11}\varphi_1 - k_{12}\varphi_2 - k_{13}\varphi_3) + \varphi_1^* - \rho\ddot{\varphi}_1]$$
$$+\nabla\times[\nabla^2(\bar{k}_{11}\mathbf{H}_1 - \bar{k}_{12}\mathbf{H}_2 - \bar{k}_{13}\mathbf{H}_3) + \mathbf{H}_1^* - \rho\ddot{\mathbf{H}}_1] = 0, \qquad (27a)$$

$$(\nabla\nabla)^D[k_{21}\varphi_1 + k_{22}\nabla^2\varphi_2 - k'_{22}\varphi_2 + k_{23}\nabla^2\varphi_3 + \tfrac{3}{2}\varphi_2^* - \tfrac{1}{2}\rho' d^2\ddot{\varphi}_2]$$
$$+(\nabla\nabla\times)^S[\bar{k}_{21}\mathbf{H}_1 + \bar{k}_{22}\nabla^2\mathbf{H}_2 - \bar{k}'_{22}\mathbf{H}_2 + \bar{k}_{23}\nabla^2\mathbf{H}_3 + 2\mathbf{H}_2^* - \tfrac{2}{3}\rho' d^2\ddot{\mathbf{H}}_2]$$
$$+(a_{10}+a_{13})\nabla^2\mathbf{\Gamma} - (b_2+b_3)\mathbf{\Gamma} + \mathbf{\Gamma}^* - \tfrac{1}{3}\rho' d^2\ddot{\mathbf{\Gamma}} = 0, \qquad (27b)$$

$$\nabla^2(k_{31}\varphi_1 + k_{32}\nabla^2\varphi_2 + k_{33}\nabla^2\varphi_3 - k'_{33}\varphi_3 + 3\varphi_3^* - \rho' d^2\ddot{\varphi}_3) = 0, \qquad (27c)$$

$$\nabla^2(\bar{k}_{31}\mathbf{H}_1 + \bar{k}_{32}\nabla^2\mathbf{H}_2 + \bar{k}_{33}\nabla^2\mathbf{H}_3 - \bar{k}'_{33}\mathbf{H}_3 + 2\mathbf{H}_3^* - \tfrac{2}{3}\rho' d^2\ddot{\mathbf{H}}_3)$$
$$+\nabla[(a_{10}-a_{13})\nabla^2 \mathrm{Y} - (b_2-b_3)\mathrm{Y} + \mathrm{Y}^* - \tfrac{1}{3}\rho' d^2\ddot{\mathrm{Y}}] = 0. \qquad (27d)$$

Finally, as solutions of (27), we may take solutions of

$$\left.\begin{aligned} k_{11}\nabla^2\varphi_1 - k_{12}\nabla^2\varphi_2 - k_{13}\nabla^2\varphi_3 + \varphi_1^* &= \rho\ddot{\varphi}_1, \\ k_{21}\varphi_1 + k_{22}\nabla^2\varphi_2 - k'_{22}\varphi_2 + k_{23}\nabla^2\varphi_3 + \tfrac{3}{2}\varphi_2^* &= \tfrac{1}{2}\rho' d^2\ddot{\varphi}_2, \\ k_{31}\varphi_1 + k_{32}\nabla^2\varphi_2 + k_{33}\nabla^2\varphi_3 - k'_{33}\varphi_3 + 3\varphi_3^* &= \rho' d^2\ddot{\varphi}_3, \end{aligned}\right\} \qquad (28)$$

$$\left.\begin{aligned} \bar{k}_{11}\nabla^2\mathbf{H}_1 - \bar{k}_{12}\nabla^2\mathbf{H}_2 - \bar{k}_{13}\nabla^2\mathbf{H}_3 + \mathbf{H}_1^* &= \rho\ddot{\mathbf{H}}_1, \\ \bar{k}_{21}\mathbf{H}_1 + \bar{k}_{22}\nabla^2\mathbf{H}_2 - \bar{k}'_{22}\mathbf{H}_2 + \bar{k}_{23}\nabla^2\mathbf{H}_3 + 2\mathbf{H}_2^* &= \tfrac{2}{3}\rho' d^2\ddot{\mathbf{H}}_2, \\ \bar{k}_{31}\mathbf{H}_1 + \bar{k}_{32}\nabla^2\mathbf{H}_2 + \bar{k}_{33}\nabla^2\mathbf{H}_3 - \bar{k}'_{33}\mathbf{H}_3 + 2\mathbf{H}_3^* &= \tfrac{2}{3}\rho' d^2\ddot{\mathbf{H}}_3, \end{aligned}\right\} \qquad (29)$$

$$(a_{10}+a_{13})\nabla^2\mathbf{\Gamma} - (b_2+b_3)\mathbf{\Gamma} + \mathbf{\Gamma}^* = \tfrac{1}{3}\rho' d^2\ddot{\mathbf{\Gamma}}, \qquad (30)$$

$$(a_{10}-a_{13})\nabla^2\mathrm{Y} - (b_2-b_3)\mathrm{Y} + \mathrm{Y}^* = \tfrac{1}{3}\rho' d^2\ddot{\mathrm{Y}}. \qquad (31)$$

These are equations analogous to (3). The solution of (28)–(31) for plane waves is identical with the complete solution of (4), for plane waves, that was given in [1]. However, the general completeness of (28)–(31) has not been ascertained.

REFERENCE

[1] R. D. MINDLIN, *Arch. Ration. Mech. Analysis*, **16**, 51 (1964).

(*Received* 28 *June* 1964)

Résumé—Dans la théorie de l'élasticité classique il est possible de réduire les équations du mouvement en déplacements à la forme de Lamé en utilisant les résolutions de Helmholtz. L'étude présente la réduction analogue pour la théorie d'un corps elastique isotrope avec micro-structure, évolue par l'auteur [1].

Абстракт—По классической теории эластичности возможно свести уравнения смещения для движения к форме Ламэ при помощи разрешения Гельмгольца.

В настоящей работе аналогичное преобразование сделано для уравнений, недавно выведенных [1], изотропного эластичного матерьяла обладающего микро-структурой.

Int. J. Solids Structures, 1965, Vol. 1, pp. 141 to 145. Pergamon Press Ltd. Printed in Great Britain

THICKNESS–TWIST VIBRATIONS OF AN INFINITE, MONOCLINIC, CRYSTAL PLATE

R. D. MINDLIN

Department of Civil Engineering, Columbia University, New York, N.Y.

Abstract—Thickness–twist modes of vibration of an infinite plate are standing waves with displacement and wave-normal at right angles to each other and parallel to the faces of the plate. In a monoclinic crystal plate, such waves are possible if the displacement is in the direction of the digonal axis. The pertinent solution of the three-dimensional equations of elasticity for a plate with free faces is given, as well as another solution with the effect of the inertia of electrode films taken into account. The results are applicable to the rotated-Y-cut quartz plates used as oscillator and filter crystals in electric circuits.

INTRODUCTION

INTEREST in the frequencies of vibration of monoclinic, crystal plates, stems from the fact that the widely used rotated-Y-cut quartz plates [1] exhibit monoclinic symmetry in the stress–strain relation when it is referred to rectangular axes in and normal to the plane of the plate. Thus, with strain–displacement relations

$$\begin{aligned} S_1 &= \frac{\partial u}{\partial x}, & S_4 &= \frac{\partial w}{\partial y}+\frac{\partial v}{\partial z}, \\ S_2 &= \frac{\partial v}{\partial y}, & S_5 &= \frac{\partial u}{\partial z}+\frac{\partial w}{\partial x}, \\ S_3 &= \frac{\partial w}{\partial z}, & S_6 &= \frac{\partial v}{\partial x}+\frac{\partial u}{\partial y}, \end{aligned} \tag{1}$$

the stress–strain relation for a rotated-Y-cut plate of quartz, with x the digonal axis, is

$$\begin{aligned} T_1 &= c_{11}S_1+c_{12}S_2+c_{13}S_3+c_{14}S_4, \\ T_2 &= c_{21}S_1+c_{22}S_2+c_{23}S_3+c_{24}S_4, \\ T_3 &= c_{31}S_1+c_{32}S_2+c_{33}S_3+c_{34}S_4, \\ T_4 &= c_{41}S_1+c_{42}S_2+c_{43}S_3+c_{44}S_4, \\ T_5 &= c_{55}S_5+c_{56}S_6, \\ T_6 &= c_{65}S_5+c_{66}S_6. \end{aligned} \tag{2}$$

The stresses, in turn, must satisfy the equations of motion

$$\frac{\partial T_1}{\partial x}+\frac{\partial T_6}{\partial y}+\frac{\partial T_5}{\partial z} = \rho\frac{\partial^2 u}{\partial t^2},$$

$$\frac{\partial T_6}{\partial x}+\frac{\partial T_2}{\partial y}+\frac{\partial T_4}{\partial z} = \rho\frac{\partial^2 v}{\partial t^2},$$
$$\frac{\partial T_5}{\partial x}+\frac{\partial T_4}{\partial y}+\frac{\partial T_3}{\partial z} = \rho\frac{\partial^2 w}{\partial t^2}. \tag{3}$$

The problem at hand is the study of the steady state solutions of the fifteen equations (1)–(3) subject to the boundary conditions

$$T_2 = T_4 = T_6 = 0 \qquad \text{on} \qquad y = \pm h. \tag{4}$$

Detailed analyses have been made [2,3] of the highly complicated solution for the case of the wave-normal parallel to the digonal axis. There is, however, a much simpler solution for waves with normal in the direction of z and displacement in the direction of x. Moreover, the simple solution is one of technological interest as it involves only the strains (S_5 and S_6) that are linked directly, through the piezoelectric relations [1], to the electric field (E_2) applied via electrodes on the surfaces $y = \pm h$.

SOLUTION

Consider the possibility of a solution of equations (1)–(3) of the form

$$u = U(y, z)e^{i\omega t}, \qquad v = w = 0. \tag{5}$$

Then, with the exponential factor omitted, equations (1) reduce to

$$S_1 = S_2 = S_3 = S_4 = 0,$$
$$S_5 = \frac{\partial U}{\partial z}, \qquad S_6 = \frac{\partial U}{\partial y}; \tag{6}$$

equations (2), with (6), reduce to

$$T_1 = T_2 = T_3 = T_4 = 0,$$
$$T_5 = c_{55}\frac{\partial U}{\partial z}+c_{56}\frac{\partial U}{\partial y},$$
$$T_6 = c_{56}\frac{\partial U}{\partial z}+c_{66}\frac{\partial U}{\partial y}; \tag{7}$$

and equations (3), with (7), reduce to

$$c_{66}\frac{\partial^2 U}{\partial y^2}+2c_{56}\frac{\partial^2 U}{\partial y\,\partial z}+c_{55}\frac{\partial^2 U}{\partial z^2} = -\rho\omega^2 U. \tag{8}$$

In particular,

$$U = A\sin\eta y\cos\zeta z+B\cos\eta y\sin\zeta z \tag{9}$$

is a solution of (8) if

$$B = \pm A, \qquad \rho\omega^2 = c_{66}\eta^2+c_{55}\zeta^2\pm 2c_{56}\eta\zeta.$$

Thus, two solutions of the form (9) are possible. With a view to satisfying the boundary conditions (4), take a sum of the two solutions:

$$U = C_1 \sin(\eta_1 y + \zeta z) + C_2 \sin(\eta_2 y - \zeta z). \tag{10}$$

Then (8) requires

$$\rho\omega^2 = c_{66}\eta_1^2 + c_{55}\zeta^2 + 2c_{56}\eta_1\zeta = c_{66}\eta_2^2 + c_{55}\zeta^2 - 2c_{56}\eta_2\zeta, \tag{11}$$

from which

$$c_{66}(\eta_1^2 - \eta_2^2) + 2c_{56}\zeta(\eta_1 + \eta_2) = 0.$$

Hence, either $\eta_1 + \eta_2 = 0$ or

$$c_{66}(\eta_2 - \eta_1) = 2c_{56}\zeta. \tag{12}$$

Now, with (5), the boundary conditions (4) reduce to

$$c_{56}\frac{\partial U}{\partial z} + c_{66}\frac{\partial U}{\partial y} = 0 \qquad \text{on} \qquad y = \pm h. \tag{13}$$

Then, with (10) and (12), (13) reduce to

$$\frac{C_2}{C_1} = -\frac{\cos \eta_1 h}{\cos \eta_2 h} = \frac{\sin \eta_1 h}{\sin \eta_2 h}, \tag{14}$$

whence

$$\sin(\eta_1 + \eta_2)h = 0, \tag{15}$$

or

$$(\eta_1 + \eta_2)h = m\pi; \qquad m = 0, 1, 2, 3, \ldots. \tag{16}$$

From (16) and (12),

$$\eta_1 = \frac{m\pi}{2h} - \frac{c_{56}}{c_{66}}\zeta, \qquad \eta_2 = \frac{m\pi}{2h} + \frac{c_{56}}{c_{66}}\zeta$$

and then (11) becomes

$$\omega^2 = \frac{c_{66}}{\rho}\left(\frac{\pi}{2h}\right)^2\left[m^2 + \frac{\gamma_{55}}{c_{66}}\left(\frac{2\zeta h}{\pi}\right)^2\right], \tag{17}$$

where $\gamma_{55} = c_{55} - c_{56}^2/c_{66}$.

The dispersion relation (17) is plotted in Fig. 1 with the ordinate the dimensionless ratio of the frequency, ω, to the frequency, ω_s, of the lowest $x-y$ thickness-shear mode:

$$\omega_s = \frac{\pi}{2h}\sqrt{\frac{c_{66}}{\rho}}.$$

The branches for real, positive ζ are quarter-rectangular hyperbolas and for imaginary, positive ζ are quarter-circles for the chosen dimensionless ordinate and abscissa.

Effect of inertia of electrodes

Suppose that each face $y = \pm h$ of the plate is coated with a thin electrode film of

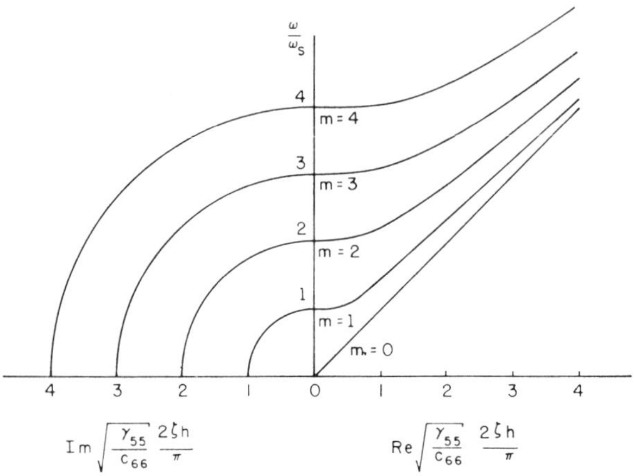

FIG. 1. Dispersion curves for thickness–twist modes of an infinite, monoclinic, crystal plate.

density ρ' and thickness $2h' \ll h$. The inertia of the electrodes is taken into account by replacing the boundary conditions (13) with

$$c_{56}\frac{\partial U}{\partial z}+c_{66}\frac{\partial U}{\partial y} = \pm 2\rho' h'\omega^2 U \qquad \text{on} \qquad y = \pm h, \tag{18}$$

which are satisfied by (10) if

$$\frac{C_2}{C_1} = -\frac{k\cos\eta_1 h-\sin\eta_1 h}{k\cos\eta_2 h-\sin\eta_2 h} = \frac{k\sin\eta_1 h+\cos\eta_1 h}{k\sin\eta_2 h+\cos\eta_2 h}, \tag{19}$$

where

$$k = \frac{c_{66}\eta_1+c_{56}\zeta}{2\rho' h'\omega^2} = \frac{c_{66}\eta_2-c_{56}\zeta}{2\rho' h'\omega^2}.$$

From (19),

$$\tan(\eta_1+\eta_2)h = 2k/(1-k^2), \tag{20}$$

which replaces (15) and, along with

$$(\eta_2-\eta_1)c_{66} = 2c_{56}\zeta, \tag{21}$$

$$\rho\omega^2 = c_{66}\eta_1^2+c_{55}\zeta^2+2c_{56}\eta_1\zeta, \tag{22}$$

constitutes the dispersion relation for thickness–twist waves in the plate with mass loading of electrodes accounted for.

Equations (20)–(22) give frequencies only slightly lower than (17) as the ratio

$$R = 2\rho' h/\rho h \tag{23}$$

of the mass per unit area of electrodes to the mass per unit area of plate is usually only of the order of 10^{-2}. For example, at cut-off ($\zeta = 0$), (20)–(22) reduce to

$$\omega/\omega_s = 2\eta h/\pi, \tag{24}$$

$$\tan 2\eta h = 2R\eta h/(R^2\eta^2h^2-1). \tag{25}$$

The roots of technological interest give frequencies near those of (17) for $\zeta = 0$ and m a small, odd integer. With

$$2\eta h = m\pi - \varepsilon, \qquad 0 < \varepsilon \ll 1,$$

equation (25) gives, approximately,

$$2\eta h = m\pi(1-R).$$

Hence, the cut-off frequencies for a plate with electrodes are approximately

$$\omega/\omega_s = m(1-R),$$

for small R and m, instead of $\omega/\omega_s = m$ which was found for the plate without electrodes.

REFERENCES

[1] W. P. Mason, *Piezoelectric crystals and their application to ultrasonics*, D. Van Nostrand Company, New York (1950).
[2] H. Ekstein, *Phys. Rev.* **68,** 11 (1945).
[3] R. K. Kaul and R. D. Mindlin, *J. acoust. Soc. Am.* **34,** 1902 (1962).

(*Received* 28 *June* 1964)

Zusammenfassung—Bei den torsionalen Dickenschwingungen einer unbegrenzten Platte entstehen parallel zu den Plattenflächen verlaufende Stehwellen, deren Auslenkung und Wellenfrontnormale rechtwinklig zueinander stehen. In einem monoklinischem Kristall sind solche Wellen dann gegeben, wenn die Auslenkung in die Richtung der Digonalachse fällt. Eine geeignete Lösung der dreidimensionalen Elastizitätsgleichungen und ebenfalls eine andere Lösung, welche den Trägheitseinfluss des Elektrodenbelags in Betracht zieht, wird für eine Platte mit freien Flächen angegeben. Die Ergebnisse sind auf die in elektrischen Schaltungen als Oszillatoren und Filterkristalle verwendeten Quarzplättchen mit verdrehtem Y-Schnitt anwendbar.

Абстракт—Формы колебаний, зависящих от толщины и кручения бесконечной плиты, являются стоячими волнами, со смещением и главной волной перпендикулярными между собой и параллельными к поверхностям плиты. В моноклинической кристаллической плите такие волны возможны, если смещение имеет место в направлении диагональной оси. Дается подходящее решение пространственных уравнений упругости для плиты со свободными поверхностями; также дается другое решение, в котором принят во внимание эффект инерции электродных фильм. Результаты могут быть применяемы для вращаемой, срезанной под острым углом кварцевой плиты, употребляемой как осциллятор и как фильтр-кристалл в электрических схемах.

Int. J. Solids Structures, 1965, Vol. 1, pp. 265 to 271. Pergamon Press Ltd. Printed in Great Britain

STRESS FUNCTIONS FOR A COSSERAT CONTINUUM

R. D. MINDLIN

Department of Civil Engineering, Columbia University, New York, N.Y.

Abstract—In this paper, a complete solution, in terms of stress functions analogous to the Papkovitch functions of classical elasticity, is obtained for the linear equations of an isotropic, elastic, Cosserat continuum. The special solutions for the concentrated force and couple are also given.

EQUATIONS OF A COSSERAT CONTINUUM

AT EACH point of a Cosserat continuum [1] there is a micro-structure which can rotate with respect to the surrounding medium. In the linear, elastic case, the potential energy-density may be expressed as a quadratic function of the classical small strain, the difference between the classical small rotation and the rotation of the micro-structure, and the gradient of the rotation of the micro-structure. Thus, employing the notations

$$A_{j,i} \equiv \frac{\partial A_j}{\partial x_i}, \qquad i = 1, 2, 3,$$

$$A_{[j,i]} \equiv \tfrac{1}{2}(A_{j,i} - A_{i,j}) = -A_{[i,j]},$$

$$B_{[ij]} \equiv \tfrac{1}{2}(B_{ij} - B_{ji}) = -B_{[ji]},$$

we may write, for the potential energy-density,

$$W = W(\varepsilon_{ij}, \gamma_{[ij]}, \kappa_{i[jk]}), \tag{1}$$

where

$$\varepsilon_{ij} \equiv \tfrac{1}{2}(u_{j,i} + u_{i,j}) = \varepsilon_{ji}, \tag{2a}$$

$$\gamma_{[ij]} \equiv u_{[j,i]} - \psi_{[ij]} = -\gamma_{[ji]}, \tag{2b}$$

$$\kappa_{i[jk]} \equiv \psi_{[jk],i} = -\kappa_{i[kj]}, \tag{2c}$$

in which the u_i are the components of displacement and the $\psi_{[ij]}$ are the components of rotation of the micro-structure.

We adopt the definitions

$$\tau_{ij} \equiv \frac{\partial W}{\partial \varepsilon_{ij}} = \tau_{ji}, \tag{3a}$$

$$\sigma_{[ij]} \equiv \frac{\partial W}{\partial \gamma_{[ij]}} = -\sigma_{[ji]}, \tag{3b}$$

$$\mu_{i[jk]} \equiv \frac{\partial W}{\partial \kappa_{i[jk]}} = -\mu_{i[kj]}. \tag{3c}$$

Here τ_{ij} is the classical (Cauchy) stress and $\mu_{i[jk]}$ is the Cosserat couple-stress. The Cosserats considered $\sigma_{[ij]}$ to be the antisymmetric part of an asymmetric tensor whose symmetric part is the Cauchy stress. However, if the Cosserat theory is extended [2] to include strain of the micro-structure, as well as rotation, $\sigma_{[ij]}$ appears naturally as the antisymmetric part of another tensor. This distinction is retained here.

The variation of W is

$$\begin{aligned}\delta W &= \tau_{ij}\delta\varepsilon_{ij} + \sigma_{[ij]}\delta\gamma_{[ij]} + \mu_{i[jk]}\delta\kappa_{i[jk]},\\ &= [(\tau_{ij}+\sigma_{[ij]})\delta u_j]_{,i} - (\tau_{ij}+\sigma_{[ij]})_{,i}\delta u_j - \sigma_{[ij]}\delta\psi_{[ij]} + (\mu_{i[jk]}\delta\psi_{[jk]})_{,i} - \mu_{i[jk],i}\delta\psi_{[jk]}\end{aligned}$$

and, with the divergence theorem,

$$\begin{aligned}\int_V \delta W\,\mathrm{d}V = \int_S n_i[(\tau_{ij}+\sigma_{[ij]})\delta u_j + \mu_{i[jk]}\delta\psi_{[jk]}]\,\mathrm{d}S\\ -\int_V [(\tau_{ij}+\sigma_{[ij]})\delta u_j + (\mu_{i[jk],i}+\sigma_{[jk]})\delta\psi_{jk}]\,\mathrm{d}V,\end{aligned} \tag{4}$$

where the n_i are the components of the unit outward normal to the surface S of a volume V.

A principle of virtual work is expressed by

$$\int_V \delta W\,\mathrm{d}V = \int_S (t_j\delta u_j + \mathrm{T}_{[jk]}\delta\psi_{[jk]})\,\mathrm{d}S + \int_V (f_j\delta u_j + \Phi_{[jk]}\delta\psi_{[jk]})\,\mathrm{d}V, \tag{5}$$

where t_j is the surface traction, $\mathrm{T}_{[jk]}$ is the surface couple, f_j is the body force and $\Phi_{[jk]}$ is the body couple.

Upon equating coefficients of like independent variations, δu_i and $\delta\psi_{[jk]}$, in the surface and volume integrals in (4) and (5), we arrive at the Cosserat stress-equations of equilibrium [1]

$$(\tau_{ij}+\sigma_{[ij]})_{,i} + f_j = 0, \tag{6a}$$

$$\mu_{i[jk],i} + \sigma_{[jk]} + \Phi_{[jk]} = 0, \tag{6b}$$

and the boundary conditions

$$t_j = n_i(\tau_{ij}+\sigma_{[ij]}), \tag{7a}$$

$$\mathrm{T}_{[jk]} = n_i\mu_{i[jk]}. \tag{7b}$$

We shall consider only isotropic materials without initial stress. Then the potential energy-density must be a linear function of such product pairs of ε_{ij}, $\gamma_{[ij]}$ and $\kappa_{i[jk]}$ as are scalars. There are two such for ε_{ij} ($\varepsilon_{ii}\varepsilon_{jj}$ and $\varepsilon_{ij}\varepsilon_{ij}$), one for $\gamma_{[ij]}$ ($\gamma_{[ij]}\gamma_{[ij]}$) and three for $\kappa_{i[jk]}$ ($\kappa_{i[ik]}\kappa_{j[kj]}$, $\kappa_{i[jk]}\kappa_{i[jk]}$ and $\kappa_{i[jk]}\kappa_{j[ki]}$) and no others. Hence*

$$W = \tfrac{1}{2}\lambda\varepsilon_{ii}\varepsilon_{jj} + \mu\varepsilon_{ij}\varepsilon_{ij} + \beta\gamma_{[ij]}\gamma_{[ij]} + \alpha_1\kappa_{i[ik]}\kappa_{j[kj]} + \alpha_2\kappa_{i[jk]}\kappa_{i[jk]} + \alpha_3\kappa_{i[jk]}\kappa_{j[ki]}. \tag{8}$$

* The relations between the constants β and α_i in (8) and those in [2] are

$$\beta = \tfrac{1}{2}(b_2 - b_3), \qquad \alpha_2 = \tfrac{1}{2}(a_{10} - a_{13}),$$

$$\alpha_1 = a_2 - \tfrac{1}{2}a_3 - \tfrac{1}{2}a_8,$$

$$\alpha_3 = a_{11} - \tfrac{1}{2}a_{14} - \tfrac{1}{2}a_{15}.$$

Then (3) become

$$\tau_{ij} = \lambda\delta_{ij}\varepsilon_{kk} + 2\mu\varepsilon_{ij}, \tag{9a}$$

$$\sigma_{[ij]} = 2\beta\gamma_{[ij]}, \tag{9b}$$

$$\mu_{i[jk]} = \alpha_1(\kappa_{l[lj]}\delta_{ik} + \kappa_{l[kl]}\delta_{ij}) + 2\alpha_2\kappa_{i[jk]} + \alpha_3(\kappa_{k[ij]} + \kappa_{j[ki]}). \tag{9c}$$

The forty-two equations (2), (6) and (9) constitute one form of the equations of the linear theory of equilibrium of an isotropic, elastic, Cosserat continuum. As noted previously, these equations have been amplified, in another paper [2], to include a homogeneous strain of the micro-structure. In [2], the complete deformation of the micro-structure is represented by an asymmetric tensor ψ_{ij}. To reduce to the Cosserat continuum, it is only necessary to set the symmetric part of ψ_{ij} equal to zero.

Passage to a limit, as $\beta \to \infty$, $\gamma_{[ij]} \to 0$, reduces the Cosserat continuum to one without micro-structure but with its potential energy-density dependent on strain and gradient of rotation. The resulting equations are those considered in [3].

Upon substituting (2) in (9) and the result in (6), we find the following equations on u_i and $\psi_{[ij]}$:

$$(\lambda+\mu-\beta)u_{j,ji} + (\mu+\beta)u_{i,jj} - 2\beta\psi_{[ji],j} + f_i = 0, \tag{10a}$$

$$(\alpha_1+\alpha_3)(\psi_{[ki],kj} + \psi_{[jk],ki}) + 2\alpha_2\psi_{[ij],kk} - 2\beta\psi_{[ij]} + \beta(u_{j,i} - u_{i,j}) + \Phi_{[ij]} = 0. \tag{10b}$$

STRESS FUNCTIONS

For the purpose of this section, it is convenient to express (10) in an invariant form. Let $\mathbf{u}$ and $\mathbf{f}$ be the vectors with components u_i and f_i; let $\boldsymbol{\psi}^A$, $\boldsymbol{\Phi}^A$ and $\mathbf{I}$ be the dyadics with components $\psi_{[ij]}$, $\Phi_{[ij]}$ and δ_{ij}; and write $\boldsymbol{\Phi}^A = -\frac{1}{2}\mathbf{I}\times\mathbf{c}$, i.e. in terms of a body couple vector $\mathbf{c}$. Then (10) become

$$(\lambda+\mu-\beta)\nabla\nabla\cdot\mathbf{u} + (\mu+\beta)\nabla^2\mathbf{u} - 2\beta\nabla\cdot\boldsymbol{\psi}^A + \mathbf{f} = 0, \tag{11a}$$

$$(\alpha_1+\alpha_3)(\nabla\cdot\boldsymbol{\psi}^A\nabla + \nabla\boldsymbol{\psi}^A\cdot\nabla) + 2\alpha_2\nabla^2\boldsymbol{\psi}^A - 2\beta\boldsymbol{\psi}^A + \beta(\nabla\mathbf{u} - \mathbf{u}\nabla) - \tfrac{1}{2}\mathbf{I}\times\mathbf{c} = 0. \tag{11b}$$

A proof will now be given that any solution $(\mathbf{u}, \boldsymbol{\psi}^A)$ of (11), in a region V bounded by a surface S, can be expressed as

$$\mathbf{u} = \nabla\times\mathbf{K} + (1 - l_3^2\nabla^2)(\mathbf{B} - l_1^2\nabla\nabla\cdot\mathbf{B}) - \tfrac{1}{2}(k_1 - l_3^2\nabla^2)\nabla[\mathbf{r}\cdot(1 - l_1^2\nabla^2)\mathbf{B} + B_0], \tag{12a}$$

$$\boldsymbol{\psi}^A = -\tfrac{1}{4}\mathbf{I}\times[\nabla^2\nabla(\mathbf{r}\cdot\mathbf{K} + K_0) + 2\nabla\times\mathbf{B}], \tag{12b}$$

where

$$\mu(1 - l_1^2\nabla^2)\nabla^2\mathbf{B} = -\mathbf{f} - \tfrac{1}{2}(1 + \mu/\beta)\nabla\times\mathbf{c}, \tag{13a}$$

$$\mu\nabla^2 B_0 = \mathbf{r}\cdot[\mathbf{f} + \tfrac{1}{2}(1 + \mu/\beta)\nabla\times\mathbf{c}], \tag{13b}$$

$$2\beta\nabla^2\mathbf{K} = \mathbf{c}, \tag{13c}$$

$$2\beta(1 - l_2^2\nabla^2)\nabla^2 K_0 = 4l_2^2\nabla\cdot\mathbf{c} - \mathbf{r}\cdot(1 - l_2^2\nabla^2)\mathbf{c}, \tag{13d}$$

$$k_1 = (\lambda+\mu)/(\lambda+2\mu),$$

$$l_1^2 = (\mu+\beta)(2\alpha_2 - \alpha_1 - \alpha_3)/2\mu\beta,$$

$$l_2^2 = \alpha_2/\beta, \quad l_3^2 = (2\alpha_2-\alpha_1-\alpha_3)/\beta,$$

and $\mathbf{r}$ is the position vector.

Consider a field point $P(x, y, z)$ and a source point $Q(\xi, \eta, \zeta)$ and define

$$4\pi\mathbf{U}_P \equiv -\int_V r_1^{-1}\mathbf{u}_Q \,\mathrm{d}V_Q,$$

where $\mathrm{d}V_Q = \mathrm{d}\xi\,\mathrm{d}\eta\,\mathrm{d}\zeta$ and r_1 is the distance between P and Q.
Then $\nabla^2\mathbf{U} = \mathbf{u}$, or

$$\mathbf{u} = \nabla\nabla\cdot\mathbf{U}-\nabla\times\nabla\times\mathbf{U}. \tag{14}$$

Define

$$\varphi \equiv \nabla\cdot\mathbf{U}, \qquad \mathbf{H} \equiv -\nabla\times\mathbf{U}, \qquad (\nabla\cdot\mathbf{H} = 0).$$

Substitution of these in (14) produces Helmholtz's resolution:

$$\mathbf{u} = \nabla\varphi+\nabla\times\mathbf{H}, \qquad \nabla\cdot\mathbf{H} = 0. \tag{15}$$

Similarly [4] a scalar function χ and a vector function $\mathbf{G}$ can be defined in terms of $\boldsymbol{\psi}^A$ in such a way that

$$\boldsymbol{\psi}^A = -\tfrac{1}{2}\mathbf{I}\times(\nabla\chi-\nabla^2\mathbf{G}), \qquad \nabla\cdot\mathbf{G} = 0. \tag{16}$$

The substitution of (15) and (16) into (11) converts the latter to*

$$\nabla^2\{(\lambda+2\mu)\nabla\varphi+\nabla\times[(\mu+\beta)\mathbf{H}-\beta\mathbf{G}]\}+\mathbf{f} = 0, \tag{17a}$$

$$\beta(1-l_2^2\nabla^2)\nabla\chi+\beta\nabla^2[\mathbf{H}-(1-l_3^2\nabla^2)\mathbf{G}]-\tfrac{1}{2}\mathbf{c} = 0. \tag{17b}$$

From (17b),

$$\beta\nabla^2\nabla\times\mathbf{H} = \beta(1-l_3^2\nabla^2)\nabla^2\nabla\times\mathbf{G}+\tfrac{1}{2}\nabla\times\mathbf{c}.$$

Upon substituting this expression for $\nabla^2\nabla\times\mathbf{H}$ into (17a), we find

$$\mu\nabla^2[k\nabla\varphi+(1-l_1^2\nabla^2)\nabla\times\mathbf{G}] = -\mathbf{f}-\tfrac{1}{2}(1+\mu/\beta)\nabla\times\mathbf{c}, \tag{18}$$

where $k = (\lambda+2\mu)/\mu$.
Now define

$$4\pi l_1^2\mathbf{B}'_P \equiv \int_V r_1^{-1}e^{-r_1/l_1}[k\nabla\varphi+(1-l_1^2\nabla^2)\nabla\times\mathbf{G}]_Q \,\mathrm{d}V_Q.$$

Then

$$(1-l_1^2\nabla^2)\mathbf{B}' = k\nabla\varphi+(1-l_1^2\nabla^2)\nabla\times\mathbf{G} \tag{19}$$

and, from (18) and (19),

$$\mu(1-l_1^2\nabla^2)\nabla^2\mathbf{B}' = -\mathbf{f}-\tfrac{1}{2}(1+\mu/\beta)\nabla\times\mathbf{c}. \tag{20}$$

Also, from (19),

$$(1-l_1^2\nabla^2)\nabla\cdot\mathbf{B}' = k\nabla^2\varphi. \tag{21}$$

Define

$$2k\varphi^* \equiv \mathbf{r}\cdot(1-l_1^2\nabla^2)\mathbf{B}'. \tag{22}$$

* Note that (17) are what (27) of Ref. [4] reduce to when $\boldsymbol{\psi} = \boldsymbol{\psi}^A$.

Then, using (20) and (21), we find

$$2k\mu\nabla^2\varphi^* = 2k\mu\nabla^2\varphi - \mathbf{r}\cdot[\mathbf{f}+\tfrac{1}{2}(1+\mu/\beta)\nabla\times\mathbf{c}]. \tag{23}$$

With the definition $B_0 \equiv 2k(\varphi-\varphi^*)$, we have, from (23) and (22),

$$\mu\nabla^2 B_0 = \mathbf{r}\cdot[\mathbf{f}+\tfrac{1}{2}(1+\mu/\beta)\nabla\times\mathbf{c}], \tag{24}$$

$$2k\varphi = \mathbf{r}\cdot(1-l_1^2\nabla^2)\mathbf{B}'+B_0. \tag{25}$$

Define

$$\mathbf{D} \equiv \mathbf{B}'-l_1^2\nabla\nabla\cdot\mathbf{B}'-k\nabla\varphi.$$

By (21), $\nabla\cdot\mathbf{D} = 0$. Hence there exists a function $\mathbf{G}^*$ such that $\nabla\times\mathbf{G}^* = \mathbf{D}$, or

$$\nabla\times\mathbf{G}^* = \mathbf{B}'-l_1^2\nabla\nabla\cdot\mathbf{B}'-k\nabla\varphi; \tag{26}$$

whence, by (21) and (19),

$$(1-l_1^2\nabla^2)\nabla\times\mathbf{G}^* = (1-l_1^2\nabla^2)\nabla\times\mathbf{G}^*.$$

With the definition $\mathbf{B}'' \equiv \nabla\times\mathbf{G}-\nabla\times\mathbf{G}^*$, we have

$$(1-l_1^2\nabla^2)\mathbf{B}'' = 0, \qquad \nabla\cdot\mathbf{B}'' = 0 \tag{27}$$

and, from (26) and (25),

$$\nabla\times\mathbf{G} = \mathbf{B}''+\mathbf{B}'-l_1^2\nabla\nabla\cdot\mathbf{B}'-\tfrac{1}{2}\nabla[\mathbf{r}\cdot(1-l_1^2\nabla^2)\mathbf{B}'+B_0]. \tag{28}$$

Define $\mathbf{B} \equiv \mathbf{B}'+\mathbf{B}''$. Then, in view of (27), (28) may be written as

$$\nabla\times\mathbf{G} = \mathbf{B}-l_1^2\nabla\nabla\cdot\mathbf{B}-\tfrac{1}{2}\nabla[\mathbf{r}\cdot(1-l_1^2\nabla^2)\mathbf{B}+B_0], \tag{29}$$

while (20) becomes

$$\mu(1-l_1^2\nabla^2)\nabla^2\mathbf{B} = -\mathbf{f}-\tfrac{1}{2}(1+\mu/\beta)\nabla\times\mathbf{c} \tag{30}$$

and (25) becomes

$$2k\varphi = \mathbf{r}\cdot(1-l_1^2\nabla^2)\mathbf{B}+B_0. \tag{31}$$

From (29), $\nabla\times\nabla\times\mathbf{G} = \nabla\times\mathbf{B}$; but $\nabla\times\nabla\times\mathbf{G} = \nabla\nabla\cdot\mathbf{G}-\nabla^2\mathbf{G}$ and, by (16), $\nabla\cdot\mathbf{G} = 0$. Hence

$$\nabla^2\mathbf{G} = -\nabla\times\mathbf{B}. \tag{32}$$

Now define

$$4\pi\chi_P^* \equiv -\int_V r_1^{-1}\chi_Q\,\mathrm{d}V_Q,$$

so that

$$\chi = \nabla^2\chi^*. \tag{33}$$

Substitution of (33) into (17b) yields

$$\beta\nabla^2[(1-l_2^2\nabla^2)\nabla\chi^*+\mathbf{H}-(1-l_3^2\nabla^2)\mathbf{G}]-\tfrac{1}{2}\mathbf{c} = 0. \tag{34}$$

Define

$$\mathbf{K} \equiv (1-l_2^2\nabla^2)\nabla\chi^*+\mathbf{H}-(1-l_3^2\nabla^2)\mathbf{G}.$$

Then

$$2\beta\nabla^2\mathbf{K} = \mathbf{c}, \tag{35}$$

$$\nabla\cdot\mathbf{K} = (1-l_2^2\nabla^2)\nabla^2\chi^*, \tag{36}$$

$$\nabla\times\mathbf{K} = \nabla\times\mathbf{H}-(1-l_3^2\nabla^2)\nabla\times\mathbf{G}. \tag{37}$$

Define

$$\chi^{**} = \tfrac{1}{2}\mathbf{r}\cdot\mathbf{K} \tag{38}$$

and find, using (35) and (36),

$$4\beta(1-l_2^2\nabla^2)\nabla^2(\chi^{**}-\chi^*) = -4l_2^2\nabla\cdot\mathbf{c}+\mathbf{r}\cdot(1-l_2^2\nabla^2)\mathbf{c}. \tag{39}$$

Thus, with the definition

$$K_0 \equiv 2(\chi^*-\chi^{**}), \tag{40}$$

we have, from (39),

$$2\beta(1-l_2^2\nabla^2)\nabla^2 K_0 = 4l_2^2\nabla\cdot\mathbf{c}-\mathbf{r}\cdot(1-l_2^2\nabla^2)\mathbf{c}. \tag{41}$$

Finally, from (33), (40) and (38),

$$\chi = \tfrac{1}{2}\nabla^2(\mathbf{r}\cdot\mathbf{K}+K_0). \tag{42}$$

The expression (12a) for $\mathbf{u}$ is given by (15) with φ expressed in terms of $\mathbf{B}$ and B_0 by (31), $\nabla\times\mathbf{H}$ expressed in terms of $\mathbf{K}$ and $\nabla\times\mathbf{G}$ by (37) and $\nabla\times\mathbf{G}$ expressed in terms of $\mathbf{B}$ and B_0 by (29). The expression (12b) for $\boldsymbol{\psi}^A$ is given by (16) with χ expressed in terms of $\mathbf{K}$ and K_0 by (42) and $\nabla^2\mathbf{G}$ expressed in terms of $\nabla\times\mathbf{B}$ by (32). Equations (13) are, respectively, (30), (24), (35) and (41). These equations are elliptic because positive definiteness of the potential energy-density (8) requires

$$\mu > 0, \qquad \beta > 0, \qquad \alpha_2 > 0, \qquad 2\alpha_2-\alpha_1-\alpha_3 > 0,$$

and, hence, l_1^2 and l_2^2 are positive.

CONCENTRATED FORCE AND COUPLE

For the problem of a concentrated force in an infinite region, $\mathbf{c} = 0$ and we may take $\mathbf{K} = 0$, $K_0 = 0$. Then of the four equations (13), there remain only the following on $\mathbf{B}$ and B_0:

$$\mu(1-l_1^2\nabla^2)\nabla^2\mathbf{B} = -\mathbf{f},$$

$$\mu\nabla^2 B_0 = \mathbf{r}\cdot\mathbf{f}.$$

These are the same as the equations encountered in [3] except for the replacement of l_1 by another constant. Hence, the solution for a concentrated force $\mathbf{P}$ at the origin in an infinite region has the same form as that found in [3], namely:

$$\mathbf{B} = \frac{\mathbf{P}}{4\pi\mu r}(1-e^{-r/l_1}), \qquad B_0 = 0.$$

If body couples are present, but body forces are absent, the equations (13) on the stress functions are

$$2\mu(1-l_1^2\nabla^2)\nabla^2\mathbf{B} = -(1+\mu/\beta)\nabla\times\mathbf{c},$$
$$2\mu\nabla^2 B_0 = (1+\mu/\beta)\mathbf{r}\cdot\nabla\times\mathbf{c},$$
$$2\beta\nabla^2\mathbf{K} = \mathbf{c},$$
$$2\beta(1-l_2^2\nabla^2)\nabla^2 K_0 = 4l_2^2\nabla\cdot\mathbf{c}-\mathbf{r}\cdot(1-l_2^2\nabla^2)\mathbf{c}.$$

These have the same forms as equations encountered in the body couple problem considered in [3] and the body force problem considered in [2]. The same methods of integration employed in those papers may be used here for the case of a concentrated couple **C** at the origin of an infinite region. After noting that, in [2], the right hand side of (48) should be multiplied by 1/2, we find:

$$\mathbf{B} = -\frac{\mu+\beta}{8\pi\mu\beta}\mathbf{C}\times\nabla\left(\frac{1}{r}-\frac{\exp(-r/l_1)}{r}\right),$$

$$B_0 = 0, \qquad \mathbf{K} = -\frac{\mathbf{C}}{8\pi\beta r},$$

$$K_0 = -\frac{l_2^2}{4\pi\beta}\,\mathbf{C}\cdot\nabla\left(\frac{1}{r}-\frac{\exp(-r/l_2)}{r}\right).$$

REFERENCES

[1] E. and F. Cosserat, *Théorie des Corps Déformables*. Hermann & Fils, Paris (1909).
[2] R. D. Mindlin, *Arch. ration. Mech. Analysis* **16,** 51–78 (1964).
[3] R. D. Mindlin and H. F. Tiersten, *Arch. ration. Mech. Analysis* **11,** 415–448 (1962).
[4] R. D. Mindlin, *Int. J. Solids Structures* **1,** 73–78 (1965).

(*Received* 28 *September* 1964)

Zusammenfassung—In diesem Beitrag werden die Lineargleichungen eines isotropen, elastischen Cosserat Kontinuums vollständig gelöst, und zwar werden diese in Spannungsfunktionen, welche den Papkovitsch Funktionen der klassischen Elastizitätstheorie analog sind, ausgedrückt. Speziallösungen für eine konzentrierte Kraft und ein Kräftepaar werden ebenfalls angegeben.

Абстракт—В настоящей бумаге выводится полное решение, в терминах функций напряжения, аналогичных функциям Папковича классической упругости, для линейных уравнений изотропного упругого континуума Коссера.

Даются также специальные решения для сосредоточенного усилия и для пары сил.

Int. J. Solids Structures, 1965, Vol. 1, pp. 417 to 438. Pergamon Press Ltd. Printed in Great Britain

SECOND GRADIENT OF STRAIN AND SURFACE-TENSION IN LINEAR ELASTICITY

R. D. MINDLIN

Department of Civil Engineering, Columbia University, New York, N.Y.

Abstract—In this paper there is formulated a linear theory of deformation of an elastic solid in which the potential energy-density is a function of the strain and its first and second gradients. This is a theory in which cohesive force and surface-tension are intrinsic. A solution is given for the strain and surface-tension, or surface-energy per unit area, resulting from separation of a solid along a plane; and a comparison is made with an analogous lattice model. Also presented are a general solution of the displacement-equation of equilibrium in terms of stress functions and the particular solution for the concentrated force. The special case of a liquid is considered and the solutions are given for the surface-tensions at plane and spherical surfaces.

1. INTRODUCTION

IF THE potential energy–density of an elastic continuum is assumed to depend on the rotation-gradient, in addition to the strain, there results a theory of elasticity with couple-stresses, i.e. couples per unit area, in addition to the usual stresses. In a paper on elastic materials with couple-stresses [1], R. A. Toupin observed that the rotation-gradient comprises only eight of the eighteen components of the strain-gradient and he proceeded to establish the mathematical theory in which all the components are accounted for. The additional components of the strain-gradient are accompanied by ten components of self-equilibrating double forces per unit area, i.e. stresses contributing neither resultant force nor couple, per unit area, across a surface in the material but, nevertheless, contributing to the potential energy and to the boundary conditions. In a subsequent paper, Toupin and Gazis [2] exhibited the correspondence between strain-gradient theory and atomic lattices with nearest neighbor and next nearest neighbor interactions; and they explored the consequences of an initial, homogeneous, self-equilibrating stress. They showed that a free surface will draw in or push out in a surface-layer; but only in non-centrosymmetric materials. Toupin pointed out, in a private conversation, that this restriction could be removed by taking into account, further, the second gradient of the strain. A start in this direction was made in 1959 by Hart [3] for liquids. Also, Green and Rivlin [4] have recently established the basis of a very general theory which includes strain-gradients of any order. It seems worthwhile, however, to formulate completely the equations of the simplest theory, incorporating the second gradient of strain, and to exhibit some typical solutions. The cohesive force and surface-tension, or surface-energy per unit area, which appear, may have some bearing on criteria of failure of solids and on a variety of surface-phenomena observed in both solids and liquids.

In what follows, equations of equilibrium, boundary conditions and constitutive equations are derived for a linear theory of elastic materials in which the potential energy-density is a 2nd degree polynomial in the infinitesimal strain and its first and second gradients. The case of the homogeneous, isotropic, centrosymmetric material is explored in some detail. Terms appear which describe the property of cohesion and a general formula is given for surface-tension: one-half the product of a modulus of cohesion and the normal gradient of the dilatation at the surface. After a formulation of the general displacement-equation of equilibrium, a solution is given for the problem of separation of a solid along a plane, including formulas for the surface-energy per unit area and for the residual strain—which decays exponentially into the interior. A decay constant of the order of magnitude of interatomic distances is estimated from electron diffraction data obtained by Germer, MacRae and Hartman [5]. The theory admits either monotonic or oscillatory decay: a property which is shown to be also the case in a solution of the analogous problem for a monatomic lattice with first, second and third neighbor interactions. Following this, a general solution of the displacement-equation of equilibrium is given in terms of stress functions. Then the particular solution is found for the concentrated force. Finally, a special theory for a liquid is formulated by restricting the strain and its first two gradients to the dilatation and its first two gradients. Solutions are given for the surface-tensions at plane, and spherical, liquid surfaces. It is found that, as the curvature of the surface increases (decreases) from zero, the surface-tension becomes algebraically smaller (larger).

2. STRESS-EQUATION OF EQUILIBRIUM AND BOUNDARY CONDITIONS

In the linear theory to be formulated, the potential energy-density, W, is assumed to be a function of three polyadics:

$$W = W(\overset{1}{\boldsymbol{\varepsilon}}, \overset{2}{\boldsymbol{\varepsilon}}, \overset{3}{\boldsymbol{\varepsilon}}), \tag{1}$$

where

$$\overset{1}{\boldsymbol{\varepsilon}} = \tfrac{1}{2}(\nabla\mathbf{u}+\mathbf{u}\nabla), \qquad \overset{2}{\boldsymbol{\varepsilon}} = \nabla\nabla\mathbf{u}, \qquad \overset{3}{\boldsymbol{\varepsilon}} = \nabla\nabla\nabla\mathbf{u}. \tag{2}$$

In (2), ∇ is the gradient operator and $\mathbf{u}$ is the displacement. The symmetric dyadic $\overset{1}{\boldsymbol{\varepsilon}}$ is the classical infinitesimal strain with six independent components. The triadic $\overset{2}{\boldsymbol{\varepsilon}}$, symmetric in the first two positions, has eighteen independent components which could be replaced [6] by the strain-gradient $\nabla\overset{1}{\boldsymbol{\varepsilon}}$, with eighteen independent components which are linear combinations of those of $\overset{2}{\boldsymbol{\varepsilon}}$, or by the rotation-gradient and a symmetric triadic with, respectively, eight and ten independent components which are linear combinations of those of $\overset{2}{\boldsymbol{\varepsilon}}$ or of $\nabla\overset{1}{\boldsymbol{\varepsilon}}$. Similarly, $\overset{3}{\boldsymbol{\varepsilon}}$, symmetric in the first three positions, has thirty independent components which could be replaced by the second gradient of strain $\nabla\nabla\overset{1}{\boldsymbol{\varepsilon}}$ or by a variety of combinations of other polyadics. Thus, the energy-density could be expressed as any one of a large number of functions of fifty-four independent variables. All of these energy functions lead to the same displacement-equation of equilibrium. The function chosen in (1) produces simpler forms of the boundary conditions and constitutive equations than most and as simple forms as any.

The variation of the total potential energy, in a volume V, with arbitrary variation of $\mathbf{u}$, is

$$\delta\int_V W\,\mathrm{d}V = \int_V (\overset{1}{\boldsymbol{\tau}} : \nabla\delta\mathbf{u} + \overset{2}{\boldsymbol{\tau}} \,\vdots\, \nabla\nabla\delta\mathbf{u} + \overset{3}{\boldsymbol{\tau}} :: \nabla\nabla\nabla\delta\mathbf{u})\,\mathrm{d}V, \tag{3}$$

where

$$\overset{1}{\boldsymbol{\tau}} = \frac{\partial W}{\partial\overset{1}{\boldsymbol{\varepsilon}}}, \qquad \overset{2}{\boldsymbol{\tau}} = \frac{\partial W}{\partial\overset{2}{\boldsymbol{\varepsilon}}}, \qquad \overset{3}{\boldsymbol{\tau}} = \frac{\partial W}{\partial\overset{3}{\boldsymbol{\varepsilon}}}. \tag{4}$$

and the rule for multiple scalar multiplication of dyads, triads, etc. is

$$\mathbf{ab} : \mathbf{cd} = (\mathbf{a} \cdot \mathbf{c})(\mathbf{b} \cdot \mathbf{d}), \qquad \mathbf{abc} \vdots \mathbf{def} = (\mathbf{a} \cdot \mathbf{d})(\mathbf{b} \cdot \mathbf{e})(\mathbf{c} \cdot \mathbf{f}), \qquad \text{etc.}$$

In the definitions (4), it is to be understood that $\overset{1}{\tau}$, $\overset{2}{\tau}$, $\overset{3}{\tau}$ have the same symmetries as $\overset{1}{\varepsilon}$, $\overset{2}{\varepsilon}$, $\overset{3}{\varepsilon}$, respectively.

By application of the chain rule of differentiation and the divergence theorem, the right-hand side of (3) is converted into the sum of

$$-\int_V [\nabla \cdot (\overset{1}{\tau} - \nabla \cdot \overset{2}{\tau} + \nabla\nabla : \overset{3}{\tau})] \cdot \delta\mathbf{u}\, \mathrm{d}V + \int_S \mathbf{n} \cdot (\overset{1}{\tau} - \nabla \cdot \overset{2}{\tau} + \nabla\nabla : \overset{3}{\tau}) \cdot \delta\mathbf{u}\, \mathrm{d}S \tag{5}$$

and

$$\int_S \mathbf{n} \cdot (\overset{2}{\tau} - \nabla \cdot \overset{3}{\tau}) : \nabla\delta\mathbf{u}\, \mathrm{d}S + \int_S \mathbf{n} \cdot \overset{3}{\tau} \vdots \nabla\nabla\delta\mathbf{u}\, \mathrm{d}S, \tag{6}$$

where S is the boundary of V and $\mathbf{n}$ is the unit outward normal to S. Now, $\nabla\delta\mathbf{u}$ is not independent of $\delta\mathbf{u}$ on S because, if $\delta\mathbf{u}$ is known on S, so is the surface-gradient of $\delta\mathbf{u}$. This is analogous to the situation in couple-stress theory [7] where, if the displacement is known on S, so is the normal component of rotation—just as, in the classical theory of flexure of plates, if the deflection at the edge is known, so is the component of rotation about an axis normal to the edge. To prepare for the formulation of a variational principle, we resolve $\nabla\delta\mathbf{u}$ into a surface-gradient and a normal gradient:

$$\nabla\delta\mathbf{u} = \overset{s}{\nabla}\delta\mathbf{u} + \mathbf{n}D\delta\mathbf{u}; \qquad \overset{s}{\nabla} = (\mathbf{I} - \mathbf{nn}) \cdot \nabla, \qquad D = \mathbf{n} \cdot \nabla, \tag{7}$$

where $\mathbf{I}$ is the unit dyadic. The term, in the integrand, with $\overset{s}{\nabla}\delta\mathbf{u}$ as coefficient can then be reduced to one with $\delta\mathbf{u}$ as coefficient by employing the chain rule and a surface integration—leaving only the independent variations $\delta\mathbf{u}$ and $D\delta\mathbf{u}$ as coefficients. A similar procedure, with some additional steps, can be used to convert the term with $\nabla\nabla\delta\mathbf{u}$ as coefficient to terms with the independent variations $\delta\mathbf{u}$, $D\delta\mathbf{u}$ and $D^2\delta\mathbf{u}$ as coefficients. Thus, if $\mathbf{\Phi}$ is a dyadic and $\mathbf{v}$ is a vector,

$$\mathbf{\Phi} : \nabla\mathbf{v} = \mathbf{\Phi} : \overset{s}{\nabla}\mathbf{v} + \mathbf{n} \cdot \mathbf{\Phi} \cdot D\mathbf{v} = \overset{s}{\nabla} \cdot (\mathbf{\Phi} \cdot \mathbf{v}) - (\overset{s}{\nabla} \cdot \mathbf{\Phi}) \cdot \mathbf{v} + \mathbf{n} \cdot \mathbf{\Phi} \cdot D\mathbf{v}.$$

By the surface divergence theorem [8] for a smooth, closed surface (which S is assumed to be),

$$\int_S \overset{s}{\nabla} \cdot (\mathbf{\Phi} \cdot \mathbf{v})\, \mathrm{d}S = \int_S (\overset{s}{\nabla} \cdot \mathbf{n})\mathbf{n} \cdot \mathbf{\Phi} \cdot \mathbf{v}\, \mathrm{d}S.$$

Hence,

$$\int_S \mathbf{\Phi} : \nabla\mathbf{v}\, \mathrm{d}S = \int_S (\mathbf{L} \cdot \mathbf{\Phi}) \cdot \mathbf{v}\, \mathrm{d}S + \int_S \mathbf{n} \cdot \mathbf{\Phi} \cdot D\mathbf{v}\, \mathrm{d}S, \tag{8}$$

where

$$\mathbf{L} = \mathbf{n}\overset{s}{\nabla} \cdot \mathbf{n} - \overset{s}{\nabla}. \tag{9}$$

Similarly, if $\mathbf{\Psi}$ is a triadic,

$$\int_S \mathbf{\Psi} \vdots \nabla\mathbf{\Phi}\, \mathrm{d}S = \int_S (\mathbf{L} \cdot \mathbf{\Psi}) : \mathbf{\Phi}\, \mathrm{d}S + \int_S \mathbf{n} \cdot \mathbf{\Psi} : D\mathbf{\Phi}\, \mathrm{d}S. \tag{10}$$

Thus, by (8), the first integral in (6) can be written as

$$\int_S \mathbf{L}\cdot[\mathbf{n}\cdot(\overset{2}{\boldsymbol{\tau}}-\nabla\cdot\overset{3}{\boldsymbol{\tau}})]\cdot\delta\mathbf{u}\,\mathrm{d}S+\int_S \mathbf{nn}:(\overset{2}{\boldsymbol{\tau}}-\nabla\cdot\overset{3}{\boldsymbol{\tau}})\cdot D\delta\mathbf{u}\,\mathrm{d}S \tag{11}$$

and, by (10), the second integral in (6) is

$$\int_S \mathbf{L}\cdot(\mathbf{n}\cdot\overset{3}{\boldsymbol{\tau}}):\nabla\delta\mathbf{u}\,\mathrm{d}S+\int_S \mathbf{nn}:\overset{3}{\boldsymbol{\tau}}:D\nabla\delta\mathbf{u}\,\mathrm{d}S. \tag{12}$$

Again by (8), the first integral in (12) can be written as

$$\int_S \mathbf{L}\cdot[\mathbf{L}\cdot(\mathbf{n}\cdot\overset{3}{\boldsymbol{\tau}})]\cdot\delta\mathbf{u}\,\mathrm{d}S+\int_S \mathbf{n}\cdot[\mathbf{L}\cdot(\mathbf{n}\cdot\overset{3}{\boldsymbol{\tau}})]\cdot D\delta\mathbf{u}\,\mathrm{d}S. \tag{13}$$

As for the second integral in (12), we suppose that S is a coordinate surface of an orthogonal, curvilinear, coordinate system and interpret $\mathbf{n}$ as one of the orthogonal unit vectors in that system. Then

$$D\nabla\mathbf{v} = \mathbf{n}\cdot\nabla\nabla\mathbf{v} = \nabla(\mathbf{n}\cdot\nabla\mathbf{v})-(\nabla\mathbf{n})\cdot\nabla\mathbf{v}.$$

But $\nabla(\mathbf{n}\cdot\nabla\mathbf{v}) = \nabla D\mathbf{v} = \overset{s}{\nabla}D\mathbf{v}+\mathbf{n}D^2\mathbf{v}$ and $(\nabla\mathbf{n})\cdot\nabla\mathbf{v} = (\overset{s}{\nabla}\mathbf{n})\cdot\overset{s}{\nabla}\mathbf{v}$. Hence:

$$D\nabla\mathbf{v} = \overset{s}{\nabla}D\mathbf{v}+\mathbf{n}D^2\mathbf{v}-(\overset{s}{\nabla}\mathbf{n})\cdot\overset{s}{\nabla}\mathbf{v}.$$

Accordingly,

$$\begin{aligned}\boldsymbol{\Phi}:D\nabla\mathbf{v} &= \boldsymbol{\Phi}:[\overset{s}{\nabla}D\mathbf{v}-(\overset{s}{\nabla}\mathbf{n})\cdot\overset{s}{\nabla}\mathbf{v}+\mathbf{n}D^2\mathbf{v}],\\ &= \overset{s}{\nabla}\cdot[\boldsymbol{\Phi}\cdot D\mathbf{v}-(\overset{s}{\nabla}\mathbf{n})\cdot\boldsymbol{\Phi}\cdot\mathbf{v}]-(\overset{s}{\nabla}\cdot\boldsymbol{\Phi})\cdot D\mathbf{v}+\overset{s}{\nabla}\cdot[(\overset{s}{\nabla}\mathbf{n})\cdot\boldsymbol{\Phi}]\cdot\mathbf{v}+\mathbf{n}\cdot\boldsymbol{\Phi}\cdot D^2\mathbf{v},\\ &= \overset{s}{\nabla}\cdot[\boldsymbol{\Phi}\cdot D\mathbf{v}-(\overset{s}{\nabla}\mathbf{n})\cdot\boldsymbol{\Phi}\cdot\mathbf{v}]-(\overset{s}{\nabla}\cdot\boldsymbol{\Phi})\cdot D\mathbf{v}-\mathbf{L}\cdot[(\overset{s}{\nabla}\mathbf{n})\cdot\boldsymbol{\Phi}]\cdot\mathbf{v}+\mathbf{n}\cdot\boldsymbol{\Phi}\cdot D^2\mathbf{v},\end{aligned}$$

where the last step is a consequence of $\mathbf{n}\cdot(\overset{s}{\nabla}\mathbf{n}) = 0$ and (9). Hence, using the surface divergence theorem and noting again that $\mathbf{n}\cdot(\overset{s}{\nabla}\mathbf{n}) = 0$, we can write the second integral in (12) as

$$-\int_S \mathbf{L}\cdot[(\overset{s}{\nabla}\mathbf{n})\cdot(\mathbf{nn}:\overset{3}{\boldsymbol{\tau}})]\cdot\delta\mathbf{u}\,\mathrm{d}S+\int_S \mathbf{L}\cdot(\mathbf{nn}:\overset{3}{\boldsymbol{\tau}})\,D\delta\mathbf{u}\,\mathrm{d}S+\int_S \mathbf{nnn}\,\vdots\,\overset{3}{\boldsymbol{\tau}}\cdot D^2\delta\mathbf{u}\,\mathrm{d}S. \tag{14}$$

Altogether, the right-hand side of (3) is the sum of (5), (11), (13) and (14); so that (3) becomes

$$\begin{aligned}\delta\int_V W\,\mathrm{d}V = &-\int_V [\nabla\cdot(\overset{1}{\boldsymbol{\tau}}-\nabla\cdot\overset{2}{\boldsymbol{\tau}}+\nabla\nabla:\overset{3}{\boldsymbol{\tau}})]\cdot\delta\mathbf{u}\,\mathrm{d}V+\int_S \mathbf{n}\cdot(\overset{1}{\boldsymbol{\tau}}-\nabla\cdot\overset{2}{\boldsymbol{\tau}}+\nabla\nabla:\overset{3}{\boldsymbol{\tau}})\cdot\delta\mathbf{u}\,\mathrm{d}S\\ &+\int_S \{\mathbf{L}\cdot[\mathbf{n}\cdot(\overset{2}{\boldsymbol{\tau}}-\nabla\cdot\overset{3}{\boldsymbol{\tau}})+\mathbf{L}\cdot(\mathbf{n}\cdot\overset{3}{\boldsymbol{\tau}})-(\overset{s}{\nabla}\mathbf{n})\cdot(\mathbf{nn}:\overset{3}{\boldsymbol{\tau}})]\}\cdot\delta\mathbf{u}\,\mathrm{d}S\\ &+\int_S \{\mathbf{nn}:(\overset{2}{\boldsymbol{\tau}}-\nabla\cdot\overset{3}{\boldsymbol{\tau}})+\mathbf{n}\cdot[\mathbf{L}\cdot(\mathbf{n}\cdot\overset{3}{\boldsymbol{\tau}})]+\mathbf{L}\cdot(\mathbf{nn}:\overset{3}{\boldsymbol{\tau}})\}\cdot D\delta\mathbf{u}\,\mathrm{d}S\\ &+\int_S (\mathbf{nnn}\,\vdots\,\overset{3}{\boldsymbol{\tau}})\cdot D^2\delta\mathbf{u}\,\mathrm{d}S.\end{aligned} \tag{15}$$

We now assume the following principle of stationary potential energy:

$$\delta\int_V W\,\mathrm{d}V = \int_V \mathbf{f}\cdot\delta\mathbf{u}\,\mathrm{d}V + \int_S (\overset{1}{\mathbf{t}}\cdot\delta\mathbf{u} + \overset{2}{\mathbf{t}}\cdot D\delta\mathbf{u} + \overset{3}{\mathbf{t}}\cdot D^2\delta\mathbf{u})\,\mathrm{d}S, \tag{16}$$

where $\mathbf{f}$ is the body force per unit volume and $\overset{1}{\mathbf{t}}$, $\overset{2}{\mathbf{t}}$, $\overset{3}{\mathbf{t}}$ are generalized surface tractions. From (15) and (16) there ensues the stress-equation of equilibrium

$$\nabla\cdot(\overset{1}{\tau} - \nabla\cdot\overset{2}{\tau} + \nabla\nabla : \overset{3}{\tau}) + \mathbf{f} = 0 \tag{17}$$

and the traction boundary conditions

$$\overset{1}{\mathbf{t}} = \mathbf{n}\cdot(\overset{1}{\tau} - \nabla\cdot\overset{2}{\tau} + \nabla\nabla : \overset{3}{\tau}) + \mathbf{L}\cdot[\mathbf{n}\cdot(\overset{2}{\tau} - \nabla\cdot\overset{3}{\tau}) + \mathbf{L}\cdot(\mathbf{n}\cdot\overset{3}{\tau}) - (\overset{s}{\nabla}\mathbf{n})\cdot(\mathbf{nn} : \overset{3}{\tau})], \tag{18a}$$

$$\overset{2}{\mathbf{t}} = \mathbf{nn} : (\overset{2}{\tau} - \nabla\cdot\overset{3}{\tau}) + \mathbf{n}\cdot[\mathbf{L}\cdot(\mathbf{n}\cdot\overset{3}{\tau})] + \mathbf{L}\cdot(\mathbf{nn} : \overset{3}{\tau}), \tag{18b}$$

$$\overset{3}{\mathbf{t}} = \mathbf{nnn}\,\vdots\,\overset{3}{\tau}. \tag{18c}$$

Thus, there are three scalar equations of equilibrium and only nine scalar boundary conditions instead of what appeared at first, in (5) and (6), to be $3+9+18 = 30$ boundary conditions. This reduction is analogous to the reduction from three to two boundary conditions in the classical theory of flexure of plates.

Other admissible sets of nine boundary conditions, in terms of $\mathbf{u}$, $D\mathbf{u}$ and $D^2\mathbf{u}$ or products of appropriate components of $\overset{1}{\mathbf{t}}$ and $\mathbf{u}$ or $\overset{2}{\mathbf{t}}$ and $D\mathbf{u}$ or $\overset{3}{\mathbf{t}}$ and $D^2\mathbf{u}$, are apparent from the form of (16) and the uniqueness theorem which would follow the steps from (1) to (17) in reverse.

3. CONSTITUTIVE EQUATIONS

We shall consider, here, only homogeneous, centrosymmetric, isotropic materials. Then, in the 2nd degree polynomial W, with constant coefficients, the variables $\overset{1}{\boldsymbol{\varepsilon}}$, $\overset{2}{\boldsymbol{\varepsilon}}$, $\overset{3}{\boldsymbol{\varepsilon}}$ can appear in the linear terms only as scalars and in the quadratic terms only as products that are scalars. This eliminates the terms linear in $\overset{2}{\boldsymbol{\varepsilon}}$ and the products of $\overset{2}{\boldsymbol{\varepsilon}}$ with $\overset{1}{\boldsymbol{\varepsilon}}$ and $\overset{3}{\boldsymbol{\varepsilon}}$. Also, there are only two quadratic products of $\overset{1}{\boldsymbol{\varepsilon}}$, five of $\overset{2}{\boldsymbol{\varepsilon}}$, seven of $\overset{3}{\boldsymbol{\varepsilon}}$ and three of $\overset{1}{\boldsymbol{\varepsilon}}$ with $\overset{3}{\boldsymbol{\varepsilon}}$ that are scalars. Furthermore, we may omit the constant term and the term linear in $\overset{1}{\boldsymbol{\varepsilon}}$ because, just as in classical elasticity, the material configuration, to which W is referred, can be chosen so that these terms do not appear. What remains, then, is

$$\begin{aligned} W = {} & \tfrac{1}{2}\lambda\varepsilon_{ii}\varepsilon_{jj} + \mu\varepsilon_{ij}\varepsilon_{ij} + a_1\varepsilon_{ijj}\varepsilon_{ikk} + a_2\varepsilon_{iik}\varepsilon_{kjj} + a_3\varepsilon_{iik}\varepsilon_{jjk} \\ & + a_4\varepsilon_{ijk}\varepsilon_{ijk} + a_5\varepsilon_{ijk}\varepsilon_{kji} + b_1\varepsilon_{iijj}\varepsilon_{kkll} + b_2\varepsilon_{ijkk}\varepsilon_{ijll} \\ & + b_3\varepsilon_{iijk}\varepsilon_{jkll} + b_4\varepsilon_{iijk}\varepsilon_{llkj} + b_5\varepsilon_{iijk}\varepsilon_{lljk} + b_6\varepsilon_{ijkl}\varepsilon_{ijkl} \\ & + b_7\varepsilon_{ijkl}\varepsilon_{jkli} + c_1\varepsilon_{ii}\varepsilon_{jjkk} + c_2\varepsilon_{ij}\varepsilon_{ijkk} + c_3\varepsilon_{ij}\varepsilon_{kkij} + b_0\varepsilon_{iijj}, \end{aligned} \tag{19}$$

where ε_{ij}, ε_{ijk}, ε_{ijkl} $(i, j, k, l = 1, 2, 3)$ are the components of $\overset{1}{\boldsymbol{\varepsilon}}$, $\overset{2}{\boldsymbol{\varepsilon}}$, $\overset{3}{\boldsymbol{\varepsilon}}$, respectively, in any orthogonal coordinate system, and the summation convention for repeated indices is employed.

The constants λ and μ are the usual Lamé constants and the five a_n are the additional constants which appear in Toupin's strain-gradient theory [1, 6].

The choice of the polynomial (19) implies that terms of higher degree are small in comparison with those retained. This, in turn, implies that certain assumptions as to order of magnitude have been made in addition to $|\nabla\mathbf{u}| \ll 1$ of the classical linear theory. While $\overset{1}{\boldsymbol{\varepsilon}}$ is dimensionless, $\overset{2}{\boldsymbol{\varepsilon}}$ has the dimension of reciprocal of length and $\overset{3}{\boldsymbol{\varepsilon}}$ has the dimension of reciprocal of the square of length. Thus, the ratios b_n/μ have the dimension of the fourth power of length and the ratios a_n/μ, b_0/μ, c_n/μ have the dimension of square of length. Suppose that l^2 is the ratio of one of the constants a_n to μ; i.e. l is a material property with the dimension of length. Then we assume that $a_n/\mu l^2$, $b_n/\mu l^4$, $c_n/\mu l^2$, $b_0/\mu l^2$ are of order of magnitude unity and

$$|\nabla\mathbf{u}| \ll 1, \qquad l|\nabla\nabla\mathbf{u}| \ll 1, \qquad l^2|\nabla\nabla\nabla\mathbf{u}| \ll 1,$$

where $|\boldsymbol{\Psi}|$ signifies the absolute values of the components of $\boldsymbol{\Psi}$. With these assumptions, all of the terms in (19) are of the same order of magnitude and terms of higher degree are negligibly small.

From (19) and (4) follow the constitutive equations:

$$\tau_{pq} = \lambda\varepsilon_{ii}\delta_{pq}+2\mu\varepsilon_{pq}+c_1\varepsilon_{iijj}\delta_{pq}+c_2\varepsilon_{pqii}+\tfrac{1}{2}c_3(\varepsilon_{iipq}+\varepsilon_{iiqp}), \tag{20a}$$

$$\begin{aligned}\tau_{pqr} = {} & a_1(\varepsilon_{pii}\delta_{qr}+\varepsilon_{qii}\delta_{pr})+\tfrac{1}{2}a_2(\varepsilon_{iip}\delta_{qr}+2\varepsilon_{rii}\delta_{qp}+\varepsilon_{iiq}\delta_{pr})\\ & +2a_3\varepsilon_{iir}\delta_{pq}+2a_4\varepsilon_{pqr}+a_5(\varepsilon_{rqp}+\varepsilon_{rpq}),\end{aligned} \tag{20b}$$

$$\begin{aligned}\tau_{pqrs} = {} & \tfrac{2}{3}b_1\varepsilon_{iijj}\delta_{pqrs}+\tfrac{2}{3}b_2\varepsilon_{jkii}\delta_{jkpqrs}+\tfrac{1}{6}b_3[(\varepsilon_{iijk}+\varepsilon_{iikj})\delta_{jkpqrs}+2\varepsilon_{jsii}\delta_{jpqr}]\\ & +\tfrac{2}{3}b_4\varepsilon_{iisj}\delta_{jpqr}+\tfrac{2}{3}b_5\varepsilon_{iijs}\delta_{jpqr}+2b_6\varepsilon_{pqrs}+\tfrac{2}{3}b_7(\varepsilon_{qrsp}+\varepsilon_{rspq}+\varepsilon_{spqr})\\ & +\tfrac{1}{3}c_1\varepsilon_{ii}\delta_{pqrs}+\tfrac{1}{3}c_2\varepsilon_{ij}\delta_{ijpqrs}+\tfrac{1}{3}c_3\varepsilon_{is}\delta_{ipqr}+\tfrac{1}{3}b_0\delta_{pqrs},\end{aligned} \tag{20c}$$

where $\tau_{pq}, \tau_{pqr}, \tau_{pqrs}$ are the components of $\overset{1}{\boldsymbol{\tau}}, \overset{2}{\boldsymbol{\tau}}, \overset{3}{\boldsymbol{\tau}}$, respectively; δ_{ij} is the Kronecker delta and

$$\delta_{ijkl} = \delta_{ij}\delta_{kl}+\delta_{ik}\delta_{jl}+\delta_{jk}\delta_{il}, \qquad \delta_{ijklmn} = \delta_{ik}\delta_{jl}\delta_{mn}+\delta_{ik}\delta_{jm}\delta_{ln}+\delta_{il}\delta_{jm}\delta_{kn}.$$

The τ_{pq} are like the components of the usual stress with the dimensions of force per unit area; but here they depend on the second gradient of strain in addition to the strain. The τ_{pqr} have the character of double forces per unit area. The combinations $\frac{1}{2}(\tau_{pqr}-\tau_{prq})$ are the eight components of the deviator of the couple-stress, i.e. couples per unit area, while the remaining ten independent combinations are self-equilibrating. The thirty τ_{pqrs} have the dimensions of force and the character of triple forces per unit area—all self-equilibrating.

4. SURFACE-TENSION

Of special interest, in the energy–density (19), is the term linear in $\overset{3}{\boldsymbol{\varepsilon}}$ which produces the homogeneous, self-equilibrating components of *cohesive force*, $\frac{1}{3}b_0\delta_{pqrs}$, in (20c). The coefficient b_0, with the dimensions of force, is a *modulus of cohesion*. It gives rise to surface-tension; or, equivalently, to the energy, per unit area, associated with the formation of a new surface.

In classical hydrostatics, surface-tension is introduced by postulating the existence of a vanishingly thin surface-membrane under uniform tension. If the liquid is under no external forces, the surface-tension is equal to the total potential energy divided by the surface-area. In a solid, the surface-tension in a body under no external forces need not be uniform; but we may adopt the same definition for the average surface-tension:

$$T_{\text{ave}} = \mathscr{W}/A, \tag{21}$$

where A is the area of the surface and

$$\mathscr{W} = \int_V W \, dV;$$

or, from (19) and (20),

$$\mathscr{W} = \tfrac{1}{2}\int_V (\overset{1}{\tau} : \nabla \mathbf{u} + \overset{2}{\tau} \vdots \nabla\nabla \mathbf{u} + \overset{3}{\tau} :: \nabla\nabla\nabla \mathbf{u}) \, dV + \tfrac{1}{2} b_0 \int_V \nabla^2 \nabla \cdot \mathbf{u} \, dV. \tag{22}$$

The second integral in (22) enters because the cohesive force is constant whereas the remaining parts of the "stresses" vary in proportion to the strain and its gradients.

Now, if there are no external forces, i.e. if $\overset{1}{\mathbf{t}}, \overset{2}{\mathbf{t}}, \overset{3}{\mathbf{t}}$ and $\mathbf{f}$ are zero, the first integral in (22) vanishes; as can be shown by the procedure employed in Section 2. Hence,

$$T_{\text{ave}} = \frac{b_0}{2A}\int_V \nabla^2 \nabla \cdot \mathbf{u} \, dV = \frac{b_0}{2A}\int_S \mathbf{n} \cdot \nabla\nabla \cdot \mathbf{u} \, dS.$$

With this result, it seems appropriate to take

$$T = \tfrac{1}{2} b_0 \mathbf{n} \cdot \nabla\nabla \cdot \mathbf{u}|_S \tag{23}$$

as the definition of the *punctual* surface-tension, or surface-energy per unit area, in a centrosymmetric, isotropic, elastic solid; i.e. one-half the product of the modulus of cohesion and the normal gradient of the dilatation at the surface.

5. DISPLACEMENT-EQUATION OF EQUILIBRIUM

Upon substituting the strain–displacement relations (2) into the stress–strain relations (20) and the resulting expressions for the stresses into the stress-equation of equilibrium (17), we find the displacement-equation of equilibrium

$$[\lambda + 2\mu - (\bar{a} - 2\bar{c})\nabla^2 + \bar{b}\nabla^4]\nabla\nabla \cdot \mathbf{u} - [\mu - (\bar{a}' - c_3)\nabla^2 + \bar{b}'\nabla^4]\nabla \times \nabla \times \mathbf{u} + \mathbf{f} = 0, \tag{24}$$

where

$$\bar{a} = 2(a_1 + a_2 + a_3 + a_4 + a_5), \quad \bar{b} = 2(b_1 + b_2 + b_3 + b_4 + b_5 + b_6 + b_7),$$

$$\bar{c} = c_1 + c_2 + c_3, \quad \bar{a}' = 2(a_3 + a_4), \quad \bar{b}' = 2(b_5 + b_6).$$

Alternatively, (24) may be written in the form

$$(\lambda + 2\mu) D_{11}^2 D_{12}^2 \nabla\nabla \cdot \mathbf{u} - \mu D_{21}^2 D_{22}^2 \nabla \times \nabla \times \mathbf{u} + \mathbf{f} = 0, \tag{25}$$

where

$$D_{ij}^2 = 1 - l_{ij}^2 \nabla^2, \qquad i = 1, 2; \qquad j = 1, 2; \tag{26}$$

and

$$2(\lambda + 2\mu) l_{1j}^2 = \bar{a} - 2\bar{c} \pm [(\bar{a} - 2\bar{c})^2 - 4\bar{b}(\lambda + 2\mu)]^{\frac{1}{2}}, \qquad j = 1, 2; \tag{27a}$$

$$2\mu l_{2j}^2 = \bar{a}' - c_3 \pm [(\bar{a}' - c_3)^2 - 4\bar{b}'\mu]^{\frac{1}{2}}, \qquad j = 1, 2. \tag{27b}$$

Thus, in general, solutions of the equation of equilibrium will contain Lamé's constants, λ and μ, and four additional material constants, l_{ij}, having the dimension of length. Application of boundary conditions may introduce some or all of the remaining material constants which appear in the energy-density (19).

The conditions for positive W do not include relations between $\bar{a}$ (or $\bar{a}'$) and $\bar{b}, \bar{c}$ (or $\bar{b}', c_3$) and, hence, supply no indication of the character, real or complex, of the l_{ij}. In what follows, the l_{ij} will be treated as if they were real and positive; but complex l_{ij} are equally admissible.

6. STRAIN AND SURFACE-TENSION AT A PLANE SURFACE OF A SOLID

In a rectangular coordinate system x, y, z, consider the half-space $x \geqslant 0$ without body force and with its plane surface traction-free:

$$\overset{1}{\mathbf{t}} = \overset{2}{\mathbf{t}} = \overset{3}{\mathbf{t}} = 0 \qquad \text{on } x = 0. \tag{28}$$

Assume

$$u_x = u_x(x), \qquad u_y = u_z = 0. \tag{29}$$

Then, with primes designating differentiations with respect to x, the stress-equation of equilibrium (17) reduces to

$$\tau'_{xx} - \tau''_{xxx} + \tau'''_{xxxx} = 0, \tag{30}$$

while the boundary conditions (18) become

$$(\tau_{xx} - \tau'_{xxx} + \tau''_{xxxx})_{x=0} = (\tau_{xxx} - \tau'_{xxxx})_{x=0} = (\tau_{xxxx})_{x=0} = 0. \tag{31}$$

Also, with the assumed form of displacement (29), the constitutive equations (20) yield

$$\tau_{xx} = (\lambda + 2\mu)\varepsilon + \bar{c}\varepsilon'', \qquad \tau_{xxx} = \bar{a}\varepsilon', \qquad \tau_{xxxx} = b_0 + \bar{c}\varepsilon + \bar{b}\varepsilon'', \tag{32}$$

where $\varepsilon = \mathrm{d}u_x/\mathrm{d}x$. Further, the displacement-equation of equilibrium (25) becomes

$$\left(1 - l_{11}^2 \frac{\mathrm{d}^2}{\mathrm{d}x^2}\right)\left(1 - l_{12}^2 \frac{\mathrm{d}^2}{\mathrm{d}x^2}\right)\frac{\mathrm{d}^2 u_x}{\mathrm{d}x^2} = 0. \tag{33}$$

One integration of the stress-equation of equilibrium (30) makes $\tau_{xx} - \tau'_{xxx} + \tau''_{xxxx}$ equal to a constant which, by the first of the boundary conditions (31), must be zero. This disposes of one of the three boundary conditions on $x = 0$. The remaining two, expressed in terms of ε through (32), are

$$[(\bar{a} - \bar{c})\varepsilon' - \bar{b}\varepsilon''']_{x=0} = 0, \qquad (\bar{c}\varepsilon + \bar{b}\varepsilon'')_{x=0} = -b_0. \tag{34}$$

The solution of (33), vanishing at infinity, is

$$u_x = A_1 e^{-x/l_{11}} + A_2 e^{-x/l_{12}}. \tag{35}$$

Inserting (35) into (34) and making use of the definitions (27a), we find the following equations for determining the constants A_1 and A_2:

$$A_1 l_{12}^2 (l_{11}^2 + l_{10}^2) + A_2 l_{11}^2 (l_{12}^2 + l_{10}^2) = 0, \tag{36a}$$

$$A_1 l_{12} (l_{12}^2 + l_{10}^2) + A_2 l_{11} (l_{11}^2 + l_{10}^2) = b_0 l_{11} l_{12}/(\lambda + 2\mu), \tag{36b}$$

where $l_{10}^2 = \bar{c}/(\lambda + 2\mu)$. From (36), the values of A_1 and A_2 follow immediately.

The formula (23), for the surface-tension yields

$$T = -\tfrac{1}{2}b_0\varepsilon'|_{x=0} = -\tfrac{1}{2}b_0(A_1 l_{11}^{-2} + A_2 l_{12}^{-2}),$$

or

$$T = \frac{b_0^2(l_{11}^2 - l_{12}^2)}{2(\lambda+2\mu)[l_{11}(l_{12}^2+l_{10}^2)^2 - l_{12}(l_{11}^2+l_{10}^2)^2]}. \tag{37}$$

This is also, for each surface, the energy per unit area required to separate the body along a plane.

As may be seen from (35), the displacement and strain decay exponentially with distance from the surface. It is known that such an effect is confined to an extremely thin surface-layer in physical materials; so that l_{11} and l_{12} are probably very small lengths. An estimate of order of magnitude can be obtained from measurements of electron diffraction at nickel surfaces by Germer, MacRae and Hartman [5]. They conclude that the displacement of the superficial layer of atoms toward the interior is five times as large as that of the next layer. The assumption of a simple exponential decay, proportional to $e^{-x/l}$, would lead to an l of about five eighths of the distance between adjacent planes of atoms. This is a ratio of the same order of magnitude as the analogous one of about three eighths of the diameter of a sphere, in a simple cubic array of contiguous elastic spheres, computed for the case when only the rotation-gradient (leading to couple-stresses) is taken into account, in the energy-density, in addition to the strain [9]. If the l_{ij} are complex, the solution (35) would take the form of products of negative exponentials and trigonometric functions. Then the decay of displacement and strain would be oscillatory and the estimate of order of magnitude would apply to the real part.

7. LATTICE MODEL

Consider a one-dimensional lattice of interacting particles distributed along the x-axis at points $x = nh$, where n is a positive or negative integer and h is a constant which is taken equal to unity. Suppose that forces between particles are limited to first, second and third neighbors and include self-equilibrating initial forces. Now separate the lattice between particles $n = 0$ and $n = -1$ and consider the semi-infinite lattice $n \geqslant 0$. The separation is effected by the addition of forces P_0, P_1, P_2, on particles 0, 1, 2, respectively, equal and opposite to the resultants of the initial forces exerted on those particles by the particles -1, -2, -3. Since the initial forces are self-equilibrating,

$$P_0 + P_1 + P_2 = 0. \tag{38}$$

Now, suppose that the change of force between two particles is proportional to their relative displacement; with constants of proportionality $\alpha_1, \alpha_2, \alpha_3$ for first, second and third neighbors, respectively. Then, if u_n is the displacement of the nth particle, the equilibrium of that particle is expressed by

$$\sum_{i=1}^{i=3} \alpha_i(u_{n+i} - 2u_n + u_{n-i}) = 0, \qquad n \geqslant 3, \tag{39}$$

$$\sum_{i=1}^{i=2} \alpha_i(u_{2+i} - 2u_2 + u_{2-i}) + \alpha_3(u_5 - u_2) = P_2, \qquad n = 2, \tag{40a}$$

$$\alpha_1(u_2-2u_1+u_0)+\alpha_2(u_3-u_1)+\alpha_3(u_4-u_1) = P_1, \qquad n = 1, \tag{40b}$$

$$\alpha_1(u_1-u_0)+\alpha_2(u_2-u_0)+\alpha_3(u_3-u_0) = P_0, \qquad n = 0. \tag{40c}$$

If we adopt the notation $\Delta u_n = u_{n+1}-u_n$, we have

$$\Delta^2 u_n = u_{n+1}-2u_n+u_{n-1},$$

$$\Delta^4 u_n = u_{n+2}-4u_{n+1}+6u_n-4u_{n-1}+u_{n-2},$$

$$\Delta^6 u_n = u_{n+3}-6u_{n+2}+15u_{n+1}-20u_n+15u_{n-1}-6u_{n-2}+u_{n-3}.$$

Then, with

$$\beta_1 = \alpha_1+4\alpha_2+9\alpha_3, \qquad \beta_2 = -\alpha_2-6\alpha_3, \tag{41}$$

equation (39) can be written as

$$(\beta_1-\beta_2\Delta^2+\alpha_3\Delta^4)\Delta^2 u_n = 0; \tag{42}$$

or

$$(1-\lambda_1^2\Delta^2)(1-\lambda_2^2\Delta^2)\Delta^2 u_n = 0, \tag{43}$$

where

$$2\beta_1\lambda_i^2 = \beta_2 \pm (\beta_2^2-4\beta_1\alpha_3)^{\frac{1}{2}}, \qquad i = 1, 2. \tag{44}$$

The solution of (43), vanishing as $n \to \infty$, is

$$u_n = A_1' e^{-n/\mu_1} + A_2' e^{-n/\mu_2}, \tag{45}$$

where

$$\cosh \mu_i^{-1} = 1+\tfrac{1}{2}\lambda_i^{-2}, \qquad i = 1, 2. \tag{46}$$

As for the boundary conditions, again there are three for the two constants A_1' and A_2'. However, following Gazis and Wallis [10], we sum the three equations (40) and obtain

$$(1-\lambda_1^2\Delta^2)(1-\lambda_2^2\Delta^2)\Delta u_2 = 0, \tag{47}$$

which is satisfied identically by the solution (45). Thus, as in Section 6, the general condition of equilibrium disposes of one of the three boundary conditions. This leaves two conditions which may be any two, independent, linear combinations of the three conditions (40) other than their sum. The resulting formulas for A_1' and A_2' are lengthy expressions of little present interest.

More important is a comparison of the solutions (45), for the lattice, and (35) for the continuum. Both are comprised of two negative exponentials. In the case of the continuum, no conclusion could be reached regarding the character, real or complex, of the decay constants. The same is true for the lattice.

The quadratic part of the potential energy of the nth particle, $n > 2$, is

$$W_n = \tfrac{1}{2}\beta_1(\Delta u_n)^2+\tfrac{1}{2}\beta_2'(\Delta^2 u_n)^2+\tfrac{1}{2}\alpha_3(\Delta^3 u_n)^2+\beta_2''\Delta u_n\Delta^3 u_n;$$

i.e. an energy leading to the equilibrium equation (42):

$$-\frac{\partial W_n}{\partial u_n} = (\beta_1-\beta_2\Delta^2+\alpha_3\Delta^4)\Delta^2 u_n = 0,$$

where $\beta_2 = \beta'_2 - 2\beta''_2$. The conditions for positive W_n supply no relation between β'_2 (corresponding to $\bar{a}$ of the continuum) and the remaining constants; and, hence, no indication of the character of the λ_i and μ_i.

For real, positive force constants, α_i, standing in the relations

$$\alpha_1 \gg \alpha_2 \gg \alpha_3,$$

(for example, going inversely as high powers of the distances between interacting particles) the roots λ_i in (44) would be complex and then (46) would require complex μ_i. However, real, positive α_i, though appealing, are not necessary for stability.

8. STRESS FUNCTIONS

In this section, it is proved that any solution, $\mathbf{u}$, of the displacement-equation of equilibrium, (24) or (25), in a region V bounded by a surface S, can be expressed as

$$\mathbf{u} = \mathbf{B} - (l_{21}^2 + l_{22}^2 - l_{21}^2 l_{22}^2 \nabla^2)\nabla\nabla\cdot\mathbf{B} - \tfrac{1}{2}\nabla(D_{11}^2 D_{12}^2 - k^{-1})(\mathbf{r}\cdot D_{21}^2 D_{22}^2 \mathbf{B} + B_0), \tag{48}$$

where $k = (\lambda + 2\mu)/\mu$, $\mathbf{r}$ is the position vector and

$$\mu D_{11}^2 D_{12}^2 \nabla^2 B_0 = \mathbf{r}\cdot D_{11}^2 D_{12}^2 \mathbf{f} - 4(l_{11}^2 + l_{12}^2 - \tfrac{3}{2} l_{11}^2 l_{12}^2 \nabla^2)\nabla\cdot\mathbf{f}, \tag{49}$$

$$\mu D_{21}^2 D_{22}^2 \nabla^2 \mathbf{B} = -\mathbf{f}. \tag{50}$$

First, by the usual proof of the Helmholtz resolution, functions φ and $\mathbf{H}$ can be defined, in terms of $\mathbf{u}$, so that

$$\mathbf{u} = \nabla\varphi + \nabla\times\mathbf{H}, \qquad \nabla\cdot\mathbf{H} = 0. \tag{51}$$

Upon substituting (51) in (25), we find

$$\mu\nabla^2[k D_{11}^2 D_{12}^2 \nabla\varphi + D_{21}^2 D_{22}^2 \nabla\times\mathbf{H}] + \mathbf{f} = 0. \tag{52}$$

Define

$$4\pi\mathbf{B}' = \int_V \psi(r_1)[k D_{11}^2 D_{12}^2 \nabla\varphi + D_{21}^2 D_{22}^2 \nabla\times\mathbf{H}]_Q \, \mathrm{d}V_Q, \tag{53}$$

where

$$\psi(r_1) = r_1^{-1}(e^{-r_1/l_{21}} - e^{-r_1/l_{22}})/(l_{21}^2 - l_{22}^2), \tag{54}$$

r_1 is the distance between a field point $P(x, y, z)$ and a source point $Q(\xi, \eta, \zeta)$ and $\mathrm{d}V_Q = \mathrm{d}\xi\,\mathrm{d}\eta\,\mathrm{d}\zeta$. Then, by a process similar to that for Poisson's equation [11],

$$D_{21}^2 D_{22}^2 \mathbf{B}' = k D_{11}^2 D_{12}^2 \nabla\varphi + D_{21}^2 D_{22}^2 \nabla\times\mathbf{H} \tag{55}$$

and, from (55) and (52),

$$\mu D_{21}^2 D_{22}^2 \nabla^2 \mathbf{B}' + \mathbf{f} = 0. \tag{56}$$

The divergence of (55) produces

$$D_{21}^2 D_{22}^2 \nabla\cdot\mathbf{B}' = k D_{11}^2 D_{12}^2 \nabla^2\varphi. \tag{57}$$

Define

$$B_0 = 2k\varphi - \mathbf{r}\cdot D_{21}^2 D_{22}^2 \mathbf{B}' \tag{58}$$

and find

$$\mu D_{11}^2 D_{12}^2 \nabla^2 B_0 = \mathbf{r} \cdot D_{11}^2 D_{12}^2 \mathbf{f} - 4(l_{11}^2 + l_{12}^2 - \tfrac{3}{2} l_{11}^2 l_{12}^2 \nabla^2) \nabla \cdot \mathbf{f} \tag{59}$$

by using (56) and (57). Now, define

$$\mathbf{B}'' = \nabla \times \mathbf{H} - \mathbf{B}' + (l_{21}^2 + l_{22}^2 - l_{21}^2 l_{22}^2 \nabla^2) \nabla\nabla \cdot \mathbf{B}' + k D_{11}^2 D_{12}^2 \nabla \varphi \tag{60}$$

and note that, by (55) and (57),

$$D_{21}^2 D_{22}^2 \mathbf{B}'' = 0, \qquad \nabla \cdot \mathbf{B}'' = 0. \tag{61}$$

Then substitute (60) and (58) into (51) to get

$$\mathbf{u} = \mathbf{B}' + \mathbf{B}'' - (l_{21}^2 + l_{22}^2 - l_{21}^2 l_{22}^2 \nabla^2) \nabla\nabla \cdot \mathbf{B}' - \tfrac{1}{2} \nabla (D_{11}^2 D_{12}^2 - k^{-1})(\mathbf{r} \cdot D_{21}^2 D_{22}^2 \mathbf{B}' + B_0). \tag{62}$$

Finally, define

$$\mathbf{B} = \mathbf{B}' + \mathbf{B}''. \tag{63}$$

In view of (61), we may write (62) and (56) in the form of (48) and (50); and (59) is already in the form of (49). Thus, (48) is a complete solution of the equation of equilibrium, (24) or (25), if the stress functions B_0 and $\mathbf{B}$ satisfy (49) and (50).

9. CONCENTRATED FORCE

In an infinite region V, let the body force be zero outside a finite region V_0 which contains the origin and a non-vanishing field of undirectional forces $\mathbf{f}$ per unit volume. A concentrated force is defined by

$$\mathbf{F} = \lim_{V_0 \to 0} \int_V \mathbf{f}_Q \, dV_Q. \tag{64}$$

To find B_0 and $\mathbf{B}$ for the concentrated force, we have to solve equations of the type (49) and (50), viz.:

$$D_{i1}^2 D_{i2}^2 \nabla^2 \psi = \rho, \qquad i = 1, 2 \text{ (not summed)}. \tag{65}$$

This may be done by constructing the pertinent Green's formula from Green's identity

$$\int_S \mathbf{n} \cdot (\varphi_i \nabla \psi - \psi \nabla \varphi_i)_Q \, dS_Q = \int_V (\varphi_i \nabla^2 \psi - \psi \nabla^2 \varphi_i)_Q \, dV_Q \tag{66}$$

in the usual way, as follows. In (66) replace φ_i and ψ, successively, by

$$(1 - l_{i1}^2 \nabla^2)(1 - l_{i2}^2 \nabla^2)(\varphi_i, \psi),$$

$$l_{i1} l_{i2} [1 - (l_{i1}^2 + l_{i2}^2) \nabla^2] \nabla^2 (\varphi_i, \psi),$$

$$[(l_{i1}^2 + l_{i2}^2)^2 - 2 l_{i1}^2 l_{i2}^2]^{\frac{1}{2}} \nabla^2 (\varphi_i, \psi),$$

$$[l_{i1}^4 l_{i2}^4 - l_{i1}^2 l_{i2}^2 (l_{i1}^2 + l_{i2}^2)^2]^{\frac{1}{2}} \nabla^4 (\varphi_i, \psi)$$

and subtract the sum of the last three from the first of the resulting four equations. Now choose

$$\varphi_i(r_1) = \frac{1}{r_1} \left(1 - \frac{l_{i1}^2 e^{-r_1/l_{i1}}}{l_{i1}^2 - l_{i2}^2} + \frac{l_{i2}^2 e^{-r_1/l_{i2}}}{l_{i1}^2 - l_{i2}^2} \right), \qquad i = 1, 2$$

and note that

$$D_{i1}^2 D_{i2}^2 \nabla^2 \varphi_i = 0$$

except at P. Exclude the point P by surrounding it with a spherical surface of radius ε and center P. Upon passing to the limit as $\varepsilon \to 0$, we find the Green's formula

$$4\pi\psi = \int_S \mathbf{n} \cdot \sum_{j=1}^{j=4} A_j [\mathscr{L}_j \varphi_i \nabla \mathscr{L}_j \psi - \mathscr{L}_j \psi \nabla \mathscr{L}_j \varphi_i]_Q \, \mathrm{d}S_Q - \int_V \rho_Q \varphi_i \, \mathrm{d}V_Q, \tag{67}$$

where

$$\mathscr{L}_1 = D_{i1}^2 D_{i2}^2, \quad \mathscr{L}_2 = [1-(l_{i1}^2+l_{i2}^2)\nabla^2]\nabla^2, \quad \mathscr{L}_3 = \nabla^2, \quad \mathscr{L}_4 = \nabla^4,$$

$$A_1 = 1, \quad A_2 = -l_{i1}^2 l_{i2}^2, \quad A_3 = 2l_{i1}^2 l_{i2}^2 - (l_{i1}^2 + l_{i2}^2)^2, \quad A_4 = l_{i1}^2 l_{i2}^2[(l_{i1}^2 + l_{i2}^2)^2 - l_{i1}^2 l_{i2}^2].$$

If B_0 and $\mathbf{B}$ vanish at infinity at least as r^{-1}, the solution for the concentrated force is, from (49), (50), (64) and (67),

$$4\pi\mu B_0 = -\lim_{V_0 \to 0} \int_V [\mathbf{r}' \cdot D_{11}^2 D_{12}^2 \mathbf{f} - 4(l_{11}^2 + l_{12}^2 - \tfrac{3}{2} l_{11}^2 l_{12}^2 \nabla^2)\nabla \cdot \mathbf{f}]_Q \varphi_1(r_1) \, \mathrm{d}V_Q, \tag{68}$$

$$4\pi\mu \mathbf{B} = \lim_{V_0 \to 0} \int_V \mathbf{f}_Q \varphi_2(r_1) \, \mathrm{d}V_Q, \tag{69}$$

where $r' = (\xi^2 + \eta^2 + \zeta^2)^{\frac{1}{2}}$. Now,

$$\lim_{V_0 \to 0} r_1 = r, \qquad \lim_{V_0 \to 0} r' = 0.$$

Hence, the solution for the concentrated force is

$$2\pi\mu B_0 = \mathbf{F} \cdot \nabla(l_{11}^2 + l_{12}^2 - l_{11}^2 l_{12}^2 \nabla^2)\varphi_1(r),$$

$$4\pi\mu \mathbf{B} = \mathbf{F}\varphi_2(r).$$

The limit for $\mathbf{B}$ follows directly from (69) while that for B_0 is obtained from (68) after successive applications of the chain rule and the divergence theorem to separate $\mathbf{f}$ as a factor, of the integrand of the volume integral, before passing to the limit. Finally, it is apparent that B_0 and $\mathbf{B}$ have the required behavior at infinity.

10. ELASTIC LIQUID

The linear theory of deformation of an elastic liquid can be treated as a special case of a solid in which the potential energy-density is a function of only the infinitesimal dilatation and its gradients instead of the full infinitesimal strain and its gradients. Thus,

$$W_\Delta = W_\Delta(\Delta, \nabla\Delta, \nabla\nabla\Delta), \tag{70}$$

where $\Delta = \nabla \cdot \mathbf{u}$ is the infinitesimal dilatation. For the centrosymmetric, isotropic liquid, the form of W_Δ is obtained from (19) by discarding all but the terms indicated in (70), leaving

$$W_\Delta = \tfrac{1}{2}\lambda\Delta^2 + a_1 \nabla\Delta \cdot \nabla\Delta + b_1(\nabla^2\Delta)^2 + b_2 \nabla\nabla\Delta : \nabla\nabla\Delta + c_1 \Delta\nabla^2\Delta + b_0 \nabla^2\Delta. \tag{71}$$

This contains the term $b_2 \nabla\nabla\Delta : \nabla\nabla\Delta$ which Hart omitted in a previous study [3].

As noted by Hart, either the dilatation or the displacement may be subject to variation in varying the potential energy:

$$\delta \int_V W_\Delta \, dV = -\int_V (p\delta\Delta + \mathbf{p} \cdot \nabla\delta\Delta + \boldsymbol{\pi} : \nabla\nabla\delta\Delta) \, dV, \tag{72}$$

or

$$\delta \int_V W_\Delta \, dV = -\int_V (p\nabla \cdot \delta\mathbf{u} + \mathbf{p} \cdot \nabla\nabla \cdot \delta\mathbf{u} + \boldsymbol{\pi} : \nabla\nabla\nabla \cdot \delta\mathbf{u}) \, dV, \tag{73}$$

where, in either case,

$$p = -\frac{\partial W_\Delta}{\partial \Delta} = -\lambda\Delta - c_1\nabla^2\Delta, \tag{74a}$$

$$\mathbf{p} = -\frac{\partial W_\Delta}{\partial \nabla\Delta} = -2a_1\nabla\Delta, \tag{74b}$$

$$\boldsymbol{\pi} = -\frac{\partial W_\Delta}{\partial \nabla\nabla\Delta} = -(b_0 + 2b_1\nabla^2\Delta + c_1\Delta)\mathbf{I} - 2b_2\nabla\nabla\Delta. \tag{74c}$$

We consider, first, $\delta\Delta$ as the independent variation. Then, by applying the chain rule and the divergence theorem, we transform the right-hand side of (72) to

$$-\int_V (p - \nabla \cdot \mathbf{p} + \nabla\nabla : \boldsymbol{\pi})\delta\Delta \, dV - \int_S \mathbf{n} \cdot (\mathbf{p} - \nabla \cdot \boldsymbol{\pi})\delta\Delta \, dS - \int_S \mathbf{n} \cdot \boldsymbol{\pi} \cdot \nabla\delta\Delta \, dS.$$

Hart did not observe that $\nabla\delta\Delta$ is not independent of $\delta\Delta$ on S; but, by the same procedure as in Section 2, we find

$$\int_S \mathbf{n} \cdot \boldsymbol{\pi} \cdot \nabla\delta\Delta \, dS = \int_S [\mathbf{L} \cdot (\mathbf{n} \cdot \boldsymbol{\pi})]\delta\Delta \, dS + \int_S \mathbf{nn} : \boldsymbol{\pi} D\delta\Delta \, dS,$$

thereby converting to independent variations $\delta\Delta$ and $D\delta\Delta$. Hence

$$\delta \int_V W_\Delta \, dV = -\int_V (p - \nabla \cdot \mathbf{p} + \nabla\nabla : \boldsymbol{\pi})\delta\Delta \, dV - \int_S \mathbf{n} \cdot (\mathbf{p} - \nabla \cdot \boldsymbol{\pi})\delta\Delta \, dS$$
$$- \int_S [\mathbf{L} \cdot (\mathbf{n} \cdot \boldsymbol{\pi})]\delta\Delta \, dS - \int_S \mathbf{nn} : \boldsymbol{\pi} D\delta\Delta \, dS. \tag{75}$$

The principle of stationary potential energy now assumed is

$$\delta \int_V W_\Delta \, dV = \int_V \varphi\delta\Delta \, dV + \int_S (\overset{1}{t}_\Delta \delta\Delta + \overset{2}{t}_\Delta D\delta\Delta) \, dS, \tag{76}$$

where φ is the potential of a body force per unit volume and $\overset{1}{t}_\Delta$, $\overset{2}{t}_\Delta$ are generalized surface-tractions. Equating the right-hand sides of (75) and (76), we find the stress-equation of equilibrium

$$p - \nabla \cdot \mathbf{p} + \nabla\nabla : \boldsymbol{\pi} + \varphi = 0 \tag{77}$$

and the traction boundary conditions

$$\overset{1}{t}_\Delta = -\mathbf{n} \cdot (\mathbf{p} - \nabla \cdot \boldsymbol{\pi}) - \mathbf{L} \cdot (\mathbf{n} \cdot \boldsymbol{\pi}), \qquad \overset{2}{t}_\Delta = -\mathbf{nn} : \boldsymbol{\pi}. \tag{78}$$

The equation of equilibrium on Δ is obtained by substituting (74) into (77), with the result:

$$\lambda\Delta - 2(a_1 - c_1)\nabla^2\Delta + 2(b_1 + b_2)\nabla^4\Delta = \varphi, \tag{79}$$

or

$$\lambda(1-l_1^2\nabla^2)(1-l_2^2\nabla^2)\Delta = \varphi, \tag{80}$$

where

$$\lambda l_i^2 = a_1 - c_1 \pm [(a_1-c_1)^2 - 2\lambda(b_1+b_2)]^{\frac{1}{2}}, \qquad i = 1, 2. \tag{81}$$

We now turn to (73), in which $\delta \mathbf{u}$ is the variation, and find, by application of the chain rule and the divergence theorem,

$$\delta \int_V W_\Delta \, dV = \int_V [\nabla(p - \nabla \cdot \mathbf{p} + \nabla\nabla : \boldsymbol{\pi})] \cdot \delta\mathbf{u} \, dV - \int_S (p - \nabla \cdot \mathbf{p} + \nabla\nabla : \boldsymbol{\pi})\mathbf{n} \cdot \delta\mathbf{u} \, dS$$
$$- \int_S [\mathbf{n} \cdot (\mathbf{p} - \nabla \cdot \boldsymbol{\pi})]\nabla \cdot \delta\mathbf{u} \, dS - \int_S \mathbf{n} \cdot \boldsymbol{\pi} \cdot \nabla\nabla \cdot \delta\mathbf{u} \, dS. \tag{82}$$

Here, again, Hart did not notice the lack of independence of $\delta\mathbf{u}$, $\nabla \cdot \delta\mathbf{u}$ and $\nabla\nabla \cdot \delta\mathbf{u}$ on S; but, by the same procedure as in Section 2, we find

$$\delta \int_V W_\Delta \, dV = \int_V [\nabla(p - \nabla \cdot \mathbf{p} + \nabla\nabla : \boldsymbol{\pi})] \cdot \delta\mathbf{u} \, dV - \int_S (p - \nabla \cdot \mathbf{p} + \nabla\nabla : \boldsymbol{\pi})\mathbf{n} \cdot \delta\mathbf{u} \, dS$$
$$- \int_S \{\mathbf{L}[\mathbf{n} \cdot (\mathbf{p} - \nabla \cdot \boldsymbol{\pi}) + \mathbf{L} \cdot (\mathbf{n} \cdot \boldsymbol{\pi})] - \mathbf{L} \cdot [(\overset{s}{\nabla}\mathbf{n})\mathbf{nn} : \boldsymbol{\pi}]\} \cdot \delta\mathbf{u} \, dS$$
$$- \int_S \{[\mathbf{n} \cdot (\mathbf{p} - \nabla \cdot \boldsymbol{\pi}) + \mathbf{L} \cdot (\mathbf{n} \cdot \boldsymbol{\pi})]\mathbf{n} + \mathbf{L}(\mathbf{nn} : \boldsymbol{\pi})\} \cdot D\delta\mathbf{u} \, dS$$
$$- \int_S (\mathbf{nn} : \boldsymbol{\pi})\mathbf{n} \cdot D^2\delta\mathbf{u} \, dS. \tag{83}$$

In this case, the principle of stationary potential energy is assumed to be:

$$\delta \int_V W_\Delta \, dV = \int_V \mathbf{f} \cdot \delta\mathbf{u} \, dV + \int_S (\overset{1}{\mathbf{t}} \cdot \delta\mathbf{u} + \overset{2}{\mathbf{t}} \cdot D\delta\mathbf{u} + \overset{3}{\mathbf{t}} \cdot D^2\delta\mathbf{u}) \, dS, \tag{84}$$

so that the stress-equation of equilibrium is

$$\nabla(p - \nabla \cdot \mathbf{p} + \nabla\nabla : \boldsymbol{\pi}) - \mathbf{f} = 0 \tag{85}$$

and the traction boundary conditions are

$$\overset{1}{\mathbf{t}} = -\mathbf{n}(p - \nabla \cdot \mathbf{p} + \nabla\nabla : \boldsymbol{\pi}) - \mathbf{L}[\mathbf{n} \cdot (\mathbf{p} - \nabla \cdot \boldsymbol{\pi}) + \mathbf{L} \cdot (\mathbf{n} \cdot \boldsymbol{\pi})] + \mathbf{L} \cdot [(\overset{s}{\nabla}\mathbf{n})\mathbf{nn} : \boldsymbol{\pi}], \tag{86a}$$

$$\overset{2}{\mathbf{t}} = -\mathbf{nn} \cdot (\mathbf{p} - \nabla \cdot \boldsymbol{\pi}) - \mathbf{nL} \cdot (\mathbf{n} \cdot \boldsymbol{\pi}) - \mathbf{L}(\mathbf{nn} : \boldsymbol{\pi}), \tag{86b}$$

$$\overset{3}{\mathbf{t}} = -\mathbf{nnn} : \boldsymbol{\pi}. \tag{86c}$$

The displacement-equation of equilibrium, obtained by substituting (74) in (85) and employing (81), is

$$\lambda(1-l_1^2\nabla^2)(1-l_2^2\nabla^2)\nabla\nabla \cdot \mathbf{u} + \mathbf{f} = 0. \tag{87}$$

The formula for the surface-tension is obtained as in Section 4:

$$T = \tfrac{1}{2}b_0 \mathbf{n} \cdot \nabla\Delta|_S \tag{88}$$

and is the same for variation of Δ as for variation of $\mathbf{u}$.

11. DILATATION AND SURFACE-TENSION AT A PLANE SURFACE OF A LIQUID

The problem of the liquid half-space $x \geqslant 0$, with traction-free boundary and no body force, is formulated, first, in terms of the equations derived from variation of Δ. The solution of (80), vanishing as $x \to \infty$, with Δ a function of x only and $\varphi = 0$, is

$$\Delta = A_1 e^{-x/l_1} + A_2 e^{-x/l_2}; \tag{89}$$

and the two boundary conditions (78) become

$$[(2a_1 - c_1)\Delta' - 2(b_1+b_2)\Delta''']_{x=0} = 0, \qquad [c_1\Delta + 2(b_1+b_2)\Delta'']_{x=0} = -b_0, \tag{90}$$

where primes signify differentiations with respect to x. Substituting (89) in (90), we have a pair of equations for A_1 and A_2—the solution of which, after the use of (81), is

$$A_1 = -\frac{b_0 l_1(l_2^2+l_0^2)}{\lambda[l_1(l_2^2+l_0^2)^2 - l_2(l_1^2+l_0^2)^2]},$$

$$A_2 = \frac{b_0 l_2(l_1^2+l_0^2)}{\lambda[l_1(l_2^2+l_0^2)^2 - l_2(l_1^2+l_0^2)^2]},$$

where $l_0^2 = c_1/\lambda$. Then (88) and (89) yield the formula for the surface-tension at a plane, liquid surface:

$$T_0 = \frac{b_0^2(l_1^2 - l_2^2)}{2\lambda[l_1(l_2^2+l_0^2)^2 - l_2(l_1^2+l_0^2)^2]}. \tag{91}$$

In the alternate case of the formulation derived from varying $\mathbf{u}$, the appropriate solution of (87) may be written in the form

$$u_x = A_0 - l_1 A_1' e^{-x/l_1} - l_2 A_2' e^{-x/l_2}. \tag{92}$$

There are now the three boundary conditions (86); but they lead to

$$A_0 = 0, \qquad A_1' = A_1, \qquad A_2' = A_2$$

so that the two solutions are the same.

12. DILATATION AND SURFACE-TENSION AT A SPHERICAL SURFACE OF A LIQUID

We consider a liquid sphere under no body force and with a traction-free surface of radius r_0. The appropriate solution of the equation of equilibrium (80), regular at the origin $r = 0$, can be written in the form

$$\Delta = A_1 l_1 r^{-1}\sinh(r/l_1) + A_2 l_2 r^{-1}\sinh(r/l_2). \tag{93}$$

Then the boundary conditions (78) yield

$$\sum_{i=1}^{i=2} A_i r_i[2(b_1+b_2) - (2a_1-c_1)l_i^2](r_i\cosh r_i - \sinh r_i) = 0, \tag{94a}$$

$$\sum_{i=1}^{i=2} A_i\{r_i[2b_2(1+2r_i^{-1}) + 2b_1 + c_1 l_i^2]\sinh r_i - 4b_2\cosh r_i\} = -b_0 r_0^2, \tag{94b}$$

where $r_i = r_0/l_i$. Again, it may be verified that the same result is obtained from the equations derived by varying the displacement.

To find a simple result for the first order effect of the curvature of the surface, assume that the radius of curvature is large in comparison with the l_i (i.e. $r_i \gg 1$). Then the solution of (94) is

$$A_1 \approx -\frac{2b_0\lambda^{-1}r_0(l_2^2+l_0^2)e^{-r_1}}{l_1(l_2^2+l_0^2)^2-l_2(l_1^2+l_0^2)^2+4b_2\lambda^{-1}r_0^{-1}(l_1^2-l_2^2)},$$

$$A_2 \approx \frac{2b_0\lambda^{-1}r_0(l_1^2+l_0^2)e^{-r_2}}{l_1(l_2^2+l_0^2)^2-l_2(l_1^2+l_0^2)^2+4b_2\lambda^{-1}r_0^{-1}(l_1^2-l_2^2)};$$

and the surface tension is, to the first order in l_i/r_0,

$$T \approx \frac{T_0}{1+(8b_2T_0/b_0^2r_0)}, \tag{95}$$

where T_0 is the surface-tension at a plane surface as given by (91). The condition of positive definiteness of the quadratic part of the potential energy-density (71) requires $b_2 > 0$. Hence, as the radius of the surface diminishes, the surface-tension becomes algebraically smaller.

A result of the classical theory of surface-tension is that the internal pressure in the sphere is $2T/r_0$. In the present type of theory, a variety of quantities could be assigned the role of internal pressure. We could choose p or $\overset{1}{t}_\theta$; but $\overset{1}{t}_\theta$ could take a variety of forms for various choices of independent variables and independent variations. Internal pressure, along with stress, appears to have lost its former significance owing to the consideration of long range forces. However, dilatation is defined uniquely and can serve as a basis of comparison of the two theories. In the classical case, the dilatation in the sphere would be uniform and proportional to T/r_0. In the present theory, the dilatation is non-uniform; but we may calculate the *average* dilatation:

$$\Delta_{\text{ave}} = \frac{3}{4\pi r_0^3}\int_V \Delta \, \mathrm{d}V,$$
$$= \frac{3}{r_0^2}\sum_{i=1}^{i=2} A_i l_i^2(\cosh r_i - r_i^{-1}\sinh r_i).$$

Assuming, again, that $r_i \gg 1$, we find, to the first order in l_i/r_0,

$$\Delta_{\text{ave}} \approx -\frac{6c_1}{b_0\lambda}\frac{T}{r_0},$$

where T is given by (95). Thus, as in the classical theory, the average dilatation in the sphere is directly proportional to the surface-tension and inversely proportional to the radius.

Now consider a spherical *cavity* of radius r_0. For this we take the solution of (80), with $\varphi = 0$, in the form

$$\Delta = A_1 l_1 r^{-1}e^{-r/l_1} + A_2 l_2 r^{-1}e^{-r/l_2}. \tag{96}$$

The conditions (78), for a free boundary, require

$$\sum_{i=1}^{i=2} A_i[2(b_1+b_2)-(2a_1-c_1)l_i^2](1+r_i)r_i e^{-r_i} = 0, \tag{97a}$$

$$\sum_{i=1}^{i=2} A_i r_i[2b_2(1+2r_i^{-1}+2r_i^{-2})+2b_1+c_1 l_i^2]e^{-r_i} = -b_0 r_0^2. \tag{97b}$$

Again we consider the case $r_i \gg 1$ and find, to the first order in l_i/r_0,

$$T = \frac{T_0}{1-(8b_2T_0/b_0^2r_0)}. \tag{98}$$

Thus, for both convex and concave spherical surfaces, the surface-tension, to the first order in l_i/r_0, is

$$T = \frac{T_0}{1+8b_0^{-2}b_2\kappa T_0} \tag{99}$$

where κ is the curvature: positive for the sphere and negative for the cavity.

Acknowledgements—The idea that the extension of R. A. Toupin's strain-gradient theory, to include the second gradient of strain, would bring in the cohesive force and its effects in centrosymmetric materials was suggested to me several years ago by Dr. Toupin. I am indebted to him for many valuable discussions of the associated mathematical theory.

This investigation was supported by the Office of Naval Research under a contract with Columbia University.

REFERENCES

[1] R. A. TOUPIN, *Arch. ration. Mech. Analysis* **11,** 385 (1962).
[2] R. A. TOUPIN and D. C. GAZIS, *Proceedings of the International Conference on Lattice Dynamics, Copenhagen, August, 1963* (Edited by R. F. WALLIS) pp. 597–602. Pergamon Press (1964).
[3] E. W. HART, *Phys. Rev.* **113,** 412 (1959); **114,** 27 (1959); *J. Chem. Phys.* **39,** 3075 (1963).
[4] A. E. GREEN and R. S. RIVLIN, *Arch. ration. Mech. Analysis* **16,** 325 (1964).
[5] L. H. GERMER, A. U. MACRAE and C. D. HARTMAN, *J. appl. Phys.* **32,** 2432 (1961).
[6] R. D. MINDLIN, *Arch. ration. Mech. Analysis* **16,** 51 (1964).
[7] R. D. MINDLIN and H. F. TIERSTEN, *Arch. ration. Mech. Analysis* **11,** 415 (1962).
[8] L. BRAND, *Vector and Tensor Analysis*, p. 222. John Wiley & Sons (1947).
[9] R. D. MINDLIN, *Expl Mech.* **3,** 1 (1963).
[10] D. C. GAZIS and R. F. WALLIS, *Surface Science* **3,** 19 (1964).
[11] H. and B. S. JEFFREYS, *Methods of Mathematical Physics*, 2nd edition, p. 210. Cambridge University Press (1950).

APPENDIX

Boundary conditions at edges and corners

In Sections 2 and 10, it was assumed that the surface S is smooth. Additional boundary conditions are required if S has edges and corners.

Suppose the surface has an edge, C, formed by the intersection of two segments, S_1 and S_2, of S. Then, for each segment, the surface divergence theorem [8] is

$$\int_{S_i} \overset{s}{\nabla} \cdot \mathbf{f}\, dS = \int_{S_i} (\overset{s}{\nabla} \cdot \mathbf{n})\mathbf{n} \cdot \mathbf{f}\, dS + \oint_C \mathbf{m}_i \cdot \mathbf{f}_i\, ds, \qquad i = 1, 2, \qquad \text{(not summed)}$$

where s is measured along C in the direction of its unit tangent $\mathbf{s}_i$; and $\mathbf{m}_i$ $(= \mathbf{s}_i \times \mathbf{n}_i)$ is the unit outward normal to C tangent to S_i. The contributions of the line integrals have to be added to the previous results. Thus, to (8) and (10), we have to add

$$\oint_C [\mathbf{m} \cdot \boldsymbol{\Phi} \cdot \mathbf{v}]\, ds, \tag{8A}$$

and

$$\oint_C [\mathbf{m} \cdot \boldsymbol{\Psi} : \boldsymbol{\Phi}]\, ds, \tag{10A}$$

where the boldfaced brackets **[ ]** denote the value of the enclosed quantity on C when approached over S_1 minus its value when approached over S_2; and ds is positive in the direction of $\mathbf{s}_1$.

Considering, first, the case of the solid, we must add to (11), (12) and (13):

$$\oint_C [\mathbf{mn} : (\overset{2}{\boldsymbol{\tau}} - \nabla \cdot \overset{3}{\boldsymbol{\tau}})] \cdot \delta\mathbf{u}\, ds, \tag{11A}$$

$$\oint_C [\mathbf{mn} : \overset{3}{\boldsymbol{\tau}} : \nabla\delta\mathbf{u}]\, ds, \tag{12A*}$$

$$\oint_C [\mathbf{m} \cdot [\mathbf{L} \cdot (\mathbf{n} \cdot \overset{3}{\boldsymbol{\tau}})]] \cdot \delta\mathbf{u}\, ds, \tag{13A}$$

respectively, while to (14) we have to add

$$-\oint_C [\mathbf{m} \cdot (\overset{s}{\nabla}\mathbf{n}) \cdot (\mathbf{nn} : \overset{3}{\boldsymbol{\tau}})] \cdot \delta\mathbf{u}\, ds \tag{14A1}$$

and

$$\oint_C [\mathbf{mnn} \vdots \overset{3}{\boldsymbol{\tau}} \cdot D\delta\mathbf{u}]\, ds. \tag{14A*}$$

The integrands in (12A*) and (14A*) require resolution into terms with coefficients that are independent variations. For (14A*) we write

$$\oint_C (\mathbf{nnm} \vdots \overset{3}{\boldsymbol{\tau}})_1 D_{n_1}\delta\mathbf{u}\, ds - \oint_C (\mathbf{nnm} \vdots \overset{3}{\boldsymbol{\tau}})_2 D_{n_2}\delta\mathbf{u}\, ds, \tag{14A2}$$

where $D_{n_i}\delta\mathbf{u}$ and $(\mathbf{nnm} \vdots \overset{3}{\boldsymbol{\tau}})_i$ are the limits at C of $\mathbf{n} \cdot \nabla\delta\mathbf{u}$ and $\mathbf{nnm} \vdots \overset{3}{\boldsymbol{\tau}}$ approached over S_i. As for (12A*), we first write

$$\nabla\delta\mathbf{u} = \mathbf{nn} \cdot \nabla\delta\mathbf{u} + \mathbf{mm} \cdot \nabla\delta\mathbf{u} + \mathbf{ss} \cdot \nabla\delta\mathbf{u} = \mathbf{n}D_n\delta\mathbf{u} + \mathbf{m}D_m\delta\mathbf{u} + \mathbf{s}D_s\delta\mathbf{u}.$$

Then (12A*) is the sum of

$$\oint_C (\mathbf{nmn} \vdots \overset{3}{\tau})_1 \cdot D_{n_1}\delta\mathbf{u}\, ds - \oint_C (\mathbf{nmn} \vdots \overset{3}{\tau})_2 \cdot D_{n_2}\delta\mathbf{u}\, ds + \oint_C (\mathbf{nmm} \vdots \overset{3}{\tau})_1 \cdot D_{m_1}\delta\mathbf{u}\, ds - \oint_C (\mathbf{nmm} \vdots \overset{3}{\tau})_2 \cdot D_{m_2}\delta\mathbf{u}\, ds \qquad \text{(12A1)}$$

and

$$\oint_C [\mathbf{nm} : \overset{3}{\tau} : \mathbf{s}D_s\delta\mathbf{u}]\, ds = \oint_C [D_s(\mathbf{nms} \vdots \overset{3}{\tau} \cdot \delta\mathbf{u})]\, ds - \oint_C [D_s(\mathbf{nms} \vdots \overset{3}{\tau})] \cdot \delta\mathbf{u}\, ds$$

$$= [|\mathbf{nms} \vdots \overset{3}{\tau}|{}^{+}_{-}] \cdot \delta\mathbf{u} - \oint_C [D_s(\mathbf{nms} \vdots \overset{3}{\tau})] \cdot \delta\mathbf{u}\, ds, \qquad \text{(12A2)}$$

where the quantity $[|\mathbf{nms} \vdots \overset{3}{\tau}|{}^{+}_{-}]$ is the difference between discontinuities in $(\mathbf{nms} \vdots \overset{3}{\tau})_1$ and $(\mathbf{nms} \vdots \overset{3}{\tau})_2$ along C or at a corner of C.

Altogether, the additional terms are (11A), (12A1), (12A2), (13A), (14A1) and (14A2), so that we have to add to (15) and (16):

$$\oint_C [\mathbf{mn} : (\overset{2}{\tau} - \nabla \cdot \overset{3}{\tau}) + \mathbf{m} \cdot [\mathbf{L} \cdot (\mathbf{n} \cdot \overset{3}{\tau})] - D_s(\mathbf{nms} \vdots \overset{3}{\tau}) - \mathbf{m} \cdot (\overset{s}{\nabla}\mathbf{n}) \cdot (\mathbf{nn} : \overset{3}{\tau})] \cdot \delta\mathbf{u}\, ds$$
$$+ \oint_C 2(\mathbf{nmn} \vdots \overset{3}{\tau})_1 \cdot D_{n_1}\delta\mathbf{u}\, ds - \oint_C 2(\mathbf{nmn} \vdots \overset{3}{\tau})_2 \cdot D_{n_2}\delta\mathbf{u}\, ds$$
$$+ \oint_C (\mathbf{mnm} \vdots \overset{3}{\tau})_1 \cdot D_{m_1}\delta\mathbf{u}\, ds - \oint_C (\mathbf{mnm} \vdots \overset{3}{\tau})_2 \cdot D_{m_2}\delta\mathbf{u}\, ds$$
$$+ [|\mathbf{mns} \vdots \overset{3}{\tau}|{}^{+}_{-}] \cdot \delta\mathbf{u} \qquad \text{(15A)}$$

and

$$\oint_C \mathbf{F} \cdot \delta\mathbf{u}\, ds + \oint_C (\mathbf{N}_1 \cdot D_{n_1}\delta\mathbf{u} + \mathbf{N}_2 \cdot D_{n_2}\delta\mathbf{u} + \mathbf{T}_1 \cdot D_{m_1}\delta\mathbf{u} + \mathbf{T}_2 \cdot D_{m_2}\delta\mathbf{u})\, ds + \mathbf{G} \cdot \delta\mathbf{u}, \qquad \text{(16A)}$$

respectively. Hence, equating (15A) and (16A), we have the edge and corner conditions

$$\mathbf{F} = [\mathbf{mn} : (\overset{2}{\tau} - \nabla \cdot \overset{3}{\tau}) + \mathbf{m} \cdot [\mathbf{L} \cdot (\mathbf{n} \cdot \overset{3}{\tau})] - D_s(\mathbf{nms} \vdots \overset{3}{\tau}) - \mathbf{m} \cdot (\overset{s}{\nabla}\mathbf{n}) \cdot (\mathbf{nn} : \overset{3}{\tau})],$$
$$\mathbf{N}_1 = 2(\mathbf{nmn} \vdots \overset{3}{\tau})_1,$$
$$\mathbf{N}_2 = -2(\mathbf{nmn} \vdots \overset{3}{\tau})_2,$$
$$\mathbf{T}_1 = (\mathbf{nmm} \vdots \overset{3}{\tau})_1,$$
$$\mathbf{T}_2 = -(\mathbf{nmm} \vdots \overset{3}{\tau})_2,$$
$$\mathbf{G} = [|\mathbf{mns} \vdots \overset{3}{\tau}|{}^{+}_{-}].$$

As in the case of the boundary conditions on a smooth surface, alternative edge and corner conditions involving displacements are apparent from the form of (16A).

Turning, now, to the boundary conditions for the liquid, we find that, when it is the dilatation that is varied, we have only to add to (75) and (76):

$$-\oint_C [\mathbf{mn} : \boldsymbol{\pi}]\delta\Delta \, \mathrm{d}s \tag{75A}$$

and

$$\oint_C \overset{0}{t}_\Delta \delta\Delta \, \mathrm{d}s, \tag{76A}$$

respectively; so that the only boundary condition to be added to (78) is

$$\overset{0}{t}_\Delta = -[\mathbf{mn} : \boldsymbol{\pi}]. \tag{78A}$$

On the other hand, when it is the displacement that is varied, the additional terms are similar to those for the solid. To (83) and (84) we have to add

$$\begin{aligned} &-\oint_C [\mathbf{n} \cdot (\mathbf{p} - \nabla \cdot \boldsymbol{\pi})\mathbf{m} + \mathbf{mL} \cdot (\mathbf{n} \cdot \boldsymbol{\pi}) - D_s(\mathbf{mn} : \boldsymbol{\pi}\mathbf{s}) - \mathbf{nn} : \boldsymbol{\pi}\mathbf{m} \cdot (\overset{s}{\nabla}\mathbf{n})] \cdot \delta\mathbf{u} \, \mathrm{d}s \\ &-\oint_C (\mathbf{nn} : \boldsymbol{\pi}\mathbf{m} + \mathbf{mn} : \boldsymbol{\pi}\mathbf{n})_1 \cdot D_{n_1}\delta\mathbf{u} \, \mathrm{d}s + \oint_C (\mathbf{nn} : \boldsymbol{\pi}\mathbf{m} + \mathbf{mn} : \boldsymbol{\pi}\mathbf{n})_2 \cdot D_{n_2}\delta\mathbf{u} \, \mathrm{d}s \\ &-\oint_C (\mathbf{mn} : \boldsymbol{\pi}\mathbf{m})_1 \cdot D_{m_1}\delta\mathbf{u} \, \mathrm{d}s + \oint_C (\mathbf{mn} : \boldsymbol{\pi}\mathbf{m})_2 \cdot D_{m_2}\delta\mathbf{u} \, \mathrm{d}s - [|\mathbf{mn} : \boldsymbol{\pi}\mathbf{s}|^+_-] \cdot \delta\mathbf{u} \end{aligned} \tag{83A}$$

and

$$\oint_C \mathbf{F} \cdot \delta\mathbf{u} \, \mathrm{d}s + \oint_C (\mathbf{N}_1 \cdot D_{n_1}\delta\mathbf{u} + \mathbf{N}_2 \cdot D_{n_2}\delta\mathbf{u} + \mathbf{T}_1 \cdot D_{m_1}\delta\mathbf{u} + \mathbf{T}_2 \cdot D_{m_2}\delta\mathbf{u}) \, \mathrm{d}s + \mathbf{G} \cdot \delta\mathbf{u}, \tag{84A}$$

respectively. Then, upon equating (83A) and (84A) we find the edge and corner conditions for the liquid:

$$\begin{aligned} \mathbf{F} &= -[\mathbf{n} \cdot (\mathbf{p} - \nabla \cdot \boldsymbol{\pi})\mathbf{m} + \mathbf{mL} \cdot (\mathbf{n} \cdot \boldsymbol{\pi}) - D_s(\mathbf{mn} : \boldsymbol{\pi}\mathbf{s}) - \mathbf{nn} : \boldsymbol{\pi}\mathbf{m} \cdot (\overset{s}{\nabla}\mathbf{n})], \\ \mathbf{N}_1 &= -(\mathbf{nn} : \boldsymbol{\pi}\mathbf{m} + \mathbf{mn} : \boldsymbol{\pi}\mathbf{n})_1, \\ \mathbf{N}_2 &= (\mathbf{nn} : \boldsymbol{\pi}\mathbf{m} + \mathbf{mn} : \boldsymbol{\pi}\mathbf{n})_2, \\ \mathbf{T}_1 &= -(\mathbf{mn} : \boldsymbol{\pi}\mathbf{m})_1, \\ \mathbf{T}_2 &= (\mathbf{mn} : \boldsymbol{\pi}\mathbf{m})_2, \\ \mathbf{G} &= -[|\mathbf{mn} : \boldsymbol{\pi}\mathbf{s}|^+_-]. \end{aligned}$$

Again, alternative edge and corner conditions involving displacements are apparent from the form of (84A).

(*Received* 29 *January* 1965)

Zusammenfassung—In diesem Aufsatz wird die lineare Deformationstheorie eines elastischen Festkörpers formuliert, bei welchem die potentielle Energiedichte eine Funktion der Beanspruchung und deren Gradienten erster und zweiter Ordnung ist. In diese Theorie sind die Kohäsionskraft und Oberflächenbeanspruchung als wesentliche Merkmale eingebaut. Eine Lösung für die Beanspruchung und Oberflächenspannung pro Flächeneinheit, wie sie bei der in einer Ebene liegenden Trennung eines Festkörpers auftreten, wird angegeben, und ein Vergleich mit dem analogen Gittermodell angestellt. Ausserdem wird die allgemeine Lösung der Gleichgewichtsverschiebegleichung angegeben, und zwar drückt man diese in Spannungsfunktionen und mit Hilfe der besonderen Lösung für die konzentrierte Kraft aus. Der Spezialfall einer Flüssigkeit wird erwogen, und die Lösungen für die Oberflächenspannung bei ebenen und runden Oberflächen werden angegeben.

Абстракт—В настоящей статье формулируется линейная теория деформации упругого твердого тела, в котором потенциальная энергия-плотность является функцией деформации и ее первым и вторым градиентом. В этой теории сила сцепления и поверхностное натяжение являются внутренними силами. Дано решение для деформации и поверхностного натяжения, или поверхностной энергии на единицу площади, полученной в разделении твердого тела по плоскости; сделано сравнение с аналогичной решетчатой моделью. Также предлагается общее решение уравнения смещения при равновесии в условиях функции напряжения и особое решение для сосредоточенной силы.

Рассматривается специальный случай жидкости и дается решение для поверхностного натяжения в плоской и сферической поверхностях.

Int. J. Solids Structures, 1966, Vol. 2, pp. 125 to 139. Pergamon Press Ltd. Printed in Great Britain

THICKNESS-SHEAR AND FLEXURAL VIBRATIONS OF PARTIALLY PLATED, CRYSTAL PLATES

R. D. MINDLIN and P. C. Y. LEE
Department of Civil Engineering, Columbia University, New York, N.Y.

Abstract—In this paper, mathematical solutions are exhibited which predict the influence of length and mass, of rectangular electrode platings, on the sharpness of the fundamental thickness-shear resonance and on the number and distribution of anharmonic overtones between the cut-off frequencies of the plated and unplated portions of AT-type quartz plates.

1. INTRODUCTION

TWO INTERESTING and useful effects of the size of electrode platings on vibrations of AT-type quartz plates have been employed, principally by Bechmann [1], to improve the performance of the crystals in filter circuits. First: if electrode platings, symmetrically disposed on the two faces of an AT-type quartz plate, cover only a central portion of the plate and if the plate is excited in a thickness-shear resonance, much of the motion of the plate is confined to the region under the electrodes. Second: as the area of the electrodes is reduced, the responses of the anharmonic overtones of thickness-shear are diminished, successively, until, below a certain length/thickness ratio (Bechmann's Number), the response of the fundamental is, essentially, all that remains.

The first of these phenomena was given a qualitative, physical explanation, by Shockley, Curran and Koneval [2], on the basis of the dispersion relations for elastic waves in quartz plates [3, 4, 5]. Consider the dispersion curve (TS_1 in Fig. 1) for a straight-crested, thickness-shear wave progressing in the direction of the axis of digonal symmetry in the plane of an AT-type quartz plate without electrodes. At zero wave-number there is a cut-off circular frequency, ω_1, below which the wave-number is imaginary; i.e. at frequencies below ω_1, the wave is non-propagating and decays exponentially with distance. Alternatively, consider the analogous curve for a plate coated with electrodes. Owing to the mass loading of the electrodes, the frequency is lowered throughout (curve $\overline{TS}_1$ in Fig. 1). In particular, the cut-off frequency $\bar{\omega}_1$ is less than ω_1. Now suppose that only a central portion of the plate is coated with electrodes across which the impressed voltage has a frequency between ω_1 and $\bar{\omega}_1$. A propagating thickness-shear wave is excited in the part of the plate under the electrodes; but the corresponding wave at the same frequency, in the remainder of the plate, decays exponentially: resulting in a total reflection phenomenon whereby thickness-shear vibrational energy is trapped in the region under the electrodes.

Now, flexural, face-shear and extensional waves can propagate at any frequency. These waves are excited, at the boundary between the plated and unplated portions of the plate, and propagate into both regions; so that, even at frequencies between $\bar{\omega}_1$ and ω_1, some of the energy escapes from the plated region—mostly in the outgoing flexural

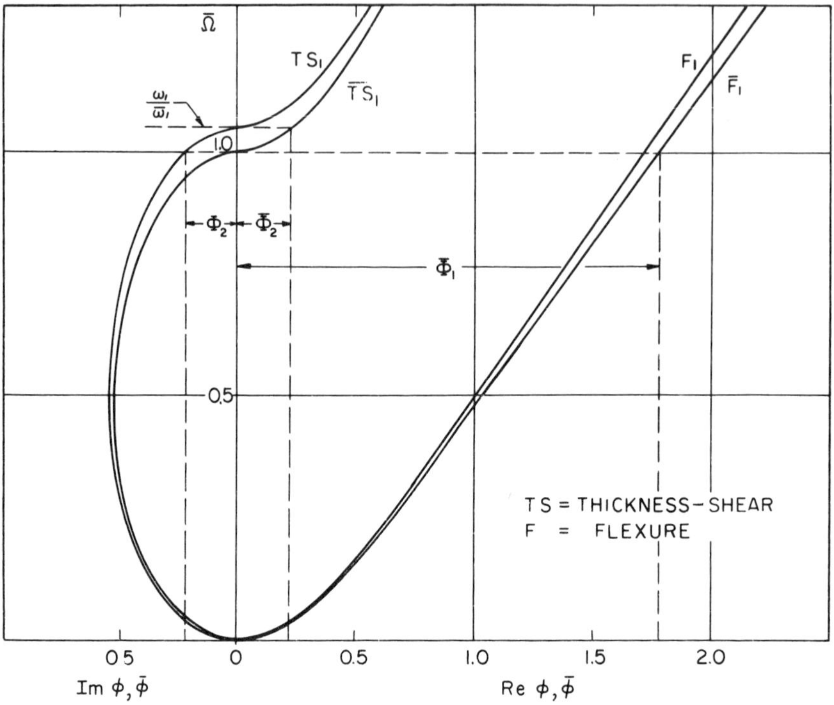

FIG. 1. Dispersion curves for straight-crested, thickness-shear and flexural waves with wave-normals in the direction of the digonal axis in unplated (TS_1 and F_1) and plated ($\overline{TS}_1$ and $\bar{F}_1$) AT-type quartz plates.

wave if the edge of the electrode is straight and perpendicular to the digonal axis in the plane of the plate. It is shown, in this paper, how the energy in the escaping flexural wave depends on the dimensions and material properties of the electrodes and the crystal in the case of rectangular electrodes. The Q of the resonance (proportional to the ratio of the energy trapped to the the energy lost) is found to be an almost periodic function of the ratio of the length of the electrode, in the digonal direction, to the thickness of the plate. The Q is a maximum when the length of the electrode is approximately an integral multiple of the wavelength of the flexural wave in the plated portion of the plate; i.e. there is little transmission of flexural energy across the boundary between the plated and unplated portions when there is a node of flexure there. The minima of Q become lower as the length of the electrodes is reduced and as the thickness of the electrodes is increased.

An explanation of the phenomenon of Bechmann's Number can be based on the observations that only the deformation in the portion of the plate under the electrodes contributes to the current through the crystal and only those anharmonic overtones with frequencies less than ω_1 are trapped under the electrodes. The anharmonic overtone modes of thickness-shear, in the plated portion of the plate, have successively shorter and shorter half-wave-lengths: approximately equal to $\frac{1}{3}, \frac{1}{5}, \frac{1}{7} \ldots$ of the length of a rectangular electrode in the digonal direction. To each successively higher anharmonic overtone mode, there corresponds an only slightly higher frequency; so that, for a long enough electrode, many anharmonic overtone modes can have frequencies less than ω_1 and, hence, can be totally trapped. A forcing frequency, equal to that of a trapped overtone mode, will produce a relatively large amplitude of thickness-shear deformation in the plated portion of the plate because much of the input energy is trapped there; but a forcing frequency equal to that of an anharmonic overtone resonance above ω_1 excites the whole plate because only a small proportion of that mode is reflected at the edge of the electrode. Consequently, a smaller proportion of the total energy is to be found in the part of the plate under the electrodes—the part which contributes to the current through the crystal. Now, if the electrode is small enough, the wave-length of the first anharmonic overtone mode will be so short that the frequency of that mode, and all the higher ones, will be greater than ω_1 and there will be no total trapping of any anharmonic overtone modes—only the fundamental can be trapped. Thus, the critical length of electrode is simply the wave-length of a thickness-shear wave, in the plated portion of the plate, at the cut-off frequency of the thickness-shear wave in the unplated portion of the plate. The ratio of this critical wave-length to the thickness of the plate is Bechmann's Number.

Curran and Koneval [6] have made a series of experiments with rectangular electrodes of successively smaller lengths. Their data of number and frequency of trapped anharmonic overtones *versus* length of electrode are compared, in the second part of this paper, with the results of a mathematical solution. Their formula for Bechmann's Number, in terms of $\bar{\omega}_1/\omega_1$ and a numerical constant, is compared with a formula, obtained mathematically, expressed in terms of the dimensions and material properties of the electrodes and the crystal.

2. THICKNESS-SHEAR AND FLEXURAL MOTIONS OF INFINITE, UNPLATED AND PLATED PLATES

It has been shown [3, 4] that the motions of crystal plates are well described, up to a frequency somewhat higher than the cut-off frequency of the lowest thickness-shear mode, by solutions of a system of five, coupled, two-dimensional, second order partial differential equations governing three components of displacement and two components of rotation. Quartz is a crystal of the trigonal trapezohedral class [7]: with one trigonal axis and, in the plane at right angles, three digonal axes. A rotated Y-cut plate [7] contains a digonal axis and is cut at an angle (35° 15′ for the AT-cut) to the trigonal axis. In the case of straight-crested waves advancing in the direction of the digonal axis in the plane of a rotated Y-cut plate, three equations of coupled thickness-shear, flexure and face-shear separate from the five coupled equations [3]. If, in these three, the coupling with face-shear is neglected, the remaining equations of coupled thickness-shear and flexure [8] have the same form as Timoshenko's equations of flexural vibrations of beams in which transverse shear deformation and rotatory inertia are taken into account [9]:

$$k_1^2 c_{66}\left(\frac{\partial^2 v}{\partial x^2} + \frac{\partial \psi_x}{\partial x}\right) = \rho \frac{\partial^2 v}{\partial t^2}, \tag{1a}$$

$$\frac{\gamma_{11} h^2}{3}\frac{\partial^2 \psi_x}{\partial x^2} - k_1^2 c_{66}\left(\frac{\partial v}{\partial x} + \psi_x\right) = \frac{\rho h^2}{3}\frac{\partial^2 \psi_x}{\partial t^2}, \tag{1b}$$

9

where x is the digonal axis in the plane of the plate, v is the deflection of the plate, ψ_x is the rotation of a line element initially normal to the middle plane of the plate, ρ is the density, $2h$ is the thickness of the plate, k_1 is the shear-correction factor, c_{66} is the shear modulus of elasticity in the plane normal to the plate through the x-axis and γ_{11} is Voigt's stretch modulus of the plate in the x-direction. In terms of the compliances, s_{pq}, and stiffnesses, c_{pq}, referred to rectangular coordinates x, y, z, with y normal to the middle plane of the plate,

$$\begin{aligned}\gamma_{11} &= s_{33}/(s_{11}s_{33} - s_{13}^2),\\ &= c_{11} - c_{12}^2/c_{22} - (c_{14} - c_{12}c_{24}/c_{22})^2/(c_{44} - c_{24}^2/c_{22}).\end{aligned}$$

For the AT-cut of quartz,

$$c_{66} = 29{\cdot}01, \qquad \gamma_{11} = 85{\cdot}93$$

in units of 10^{10} dyne/cm^2, as calculated from Bechmann's values of the principal constants [10].

As in Timoshenko's equations, the transverse shear force, V_x, and the bending moment, M_x, both per unit width of plate, are given by

$$V_x = 2hk_1^2c_{66}\left(\frac{\partial v}{\partial x} + \psi_x\right), \qquad M_x = \frac{2h^3\gamma_{11}}{3}\frac{\partial \psi_x}{\partial x}. \tag{2}$$

Consider, now, the same plate with electrode platings, on each face, of like thicknesses $2h'$ and density ρ'. An alternating voltage, impressed across the electrodes, is mechanically equivalent to a couple, C per unit area, distributed uniformly over the plate [11, 12]. If the remaining coupling with the electric field is neglected and if the stiffnesses of the platings are neglected, the equations analogous to (1) are [13]

$$\bar{k}_1^2c_{66}\left(\frac{\partial^2 \bar{v}}{\partial x^2} - \frac{\partial \bar{\psi}_x}{\partial x}\right) = \rho(1 + R)\frac{\partial^2 \bar{v}}{\partial t^2}, \tag{3a}$$

$$\frac{\gamma_{11}h^2}{3}\frac{\partial^2 \bar{\psi}_x}{\partial x^2} - \bar{k}_1^2c_{66}\left(\frac{\partial \bar{v}}{\partial x} + \bar{\psi}_x\right) + \frac{C}{2h} = \frac{\rho h^2(1 + 3R)}{3}\frac{\partial^2 \bar{\psi}_x}{\partial t^2}, \tag{3b}$$

where barred symbols pertain to a plated plate and

$$R = 2\rho' h'/\rho h, \tag{4}$$

i.e. R is the ratio of the mass per unit area of both electrodes to the mass per unit area of the plate. The transverse shear force and bending moment, per unit width of plated plate, are

$$\bar{V}_x = 2h\bar{k}_1^2c_{66}\left(\frac{\partial \bar{v}}{\partial x} + \bar{\psi}_x\right), \qquad \bar{M}_x = \frac{2h^3\gamma_{11}}{3}\frac{\partial \bar{\psi}_x}{\partial x}. \tag{5}$$

The shear correction factor $\bar{k}_1$ is expressed in terms of R by equating the thickness-shear cut-off frequency, obtained from (3), with that obtained from a solution of the three-dimensional equations of elasticity. In the case of (3), we set

$$C = \bar{v} = 0, \qquad \bar{\psi}_x = \bar{A}\exp(i\bar{\omega}_1 t),$$

where $\bar{A}$ is a constant. Then

$$\bar{\omega}_1^2 = 3\bar{k}_1^2 c_{66}/\rho h^2(1 + 3R). \tag{6}$$

For simple thickness-shear motion, the three-dimensional equations reduce to

$$c_{66}\frac{\partial^2 u}{\partial y^2} = \rho\frac{\partial^2 u}{\partial t^2}, \qquad T_{xy} = c_{66}\frac{\partial u}{\partial y}, \tag{7}$$

where y is the coordinate normal to the middle plane of the plate, u is the displacement in the direction of x and T_{xy} is the shear stress on planes normal to x or y. For

$$u = A \sin \eta y \exp(i\omega t), \tag{8}$$

the first of (7) yields

$$\omega^2 = \eta^2 c_{66}/\rho. \tag{9}$$

As boundary conditions we take

$$T_{xy} = \mp 2\rho' h' \frac{\partial^2 u}{\partial t^2} \quad \text{on} \quad y = \pm h, \tag{10}$$

i.e. the inertia of the electrode plating is balanced by a shear traction on the surface of the plate. With (8) and the second of (7), (10) becomes

$$c_{66}\eta \cos \eta h = 2h'\rho'\omega^2 \sin \eta h;$$

or, with (4) and (9),

$$R\eta h \tan \eta h = 1. \tag{11}$$

For $R \ll 1$, the smallest root of (11) is given, approximately, by

$$(\eta h)_1 = \pi/2(1 + R). \tag{12}$$

Upon substituting (12) into (9) and equating the resulting expression for ω^2 to (6), we find

$$\bar{k}_1^2 = \pi^2(1 + 3R)/12(1 + R)^2, \qquad \bar{\omega}_1^2 = \pi^2 c_{66}/4\rho h^2(1 + R)^2. \tag{13}$$

The corresponding shear correction factor and cut-off frequency for the unplated plate ($R = 0$) are

$$k_1^2 = \pi^2/12, \qquad \omega_1^2 = \pi^2 c_{66}/4\rho h^2. \tag{14}$$

Thus, the effect of the mass of the electrodes is equivalent to increases of transverse inertia, rotatory inertia and shear stiffness by factors $1 + R$, $1 + 3R$ and $(1 + 3R)/(1 + R)^2$, respectively.

From (13) and (14),

$$\omega_1/\bar{\omega}_1 = 1 + R. \tag{15}$$

As shown in [14], inclusion of piezoelectric properties would make $\omega_1/\bar{\omega}_1$ slightly greater than $1 + R$; but this effect is omitted here.

Two fundamental solutions of the equations of motion, (1) and (3), are recorded here for later use. First: for an applied voltage mechanically equivalent to $C_0 \cos \omega t$, with C_0 a constant, a particular solution of (3) is $\bar{v} = 0$ and

$$\bar{\psi}_x = \bar{\psi}_0 \cos \omega t, \qquad \bar{\psi}_0 = C_0/2h\bar{k}_1^2 c_{66}(1 - \omega^2/\bar{\omega}_1^2). \tag{16}$$

Second: (1) and (3) admit the free wave solutions

$$\psi_x = A \exp[i(\xi x - \omega t)], \qquad v = i\alpha A h \exp[i(\xi x - \omega t)],$$
$$\bar{\psi}_x = \bar{A} \exp[i(\bar{\xi} x - \omega t)], \qquad \bar{v} = i\bar{\alpha}\bar{A} h \exp[i(\bar{\xi} x - \omega t)],$$

if the amplitude ratios, α and $\bar{\alpha}$, satisfy the equations

$$\alpha = \frac{\xi h}{\xi^2 h^2 - 3\Omega^2} = \frac{1 + \hat{\gamma}_{11}\xi^2 h^2 - \Omega^2}{\xi h}, \tag{17}$$

$$\bar{\alpha} = \frac{\bar{\xi} h}{\bar{\xi}^2 h^2 - 3r\bar{\Omega}^2} = \frac{1 + \tilde{\gamma}_{11}\bar{\xi}^2 h^2 - \bar{\Omega}^2}{\bar{\xi} h}, \tag{18}$$

where

$$\Omega = \omega/\omega_1, \qquad \hat{\gamma}_{11} = \gamma_{11}/3k_1^2 c_{66},$$
$$\bar{\Omega} = \omega/\bar{\omega}_1, \qquad \tilde{\gamma}_{11} = \gamma_{11}/3\bar{k}_1^2 c_{66},$$
$$r = (1 + R)/(1 + 3R).$$

The second of (17) and (18) yield the dispersion relations depicted in Fig. 1 with dimensionless frequency, $\bar{\Omega}$, as ordinate and dimensionless wave-number, $\varphi = \xi h$ or $\bar{\varphi} = \bar{\xi} h$, as abscissa. Points on the curves marked TS_1 and $\overline{TS_1}$ correspond to motions in the unplated and plated plates, respectively, which are predominantly thickness-shear ($|\alpha| < 1$, $|\bar{\alpha}| < 1$). The curves marked F_1 and $\bar{F}_1$ give the dispersion relations for predominantly flexural motion ($|\alpha| > 1$, $|\bar{\alpha}| > 1$). It will be observed that, in the frequency range $\bar{\omega}_1 < \omega < \omega_1$, i.e. $1 < \bar{\Omega} < \omega_1/\bar{\omega}_1$, the thickness-shear waves are propagating waves ($\bar{\varphi}$ real) in the plated plate but non-propagating waves (φ imaginary) in the unplated plate; whereas flexural waves in both plates are propagating waves at all frequencies.

3. EFFECT OF SIZE OF ELECTRODES ON Q

Suppose that the strip $-a \leqslant x \leqslant a$, on each face of an infinite plate, is coated with electrodes, Fig. 2, across which is impressed a voltage mechanically equivalent to $C_0 \cos \omega t$ with $\bar{\omega}_1 \leqslant \omega \leqslant \omega_1$. A suitable form of solution of (1), for the unplated portion $x \geqslant a$, is

$$\psi_x = A_1 \sin(\xi_1 x - \xi_1 a - \omega t) + A_2 \exp[-\xi_2(x - a)] \cos \omega t$$
$$+ B_1 \cos(\xi_1 x - \xi_1 a - \omega t) + B_2 \exp[-\xi_2(x - a)] \sin \omega t, \tag{19a}$$

$$v = -\alpha_1 A_1 h \cos(\xi_1 x - \xi_1 a - \omega t) + \alpha_2 A_2 h \exp[-\xi_2(x - a)] \cos \omega t$$
$$+ \alpha_1 B_1 h \sin(\xi_1 x - \xi_1 a - \omega t) + \alpha_2 B_2 h \exp[-\xi_2(x - a)] \sin \omega t. \tag{19b}$$

For the unplated portion $x \leqslant -a$, x and v, in (19), are changed to $-x$ and $-v$. For the plated portion $-a \leqslant x \leqslant a$, we take

$$\bar{\psi}_x = \bar{\psi}_0 \cos \omega t + (\bar{A}_1 \cos \bar{\xi}_1 x + \bar{A}_2 \cos \bar{\xi}_2 x) \cos \omega t$$
$$+ (\bar{B}_1 \cos \bar{\xi}_1 x + \bar{B}_2 \cos \bar{\xi}_2 x) \sin \omega t, \tag{20a}$$

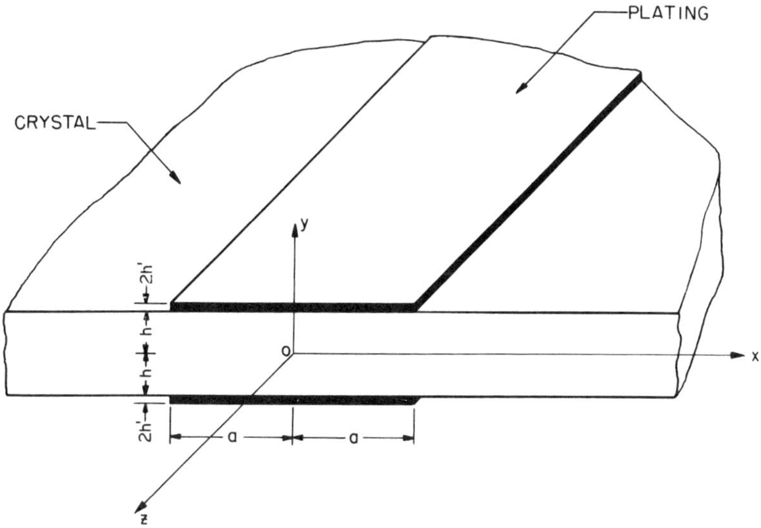

FIG. 2. Partially plated, infinite plate.

$$\bar{v} = (\bar{\alpha}_1 \bar{A}_1 h \sin \bar{\xi}_1 x + \bar{\alpha}_2 \bar{A}_2 h \sin \bar{\xi}_2 x) \cos \omega t + (\bar{\alpha}_1 \bar{B}_1 h \sin \bar{\xi}_1 x + \bar{\alpha}_2 \bar{B}_2 h \sin \bar{\xi}_2 x) \sin \omega t, \qquad (20b)$$

where $\bar{\psi}_0$ is given by (16).

The functions (19) and (20) are solutions of (1) and (3), respectively, if

$$\alpha_i = \frac{\varphi_i}{3\Omega^2 + (-1)^i \varphi_i^2} = \frac{(-1)^i(1 - \Omega^2) - \hat{\gamma}_{11}\varphi_i^2}{\varphi_i}, \qquad i = 1, 2, \qquad (21)$$

$$\bar{\alpha}_i = \frac{\bar{\varphi}_i}{3r\bar{\Omega}^2 - \bar{\varphi}_i^2} = \frac{\bar{\Omega}^2 - 1 - \tilde{\gamma}_{11}\bar{\varphi}_i^2}{\bar{\varphi}_i}, \qquad i = 1, 2, \qquad (22)$$

where

$$\varphi_i = \xi_i h, \qquad \bar{\varphi}_i = \bar{\xi}_i h.$$

The second of (21) and (22) give the dispersion relations in the form of biquadratic equations from which we select, as roots, the positive roots of

$$\varphi_i^2 = \tfrac{1}{2}\hat{\gamma}_{11}^{-1}\Omega^2\{[(1 + 3\hat{\gamma}_{11})^2 + 12\hat{\gamma}_{11}(\Omega^{-2} - 1)]^{\frac{1}{2}} - (-1)^i(1 + 3\hat{\gamma}_{11})\}, \qquad (23)$$

$$\bar{\varphi}_i^2 = \tfrac{1}{2}\tilde{\gamma}_{11}^{-1}\bar{\Omega}^2\{1 + 3r\tilde{\gamma}_{11} - (-1)^i[(1 + 3r\tilde{\gamma}_{11})^2 - 12r\tilde{\gamma}_{11}(1 - \bar{\Omega}^{-2})]^{\frac{1}{2}}\}. \qquad (24)$$

With these choices, ξ_1, ξ_2, $\bar{\xi}_1$ and $\bar{\xi}_2$ are all real and positive in the frequency range $\bar{\omega}_1 < \omega < \omega_1$, i.e. $\Omega < 1$ and $\bar{\Omega} > 1$. Hence, in that range, the solutions (19) and (20) have the following properties. In the unplated portion, there are two flexural waves (A_1, B_1) ninety degrees out of phase and propagating away from the electrodes; and two thickness-shear vibrations (A_2, B_2) ninety degrees out of phase and decaying away from the electrodes. In the plated portion, there are a forced thickness-shear vibration $\bar{\psi}_0 \cos \omega t$

and four additional vibrational motions: two of flexure ($\bar{A}_1, \bar{B}_1$) ninety degrees out of phase and two of thickness-shear ($\bar{A}_2, \bar{B}_2$) ninety degrees out of phase.

The conditions of continuity at the boundaries of the plated and unplated portions of the plate require that, on $x = \pm a$,

$$\psi_x = \bar{\psi}_x, \qquad v = \bar{v}, \qquad M_x = \bar{M}_x, \qquad V_x = \bar{V}_x.$$

These four conditions lead to eight equations on the eight constants $A_i, B_i, \bar{A}_i, \bar{B}_i$:

$$\begin{bmatrix} a_{11} & a_{12} & 1 & 0 & 1 & 0 & 0 & 0 \\ a_{21} & a_{22} & \beta_2 & 0 & \beta_1 & 0 & 0 & 0 \\ a_{31} & a_{32} & -\alpha_2 & \alpha_1 & 0 & 0 & 0 & 0 \\ a_{41} & a_{42} & -\varphi_2 & \varphi_1 & 0 & 0 & 0 & 0 \\ 0 & 0 & 0 & -1 & 0 & 1 & b_{11} & b_{12} \\ 0 & 0 & 0 & -\beta_1 & 0 & \beta_2 & b_{21} & b_{22} \\ 0 & 0 & 0 & 0 & \alpha_1 & -\alpha_2 & b_{31} & b_{32} \\ 0 & 0 & 0 & 0 & \varphi_1 & -\varphi_2 & b_{41} & b_{42} \end{bmatrix} \begin{bmatrix} \bar{A}_1 \\ \bar{A}_2 \\ A_2 \\ A_1 \\ B_1 \\ B_2 \\ \bar{B}_1 \\ \bar{B}_2 \end{bmatrix} = \bar{\psi}_0 \begin{bmatrix} 1 \\ k \\ 0 \\ 0 \\ 0 \\ 0 \\ 0 \\ 0 \end{bmatrix} \tag{25}$$

where

$$\begin{aligned} a_{11} &= b_{11} = -\cos \bar{\varphi}_1\lambda & a_{31} &= b_{31} = \bar{\alpha}_1 \sin \bar{\varphi}_1\lambda \\ a_{12} &= b_{12} = -\cos \bar{\varphi}_2\lambda & a_{32} &= b_{32} = \bar{\alpha}_2 \sin \bar{\varphi}_2\lambda \\ a_{21} &= b_{21} = -k\bar{\beta}_1 \cos \bar{\varphi}_1\lambda & a_{41} &= b_{41} = \bar{\varphi}_1 \sin \bar{\varphi}_1\lambda \\ a_{22} &= b_{22} = -k\bar{\beta}_2 \cos \bar{\varphi}_2\lambda & a_{42} &= b_{42} = \bar{\varphi}_2 \sin \bar{\varphi}_2\lambda \\ \beta_i &= 1 - (-1)^i\alpha_i\varphi_i & \bar{\beta}_i &= 1 + \bar{\alpha}_i\bar{\varphi}_i \\ \lambda &= a/h & k &= \bar{k}_1^2/k_1^2 \end{aligned}$$

For given forcing frequency ω, material properties of plate and electrodes, mass ratio R and dimensional ratio λ, all of the coefficients in (25) are known and the amplitudes A_i, $\bar{A}_i$, B_i, $\bar{B}_i$ can be computed.

The quantity of present interest is the Q of the partially plated plate. This is defined as 2π times the ratio of the energy per cycle in the plated portion of the plate to the energy per cycle in the escaping flexural waves—at resonance. Hence, to compute Q, the frequency at resonance must be known. Now, resonance is identified by a maximum of the current through the crystal. The current is proportional to the integral of the surface charge over the electrodes; and the surface charge, in turn, is a linear function of the thickness-shear strain $\partial\bar{v}/\partial x + \bar{\psi}_x$ [12]. Hence the criterion for resonance is a maximum of the amplitude of

$$\int_{-a}^{a} (\partial\bar{v}/\partial x + \bar{\psi}_x)\, dx. \tag{26}$$

Upon substituting (20) into (26) and performing the integration, we find that the amplitude of (26) is proportional to

$$S = (\bar{\Omega}^2 - 1)^{-1}(S_c^2 + S_s^2)^{\frac{1}{2}} \tag{27}$$

where

$$S_c = \lambda + \bar{A}_1\bar{\beta}_1\bar{\varphi}_1^{-1}\sin\bar{\varphi}_1\lambda + \bar{A}_2\bar{\beta}_2\bar{\varphi}_2^{-1}\sin\bar{\varphi}_2\lambda,$$
$$S_s = \bar{B}_1\bar{\beta}_1\bar{\varphi}_1^{-1}\sin\bar{\varphi}_1\lambda + \bar{B}_2\bar{\beta}_2\bar{\varphi}_2^{-1}\sin\bar{\varphi}_2\lambda.$$

For given material properties and given R and λ, the amplitudes $A_i, B_i, \bar{A}_i, \bar{B}_i$ can be computed, from (25), for trial values of dimensionless frequency $\bar{\Omega}$; and the values of $\bar{\Omega}$ that give maxima of S can be identified. The process can then be repeated over a range of values of λ. The result is a spectrum of resonance frequencies *versus* λ. The range of trial frequencies can be narrowed by centering on the approximate values obtained by the simplified procedure described in Section 4.

Next, the energies must be computed. The kinetic and potential energies, per cycle, in the plated portion of the plate are given by [13]

$$K = \frac{1}{T}\int_0^T\int_{-a}^{a}\left[\rho h(1+R)\left(\frac{\partial\bar{v}}{\partial t}\right)^2 + \tfrac{1}{3}\rho h^3(1+3R)\left(\frac{\partial\bar{\psi}_x}{\partial t}\right)^2\right]\mathrm{d}x\,\mathrm{d}t,$$

$$U = \frac{1}{T}\int_0^T\int_{-a}^{a}\left[\bar{k}_1^2 c_{66} h\left(\frac{\partial\bar{v}}{\partial x} + \bar{\psi}_x\right)^2 + \tfrac{1}{3}\gamma_{11}h^3\left(\frac{\partial\bar{\psi}_x}{\partial x}\right)^2\right]\mathrm{d}x\,\mathrm{d}t,$$

where T $(=2\pi/\omega)$ is the period. Upon performing the integrations, we find

$$\begin{aligned}(\tfrac{1}{2}h_1^2\bar{k}_1^2 c_{66}\bar{\Omega}^2)^{-1}K &= (\bar{A}_1^2 + \bar{B}_1^2)(\mu_1\lambda + \mu_2\sin 2\bar{\varphi}_1\lambda)\\ &\quad + (\bar{A}_1\bar{A}_2 + \bar{B}_1\bar{B}_2)[\mu_3\sin(\bar{\varphi}_1 - \bar{\varphi}_2)\lambda + \mu_4\sin(\bar{\varphi}_1 + \bar{\varphi}_2)\lambda]\\ &\quad + (\bar{A}_2^2 + \bar{B}_2^2)(\mu_5\lambda + \mu_6\sin 2\bar{\varphi}_2\lambda)\\ &\quad + \mu_7\bar{\psi}_0\bar{A}_1\sin\bar{\varphi}_1\lambda + \mu_8\bar{\psi}_0\bar{A}_2\sin\bar{\varphi}_2\lambda + 2\bar{\psi}_0^2\lambda,\end{aligned}$$

$$\begin{aligned}(\tfrac{1}{2}h^2\bar{k}_1^2 c_{66})^{-1}U &= (\bar{A}_1^2 + \bar{B}_1^2)(\nu_1\lambda + \nu_2\sin 2\bar{\varphi}_1\lambda)\\ &\quad + (\bar{A}_1\bar{A}_2 + \bar{B}_1\bar{B}_2)[\nu_3\sin(\bar{\varphi}_1 - \bar{\varphi}_2)\lambda + \nu_4\sin(\bar{\varphi}_1 + \bar{\varphi}_2)\lambda]\\ &\quad + (\bar{A}_2^2 + \bar{B}_2^2)(\nu_5\lambda + \nu_6\sin 2\bar{\varphi}_2\lambda)\\ &\quad + \nu_7\bar{\psi}_0\bar{A}_1\sin\bar{\varphi}_1\lambda + \nu_8\bar{\psi}_0\bar{A}_2\sin\bar{\varphi}_2\lambda + 2\bar{\psi}_0^2\lambda,\end{aligned}$$

where

$$\begin{aligned}
\mu_1 &= 1 + 3r\bar{\alpha}_1^2 & \nu_1 &= \bar{\beta}_1^2 + \tilde{\gamma}_{11}\bar{\varphi}_1^2\\
\mu_2 &= (1 - 3r\bar{\alpha}_1^2)/2\bar{\varphi}_1 & \nu_2 &= (\bar{\beta}_1^2 - \tilde{\gamma}_{11}\bar{\varphi}_1^2)/2\bar{\varphi}_1\\
\mu_3 &= 2(1 + 3r\bar{\alpha}_1\bar{\alpha}_2)/(\bar{\varphi}_1 - \bar{\varphi}_2) & \nu_3 &= 2(\bar{\beta}_1\bar{\beta}_2 + \tilde{\gamma}_{11}\bar{\varphi}_1\bar{\varphi}_2)/(\bar{\varphi}_1 - \bar{\varphi}_2)\\
\mu_4 &= 2(1 - 3r\bar{\alpha}_1\bar{\alpha}_2)/(\bar{\varphi}_1 + \bar{\varphi}_2) & \nu_4 &= 2(\bar{\beta}_1\bar{\beta}_2 - \tilde{\gamma}_{11}\bar{\varphi}_1\bar{\varphi}_2)/(\bar{\varphi}_1 + \bar{\varphi}_2)\\
\mu_5 &= 1 + 3r\bar{\alpha}_2^2 & \nu_5 &= \bar{\beta}_2^2 + \tilde{\gamma}_{11}\bar{\varphi}_2^2\\
\mu_6 &= (1 - 3r\bar{\alpha}_1^2)/2\bar{\varphi}_2 & \nu_6 &= (\bar{\beta}_2^2 - \tilde{\gamma}_{11}\bar{\varphi}_2^2)/2\bar{\varphi}_2\\
\mu_7 &= 4/\bar{\varphi}_1 & \nu_7 &= 4\bar{\beta}_1/\bar{\varphi}_1\\
\mu_8 &= 4/\bar{\varphi}_2 & \nu_8 &= 4\bar{\beta}_2/\bar{\varphi}_2.
\end{aligned}$$

The energy, per cycle, that escapes in the form of flexural waves in the unplated portions of the plate is equal to the work done, per cycle, by the transverse shear force and bending moment on sections of the unplated plate perpendicular to the x-axis:

$$E_l = -2\int_0^T \left(V_x\frac{\partial v}{\partial t} + M_x\frac{\partial \psi_x}{\partial t}\right) dt,$$

$$= 4\pi h^2 k_1^2 c_{66}(A_1^2 + B_1^2)(\hat{\gamma}_{11}\varphi_1 + \alpha_1\beta_1).$$

As the material is assumed to be lossless, E_l must be equal to the energy, per cycle, entering the plated portion of the plate:

$$E_a = \int_0^T\int_{-a}^{a} C\frac{\partial \bar{\psi}_x}{\partial t}\, dx\, dt,$$

$$= 4\pi h^2 \bar{k}_1^2 c_{66}(1 - \bar{\Omega}^2)\bar{\psi}_0(\bar{B}_1\bar{\varphi}_1^{-1}\sin\bar{\varphi}_1\lambda + \bar{B}_2\bar{\varphi}_2^{-1}\sin\bar{\varphi}_2\lambda).$$

A comparison of independent computations of E_l and E_a serves as a check on a major part of the entire computation.

The results of a series of computations of

$$Q = 2\pi(K + U)/E_l \tag{28}$$

for a partially plated AT-type plate for which $\omega_1/\bar{\omega}_1 = 1{\cdot}0354$ are illustrated in Fig. 3, where Q is plotted against a/h. As may be seen, Q is an almost periodic function of a/h with maxima exceeding 10^8 and minima ranging from about 10^3, for very short electrodes, to about 10^5 for a/h near Bechmann's Number which, as shown in the next section, is 14·3 for this case.

The maxima of Q occur, approximately, when the flexural mode in the plated portion has nodes at the boundaries; i.e. whenever an integral number of wave-lengths of the flexural mode just fits across the electrode:

$$a/h = m\pi/\bar{\Phi}_1, \qquad m = 1, 2, 3\ldots,$$

where $\bar{\Phi}_1$ is given by (24) with $i = 1$ and the frequency that of resonance. The latter is given closely enough, for the present purpose, by $\bar{\Omega} = 1$. (The corresponding $\bar{\Phi}_1$ is illustrated in Fig. 1). Then, in terms of material properties and mass loading, the maxima of Q occur, approximately, when

$$a/h = m\pi/(\tilde{\gamma}_{11}^{-1} + 3r)^{\frac{1}{2}}, \qquad m = 1, 2, 3\ldots$$

This is insensitive to R; so that, if we take $R = 0$, we find

$$a/h \approx m\pi/(3 + \pi^2 c_{66}/4\gamma_{11})^{\frac{1}{2}},\ m = 1, 2, 3\ldots \tag{29}$$

For the AT-cut of quartz, a/h, in (29), is about 1·6m. Thus, for the AT-cut, the maxima of Q occur at increments of electrode length of about 1·6 times the thickness of the plate.

Additional computations, with other mass ratios R, reveal that the locus of the minima of Q is raised as the mass per unit area of the electrodes is reduced.

In a physical plate, there would be additional losses which would affect the Q in a variety of ways. Losses through internal friction and the thermoelastic effect would probably have a quantitative, rather than a qualitative, influence—resulting in a great

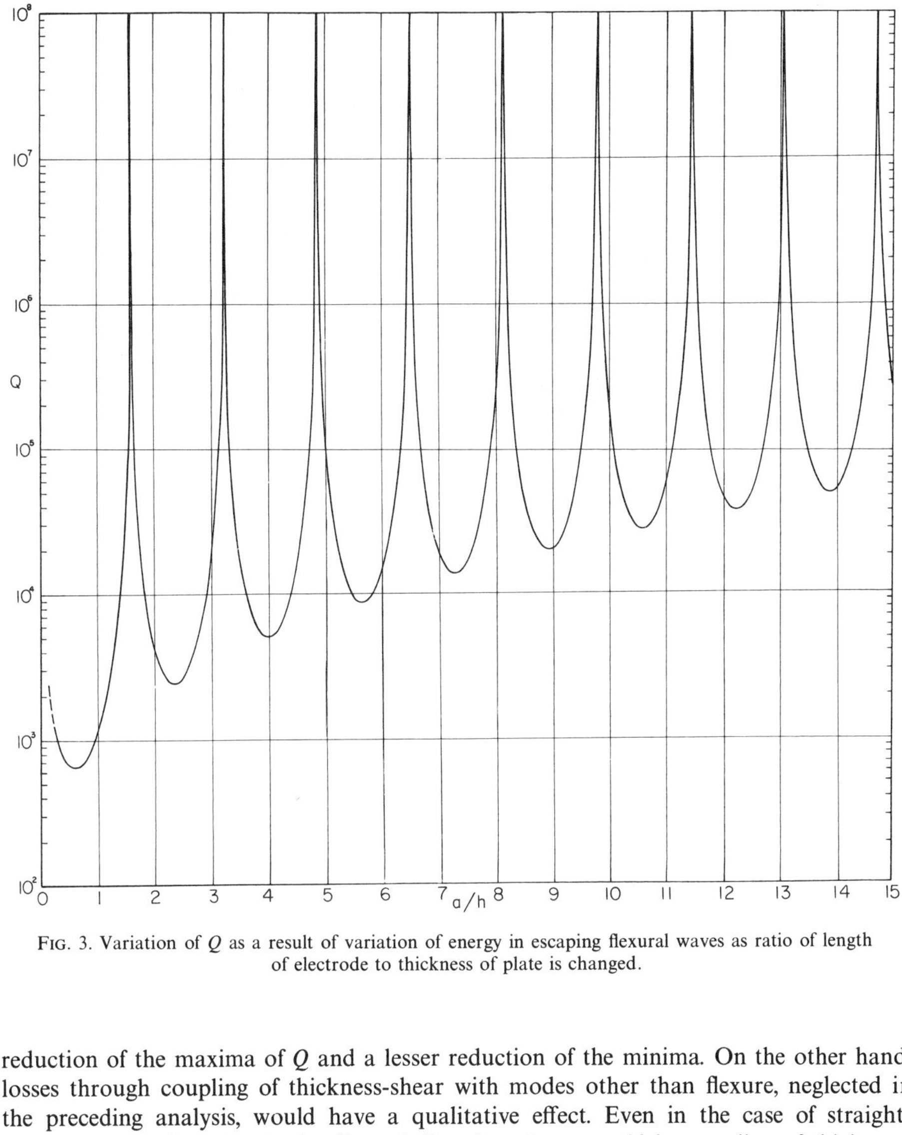

FIG. 3. Variation of Q as a result of variation of energy in escaping flexural waves as ratio of length of electrode to thickness of plate is changed.

reduction of the maxima of Q and a lesser reduction of the minima. On the other hand, losses through coupling of thickness-shear with modes other than flexure, neglected in the preceding analysis, would have a qualitative effect. Even in the case of straight-crested waves advancing in the digonal direction, there would be coupling of thickness-shear and flexure with face-shear. The energy in the escaping face-shear wave would have a periodicity different from that of the flexural wave— with a consequent destruction of the simple periodicity exhibited in Fig. 3. Moreover, the actual finite dimensions of the electrodes and plate would not permit simple, straight-crested waves. There would be some excitation of flexural, extensional and face-shear waves radiating laterally and extensional waves radiating longitudinally—each with its own periodicity of energy loss *versus* dimensional ratios. What might be expected in a series of experiments, then, would be an apparently erratic fluctuation of Q about a general diminution with reduction of areal dimensions (and increase of thickness) of electrodes.

4. BECHMANN'S NUMBER

In long, thin plates, the coupling of thickness-shear and flexural waves at a free boundary does not materially affect the frequencies of the fundamental thickness-shear mode and its anharmonic overtones in the neighborhood of the cut-off frequency [8]. Also, the flexural rigidity of the plate has little influence on the dispersion of thickness-shear waves at long wave-lengths. As a consequence of these two properties, the frequencies of the thickness-shear mode and its anharmonic overtones, in long, thin plates, are governed, to a good approximation, by equations of motion in which the term arising from the flexural rigidity of the plate is omitted—especially if, by the introduction of a suitable correction factor, the resulting dispersion relation is adjusted to nearly its original form: an adjustment which can be made very closely in a narrow frequency range [15].

The contribution of flexural rigidity, in (1) and (3), may be removed by omitting the first term in (1b) and in (3b). Then the equations of free vibration are

$$k_1^2 c_{66}\left(\frac{\partial^2 v}{\partial x^2}+\frac{\partial \psi_x}{\partial x}\right) = -k_2\rho\omega^2 v, \tag{30a}$$

$$k_1^2 c_{66}\left(\frac{\partial v}{\partial x}+\psi_x\right) = \tfrac{1}{3}\rho\omega^2 h^2\psi_x, \tag{30b}$$

for the unplated plate and, for the plated plate:

$$\bar{k}_1^2 c_{66}\left(\frac{\partial^2 \bar{v}}{\partial x^2}+\frac{\partial \bar{\psi}_x}{\partial x}\right) = -\bar{k}_2\rho\omega^2(1+R)\bar{v}, \tag{31a}$$

$$\bar{k}_1^2 c_{66}\left(\frac{\partial \bar{v}}{\partial x}+\bar{\psi}_x\right) = \tfrac{1}{3}\rho\omega^2 h^2(1+3R)\bar{\psi}_x, \tag{31b}$$

where k_2 and $\bar{k}_2$ are the factors introduced to correct the dispersion relations.

Equations (30) and (31) may be written in the forms

$$v = -\frac{h^2}{3k_2}\frac{\partial \psi_x}{\partial x}, \qquad \frac{h^2}{3k_2}\frac{\partial^2 \psi_x}{\partial x_2} - (1-\omega^2/\omega_1^2)\psi_x = 0, \tag{32}$$

$$\bar{v} = -\frac{h^2}{3r\bar{k}_2}\frac{\partial \bar{\psi}_x}{\partial x}, \qquad \frac{h^2}{3r\bar{k}_2}\frac{\partial^2 \bar{\psi}_x}{\partial x^2} + (\omega^2/\bar{\omega}_1^2 - 1)\bar{\psi}_x = 0. \tag{33}$$

To fix the correction factors k_2 and $\bar{k}_2$, we first find the dispersion relations from the approximate equations (32) and (33) by setting

$$\psi_x = A\exp(-\xi x), \qquad \bar{\psi}_x = \bar{A}\cos\bar{\xi}x,$$

with the results

$$\xi^2 h^2 = 3k_2(1-\omega^2/\omega_1^2), \qquad \bar{\xi}^2 h^2 = 3r\bar{k}_2(\omega^2/\bar{\omega}_1^2 - 1). \tag{34}$$

These give the correct frequencies of zero wave-number. As we are interested only in the frequency range $\bar{\omega}_1 \leqslant \omega \leqslant \omega_1$, we choose k_2 and $\bar{k}_2$ to give the correct wave-number ξ at $\omega = \bar{\omega}_1$ and the correct wave-number $\bar{\xi}$ at $\omega = \omega_1$:

$$k_2 = \Phi_2^2/3(1-\bar{\omega}_1^2/\omega_1^2), \qquad \bar{k}_2 = \bar{\Phi}_2^2/3r(\omega_1^2/\bar{\omega}_1^2-1), \tag{35}$$

where

$$\Phi_2 = [\varphi_2]_{\omega=\bar{\omega}_1}, \qquad \bar{\Phi}_2 = [\bar{\varphi}_2]_{\omega=\omega_1} \tag{36}$$

as calculated from (23) and (24), respectively, and identified in Fig. 1. Thus, k_2 and $\bar{k}_2$ are expressed in terms of material constants and mass loading. By this device, the dispersion relations for thickness-shear waves are returned, in the range $\bar{\omega}_1 \leqslant \omega \leqslant \omega_1$, to almost exactly their forms before omission of the flexural rigidities.

Turning, now, to the problem of vibrations of an infinite plate coated with electrodes over $-a \leqslant x \leqslant a$ (Fig. 2) we take

$$\psi_x|_{x>a} = A\exp[\xi(a-x)], \qquad \psi_x|_{x<-a} = A\exp[\xi(a+x)], \qquad \bar{\psi}_x = \bar{A}\cos\bar{\xi}x. \tag{37}$$

Upon substituting (37) into (32) and (33), we find

$$\omega^2 = \omega_1^2(1-\varphi^2/3k_2) = \bar{\omega}_1^2(1+\bar{\varphi}^2/3r\bar{k}_2), \tag{38}$$

where

$$\varphi = \xi h, \qquad \bar{\varphi} = \bar{\xi}h.$$

With the flexural rigidity omitted, the conditions of continuity at the boundaries of the plated and unplated portions of the plate reduce to

$$v = \bar{v}, \qquad V_x = \bar{V}_x \quad \text{on} \quad x = \pm a. \tag{39}$$

Upon substituting (37) into (39), and using (38), we find

$$\frac{A}{\bar{A}} = \frac{\bar{\varphi}k_2\sin\bar{\varphi}\lambda}{\varphi r\bar{k}_2} = \frac{\bar{k}_1^2\omega_1^2\cos\bar{\varphi}\lambda}{k_1^2\bar{\omega}_1^2}, \tag{40}$$

where, as before, $\lambda = a/h$. The second of (40) yields the frequency equation

$$\frac{a}{h} = \frac{1}{\bar{\varphi}}\tan^{-1}\left(\frac{r\bar{k}_1^2\bar{k}_2\omega_1^2\varphi}{k_1^2k_2\bar{\omega}_1^2\bar{\varphi}}\right). \tag{41}$$

The quantities $k_1, \bar{k}_1, k_2, \bar{k}_2, \omega_1, \bar{\omega}_1$ and r are determined solely by the material properties of the crystal and the mass ratio R; while φ and $\bar{\varphi}$ depend on the frequency through (38). Hence, for a given frequency, there corresponds a sequence of roots a/h equal to a constant multiplied by the integers $1, 2, 3, \ldots$; and only the constant changes for a different frequency. The computation of a frequency spectrum is thus a simple matter. An example, for $\omega_1/\bar{\omega}_1 = 1{\cdot}0354$, is illustrated in Fig. 4 along with experimental data obtained by Curran and Koneval [6]. (In the experiments, the width of the electrodes was small, in comparison with the length $2a$, instead of infinite as assumed in the mathematical solution. However, it was shown by Sykes, many years ago [16], that the thickness-shear frequencies are insensitive to width).

It may be seen, by inspection of (41) or Fig. 4, that the number of anharmonic overtones between $\bar{\omega}_1$ and ω_1 increases with increasing a/h. To find the values of a/h for which an additional overtone is included, it is only necessary to set $\omega = \omega_1$ in (41). Then,

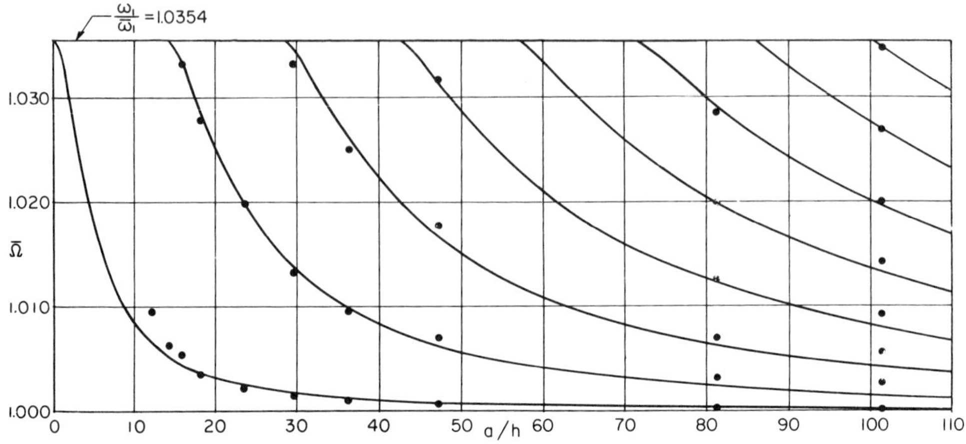

FIG. 4. Variation of number and distribution of resonances between cut-off frequencies. Comparison of theory with experiments by Curran and Koneval.

from (38), $\varphi = 0$ and, from (36), $\bar{\varphi} = \bar{\Phi}_2$; so that

$$a/h = m\pi/\bar{\Phi}_2, \qquad m = 1, 2, 3 \ldots$$

where, from (24),

$$\bar{\Phi}_2^2 = \frac{\omega_1^2(1+3r\tilde{\gamma}_{11})}{2\bar{\omega}_1^2\tilde{\gamma}_{11}}\left[1-\left(1-\frac{12r\tilde{\gamma}_{11}(1-\bar{\omega}_1^2/\omega_1^2)}{(1+3r\tilde{\gamma}_{11})^2}\right)^{\frac{1}{2}}\right]. \tag{42}$$

The value of a/h, below which there are no anharmonic overtones, is, then,

$$a/h = \pi/\bar{\Phi}_2. \tag{43}$$

For the AT-type plate with $\omega_1/\bar{\omega}_1 = 1{\cdot}0354$, (43) yields $a/h = 14{\cdot}3$.

Equation (43) is to be compared with Curran and Koneval's

$$\frac{a}{h} = M\left(\frac{\bar{\omega}_1}{\omega_1-\bar{\omega}_1}\right)^{\frac{1}{2}}, \tag{44}$$

where M is a numerical constant estimated to be about 2·8. Equation (44) may be shown to be an asymptotic form of (43) for $R \ll 1$. In that case,

$$1-\bar{\omega}_1^2/\omega_1^2 \approx 2(1-\bar{\omega}_1/\omega_1) \ll 1, \qquad r \approx 1, \qquad \tilde{\gamma}_{11} \approx \hat{\gamma}_{11} = 4\gamma_{11}/\pi^2 c_{66}.$$

With these assumptions, (43) becomes

$$\frac{a}{h} \approx \pi\left(\frac{1}{6}+\frac{2\gamma_{11}}{\pi^2 c_{66}}\right)^{\frac{1}{2}}\left(\frac{\bar{\omega}_1}{\omega_1-\bar{\omega}_1}\right)^{\frac{1}{2}}. \tag{45}$$

Hence, in terms of the elastic properties of the plate, Curran and Koneval's constant M is

$$M = \pi\left(\frac{1}{6}+\frac{2\gamma_{11}}{\pi^2 c_{66}}\right)^{\frac{1}{2}}. \tag{46}$$

For the AT-cut of quartz, this formula gives $M = 2{\cdot}75$ and (45) gives $a/h = 14{\cdot}6$ for the plate with $\omega_1/\bar{\omega}_1 = 1{\cdot}0354$.

It may be remarked that the frequencies of the lowest branch in Fig. 4, although computed by the approximate method described in this section, match the frequencies computed in the preceding section extremely closely—even down to very small values of a/h. This is due to the fact that the deformation in the plated portion is almost entirely thickness-shear; i.e. very little of the reflection is in the form of flexure.

REFERENCES

[1] R. Bechmann, *Proc. IRE*, **49**, 523–524 (1961). U.S. Patent No. 2,249,933 (1941).
[2] W. Shockley, D. R. Curran and D. J. Koneval, *Proc. 17th Symp. on Frequency Control*, 88–126 (1963).
[3] R. D. Mindlin, *Quart. Appl. Math.*, **19**, 51–61 (1961).
[4] R. D. Mindlin and D. C. Gazis, *Proc. 4th U.S. Nat'l. Cong. Appl. Mechanics*, 305–310 (1962).
[5] R. K. Kaul and R. D. Mindlin, *J. Acoust. Soc. Amer.*, **34**, 1902–1910 (1962).
[6] D. R. Curran and D. J. Koneval, *Proc. 18th Symp. Freq. Control*, 93–119 (1964).
[7] W. P. Mason, *Piezoelectric Crystals and their Application to Ultrasonics*, D. Van Nostrand, New York (1950).
[8] R. D. Mindlin, *J. Appl. Phys.*, **22**, 316–323 (1951).
[9] S. Timoshenko, *Phil. Mag.* Ser. 6, **41**, 744 (1921).
[10] R. Bechmann, *Phys. Rev.* **110**, 1060–1061 (1958).
[11] R. D. Mindlin, *J. Appl. Phys.* **23**, 83–88 (1952).
[12] H. F. Tiersten and R. D. Mindlin, *Quart. Appl. Math.*, **20**, 107–119 (1962).
[13] R. D. Mindlin, *Progress in Applied Mechanics*, pp. 73–84, Macmillan Co., (1963).
[14] R. D. Mindlin and H. Deresiewicz, *J. Appl. Phys.*, **25**, 21–24 (1954).
[15] R. D. Mindlin and M. Forray, *J. Appl. Phys.*, **25**, 12–20 (1954).
[16] R. A. Sykes, *Quartz Crystals for Electrical Circuits*, R. A. Heising, Editor, D. Van Nostrand, New York (1946).

(*Received* 28 *June* 1965)

Résumé—Dans cette étude, des solutions mathématiques sont présentées, et celles-ci prédisent l'influence de la longueur et de la masse de revêtements d'électrodes rectangulaires sur l'acuité de la résonance fondamentale "épaisseur-cisaillement" et sur le nombre et la distribution des vibrations anharmoniques entre les fréquences de détente des portions recouvertes et non recouvertes de plaques de quartz du type -AT.

Zusammenfassung—Mathematische Lösungen sind in dieser Abhandlung aufgezeigt welche den Einfluss der Länge und Masse von rechteckigen Elektrodenplatten auf die Schärfe der grundlegenden Scherungs Resonanz und auf die Anzahl und Verteilung von unharmonischen Obertönen zwischen den abgeschnittenen Frequenzen der gedeckfen und nicht-gedeckfen Teilen der AT-artigen Quarzplatten vorhersagen.

Абстракт—Настоящая работа демонстрирует математические решения, указывающие на влияние длины и массы прямоугольных гальванопокрытий на остроту основного резонанса поперечниа сдвигов и на количество и распределение ангармонических обертонов между граничными частотами покрытых и непокрытых частей кварцевых пластинок вида "AT".

Reprinted from The Journal of the Acoustical Society of America, Vol. 41, No. 4 (Part 2), 969–973, April 1967

Bechmann's Number for Harmonic Overtones of Thickness/Twist Vibrations of Rotated-Y-Cut Quartz Plates

R. D. Mindlin

Department of Civil Engineering, Columbia University, New York, New York 10027

Two methods are described for calculating Bechmann's number for harmonic overtones of thickness/twist modes in rotated Y-cut quartz plates: one based on an exact solution of the equations of elasticity and the other based on an asymptotic form of the exact solution. The latter leads to an explicit formula for Bechmann's number and it is shown to give results very close to the former—which requires extensive computations.

INTRODUCTION

Thickness/twist waves in a plate are those in which the displacement is parallel to the middle plane of the plate and perpendicular to the direction of propagation. In rotated-Y-cut quartz plates,[1] such waves are possible if the displacement is parallel to the digonal axis of elastic symmetry that lies in the plane of the plate.[2] At long wavelengths, the frequencies of thickness/twist waves are approximately proportional to the number of nodes of displacement across the thickness of the plate; i.e., the waves occur in harmonic overtones. Each harmonic overtone has a low-frequency cutoff—a frequency, say ω_m, below which the wavenumber is imaginary so that the wave is nonpropagating. In a portion of a plate coated with electrodes, the cutoff frequency is lowered slightly, say to $\bar{\omega}_m$, owing to the added mass. Consequently, there is a narrow frequency range, between $\bar{\omega}_m$ and ω_m, in which waves can propagate in the plated portion but not in the unplated portion. This results in the trapping of vibrational motion in the coated portion of the plate.[3] These vibrational modes have closely spaced anharmonic overtones corresponding to the number of nodes of displacement *along* the plate between the edges of the electrode. Thus, each harmonic-overtone mode can have successively higher anharmonic overtones with only slightly higher frequencies; so that, for a long enough electrode, many anharmonic overtones of a single harmonic overtone can have frequencies between $\bar{\omega}_m$ and ω_m and can be trapped under the electrodes. However, if the length of electrode is short enough, only one such mode can be trapped because the next anharmonic overtone and all the higher ones will have frequencies above ω_m and will propagate into the unplated portion of the plate. For this situation to obtain, the length of electrode must be less than the wavelength, along the plate, of that thickness/twist wave, in the plated portion of the plate, which propagates at the cutoff frequency of the same harmonic overtone mode in the unplated portion of the plate. The ratio of the critical length of electrode to the thickness of the plate is Bechmann's number—named after the discoverer of the effect.[4]

Calculations of Bechmann's number for harmonic overtones of thickness/twist modes are described in this paper. Two methods are presented: one based on an exact solution, of the three-dimensional equations of the theory of elasticity, for thickness/twist waves in plates with and without electrode coatings; and a second based on an asymptotic form of the exact solution—a form that can more readily be applied to problems of greater difficulty. The first method requires extensive computations, whereas the asymptotic form leads to an explicit formula for Bechmann's number in terms of the material properties and dimensions of the plate and electrodes and the order of the harmonic overtone. A comparison of the two solutions reveals, in the case of AT-type plates, that the error in the

[1] W. P. Mason, *Piezoelectric Crystals and Their Application to Ultrasonsics* (D. Van Nostrand Co., Inc., Princeton, N. J., 1950).
[2] R. D. Mindlin, Intern. J. Solids & Structures **1**, 141–145 (1965).
[3] W. S. Mortley, Proc. Inst. Elec. Engrs. (London) **104B**, 239–253 (1957).
[4] R. Bechmann, U. S. Patent No. 2,249,933 (26 Sept. 1939).

approximation for the fundamental mode is about 1% for 3% plateback and diminishes as the order of the harmonic overtone increases and the percent plateback decreases.

I. EQUATIONS OF ELASTICITY

The plate is referred to rectangular coordinate axes x, y, z, with the faces at $y=\pm h$ and x a digonal axis of elastic symmetry of the quartz crystal. The stress/strain relations, referred to these axes, exhibit monoclinic symmetry:

$$\begin{aligned}
T_1&=c_{11}S_1+c_{12}S_2+c_{13}S_3+c_{14}S_4,\\
T_2&=c_{21}S_1+c_{22}S_2+c_{23}S_3+c_{24}S_4,\\
T_3&=c_{31}S_1+c_{32}S_2+c_{33}S_3+c_{34}S_4,\\
T_4&=c_{41}S_1+c_{42}S_2+c_{43}S_3+c_{44}S_4,\\
T_5&=c_{55}S_5+c_{56}S_6,\\
T_6&=c_{65}S_5+c_{66}S_6, \quad c_{pq}=c_{qp},
\end{aligned}\tag{1}$$

where the components of strain S_q are related to the components of displacement u, v, w by

$$\begin{aligned}
S_1&=\frac{\partial u}{\partial x}, & S_4&=\frac{\partial w}{\partial y}+\frac{\partial v}{\partial z},\\
S_2&=\frac{\partial v}{\partial y}, & S_5&=\frac{\partial u}{\partial z}+\frac{\partial w}{\partial x},\\
S_3&=\frac{\partial w}{\partial z}, & S_6&=\frac{\partial v}{\partial x}+\frac{\partial u}{\partial y},
\end{aligned}\tag{2}$$

and the components of stress T_p must satisfy the stress equations of motion

$$\begin{aligned}
\frac{\partial T_1}{\partial x}+\frac{\partial T_6}{\partial y}+\frac{\partial T_5}{\partial z}&=\rho\frac{\partial^2 u}{\partial t^2},\\
\frac{\partial T_6}{\partial x}+\frac{\partial T_2}{\partial y}+\frac{\partial T_4}{\partial z}&=\rho\frac{\partial^2 v}{\partial t^2},\\
\frac{\partial T_5}{\partial x}+\frac{\partial T_4}{\partial y}+\frac{\partial T_3}{\partial z}&=\rho\frac{\partial^2 w}{\partial t^2}.
\end{aligned}\tag{3}$$

The 15 equations (1–3) constitute the equations of the linear theory of elasticity solutions of which for thickness/twist vibrations of infinite plates without and with electrode coatings are given in the next two sections.

II. THICKNESS/TWIST VIBRATIONS OF A PLATE WITHOUT ELECTRODES

The thickness/twist modes that are to be studied have the general form

$$u=U(y,z)e^{i\omega t}, \quad v=w=0. \tag{4}$$

Then, with the exponential factor omitted, Eqs. 2 reduce to

$$\begin{aligned}
&S_1=S_2=S_3=S_4=0,\\
&S_5=\partial U/\partial z, \quad S_6=\partial U/\partial y;
\end{aligned}\tag{5}$$

Eqs. 1, with Eqs. 5, reduce to

$$\begin{aligned}
&T_1=T_2=T_3=T_4=0,\\
&T_5=c_{55}\frac{\partial U}{\partial z}+c_{56}\frac{\partial U}{\partial y},\\
&T_6=c_{65}\frac{\partial U}{\partial z}+c_{66}\frac{\partial U}{\partial y};
\end{aligned}\tag{6}$$

and Eqs. 3, with Eqs. 6 and 4 reduce to

$$c_{66}\frac{\partial^2 U}{\partial y^2}+2c_{56}\frac{\partial^2 U}{\partial y\partial z}+c_{55}\frac{\partial^2 U}{\partial z^2}=-\rho\omega^2 U. \tag{7}$$

Now, the main interest is in modes with displacement even in z and odd in y: even in z because the modes will then include the lowest of the anharmonic series, which is the only one of present interest; odd in y because modes even in y do not contribute to current through the plate. The appropriate form of U is

$$U=A\sin\eta y\cos\zeta\left(\frac{c_{56}}{c_{66}}y-z\right). \tag{8}$$

It will be observed that Eq. 8 is purely antisymmetric in y only at $\zeta=0$; whereas, at small ζy, there is a small symmetric contribution. A mode purely antisymmetric in y, with $\zeta\neq 0$, is not possible (i.e., cannot satisfy the equations) because of the presence of the elastic coupling constant c_{56}.

Upon substituting Eq. 8 in Eq. 7, we find that

$$\rho\omega^2=c_{66}\eta^2+\gamma_{55}\zeta^2, \tag{9}$$

where $\gamma_{55}=c_{55}-c_{56}^2/c_{66}$, i.e., Voigt's face/shear constant.

For a plate without electrode coatings, the boundary conditions on the faces are

$$T_2=T_4=T_6=0 \quad \text{on} \quad y=\pm h. \tag{10}$$

With Eqs. 6 and 8, the first two of these are satisfied identically and the third is satisfied if

$$2\eta h=m\pi, \quad m=1, 3, 5, \cdots. \tag{11}$$

Hence Eq. 9 becomes

$$\omega^2=\frac{c_{66}}{\rho}\left(\frac{\pi}{2h}\right)^2\left[m^2+\frac{\gamma_{55}}{c_{66}}\left(\frac{2\zeta h}{\pi}\right)^2\right], \tag{12}$$

which is the dispersion relation for thickness/twist waves in an infinite, unplated, rotated-**Y**-cut, quartz plate.[3] At $\zeta=0$, the frequencies, say ω_m, are proportional to the odd integers m; i.e., the overtones of an infinite plate are harmonic. For a given m, the wavenumbers ζ are real for $\omega>\omega_m$ and imaginary for $\omega<\omega_m$. Thus, a given harmonic overtone is a propagating wave above ω_m and a nonpropagating wave below ω_m.

III. PLATE WITH ELECTRODE COATINGS

In the case of a plate with like electrode films on its two faces, we take

$$U=\bar{A}\sin\bar{\eta}y\cos\bar{\zeta}\left(\frac{c_{56}}{c_{66}}y-z\right). \tag{13}$$

Then Eq. 7 becomes

$$\rho\bar{\omega}^2=c_{66}\bar{\eta}^2+\gamma_{55}\bar{\zeta}^2, \tag{14}$$

where $\bar{\omega}$ is the frequency of vibration of the plated plate.

Again, the first two of the boundary conditions in Eqs. 10 are satisfied identically. Instead of using the third of Eqs. 10, we take the inertia of the electrodes into account by setting

$$(T_6)_{y=\pm h}=\pm 2\rho'h'\bar{\omega}^2(U)_{y=\pm h}, \tag{15}$$

where ρ' is the density of the electrode material and h' is the thickness of each electrode. Upon substituting Eq. 13 into the last of Eqs. 6 and the result into Eq. 15, we find that

$$\tan\bar{\eta}h=4\bar{\eta}h/\pi^2\bar{\Omega}^2R, \tag{16}$$

where

$$\bar{\Omega}^2=\frac{\bar{\omega}^2}{(\pi^2c_{66}/4h^2\rho)},\quad R=2\rho'h'/\rho h. \tag{17}$$

$\bar{\Omega}$ is the ratio of the frequency of vibration of the plated plate to the cutoff frequency of the lowest mode of the unplated plate. R is the ratio of the mass per unit area of both electrodes to the mass per unit area of the plate.

Solving Eq. 14 for $\bar{\zeta}h$ in terms of $\bar{\eta}h$ and $\bar{\Omega}$, we have

$$\bar{\zeta}h=\frac{\pi}{2}\left(\frac{c_{66}}{\gamma_{55}}\right)^{\frac{1}{2}}\left[\bar{\Omega}^2-\left(\frac{2\bar{\eta}h}{\pi}\right)^2\right]^{\frac{1}{2}}. \tag{18}$$

The dispersion relation for thickness/twist waves in an infinite, rotated-**Y**-cut quartz plate, coated with electrodes, is given by Eqs. 18 and 16. As R is usually small in comparison with unity, the dispersion curves are similar to those for the unplated plate but the frequencies are slightly lower. The computation of the dispersion curves is simplified by the substitution

$$\bar{\eta}h=\tfrac{1}{2}m\pi-\phi. \tag{19}$$

Then Eq. 16 becomes

$$\operatorname{ctn}\phi=(2m\pi-4\phi)/m^2\pi^2R. \tag{20}$$

For $2mR\ll\pi$, the cutoff frequencies (those at $\bar{\zeta}=0$) are found, from Eqs. 18–20, to be approximately

$$\bar{\omega}_m=\frac{m\pi(1-R)}{2h}\left(\frac{c_{66}}{\rho}\right)^{\frac{1}{2}}, \tag{21}$$

whereas the cutoff frequencies for the unplated plate are given by the same formula with R set equal to zero. Thus, for small mR, $100R$ is approximately the percent reduction in cutoff frequency (the percent plateback) due to the inertia of the electrodes.

IV. BECHMANN'S NUMBER

Consider, now, a plate coated with a pair of electrode strips symmetrically disposed on the two faces of the plate and aligned parallel to the x axis. As explained previously,[5] Bechmann's number is the ratio, to the thickness of the plate, of the wavelength of a propagating wave, in the plated portion of the plate, at the cutoff frequency of the same type of wave in the unplated portion of the plate. (This definition involves neglect of the slight difference between the distributions of displacements, in the two waves, across the thickness of the plate.)

Now, the wavelength in the plated portion of the plate is $2\pi/\bar{\zeta}$, the cutoff frequency of the mth harmonic in the unplated plate is m times the cutoff frequency of the first mode, and the thickness of the plate is $2h$. Hence, Bechmann's number for thickness/twist waves is

$$B_{\mathrm{TT}}=\pi/(\bar{\zeta}h)_{\bar{\Omega}=m}. \tag{22}$$

To calculate B_{TT} for a given order m of harmonic overtone and a given plateback fraction R, insert the chosen m and R in Eq. 20 and calculate the first root ϕ. With this value of ϕ, Eqs. 19 and 18, with $\bar{\Omega}=m$, give the denominator in Eq. 22. The results of such computations are shown in Fig. 1 for **AT**-type quartz plates[1] for which, in units of 10^{10} dyn/cm^2,

$$c_{55}=68.81,\quad c_{66}=29.01,\quad c_{56}=2.53,\quad \gamma_{55}=68.59, \tag{23}$$

as calculated from Bechmann's values[6] of the principal constants of quartz. As may be seen from the Figure, Bechmann's number varies approximately inversely as the product of the order of the harmonic overtone and the square root of the percent plateback. For small mR, the curves in Fig. 1 are very nearly straight lines and, in fact, a simple, approximate, explicit formula for B_{TT} may be obtained as follows. The first root of Eq. 20 is approximately $\phi=m\pi R/2$ for small mR, so that, from Eq. 19,

$$\bar{\eta}h\approx m\pi(1-R)/2. \tag{24}$$

---

[5] R. D. Mindlin and P. C. Y. Lee, Intern. J. Solids & Structures 2, 125–139 (1966).
[6] R. Bechmann, Phys. Rev. 110, 1060–1061 (1958).

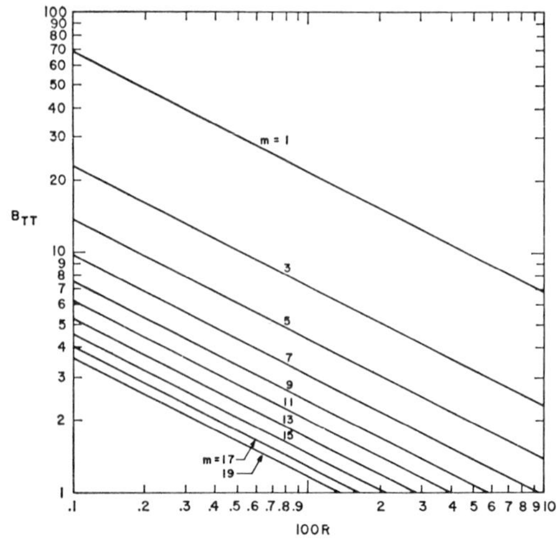

FIG. 1. Bechmann's number (B_{TT}) for thickness/twist modes of order m, in **AT**-type quartz plates, versus percent plateback ($100R$).

Inserting this in Eq. 18, with $\bar{\Omega}=m$, and the result in Eq. 22, we have

$$B_{TT}\approx(1/m)(2\gamma_{55}/Rc_{66})^{\frac{1}{2}}. \qquad (25)$$

This is to be compared with Curran and Koneval's empirical formula[7] for B_{TT} which, in the present notation, is

$$B_{TT}=\frac{M}{m}[\bar{\omega}_m/(\omega_m-\bar{\omega}_m)]^{\frac{1}{2}}, \qquad (26)$$

where M is an empirical constant to which they assign the value 2.4. Now, for small mR, we have, from Eq. 21,

$$\frac{\bar{\omega}_m}{\omega_m-\bar{\omega}_m}=\frac{1-R}{1-(1-R)}\approx\frac{1}{R}. \qquad (27)$$

Hence, Curran and Koneval's constant M is

$$M=(2\gamma_{55}/c_{66})^{\frac{1}{2}}. \qquad (28)$$

For the **AT** cut of quartz, Eqs. 28 and 23 give $M=2.17$.

V. ASYMPTOTIC APPROXIMATION

The solution described in the preceding sections is the simplest of all two-dimensional solutions of the equations of elasticity for vibrations of rotated-**Y**-cut quartz plates. In the order of increasing complexity, the next is the solution for thickness/shear and flexural vibrations[8,9] in which Eq. 12 is replaced by a system of algebraic and complicated transcendental equations and Eqs. 16 and 18 are replaced by an even more complicated system of equations. The three-dimensional solutions appear to be almost hopelessly unmanageable.

Now, in the applications, use is made, in the end, only of the long wavelength portions of individual dispersion curves; and it is not difficult to write simple, asymptotic expressions for these portions. For example, such expressions can be constructed from the ordinates, slopes, and curvatures of the exact dispersion curves at zero wavenumber and the results can be substituted for the complicated transcendental equations. To test this method, Bechmann's number for thickness/twist harmonics can be calculated by means of it and compared with the solution in the preceding Section.

The dispersion relation to be approximated is given by Eqs. 16 and 18. At $\bar{\xi}=0$, these reduce to the single equation

$$R(\tfrac{1}{2}\pi\bar{\Omega}_m)\tan(\tfrac{1}{2}\pi\bar{\Omega}_m)=1, \qquad (29)$$

where $\bar{\Omega}_m=(\bar{\Omega})_{\bar{\xi}=0}$. The roots of the simple Eq. 29 give the ordinates of the individual dispersion curves at $\bar{\xi}=0$, i.e., the cutoff frequencies of the harmonic-overtone modes in the plated plate.

Returning to Eqs. 16 and 18, we calculate

$$(d\bar{\Omega}/hd\bar{\xi})_{\bar{\xi}=0}=0, \qquad (30)$$

$$\left(\frac{d^2\bar{\Omega}}{h^2d\bar{\xi}^2}\right)_{\bar{\xi}=0}=\frac{4\gamma_{55}[4(1-R)+\pi^2R^2\bar{\Omega}_0^2]}{\pi^2\bar{\Omega}_m c_{66}[4(1+R)+\pi^2R^2\bar{\Omega}_m^2]}=\kappa_m,\ \text{say}. \qquad (31)$$

[7] D. R. Curran and D. J. Koneval, Proc. Ann. Symp. Frequency Control, 19th, U. S. Army Electron. Labs., Ft. Monmouth, N. J., 213–268 (1965).

[8] H. Ekstein, Phys. Rev. **68**, 11-23 (1945).

[9] R. K. Kaul and R. D. Mindlin, J. Acoust. Soc. Am. **34**, 1895–1910 (1962).

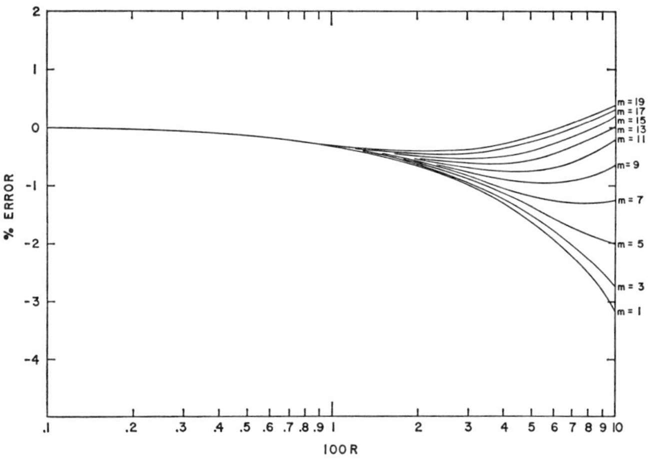

FIG. 2. Error in the computation of Bechmann's number from an asymptotic formula for the dispersion relation of thickness/twist waves in AT-type quartz plates.

Hence, an asymptotic expression for the dispersion curve of order m, valid for $\bar{\xi}h \ll 1$, is

$$\bar{\Omega}=\bar{\Omega}_m+\tfrac{1}{2}\kappa_m(\bar{\xi}h)^2, \tag{32}$$

where $\bar{\Omega}_m$, in Eqs. 31 and 32 is the $\frac{1}{2}(m+1)$th root of Eq. 29.

The approximate wavenumber of the mth harmonic wave in the plated portion of the plate, at the cutoff frequency of the mth harmonic wave in the unplated portion, is, from Eq. 32,

$$(\bar{\xi}h)^2{}_{\Omega=m}=2(m-\bar{\Omega}_m)/\kappa_m. \tag{33}$$

Hence, the approximate value of Bechmann's number is

$$(B_{\mathrm{TT}})_{\mathrm{approx}}=\pi(\kappa_m/2(m-\bar{\Omega}_m))^{\frac{1}{2}}. \tag{34}$$

For small mR, this reduces, again, to Eq. 25.

The quantity $100(B_{\mathrm{TT}})_{\mathrm{approx}}/B_{\mathrm{TT}}$ is plotted in Fig. 2 against percent plateback for a range of harmonic overtones. It may be seen that the approximation is quite good. At 3% plateback, the error is 1% for the lowest mode and diminishes for smaller platebacks and higher modes. These results give promise of possible successful application of the asymptotic approximation to more-complicated modes of motion and, possibly, to problems of vibration of bounded plates.

Reprinted from THE JOURNAL OF THE ACOUSTICAL SOCIETY OF AMERICA, Vol. 42, No. 6, 1268–1277, December 1967

Anharmonic, Thickness-Twist Overtones of Thickness-Shear and Flexural Vibrations of Rectangular, *AT*-Cut Quartz Plates

R. D. MINDLIN

Department of Civil Engineering, Columbia University, New York, New York 10027

W. J. SPENCER

Bell Telephone Laboratories, Allentown, Pennsylvania 18103

Equations of motion accommodating thickness-twist overtones of thickness-shear and flexural vibrations of rectangular, *AT*-cut, quartz plates are obtained. The equations are solved in closed form for plates with all four edges free. The predicted frequencies are compared with measured resonances. X-ray topographs of typical thickness-twist overtone modes are exhibited which verify that the mode shapes correspond to those predicted.

INTRODUCTION

MOST of the piezoelectric crystal resonators employed in oscillator and filter circuits, at present, are rotated-*Y*-cut quartz plates. Up to and including the fundamental thickness-shear frequencies of such plates, their motions fall within the range of applicability of the classical equations of generalized plane stress, coupled, through anisotropy, with the classical equations of flexure of plates—extended to include the effects of rotatory inertia and transverse shear deformation.[1] There is no difficulty in finding relatively simple solutions of these equations for infinite strips with free edges, and solutions of this type have been shown to predict with great accuracy the frequencies of modes of vibration that are associated with reflections from one pair of parallel edges.[2] However, in the case of rectangular plates with all four edges free, the same difficulty arises that is encountered even in the simplest of the classical equations of generalized plane stress or of low frequency flexure for isotropic plates: solutions are not expressible in terms of a finite number of elementary functions. The additional complications of anisotropy and high frequency make general solutions even more intractable. However, in the frequency range of interest, the special anisotropy of the widely used *AT* cut of quartz[3] serves to reduce the coupling of certain of the modes of motion, and to minimize the piezoelectric contributions of others; these include the modes which cause the difficulties with the additional boundaries. Consequently, a judicious pruning of the equations results in a set that can be solved, in closed form, for rectangular plates with all four edges free and that contains an additional family of modes—the thickness-twist overtones. These are modes, associated with reflections from all four edges of the plate, which have not heretofore been included in mathematical descriptions. The mode shapes comprise phase reversals, across the width of the plate, of thickness-shear and flexural deformations along the length—resulting in twisting deformations across the width.

In this paper, the approximate equations are deduced along with the appropriate boundary conditions. The problem of vibrations of a rectangular plate with all four edges free is solved, and the calculated frequencies are compared with the results of a series of measurements. Finally, x-ray-diffraction topographs[4] of several typical modes are exhibited which verify that pairs of predicted and measured frequencies of thickness-twist

[1] R. D. Mindlin, Quart. Appl. Math. 19, 51–61 (1961).
[2] R. D. Mindlin and D. C. Gazis, Proc. U. S. Natl. Congr. Appl. Mech. 4th, 305–310 (1962).
[3] W. P. Mason, *Piezoelectric Crystals and their Application to Ultrasonics* (D. Van Nostrand Co., Inc., New York, 1950).
[4] W. J. Spencer and R. M. Hunt, J. Acoust. Soc. Am. 39, 929–935 (1966).

overtones correspond to the same predicted and observed modes.

I. TWO-DIMENSIONAL EQUATIONS OF MOTION

The equations considered in this Section are two-dimensional equations of motion, of anisotropic plates, with rotatory inertia and transverse shear deformation taken into account.[1] The plates are of uniform thickness $2b$ and lie in the plane of rectangular coordinates x_1 and x_3. Then the stress equations of motion can be written as[5]

$$N_{1,1}+N_{5,3}=2b\rho\ddot{u}_1,\qquad N_{5,1}+N_{3,3}=2b\rho\ddot{u}_3, \tag{1}$$

and

$$Q_{1,1}+Q_{3,3}=2b\rho\ddot{u}_2,\qquad M_{1,1}+M_{5,3}-Q_1=\tfrac{2}{3}b^3\rho\ddot{\psi}_1,\qquad M_{5,1}+M_{3,3}-Q_3=\tfrac{2}{3}b^3\rho\ddot{\psi}_3. \tag{2}$$

Equations 1 are the stress equations of motion of generalized plane stress and Eqs. 2 are those of flexure. In the rotated-Y cuts of quartz,[3] the two sets are coupled through the stress-displacement relations[1]

$$\begin{aligned}
N_1&=2b[\bar{c}_{11}u_{1,1}+\bar{c}_{13}u_{3,3}+\mathcal{K}_3\bar{c}_{14}(u_{2,3}+\psi_3)],\\
N_3&=2b[\bar{c}_{13}u_{1,1}+\bar{c}_{33}u_{3,3}+\mathcal{K}_3\bar{c}_{34}(u_{2,3}+\psi_3)],\\
N_5&=2b[c_{55}(u_{3,1}+u_{1,3})+\mathcal{K}_1c_{56}(u_{2,1}+\psi_1)],\\
Q_1&=2b[\mathcal{K}_1c_{56}(u_{3,1}+u_{1,3})+\mathcal{K}_1^2c_{66}(u_{2,1}+\psi_1)],\\
Q_3&=2b[\mathcal{K}_3\bar{c}_{14}u_{1,1}+\mathcal{K}_3\bar{c}_{34}u_{3,3}+\mathcal{K}_3^2\bar{c}_{44}(u_{2,3}+\psi_3)],\\
M_1&=\tfrac{2}{3}b^3(\gamma_{11}\psi_{1,1}+\gamma_{13}\psi_{3,3}),\\
M_3&=\tfrac{2}{3}b^3(\gamma_{13}\psi_{1,1}+\gamma_{33}\psi_{3,3}),\\
M_5&=\tfrac{2}{3}b^3\gamma_{55}(\psi_{3,1}+\psi_{1,3}).
\end{aligned} \tag{3}$$

In these equations, x_1 is the axis of digonal elastic symmetry in the plane of the plate, and

$$\begin{aligned}
\mathcal{K}_1^2&=\pi^2/12,\\
\mathcal{K}_3^2&=\pi^2\{c_{22}+c_{44}-[(c_{22}-c_{44})^2+4c_{24}^2]^{\frac{1}{2}}\}/24\bar{c}_{44},\\
\bar{c}_{pq}&=c_{pq}-c_{2p}c_{q2}/c_{22},\qquad \gamma_{pq}=\bar{c}_{pq}-\bar{c}_{4p}\bar{c}_{q4}/\bar{c}_{44},\\
\gamma_{55}&=c_{55}-c_{56}^2/c_{66},
\end{aligned}$$

where the c_{pq} are the elastic stiffnesses referred to the coordinate axes x_1, x_2, x_3. In terms of the compliances s_{pq}, referred to the same axes,

$$\gamma_{11}=\frac{s_{33}}{s_{11}s_{33}-s_{13}^2},\qquad \gamma_{33}=\frac{s_{11}}{s_{11}s_{33}-s_{13}^2},$$

$$\gamma_{13}=-\frac{s_{13}}{s_{11}s_{33}-s_{13}^2},\qquad \gamma_{55}=\frac{1}{s_{55}}.$$

Insertion of Eqs. 3 in Eqs. 1 and 2 results in the displacement equations of motion:

$$\begin{aligned}
&\bar{c}_{11}u_{1,11}+\bar{c}_{13}u_{3,13}+\mathcal{K}_3\bar{c}_{14}(u_{2,13}+\psi_{3,1})\\
&\qquad+c_{55}(u_{3,13}+u_{1,33})+\mathcal{K}_1c_{56}(u_{2,13}+\psi_{1,3})=\rho\ddot{u}_1,\\
&\bar{c}_{13}u_{1,13}+\bar{c}_{33}u_{3,33}+\mathcal{K}_3\bar{c}_{34}(u_{2,33}+\psi_{3,3})\\
&\qquad+c_{55}(u_{3,11}+u_{1,13})+\mathcal{K}_1c_{56}(u_{2,11}+\psi_{1,1})=\rho\ddot{u}_3,
\end{aligned} \tag{4}$$

$$\begin{aligned}
&\mathcal{K}_3\bar{c}_{14}u_{1,13}+\mathcal{K}_3\bar{c}_{34}u_{3,33}+\mathcal{K}_3^2\bar{c}_{44}(u_{2,33}+\psi_{3,3})\\
&\qquad+\mathcal{K}_1c_{56}(u_{3,11}+u_{1,13})+\mathcal{K}_1^2c_{66}(u_{2,11}+\psi_{1,1})=\rho\ddot{u}_2,\\
&\gamma_{11}\psi_{1,11}+\gamma_{13}\psi_{3,13}+\gamma_{55}(\psi_{3,13}+\psi_{1,33})\\
&\qquad-3b^{-2}[\mathcal{K}_1c_{56}(u_{3,1}+u_{1,3})+\mathcal{K}_1^2c_{66}(u_{2,1}+\psi_1)]=\rho\ddot{\psi}_1,\\
&\gamma_{55}(\psi_{3,11}+\psi_{1,13})+\gamma_{13}\psi_{1,13}+\gamma_{33}\psi_{3,33}\\
&\qquad-3b^{-2}[\mathcal{K}_3\bar{c}_{14}u_{1,1}+\mathcal{K}_3\bar{c}_{34}u_{3,3}+\mathcal{K}_3^2\bar{c}_{44}(u_{2,3}+\psi_3)]=\rho\ddot{\psi}_3.
\end{aligned} \tag{5}$$

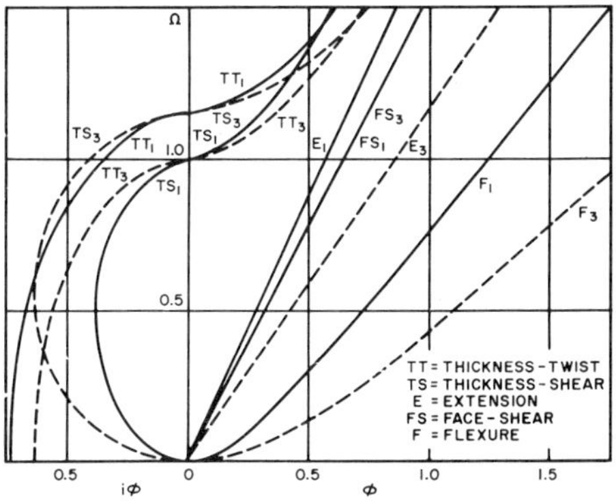

FIG. 1. Dispersion curves for straight crested waves in the x_1 and x_3 directions in an infinite, AT-cut quartz plate. Equations 6–9.

For modes of motion independent of x_3 or x_1 (cylindrical modes) the displacement equations of motion separate into four groups:

A. Flexure (F_1), *face-shear* (FS_1) *and thickness-shear* (TS_1) *varying along* x_1:

$$\begin{aligned}
&\mathcal{K}_1c_{56}u_{3,11}+\mathcal{K}_1^2c_{66}(u_{2,11}+\psi_{1,1})=\rho\ddot{u}_2\ \cdots\ (F_1),\\
&c_{55}u_{3,11}+\mathcal{K}_1c_{56}(u_{2,11}+\psi_{1,1})=\rho\ddot{u}_3\ \cdots\ (FS_1),\\
&\gamma_{11}\psi_{1,11}-3b^{-2}\\
&\qquad\times[\mathcal{K}_1c_{56}u_{3,1}+\mathcal{K}_1^2c_{66}(u_{2,1}+\psi_1)]=\rho\ddot{\psi}_1\ \cdots\ (TS_1);
\end{aligned} \tag{6}$$

B. Face-shear (FS_3) *and thickness-twist* (TT_3) *varying along* x_3:

$$\begin{aligned}
&c_{55}u_{1,33}+\mathcal{K}_1c_{56}\psi_{1,3}=\rho\ddot{u}_1\ \cdots\ (FS_3),\\
&\gamma_{55}\psi_{1,33}-3b^{-2}(\mathcal{K}_1c_{56}u_{1,3}+\mathcal{K}_1^2c_{66}\psi_1)=\rho\ddot{\psi}_1\ \cdots\ (TT_3);
\end{aligned} \tag{7}$$

C. Extension (E_1) *and thickness-twist* (TT_1) *varying along* x_1:

$$\begin{aligned}
&\bar{c}_{11}u_{1,11}+\mathcal{K}_3\bar{c}_{14}\psi_{3,1}=\rho\ddot{u}_1\ \cdots\ (E_1),\\
&\gamma_{55}\psi_{3,11}-3b^{-2}(\mathcal{K}_3\bar{c}_{14}u_{1,1}+\mathcal{K}_3^2\bar{c}_{44}\psi_3)\\
&\qquad=\rho\ddot{\psi}_3\ \cdots\ (TT_1);
\end{aligned} \tag{8}$$

[5] u_1, u_2, u_3, ψ_1, ψ_3, N_1, N_3, N_5, Q_1, Q_3, M_1, M_3, M_5 are, respectively, $u_1^{(0)}$, $u_2^{(0)}$, $u_3^{(0)}$, $u_1^{(1)}$, $u_3^{(1)}$, $T_{11}^{(0)}$, $T_{33}^{(0)}$, $T_{13}^{(0)}$, $T_{12}^{(0)}$, $T_{23}^{(0)}$, $T_{11}^{(1)}$, $T_{33}^{(1)}$, $T_{13}^{(1)}$ of Ref. 1.

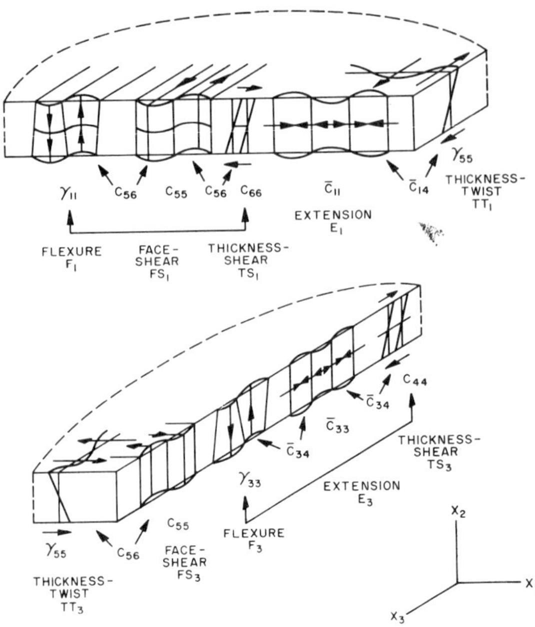

FIG. 2. Mode shapes associated with dispersion curves of Fig. 1.

D. Flexure (F_3), *extension* (E_3) *and thickness-shear* (TS_3) *varying along* x_3:

$$\begin{aligned} &\mathcal{K}_3\bar{c}_{34}u_{3,33}+\mathcal{K}_3^2\bar{c}_{44}(u_{2,33}+\psi_{3,3})=\rho\ddot{u}_2 \cdots (F_3),\\ &\bar{c}_{33}u_{3,33}+\mathcal{K}_3\bar{c}_{34}(u_{2,33}+\psi_{3,3})=\rho\ddot{u}_3 \cdots (E_3),\\ &\gamma_{33}\psi_{3,33}-3b^{-2}\\ &\quad\times[\mathcal{K}_3\bar{c}_{34}u_{3,3}+\mathcal{K}_3^2\bar{c}_{44}(u_{2,3}+\psi_3)]=\rho\ddot{\psi}_3 \cdots (TS_3). \end{aligned} \tag{9}$$

The dispersion curves for wave solutions of Eqs. 6–9 are shown in Fig. 1, and the associated mode shapes are shown in Fig. 2. The ordinate in Fig. 1 is proportional to frequency, and the abscissa is proportional to the wavenumber along x_1 or x_3.

Solutions of the first set of equations ($F_1-FS_1-TS_1$) for an infinite strip with free edges at $x_1=\pm a$, and solutions of the second set (FS_3-TT_3) for an infinite strip with free edges at $x_3=\pm c$ have been shown,[2] by comparison with measurements by Koga and Fukuyo, to predict accurately the frequencies of the "strong" resonances of rectangular, AT-cut, quartz plates, i.e. the frequencies of modes that are coupled most strongly with the component of the electric field, E_2, normal to the plane of the plate. The strongest coupling is between E_2 and the thickness-shear strain S_6 which, in these equations, is represented by $u_{2,1}+\psi_1$. There is also a weaker coupling between E_2 and the face-shear strain $S_5(=u_{3,1}+u_{1,3})$. These two shear strains, it may be seen, appear only in the first two sets, A and B, of the equations governing cylindrical modes. Now, in these, the face-shear modes are coupled to the flexural, thickness-shear and thickness-twist modes through the constant c_{56} which is small in comparison with the other constants in the equations. Omission of the coupling terms permits the face-shear modes to be separated, leaving

$$\begin{aligned} \mathcal{K}_1^2c_{66}(u_{2,11}+\psi_{1,1})&=\rho\ddot{u}_2 \cdots (F_1),\\ \gamma_{11}\psi_{1,11}-3b^{-2}\mathcal{K}_1^2c_{66}(u_{2,1}+\psi_1)&=\rho\ddot{\psi}_1 \cdots (TS_1), \end{aligned} \tag{10}$$

and

$$\gamma_{55}\psi_{1,33}-3b^{-2}\mathcal{K}_1^2c_{66}\psi_1=\rho\ddot{\psi}_1 \cdots (TT_3) \tag{11}$$

as the equations governing cylindrical modes of motion most strongly coupled to E_2. Owing to the smallness of c_{56}, it may be expected that solutions of Eqs. 10 and 11 will predict the frequencies of the flexural, thickness-shear and thickness-twist modes almost as well as when the influence of face-shear is included. That this is so has been confirmed, for the coupled thickness-shear and flexure equations, by comparison with measurements by Sykes.[6]

In solutions of Eqs. 10, u_2 and ψ_1 are independent of x_3, and conditions of free edges can be satisfied only on $x_1=\pm a$. Similarly, in solutions of Eq. 11, ψ_1 is independent of x_1 and conditions of free edges can be

[6] R. D. Mindlin, J. Appl. Phys. 22, 316–323 (1951).

satisfied only on $x_3=\pm c$. However, in a rectangular plate with all four edges free, there will be additional modes of motion in which u_2 and ψ_1 are functions of both x_1 and x_3 owing to reflections of waves from four boundaries instead of two. The construction of approximate equations that have simple solutions for the additional modes is described in the following section.

II. FLEXTURE–TWIST APPROXIMATION

General solutions of the five displacement equations of motion (Eqs. 4 and 5), for a rectangular plate with all four edges free, cannot be expressed in closed form, i.e., in terms of a finite number of elementary functions. However, the greater part of such a general solution would contribute little to u_2 and ψ_1, and it is possible to pare down the equations beforehand so as to eliminate the portions that generate the small effects.

In the AT cut of quartz, the cutoff frequency of thickness-shear vibrations (TS_3) in the x_2–x_3 plane is about 15% higher than that of thickness-shear vibrations (TS_1) in the x_2–x_1 plane (see Fig. 1), so that it is appropriate to remove the former vibrations. This may be done by dropping the inertia term $\frac{2}{3}\rho b^2\ddot{\psi}_3$ in the last of the stress equations of flexural motion (Eqs. 2) and passing to a limit, in the expression for Q_3 in the fifth of Eqs. 3, as the pertinent shear modulus $\bar{c}_{44}$ approaches infinity and the shear strain $u_{2,3}+\psi_3$ approaches zero, while Q_3 remains finite. The limiting process leaves Q_3 indeterminate, but it can be calculated from what remains of the last of Eqs. 2. In that equation, there still remains the unimportant stress couple M_3, associated with the flexural curvature $\psi_{3,3}$ in the x_2–x_3 plane. The contributions of M_3 and $\psi_{3,3}$ may be removed by setting γ_{13} and γ_{33} equal to zero in the sixth and seventh of the stress-displacement relations (Eqs. 3). Finally, coupling with face–shear, through c_{56}, may be eliminated by omitting the coupling terms from the expressions for N_5 and Q_1 in Eqs. 3.

As a result of the simplifications, the stress resultants N_1, N_3, and N_5 are not coupled with u_2 and ψ_1, so that the stress equations of motion of generalized plane stress (Eqs. 1) may be omitted. The last of the stress equations of flexural motion (Eqs. 2) is reduced to $Q_3=M_{5,1}$, by means of which Q_3 may be eliminated from the first of Eqs. 2. What remains of the stress equations of motion is the pair:

$$\begin{aligned} Q_{1,1}+M_{5,13}&=2b\rho\ddot{u}_2,\\ M_{1,1}+M_{5,3}-Q_1&=\tfrac{2}{3}b^3\rho\ddot{\psi}_1. \end{aligned} \tag{12}$$

The expressions for the surviving stress resultant Q_1 and stress couples M_1 and M_5, in terms of the displacements, are also simplified:

$$\begin{aligned} Q_1&=2b\mathcal{K}_1^2c_{66}(u_{2,1}+\psi_1),\\ M_1&=\tfrac{2}{3}b^3\gamma_{11}\psi_{1,1},\\ M_5&=-\tfrac{2}{3}b^3\gamma_{55}(u_{2,13}-\psi_{1,3}). \end{aligned} \tag{13}$$

Thus, the transverse shear stress resultant Q_1, on $x_1=$ constant, is proportional to the shear strain $u_{2,1}+\psi_1$, in the x_1–x_2 plane and the bending stress couple M_1, on $x_1=$ constant, is proportional to $\psi_{1,1}$—the gradient, along x_1, of the rotation, about x_3, of the normal to the plane of the plate. The new form for the twisting stress couple, M_5, stems from the suppression of the shear strain $u_{2,3}+\psi_3$; as a result of which $\psi_{3,1}$ in the expression for M_5 in Eqs. 3 is replaced by $-u_{2,13}$ and M_5 becomes proportional to $(u_{2,13}-\psi_{1,3})$. Now $(u_{2,1}-\psi_1)$ is twice the rigid rotation, about x_3, of an *element* of the *plate*, rather than of the normal to the plane of the plate. Hence $(u_{2,13}-\psi_{1,3})$ is twice the gradient, along x_3, of the rigid rotation of an element of the plate about x_3; i.e. it is the twist about x_3 per unit length along x_3. The first minus sign in the third of Eqs. 13 arises from the definitions of positive stress couple and rotation in opposite senses.

It may be verified that the stress-displacement relations, Eqs. 13, are derivable from a positive definite potential energy density

$$W=b\mathcal{K}_1^2c_{66}(u_{2,1}+\psi_1)^2+\tfrac{1}{3}b^3\gamma_{11}(\psi_{1,1})^2+\tfrac{1}{3}b^3\gamma_{55}(u_{2,13}-\psi_{1,3})^2 \tag{14}$$

through

$$Q_1=\frac{\partial W}{\partial(u_{2,1}+\psi_1)},\quad M_1=\frac{\partial W}{\partial\psi_{1,1}},\quad M_5=-\frac{\partial W}{\partial(u_{2,13}-\psi_{1,3})}. \tag{15}$$

The displacement equations of motion are obtained by substituting Eqs. 13 in Eqs. 12, with the result:

$$\begin{aligned} &\mathcal{K}_1^2c_{66}(u_{2,11}+\psi_{1,1})-\tfrac{1}{3}b^2\gamma_{55}(u_{2,1133}-\psi_{1,133})=\rho\ddot{u}_2,\\ &\gamma_{11}\psi_{1,11}-\gamma_{55}(u_{2,133}-\psi_{1,33})-3b^{-2}\mathcal{K}_1^2c_{66}(u_{2,1}+\psi_1)=\rho\ddot{\psi}_1. \end{aligned} \tag{16}$$

These equations reduce to Eqs. 10 when u_2 and ψ_1 are independent of x_3 and to Eq. 11 when u_2 and ψ_1 are independent of x_1.

III. UNIQUENESS OF SOLUTIONS

The boundary conditions, appropriate to the system of Eqs. 12 and 13 and a rectangular plate with edges at $x_1=\pm a$, $x_3=\pm c$, can be obtained from a theorem of uniqueness of solutions analogous to Neumann's [Ref. 7, p. 176].

Consider two sets, Q_1', M_1', M_5', u_2', ψ_1' and Q_1'', M_1'', M_5'', u_2'', ψ_1'', of the five dependent variables and form their differences $Q_1^*=Q_1'-Q_1''$, etc. Since Eqs. 12 are linear, they are satisfied by the difference variables.

[7] A. E. H. Love, *A Treatise on the Mathematical Theory of Elasticity* (Cambridge University Press, Cambridge, England, 1927), 4th ed.

Accordingly, the equation

$$\int_{t_0}^{t} dt \int_{-a}^{a} \int_{-c}^{c} (Q_{1,1}^* + M_{5,13}^* - 2b\rho\ddot{u}_2^*)\dot{u}_2^* dx_1 dx_3$$
$$+ \int_{t_0}^{t} dt \int_{-a}^{a} \int_{-c}^{c} (M_{1,1}^* + M_{5,3}^* - Q_1^* - \tfrac{2}{3}b^3\rho\ddot{\psi}_1^*) \times \dot{\psi}_1^* dx_1 dx_3 = 0, \quad (17)$$

where t_0 is an initial time, is satisfied identically.

By application of the chain rule of differentiation,

$$\int_{-a}^{a} \int_{-c}^{c} [(Q_{1,1}^* + M_{5,13}^*)\dot{u}_2^* + (M_{1,1}^* + M_{5,3}^* - Q_1^*)\dot{\psi}_1^*] dx_1 dx_3$$
$$= -\int_{-a}^{a} \int_{-c}^{c} [Q_1^*(\dot{u}_{2,1}^* + \dot{\psi}_1^*) + M_1^*\dot{\psi}_{1,1}^* - M_5^*(\dot{u}_{2,13}^* - \dot{\psi}_{1,3}^*)] dx_1 dx_3$$
$$+ \int_{-a}^{a} \int_{-c}^{c} (Q_1^*\dot{u}_2^* - M_5^*\dot{u}_{2,3}^* + M_1^*\dot{\psi}_1^*)_{,1} dx_1 dx_3$$
$$+ \int_{-a}^{a} \int_{-c}^{c} (M_{5,1}^*\dot{u}_2^* + M_5^*\dot{\psi}_1^*)_{,3} dx_1 dx_3. \quad (18)$$

Now,

$$\int_{-a}^{a} \int_{-c}^{c} (Q_1^*\dot{u}_2^* - M_5^*\dot{u}_{2,3}^* + M_1^*\dot{\psi}_1^*)_{,1} dx_1 dx_3$$
$$= \int_{-c}^{c} [Q_1^*\dot{u}_2^* - M_5^*\dot{u}_{2,3}^* + M_1^*\dot{\psi}_1^*]_{-a}^{a} dx_3 \quad (19a)$$

and

$$-\int_{-c}^{c} [M_5^*\dot{u}_{2,3}^*]_{-a}^{a} dx_3$$
$$= -\int_{-c}^{c} [(M_5^*\dot{u}_2^*)_{,3} - M_{5,3}^*\dot{u}_2^*]_{-a}^{a} dx_3$$
$$= -\{[M_5^*\dot{u}_2^*]_{-a}^{a}\}_{-c}^{c} + \int_{-c}^{c} [M_{5,3}^*\dot{u}_2^*]_{-a}^{a} dx_3. \quad (19b)$$

Also,

$$\int_{-a}^{a} \int_{-c}^{c} (M_{5,1}^*\dot{u}_2^* + M_5^*\dot{\psi}_1^*)_{,3} dx_1 dx_3$$
$$= \int_{-a}^{a} [M_{5,1}^*\dot{u}_2^* + M_5^*\dot{\psi}_1^*]_{-c}^{c} dx_1 \quad (20a)$$

and

$$\int_{-a}^{a} [M_{5,1}^*\dot{u}_2^*]_{-c}^{c} dx_1$$
$$= \int_{-a}^{a} [(M_5^*\dot{u}_2^*)_{,1} - M_5^*\dot{u}_{2,1}^*]_{-c}^{c} dx_1$$
$$= \{[M_5^*\dot{u}_2^*]_{-c}^{c}\}_{-a}^{a} - \int_{-a}^{a} [M_5^*\dot{u}_{2,1}^*]_{-c}^{c} dx_1. \quad (20b)$$

Further, Eqs. 13 are also satisfied by the difference variables and, hence,

$$Q_1^*(\dot{u}_{2,1}^* + \dot{\psi}_1^*) + M_1^*\dot{\psi}_{1,1}^* - M_5^*(\dot{u}_{2,13}^* - \dot{\psi}_{1,3}^*) = \dot{W}^*, \quad (21)$$

where W^* is the potential energy density of the difference system; i.e., Eq. 14 in terms of u_2^* and ψ_1^*.

Inerting Eqs. 19, 20, and 21 in Eq. 18, the result in Eq. 17, and integrating $\dot{W}^*$, $\ddot{u}_2^*\dot{u}_2^*$ and $\ddot{\psi}_1^*\dot{\psi}_1^*$ with respect to t, we find

$$\int_{-a}^{a} \int_{-c}^{c} [K^* + W^*]_{t_0}^{t} dx_1 dx_3$$
$$= \int_{t_0}^{t} \int_{-c}^{c} [(Q_1^* + M_{5,3}^*)\dot{u}_2^* + M_1^*\dot{\psi}_1^*]_{-a}^{a} dx_3 dt$$
$$- \int_{t_0}^{t} \int_{-a}^{a} [M_5^*(\dot{u}_{2,1}^* - \dot{\psi}_1^*)]_{-c}^{c} dx_1 dt$$
$$- \int_{t_0}^{t} \{[M_5^*\dot{u}_2^*]_{-a}^{a}\}_{-c}^{c} dt + \int_{t_0}^{t} \{[M_5^*\dot{u}_2^*]_{-c}^{c}\}_{-a}^{a} dt. \quad (22)$$

where $K^* = \rho b(\dot{u}_2^*)^2 + \frac{1}{3}b^3\rho(\dot{\psi}_1^*)^2$; i.e., K^* is the kinetic energy density of the difference system.

By the usual argument based on the positive definiteness of K^* and W^* (Ref. 7, p. 176), Eq. 22 leads to the following conditions sufficient for a unique solution:

- At time $t = t_6$, specification of u_2, $\dot{u}_2$, ψ_1 and $\dot{\psi}_1$.
- On $x_1 = \pm a$, either $Q_1 + M_{5,3}$ or u_2 and either M_1 or ψ_1.
- On $x_3 = \pm c$, either M_5 or $u_{2,1} - \psi_1$.
- At the corners, either u_2 or the corner forces (Ref. 7, p. 458)

$$\{[M_5]_{-c}^{c}\}_{-a}^{a} - \{[M_5]_{-a}^{a}\}_{-c}^{c}.$$

Thus, a rectangular plate with all four edges free has boundary conditions

$$[Q_1 + M_{5,3}]_{x_1=\pm a} = 0, \quad [M_1]_{x_1=\pm a} = 0, \quad [M_5]_{x_3=\pm c} = 0, \quad (23)$$

and the additional requirement that M_5 be continuous at the corners.

IV. VIBRATIONS OF A PLATE WITH FREE EDGES

If the impressed electric field is uniform over the plate, the modes excited will have ψ_1 even in x_1 and x_3 and u_2 odd in x_1 and even in x_3. Furthermore, from the forms of the displacement equations of motion, Eqs. 16, and the boundary conditions, Eq. 23, it can be anticipated that the solution for free vibrations will involve two wave numbers along x_1 and one along x_3. Hence the appropriate forms of u_2 and ψ_1 are

$$u_2=\sum_{i=1}^{i=2} A_i b \sin\xi_i x_1 \cos\zeta x_3 e^{j\omega t},$$
$$\psi_1=\sum_{i=1}^{i=2} B_i \cos\xi_i x_1 \cos\zeta x_3 e^{j\omega t}. \quad (24)$$

Substitution of Eqs. 24 into Eqs. 16 yields

$$(\bar\xi_i^2+\hat\gamma_{55}\bar\xi_i^2\bar\zeta^2-3\Omega^2)A_i+(1-\hat\gamma_{55}\bar\zeta^2)\bar\xi_i B_i=0,$$
$$(1-\hat\gamma_{55}\bar\zeta^2)\bar\xi_i A_i+(1+\hat\gamma_{11}\bar\xi_i^2+\hat\gamma_{55}\bar\zeta^2-\Omega^2)B_i=0, \quad (25)$$

where $\bar\xi_i=\xi_i b$, $\bar\zeta=\zeta b$ and

$$\hat\gamma_{11}=\gamma_{11}/3\mathcal{K}_1^2c_{66},\quad \hat\gamma_{55}=\gamma_{55}/3\mathcal{K}_1^2c_{66},$$
$$\Omega^2=b^2\rho\omega^2/3\mathcal{K}_1^2c_{66}. \quad (26)$$

From Eqs. 25,

$$\frac{A_i}{B_i}=-\frac{(1-\hat\gamma_{55}\bar\zeta^2)\bar\xi_i}{(1+\hat\gamma_{55}\bar\zeta^2)\bar\xi_i^2-3\Omega^2}$$
$$=-\frac{1+\hat\gamma_{11}\bar\xi_i^2+\hat\gamma_{55}\bar\zeta^2-\Omega^2}{(1-\hat\gamma_{55}\bar\zeta^2)\bar\xi_i}\equiv\alpha_i. \quad (27)$$

The second of Eqs. 27 is a quadratic equation on $\bar\xi_i^2$, the roots of which are

$$\bar\xi_i^2=[\mathcal{B}-(-1)^i(\mathcal{B}^2+\mathcal{C})^{\frac{1}{2}}]/2\hat\gamma_{11}(1+\hat\gamma_{55}\bar\zeta^2),\ i=1,2, \quad (28)$$

where

$$\mathcal{B}=(1+\hat\gamma_{55}\bar\zeta^2)(\Omega^2-1-\hat\gamma_{55}\bar\zeta^2)+(1-\hat\gamma_{55}\bar\zeta^2)^2+3\hat\gamma_{11}\Omega^2, \quad (29)$$
$$\mathcal{C}=12\hat\gamma_{11}\Omega^2(1+\hat\gamma_{55}\bar\zeta^2)(1+\hat\gamma_{55}\bar\zeta^2-\Omega^2). \quad (30)$$

Equation 28 gives the two branches of the dispersion relation (Ω vs $\bar\xi_i$) for waves in an infinite plate; ξ_1 is the wavenumber for waves which are predominantly flexural, and ξ_2 is the wavenumber for waves which are predominantly thickness-twist or, for $\zeta=0$, thickness-shear. To determine the character (real or imaginary) of the $\bar\xi_i$, note first that $\mathcal{B}^2+\mathcal{C}>0$, as it can be written in the form

$$[4\hat\gamma_{55}\bar\zeta^2+3\hat\gamma_{11}\Omega^2-(1+\hat\gamma_{55}\bar\zeta^2)\Omega^2]^2+12\hat\gamma_{11}\Omega^2(1-\hat\gamma_{55}\bar\zeta^2)^2.$$

Hence $\bar\xi_1^2$ and $\bar\xi_2^2$ are real. The boundary conditions to be set require ζ to be real, so that $1+\hat\gamma_{55}\bar\zeta^2>0$. Then, from Eqs. 29 and 30:

- If $\Omega^2>1+\hat\gamma_{55}\bar\zeta^2$: $\mathcal{B}>0$, $\mathcal{C}<0$; hence $(\mathcal{B}^2+\mathcal{C})^{\frac{1}{2}}<\mathcal{B}$ and, from Eq. 28, $\bar\xi_1$ and $\bar\xi_2$ are real;
- If $\Omega^2<1+\hat\gamma_{55}\bar\zeta^2$: $\mathcal{C}>0$; hence $(\mathcal{B}^2+\mathcal{C})^{\frac{1}{2}}>\mathcal{B}$ and, from Eq. 28, $\bar\xi_1$ is real, and $\bar\xi_2$ is imaginary.

The boundary conditions, for free edges $x_1=\pm a$, $x_3=\pm c$, become, from Eqs. 23, 13 and 24,

$$\sum_{i=1}^{i=2} B_i[1-\hat\gamma_{55}\bar\zeta^2+\alpha_i\bar\xi_i(1+\hat\gamma_{55}\bar\zeta^2)]\cos\xi_i a=0,$$
$$\sum_{i=1}^{i=2} B_i\bar\xi_i\sin\xi_i a=0, \quad (31)$$
$$\sin\zeta c=0.$$

and the corner conditions are satisfied identically.

From the last of Eqs. 31,

$$\bar\zeta=n\pi b/2c,\quad n=0,2,4\cdots \quad (32)$$

and, from the first two of Eqs. 31,

$$\frac{B_1}{B_2}=-\frac{[1-\hat\gamma_{55}\bar\zeta^2+\alpha_2\bar\xi_2(1+\hat\gamma_{55}\bar\zeta^2)]\cos\xi_2 a}{[1-\hat\gamma_{55}\bar\zeta^2+\alpha_1\bar\xi_1(1+\hat\gamma_{55}\bar\zeta^2)]\cos\xi_1 a}$$
$$=-\frac{\bar\xi_2\sin\xi_2 a}{\bar\xi_1\sin\xi_1 a}. \quad (33)$$

For real ξ_2, the second of Eqs. 33 gives the frequency equation

$$\frac{\tan\xi_1 a}{\tan\xi_2 a}=\frac{\bar\xi_2[1-\hat\gamma_{55}\bar\zeta^2+\alpha_1\bar\xi_1(1+\hat\gamma_{55}\bar\zeta^2)]}{\bar\xi_1[1-\hat\gamma_{55}\bar\zeta^2+\alpha_2\bar\xi_2(1+\hat\gamma_{55}\bar\zeta^2)]},$$
$$\Omega^2>1+\hat\gamma_{55}\bar\zeta^2. \quad (34)$$

For imaginary ξ_2, it is convenient to have Eq. 34 in real form. Set $\bar\xi_2=i\bar\xi_2'$ and $\alpha_2=i\alpha_2'$ so that, from Eqs. 27,

$$\alpha_2'=\frac{(1-\hat\gamma_{55}\bar\zeta^2)\bar\xi_2'}{(1+\hat\gamma_{55}\bar\zeta^2)\bar\xi_2'^2+3\Omega^2}=\frac{1-\hat\gamma_{11}\bar\xi_2'^2+\hat\gamma_{55}\bar\zeta^2-\Omega^2}{(1-\gamma_{55}\bar\zeta^2)\bar\xi_2'}. \quad (35)$$

Equation 34 then becomes

$$\frac{\tan\xi_1 a}{\tanh\xi_2' a}=-\frac{\bar\xi_2'[1-\hat\gamma_{55}\bar\zeta^2+\alpha_1\bar\xi_1(1+\hat\gamma_{55}\bar\zeta^2)]}{\bar\xi_1[1-\hat\gamma_{55}\bar\zeta^2-\alpha_2'\bar\xi_2'(1+\hat\gamma_{55}\bar\zeta^2)]},$$
$$\Omega^2<1+\hat\gamma_{55}\bar\zeta^2. \quad (36)$$

In the range $\Omega^2>1+\hat\gamma_{55}\bar\zeta^2$, there are some useful, simple solutions of Eqs. 33:

- $\cos\xi_1 a=0$, $\cos\xi_2 a=0$; i.e.,

$$\xi_1=m_1\pi/2a,\quad \xi_2=m_2\pi/2a,\quad m_1 \text{ and } m_2 \text{ odd}, \quad (37)$$
$$B_1/B_2=-(m_2/m_1)(-1)^{(m_2-m_1)/2}. \quad (38)$$

These are predominantly thickness-twist modes or, for $\zeta=0$, thickness-shear modes ($|B_2/B_1|\gg1$ for the usual case of $m_1\gg m_2$).

- $\sin\xi_1 a=0$, $\sin\xi_2 a=0$; i.e.,

$$\xi_1=m_1\pi/2a,\quad \xi_2=m_2\pi/2a,\quad m_1 \text{ and } m_2 \text{ even}, \quad (39)$$
$$\frac{B_1}{B_2}=-\frac{[1-\hat\gamma_{55}\bar\zeta^2+\alpha_2\bar\xi_2(1+\hat\gamma_{55}\bar\zeta^2)](-1)^{(m_2-m_1)/2}}{1-\hat\gamma_{55}\bar\zeta^2+\alpha_1\bar\xi_1(1+\gamma_{55}\bar\zeta^2)}. \quad (40)$$

These are predominantly flexural modes ($|B_1/B_2|>1$ for $m_1\gg m_2$).

In both classes of simple solutions, the frequency is obtained by inserting the special values of a pair of ξ_i in the two Eqs. 28, and solving simultaneously for a given set of a/b, c/b, and n. The two equations may be written in the form

$$a/b=F_i(\Omega,c/b,n,m_i),\quad i=1,2; \quad (41)$$

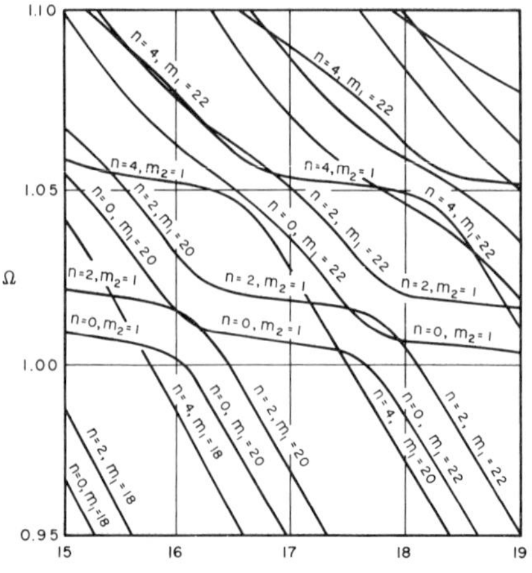

FIG. 3. Resonant frequencies of thickness–shear ($n=0$; $m_2=1$) and flexure ($n=0$; $m_1=\cdots 18, 20, 22\cdots$), and thickness–twist overtones of thickness–shear ($n=2, 4\cdots$; $m_2=1$) and flexure ($n=2, 4, \cdots$; $m_1=\cdots 18, 20, 22\cdots$) in a rectangular AT-cut quartz plate as computed from Eqs. 28, 34 and 36 for $c/b=20.0$.

where

$$F_i(\Omega,c/b,n,m_i)=\frac{m_i\pi}{2}\left[\frac{2\hat{\gamma}_{11}(1+\hat{\gamma}_{55}\bar{\zeta}^2)}{\mathcal{B}-(-1)^i(\mathcal{B}^2+\mathcal{C})^{\frac{1}{2}}}\right]^{\frac{1}{2}}.$$

For given c/b and n, Eqs. 41 represent two families of curves in the Ω, a/b plane. The intersections of curves m_1 even and m_2 even and the intersections of curves m_1 odd and m_2 odd are roots of Eqs. 28 and 34. Points on the curves not at such intersections are *not* roots. Hence, between such intersections, the curves are bounds of the branches of the frequency spectrum.

Detailed computations for the AT cut of quartz have been made of the roots of Eqs. 28 and 34 or 36 in the ranges $0.95<\Omega<1.10$ and $15<a/b<19$ for $n=0$, 2, 4 and $c/b=20$; and the results are shown in Fig. 3.

The designations n, m_i, on portions of the branches in Fig. 3, indicate that those portions are near the bounds, Eqs. 41, with the corresponding n and m_i, and have similar mode shapes. The steep portions of the branches, identified by m_1, are close to $a/b=F_1$ in Eqs. 41; i.e., they are predominantly flexural. The flat portions, identified by m_2, are close to $a/b=F_2$, and are predominantly thickness–shear for $n=0$, and thickness–twist for $n\neq 0$ (n is the number of nodes across the width, $2c$, of the plate).

Below $\Omega^2=1+\hat{\gamma}_{55}\bar{\zeta}^2$, which, for each n, is slightly below the flat portions in Fig. 1 marked $m_2=1$, the thickness–shear or thickness–twist branch, ξ_2, of the dispersion relation given in Eqs. 28, has an imaginary wave-number and its contribution is given by a hyperbolic function with small amplitude. The main contribution to the mode shape is from the flexural branch of Eqs. 28 with approximately m_1 (even) antinodes of u_2 and nodes of ψ_1 along the length $2a$, since the branch of the frequency equation, 36, is near the bound $a/b=F_1$.

Above $\Omega^2=1+\hat{\gamma}_{55}\bar{\zeta}^2$, the centers of the flat portions in Fig. 3 correspond to modes with maximum content of thickness–shear ($n=0$) or thickness–twist ($n\neq 0$). For example, a little below the center of Fig. 3, there is a flat portion marked $n=0$, $m_2=1$. Near the center of this portion is the intersection of the two bounds

$$a/b=F_1(\Omega,c/b,0,21),\quad a/b=F_2(\Omega,c/b,0,1),$$

for which, from Eq. 38, $B_2/B_1=-21$; i.e., the contribution of the thickness–shear branch of Eqs. 28 to ψ_1 is 21 times that of the flexural branch. The shear branch $m_2=1$ has $m_2-1=0$ nodes of ψ_1 and the flexural branch has $m_1=21$ antinodes of ψ_1 along the length $2a$. Moving along the flat portion to the right, $|B_2/B_1|$ diminishes, and m_1 increases toward 22; toward the left, $|B_2/B_1|$ also diminishes, and m_1 approaches 20. Upon rounding the bends at the right and left ends of the flat portion, the modes become predominantly flexural and approach 22 and 20 antinodes of ψ_1, respectively.

Above and to the right of the flat portion under discussion is another flat portion marked $n=2$, $m_2=1$, and still farther up and again to the right is yet another flat portion, marked $n=4$, $m_2=1$. These portions are close to the bounds

$$a/b=F_2(\Omega,c/b,2,1)\quad\text{and}\quad a/b=F_2(\Omega,c/b,4,1)$$

respectively. At their centers they intersect with

$$a/b=F_1(\Omega,c/b,2,21)\quad\text{and}\quad a/b=F_1(\Omega,c/b,4,21),$$

and so, from Eq. 38, they again have $B_2/B_1=-21$, but now they are thickness–twist modes with two and four nodes across the width. Moving to the right along these flat portions, the mode shapes change, upon turning the bends, to flexural modes with $m_1=22$ antinodes of ψ_1. For a given a/b, say 19, there is, thus, a sequence of flexural modes all with the same number of phase reversals in ψ_1 (or u_2) along the length, but with $n=0$, 2, 4 phase reversals across the width, and successively higher frequencies. It is apparent that the coupling of flexure and thickness–twist introduces additional modes; these may be termed anharmonic overtones of coupled flexure and thickness–twist or, simply, flexure–twist overtones.

The usual anharmonic overtones of thickness–shear ($m_2=3, 5\cdots$) are also present, but they do not appear in the form of pronounced flat portions of branches, in Fig. 3, because of the low range of length-to-thickness ratios shown there.

V. DETERMINATION OF RESONANCES IN QUARTZ PLATES

The curves shown in Fig. 3 give the frequencies of the twist overtones of thickness–shear and flexural vibrations of an *AT*-cut plate with a constant width as a function of the length-to-thickness ratio. The curves were obtained from Eqs. 28 and 34 or 36 using the elastic constants given by Bechman,[8] from which, for the *AT* cut,

$$c_{66}=29.01,\quad \gamma_{11}=85.93,\quad \gamma_{55}=68.59$$

in units of 10^{10} dyn/cm^2. These elastic constants were chosen since they coincide most closely with those previously determined in matching the Koga–Fukuyo data to curves computed from the results of Section I. McSkimin[9] has made a careful study of elastic constants of quartz at 25°C by a pulse superposition method. His analysis is based on a plane wave approximation, and he considers the uncertainties due to various experimental procedures. His constants generally agree (to better than 1%) with those used here. However, many of the mechanical coupling effects shown in Fig. 3 produce frequency changes of less than 1%. This necessitates the use of a set of elastic constants that are consistent with the magnitude of the effects to be predicted. The equations used to generate the curves in Fig. 3 do not include piezoelectric or dissipative properties of the quartz. The effects of both of these should be quite small owing to the low electromechanical coupling and high intrinsic Q of quartz, but could introduce changes of a few tenths of a percent in the effective elastic constants.

The modes in piezoelectric single crystal quartz plates can be excited by placing the plate in an alternating electric field. The resonant frequency may be determined by measuring the charge on the plate surfaces as a function of the frequency of the driving field. In the present experiment, the same electrodes were used to generate the electric field and to detect the various resonances. The measuring technique is shown schematically in Fig. 4. The bridge circuit is balanced by adjusting the variable capacitor to account for the static capacitance of the quartz plate and its electrodes. The bridge output is plotted on an x–y recorder which is driven by a voltage obtained from the variable frequency oscillator.

The resulting frequency scan of a quartz plate produces many more resonances than can be accounted for by present mathematical solutions. To sort out the various modes and their dependence on plate dimensions, the quartz plates have traditionally been reduced, step by step, in one dimension, and the change in frequency of the various resonances has been recorded. It is usually possible to choose, from such a plot, families of modes which depend primarily on the dimension which has been changed. The experimentally measured modes which correspond to the coupled thickness–shear and flexure and their twist overtones are shown in Fig. 5. For simplicity, only the resonances which correspond to the theoretical curves shown in Fig. 3 are included.

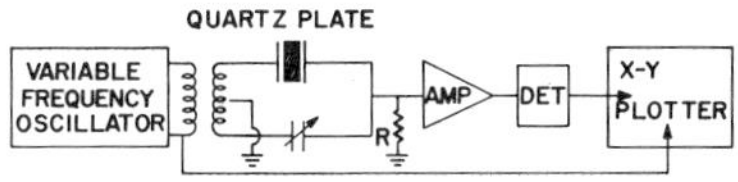

FIG. 4. Schematic diagram of bridge circuit used for measuring responses in quartz plates.

The data shown in Fig. 5 are for a rectangular rotated-Y-cut quartz plate, $\Phi=35°\ 15'$,[3] with original dimensions of $a/b=19$, $c/b=20$, and for the frequency range $0.95\leq\Omega\leq1.10$. The plate thickness was 0.750 mm, making the fundamental thickness–shear frequency about 2.220 MHz. The plate length was carefully reduced in steps of approximately 0.10 mm. The plate dimensions and its principal resonances, within ±10% of the fundamental thickness–shear mode, were recorded after each reduction in plate length. The plate was checked after each reduction for parallelism of its sides, and for small conchoidal fractures which might affect its resonance spectrum.

The twist overtones are extremely weak responses, when the quartz plate is completely covered with an electrode, owing to charge cancellation. This is related to the mode shape in the x_3 direction given by Eqs. 24 and 32. The electrical strength of these resonances, which is what is recorded, may be enhanced by the use of an electrode strip in place of a complete electrode along the plate length. The electrical symmetry of many of the resonances may be determined by noting the enhancement of the various modes by electrode strips of different width. The flexure modes are always much weaker, owing to their short wave length along the plate length, than are the thickness–shear and twist modes.

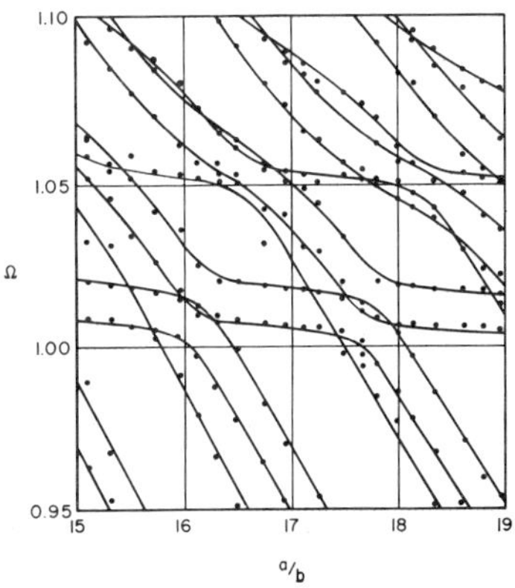

FIG. 5. Comparison of computed and measured resonant frequencies for a rectangular *AT*-cut quartz plate.

[8] R. Bechmann, Phys. Rev. **100**, 1060–1061 (1958).
[9] H. J. McSkimin, J. Acoust. Soc. Am. **34**, 1271–1274 (1962).

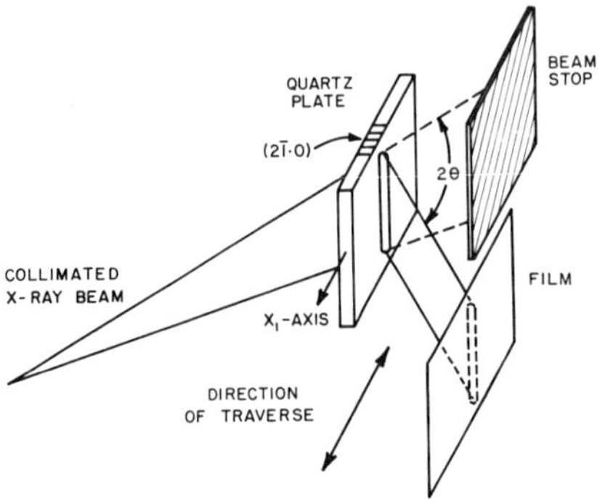

FIG. 6. Schematic of camera used to obtain x-ray-diffraction topographs.

The curves computed from the equations in Sec. IV predict quite well the branches which are predominately thickness–shear and twist, including the coupling with flexure. In regions where the modes are predominately flexure, the experimental points generally show more scatter. This is most probably due to the difficulty in the measurement of these modes, since they are relatively weak. Scatter at other points can also be due to coupling with modes not included in this analysis.

The abundance of resonances in quartz plates might lead one to believe that any theory could be verified by making frequency measurements alone. To identify more completely the measured resonances, a number of x-ray-diffraction topographs were made to show the actual mode displacement. The x-ray topographs may be used to map the rotation ψ_1 throughout the entire plate.[4] The technique is shown schematically in Fig. 6. A narrow beam of x rays is diffracted from Bragg planes normal to the plate surface and the x_1 axis on to an x-ray-sensitive film. The crystal and film are locked together and translated back and forth through the diffracted beam to obtain the proper film exposure. The diffracted intensity may be shown to be proportional to ψ_1. The film record is then a two-dimensional projection, of the rotation ψ_1 with respect to the diffracting planes, on the x-ray film.

A series of topographs taken of modes in a square quartz plate with $a/b=c/b=29.8$ is shown in Fig. 7. Each mode has displacements which are made up of components of both flexure and thickness–shear and three of the modes show twist overtones of the thickness–shear modes. The x_1 axis is horizontal and the x_3 axis vertical in all the topographs. The lowest frequency mode is predominately the 36th overtone of flexure. This mode lies near the bound $a/b=F_1(\Omega,c/b,0,36)$. The displacements are nearly straight crested in the x_3 direction although some "beading" of the antinodes of ψ_1 (dark regions) is visible. There are only 35 antinodes of ψ_1 in this topograph instead of the expected 36. The discrepancy is not entirely resolved, but may be due to limitations of the experimental technique.

The fundamental thickness–shear mode, $n=0$, $m_2=1$, is not shown. Directly above this mode in frequency is the $n=2$ twist overtone of the thickness–shear mode, for which a topograph is shown. The flexure components of displacement are still visible as banding of the shear displacement. This banding is now related to a full wavelength of the flexure component rather than the half-wavelength observed when the motion is predominately flexural. The diffracted intensity is proportional to the absolute value of ψ_1 and the number of fringes will depend on the relative contribution to the rotation of both the flexural and shear components.[4] Generally when $\Omega>1+\hat{\gamma}_{55}\bar{\zeta}^2$ the spacing between the fringes will correspond to a full wavelength of the flexure, while below this frequency the spacing corresponds to a half-wavelength. As pointed out in the preceding section, there are two nodes of ψ_1 along the width of the plate. Directly above the $n=2$ $m_2=1$ branch are the first symmetric anaharmonic overtone of the $n=0$, $m_2=3$ thickness-shear mode and the $n=2$, $m_2=3$ twist overtone. The latter mode has nine dark areas with two nodes of ψ_1 along both the length and the width of the plate. These nodes are not equally spaced in the two directions. The last topographs is the $n=4$ twist overtone of the thickness–shear mode. Again the flexure components are noticeable, although the displacement is predominately thickness–shear and thickness–twist. There are four nodes of ψ_1 along the plate width.

Topographs were also made of many of the modes shown in Fig. 5. The flexure components are much more apparent in these lower length-to-thickness ratio plates and some of the displacement patterns are complicated by contributions of the face–shear modes. Numerous relatively strong modes were also found which correspond to the antisymmetric solutions of the equations in Sec. IV, but they are not included in the results given in Fig. 5.

The results presented here are for rotated Y-cut quartz plates without electrodes. The results could also be applied to plates with complete adherent electrodes by making the following replacements

$$
\begin{aligned}
(d^2/dt^2)u_2 &\rightarrow (d^2/dt^2)\bar{u}_2(1+R)\\
(d^2/dt^2)\psi_1 &\rightarrow (d^2/dt^2)\bar{\psi}_1(1+3R)\\
\mathcal{K} &\rightarrow \bar{\mathcal{K}}=\frac{\pi^2(1+3R)}{12(1+R)^2}
\end{aligned}
\tag{42}
$$

where R is the ratio of the mass per unit area of both electrodes to the mass per unit area of the plate.[10]

The equations in Sec. IV and the curves shown in

[10] R. D. Mindlin and P. C. Y. Lee, Intern. J. Solids Struct. **2**, 125–139 (1966).

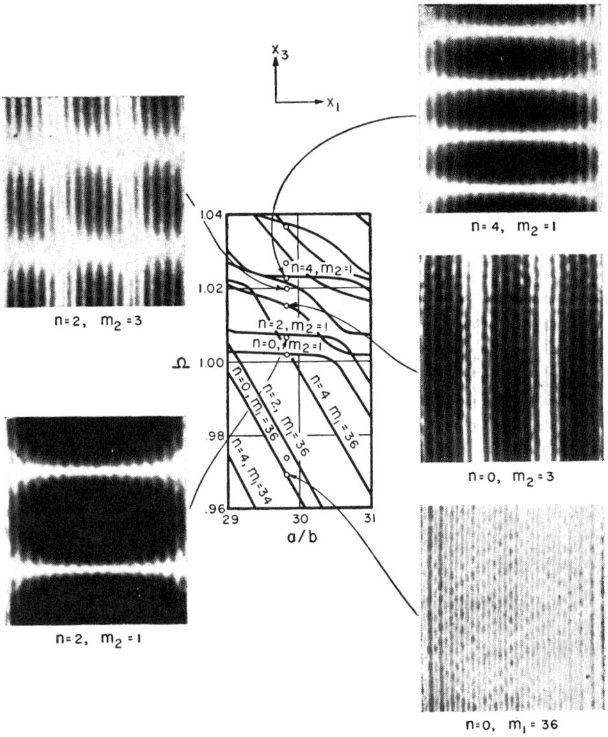

FIG. 7. X-ray-diffraction topographs of thickness–shear, thickness–twist, and flexural modes in a rectangular *AT*-cut quartz plate. $c/b=29.8$.

Figs. 3 and 5 may also be used to determine approximately the points at which the flexure–twist modes will coincide with the fundamental thickness–shear mode in quartz plates with partial electrodes. Partial electrodes will lower the frequency of the thickness–shear modes, but will have very little effect on the flexural branches. By judicious choice of plate dimensions, it is possible to make a partially electroded quartz resonator with a given mass lowering of the thickness–shear frequency to a value which will not coincide with a flexure or flexure–twist frequency.

ACKNOWLEDGMENT

The authors are indebted to Dr. K. Haruta for his programming help. The work of one of us (R.D.M.) was supported by the U. S. Army Electronics Laboratories and the Office of Naval Research, through contracts with Columbia University.

Int. J. Solids Structures, 1968, Vol. 4, pp. 109 to 124. Pergamon Press. Printed in Great Britain

ON FIRST STRAIN-GRADIENT THEORIES IN LINEAR ELASTICITY

R. D. MINDLIN

Department of Civil Engineering, Columbia University

and

N. N. ESHEL

Department of Civil Engineering, The City College,
University of the City of New York

Abstract—This paper contains a study of the linear theory of elasticity in which the potential energy-density depends on the gradient of the strain in addition to the strain.

In the first part of the paper, three forms of the theory are compared and the relations connecting the stresses in the three forms and the boundary conditions in the three forms are derived. In the second part of the paper, ambiguities in the form of the moment-equation of equilibrium and the definition of couple-stress are resolved by a derivation based on conservation principles rather than the variational principles employed previously.

INTRODUCTION

THE purpose of this paper is to clarify certain aspects of the linear theory of elasticity in which the potential energy of deformation is a function of the six components of the strain and the eighteen components of the gradient of the strain.

Stress-equations of equilibrium, constitutive equations and boundary conditions of the "strain-gradient theory" were first given in a general, non-linear form by Toupin [1, §7]. Subsequently [2], linear versions of the theory were given in three forms—distinguished by different groupings of the eighteen additional variables in the potential energy-density: I, the eighteen components of the second gradient of the displacement; II, the eighteen components of the first gradient of the strain; III, the eight components of the gradient of the rotation and the ten components of the fully symmetric part of the second gradient of the displacement (or of the gradient of the strain). The components in the second and third sets are simply linear combinations of those in the first. The third form of the theory is the most convenient one for reduction to the theory in which the potential energy-density is a function of the strain and the gradient of the rotation [1, 3–6].

In [2], the three forms of the strain-gradient theory were shown to lead to the same displacement-equations of motion for isotropic materials. However, the general identity of the stress-equations of motion and the general relations among the stresses in the three forms and among the traction boundary conditions for the three forms were not exhibited. These results are derived in the first part of the present paper.

The second part of the paper is concerned with the moment equation and the couple-stress. In the derivation of the equations of the strain-gradient theory by the variational methods employed previously, the moment equation does not appear explicitly. Although a moment equation can be deduced subsequently from the condition of invariance of the potential energy-density in a rigid rotation of the deformed body [7], the equation can be produced in a variety of forms. As a result, the identification of the couple-stress is uncertain to a constant factor. To clarify the situation, the complete equations of the linear strain-gradient theory are rederived, here, starting from principles of conservation of linear momentum, angular momentum and energy. The moment equation and couple-stress thereby are displayed without ambiguity. A theorem of uniqueness of solutions leaves the spherical part of the couple-stress undetermined just as in the theory in which the potential energy-density depends on the strain and the gradient of the rotation.

1. KINEMATIC VARIABLES

The kinematic variables to be employed are defined in terms of derivatives of components of displacement as follows:

$$
\begin{aligned}
\varepsilon_{ij} &= \tfrac{1}{2}(u_{j,i}+u_{i,j}) = u_{(j,i)} = \varepsilon_{ji} = \text{strain},\\
\omega_{ij} &= \tfrac{1}{2}(u_{j,i}-u_{i,j}) = u_{[j,i]} = -\omega_{ji} = \text{rotation},\\
w_i &= \tfrac{1}{2}e_{ijk}u_{k,j} = \text{vector rotation},\\
\tilde{\kappa}_{ijk} &= u_{k,ij} = \tilde{\kappa}_{jik} = \text{second gradient of displacement},\\
\hat{\kappa}_{ijk} &= \tfrac{1}{2}(u_{k,ji}+u_{j,ki}) = u_{(k,j)i} = \hat{\kappa}_{ikj} = \text{gradient of strain},\\
\bar{\kappa}_{ij} &= \tfrac{1}{2}e_{jlk}u_{k,li} = \text{gradient of rotation } (\bar{\kappa}_{ii}=0),\\
\bar{\bar{\kappa}}_{ijk} &= \tfrac{1}{3}(u_{k,ij}+u_{i,jk}+u_{j,ki}) = u_{(k,ij)}\\
&= \bar{\bar{\kappa}}_{jki} = \bar{\bar{\kappa}}_{kij} = \bar{\bar{\kappa}}_{kji} = \text{symmetric part of } \tilde{\kappa}_{ijk} \text{ or } \hat{\kappa}_{ijk},
\end{aligned}
\tag{1.1}
$$

where e_{ijk} is the alternator.

The following relations among the variables are found by eliminating the displacement from the definitions (1.1):

$$
\begin{aligned}
\omega_{ij} &= e_{ijk}w_k,\\
w_i &= \tfrac{1}{2}e_{ijk}\omega_{jk},
\end{aligned}
\tag{1.2}
$$

$$
\begin{aligned}
\tilde{\kappa}_{ijk} &= \hat{\kappa}_{ijk}+\hat{\kappa}_{jki}-\hat{\kappa}_{kij} = \bar{\bar{\kappa}}_{ijk}+\tfrac{2}{3}\bar{\kappa}_{il}e_{ljk}+\tfrac{2}{3}\bar{\kappa}_{jl}e_{lik},\\
\hat{\kappa}_{ijk} &= \bar{\bar{\kappa}}_{ijk}-\tfrac{1}{3}\bar{\kappa}_{jl}e_{kil}-\tfrac{1}{3}\bar{\kappa}_{kl}e_{jil} = \tfrac{1}{2}(\tilde{\kappa}_{ijk}+\tilde{\kappa}_{ikj}),\\
\bar{\kappa}_{ij} &= w_{j,i} = \tfrac{1}{2}e_{jkl}\omega_{kl,i} = \tfrac{1}{2}\tilde{\kappa}_{ilk}e_{jlk} = \hat{\kappa}_{lik}e_{jlk},\\
\bar{\bar{\kappa}}_{ijk} &= \tfrac{1}{3}(\tilde{\kappa}_{ijk}+\tilde{\kappa}_{jki}+\tilde{\kappa}_{kij}) = \tfrac{1}{3}(\hat{\kappa}_{ijk}+\hat{\kappa}_{jki}+\hat{\kappa}_{kij}).
\end{aligned}
\tag{1.3}
$$

2. EULER EQUATIONS AND NATURAL BOUNDARY CONDITIONS

The derivations of the three forms of the strain-gradient theory, given in [2], are summarized, here, with some minor alterations.

Hamilton's principle is written for independent variations δu_i between fixed limits of u_i at times t_0 and t_1:

$$\delta\int_{t_0}^{t_1}(\mathscr{T}-\mathscr{W})\mathrm{d}t+\int_{t_0}^{t_1}\delta\mathscr{W}_1\,\mathrm{d}t = 0, \tag{2.1}$$

where $\mathscr{T}$ and $\mathscr{W}$ are the total kinetic and potential energies in a volume V:

$$\mathscr{T} = \int_{\mathrm{V}}\mathrm{T}\mathrm{d}\mathrm{V}, \qquad \mathscr{W} = \int_{\mathrm{V}}\mathrm{W}\mathrm{d}\mathrm{V}$$

and $\delta\mathscr{W}_1$ is the variation of work done by external forces.

The kinetic energy-density is taken to be

$$\mathrm{T} = \tfrac{1}{2}\rho\dot{u}_i\dot{u}_i. \tag{2.2}$$

In [2], velocity-gradient terms were included in the kinetic energy-density. Such terms are appropriate if the strain-gradient equations are regarded as a low frequency approximation to the equations of a certain elastic material with a deformable micro-structure [2]. However, if the strain-gradient equations are viewed as a moderately long wavelength limit of the finite difference equations of a simple, crystal lattice, the velocity-gradient terms must be omitted. The latter view is adopted, here, in order to avoid carrying along complicated terms which are not germane to the present study.

Three forms are considered for the potential energy-density:

$$\mathrm{W} = \tilde{\mathrm{W}}(\varepsilon_{ij}, \tilde{\kappa}_{ijk}) = \hat{\mathrm{W}}(\varepsilon_{ij}, \hat{\kappa}_{ijk}) = \overline{\mathrm{W}}(\varepsilon_{ij}, \bar{\kappa}_{ij}, \bar{\bar{\kappa}}_{ijk}) \tag{2.3}$$

—all for the same displacement field. In the case of isotropic materials,

$$\begin{aligned}\tilde{\mathrm{W}} = {} & \tfrac{1}{2}\lambda\varepsilon_{ii}\varepsilon_{jj}+\mu\varepsilon_{ij}+\tilde{a}_1\tilde{\kappa}_{iik}\tilde{\kappa}_{kjj}+\tilde{a}_2\tilde{\kappa}_{ijj}\tilde{\kappa}_{ikk}\\ & +\tilde{a}_3\tilde{\kappa}_{iik}\tilde{\kappa}_{jjk}+\tilde{a}_4\kappa_{ijk}\tilde{\kappa}_{ijk}+\tilde{a}_5\tilde{\kappa}_{ijk}\kappa_{kji},\end{aligned} \tag{2.4}$$

$$\begin{aligned}\hat{\mathrm{W}} = {} & \tfrac{1}{2}\lambda\varepsilon_{ii}\varepsilon_{jj}+\mu\varepsilon_{ij}\varepsilon_{ij}+\hat{a}_1\hat{\kappa}_{iik}\hat{\kappa}_{kjj}+\hat{a}_2\hat{\kappa}_{ijj}\hat{\kappa}_{ikk}\\ & +\hat{a}_3\hat{\kappa}_{iik}\hat{\kappa}_{jjk}+\hat{a}_4\hat{\kappa}_{ijk}\hat{\kappa}_{ijk}+\hat{a}_5\hat{\kappa}_{ijk}\hat{\kappa}_{kji},\end{aligned} \tag{2.5}$$

where

$$\begin{aligned}\hat{a}_1 = 2\tilde{a}_1-4\tilde{a}_3, \quad \hat{a}_2 = -\tilde{a}_1+\tilde{a}_2+\tilde{a}_3,\\ \hat{a}_3 = 4\tilde{a}_3, \quad \hat{a}_4 = 3\tilde{a}_4-\tilde{a}_5, \quad \hat{a}_5 = -2\tilde{a}_4+2\tilde{a}_5,\end{aligned} \tag{2.6}$$

and

$$\begin{aligned}\overline{\mathrm{W}} = {} & \tfrac{1}{2}\lambda\varepsilon_{ii}\varepsilon_{jj}+\mu\varepsilon_{ij}\varepsilon_{ij}+2\bar{d}_1\bar{\kappa}_{ij}\bar{\kappa}_{ij}+2\bar{d}_2\bar{\kappa}_{ij}\bar{\kappa}_{ji}\\ & +\tfrac{3}{2}\bar{a}_1\bar{\bar{\kappa}}_{iij}\bar{\bar{\kappa}}_{kkj}+\bar{a}_2\bar{\bar{\kappa}}_{ijk}\bar{\bar{\kappa}}_{ijk}+\bar{f}e_{ijk}\bar{\kappa}_{ij}\bar{\bar{\kappa}}_{kll},\end{aligned} \tag{2.7}$$

where

$$\begin{aligned}18\bar{d}_1 = -2\hat{a}_1+4\hat{a}_2+\hat{a}_3+6\hat{a}_4-3\hat{a}_5, \quad 18\bar{d}_2 = 2\hat{a}_1-4\hat{a}_2-\hat{a}_3,\\ 3\bar{a}_1 = 2(\hat{a}_1+\hat{a}_2+\hat{a}_3), \quad \bar{a}_2 = \hat{a}_4+\hat{a}_5, \quad 3\bar{f} = \hat{a}_1+4\hat{a}_2-2\hat{a}_3,\end{aligned} \tag{2.8}$$

The variation of the work done by external forces is given a separate form for each form of W:

$$\delta\mathscr{W}_1 = \delta\tilde{\mathscr{W}}_1 = \delta\hat{\mathscr{W}}_1 = \delta\overline{\mathscr{W}}_1, \tag{2.9}$$

where, for $\tilde{W}$,

$$\delta\tilde{\mathscr{W}}_1 = \int_V F_k\delta u_k dV + \int_S (\tilde{P}_k\delta u_k + \tilde{R}_k D\delta u_k) dS + \oint_C \tilde{E}_k\delta u_k ds; \tag{2.10}$$

for $\hat{W}$,

$$\delta\hat{\mathscr{W}}_1 = \int_V F_k\delta u_k dV + \int_S (\hat{P}_k\delta u_k + \hat{R}_k D\delta u_k) dS + \oint_C \hat{E}_k\delta u_k ds; \tag{2.11}$$

and, for $\overline{W}$,

$$\delta\overline{\mathscr{W}}_1 = \int_V F_k\delta u_k dV + \int_S [\bar{P}_k\delta u_k + \bar{Q}_k(\delta_{kj} - n_k n_j)\delta w_j + \bar{R}\delta\varepsilon_{nn}] dS + \oint_C \bar{E}_k\delta u_k ds; \tag{2.12}$$

where S is the bounding surface, n_i are the components of the outward unit normal to S, C is an edge of S, s is the coordinate along C, D is the normal derivative:

$$D\varphi = n_i\varphi_{,i}; \tag{2.13}$$

and

$$\varepsilon_{nn} = n_i n_j \varepsilon_{ij}, \tag{2.14}$$

In (2.12), $\bar{Q}_k(\delta_{kj} - n_k n_j)\delta w_j$ replaces $\bar{Q}_k n_j \delta w_i e_{ijk}$ which was employed inadvertently in [2, (12.12)]. The change constitutes a rotation of ninety degrees about the normal to S.

As shown in [2], the three forms of the variational principle lead to the following three sets of Euler equations (stress-equations of motion) and natural (traction) boundary conditions:

I:

$$\tilde{\tau}_{jk,j} - \tilde{\mu}_{ijk,ij} + F_k = \rho\ddot{u}_k, \tag{2.15}$$

$$\begin{aligned} \tilde{P}_k &= n_j(\tilde{\tau}_{jk} - \tilde{\mu}_{ijk,i}) - D_j(n_i\tilde{\mu}_{ijk}) + (D_l n_l) n_i n_j \tilde{\mu}_{ijk}, \\ \tilde{R}_k &= n_i n_j \tilde{\mu}_{ijk}, \\ \tilde{E}_k &= s_p[n_l n_i \tilde{\mu}_{ijk}] e_{plj}, \end{aligned} \tag{2.16}$$

where the components of stress, $\tilde{\tau}_{ij}$, and double stress, $\tilde{\mu}_{ijk}$, are defined by

$$\tilde{\tau}_{ij} = \frac{\partial \tilde{W}}{\partial \varepsilon_{ij}} = \tilde{\tau}_{ji}, \tag{2.17}$$

$$\tilde{\mu}_{ijk} = \frac{\partial \tilde{W}}{\partial \tilde{\kappa}_{ijk}} = \tilde{\mu}_{jik}, \tag{2.18}$$

the D_i are the components of the surface gradient:

$$D_i\varphi = \varphi_{,i} - n_i D\varphi,$$

the s_p are the components of the unit vector tangent to C and the bold face brackets in the third of (2.16) indicate that the enclosed quantity is the difference of its values, at C, on the two portions of S that intersect at C. Equations (2.15)–(2.18), without the acceleration term, are linear forms of Toupin's results [1, §7].

II:
$$\hat{\tau}_{jk,j}-\hat{\mu}_{ijk,ij}+F_k = \rho\ddot{u}_k, \tag{2.19}$$

$$\begin{aligned}\hat{P}_k &= n_j(\hat{\tau}_{jk}-\hat{\mu}_{ijk,i})-\mathrm{D}_j(n_i\hat{\mu}_{ijk})+(\mathrm{D}_l n_l)n_i n_j\hat{\mu}_{ijk},\\ \hat{R}_k &= n_i n_j\hat{\mu}_{ijk},\\ \hat{E}_k &= s_p[n_l n_i\hat{\mu}_{ijk}]e_{plj},\end{aligned} \tag{2.20}$$

where

$$\hat{\tau}_{ij} = \frac{\partial\hat{\mathrm{W}}}{\partial\varepsilon_{ij}} = \hat{\tau}_{ji}, \tag{2.21}$$

$$\hat{\mu}_{ijk} = \frac{\partial\hat{\mathrm{W}}}{\partial\hat{\kappa}_{ijk}} = \hat{\mu}_{ikj}, \tag{2.22}$$

III:
$$\bar{\tau}_{jk,j}-\tfrac{1}{2}\bar{\mu}_{il,ij}e_{ljk}-\bar{\bar{\mu}}_{ijk,ij}+F_k = \rho\ddot{u}_k, \tag{2.23}$$

$$\begin{aligned}\bar{P}_k &= n_j[\bar{\tau}_{jk}+\tfrac{1}{2}(\bar{\mu}_{li,l}-\bar{\mu}_{nn,i})e_{jik}-\bar{\bar{\mu}}_{ijk,i}]-(\mathrm{D}_j-n_j\mathrm{D}_l n_l)(n_i\bar{\bar{\mu}}_{ijk}+n_p n_q n_k\bar{\bar{\mu}}_{pqj}),\\ \bar{Q}_k &= n_i\bar{\mu}_{ij}(\delta_{jk}-n_j n_k)+2n_q n_i n_j\bar{\bar{\mu}}_{ijp}e_{qpk},\\ \bar{R} &= n_i n_j n_k\bar{\bar{\mu}}_{ijk},\\ \bar{E}_k &= s_p[\tfrac{1}{2}\delta_{pk}\bar{\mu}_{nn}+n_q n_i(\bar{\bar{\mu}}_{ijk}+n_k n_l\bar{\bar{\mu}}_{lij})e_{jpq}],\end{aligned} \tag{2.24}$$

where

$$\bar{\tau}_{ij} = \frac{\partial\overline{\mathrm{W}}}{\partial\varepsilon_{ij}} = \bar{\tau}_{ji}, \tag{2.25}$$

$$\bar{\mu}_{ij} = \frac{\partial\overline{\mathrm{W}}}{\partial\bar{\kappa}_{ij}}, \quad \bar{\mu}_{ii} = 0, \tag{2.26}$$

$$\bar{\bar{\mu}}_{ijk} = \frac{\partial\overline{\mathrm{W}}}{\partial\bar{\bar{\kappa}}_{ijk}} = \bar{\bar{\mu}}_{jki} = \bar{\bar{\mu}}_{kij} = \bar{\bar{\mu}}_{kji} \tag{2.27}$$

and $\bar{\mu}_{nn} = n_i n_j\bar{\mu}_{ij}$.

In [2], $\bar{\mu}_{ij}$ was designated the "deviator of the couple-stress". If $u_{(k,ij)}$ is set equal to zero in $\overline{\mathrm{W}}$, $\bar{\mu}_{ij}$ does, in fact, reduce to the deviator of the couple-stress tensor μ_{ij} which was defined in [6] by means of the angular momentum principle and appears in the angular momentum equation [6, (1.9)]. That $\bar{\mu}_{ij}$ remains the deviator of the couple-stress when $\overline{\mathrm{W}}$ depends on $u_{(k,ij)}$ is verified in Section 5.

3. RELATIONS AMONG THE THREE FORMS

From (2.3) and the definitions (2.17), (2.21) and (2.25), it follows that the stresses, in the three forms of the theory, are the same:

$$\tilde{\tau}_{ij} = \hat{\tau}_{ij} = \bar{\tau}_{ij}. \tag{3.1}$$

To find the relations among the double stresses, we first form the relations

$$\begin{aligned}
\tilde{\mu}_{ijk} &= \frac{\partial \hat{W}}{\partial \hat{\kappa}_{pqr}} \frac{\partial \hat{\kappa}_{pqr}}{\partial \tilde{\kappa}_{ijk}} = \frac{\partial \overline{W}}{\partial \bar{\bar{\kappa}}_{pqr}} \frac{\partial \bar{\bar{\kappa}}_{pqr}}{\partial \tilde{\kappa}_{ijk}} + \frac{\partial \overline{W}}{\partial \bar{\kappa}_{pq}} \frac{\partial \bar{\kappa}_{pq}}{\partial \tilde{\kappa}_{ijk}}, \\
\hat{\mu}_{ijk} &= \frac{\partial \tilde{W}}{\partial \tilde{\kappa}_{pqr}} \frac{\partial \tilde{\kappa}_{pqr}}{\partial \hat{\kappa}_{ijk}} = \frac{\partial \overline{W}}{\partial \bar{\bar{\kappa}}_{pqr}} \frac{\partial \bar{\bar{\kappa}}_{pqr}}{\partial \hat{\kappa}_{ijk}} + \frac{\partial \overline{W}}{\partial \bar{\kappa}_{pq}} \frac{\partial \bar{\kappa}_{pq}}{\partial \hat{\kappa}_{ijk}}, \\
\bar{\mu}_{ij} &= \frac{\partial \tilde{W}}{\partial \tilde{\kappa}_{pqr}} \frac{\partial \tilde{\kappa}_{pqr}}{\partial \bar{\kappa}_{ij}} = \frac{\partial \hat{W}}{\partial \hat{\kappa}_{pqr}} \frac{\partial \hat{\kappa}_{pqr}}{\partial \bar{\kappa}_{ij}}, \\
\bar{\bar{\mu}}_{ijk} &= \frac{\partial \tilde{W}}{\partial \tilde{\kappa}_{pqr}} \frac{\partial \tilde{\kappa}_{pqr}}{\partial \bar{\bar{\kappa}}_{ijk}} = \frac{\partial \hat{W}}{\partial \hat{\kappa}_{pqr}} \frac{\partial \hat{\kappa}_{pqr}}{\partial \bar{\bar{\kappa}}_{ijk}}.
\end{aligned} \tag{3.2}$$

Then we replace the derivatives of the energy, in (3.2), with the definitions (2.18), (2.22), (2.26) and (2.27) and we use the relations (1.3) in evaluating the derivatives of the kinematic quantities. The results are

$$\tilde{\mu}_{ijk} = \tfrac{1}{2}(\hat{\mu}_{ijk} + \hat{\mu}_{jik}) = \bar{\bar{\mu}}_{ijk} + \tfrac{1}{4}\bar{\mu}_{il}e_{ljk} + \tfrac{1}{4}\bar{\mu}_{jl}e_{lik}, \tag{3.3}$$

$$\hat{\mu}_{ijk} = \tilde{\mu}_{ijk} + \tilde{\mu}_{kij} - \tilde{\mu}_{jki} = \bar{\bar{\mu}}_{ijk} + \tfrac{1}{2}\bar{\mu}_{jl}e_{lik} + \tfrac{1}{2}\bar{\mu}_{kl}e_{lij}, \tag{3.4}$$

$$\bar{\mu}_{ij} = \tfrac{4}{3}\tilde{\mu}_{ipq}e_{jpq} = \tfrac{2}{3}(\hat{\mu}_{ipq} + \hat{\mu}_{piq})e_{jpq}, \tag{3.5}$$

$$\bar{\bar{\mu}}_{ijk} = \tfrac{1}{3}(\tilde{\mu}_{ijk} + \tilde{\mu}_{jki} + \tilde{\mu}_{kij}) = \tfrac{1}{3}(\hat{\mu}_{ijk} + \hat{\mu}_{jki} + \hat{\mu}_{kij}). \tag{3.6}$$

A useful, alternative form of the second of (3.3) is

$$\tilde{\mu}_{ijk} = \bar{\bar{\mu}}_{ijk} + \tfrac{1}{2}\bar{\mu}_{il}e_{ljk} + \tfrac{1}{4}\bar{\mu}_{kl}e_{lij}. \tag{3.7}$$

From (3.7) and the first of (3.3), we find

$$\tilde{\mu}_{ijk,ij} = \hat{\mu}_{ijk,ij} = \tfrac{1}{2}\bar{\mu}_{il,ij}e_{ljk} + \bar{\bar{\mu}}_{ijk,ij}. \tag{3.8}$$

Accordingly, the three stress-equations of motion, (2.15), (2.19) and (2.23), are the same and they lead to the same displacement-equations of motion—as shown in [2] for isotropic materials.

The relations among the boundary conditions may be obtained by equating coefficients of like variations in (2.10), (2.11) and (2.12). From (2.10) and (2.11), it follows immediately that

$$\tilde{P}_k = \hat{P}_k, \quad \tilde{R}_k = \hat{R}_k, \quad \tilde{E}_k = \hat{E}_k. \tag{3.9}$$

To find the corresponding relations between the first and third forms, we can proceed by noting that, in (2.10),

$$\tilde{R}_k \mathrm{D}\delta u_k = 2\tilde{R}_k n_j \delta w_l e_{ljk} + \tilde{R}_k n_j \mathrm{D}_k \delta u_j + \tilde{R}_k n_k \delta\varepsilon_{nn},$$

$$2\tilde{R}_k n_j \delta w_l e_{ljk} = 2n_j \tilde{R}_k(\delta_{li} - n_l n_i)\delta w_i e_{ljk},$$

and

$$\begin{aligned}
\int_S \tilde{R}_k n_j \mathrm{D}_k \delta u_j \mathrm{dS} &= \int_S [\mathrm{D}_k(\tilde{R}_k n_j \delta u_j) - \mathrm{D}_k(\tilde{R}_k n_j)\delta u_j]\mathrm{dS} \\
&= \int_S [(\mathrm{D}_l n_l) n_k \tilde{R}_k n_j - \mathrm{D}_k(\tilde{R}_k n_j)]\delta u_j \mathrm{dS} + \oint_C s_l[n_i \tilde{R}_j n_k] e_{lij} \delta u_k \mathrm{ds}
\end{aligned}$$

by the surface divergence theorem. Thus, (2.10) takes the form

$$\begin{aligned}\delta\tilde{\mathscr{W}}_1 = \int_V F_k\delta u_k \mathrm{dV} + \int_S [\tilde{P}_k - (\mathrm{D}_j - n_j\mathrm{D}_l n_l)(\tilde{R}_j n_k)]\delta u_k \mathrm{dS} \\ + \int_S 2n_j\tilde{R}_k e_{ljk}(\delta_{li} - n_l n_i)\delta w_i \mathrm{dS} + \int_S \tilde{R}_k n_k \delta\varepsilon_{nn}\mathrm{dS} \\ + \oint_C (\tilde{E}_k + s_l[n_i\tilde{R}_j n_k]e_{lij})\delta u_k \mathrm{ds}.\end{aligned} \tag{3.10}$$

Equating coefficients of like variations in (3.10) and (2.12), we have

$$\begin{aligned}\bar{P}_k &= \tilde{P}_k - (\mathrm{D}_j - n_j\mathrm{D}_i n_i)(\tilde{R}_j n_k),\\ \bar{Q}_k &= 2n_i\tilde{R}_j e_{ijk},\\ \bar{R} &= n_i\tilde{R}_i,\\ \bar{E}_k &= \tilde{E}_k + s_l[n_i\tilde{R}_j n_k]e_{lij}\end{aligned} \tag{3.11}$$

or, in vector notation,

$$\begin{aligned}\bar{\mathbf{P}} &= \tilde{\mathbf{P}} - \mathbf{n}\cdot\nabla\times(\mathbf{n}\times\tilde{\mathbf{R}}\mathbf{n}),\\ \bar{\mathbf{Q}} &= 2\mathbf{n}\times\mathbf{R},\\ \bar{\mathbf{R}} &= \tilde{\mathbf{R}}\cdot\mathbf{n},\\ \bar{\mathbf{E}} &= \tilde{\mathbf{E}} + \mathbf{s}\cdot[\mathbf{n}\times\tilde{\mathbf{R}}\mathbf{n}].\end{aligned} \tag{3.12}$$

Conversely,

$$\begin{aligned}\tilde{\mathbf{P}} &= \bar{\mathbf{P}} + \tfrac{1}{2}\mathbf{n}\cdot\nabla\times(\bar{\mathbf{Q}}\mathbf{n}) = \hat{\mathbf{P}},\\ \tilde{\mathbf{R}}(&= \mathbf{n}\times\tilde{\mathbf{R}}\times\mathbf{n} + \mathbf{n}\cdot\tilde{\mathbf{R}}\mathbf{n}) = \tfrac{1}{2}\bar{\mathbf{Q}}\times\mathbf{n} + \bar{\mathbf{R}}\mathbf{n} = \hat{\mathbf{R}},\\ \tilde{\mathbf{E}} &= \bar{\mathbf{E}} - \tfrac{1}{2}\mathbf{s}\cdot[\bar{\mathbf{Q}}\mathbf{n}] = \hat{\mathbf{E}}.\end{aligned} \tag{3.13}$$

The formulas (3.12) and (3.13) show how the same traction boundary conditions can be set in each of the three forms of the theory. The same results may also be found by substituting (3.3)–(3.8) in (2.16), (2.20) and (2.24).

4. MOMENT EQUATIONS

The assumption that the first gradient of the displacement enters the potential energy-density only in the symmetric form ε_{ij} makes W *ab initio* invariant in a rigid rotation of the deformed body and, in the variational derivation, precludes the display of a differential equation expressing the equilibrium of moments. Toupin [7] has shown how a moment equation can be found by assuming an energy function

$$\mathrm{W} = \mathrm{W}'(u_{j,i}, \tilde{\kappa}_{ijk}) \tag{4.1}$$

and applying the requirement of rotational invariance subsequently. It is shown, here, how such a procedure can lead to a variety of equivalent forms of the moment equation. To exhibit this result in a simple context, we shall omit time dependence and edges.

Hamilton's principle then reduces to

$$\int_V \delta W' \, dV = \int_V F_k \delta u_k \, dV + \int_S (\tilde{P}'_k \delta u_k + \tilde{R}'_k D\delta u_k) \, dS. \tag{4.2}$$

Starting with

$$\delta W' = \frac{\partial W'}{\partial u_{j,i}} \delta u_{j,i} + \frac{\partial W'}{\partial \tilde{\kappa}_{ijk}} \delta \tilde{\kappa}_{ijk},$$

we find

$$\int_V \delta W' \, dV = -\int_V \left[\left(\frac{\partial W'}{\partial u_{k,i}} \right)_{,i} - \left(\frac{\partial W'}{\partial \tilde{\kappa}_{ijk}} \right)_{,ij} \right] \delta u_k \, dV$$
$$+ \int_V n_j \left[\frac{\partial W'}{\partial u_{k,j}} - \left(\frac{\partial W'}{\partial \tilde{\kappa}_{ijk}} \right)_{,i} \right] \delta u_k \, dS + \int_S n_i \frac{\partial W'}{\partial \tilde{\kappa}_{ijk}} \delta u_{k,j} \, dS \tag{4.3}$$

after application of the chain rule and the divergence theorem.

Define

$$\tilde{\mu}'_{ijk} = \frac{\partial W'}{\partial \tilde{\kappa}_{ijk}} = \tilde{\mu}'_{jik}, \tag{4.4}$$

$$\tau'_{jk} = \frac{\partial W'}{\partial u_{k,j}} - \tilde{\mu}'_{ijk,i}. \tag{4.5}$$

Then

$$\int_V \delta W' \, dV = -\int_V \tau'_{jk,j} \delta u_k \, dV + \int_S n_j \tau'_{jk} \delta u_k \, dS + \int_S n_i \tilde{\mu}'_{ijk} \delta u_{k,j} \, dS.$$

Now,

$$\int_S n_i \tilde{\mu}'_{ijk} \delta u_{k,j} \, dS = \int_S n_i \tilde{\mu}'_{ijk} D_j \delta u_k \, dS + \int_S n_i n_j \tilde{\mu}'_{ijk} D\delta u_k \, dS$$

and

$$\int_S n_i \tilde{\mu}'_{ijk} D_j \delta u_k \, dS = \int_S D_j(n_i \tilde{\mu}'_{ijk} \delta u_k) \, dS - \int_S D_j(n_i \tilde{\mu}'_{ijk}) \delta u_k \, dS.$$

Also,

$$\int_S D_j(n_i \tilde{\mu}'_{ijk} \delta u_k) \, dS = \int_S (D_l n_l) n_i n_j \tilde{\mu}'_{ijk} \delta u_k \, dS$$

by the surface divergence theorem for a smooth surface. Assembling these results, we have

$$\int_V \delta W' \, dV = -\int_V \tau'_{jk,j} \delta u_k \, dV + \int_S n_i n_j \tilde{\mu}'_{ijk} D\delta u_k \, dS$$
$$+ \int_S [n_j \tau'_{jk} - D_j(n_i \tilde{\mu}_{ijk}) + (D_l n_l) n_i n_j \tilde{\mu}'_{ijk}] \delta u_k \, dS. \tag{4.6}$$

Upon equating coefficients of like variations in (4.6) and (4.2), we find

$$\tau'_{jk,j}+F_k = 0, \tag{4.7}$$

$$\begin{aligned} P'_k &= n_j\tau'_{jk}-D_j(n_i\tilde{\mu}'_{ijk})+(D_l n_l)n_i n_j\tilde{\mu}'_{ijk}, \\ R'_k &= n_i n_j\tilde{\mu}'_{ijk}, \end{aligned} \tag{4.8}$$

which are equivalent to Toupin's results (10.13) and (10.14) in [7].

The conditions of invariance of the potential energy-density in a rigid rotation of the deformed body are, in the present linear case,

$$\frac{\partial W'}{\partial u_{[j,i]}} = 0. \tag{4.9}$$

With (4.9), $\tilde{\mu}'_{ijk}$ becomes $\tilde{\mu}_{ijk}$ and the definition (4.5) yields

$$\tau'_{[jk]}+\tilde{\mu}_{i[jk],i} = 0, \tag{4.10}$$

which, allowing for differences in notation, is Toupin's moment equation (10.20) in [7].

Now, return to the definition (4.5) and replace it with

$$\tau''_{jk} = \frac{\partial W'}{\partial u_{k,j}}-\tilde{\mu}'_{ijk,i}+2A\tilde{\mu}'_{k[ij],i}, \tag{4.11}$$

where A is a constant, and note that $\tilde{\mu}_{k[ij],ij} = 0$. Then, instead of (4.7) and (4.8), we have

$$\tau''_{jk,j}+F_k = 0, \tag{4.12}$$

$$\begin{aligned} \tilde{P}'_k &= n_j(\tau''_{jk}-2A\tilde{\mu}'_{k[ij],i})-D_j(n_i\tilde{\mu}'_{ijk})+(D_l n_l)n_i n_j\tilde{\mu}'_{ijk}, \\ \tilde{R}'_k &= n_i n_j\tilde{\mu}'_{ijk}. \end{aligned} \tag{4.13}$$

In terms of W′, (4.12) and (4.13) are the same as (4.7) and (4.8). The invariance conditions (4.9), applied to the definition (4.11), produce

$$\tau''_{[jk]}+(1+A)\tilde{\mu}_{i[jk],i} = 0. \tag{4.14}$$

Again, (4.14) and (4.10) are equivalent equations; but derivatives of the potential energy-density are distributed differently between the two terms in each equation.

By the first of (3.5),

$$\tilde{\mu}_{i[jk]} = \tfrac{3}{8}\bar{\mu}_{il}e_{jkl}. \tag{4.15}$$

Hence (4.14) can be written as

$$\tau''_{[jk]}+\tfrac{3}{8}(1+A)\bar{\mu}_{il,i}e_{jkl} = 0. \tag{4.16}$$

This is to be compared with the moment equation derived from the principle of angular momentum [6, (1.9)]:

$$\tau_{[jk]}+\tfrac{1}{2}\mu_{il,i}e_{jkl} = 0. \tag{4.17}$$

Noting that A is arbitrary and that $\bar{\mu}_{ij}$ is a deviator whereas μ_{ij} is not, it is apparent that the invariance conditions (4.9) do not contribute to the identification of the couple-stress.

5. DERIVATION BASED ON CONSERVATION PRINCIPLES

In this section, principles of conservation of linear and angular momentum and of energy are employed in the derivation of the equations of strain-gradient theory. For completeness, body double forces, with and without moment, are included along with body forces, edges and time dependence.

Let t_i and m_i be the components of force and couple, per unit area, acting on the surface S of a body occupying a volume V; and let F_i and C_i be the components of force and couple, per unit volume, in V. Then principles of linear and angular momentum are expressed by

$$\int_S t_i \,\mathrm{dS} + \int_V F_i \,\mathrm{dV} = \int_V \rho \ddot{u}_i \,\mathrm{dV}, \tag{5.1}$$

$$\int_S (x_j t_k e_{ijk} + m_i)\,\mathrm{dS} + \int_V (x_j F_k e_{ijk} + C_i)\,\mathrm{dV} = \int_V \rho x_j \ddot{u}_k e_{ijk}\,\mathrm{dV}. \tag{5.2}$$

Application of (5.1) to an elementary tetrahedron leads, in the limit as the tetrahedron shrinks to zero, to the definition of stress, $\boldsymbol{\tau}$, such that

$$t_j = n_i \tau_{ij}. \tag{5.3}$$

Substitution of (5.3) into (5.1) and application of the divergence theorem lead to the stress-equations of motion:

$$\tau_{jk,j} + F_k = \rho \ddot{u}_k. \tag{5.4}$$

Similarly, application of (5.2) to an elementary tetrahedron leads to the definition of couple-stress, $\boldsymbol{\mu}$, such that

$$m_j = n_i \mu_{ij} \tag{5.5}$$

which, with (5.2), yields the moment-equation

$$\mu_{ij,i} + \tau_{ki} e_{ijk} + C_j = 0 \tag{5.6}$$

or

$$\tau_{[jk]} + \tfrac{1}{2}\mu_{il,i} e_{jkl} + \tfrac{1}{2} C_l e_{jkl} = 0. \tag{5.7}$$

Now, in (5.4), write $\tau_{jk} = \tau_{(jk)} + \tau_{[jk]}$. Then, with (5.7), (5.4) becomes

$$\tau_{(jk),j} - \tfrac{1}{2}\mu_{il,ij} e_{jkl} + F_k - \tfrac{1}{2} C_{l,j} e_{jkl} = \rho \ddot{u}_k. \tag{5.8}$$

In (5.8), separate μ_{ij} into its deviatoric and spherical parts:

$$\mu_{ij} = \mu_{ij}^{\mathrm{D}} + \tfrac{1}{3}\delta_{ij}\mu_{kk}. \tag{5.9}$$

But the curl of the divergence of a spherical tensor vanishes. Hence (5.8) is

$$\tau_{(jk),j} - \tfrac{1}{2}\mu_{il,ij}^{\mathrm{D}} e_{jkl} + F_k - \tfrac{1}{2} C_{l,j} e_{jkl} = \rho \ddot{u}_k. \tag{5.10}$$

As for energy densities, we suppose, again, that the kinetic energy-density is

$$T = \tfrac{1}{2}\rho \dot{u}_i \dot{u}_i. \tag{5.11}$$

Then

$$\dot{T} = \rho \ddot{u}_i \dot{u}_i. \tag{5.12}$$

For the potential energy-density, we assume, again,

$$W = \overline{W}(\varepsilon_{ij}, \bar{\kappa}_{ij}, \bar{\bar{\kappa}}_{ijk});$$

so that

$$\dot{W} = \bar{\tau}_{ij}\dot{\varepsilon}_{ij} + \bar{\mu}_{ij}\dot{\bar{\kappa}}_{ij} + \bar{\bar{\mu}}_{ijk}\dot{\bar{\bar{\kappa}}}_{ijk}, \tag{5.13}$$

where $\bar{\tau}_{ij}$, $\bar{\mu}_{ij}$ and $\bar{\bar{\mu}}_{ijk}$ are defined, as before, by (2.25), (2.26) and (2.27).

We now adopt the following principle of conservation of energy:

$$\int_V (\dot{T} + \dot{W})\, dV = \int_V (F_j\dot{u}_j + C_j\dot{w}_j + \Phi_{(ij)}\dot{\varepsilon}_{ij})\, dV + \int_S (t_j\dot{u}_j + m_j\dot{w}_j + n_i\bar{\bar{\mu}}_{ijk}\dot{\varepsilon}_{jk})\, dS, \tag{5.14}$$

where the symmetric components $\Phi_{(ij)}$ are the densities of body double forces without moment. The antisymmetric part $\Phi_{[ij]}$ is the body couple per unit volume: $\frac{1}{2}e_{ijk}C_k$.

With (5.3), (5.5), the divergence theorem and the chain rule, the surface integral in (5.14) becomes

$$\int_V (\tau_{ij,i}\dot{u}_j + \mu_{ij,i}\dot{w}_j + \bar{\bar{\mu}}_{ijk,i}\dot{\varepsilon}_{jk} + \tau_{ij}\dot{u}_{j,i} + \mu_{ij}\dot{w}_{j,i} + \bar{\bar{\mu}}_{ijk}\dot{\varepsilon}_{jk,i})\, dV. \tag{5.15}$$

But

$$\tau_{ij}\dot{u}_{j,i} = \tau_{(ij)}\dot{u}_{(j,i)} + \tau_{[ij]}\dot{u}_{[j,i]} = \tau_{(ij)}\dot{\varepsilon}_{ij} + \tau_{ki}e_{ijk}\dot{w}_j$$

and

$$\mu_{ij}\dot{w}_{j,i} = \mu^D_{ij}\dot{\bar{\kappa}}_{ij}, \qquad \bar{\bar{\mu}}_{ijk}\dot{\varepsilon}_{jk,i} = \bar{\bar{\mu}}_{ijk}\dot{\bar{\bar{\kappa}}}_{ijk};$$

so that (5.15) becomes

$$\int_V [\tau_{ij,i}\dot{u}_j + (\mu_{ij,i} + \tau_{ki}e_{ijk})\dot{w}_j + (\tau_{(jk)} + \bar{\bar{\mu}}_{ijk,i})\dot{\varepsilon}_{jk} + \mu^D_{ij}\dot{\bar{\kappa}}_{ij} + \bar{\bar{\mu}}_{ijk}\dot{\bar{\bar{\kappa}}}_{ijk}]\, dV.$$

Hence, with (5.4), (5.6) and (5.12), the principle of conservation of energy (5.14) is converted to

$$\int_V \dot{W}\, dV = \int_V [(\tau_{(jk)} + \bar{\bar{\mu}}_{ijk,i} + \Phi_{(jk)})\dot{\varepsilon}_{jk} + \mu^D_{ij}\dot{\bar{\kappa}}_{ij} + \bar{\bar{\mu}}_{ijk}\dot{\bar{\bar{\kappa}}}_{ijk}]\, dV. \tag{5.16}$$

Finally, inserting (5.13) in the left hand side of (5.16) and equating coefficients of like kinematic variables on both sides of the equation, we find

$$\tau_{(jk)} = \bar{\tau}_{jk} - \bar{\bar{\mu}}_{ijk,i} - \Phi_{(jk)} \tag{5.17}$$

$$\mu^D_{ij} = \bar{\mu}_{ij}. \tag{5.18}$$

Thus, $\bar{\mu}_{ij}$ is indeed the deviator of the couple-stress.

Upon substituting (5.17) and (5.18) into (5.10), we recover (with the addition of body double forces) the stress-equations of motion (2.23) which were obtained from Hamilton's principle:

$$\bar{\tau}_{jk,j} - \tfrac{1}{2}\mu^D_{il,ij}e_{jkl} - \bar{\bar{\mu}}_{ijk,ij} + F_k - \tfrac{1}{2}C_{l,j}e_{jkl} - \Phi_{(jk),j} = \rho\ddot{u}_k. \tag{5.19}$$

In the case of an isotropic material, $\overline{W}$ is given by (2.7). Then, from (2.25), (2.26), (2.27) and (5.18),

$$\bar{\tau}_{pq} = \lambda\delta_{pq}\varepsilon_{ii} + 2\mu\varepsilon_{pq}, \tag{5.20}$$

$$\mu^{D}_{pq} = 4\bar{d}_1\bar{\kappa}_{pq} + 4\bar{d}_2\bar{\kappa}_{qp} + \bar{f}e_{pqi}\bar{\bar{\kappa}}_{ijj}, \tag{5.21}$$

$$\begin{aligned}\bar{\bar{\mu}}_{pqr} = {} & \bar{a}_1(\bar{\bar{\kappa}}_{iir}\delta_{pq} + \bar{\bar{\kappa}}_{iip}\delta_{qr} + \bar{\bar{\kappa}}_{iiq}\delta_{rp}) + 2\bar{a}_2\bar{\bar{\kappa}}_{pqr} \\ & + \tfrac{1}{3}\bar{f}\bar{\kappa}_{ij}(\delta_{pq}e_{ijr} + \delta_{qr}e_{ijp} + \delta_{rp}e_{ijq}).\end{aligned} \tag{5.22}$$

[Note that, in (5.21), $\bar{\bar{\kappa}}_{ijk}$ contributes to the couple-stress—contrary to the statement in [2] following (12.1)].

When (5.20), (5.21) and (5.22) are inserted in (5.19) and ε_{ij}, $\bar{\kappa}_{ij}$ and $\bar{\bar{\kappa}}_{ijk}$ are replaced by their expressions in terms of u_i, we find the displacement-equation of motion

$$(\lambda+2\mu)(1-l_1^2\nabla^2)\nabla\nabla\cdot\mathbf{u} - \mu(1-l_2^2\nabla^2)\nabla\times\nabla\times\mathbf{u} + \mathbf{F} + \tfrac{1}{2}\nabla\times\mathbf{C} - \nabla\cdot\mathbf{\Phi} = \rho\ddot{\mathbf{u}}, \tag{5.23}$$

where

$$l_1^2 = (3\bar{a}_1 + 2\bar{a}_2)/(\lambda+2\mu), \quad l_2^2 = (3\bar{d}_1 + \bar{a}_1 + 2\bar{a}_2 - \bar{f})/3\mu. \tag{5.24}$$

Necessary and sufficient conditions for positive definiteness of W are

$$\begin{aligned}&\mu > 0, \quad 3\lambda + 2\mu > 0, \quad -\bar{d}_1 < \bar{d}_2 < \bar{d}_1, \\ &\bar{a}_2 > 0, \quad 5\bar{a}_1 + 2\bar{a}_2 > 0, \quad 5\bar{f}^2 < 6(\bar{d}_1 - \bar{d}_2)(5\bar{a}_1 + 2\bar{a}_2),\end{aligned} \tag{5.25}$$

which replace (12.18) of [2]. From (5.25), it follows immediately that $l_1^2 > 0$. To show that $l_2^2 > 0$, note first that

$$\begin{aligned}&[2(\bar{a}_1 + 2\bar{a}_2) - 3\bar{d}_1 - \tfrac{1}{2}\bar{f}]^2 + \tfrac{9}{20}[6(\bar{d}_1 - \bar{d}_2)(5\bar{a}_1 + 2\bar{a}_2) - 5\bar{f}^2] + \tfrac{216}{5}\bar{d}_1\bar{a}_2 \\ &+ \tfrac{27}{10}(d_1 + d_2)(5\bar{a}_1 + 2\bar{a}_2) = (3\bar{d}_1 + \bar{a}_1 + 2\bar{a}_2 - \bar{f})(3\bar{d}_1 + 4\bar{a}_1 + 8\bar{a}_2 + 2\bar{f}).\end{aligned} \tag{5.26}$$

Since, by (5.25), the left hand side of (5.26) is the sum of positive terms, the right hand side is positive. The two factors on the right must both be positive because, if they were negative, the sum of the second factor and twice the first would be negative—in violation of (5.25). Hence

$$3\bar{d}_1 + \bar{a}_1 + 2\bar{a}_2 - \bar{f} > 0$$

and, consequently, $l_2^2 > 0$.

6. UNIQUENESS OF SOLUTIONS

As deduced from conservation principles in the preceding section, the equations of the strain-gradient theory for isotropic materials are sixty-three in number:

{3} $$\tau_{ij,i} + F_j = \rho\ddot{u}_j, \tag{6.1}$$

{3} $$\mu_{ij,i} + \tau_{ki}e_{ijk} + C_j = 0, \tag{6.2}$$

{6} $$\tau_{(jk)} = \bar{\tau}_{jk} - \bar{\bar{\mu}}_{ijk,i} - \Phi_{(jk)}, \tag{6.3}$$

{6} $$\varepsilon_{ij} = u_{(j,i)}, \tag{6.4}$$

{3} $$w_j = \tfrac{1}{2}u_{l,k}e_{jkl}, \tag{6.5}$$

{8} $$\bar{\kappa}_{ij} = w_{j,i}, \tag{6.6}$$

{10} $$\bar{\bar{\kappa}}_{ijk} = u_{(k,ij)}, \tag{6.7}$$

{6} $$\bar{\tau}_{ij} = \lambda\delta_{ij}\varepsilon_{kk}+2\mu\varepsilon_{ij}, \tag{6.8}$$

{8} $$\mu^{\mathrm{D}}_{ij} = 4\bar{d}_1\bar{\kappa}_{ij}+4\bar{d}_2\bar{\kappa}_{ji}+\bar{f}e_{ijp}\bar{\bar{\kappa}}_{pqq}, \tag{6.9}$$

{10} $$\begin{aligned}\bar{\bar{\mu}}_{ijk} = {} & \bar{a}_1(\bar{\bar{\kappa}}_{ppk}\delta_{ij}+\bar{\bar{\kappa}}_{ppi}\delta_{jk}+\bar{\bar{\kappa}}_{ppj}\delta_{ki})+2\bar{a}_2\bar{\bar{\kappa}}_{ijk}\\ & +\tfrac{1}{3}\bar{f}\bar{\kappa}_{pq}(\delta_{ij}e_{pqk}+\delta_{jk}e_{pqi}+\delta_{ki}e_{pqj}).\end{aligned} \tag{6.10}$$

Whereas there are sixty-three equations, there are sixty-four dependent variables: 3 of u_i, 3 of w_i, 6 of ε_{ij}, 8 of $\bar{\kappa}_{ij}$, 10 of $\bar{\bar{\kappa}}_{ijk}$, 9 of τ_{ij}, 9 of μ_{ij}, 6 of $\bar{\tau}_{ij}$ and 10 of $\bar{\bar{\mu}}_{ijk}$. The additional variable is the spherical part of the couple-stress which, because it does not contribute to the change of potential energy-density (5.13), is indeterminate within the framework of the theory as represented by the sixty-three equations. With this understanding, we proceed to the proof of a theorem of uniqueness of solutions in the usual manner.

Consider two sets of the sixty-four dependent variables $u'_i \ldots$ and $u''_i \ldots$ (with μ'_{ii} and μ''_{ii} arbitrary) and their differences $u_i = u'_i - u''_i \ldots$ Similarly, define body force and body double force differences: $F_i = F'_i - F''_i \ldots$ If each set of variables and body forces is a solution of (6.1)–(6.10), so is the difference set and, from (6.1), we may form the equation

$$\int_{t_0}^{t} \mathrm{d}t \int_{\mathrm{V}} (\tau_{ij,i}+F_j-\rho\ddot{u}_j)\dot{u}_j\,\mathrm{dV} = 0, \tag{6.11}$$

where t_0 is an initial time and V is a volume bounded by a surface S with an edge C. Now

$$\tau_{ij,i}\dot{u}_j = (\tau_{ij}\dot{u}_j)_{,i}-\tau_{ij}\dot{u}_{j,i} = (\tau_{ij}\dot{u}_j)_{,i}-\tau_{[ij]}\dot{u}_{[j,i]}-\tau_{(ij)}\dot{u}_{(j,i)}.$$

Hence, with (6.2) and (6.3),

$$\tau_{ij,i}\dot{u}_j = (\tau_{ij}\dot{u}_j)_{,i}+\tfrac{1}{2}e_{ijk}(\mu_{lk,l}+C_k)\dot{u}_{[j,i]}-(\bar{\tau}_{ij}-\bar{\bar{\mu}}_{kij,k}-\Phi_{(ij)})\dot{u}_{(j,i)};$$

or, with (6.4) and (6.5),

$$\tau_{ij,i}\dot{u}_j = (\tau_{ij}\dot{u}_j)_{,i}+(\mu_{lk,l}+C_k)\dot{w}_k-(\bar{\tau}_{ij}-\bar{\bar{\mu}}_{kij,k}-\Phi_{(ij)})\dot{\varepsilon}_{ij}.$$

Further,

$$\mu_{lk,l}\dot{w}_k+\bar{\bar{\mu}}_{kij,k}\dot{\varepsilon}_{ij} = (\mu_{ij}\dot{w}_j+\bar{\bar{\mu}}_{ijk}\dot{\varepsilon}_{jk})_{,i}-\mu_{ij}\dot{w}_{j,i}-\bar{\bar{\mu}}_{ijk}\dot{\varepsilon}_{jk,i};$$

or, with (6.6) and (6.7),

$$\mu_{lk,l}\dot{w}_k+\bar{\bar{\mu}}_{kij,k}\dot{\varepsilon}_{ij} = (\mu_{ij}\dot{w}_j+\bar{\bar{\mu}}_{ijk}\dot{\varepsilon}_{jk})_{,i}-\mu^{\mathrm{D}}_{ij}\dot{\bar{\kappa}}_{ij}-\bar{\bar{\mu}}_{ijk}\dot{\bar{\bar{\kappa}}}_{ijk},$$

where we have used $\bar{\kappa}_{ii} = 0$ and the symmetry of $\bar{\bar{\mu}}_{ijk}$. Also, by (6.8), (6.9) and (6.10),

$$\bar{\tau}_{ij}\dot{\varepsilon}_{ij}+\mu^{\mathrm{D}}_{ij}\dot{\bar{\kappa}}_{ij}+\bar{\bar{\mu}}_{ijk}\dot{\bar{\bar{\kappa}}}_{ijk} = \dot{\mathrm{W}},$$

where W is defined by the same form as (2.7) but in terms of the difference variables. Thus, the integrand of (6.11) becomes

$$F_j\dot{u}_j + C_j\dot{w}_j + \Phi_{(ij)}\dot{\varepsilon}_{ij} + (\tau_{ij}\dot{u}_j + \mu_{ij}\dot{w}_i + \bar{\bar{\mu}}_{ijk}\dot{\varepsilon}_{jk})_{,i} - \dot{\mathrm{T}} - \dot{\mathrm{W}},$$

where T is given by (2.2). Accordingly, application of the divergence theorem and an integration with respect to time convert (6.11) to

$$\int_{\mathrm{V}} [\mathrm{T}+\mathrm{W}]_{t_0}^{t}\,\mathrm{dV} = \int_{t_0}^{t} \mathrm{d}t \int_{\mathrm{V}} (F_j\dot{u}_j + C_j\dot{w}_j + \Phi_{(ij)}\dot{\varepsilon}_{ij})\,\mathrm{dV} + \int_{t_0}^{t} \mathrm{d}t \int_{\mathrm{S}} n_i(\tau_{ij}\dot{u}_j + \mu_{ij}\dot{w}_j + \bar{\bar{\mu}}_{ijk}\dot{\varepsilon}_{jk})\,\mathrm{dS}. \tag{6.12}$$

Thus, starting with the system of sixty-three equations (6.1)–(6.10), we have recovered the principle of conservation of energy for the difference system. We have now to reduce the twelve variables $\dot{u}_j$, $\dot{w}_j$, $\dot{\varepsilon}_{jk}$, in the surface integral of (6.12), to six as only six are independent of each other on S. For example, if the u_j are known on S, so is the normal component, n_jw_j, of w_j; and, if the u_j and the tangential component, $(\delta_{ij}-n_in_j)w_j$, of w_j are known on S, so are all the components of ε_{ij} except the normal component $n_in_j\varepsilon_{ij}$. It is also convenient to express the coefficients of $\dot{u}_j$, $(\delta_{ij}-n_in_j)\dot{w}_j$ and $n_in_j\dot{\varepsilon}_{ij}$ in forms that are independent of $\tau_{[ij]}$ and μ_{ii} so that the coefficients can be computed directly from the constitutive equations (6.8)–(6.10).

From (6.2) and (6.3),

$$\tau_{jk} = \bar{\tau}_{jk} - \tfrac{1}{2}\mu_{li,l}e_{ijk} - \bar{\bar{\mu}}_{ijk,i} - \tfrac{1}{2}C_ie_{ijk} - \Phi_{(jk)}. \tag{6.13}$$

Also,

$$\begin{aligned} n_i\mu_{ij}\dot{w}_j &= n_i\mu_{ij}n_jn_k\dot{w}_k + n_i\mu_{ij}(\delta_{jk}-n_jn_k)\dot{w}_k \\ &= \tfrac{1}{2}\mu_{nn}n_ke_{kij}\dot{u}_{j,i} + n_i\mu_{ij}(\delta_{jk}-n_jn_k)\dot{w}_k \\ &= \tfrac{1}{2}n_ke_{kij}(\mu_{nn}\dot{u}_j)_{,i} - \tfrac{1}{2}n_ke_{kij}\mu_{nn,i}\dot{u}_j + n_i\mu_{ij}(\delta_{jk}-n_jn_k)\dot{w}_k \end{aligned}$$

and

$$\int_{\mathrm{S}} n_ke_{kij}(\mu_{nn}\dot{u}_j)_{,i}\,\mathrm{dS} = \oint_{\mathrm{C}} [\mu_{nn}^{\mathrm{D}}]s_j\dot{u}_j\,\mathrm{ds}.$$

Hence

$$\int_{\mathrm{S}} (n_j\tau_{jk}\dot{u}_k + n_i\mu_{ij}\dot{w}_j)\,\mathrm{dS} = \int_{\mathrm{S}} n_j[\bar{\tau}_{jk} + \tfrac{1}{2}(\mu_{li,l}^{\mathrm{D}} - \mu_{nn,i}^{\mathrm{D}})e_{jik} - \bar{\bar{\mu}}_{ijk,i}]\dot{u}_k\,\mathrm{dS} + \int_{\mathrm{S}} n_j(\tfrac{1}{2}C_ie_{jik} - \Phi_{(jk)})\dot{u}_k\,\mathrm{dS} + \int_{\mathrm{S}} n_i\mu_{ij}^{\mathrm{D}}(\delta_{jk}-n_jn_k)\dot{w}_k\,\mathrm{dS} + \oint_{\mathrm{C}} \tfrac{1}{2}[\mu_{nn}^{\mathrm{D}}]s_k\dot{u}_k\,\mathrm{ds}, \tag{6.14}$$

in which we have used

$$\mu_{ij}(\delta_{jk}-n_jn_k) = \mu_{ij}^{D}(\delta_{jk}-n_jn_k), \quad \mu_{li,l}-\mu_{nn,i} = \mu_{li,l}^{D}-\mu_{nn,i}^{D}.$$

As for the last term in (6.12), we write

$$n_i\bar{\bar{\mu}}_{ijk}\dot{\varepsilon}_{jk} = n_i\bar{\bar{\mu}}_{ijk}\dot{u}_{k,j} = n_i\bar{\bar{\mu}}_{ijk}\mathrm{D}_j\dot{u}_k+n_in_j\bar{\bar{\mu}}_{ijk}\mathrm{D}\dot{u}_k.$$

But

$$\mathrm{D}\dot{u}_k = 2\dot{w}_ln_je_{ljk}+n_j\mathrm{D}_k\dot{u}_j+n_k\dot{\varepsilon}_{nn}$$

Hence

$$\begin{aligned} n_i\bar{\bar{\mu}}_{ijk}\dot{\varepsilon}_{jk} &= (n_i\bar{\bar{\mu}}_{ijk}+n_pn_q\bar{\bar{\mu}}_{pqj}n_k)\mathrm{D}_j\dot{u}_k+2n_qn_in_j\bar{\bar{\mu}}_{ijp}e_{qpk}\dot{w}_k+n_in_jn_k\bar{\bar{\mu}}_{ijk}\dot{\varepsilon}_{nn} \\ &= \mathrm{D}_j[(n_i\bar{\bar{\mu}}_{ijk}+n_pn_q\bar{\bar{\mu}}_{pqj}n_k)\dot{u}_k]-\mathrm{D}_j(n_i\bar{\bar{\mu}}_{ijk}+n_pn_q\bar{\bar{\mu}}_{pqj}n_k)\dot{u}_k \\ &\quad +2n_qn_in_j\bar{\bar{\mu}}_{ijp}e_{qpk}\dot{w}_k+n_in_jn_k\bar{\bar{\mu}}_{ijk}\dot{\varepsilon}_{nn}. \end{aligned}$$

Then, with the surface divergence theorem,

$$\begin{aligned} \int_S n_i\bar{\bar{\mu}}_{ijk}\dot{\varepsilon}_{jk}\,\mathrm{dS} &= \int_S [(n_j\mathrm{D}_ln_l-\mathrm{D}_j)(n_i\bar{\bar{\mu}}_{ijk}+n_pn_q\bar{\bar{\mu}}_{pqj}n_k)]\dot{u}_k\,\mathrm{dS} \\ &\quad +\int_S n_in_j(2n_q\bar{\bar{\mu}}_{ijp}e_{qpk}\dot{w}_k+\bar{\bar{\mu}}_{ijk}n_k\dot{\varepsilon}_{nn})\,\mathrm{dS} \\ &\quad +\oint_C s_p[\![n_qe_{jpq}(n_i\bar{\bar{\mu}}_{ijk}+n_ln_in_k\bar{\bar{\mu}}_{lij})]\!]\dot{u}_k\,\mathrm{ds}. \end{aligned} \tag{6.15}$$

Upon inserting (6.14) and (6.15) in (6.12), we obtain, finally,

$$\begin{aligned} \int_V [\mathrm{T}+\mathrm{W}]_{t_0}^{t}\,\mathrm{dV} &= \int_{t_0}^{t}\mathrm{d}t\int_V (F_j\dot{u}_j+C_j\dot{w}_j+\Phi_{(ij)}\dot{\varepsilon}_{ij})\,\mathrm{dV} \\ &\quad +\int_{t_0}^{t}\mathrm{d}t\int_S [\bar{P}_k\dot{u}_k+\bar{Q}_k(\delta_{kj}-n_kn_j)\dot{w}_j+\bar{R}\dot{\varepsilon}_{nn}]\,\mathrm{dS} \\ &\quad +\int_{t_0}^{t}\mathrm{d}t\int_C \bar{E}_k\dot{u}_k\,\mathrm{ds}, \end{aligned} \tag{6.16}$$

where

$$\begin{aligned} \bar{P}_k &= n_j[\bar{\tau}_{jk}+\tfrac{1}{2}(\mu_{li,l}^{D}-\mu_{nn,i}^{D})e_{jik}-\bar{\bar{\mu}}_{ijk,i}]-(\mathrm{D}_j-n_j\mathrm{D}_ln_l)(n_i\bar{\bar{\mu}}_{ijk}+n_pn_qn_k\bar{\bar{\mu}}_{pqj})+n_j(\tfrac{1}{2}C_ie_{jik}-\Phi_{(jk)}), \\ \bar{Q}_k &= n_i\mu_{ij}^{D}(\delta_{jk}-n_jn_k)+2n_in_jn_q\bar{\bar{\mu}}_{ijp}e_{qpk}, \\ \bar{R} &= n_in_jn_k\bar{\bar{\mu}}_{ijk}, \\ \bar{E}_k &= s_p[\![\tfrac{1}{2}\delta_{pk}\mu_{nn}^{D}+n_qn_i(\bar{\bar{\mu}}_{ijk}+n_kn_l\bar{\bar{\mu}}_{lij})e_{jpq}]\!]. \end{aligned} \tag{6.17}$$

It will be observed that (6.17) have the same form as (2.24) with the addition of the body double forces. Note that the body couple **C** has been placed in the expression for **P**, following Koiter [8].

Conditions sufficient for a unique solution of (6.1)–(6.10) are now obtained from (6.16) in the usual manner—based on the assumption of positive definiteness of T and W:

1. At each point in V: **F**, **C**, **Φ** and initial values of **u** and $\dot{\mathbf{u}}$.

2. At each point on S: (a) a component of **u** (or $\bar{\mathbf{P}}$), in any direction, and the resultant of $\bar{\mathbf{P}}$ (or **u**) in the plane at right angles; (b) in the tangent plane, a component of **w** (or $\bar{\mathbf{Q}}$) and the component of $\bar{\mathbf{Q}}$ (or **w**) at right angles; (c) ε_{nn} or $\bar{R}$.

3. At each point on C: a component of **u** (or $\bar{\mathbf{E}}$), in any direction, and the resultant of $\bar{\mathbf{E}}$ (or **u**) in the plane at right angles.

Note that, according to the linear momentum principle, $n_j\tau_{jk}$ and not $\bar{P}_k$ are the components of force per unit area. Also, by the angular momentum principle, $n_i\mu_{ij}(\delta_{jk}-n_jn_k)$ and not $\bar{Q}_k$ are the components of tangential couple per unit area.

As in classical elasticity, the uniqueness theorem is subject to the regularity restrictions implied by the forms of the divergence theorems employed in the proof. In particular, if a singularity is present, an additional condition, generally obtained through a limit process, is required. Sternberg and Muki [9] have shown how failure to observe such a requirement can lead to physically irrelevant "pseudo-solutions".

Acknowledgements—During this investigation, one of the authors (N.N.E.) was a Visiting Research Scientist at the Courant Institute of Mathematical Sciences, New York University. The work was supported by the Office of Naval Research, under a contract with Columbia University, and by the National Science Foundation, under a contract with New York University.

REFERENCES

[1] R. A. TOUPIN, Elastic materials with couple-stresses. *Archs ration. Mech. Analysis* **11**, 385–414 (1962).
[2] R. D. MINDLIN, Micro-structure in linear elasticity. *Archs ration. Mech. Analysis* **16**, 51–78 (1964).
[3] G. GRIOLI, Elasticità asimmetrica. *Annali Mat. pura appl.* Ser. IV, **50**, 389–417 (1960).
[4] E. L. AERO and E. V. KUVSHINSKII, Fundamental equations of the theory of elastic materials with rotationally interacting particles. *Fizika tverd. Tela* **2**, 1399–1409 (1960); Translation: *Soviet Phys. solid St.* **2**, 1272–1281 (1961).
[5] E. S. RAJAGOPAL, The existence of interfacial couples in infinitesimal elasticity. *Annln Phys.* **6**, 192–201 (1960).
[6] R. D. MINDLIN and H. F. TIERSTEN, Effects of couple-stresses in linear elasticity. *Archs ration. Mech. Analysis* **11**, 415–448 (1962).
[7] R. A. TOUPIN, Theories of elasticity with couple-stress. *Archs ration. Mech. Analysis* **17**, 85–112 (1964).
[8] W. T. KOITER, Couple-stresses in the theory of elasticity. *Proc. K. ned. Akad. Wet.* **B67**, 17–44 (1964).
[9] E. STERNBERG and R. MUKI, The effect of couple-stresses on the stress concentration around a crack. *Int. J. Solids Struct.* **3**, 69–96 (1967).

(*Received* 25 *July* 1967)

Абстракт—В работе исследуется линейная теория упругости, в которой плотность потенциальной энергии зависит от градиента дефоямации и от дефоямации.

В первой части работы сравниваются три вида теории и выводятся зависимости, касающиеся напряжений и граничных условий в этих трех видах.

Во второй части решаются неоднозначности выражения для моментного уравнения и определяется моментное напряжение, используя вывод основывающийся в большей степени на принципах сохранения, чем на вариационных принципах, которые использованы ранее.

Int. J. Solids Structures, 1968, Vol. 4, pp. 637 to 642. Pergamon Press. Printed in Great Britain

POLARIZATION GRADIENT IN ELASTIC DIELECTRICS

R. D. MINDLIN

Department of Civil Engineering, Columbia University, New York, N.Y.

Abstract—By inclusion of the polarization gradient in the stored energy function of elastic dielectrics, the classical theory of piezoelectricity is extended to accommodate an electro-mechanical interaction in centro-symmetric (including isotropic) materials and a surface energy of deformation and polarization.

1. TOUPIN'S VARIATIONAL PRINCIPLE

As a preliminary to an extension of the classical theory of electro-mechanical interaction in elastic, dielectric continua, a review is given, in this section, of a linear version of Toupin's [1] variational principle for the equilibrium equations of classical piezoelectricity.

In a body occupying a volume V bounded by a surface S, separating V from an outer vacuum V', it is assumed that

$$-\delta\int_{V^*} H\,\mathrm{d}V+\int_V (f_i\delta u_i+E_i^0\delta P_i)\,\mathrm{d}V+\int_S t_i\delta u_i\,\mathrm{d}S = 0, \tag{1.1}$$

where H is the electric enthalpy density, $V^* = V+V'$, u_i is the displacement, P_i is the polarization and f_i, E_i^0 and t_i are the external body force, electric field and surface traction, respectively. Toupin separates the electric enthalpy density into an energy density of deformation and polarization, say W^{L}, and a remainder:

$$H = W^{\mathrm{L}}(S_{ij}, P_i)-\tfrac{1}{2}\varepsilon_0\varphi_{,i}\varphi_{,i}+\varphi_{,i}P_i, \tag{1.2}$$

where S_{ij} is the strain,

$$S_{ij} = \tfrac{1}{2}(u_{j,i}+u_{i,j}), \tag{1.3}$$

φ is the potential of the Maxwell self-field,

$$E_i^{\mathrm{MS}} = -\varphi_{,i}, \tag{1.4}$$

and ε_0 is the permittivity of a vacuum.

For independent variations of u_i, P_i and φ,

$$\delta H = T_{ij}\delta S_{ij}-\bar{E}_i\delta P_i-\varepsilon_0\varphi_{,i}\delta\varphi_{,i}+\varphi_{,i}\delta P_i+P_i\delta\varphi_{,i}, \tag{1.5}$$

where T_{ij} is the stress and $\bar{E}_i$ is the effective local electric force:

$$T_{ij} \equiv \frac{\partial W^{\mathrm{L}}}{\partial S_{ij}} = T_{ji}, \qquad \bar{E}_i \equiv -\frac{\partial W^{\mathrm{L}}}{\partial P_i}. \tag{1.6}$$

By the chain rule of differentiation,

$$\delta H = -T_{ij,i}\delta u_i-(\bar{E}_i-\varphi_{,i})\delta P_i-(-\varepsilon_0\varphi_{,ii}+P_{i,i})\delta\varphi+(T_{ij}\delta u_j)_{,i}+[(-\varepsilon_0\varphi_{,i}+P_i)\delta\varphi]_{,i}. \tag{1.7}$$

Inserting (1.7) in (1.1) and applying the divergence theorem, we have

$$\int_{V^*} [(T_{ij,i}+f_j)\delta u_j+(\bar{E}_i-\varphi_{,i}+E_i^0)\delta P_i+(-\varepsilon_0\varphi_{,ii}+P_{i,i})\delta\varphi]\,\mathrm{d}V$$
$$+\int_S [(t_j-n_iT_{ij})\delta u_j+n_i(\varepsilon_0[\![\varphi_{,i}]\!]-P_i)\delta\varphi]\,\mathrm{d}S = 0; \tag{1.8}$$

whence follow the Euler equations

$$\begin{aligned} T_{ij,i}+f_j &= 0,\\ \bar{E}_i-\varphi_{,i}+E_i^0 &= 0,\\ -\varepsilon_0\varphi_{,ii}+P_{i,i} &= 0, \text{ in } V;\\ \varphi_{,ii} &= 0, \text{ in } V'; \end{aligned} \tag{1.9}$$

and the boundary conditions

$$\begin{aligned} n_iT_{ij} &= t_j,\\ n_i(-\varepsilon_0[\![\varphi_{,i}]\!]+P_i) &= 0. \end{aligned} \tag{1.10}$$

where $[\![\varphi_{,i}]\!]$ is the jump in $\varphi_{,i}$ across S.

The energy density of deformation and polarization is taken to be

$$W^{\mathrm{L}} = \tfrac{1}{2}a_{ij}P_iP_j+\tfrac{1}{2}c_{ijkl}S_{ij}S_{kl}+f_{ijk}S_{ij}P_k \tag{1.11}$$

so that, from (1.6),

$$\begin{aligned} -\bar{E}_j &= a_{jk}P_k+f_{klj}S_{kl},\\ T_{ij} &= f_{ijk}P_k+c_{ijkl}S_{kl}. \end{aligned} \tag{1.12}$$

Equations (1.3), (1.9) and (1.12), with boundary conditions (1.10), constitute the classical theory of piezoelectricity in the form given by Toupin.

2. POLARIZATION GRADIENT

The extension of the classical theory, to be considered, is obtained simply by adding a functional dependence of W^{L} on the polarization gradient. This addition may be justified on several grounds. First of all, the order of the differential equations is thereby not raised, so that we are not adding effects of higher order than those already included. Alternatively, suppose we were to start by assuming dependence of W^{L} on the displacement and polarization and their gradients and truncate after the first gradients. The requirement of invariance of W^{L}, in a rigid translation and rotation of the deformed and polarized body, eliminates dependence on the displacement and rotation only; leaving the strain, polarization and polarization gradient. Finally, from the point of view of lattice theories of crystals [2, 3], based on the "shell model" [4] of the atom, the polarization gradient represents the long wave approximation to the shell–shell and core–shell interactions between atoms which logically should accompany the core–core interaction between atoms, despite their possibly small magnitude.

In addition to taking the polarization gradient into account, we shall also include the kinetic energy and employ Hamilton's principle, so that (1.1) is replaced by

$$\delta\int_{t_0}^{t_1}\mathrm{d}t\int_{V^*}(\tfrac{1}{2}\rho\dot{u}_i\dot{u}_i-H)\,\mathrm{d}V+\int_{t_0}^{t_1}\mathrm{d}t\left[\int_V(f_i\delta u_i+E_i^0\delta P_i)\delta V+\int_S t_i\delta u_i\,\mathrm{d}S\right]=0, \tag{2.1}$$

where

$$H = W^{\mathrm{L}}(S_{ij}, P_i, P_{j,i}) - \tfrac{1}{2}\varepsilon_0 \varphi_{,i}\varphi_{,i} + \varphi_{,i}P_i. \tag{2.2}$$

As usual [5],

$$\delta \int_{t_0}^{t_1} \mathrm{d}t \int_V \tfrac{1}{2}\rho \dot{u}_i \dot{u}_i \,\mathrm{d}V = -\int_{t_0}^{t_1} \mathrm{d}t \int_V \rho \ddot{u}_j \delta u_j \,\mathrm{d}V.$$

Then, by the same procedure as that employed in the preceding section,

$$\begin{aligned} T_{ij,i} + f_j &= \rho \ddot{u}_j, \\ \bar{E}_j + E_{ij,i} - \varphi_{,j} + E_j^0 &= 0, \\ -\varepsilon_0 \varphi_{,ii} + P_{i,i} &= 0, \text{ in } V, \\ \varphi_{,ii} &= 0, \text{ in } V'; \end{aligned} \tag{2.3}$$

and, on S,

$$\begin{aligned} n_i T_{ij} &= t_j, \\ n_i(-\varepsilon_0[\varphi_{,i}] + P_i) &= 0, \\ n_i E_{ij} &= 0, \end{aligned} \tag{2.4}$$

where

$$\bar{E}_i \equiv -\frac{\partial W^{\mathrm{L}}}{\partial P_i}, \qquad E_{ij} \equiv \frac{\partial W^{\mathrm{L}}}{\partial P_{j,i}}, \qquad T_{ij} \equiv \frac{\partial W^{\mathrm{L}}}{\partial S_{ij}} = T_{ji}. \tag{2.5}$$

For W^{L}, we take

$$W^{\mathrm{L}} = b_{ij}^0 P_{j,i} + \tfrac{1}{2}a_{ij}P_iP_j + \tfrac{1}{2}b_{ijkl}P_{j,i}P_{l,k} + \tfrac{1}{2}c_{ijkl}S_{ij}S_{kl} + d_{ijkl}P_{j,i}S_{kl} + f_{ijk}S_{ij}P_k + g_{ijk}P_iP_{k,j}. \tag{2.6}$$

Then, from (2.5),

$$\begin{aligned} -\bar{E}_j &= a_{jk}P_k + g_{jkl}P_{l,k} + f_{klj}S_{kl}, \\ E_{ij} &= g_{kij}P_k + b_{ijkl}P_{l,k} + d_{ijkl}S_{kl} + b_{ij}^0, \\ T_{ij} &= f_{ijk}P_k + d_{klij}P_{l,k} + c_{ijkl}S_{kl}. \end{aligned} \tag{2.7}$$

Equations (1.3), (2.3) and (2.7), with boundary conditions (2.4), form the equations of the extended theory. The significance of the linear term $b_{ij}^0 P_{j,i}$, in W^{L}, is considered in Section 4.

3. CENTROSYMMETRIC MATERIALS

One of the properties of the extended theory is its accommodation of an electromechanical interaction, through the coefficient d_{ijkl}, even for materials with centrosymmetry. For example, for centrosymmetric cubic symmetry [6],

$$\begin{aligned} f_{ijk} &= 0, \qquad g_{ijk} = 0, \qquad b_{ij}^0 = b_0\delta_{ij}, \qquad a_{ij} = a\delta_{ij}, \\ b_{ijkl} &= b\delta_{ijkl} + b_{12}\delta_{ij}\delta_{kl} + b_{44}(\delta_{ik}\delta_{jl} + \delta_{il}\delta_{jk}) + b_{77}(\delta_{ik}\delta_{jl} - \delta_{il}\delta_{jk}), \\ c_{ijkl} &= c\delta_{ijkl} + c_{12}\delta_{ij}\delta_{kl} + c_{44}(\delta_{ik}\delta_{jl} + \delta_{il}\delta_{jk}), \\ d_{ijkl} &= d\delta_{ijkl} + d_{12}\delta_{ij}\delta_{kl} + d_{44}(\delta_{ik}\delta_{jl} + \delta_{il}\delta_{jk}), \end{aligned} \tag{3.1}$$

where

$$b = b_{11} - b_{12} - 2b_{44}, \qquad c = c_{11} - c_{12} - 2c_{44}, \qquad d = d_{11} - d_{12} - 2d_{44}, \tag{3.2}$$

δ_{ij} is the Kronecker delta and δ_{ijkl} is unity if all indices are alike and zero otherwise. Then, from (2.5), (2.6) and (3.1),

$$\begin{aligned} -\bar{E}_i &= aP_i, \\ E_{ij} &= b\delta_{ijkl}P_{l,k} + b_{12}\delta_{ij}P_{k,k} + b_{44}(P_{j,i} + P_{i,j}) + b_{77}(P_{j,i} - P_{i,j}) \\ &\quad + d\delta_{ijkl}S_{kl} + d_{12}\delta_{ij}S_{kk} + 2d_{44}S_{ij} + b_0\delta_{ij}, \\ T_{ij} &= d\delta_{ijkl}P_{l,k} + d_{12}\delta_{ij}P_{k,k} + d_{44}(P_{j,i} + P_{i,j}) \\ &\quad + c\delta_{ijkl}S_{kl} + c_{12}\delta_{ij}S_{kk} + 2c_{44}S_{ij}. \end{aligned} \tag{3.3}$$

Upon substituting (3.3) in (2.3) and employing (1.3), we find the "displacement" equations of motion:

$$\begin{aligned} &\tfrac{1}{2}c\delta_{ijkl}(u_{l,ki} + u_{k,li}) + c_{12}\delta_{ij}u_{k,ki} \\ &\qquad + c_{44}(u_{j,ii} + u_{i,ji}) + d\delta_{ijkl}P_{l,ki} + d_{12}\delta_{ij}P_{k,ki} + d_{44}(P_{j,ii} + P_{i,ji}) + f_j = \rho\ddot{u}_j \\ &\tfrac{1}{2}d\delta_{ijkl}(u_{l,ki} + u_{k,li}) + d_{12}\delta_{ij}u_{k,ki} + d_{44}(u_{j,ii} + u_{i,ji}) - aP_j - \varphi_{,j} \\ &\qquad + b\delta_{ijkl}P_{l,ki} + b_{12}\delta_{ij}P_{k,ki} + b_{44}(P_{j,ii} + P_{i,ji}) + b_{77}(P_{j,ii} - P_{i,ji}) + E_j^0 = 0, \\ &\qquad\qquad -\varepsilon_0\varphi_{,ii} + P_{i,i} = 0. \end{aligned} \tag{3.4}$$

For isotropic materials, it is only necessary to set

$$b = c = d = 0.$$

Then (3.4) become, in vector notation,

$$\begin{aligned} c_{44}\nabla^2\mathbf{u} + (c_{12} + c_{44})\nabla\nabla\cdot\mathbf{u} + d_{44}\nabla^2\mathbf{P} + (d_{12} + d_{44})\nabla\nabla\cdot\mathbf{P} + \mathbf{f} &= \rho\ddot{\mathbf{u}}, \\ d_{44}\nabla^2\mathbf{u} + (d_{12} + d_{44})\nabla\nabla\cdot\mathbf{u} + (b_{44} + b_{77})\nabla^2\mathbf{P} + (b_{12} + b_{44} - b_{77})\nabla\nabla\cdot\mathbf{P} - a\mathbf{P} - \nabla\varphi + \mathbf{E}^0 &= 0, \\ -\varepsilon_0\nabla^2\varphi + \nabla\cdot\mathbf{P} &= 0. \end{aligned} \tag{3.5}$$

It is apparent that, in both cases, the displacement and polarization fields are coupled through the constants d_{ijkl}.

4. SURFACE ENERGY OF DEFORMATION AND POLARIZATION

It will be observed that the energy density W^{L}, in (2.6), contains a linear term, $b_{ij}^0 P_{j,i}$, in the polarization gradient. The removal of the resulting $n_i E_{ij}$ from a boundary results in a polarization and strain localized at the surface. The accompanying surface energy can be found as follows.

The energy density W^{L}, given by (2.6), can be written in the alternative form

$$W^{\mathrm{L}} = \tfrac{1}{2}b_{ij}^0 P_{j,i} + \tfrac{1}{2}T_{ij}S_{ij} - \tfrac{1}{2}\bar{E}_i P_i + \tfrac{1}{2}E_{ij}P_{j,i} \tag{4.1}$$

through the use of (2.7). Then, in the case of equilibrium, application of the chain rule, the divergence theorem, the boundary conditions and the equations of equilibrium results in the following expression for the *total* energy:

$$\int_{V*} W\,\mathrm{d}V = \frac{1}{2}\int_S n_i b_{ij}^0 P_j\,\mathrm{d}S + \frac{1}{2}\int_V (f_i u_i + E_i^0 P_i)\,\mathrm{d}V + \frac{1}{2}\int_S t_i u_i\,\mathrm{d}S. \tag{4.2}$$

Hence, in the absence of external forces, we have

$$\int_{V*} W\,\mathrm{d}V = \frac{1}{2}\int_S n_i b_{ij}^0 P_j\,\mathrm{d}S. \tag{4.3}$$

Accordingly, the surface energy of deformation and polarization per unit area (the surface tension) is

$$T = \tfrac{1}{2}[n_i b_{ij}^0 P_j]_S. \tag{4.4}$$

This energy is to be added to the bond energy, per unit area, to obtain the total energy per unit area required to separate the material into two parts along a surface S [7].

As an example, consider the case of a free (100) surface of a semi-infinite, centrosymmetric, cubic crystal. In this case, the resulting fields are one-dimensional and the equations of equilibrium, (3.4), for the half space $x_1 \geqslant 0$, reduce to

$$\begin{aligned} c_{11}u_{1,11} + d_{11}P_{1,11} &= 0, \\ d_{11}u_{1,11} + b_{11}P_{1,11} - aP_1 - \varphi_{,1} &= 0, \\ -\varepsilon_0\varphi_{,11} + P_{1,1} &= 0, \end{aligned} \tag{4.5}$$

while the boundary conditions, on $x_1 = 0$, become

$$\begin{aligned} c_{11}u_{1,1} + d_{11}P_{1,1} &= 0, \\ d_{11}u_{1,1} + b_{11}P_{1,1} &= -b_0, \\ -\varepsilon_0\varphi_{,1} + P_1 &= 0. \end{aligned} \tag{4.6}$$

Consider the functions

$$u_1 = A_1\,\mathrm{e}^{-x_1/l}, \qquad P_1 = A_2\,\mathrm{e}^{-x_1/l}, \qquad \varphi = A_3\,\mathrm{e}^{-x_1/l}. \tag{4.7}$$

Upon substituting (4.7) into the equilibrium equations (4.5), we find

$$A_3 = -lA_2/\varepsilon_0 = lc_{11}A_1/\varepsilon_0 d_{11} \tag{4.8}$$

$$l = \left(\frac{c_{11}b_{11} - d_{11}^2}{c_{11}(a + \varepsilon_0^{-1})}\right)^{\frac{1}{2}}. \tag{4.9}$$

Positive definiteness of W^{L} requires the radicand in (4.9) to be positive.

With (4.8), the first and third of the boundary conditions (4.6) are satisfied identically and the second yields

$$A_1 = -\frac{b_0 d_{11}}{c_{11}l(a + \varepsilon_0^{-1})}. \tag{4.10}$$

Whence, from (4.8),

$$A_2 = \frac{b_0}{l(a + \varepsilon_0^{-1})}, \qquad A_3 = -\frac{b_0}{a\varepsilon_0 + 1}. \tag{4.11}$$

Thus, there are a strain and a polarization, at the surface, which decay exponentially into the interior with decay constant l. The surface energy of deformation and polarization per unit area (the surface tension) is, from (4.4) and (4.11),

$$T = -\frac{b_0^2}{2l(a+\varepsilon_0^{-1})}. \tag{4.12}$$

REFERENCES

[1] R. A. TOUPIN, *J. rat. Mech Analysis* **5**, 849 (1956).
[2] K. B. TOLPYGO, *J. exp. theor. Phys.* (*USSR*) **20**, 497 (1950).
[3] W. COCHRAN, *Rep. Prog. Phys.* **26**, 1 (1963).
[4] B. J. DICK and A. W. OVERHAUSER, *Phys. Rev.* **112**, 90 (1958).
[5] A. E. H. LOVE, *The Mathematical Theory of Elasticity*, p. 166. Cambridge University Press (1927).
[6] W. P. MASON, *Crystal Physics of Interaction Processes*. Academic Press (1966).
[7] M. P. TOSI, *Solid St. Phys.* **16**, 92 (1964).

(*Received* 17 *October* 1967)

Абстракт—Учитывая включение градиента поляризаций в функцию сохранения энергии упругих диэлектриков расширается классическая теория пьезоэлектричества путем учета электро-механической реакции в осесимметрических /включая изотропные/ материалах и поверхностной энергии деформации и поляризации.

Reprinted from THE JOURNAL OF THE ACOUSTICAL SOCIETY OF AMERICA, Vol. 43, No. 6, 1329–1331, June 1968

Optimum Sizes and Shapes of Electrodes for Quartz Resonators

R. D. MINDLIN

Department of Civil Engineering, Columbia University, New York, New York 10027

In this paper, formulas are given for sizes and shapes of electrode coatings, on quartz-crystal resonator plates, satisfying the criterion for Bechmann's number in all directions.

INTRODUCTION

BECHMANN'S number for coupled thickness–shear and flexural vibrations is the ratio of the length L of electrode coating, to the thickness $2h$ of a resonator plate, below which there are no anharmonic overtones of thickness–shear vibration with frequencies between the thickness–shear cutoff frequency ω_e of the coated portion of the plate and the thickness–shear cutoff frequency ω_0 of the adjacent portion of the plate. For a rotated-Y-cut quartz plate, coated only with a pair of electrode strips symmetrically disposed on the two faces of the plate with the edges of the strips perpendicular to the digonal axis of elastic symmetry in the plane of the plate, Bechmann's number for the digonal direction is,[1] neglecting the effect of piezoelectric coupling,

$$\frac{L_{TS}}{2h}=\frac{\pi}{R^{\frac{1}{2}}}\left(\frac{1}{6}+\frac{2\gamma_{11}}{\pi^2 c_{66}}\right)^{\frac{1}{2}}, \tag{1}$$

where $R(=\omega_0/\omega_e-1\ll 1)$ is the ratio of the mass per unit area of both electrodes to the mass per unit area of the plate, c_{66} is the elasticity modulus in thickness-shear, and γ_{11} is Voigt's stretch modulus of the plate in the direction of the digonal axis. Bleustein and Tiersten[2] have extended this result to include: (1) the effect of piezoelectric coupling; (2) the effect of dielectric coatings on the portions of the plate adjacent to the electrodes (the "inverted mesa" of Curran and Koneval[3]); and (3) the effect of conducting coatings replacing the dielectric coatings.

If the electrode strips are parallel to the digonal axis and thickness–twist modes are excited, the result corresponding to Eq. 1 is Bechmann's number for the direction at right angles to the digonal axis[4]:

$$L_{TT}/2h=(2\gamma_{55}/Rc_{66})^{\frac{1}{2}}, \tag{2}$$

where γ_{55} is Voigt's face-shear modulus for the plate.

In the present paper, all of these results are extended to apply to an electrode patch, rather than an electrode strip. This is accomplished by taking into account the coupling of the thickness–shear and flexure modes with the thickness–twist mode. The new results are formulas for the sizes and shapes of electrodes that satisfy the criterion for Bechmann's number in all directions.

I. DISPERSION RELATION

The equations of motion to be used for the coupled thickness–shear, flexure, and thickness–twist vibrations are[5]

$$\begin{aligned}
&\kappa^2 c_{66}(u_{2,11}+\psi_{1,1})-\tfrac{1}{3}h^2\gamma_{55}(u_{2,1}-\psi_1)_{,133} \\
&\qquad\qquad +(1+R)\rho\omega^2 u_2=0, \\
&\gamma_{11}\psi_{1,11}-\gamma_{55}(u_{2,1}-\psi_1)_{,33}-3h^{-2}\kappa^2 c_{66}(u_{2,1}+\psi_1) \\
&\qquad\qquad +(1+3R)\rho\omega^2\psi_1=0,
\end{aligned} \tag{3}$$

where u_2 is the displacement normal to the plane of the plate; ψ_1 is the rotation of a line of particles, originally normal to the plane of the plate, about the axis x_3 normal to the digonal axis x_1; ρ is the density of the

[1] R. D. Mindlin and P. C. Y. Lee, Intern. J. Solids Struct. 2, 125–139 (1966).
[2] J. L. Bleustein and H. F. Tiersten, J. Acoust. Soc. Am. 43 1311–1318 (1968).
[3] D. R. Curran and D. J. Koneval, Proc. Ann. Symp. Freq. Control, 19th, U. S. Army Electronics Command, Fort Monmouth, New Jersey, 213–268 (1965).
[4] R. D. Mindlin, J. Acoust. Soc. Am. 41, 969–973 (1967).
[5] R. D. Mindlin and W. J. Spencer, J. Acoust. Soc. Am. 42, 1268–1277 (1967).

plate; c_{66} is the thickness-shear modulus at constant electric field; and κ is the shear correction factor including the contributions of both the mass of the electrodes and the piezoelectric effect, as given by Bleustein and Tiersten[2]:

$$\kappa^2=(1+R-8k_{26}^2/\pi^2)(\pi^2\bar{c}_{66}/12c_{66}). \quad (4)$$

In Eq. 4, k_{26} is the electromechanical coupling coefficient[6]

$$k_{26}^2=e_{26}^2/c_{66}\epsilon_{22},$$

where e_{26} and ϵ_{22} are components of the piezoelectric and dielectric constants, respectively. Also, in Eq. 4,

$$\bar{c}_{66}=(1+k_{26}^2)c_{66}.$$

The shear correction factor was chosen by Bleustein and Tiersten in such a way that Eqs. 3 and 4 give the cutoff frequency

$$\omega_e=(\pi/2h)(1-R-4k_{26}^2/\pi^2)(\bar{c}_{66}/\rho)^{\frac{1}{2}}, \quad (5)$$

which they obtained from the three-dimensional equations of piezoelectricity. The difference between $\bar{c}_{66}$ and c_{66} is only a few hundredths of 1%, so that it may be neglected in the application contemplated. However, as pointed out by Bleustein and Tiersten, R and k_{26}^2 can be of the same order of magnitude, and so can have comparable influences on ω_e.

For u_2 and ψ_1, we take

$$\begin{aligned} u_2&=Ah\exp[i(2\pi/\lambda)(x_1\cos\varphi+x_3\sin\varphi)],\\ \psi_1&=B\exp[i(2\pi/\lambda)(x_1\cos\varphi+x_3\sin\varphi)]; \end{aligned} \quad (6)$$

i.e., vibrations of wavelength λ and wave normal that makes an angle φ with the digonal axis x_1. Upon substitution of these expressions in Eqs. 3, there results a pair of homogeneous, linear equations for the amplitudes A and B, the determinant of the coefficients of which, when set equal to zero, yields the dispersion relation. This relation can be represented geometrically as a surface of two sheets in ω, h/λ, φ space. For real λ, the upper sheet (the sheet of higher frequency) gives the dispersion relation for the predominantly thickness-shear and thickness-twist mode (TS–TT mode). The asymptotic behavior of this sheet, for small h/λ, is given by

$$\omega=\omega_e\{1+(2\pi h/\lambda)^2[(\tfrac{1}{3}\alpha+\hat{\gamma}_{11})\cos^2\varphi+\hat{\gamma}_{55}\sin^2\varphi]\}^{\frac{1}{2}}, \quad (7)$$

where

$$\hat{\gamma}_{11}=\gamma_{11}/3\kappa^2c_{66},\quad \hat{\gamma}_{55}=\gamma_{55}/3\kappa^2c_{66},\quad \alpha=(1+3R)/(1+R).$$

Thus, the intersection of a plane

$$\omega/\omega_e=1+\epsilon,\quad 0<\epsilon=\text{constant}\ll 1,$$

with the upper sheet, is an elliptic curve with principal axes along $\varphi=0$ and $\varphi=\pi/2$.

II. SIZES AND SHAPES OF ELECTRODES

The critical length of electrode, below which there are no anharmonic overtones of TS–TT modes between ω_e and the cutoff frequency ω_0 of these modes in the portion of the plate outside the electrodes, is[1] the wavelength of a TS–TT wave in the electrode-coated portion of the plate at the frequency ω_0. The values of ω_0 for an uncoated plate, a plate with dielectric coatings of mass ratio R_d and a plate with conducting coatings of mass ratio R' are, respectively,

$$\omega_u=(\pi/2h)(\bar{c}_{66}/\rho)^{\frac{1}{2}}, \quad (8)$$

$$\omega_d=(\pi/2h)(1-R_d)(\bar{c}_{66}/\rho)^{\frac{1}{2}}, \quad (9)$$

$$\omega_e'=(\pi/2h)(1-R'-4k_{26}^2/\pi^2)(\bar{c}_{66}/\rho)^{\frac{1}{2}}, \quad (10)$$

as given by Bleustein and Tiersten.[2] Upon successively substituting these frequencies for ω, in Eq. 7, and setting $\lambda=L_u$, L_d, L_e, respectively, we obtain the following equations for the sizes and shapes of electrodes satisfying the criterion for Bechmann's number:

$$\frac{L_u}{2h}=\pi\left[\frac{(\tfrac{1}{3}\alpha+\hat{\gamma}_{11})\cos^2\varphi+\hat{\gamma}_{55}\sin^2\varphi}{2(R+4k_{26}^2/\pi^2)}\right]^{\frac{1}{2}}, \quad (11)$$

$$\frac{L_d}{2h}=\pi\left[\frac{(\tfrac{1}{3}\alpha+\hat{\gamma}_{11})\cos^2\varphi+\hat{\gamma}_{55}\sin^2\varphi}{2(R-R_d+k_{26}^2/\pi^2)}\right]^{\frac{1}{2}}, \quad (12)$$

$$\frac{L_e}{2h}=\pi\left[\frac{(\tfrac{1}{3}\alpha+\hat{\gamma}_{11})\cos^2\varphi+\hat{\gamma}_{55}\sin^2\varphi}{2(R-R')}\right]^{\frac{1}{2}}. \quad (13)$$

Thus, the shape of the electrode is an invert of an ellipse. The diameter along the digonal axis is obtained by setting $\varphi=0$, in which case Eqs. 11, 12, and 13 reduce to the formulas for Bechmann's number for thickness-shear and flexural vibrations obtained by Bleustein and Tiersten.[2] The diameter at right angles to the digonal axis is obtained by setting $\varphi=\pi/2$. Then, if the piezoelectric effect is neglected and the plate is uncoated outside the electrodes, the result is the Bechmann's number for thickness-twist waves given in Eq. 2.

The ratios of the principal diameters of the electrodes are the same for all three of Eqs. 11, 12, and 13:

$$\frac{L_{\varphi=0}}{L_{\varphi=\pi/2}}=\left[\frac{(\tfrac{1}{3}\alpha+\hat{\gamma}_{11})}{\hat{\gamma}_{55}}\right]^{\frac{1}{2}}$$

$$\approx\left[\frac{\pi^2/12+\gamma_{11}/c_{66}}{\gamma_{55}/c_{66}}\right]^{\frac{1}{2}}=\frac{L_{TS}}{L_{TT}}; \quad (14)$$

[6] W. P. Mason, *Piezoelectric Crystals and their Application to Ultrasonics* (D. Van Nostrand Co., Inc., New York, 1950).

i.e., it is almost independent of the piezoelectric effect and the thickness of the electrode. For the AT cut of quartz,

$$c_{66}=29.01, \quad \gamma_{11}=85.93, \quad \gamma_{55}=68.59,$$

in units of 10^{10} dyn/cm^2, as calculated from Bechmann's constants.[7] Then Eq. 14 yields the result that the major diameter of the electrode is the one along the digonal direction and it exceeds the minor diameter by about 25%. The shape of the electrode differs little from that of an ellipse with the same ratio of principal diameters.

[7] R. Bechmann, Phys. Rev. **100**, 1060–1061 (1958).

Reprinted from *Mechanics of the Solid State*
F.P.J. Rimrott and J. Schwaighofer (eds.)
University of Toronto Press, (1968) 143-145

Face-Shear Waves in Rotated Y-Cut Quartz Plates

R. D. MINDLIN

IN THE LINEAR THEORY of elasticity, the propagation of waves is governed by a system of fifteen equations:

(*a*) *Stress equations of motion:*

$$\begin{aligned}
\frac{\partial \sigma_x}{\partial x} + \frac{\partial \tau_{xy}}{\partial y} + \frac{\partial \tau_{xz}}{\partial z} &= \rho \frac{\partial^2 u}{\partial t^2}, \\
\frac{\partial \tau_{xy}}{\partial x} + \frac{\partial \sigma_y}{\partial y} + \frac{\partial \tau_{yz}}{\partial z} &= \rho \frac{\partial^2 v}{\partial t^2}, \\
\frac{\partial \tau_{xz}}{\partial x} + \frac{\partial \tau_{yz}}{\partial y} + \frac{\partial \sigma_z}{\partial z} &= \rho \frac{\partial^2 w}{\partial t^2}.
\end{aligned} \tag{1}$$

(*b*) *Stress–strain equations:*

$$\begin{aligned}
\sigma_x &= c_{11}\varepsilon_x + c_{12}\varepsilon_y + c_{13}\varepsilon_z + c_{14}\gamma_{yz} + c_{15}\gamma_{xz} + c_{16}\gamma_{xy}, \\
\sigma_y &= c_{12}\varepsilon_x + c_{22}\varepsilon_y + c_{23}\varepsilon_z + c_{24}\gamma_{yz} + c_{25}\gamma_{xz} + c_{26}\gamma_{xy}, \\
\sigma_z &= c_{13}\varepsilon_x + c_{23}\varepsilon_y + c_{33}\varepsilon_z + c_{34}\gamma_{yz} + c_{35}\gamma_{xz} + c_{36}\gamma_{xy}, \\
\tau_{yz} &= c_{14}\varepsilon_x + c_{24}\varepsilon_y + c_{34}\varepsilon_z + c_{44}\gamma_{yz} + c_{45}\gamma_{xz} + c_{46}\gamma_{xy}, \\
\tau_{xz} &= c_{15}\varepsilon_x + c_{25}\varepsilon_y + c_{35}\varepsilon_z + c_{45}\gamma_{yz} + c_{55}\gamma_{xz} + c_{56}\gamma_{xy}, \\
\tau_{xy} &= c_{16}\varepsilon_x + c_{26}\varepsilon_y + c_{36}\varepsilon_z + c_{46}\gamma_{yz} + c_{56}\gamma_{xz} + c_{66}\gamma_{xy}.
\end{aligned} \tag{2}$$

(*c*) *Strain-displacement equations:*

$$\begin{gathered}
\varepsilon_x = \frac{\partial u}{\partial x}, \qquad \varepsilon_y = \frac{\partial v}{\partial y}, \qquad \varepsilon_z = \frac{\partial w}{\partial z}, \\
\gamma_{xz} = \frac{\partial w}{\partial y} + \frac{\partial v}{\partial z}, \qquad \gamma_{xz} = \frac{\partial u}{\partial z} + \frac{\partial w}{\partial x}, \qquad \gamma_{xy} = \frac{\partial v}{\partial x} + \frac{\partial u}{\partial y}.
\end{gathered} \tag{3}$$

FACE-SHEAR WAVE IN AN ISOTROPIC PLATE

If the material is isotropic, eqs. (2) reduce to

$$
\begin{aligned}
\sigma_x &= c_{11}\varepsilon_x + c_{12}\varepsilon_y + c_{12}\varepsilon_z, \\
\sigma_y &= c_{12}\varepsilon_x + c_{11}\varepsilon_y + c_{12}\varepsilon_z, \\
\sigma_z &= c_{12}\varepsilon_x + c_{12}\varepsilon_y + c_{11}\varepsilon_z, \\
\tau_{yz} &= c_{55}\gamma_{yz}, \\
\tau_{xz} &= c_{55}\gamma_{xz}, \\
\tau_{xy} &= c_{55}\gamma_{xy},
\end{aligned}
\tag{4}
$$

where

$$c_{55} = \tfrac{1}{2}(c_{11} - c_{12}).$$

Equations (1), (3), and (4) admit simple solutions. In particular, the shear wave

$$u = A \sin \zeta(z - ct), \qquad v = 0, \qquad w = 0, \tag{5}$$

with amplitude A, wave number ζ, and velocity c, produces strains, from (3):

$$
\begin{aligned}
\gamma_{xz} &= A\zeta \cos \zeta(z - ct), \\
\varepsilon_x &= \varepsilon_y = \varepsilon_z = y_{yz} = \gamma_{xy} = 0,
\end{aligned}
\tag{6}
$$

and stresses, from (2):

$$
\begin{aligned}
\tau_{xz} &= Ac_{55}\zeta \cos \zeta(z - ct), \\
\sigma_x &= \sigma_y = \sigma_z = \tau_{yz} = \tau_{xy} = 0.
\end{aligned}
\tag{7}
$$

The stresses (7) and the displacements (5) satisfy the stress equations of motion (1) if

$$c = \sqrt{(c_{55}/\rho)}. \tag{8}$$

In this classical solution, all planes $y =$ constant are traction-free because σ_y, τ_{xy}, and τ_{yz} are identically zero. The solution thus applies to an infinite, isotropic plate bounded by traction-free faces $y = \pm h$ where $2h$ can be any thickness. Because the displacement is everywhere parallel to the faces of the plate, the wave is known as a face-shear wave.

FACE-SHEAR WAVE IN A ROTATED Y-CUT QUARTZ PLATE

At the present time, many millions of quartz crystal plates are manufactured each year for use in electrical oscillator and filter circuits. Most of the plates are "rotated

Y-cuts" for which the stress–strain relations (2), referred to axes in and normal to the middle plane of the plate, reduce to

$$\begin{aligned}
\sigma_x &= c_{11}\varepsilon_x + c_{12}\varepsilon_y + c_{13}\varepsilon_z + c_{14}\gamma_{yz},\\
\sigma_y &= c_{12}\varepsilon_x + c_{22}\varepsilon_y + c_{23}\varepsilon_z + c_{24}\gamma_{yz},\\
\sigma_z &= c_{13}\varepsilon_x + c_{23}\varepsilon_y + c_{33}\varepsilon_z + c_{34}\gamma_{yz},\\
\tau_{yz} &= c_{14}\varepsilon_x + c_{24}\varepsilon_y + c_{34}\varepsilon_z + c_{44}\gamma_{yz},\\
\tau_{xz} &= c_{55}\gamma_{xz} + c_{56}\gamma_{xy},\\
\tau_{xy} &= c_{56}\gamma_{xz} + c_{66}\gamma_{xy}.
\end{aligned} \tag{9}$$

It is of interest to find a solution for the quartz plate analogous to the classical one for face-shear waves in an isotropic plate. The displacement field (5), with velocity (8), produces stresses, through (3) and (9), which satisfy the stress equations of motion; but planes $y = \text{constant}$ are not traction-free. The shear stress τ_{xy} is not zero because of the contribution from $c_{56}\gamma_{xz}$ in (9).

Instead of (5), consider the displacements

$$u = A \sin \zeta\left(z - \frac{c_{56}}{c_{66}}\, y - ct\right), \qquad v = 0, \qquad w = 0. \tag{10}$$

These displacements produce strains, from (3):

$$\begin{aligned}
\gamma_{xz} &= A\zeta \cos \zeta\left(z - \frac{c_{56}}{c_{66}}\, y - ct\right),\\
\gamma_{xy} &= -A\zeta \frac{c_{56}}{c_{66}} \cos \zeta\left(z - \frac{c_{56}}{c_{66}}\, y - ct\right),\\
\varepsilon_x &= \varepsilon_y = \varepsilon_z = \gamma_{yz} = 0,
\end{aligned} \tag{11}$$

and stresses, from (9):

$$\begin{aligned}
\tau_{xz} &= A\zeta\left(\frac{c_{55}c_{66} - c_{56}^2}{c_{66}}\right)\cos \zeta\left(z - \frac{c_{56}}{c_{66}}\, y - ct\right),\\
\sigma_x &= \sigma_y = \sigma_z = \tau_{yz} = \tau_{xy} = 0.
\end{aligned} \tag{12}$$

Furthermore, the stress equations of motion are satisfied by (12) and (10) if

$$c = \sqrt{\left(\frac{c_{55}c_{66} - c_{56}^2}{\rho c_{66}}\right)}. \tag{13}$$

Thus, we have a solution, (10), (11), (12), (13), which, again, has τ_{xz} as the only non-zero component of stress. Accordingly, the solution is that for a face-shear wave in a rotated Y-cut quartz plate with traction-free faces. A special property of the solution is that, although the direction of propagation of the energy is parallel to the axis of z, the wave normal makes an angle, whose tangent is c_{56}/c_{66}, with the axis of z.

Theories of Elastic Continua and Crystal Lattice Theories[1]

By

R. D. Mindlin

Department of Civil Engineering
Columbia University, New York

Introduction

This paper is concerned with the relations between discrete and continuum theories of the elastic behavior of perfect crystals. The point of view adopted is that the validity of any extension of the classical theory of elasticity, intended to accommodate effects of the atomic structure of crystalline solids, can be tested by comparison with an appropriate lattice theory. The particular test to be applied is how well and to how short wave lengths the dispersion relation for plane waves, deduced from the continuum theory, reproduces that for the lattice.

In the case of a simple Bravais lattice of mass particles, the difference equations and dispersion relation reduce, in the long wave limit, to the differential equations and dispersionless velocities of classical elasticity. Extensions of the continuum theory to accommodate short wave lengths and, consequently, effects of the atomic structure of solids, began with Cauchy in 1851 [1]. Although Cauchy did not carry his work very far, it is now known that his theory, if completed, would correspond to an augmentation of classical linear elasticity through the inclusion of all the gradients of strain, in addition to the strain, in the potential energy density. Interest in this type of "gradient theory" was stimulated, in 1960, by Aero and Kuvshinskii [2], Grioli [3], Rajagopal [4] and Truesdell and Toupin [5] who took into account the first gradient of the rotation, i.e. eight of the eighteen components of the first gradient of the strain. This is what Toupin [6] has termed the "Cosserat theory with constrained rotations". Further extension to the complete first gradient was accomplished by Toupin [7], to the second gradient by the writer [8] and to gradients of all orders by Green and Rivlin [9].

[1] Paper prepared for IUTAM "Symposium on the Generalized Cosserat Continuum ...", August, 1967.

In 1963, KRUMHANSL [10] showed, by expansion of the differences in the expression for the potential energy of a general Bravais lattice, that the second order terms, in the resulting series of derivatives, contain the rotation gradient and, hence, the essence of the Cosserat theory with constrained rotations. In the first part of the present paper, it is shown in what way successively higher order gradient theories correspond to successively shorter-wave approximations to the equations of a simple Bravais lattice and how the additional material constants, which appear in the gradient theories, are related to the force constants of the lattice model. It is also shown why the requirement of positive definiteness of the potential energy density should not be applied to the strain-gradient terms when the theory is viewed as an approximation to a lattice model. The example chosen for illustration is a GAZIS-HERMAN-WALLIS [11] monatomic simple cubic lattice of mass particles.

In the case of a lattice with a basis, the dispersion relation exhibits optical branches in addition to the acoustic branches. It has been conjectured [12] that the differential equations of a Cosserat continuum [13] with a deformable Cosserat *trièdre* (micro-structure) may be a long wave to moderately long wave approximation to the difference equations of a lattice with a basis—inasmuch as such a Cosserat-type theory does exhibit both transverse and longitudinal optical branches in the dispersion relation. However, it is shown, in the second part of this paper, that the long wave limit of the difference equations of a lattice with a basis corresponds to an elastic continuum of a different type. It is also indicated how to construct higher order continuum approximations, to such lattice theories, accommodating shorter wave lengths. The lattice chosen for illustration is a GAZIS-WALLIS [14] NaCl-type lattice.

In the final part of the paper, it is shown how the difference equations of motion of lattices may be converted to differential equations which yield the same dispersion relations, as do the difference equations, for all wave lengths.

To simplify the exposition and to cope with space limitation, the lattices chosen to illustrate the methods presented are simple ones which do not correspond to real crystals. However, the same methods have been applied to face centered and body centered cubic lattices and to the NaCl-type lattice with polarizable atoms for which there are dispersion data from neutron diffraction measurements. These results will be reported elsewhere.

Simple Bravais Lattice

For one particle of mass M at each point $(l\,h, m\,h, n\,h)$ of a cubic lattice having central force interactions between nearest and next nearest neighbor particles with force constants α and β, respectively, and angular interactions, with force constant γ, between three consecutive non-collinear nearest neighbors, GAZIS, HERMAN and WALLIS [11] give three difference equations of motion of the type

$$M\,\ddot{u}_1^{l,m,n} = \alpha\Big(\sum_{\lambda} u_1^{l+\lambda,m,n} - 2u_1^{l,m,n}\Big) + \beta\sum_{\lambda\mu}\big(u_1^{l+\lambda,m+\mu,n} + u_1^{l+\lambda,m,n+\mu} - 8u_1^{l,m,n}\big) + (\beta+\gamma)\sum_{\lambda\mu}\lambda\,\mu\,\big(u_2^{l+\lambda,m+\mu,n} + u_3^{l+\lambda,m,n+\mu}\big) + 4\gamma\sum_{\lambda}\big(u_1^{l,m+\lambda,n} + u_1^{l,m,n+\lambda} - 4u_1^{l,m,n}\big), \quad (1)$$

where $\lambda,\mu = \pm 1$ and the $u_i^{l,m,n}$ are the three rectangular components of displacement of the particle at $(l\,h, m\,h, n\,h)$. Displacements

$$u_j^{l,m,n} = A_j \exp[i(l\,\theta_1 + m\,\theta_2 + n\,\theta_3 - \omega\,t)], \quad (2)$$

when substituted in (1), yield the dispersion relation

$$|d_{ij}| = 0, \quad (i,j = 1,2,3) \quad (3)$$

where

$$d_{ij} = M\,\omega^2 - 2\alpha(1-\cos\theta_i) - 4\beta\Big(2 - \cos\theta_i\sum_{k\neq i}\cos\theta_k\Big) - 8\gamma\Big(2 - \sum_{k\neq i}\cos\theta_k\Big) \quad i=j,$$

$$d_{ij} = -(\beta+\gamma)\cos\theta_i\cos\theta_j \quad (i\neq j).$$

For the purpose, it is convenient to rewrite (1) in terms of difference operators defined as follows:

$$\left.\begin{aligned}
\Delta_1^+ u_i^{l,m,n} &\equiv h^{-1}(u_i^{l+1,m,n} - u_i^{l,m,n}),\\
\Delta_1^- u_i^{l,m,n} &\equiv h^{-1}(u_i^{l,m,n} - u_i^{l-1,m,n}),\\
\Delta_1^2 u_i^{l,m,n} &\equiv \Delta_1^+\Delta_1^- u_i^{l,m,n} = h^{-2}\sum_{\lambda}(u_i^{l+\lambda,m,n} - 2u_i^{l,m,n}),\\
\Delta_1\Delta_2 u_i^{l,m,n} &\equiv \tfrac{1}{4}(\Delta_1^+\Delta_2^+ + \Delta_1^-\Delta_2^- + \Delta_1^+\Delta_2^- + \Delta_1^-\Delta_2^+)\,u_i^{l,m,n},
\end{aligned}\right\} \quad (4)$$

and analogous definitions, e.g. $\Delta_2^+ u_i^{l,m,n} \equiv h^{-1}(u_i^{l,m+1,n} - u_i^{l,m,n})$, for forward, backward, second central and cross differences in the remaining coordinate directions.

Then, noting that

$$\left.\begin{aligned}
\sum_{\lambda\mu} u_i^{l+\lambda,m+\mu,n} - 4u_i^{l,m,n} &= h^2(h^2\Delta_1^2\Delta_2^2 + 2\Delta_1^2 + 2\Delta_2^2)\,u_i^{l,m,n},\\
\sum_{\lambda\mu}\lambda\,\mu\,u_i^{l+\lambda,m+\mu,n} &= 4h^2\Delta_1\Delta_2 u_i^{l,m,n},
\end{aligned}\right\} \quad (5)$$

we can write (1) in the form

$$\varrho\,\ddot{u}_1^{l,m,n} = [c_{11}\Delta_1^2 + c_{44}(\Delta_2^2+\Delta_3^2) + \tfrac{1}{2}h^2 c_{12}\Delta_1^2(\Delta_2^2+\Delta_3^2)]\,u_1^{l,m,n} + (c_{12}+c_{44})(\Delta_1\Delta_2 u_2^{l,m,n} + \Delta_1\Delta_3 u_3^{l,m,n}), \quad (6)$$

where

$$\varrho = M/h^3, \quad h\, c_{11} = \alpha + 4\beta, \quad h\, c_{12} = 2\beta, \quad h\, c_{44} = 2(\beta + 2\gamma). \quad (7)$$

Now, expand the difference operations, on the $u_i^{l,m,n}$, in Taylor series of partial derivatives, ∂_i, of continuous displacement functions $u_i(x_j)$:

$$\left.\begin{aligned} \Delta_i^2 u_j^{l,m,n} &= \left(1 + \frac{1}{12} h^2 \partial_i^2 + \cdots\right) \partial_i^2 u_j, \\ \Delta_i \Delta_j u_k^{l,m,n} &= \left(1 + \tfrac{1}{6} h^2 \partial_i^2 + \tfrac{1}{6} h^2 \partial_j^2 + \cdots\right) \partial_i \partial_j u_k, \\ \Delta_i^2 \Delta_j^2 u_k^{l,m,n} &= \left(1 + \frac{1}{12} h^2 \partial_i^2 + \frac{1}{12} h^2 \partial_j^2 + \cdots\right) \partial_i^2 \partial_j^2 u_k. \end{aligned}\right\} \quad (8)$$

If only second derivatives are retained, (6) reduces to

$$\varrho\, \ddot{u}_1 = [c_{11} \partial_1^2 + c_{44}(\partial_2^2 + \partial_3^2)]\, u_1 + (c_{12} + c_{44})\, (\partial_1 \partial_2 u_2 + \partial_1 \partial_3 u_3), \quad (9)$$

which is the equation of motion of classical elasticity for materials with the constants c_{11}, c_{12}, c_{44} of cubic symmetry. Retaining up to fourth derivatives we find

$$\begin{aligned} \varrho\, \ddot{u}_1 = {} & [c_{11} \partial_1^2 + c_{44}(\partial_2^2 + \partial_3^2)]\, u_1 + (c_{12} + c_{44})\, (\partial_1 \partial_2 u_2 + \partial_1 \partial_3 u_3) + \\ & + \frac{1}{12} h^2 [c_{11} \partial_1^4 + c_{44}(\partial_2^4 + \partial_3^4) + 6 c_{12} \partial_1^2 (\partial_2^2 + \partial_3^2)]\, u_1 + \\ & + \tfrac{1}{6} h^2 (c_{12} + c_{44})\, [(\partial_1^3 \partial_2 + \partial_1 \partial_2^3)\, u_2 + (\partial_1^3 \partial_3 + \partial_1 \partial_3^3)\, u_3], \end{aligned} \quad (10)$$

which has the fourth derivatives characteristic of the equations of motion of the first-strain-gradient theory. In general, the potential energy density of the first-gradient theory for the crystal class $m3m$ [15], to which the simple cubic lattice belongs, has the form

$$W = \tfrac{1}{2} c_{ijkl}\, \varepsilon_{ij}\, \varepsilon_{kl} + \tfrac{1}{2} c_{ijkpqr}\, \varepsilon_{ijk}\, \varepsilon_{pqr}, \quad (11)$$

where

$$\varepsilon_{ij} = \tfrac{1}{2}(\partial_i u_j + \partial_j u_i), \qquad \varepsilon_{ijk} = \partial_i \partial_j u_k, \quad (12)$$

$$c_{ijkl} = (c_{11} - c_{12} - 2c_{44})\, \delta_{ijkl} + c_{12}\, \delta_{ij}\, \delta_{kl} + c_{44}(\delta_{ik}\, \delta_{jl} + \delta_{il}\, \delta_{jk}), \quad (13)$$

$$\begin{aligned} c_{ijkpqr} = {} & a_1\, \delta_{jk}\, \delta_{ip}\, \delta_{qr} + a_2\, \delta_{ij}\, \delta_{kp}\, \delta_{qr} + a_3\, \delta_{kr}\, \delta_{ij}\, \delta_{pq} + \\ & + a_4\, \delta_{ip}\, \delta_{jq}\, \delta_{kr} + a_5\, \delta_{kp}\, \delta_{ir}\, \delta_{jq} + a_1'\, \delta_{jk}\, \delta_{ipqr} + \\ & + a_2'\, \delta_{ij}\, \delta_{kpqr} + a_3'\, \delta_{kr}\, \delta_{ijpq} + a_4'\, \delta_{ip}\, \delta_{jqkr} + \\ & + a_5'\, \delta_{kp}\, \delta_{irjq} + a''\, \delta_{ijkpqr}, \end{aligned} \quad (14)$$

in which the $\delta_{i\ldots}$ are unity if all indices are alike and zero otherwise. The $a_1, a_2 \ldots, a_1', a_2' \ldots$ and a'' are eleven additional material constants. If

$$\left.\begin{aligned} a'' = -\frac{1}{12} h^2 (c_{11} - 10 c_{12} - 5 c_{44}) &= -\frac{1}{12} h (\alpha - 26\beta - 20\gamma), \\ a_1' = a_5' = -\tfrac{1}{6} h^2 (c_{12} + c_{44}) &= -\tfrac{2}{3} h (\beta + \gamma), \\ a_3' = -\frac{1}{12} h^2 c_{44} = -\frac{1}{6} h (\beta + 2\gamma), \quad & a_4' = -\frac{1}{2} h^2 c_{12} = -h\beta, \\ a_2' = a_1 = a_2 = a_3 &= a_4 = a_5 = 0, \end{aligned}\right\} \quad (15)$$

then (10) is the Euler equation of TOUPIN's ([7], § 7) variational principle for the first-gradient theory. It should be observed that (15) are not unique as there are combinations of terms in (14) which do not contribute to the equations of motion.

Whereas the equations of motion (9), of classical elasticity, give ω^2, in the dispersion relation (7), to the order θ_i^2, the first-gradient Eq. (10) gives ω^2 to the order θ_i^4, i.e. to shorter wave lengths. However, the latter is true only if the gradient terms in the energy (11) are not required to conform to positive definiteness of W. Consider, for example, a longitudinal wave in the [100] direction. The lattice equation, (1) or (6), and the first-gradient approximation (10) give

$$\omega^2 = \frac{2c_{11}}{\varrho h^2}(1-\cos\theta_1) = \frac{2c_{11}}{\varrho h^2}\left(\frac{\theta_1^2}{2!} - \frac{\theta_1^4}{4!} + \frac{\theta_1^6}{6!} - \cdots\right), \tag{16}$$

$$\omega^2 = \frac{2c_{11}}{\varrho h^2}\left(\frac{\theta_1^2}{2!} - \frac{\theta_1^4}{4!}\right), \tag{17}$$

respectively, for the dispersion relation. Complete positive definiteness of W would require the sign of the θ_i^4 term in (17) to be positive so as to prevent instability, i.e. imaginary ω. However, with the negative sign, instability occurs only at wave numbers far beyond the range in which (17) approximates (16). Thus, if the strain-gradient equations are regarded as approximations to the lattice equations, the requirement of positive definiteness need not be applied to the gradient terms in the potential energy as long as solutions are restricted to the range of wave lengths in which the approximation is valid. The requirement of positive definiteness must, of course, be applied to the strain terms in the energy—with the usual result

$$c_{11} > 0, \quad c_{11} - c_{12} > 0, \quad c_{11} + 2c_{12} > 0, \quad c_{44} > 0.$$

The same method as that employed for the first-gradient theory can be applied to obtain shorter-wave approximations by retaining sixth, eighth, etc. derivatives in the expansions of the difference operators, resulting in second-, third-, etc. gradient approximations.

Lattice with a Basis

We now consider a lattice of NaCl-type with particles at points (lh, mh, nh)—of mass $\overset{1}{M}$ for $l+m+n$ even and mass $\overset{2}{M}$ for $l+m+n$ odd. The interactions are taken to be the same as for the simple cubic lattice considered in the preceding section. Next nearest neighbor force constants, $\overset{1}{\beta}$ and $\overset{2}{\beta}$ are assumed to be different for particles with mass $\overset{1}{M}$ and $\overset{2}{M}$. The angular force constant, γ, is the same whether a particle of mass $\overset{1}{M}$ or $\overset{2}{M}$ is at the apex. Then GAZIS and WALLIS [14] find, for the particles at $l + m + n$ even, three equations of motion of the type (1) with M and β replaced by $\overset{1}{M}$ and $\overset{2}{\beta}$; and, for particles at $l + m + n$ odd, three equations of motion of type (1) with M and β replaced by $\overset{2}{M}$ and $\overset{2}{\beta}$. When the displacements

$$u_j^{l,m,n} = \overset{\nu}{A}_j \exp[i(l\,\theta_1 + m\,\theta_2 + n\,\theta_3 - \omega\,t)],$$

with $\nu = 1, 2$ for $l + m + n$ even and odd, respectively, are substituted in the six difference equations of motion, there results the dispersion relation

$$\begin{vmatrix} \overset{1}{d}_1 & \overset{1}{d}_{12} & \overset{1}{d}_{13} & d_1 & 0 & 0 \\ \overset{1}{d}_{21} & \overset{1}{d}_2 & \overset{1}{d}_{23} & 0 & d_2 & 0 \\ \overset{1}{d}_{31} & \overset{1}{d}_{32} & \overset{1}{d}_3 & 0 & 0 & d_3 \\ d_1 & 0 & 0 & \overset{2}{d}_1 & \overset{2}{d}_{12} & \overset{2}{d}_{13} \\ 0 & d_2 & 0 & \overset{2}{d}_{21} & \overset{2}{d}_2 & \overset{2}{d}_{23} \\ 0 & 0 & d_3 & \overset{2}{d}_{31} & \overset{2}{d}_{32} & \overset{2}{d}_3 \end{vmatrix} = 0, \tag{18}$$

where

$$\overset{\nu}{d}_i = \overset{\nu}{M}\,\omega^2 - 2(\alpha + 8\gamma) - 4\overset{\nu}{\beta}(2 - \cos\theta_i \sum_{k \neq i} \cos\theta_k),$$

$$\overset{\nu}{d}_{ij} = -4(\overset{\nu}{\beta} + \gamma)\sin\theta_i \sin\theta_j, \qquad d_i = 2\alpha\cos\theta_i + 8\gamma \sum_{k \neq i} \cos\theta_k .$$

If we designate $\overset{1}{u}_i$ and $\overset{2}{u}_i$ as the displacements of particles of mass $\overset{1}{M}$ and $\overset{2}{M}$, respectively, and write

$$\overset{\nu}{u}{}_j^{l,m,n} = \overset{\nu}{A}_j \exp[i(l\,\theta_1 + m\,\theta_2 + n\,\theta_3 - \omega\,t)] \tag{19}$$

for all l, m, n, the same dispersion relation as (18) results from three equations of the type

$$\begin{aligned} \overset{1}{M}\,\overset{1}{u}{}_{1,tt}^{l,m,n} = {} & 2(\alpha + 8\gamma)\left(\overset{2}{u}{}_1^{l,m,n} - \overset{1}{u}{}_1^{l,m,n}\right) + 2h^2\,\overset{1}{\beta}(2\Delta_1^2 + \Delta_2^2 + \Delta_3^2)\,\overset{1}{u}{}_1^{l,m,n} + \\ & + 4(\overset{1}{\beta} + \gamma)\,h^2(\Delta_1\Delta_2\,\overset{1}{u}{}_2^{l,m,n} + \Delta_1\Delta_3\,\overset{1}{u}{}_3^{l,m,n}) + \\ & + h^4\,\overset{1}{\beta}(\Delta_1^2\Delta_2^2 + \Delta_1^2\Delta_3^2)\,\overset{1}{u}{}_1^{l,m,n} + \\ & + h^2\left[\alpha\,\Delta_1^2\,\overset{2}{u}{}_1^{l,m,n} + 4\gamma(\Delta_2^2 + \Delta_3^2)\,\overset{2}{u}{}_1^{l,m,n}\right] \end{aligned} \tag{20}$$

and three more obtained by interchange of superscripts 1 and 2. This form is more convenient for passing to continuum approximations since the necessity for distinguishing between $l + m + n$ even and odd is dispensed with.

At long wave lengths and low frequency, $\overset{1}{u}_i = \overset{2}{u}_i$ ($= u_i$, say). Then, employing the expansions (8), retaining only second derivatives and adding corresponding members of the two sets of three equations of motion, we recover the equations of classical elasticity (9) with $\varrho = (\overset{1}{M} + \overset{2}{M})/2h^3$ and stiffness constants given by (7) with $\beta = \frac{1}{2}(\overset{1}{\beta} + \overset{2}{\beta})$.

At long wave lengths but not necessarily low frequency, (20) becomes, (with $\overset{1}{\varrho} = \overset{1}{M}/2h^3$)

$$h\,\overset{1}{\varrho}\,\overset{1}{u}_{1,tt} = h^{-2}(\alpha + 8\gamma)\left(\overset{2}{u}_1 - \overset{1}{u}_1\right) + \overset{1}{\beta}(2\partial_1^2 + \partial_2^2 + \partial_3^2)\,\overset{1}{u}_1 + \\ + 2\left(\overset{1}{\beta} + \gamma\right)\left(\partial_1\,\partial_2\,\overset{1}{u}_2 + \partial_1\,\partial_3\,\overset{1}{u}_3\right) + \\ + \tfrac{1}{2}\alpha\,\partial_1^2\,\overset{2}{u}_1 + 2\gamma(\partial_1^2 + \partial_2^2)\,\overset{2}{u}_1. \tag{21}$$

Thus, for the long wave approximation, there are three equations of the type (21) and three more obtained by interchange of superscripts 1 and 2. The six equations yield the long wave region of the dispersion relation (18) for both the acoustic and optical branches. The equations do not have the same form as those for a Cosserat continuum with a deformable *trièdre* [12]. However, the appropriate continuum theory may be constructed without difficulty. We consider two interpenetrating continua representing the two face centered cubic sub-lattices of the NaCl structure. The potential energy density is taken as a quadratic function of the strains of the two continua and their relative displacement and rotation. For crystal class $m3m$, this is (with $\varkappa = 1, 2$; $\lambda = 1, 2$)

$$W = c_0^*\left(\overset{2}{\varepsilon}_{ii} - \overset{1}{\varepsilon}_{ii}\right) + \tfrac{1}{2}a^*\,u_i^*\,u_i^* + c^*\,\omega_{ij}^*\,\omega_{ij}^* + \tfrac{1}{2}\sum_{\varkappa\lambda} c_{ijkl}^{\varkappa\lambda}\,\overset{\varkappa}{\varepsilon}_{ij}\,\overset{\lambda}{\varepsilon}_{kl}, \tag{22}$$

where

$$u_i^* \equiv \left(\overset{2}{u}_i - \overset{1}{u}_i\right), \qquad \overset{\varkappa}{\varepsilon}_{ij} \equiv \tfrac{1}{2}\left(\partial_i\,\overset{\varkappa}{u}_j + \partial_j\,\overset{\varkappa}{u}_i\right), \qquad \omega_{ij}^* \equiv \tfrac{1}{2}(\partial_i\,u_j^* - \partial_j\,u_i^*),$$

$$c_{ijkl}^{\varkappa\lambda} = c_{ijkl}^{\lambda\varkappa} = (c_{11}^{\varkappa\lambda} - c_{12}^{\varkappa\lambda} - 2c_{44}^{\varkappa\lambda})\,\delta_{ijkl} + c_{12}^{\varkappa\lambda}\,\delta_{ij}\,\delta_{kl} + \\ + c_{44}^{\varkappa\lambda}(\delta_{ik}\,\delta_{jl} + \delta_{il}\,\delta_{jk}).$$

With kinetic energy density $\frac{1}{2}\sum_{\varkappa}\overset{\varkappa}{\varrho}\,\overset{\varkappa}{u}_{i,t}\,\overset{\varkappa}{u}_{i,t}$ and the potential energy density (22), Hamilton's variational principle, for independent variations $\delta \overset{1}{u}_i$ and $\delta \overset{2}{u}_\iota$, yields the six Euler equations

$$\begin{aligned}\overset{\varkappa}{\varrho}\,\overset{\varkappa}{u}_{j,tt} = &-(-1)^{\varkappa}\big[a^*(\overset{2}{u}_j - \overset{1}{u}_j) + c^* \sum_{\lambda} \partial_i(\partial_i \overset{\lambda}{u}_j - \partial_j \overset{\lambda}{u}_i)\big] + \\ &+ \sum_{\lambda}\big[(c_{11}^{\varkappa\lambda} - c_{12}^{\varkappa\lambda} - 2c_{44}^{\varkappa\lambda})\,\delta_{ijkl}\,\partial_i\,\partial_k\,\overset{\lambda}{u}_l + \\ &+ c_{12}^{\varkappa\lambda}\,\partial_i\,\partial_j\,\overset{\lambda}{u}_i + c_{44}^{\varkappa\lambda}\,\partial_i(\partial_i \overset{\lambda}{u}_j + \partial_j \overset{\lambda}{u}_i)\big].\end{aligned} \tag{23}$$

These become the six equations of the type (21) if

$$\begin{aligned}&h\,c_{11}^{\varkappa\varkappa} = 2h\,c_{44}^{\varkappa\varkappa} = 2\overset{\varkappa}{\beta}, \qquad h\,c_{12}^{\varkappa\varkappa} = \overset{\varkappa}{\beta} + 2\gamma, \qquad c^* = 0, \\ &h\,c_{44}^{12} = -h\,c_{12}^{12} = 2\gamma, \qquad h\,c_{11}^{12} = \tfrac{1}{2}\,\alpha, \qquad h^3\,a^* = \alpha + 8\gamma.\end{aligned} \tag{24}$$

It may be noted that the linear term in (22) is the energy density of a self equilibrated initial stress which produces a localized strain and a surface energy of deformation at a free surface.

The next higher order approximation is constructed in the same way as in the preceding section. First, the fourth order derivatives are retained in the series expansions of the difference operators in the six equations of the type (20). The resulting differential equations are then identified as the Euler equations of Hamilton's variational principle with the potential energy density taken as a quadratic function of the relative displacement and rotation of the two continua and of their strains and first strain-gradients.

Exact, Continuum Form of Lattice Equations

For exponential functions

$$u_j = A_j \exp[i(\theta_1 x_1 + \theta_2 x_2 + \theta_3 x_3)/h], \tag{25}$$

the infinite series expansions of difference operators in terms of derivatives can be summed. Thus:

$$\Delta_j^{\pm}\, u_k^{l,m,n} = \left(1 \pm \frac{h\,\partial_j}{2!} + \frac{h^2\,\partial_j^2}{3!} \pm \frac{h^3\,\partial_j^3}{4!} + \cdots\right)\partial_j\,u_k = \varkappa_j^{\pm}\,\partial_j\,u_k, \tag{26}$$

where

$$\varkappa_j^{\pm} \equiv \pm i\,\theta_j^{-1}(1 - e^{\pm i\theta_j}) \tag{27}$$

and $\varkappa_j^{\pm}\,\partial_j$ is not summed over the repeated index j. We also adopt the definitions

$$\varkappa_i = \tfrac{1}{2}(\varkappa_i^{+} + \varkappa_i^{-}), \qquad \varkappa_i^2 \equiv \varkappa_i^{+}\,\varkappa_i^{-}, \quad \text{(not summed)}. \tag{28}$$

Then, for exponential functions

$$\Delta_i^2 = \varkappa_i^2\,\partial_i^2; \qquad \Delta_i\,\Delta_j = \varkappa_i\,\varkappa_j\,\partial_i\,\partial_j, \ (i \neq j) \quad \text{(not summed)}. \tag{29}$$

Inserting (29) in the difference Eq. (6), we have three equations of the type

$$\begin{aligned}\varrho \ddot{u}_1 = [c_{11} \varkappa_1^2 \partial_1^2 + c_{44}(\varkappa_2^2 \partial_2^2 + \varkappa_3^2 \partial_3^2) + \\ + \tfrac{1}{2} c_{12} h^2 (\varkappa_1^2 \varkappa_2^2 \partial_1^2 \partial_2^2 + \varkappa_1^2 \varkappa_3^2 \partial_1^2 \partial_3^2)] u_1 + \\ + (c_{12} + c_{44}) (\varkappa_1 \varkappa_2 \partial_1 \partial_2 u_2 + \varkappa_1 \varkappa_3 \partial_1 \partial_3 u_3).\end{aligned} \tag{30}$$

These are differential equations which yield the complete dispersion relation (3). Similarly, substitution of (29) in the difference equations of the type (20), for the NaCl-type lattice, converts them to differential equations which produce the complete dispersion relation (18).

References

[1] CAUCHY, A. L.: C. R. Acad. Sci. Paris **32**, 323 (1851).
[2] AERO, E. L., and E. V. KUVSHINSKII: Fiz. Tverdogo Tela **2**, 1399 (1960).
[3] GRIOLI, G.: Ann. Mat. pura appl., Ser. IV, **50**, 389 (1960).
[4] RAJAGOPAL, E. S.: Ann. der Phys. **6**, 192 (1960).
[5] TRUESDELL, C. A., and R. A. TOUPIN: Encyclopedia of Physics, Vol. III/1. Berlin/Göttingen/Heidelberg: Springer 1960.
[6] TOUPIN, R. A.: Arch. Rat. Mech. Anal. **17**, 85 (1964).
[7] TOUPIN, R. A.: Arch. Rat. Mech. Anal. **11**, 385 (1962).
[8] MINDLIN, R. D.: Int. J. Solids Struct. **1**, 417 (1965).
[9] GREEN, A. E., and R. S. RIVLIN: Arch. Rat. Mech. Anal. **16**, 325 (1964).
[10] KRUMHANSL, J. A.: Lattice dynamics; Proc. Int. Conf., Copenhagen 1963. Pergamon Press 1964.
[11] GAZIS, D. C., R. HERMAN and R. F. WALLIS: Phys. Rev. **119**, 533 (1960).
[12] MINDLIN, R. D.: Arch. Rat. Mech. Anal. **16**, 51 (1964).
[13] COSSERAT, E., and F. COSSERAT: Théorie des Corps Déformables. Paris: Hermann 1909.
[14] GAZIS, D. C., and R. F. WALLIS: Private communication.
[15] NYE, J. F.: Physical properties of crystals. Oxford 1960.

Reprinted from
Mechanics of Generalized Continua
E. Kröner (ed.)
IUTAM Symposium, Freudenstadt-Stuttgart, 1967
Springer-Verlag, Berlin, (1968) 312-320

LATTICE AND CONTINUUM THEORIES OF SIMPLE MODES OF VIBRATION IN CUBIC CRYSTAL PLATES AND BARS

R. D. Mindlin
Department of Civil Engineering, Columbia University

Abstract

With a view toward helping to bridge the gap, from the discrete side, between discrete and continuum models of crystalline, elastic solids, analytic solutions, in closed form, are obtained of the Gazis-Herman-Wallis finite difference equations of a simple cubic, crystal lattice for the cases of thickness-shear vibrations of a plate, face-shear and thickness-twist waves in a plate and axial shear vibrations of a rectangular bar. The simple character of the solutions facilitates detailed studies of frequencies and mode-shapes as the dimensions of the bodies and the wave lengths increase from atomic to the macroscopic sizes at which the classical continuum theory may be used.

Introduction

It is known that the description of the motions of elastic solids, by means of the classical theory of the elastic continuum, is limited to bodily dimensions and wave lengths large in comparison with a representative dimension of the structure of the material. To ascetain the magnitude of the errors, as wave lengths and sizes diminish to structural dimensions, requires data outside the compass of classical elasticity. One source of such data, applicable to crystalline solids, is the dynamical theory of crystal lattices. The difference equations and boundary conditions of lattice theory have, as their long wave, low frequency limit, the differential equations and boundary conditions of classical elasticity. Hence, analogous solutions of vibration or wave propagation problems, for bodies with at least one finite dimension, can be used to calculate the errors in the continuum solution as dimensions or wave lengths approach interatomic distances. Comparisons between the two types of solution are simplified if both are analytic and closed in terms of elementary functions. This requires simple difference equations and simple modes of motion.

The difference equations employed in this study are those formulated by Gazis, Herman and Wallis [1] for a simple cubic lattice with nearest and next nearest neighbor central force interactions and angular interactions between three, successive, non-collinear atoms. These are the simplest lattice equations that do not require a relation among the elastic constants. Although no natural crystals with simple cubic structure are known to exist, the idea is a suitable one for the present study.

Three problems are considered: thickness-shear vibrations of a plate, face-shear and thickness-twist waves in a plate and axial shear vibrations of a rectangular bar. The solutions are known, simple ones in classical elasticity and almost equally simple, analogous solutions of the lattice equations are derived, here, for comparison.

Difference Equations of Motion and Boundary Conditions

In rectangular coordinates x, y, z, the atoms of a simple cubic, crystal lattice are taken to be at points $x = la$, $y = ma$, $z = na$, where l, m, n are positive or negative integers and a is the distance between nearest neighbor atoms. The force constants between nearest neighbors and between next nearest neighbors are designated by α and β, respectively; while γ is the angular force constant between three, successive, non-collinear atoms Then, if $u_{l,m,n}$, $v_{l,m,n}$, $w_{l,m,n}$ are the rectangular components of displacement of the atom at point l, m, n, Gazis, Herman and Wallis [1] find three equations of motion of the type

$$\begin{aligned}
&\alpha\,(u_{l+1,m,n} + u_{l-1,m,n} - 2u_{l,m,n}) \\
&+ \beta\,(u_{l+1,m+1,n} + u_{l-1,m-1,n} + u_{l+1,m-1,n} + u_{l-1,m+1,n} + u_{l+1,m,n+1} + u_{l-1,m,n-1} \\
&\qquad + u_{l+1,m,n-1} + u_{l-1,m,n+1}) \\
&+ (\beta+\gamma)\,(v_{l+1,m+1,n} + v_{l-1,m-1,n} - v_{l+1,m-1,n} - v_{l-1,m+1,n} + w_{l+1,m,n+1} \\
&\qquad + w_{l-1,m,n-1} - w_{l+1,m,n-1} - w_{l-1,m,n+1}) \\
&+ 4\gamma\,(u_{l,m+1,n} + u_{l,m-1,n} + u_{l,m,n+1} + u_{l,m,n-1} - 4u_{l,m,n} = M\ddot{u}_{l,m,n}\,,
\end{aligned} \tag{1}$$

where M is the mass of an atom. The remaining two equations of motion are obtained by cyclical permutation of u, v, w and l, m, n.

At free boundaries $l = \pm L$, the conditions to be satisfied are:

$$\begin{aligned}
&\pm\alpha\,(u_{\pm(L+1),m,n} - u_{\pm L,m,n}) \\
&\pm\beta\,(u_{\pm(L+1),m+1,n} + u_{\pm(L+1),m-1,n} + u_{\pm(L+1),m,n+1} + u_{\pm(L+1),m,n-1} - 4u_{\pm L,m,n}) \\
&+\beta\,(v_{\pm(L+1),m+1,n} - v_{\pm(L+1),m-1,n} + w_{\pm(L+1),m,n+1} - w_{\pm(L+1),m,n-1}) \\
&\pm 2\gamma\,(u_{\pm L,m+1,n} + u_{\pm L,m-1,n} + u_{\pm L,m,n+1} + u_{\pm L,m,n-1} - 4u_{\pm L,m,n}) \\
&+\gamma\,(v_{\pm(L+1),m+1,n} - v_{\pm(L+1),m-1,n} - v_{\pm L,m+1,n} + v_{\pm L,m-1,n}) \\
&+\gamma\,(w_{\pm(L+1),m,n+1} - w_{\pm(L+1),m,n-1} - w_{\pm L,m,n+1} + w_{\pm L,m,n-1}) = 0,
\end{aligned}$$

$$\begin{aligned}
&\pm\beta\,(v_{\pm(L+1),m+1,n} + v_{\pm(L+1),m-1,n} - 2v_{\pm L,m,n}) \\
&+\beta\,(u_{\pm(L+1),m+1,n} - u_{\pm(L+1),m-1,n}) \\
&\pm 4\gamma\,(v_{\pm(L+1),m,n} - v_{\pm L,m,n}) \\
&+\gamma\,(u_{\pm(L+1),m+1,n} - u_{\pm(L+1),m-1,n} + u_{\pm L,m+1,n} - u_{\pm L,m-1,n}) = 0,
\end{aligned} \tag{2}$$

$$\begin{aligned}
&\pm\beta\,(w_{\pm(L+1),m,n+1} + w_{\pm(L+1),m,n-1} - 2w_{\pm L,m,n}) \\
&+\beta\,(u_{\pm(L+1),m,n+1} - u_{\pm(L+1),m,n-1}) \\
&\pm 4\gamma\,(w_{\pm(L+1),m,n} - w_{\pm L,m,n}) \\
&+\gamma\,(u_{\pm(L+1),m,n+1} - u_{\pm(L+1),m,n-1} + u_{\pm L,m,n+1} - u_{\pm L,m,n-1}) = 0.
\end{aligned}$$

Although solutions are given, in the sequel, only for odd numbers of atoms between boundaries, the solutions for even numbers can be obtained in a similar manner.

Thickness-shear Vibrations of a Plate

In a plate bounded by $l = \pm L$, consider displacements

$$\begin{aligned}
&u_{l,m,n} = v_{l,m,n} = 0, \\
&w_{l,m,n} = (A_1 \cos \xi la + A_2 \sin \xi la)\, e^{i\omega t}, \quad 0 \le \xi a \le \pi.
\end{aligned} \tag{3}$$

With these displacements, the first two equations of the type (1) are satisfied identically and the third is satisfied if

$$\rho a^2 \omega^2 = 4\mu \sin^2 \tfrac{1}{2}\,\xi a\,, \tag{4}$$

3

where

$$\rho = M/a^3 , \qquad \mu = 2(\beta + 2\gamma)/a . \tag{5}$$

Upon substituting the displacements (3) in the boundary conditions (2), we find that the first two conditions are satisfied identically and the third is satisfied if

$$\xi a = p\pi/(2L+1), \quad p = 0, 1, 2 \ldots < 2L+1, \tag{6}$$

where even and odd p apply to symmetric and antisymmetric modes, respectively. Thus, the frequencies are

$$\omega = \frac{2}{a}\left(\frac{\mu}{\rho}\right)^{\frac{1}{2}} \sin \frac{p\pi}{2(2L+1)} \tag{7}$$

and the displacement is

$$w_{l,m,n} = \left(A_1 \cos \frac{p\pi l}{2L+1} + A_2 \sin \frac{p\pi l}{2L+1}\right) e^{i\omega t} . \tag{8}$$

To compare the solution with the continuum one, we consider a continuum plate of thickness $2h$, where

$$2h = (2L+1)\,a , \tag{9}$$

i.e., each layer of atoms is replaced by a continuum layer of thickness a. The frequencies of the thickness-shear modes of such a plate are given by

$$\omega = (p\pi/2h)(\mu/\rho)^{\frac{1}{2}} , \; p = 0, 1, 2 \ldots \infty \tag{10}$$

and the displacement is

$$w = [A_1 \cos(p\pi x/2h) + A_2 \sin(p\pi x/2h)]\, e^{i\omega t}. \tag{11}$$

In the long wave, low frequency approximation ($la \to x$, $p \ll 2L+1$), it may be seen that (7) and (8) approach (10) and (11), respectively.

In the frequency for $p = 1$, in (10), i.e.,

$$\omega_1 = \frac{\pi}{2h}\left(\frac{\mu}{\rho}\right)^{\frac{1}{2}} = \frac{\pi}{(2L+1)a}\left(\frac{\mu}{\rho}\right)^{\frac{1}{2}}, \tag{12}$$

is used as a reference frequency, the normalized frequencies for the lattice are

$$\Omega = \frac{\omega}{\omega_1} = \frac{2(2L+1)}{\pi} \sin \frac{p\pi}{2(2L+1)} , \quad p = 0, 1, 2 \ldots < 2L+1 \tag{13}$$

and these are to be compared with the normalized frequencies, p, for the continuum.

The normalized frequencies of the lattice are given in Table I for plates 1, 3, ... 15 atomic layers thick. It may be seen how, for a mode of any order $p > 0$, the normalized frequencies approach the normalized frequency, p, of the continuum as the number of layers increases. Also, as the number of layers increases, the number of lower modes that have frequencies nearly equal to those of the corresponding modes of the continuum increases. For a given number of layers, the discrepancy between the frequencies of the lattice and the continuum increases as the order of the mode increases. The corresponding departure of the lattice mode-shape from the sinusoidal form of the continuum mode is illustrated, for the case of fifteen layers, in Fig. 1. It may be seen that, for the first few modes, where the lattice mode-shapes are nearly sinusoidal, the normalized frequencies p and Ω are nearly the same; but, as the mode shape departs from sinusoidal for increasing orders, the frequencies separate – by almost 50 percent for the highest mode, in this case. As the number of layers increases, the discrepancy between the frequencies of the highest lattice mode and the corresponding continuum mode approaches ($50\pi - 100$) percent. It should be noted that, whereas the number of modes in the continuum is unlimited, the number of modes in the lattice is equal to the number of layers. A wave described by N points can have no more than $N-1$ nodes.

TABLE I

Normalized Frequencies, Ω, of Thickness-shear Vibrations in Plates $2L + 1$ Atomic Layers Thick

p \ L	0	1	2	3	4	5	6	7
0	0	0	0	0	0	0	0	0
1		0.955	0.984	0.990	0.995	0.997	0.998	0.998
2		1.654	1.871	1.934	1.960	1.972	1.978	1.985
3			2.575	2.778	2.865	2.909	2.935	2.951
4			3.027	3.484	3.683	3.786	3.845	3.884
5				4.015	4.389	4.586	4.701	4.775
6				4.344	4.962	5.292	5.488	5.613
7					5.385	5.891	6.196	6.390
8					5.643	6.370	6.811	7.097
9						6.719	7.328	7.726
10						6.932	7.738	8.270
11							8.038	8.724
12							8.216	9.082
13								9.340
14								9.497

Face-shear and Thickness-twist Waves in a Plate

Waves, in a plate, with displacement and wave normal at right angles to each other and parallel to the faces of the plate, are called thickness-twist waves except for the wave of zero order (in which the displacement does not vary through the thickness of the plate) which is called a face-shear wave. In the lattice plate considered in the preceding section, such waves are represented by displacements of the form

$$u_{l,m,n} = v_{l,m,n} = 0,$$
$$w_{l,m,n} = (A_1 \cos \xi la + A_2 \sin \xi la)\, e^{i(\eta ma - \omega t)}, \quad 0 \leq \xi a,\ \eta a \leq \pi . \tag{14}$$

This time, the equations of motion require

$$\rho a^2 \omega^2 = 4\mu \left(\sin^2 \tfrac{1}{2}\xi a + \sin^2 \tfrac{1}{2}\eta a\right) \tag{15}$$

and the boundary conditions again require (6). Thus, the normalized frequencies are

$$\Omega = \frac{\omega}{\omega_1} = \frac{2\,(2L+1)}{\pi}\left(\sin^2 \frac{p\pi}{2\,(2L+1)} + \sin^2 \tfrac{1}{2}\eta a\right)^{\frac{1}{2}} \tag{16}$$

and the displacement is

$$w_{l,m,n} = \left(A_1 \cos \frac{p\pi l}{2L+1} + A_2 \sin \frac{p\pi l}{2L+1}\right) e^{i(\eta ma - \omega t)} . \tag{17}$$

In the long wave, low frequency limit ($la \to x$, $ma \to y$, $p \ll 2L + 1$), these become the known results from elasticity theory:

$$\Omega = \left[p^2 + (2\eta h/\pi)^2\right]^{\frac{1}{2}}, \tag{18}$$
$$w = \left[A_1 \cos (p\pi x/2h) + A_2 \sin (p\pi x/2h)\right] e^{i(\eta y - \omega t)}. \tag{19}$$

The real branches of the dispersion relation for the lattice are illustrated in Fig. 2, for

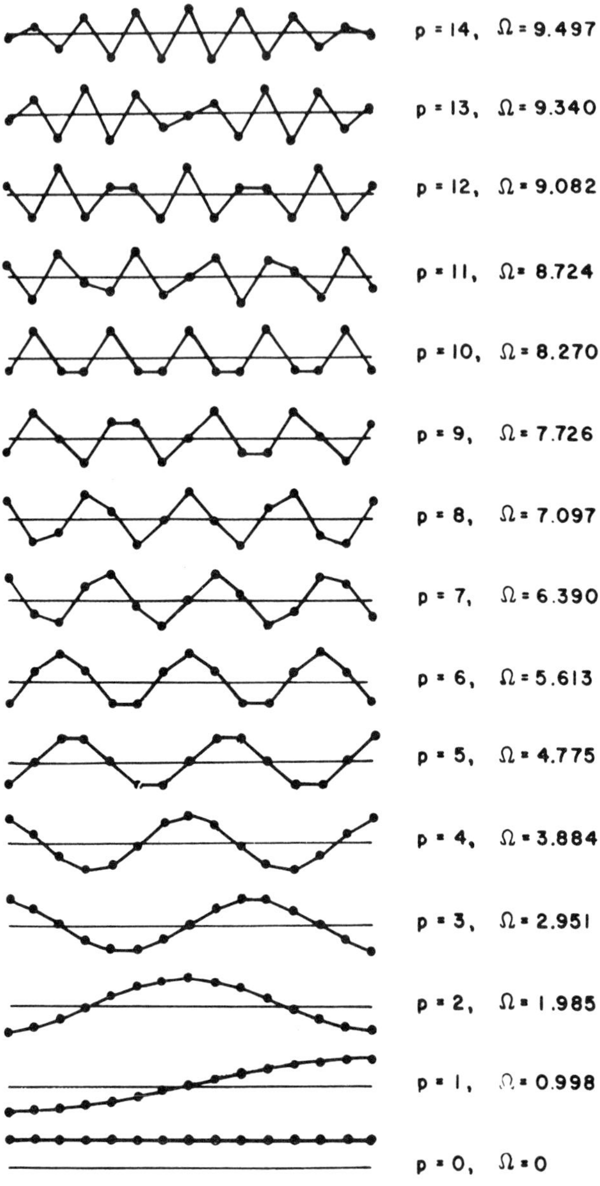

FIG. 1 MODES AND NORMALIZED FREQUENCIES OF THICKNESS-SHEAR VIBRATIONS OF A PLATE FIFTEEN ATOMIC LAYERS THICK.

6

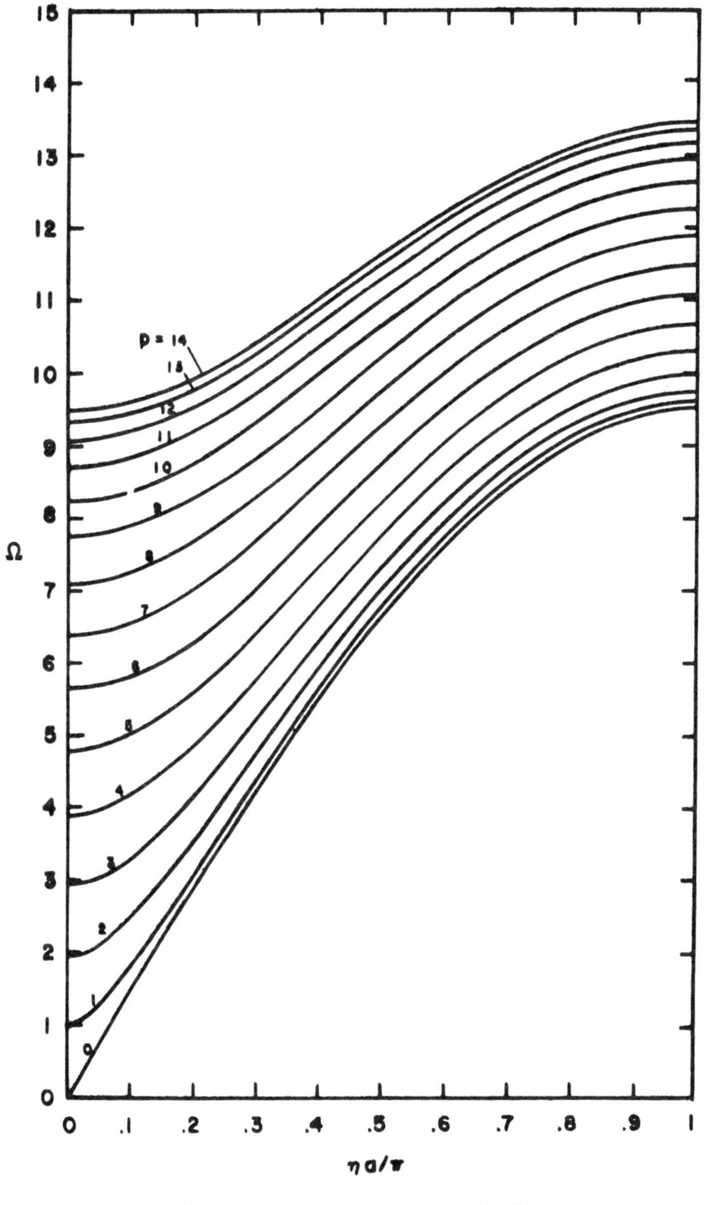

FIG. 2 REAL BRANCHES OF THE DISPERSION RELATION FOR FACE-SHEAR AND THICKNESS-TWIST WAVES IN A PLATE FIFTEEN ATOMIC LAYERS THICK.

7

a plate fifteen layers thick ($L = 7$). At infinite wave length along the plate ($\eta = 0$), the modes reduce to the thickness-shear modes illustrated in Fig. 1. These variations across the thickness of the plate are maintained for all wave lengths, $2\pi/\eta$, along the plate. The major difference between the dispersion relations for the lattice and the continuum is: each of the finite number, $2L + 1$, of branches of the lattice dispersion relation has a high frequency cutoff, in addition to a low frequency cutoff, and all have the same upper cutoff, $\eta = \pi/a$, of wave number, i.e. wave length equal to $2a$, whereas all the infinity of real branches of the continuum dispersion relation are hyperbolic curves extending from low frequency cutoffs, at $\eta = 0$, $\Omega = 1, 2 \ldots \infty$, to infinite frequencies and wave numbers – asymptotic to

$$\Omega = (2L + 1)\,\eta a/\pi, \tag{20}$$

which is a straight line starting from the lower left corner and passing through the upper right corner of Fig. 2 and is, in fact, the face-shear branch of the continuum dispersion relation. Thus, the continuum is a good approximation to the lattice only in the lower left hand region of the figure, i.e., for long wave lengths, in comparison with a, along and across the plate. For any given order of mode ($p < 2L + 1$), wave length ($2\pi/\eta$) and thickness or number of layers ($2h = [2L + 1]a$), the error in the continuum approximation may be found, readily, by comparing (18) and (19) with (16) and (17).

As in the case of the continuum, the lattice dispersion relation has branches for real frequency and imaginary wave number which are not shown in Fig. 2.

Axial Shear Vibrations of a Rectangular Bar

We consider a bar bounded by $l = \pm L$ and $m = \pm M$, with displacements

$$\begin{aligned} &u_{l,m,n} = 0, \quad v_{l,m,n} = 0, \\ &w_{l,m,n} = (A_1 \cos \xi la \sin \eta ma + A_2 \sin \xi la \sin \eta ma)\, e^{i\omega t} \\ &\qquad + (A_3 \cos \xi la \cos \eta ma + A_4 \sin \xi la \cos \eta ma)\, e^{i\omega t}. \end{aligned} \tag{21}$$

Upon substituting these displacements in the equations of motion (1) and the boundary conditions (2), we again find (15) and (6). The boundary conditions at $m = \pm M$ are obtained from (2) by one cyclical permutation of u, v, w and l, m, n. Substitution of (21) in the result yields

$$\eta a = q\pi/(2M + 1), \quad q = 0, 1, 2 \ldots < 2M + 1. \tag{22}$$

Hence, the normalized frequencies are given by

$$\Omega = \frac{2(2L + 1)}{\pi}\left(\sin^2 \frac{p\pi}{2(2L + 1)} + \sin^2 \frac{q\pi}{2(2M + 1)}\right)^{\frac{1}{2}}, \tag{23}$$

i.e., simply by the replacement of ηa, in (16), by its values (22). For a bar with a cross section $2L + 1$ atoms deep and $2M + 1$ atoms wide, the frequency of any mode with p nodes across the depth and q nodes across the width is readily calculated. The total number of modes is equal to the number of atoms in the cross section.

If $p/(2L + 1)$ and $q/(2M + 1)$ become small enough, the frequencies approach the continuum limit

$$\Omega = (p^2 + q^2 h_1^2/h_2^2)^{\frac{1}{2}}, \tag{24}$$

where h_1 and h_2 are the half-width and half-depth, respectively, of the cross section.

To see how the frequencies vary in a bar fifteen atoms deep and any number of atoms wide, enter Fig. 2 at abscissa $q/(2M + 1)$ and note the frequencies for the branches $p = 0 \ldots 14$ at that abscissa. As the ratio of the number, q, of nodes across the width to the number, $2M + 1$, of atoms across the width diminishes, the abscissa at entry is displaced to the left and the frequencies become lower. If, simultaneously, the number, p, of nodes across the depth to the number, $2L + 1$, of atoms across the depth diminishes, the frequencies approach the continuum limit (24).

Reference

[1] Gazis, D. C., R. Herman and R. F. Wallis, Physical Review, Vol. 119, 1960, pp. 533–544.

Reprinted from
Dynamics of Structured Solids
Published by
THE AMERICAN SOCIETY OF MECHANICAL ENGINEERS
345 East 47th Street, New York, New York 10017
Printed in the U.S.A.

Crystal Lattice Theory of Torsion of a Rectangular Bar of Simple Cubic Structure

R. D. MINDLIN
Department of Civil Engineering,
COLUMBIA UNIVERSITY, NEW YORK, U.S.A.

ABSTRACT: In this paper, there is described an exact closed solution of the Gazis-Herman-Wallis difference equations of a simple cubic crystal lattice for the case of torsion of an elastic bar of rectangular cross section. Examples are worked out for increasing numbers of particles in the cross section and various ratios of width to depth of cross section until the pattern of the warping function from de St. Venant's continuum solution is established. The displacement of each particle, characterizing the warping function, is given as a simple ratio of integers times the product of the angle of twist per unit length and the square of the distance between nearest neighbor particles.

1. Introduction

In recent years, there have been renewed attempts to bridge the gap between theories of elastic deformation of solids based on continuum and discrete models. Much of the effort has been directed toward extending the classical theory of elasticity to take into account effects of molecular or atomic structure of bodies as dimensions *decrease*. Examples are: addition of functional dependence of the potential energy of deformation on gradients of the strain; introduction of various extensions of the Cosserat-type microstructure. As an alternative, one may proceed in the reverse direction: starting with solutions of the difference eqations of a discrete model of an elastic solid, one may study the approach to the analogous continuum solutions as dimensions *increase*. For this purpose, it is convenient to have at hand analytic solutions of the difference equations for bodies with finite dimensions. Such a solution is described in the present paper.

The difference equations employed are those formulated by Gazis, Herman and Wallis [1] for a cubic crystal modeled as a simple cubic lattice of mass particles with central force interactions between nearest neighbor particles and between next nearest neighbor particles and angular interactions between three successive non-collinear particles. These are the simplest crystal lattice equations that do not require a relation among the elastic constants.

The problem considered is that of de St. Venant torsion of a bar of rectangular cross section. An exact, closed solution is found. At an intermediate stage, there is encountered a system of simultaneous, linear equations for the coefficients of a finite series representation of the warping function. The number of equations is equal to half the number of particles or spaces along one side of the section. However, an explicit solution is obtained for the typical unknown of the system of equations so that the exact solution of the difference equations and boundary conditions is also an explicit one.

The size of the cross section, that is, the number of particles across the width and depth, does not have to be very large before the character of the warping function, found by de St. Venant as a solution of the elasticity equations, is established. Five particles along the shorter dimensions are sufficient. The axial displacement of each particle is proportional to a simple ratio of integers, with factor of proportionality equal to the product of the angle of twist per unit length and the square of the distance between nearest neighbor particles.

2. Solution

In rectangular coordinates x, y, z, the particles of the lattice are at points $x = la$, $y = ma$, $z = na$, where l, m, n are positive or negative integers and a is the distance between nearest neighbor particles. The force constants between nearest neighbors and between next nearest neighbors are designated by α and β respectively; while γ is the angular force constant. Then, if $u_{l,m,n}$, $v_{l,m,n}$, $w_{l,m,n}$ are the rectangular components of the displacement at point l, m, n, Gazis, Herman and Wallis [1] find three equations of equilibrium of the type

$$\begin{aligned}
&\alpha(u_{l+1,m,n} - 2u_{l,m,n} + u_{l-1,m,n}) \\
&\quad + \beta(u_{l+1,m+1,n} + u_{l-1,m+1,n} + u_{l+1,m-1,n} + u_{l-1,m-1,n} \\
&\quad + u_{l+1,m,n+1} + u_{l-1,m,n+1} + u_{l+1,m,n-1} + u_{l-1,m,n-1} - 8u_{l,m,n}) \\
&\quad +(\beta + \gamma)(v_{l+1,m+1,n} + v_{l-1,m-1,n} - v_{l-1,m+1,n} - v_{l+1,m-1,n} \\
&\quad + w_{l+1,m,n+1} + w_{l-1,m,n-1} - w_{l-1,m,n-1} - w_{l+1,m,n-1}) \\
&\quad + 4\gamma(u_{l,m+1,n} + u_{l,m-1,n} + u_{l,m,n+1} + u_{l,m,n-1} - 4u_{l,m,n}) = 0. \ldots \qquad (1)
\end{aligned}$$

The remaining two equations are obtained by cyclical permutation of u, v, w and l, m, n.

The cross section of the rectangular bar is taken to be bounded by the particles at $l = \pm L$ and $m = \pm M$ and the bar is twisted about the axis of z with angle of twist ω per unit length. Although the solution is given, here, only for odd numbers of particles along a side, the solutions for even numbers can be obtained in a similar manner.

By analogy with the de St. Venant solution of the equations of classical elasticity, Todhunter and Pearson [2], it is assumed that

$$\begin{aligned} u_{l,m,n} &\equiv u_{m,n} = -\omega a^2 mn, \\ v_{l,m,n} &\equiv v_{l,n} = \omega a^2 ln \qquad \cdots \qquad (2) \\ w_{l,m,n} &\equiv w_{l,m} = \omega a^2 (lm + A \sin l\theta \sinh m\varphi), \quad 0 < \theta < \pi. \end{aligned}$$

With these displacements, the first two equations of the type (1) are satisfied identically and the third equation is satisfied if

$$\cosh \varphi = 2 - \cos \theta. \qquad \cdots \qquad (3)$$

The conditions given by Gazis, Herman and Wallis [1] for a free boundary, reduce, in view of (2), to

$$\pm 2(w_{\pm(L+1),m} - w_{\pm L,m}) + u_{m,n-1} - u_{m,n-1} = 0 \quad \text{on } l = \pm L, \qquad \cdots \qquad (4)$$

$$\pm 2(w_{l,\pm(M+1)} - w_{l \pm M}) + v_{l,n+1} - v_{l,n-1} = 0 \quad \text{on } m = \pm M. \qquad \cdots \qquad (5)$$

Upon substituting (2) in (4), we find

$$\theta = \theta_p = (2p - 1)\pi/(2L + 1), \quad p = 1, 2 \ldots L. \qquad \cdots \qquad (6)$$

Hence, the third of (2) may be written as a finite series:

$$w_{l,m} = \omega a^2 \left(lm + \sum_{p=1,2\ldots}^{L} A_p \sin l\theta_p \sinh m\varphi_p \right). \qquad \cdots \qquad (7)$$

Substitution of (7) in the boundary conditions (5) yields

$$\sum_{p=1,2\ldots}^{L} A_p \sin l\theta_p \sinh \tfrac{1}{2}\varphi_p \cosh(M + \tfrac{1}{2})\varphi_p = -l, \quad l = 1, 2 \ldots L, \qquad \cdots \qquad (8)$$

that is, a set of L simultaneous, linear, algebraic equations on the coefficients A_p. The system of equations can be solved explicitly for the A_p by a method analogous to that for determining Fourier coefficients. Multiply both sides of (8) by $\sin l\theta_q$ and sum over l from $l = 1$ to $l = L$. Now,

$$\sum_{l=1,2\ldots}^{L} \sin l\theta_q \sin l\theta_p = \begin{cases} 0, & q \neq p \\ \tfrac{1}{4}(2L + 1), & q = p \end{cases} \qquad \cdots \qquad (9)$$

and

$$\sum_{l=1,2\ldots}^{L} l \sin l\theta_p = \frac{\sin L\theta_p}{4 \sin^2 \frac{1}{2}\theta_p} - \frac{L \cos(L + \frac{1}{2})\theta_p}{2 \sin \frac{1}{2}\theta_p}. \qquad \cdots \qquad (10)$$

Hence

$$A_p = \frac{2L \sin \frac{1}{2}\theta_p \cos(L + \frac{1}{2})\theta_p - \sin L\theta_p}{(2L + 1) \sinh^3 \frac{1}{2}\varphi_p \cosh(M + \frac{1}{2})\varphi_p}. \qquad \cdots \qquad (11)$$

Substitution of (11) in (7) completes the solution for the warping function $w_{l,m}$.

3. Warping Functions

A few examples of $w_{l,m}$ for successively larger cross sections illustrate the approach of the warping from degenerate shapes to the shapes found in the de St. Venant solution for the continuum:
$L = 1$, $M = 1$:

$$w_{0,0} = w_{0,\pm 1} = w_{\pm 1,0} = 0$$

$$w_{1,1} = w_{-1,-1} = w_{1,-1} = w_{-1,1} = 0$$

$L = 1$, $M = 2$:

$$w_{0,0} = w_{0,\pm 1} = w_{0,\pm 2} = w_{\pm 1,0} = 0$$

$$w_{1,1} = w_{-1,-1} = -w_{1,-1} = -w_{-1,1} = 3\omega a^2/5$$

$$w_{1,2} = w_{-1,-2} = -w_{1,-2} = -w_{-1,2} = 4\omega a^2/5$$

$L = 1$, $M = 3$:

$$w_{0,0} = w_{0,\pm 1} = w_{0,\pm 2} = w_{0,\pm 3} = w_{\pm 1,0} = 0$$

$$w_{1,1} = w_{-1,-1} = w_{1,-1} = -w_{-1,1} = 11\omega a^2/13$$

$$w_{1,2} = w_{-1,-2} = -w_{1,-2} = -w_{-1,2} = 20\omega a^2/13$$

$$w_{1,3} = w_{-1,-3} = -w_{1,-3} = -w_{-1,3} = 23\omega a^2/13$$

$L = 1$, $M = 4$:

$$w_{0,0} = w_{0,\pm 1} = w_{0,\pm 2} = w_{0,\pm 3} = w_{0,\pm 4} = w_{\pm 1,0} = 0$$

$$w_{1,1} = w_{-1,-1} = -w_{1,-1} = -w_{-1,1} = 16\omega a^2/17$$

$$w_{1,2} = w_{-1,-2} = -w_{1,-2} = -w_{-1,2} = 31\omega a^2/17$$

$$w_{1,3} = w_{-1,-3} = -w_{1,-3} = -w_{-1,3} = 43\omega a^2/17$$

$$w_{1,4} = w_{-1,-4} = -w_{1,-4} = -w_{-1,4} = 47\omega a^2/17.$$

It may be seen that, with one dimension of the cross section restricted to three layers of particles ($L = 1$), the warping bears little resemblance to that found by de St. Venant; namely, there is no warping of the square section ($L = 1$, $M = 1$) and in each succeeding section the displacements at a long side ($l \pm 1, m = 0, \pm 1, \pm 2, \ldots$) do not have a maximum and a minimum between the center and the corners, as they have in the continuum solution. In fact, it may be shown that, for $L = 1$ and $l = 1$,

$$\frac{w_{1,m}}{\omega a^2} = m - \frac{2^{M-m+2}[(3+\sqrt{5})^m - (3-\sqrt{5})^m]}{(1+\sqrt{5})(3+\sqrt{5})^M - (1-\sqrt{5})(3-\sqrt{5})^M},$$

which is a monotonically increasing function of m from zero to M for any M.

However, for $L = 2$, $M = 2, 3, \ldots$, the dissimilarity disappears. We have:

$L = 2$, $M = 2$:

$$w_{0,0} = w_{0,\pm 1} = w_{0,\pm 2} = w_{\pm 1,0} = w_{\pm 2,0} = 0$$

$$w_{1,\pm 1} = w_{-1,\pm 1} = w_{2,\pm 2} = w_{-2,\pm 2} = 0$$

$$w_{2,1} = w_{-1,2} = w_{-2,-1} = w_{1,-2} = \omega a^2/3$$

$$w_{1,2} = w_{-2,1} = w_{-1,-2} = w_{2,-1} = -\omega a^2/3$$

$L = 2$, $M = 3$:

$$w_{0,0} = w_{0,\pm 1} = w_{0,\pm 2} = w_{0,\pm 3} = w_{\pm 1,0} = w_{\pm 2,0} = 0$$

$$w_{1,1} = w_{-1,-1} = -w_{1,-1} = -w_{-1,1} = 121\omega a^2/281$$

$$w_{1,2} = w_{-1,-2} = -w_{1,-2} = -w_{-1,2} = 184\omega a^2/281$$

$$w_{1,3} = w_{-1,-3} = -w_{1,-3} = -w_{-1,3} = 117\omega a^2/281$$

$$w_{2,1} = w_{-2,-1} = -w_{2,-1} = -w_{-2,1} = 300\omega a^2/281$$

$$w_{2,2} = w_{-2,-2} = -w_{2,-2} = -w_{-2,2} = 498\omega a^2/281$$

$$w_{2,3} = w_{-2,-3} = -w_{2,-3} = -w_{-2,3} = 448\omega a^2/281$$

$L = 2$, $M = 4$:

$$w_{0,0} = w_{0,\pm 1} = w_{0,\pm 2} = w_{0,\pm 3} = w_{0,\pm 4} = w_{\pm 1,0} = w_{\pm 2,0} = 0$$

$$w_{1,1} = w_{-1,-1} = -w_{1,-1} = -w_{-1,1} = 1539\omega a^2/2245$$

$$w_{1,2} = w_{-1,-2} = -w_{1,-2} = -w_{-1,2} = 2814\omega a^2/2245$$

$$w_{1,3} = w_{-1,-3} = -w_{1,-3} = -w_{-1,3} = 3475\omega a^2/2245$$

$$w_{1,4} = w_{-1,-4} = -w_{1,-4} = -w_{-1,4} = 3186\omega a^2/2245$$

$$w_{2,1} = w_{-2,-1} = -w_{2,-1} = -w_{-2,1} = 3342\omega a^2/2245$$

$$w_{2,2} = w_{-2,-2} = -w_{2,-2} = -w_{-2,2} = 6242\omega a^2/2245$$

$$w_{2,3} = w_{-2,-3} = -w_{2,-3} = -w_{-2,3} = 8080\omega a^2/2245$$

$$w_{2,4} = w_{-2,-4} = -w_{2,-4} = -w_{-2,4} = 7788\omega a^2/2245$$

For the square ($L = 2$, $M = 2$), it may be seen that the section is divided into eight sectors, with alternating signs of displacement, instead of the usual four sectors for a long, rectangular section. This is precisely the result found by de St. Venant for the continuum. For the rectangular sections ($L = 2$, $M = 3$; $L = 2$, $M = 4$), the displacements at a long side reach a maximum and a minimum near the corners; while, at a short side, the displacements vary monotonically. Again, this is the same as it is in de St. Venant's solution. Thus, the section need have only as many as five particles along the shorter side for the warping function to have all the qualitative properties found in the continuum solution.

REFERENCES

1. Gazis, Herman and Wallis (1960), Phys. Rev. Vol. 119, 533.
2. Todhunter and Pearson (1893), A History of the Theory of Elasticity, Vol. II, Pt. I, Cambridge University Press.

Reset from
Indian Journal of Mechanics and Mathematics
B. Sen 70th Birthday Volume
part I (1968) 35-40

Note on Vibrations of Prisms and Polygonal Plates

R. D. Mindlin,
Columbia University

Coupled dilatational and equivoluminal modes of vibration are found for certain elastic prisms and polygonal plates for special values of Poisson's ratio.

1. INTRODUCTION

The motions of an elastic body in free vibration may often be resolved into systems of plane waves traveling through the body and reflecting at traction-free surfaces so as to form closed, cyclic paths. Conversely, modes of free vibration may be found by constructing such paths.

Let β be the common angle of incidence and reflection (at a plane, free surface) of a plane, equivoluminal wave whose displacement is in the plane of incidence (SV-wave) and let α be the angle of reflection of the reflected dilatational (P-) wave, as shown in Fig. 1(a). If the amplitude of the incident wave is unity, thc amplitudes of the reflected SV- and P-waves are (see [1])

$$S = \frac{\sin 2\alpha \sin 2\beta - k^2 \cos^2 2\beta}{\sin 2\alpha \sin 2\beta + k^2 \cos^2 2\beta}, \tag{1.1}$$

$$P = \frac{k \sin 4\beta}{\sin 2\alpha \sin 2\beta + k^2 \cos^2 2\beta}, \tag{1.2}$$

where $\sin \alpha = k \sin \beta$ and k is the ratio of the velocity of dilatational waves to that of equivoluminal waves. For plane strain,

$$k^2 = \frac{\lambda + 2\mu}{\mu} = 2\frac{1 - \nu}{1 - 2\nu}, \tag{1.3}$$

where ν is Poisson's ratio. For generalized plane stress, i.e., for waves governed by Poisson's equations of extensional motion of thin plates [2, p. 497],

$$k^2 = \frac{\lambda' + 2\mu}{\mu} = \frac{2}{1 - \nu}, \tag{1.4}$$

where λ' is the plane stress constant $2\lambda\mu/(\lambda + 2\mu)$.

A well-known special case is $\beta = 45°$, for which $P = 0$; i.e., only an equivoluminal wave is reflected. Since the angles of incidence and reflection are equal and their sum is 90°, such a wave can reflect at the faces of a prism of

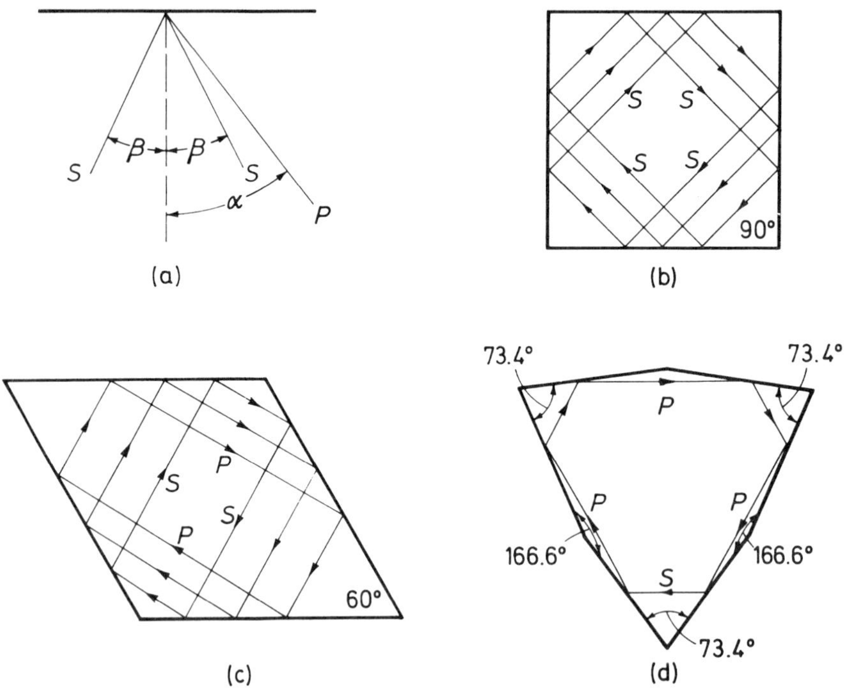

FIG. 1

square cross-section, or at the sides of a square plate, to form a closed, cyclic path, as shown in Fig. 1(b). The corresponding mode of vibration was described by Lamé [3, p. 170]. Its period T is the transit time of an S-wave across a diagonal of the square:

$$T = \frac{2L}{v_2} \sin \beta = \frac{\sqrt{2}L}{v_2}, \tag{1.5}$$

where L is the length of a side and v_2 is the velocity of S-waves.

2. COUPLED MODES

There are certain combinations of β and k for which $S = 0$ so that only a P-wave is reflected. If the angle of reflection α is such that $\alpha + \beta$ is the interior angle of an equiangular polygon with an even number of sides, that polygon is the cyclic, closed path of alternate P- and SV-waves reflecting within an equilateral polygonal boundary of the same number of sides. This is because, if an SV-wave is incident at a boundary at an angle β and reflects solely as a P-wave at an angle of reflection α, then, at the next boundary, the P-wave incident at angle α reflects solely as an SV-wave at the angle of reflection β; and so on around the circuit. Thus we should seek simultaneous solutions of the three equations:

$$\begin{cases} \sin 2\alpha \sin 2\beta = k^2 \cos^2 2\beta, \\ \sin \alpha = k \sin \beta, \\ \alpha + \beta = \dfrac{n-2}{n}\pi, \end{cases} \tag{2.1}$$

where n is the even number of sides. These equations have two solutions in the range $0 < \nu < \frac{1}{2}$.

The first solution is

$$n = 4, \qquad k^2 = 3, \qquad \alpha = 60^\circ, \qquad \beta = 30^\circ. \tag{2.2}$$

Thus, the equiangular path of the alternating P- and SV-waves is a rectangle and the equilateral boundary is a rhombus with interior angles of 60° and 120°, as illustrated in Fig. 1(c). The displacement is everywhere parallel to the long diagonal of the rhombus and the period of vibration is the transit time of an S-wave across the short diagonal of the rhombus or the transit time of a P-wave across the long diagonal:

$$T = \frac{2L}{v_2} \sin \beta = \frac{L}{v_2}, \tag{2.3}$$

where L is the length of a side of the rhombus. Since $k^2 = 3$, Poisson's ratio must be $\frac{1}{4}$ for a state of plane strain (prism) and $\frac{1}{3}$ for a state of generalized plane stress (thin plate). The body may be any shape for which the rhombus is the module.

The second solution of (2.1) is

$$n = 6, \qquad k^2 = 2.75, \qquad \alpha = 83.3^\circ, \qquad \beta = 36.7^\circ. \tag{2.4}$$

The path is an equiangular hexagon and the boundary is an equilateral hexagon whose interior angles are alternately 73.4 and 166.6 degrees, as shown in Fig. 1(d).

Poisson's ratio must be 0.216 for the prism and 0.274 for the thin plate. The period of vibration is the transit time of an S-wave across the shortest diagonal of the equilateral boundary or the transit time of a P-wave across the longest diagonal:

$$T = \frac{2L}{v_2} \sin \beta = \frac{1.2L}{v_2}, \tag{2.5}$$

where L is the length of a side of the boundary.

REFERENCES

[1] C. G. KNOTT, *Reflection and refraction of elastic waves*, Phil. Mag., 48:290 (1899), (5), pp. 64–97.
[2] A. E. H. LOVE, *A Treatise on the Mathematical Theory of Elasticity*, 4th ed., Cambridge University Press, London, 1927.
[3] G. LAMÉ, *Leçons sur la théorie mathématique de l'élasticité des corps solides*, 2ème éd., Paris, 1866.

Reprinted from
Problems of Hydrodynamics and Continuum Mechanics
L.I. Sedov 60th Birthday Volume
Society for Industrial and Applied Mathematics
(1969) 485-488

Int. J. Solids Structures, 1969, Vol. 5, pp. 1197 to 1208. Pergamon Press. Printed in Great Britain

CONTINUUM AND LATTICE THEORIES OF INFLUENCE OF ELECTROMECHANICAL COUPLING ON CAPACITANCE OF THIN DIELECTRIC FILMS

R. D. MINDLIN

Department of Civil Engineering, Columbia University

Abstract—In this paper, it is suggested that an anomaly, observed by C. A. Mead in measurements of the capacitance of thin dielectric films, may be due to an electromechanical effect not accounted for in the classical, continuum theory of piezoelectricity. A solution of equations that include the additional effect is shown to be capable of accommodating the experimental data. It is demonstrated that the augmented equations, rather than the classical ones, are a continuum approximation to the equations of a lattice of shell-model atoms. Analogous equilibrium solutions of the lattice and continuum equations are almost identical.

INTRODUCTION

ACCORDING to the classical theory of electrostatics, the capacitance of a metal–dielectric–metal sandwich is inversely proportional to the thickness of the dielectric, so that a graph of inverse capacitance vs. thickness is a straight line through the origin. In a series of experiments with a variety of thin, dielectric films, Mead [1, 2, 3] found a different result. His experimental data fall on straight lines which, if extended to zero thickness, have positive intercepts of inverse capacitance, as illustrated in Fig. 1. Initially [1], Mead suggested that the anomaly might be due to penetration of the electric field into the electrodes; but subsequently [3, 4] he abandoned that view, although it has been supported by Ku

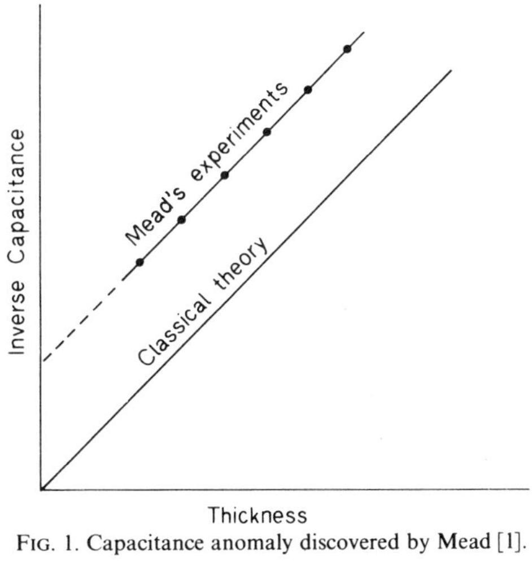

FIG. 1. Capacitance anomaly discovered by Mead [1].

and Ullman [5]. One purpose of the present paper is to show that a linear, electromechanical interaction, not accounted for in the classical, continuum theory of piezoelectricity, can produce an apparent anomaly of the same character.

The equations of classical piezoelectricity are based, in part, on the assumption that the stored energy of deformation and polarization is a function of only the strain and polarization. If the functional dependence is extended to include the polarization gradient, there can be an additional electromechanical interaction represented, in the energy, by a product of strain and polarization gradient. Since the material coefficient of such a product is a tensor of even rank, the term survives even for centrosymmetric materials—in which there is no classical piezoelectric effect. The extended equations accommodate a surface phenomenon, with a spatial exponential decay, which can become important when a dimension of the dielectric solid is small. In the first part of this paper, a solution of the extended equations, for a boundary value problem representative of Mead's experiments, is described and it is shown that the surface effect can contribute to an apparent anomaly of the type found by Mead. The appearance of the surface effect, in the solution, depends on a phenomenological constant introduced into a boundary condition to account for the influence of the properties of the metal and the metal–dielectric interface on the polarization of the dielectric at the interface.

In the second part of the paper, the same problem is treated from the point of view of a one-dimensional lattice theory, of the Cochran [6] type, based on the Dick–Overhauser [7] shell-model of the atom: a core, comprising the nucleus and inner electrons, surrounded by a shell of outer electrons. In the lattice, there are intra-atomic core–shell interactions and interatomic core–core, core–shell and shell–shell interactions. It is shown that the equations of the lattice have, as a continuum approximation, not the equations of classical piezoelectricity, but the extended equations including the contribution of the polarization gradient to the stored energy. The additional electromechanical and surface effects stem primarily from the shell–shell interaction. The solution of the difference equations of the lattice, analogous to the boundary value problem solved in the first part, yields a relation between capacitance and thickness of exactly the same form as that found in the continuum solution. The only difference in detail is in the magnitude of a length parameter characteristic of the material; and even this difference effectively disappears if the length parameter is at least as large as the distance between nearest neighbor atoms. The length parameter is the magnitude of the reciprocal wave number, at zero frequency, of the imaginary branch of the dispersion relation for longitudinal waves.

CONTINUUM THEORY

The linear equations of an elastic, dielectric continuum, including the contribution of the polarization gradient to the stored energy, have been derived elsewhere [8] by means of a simple extension of a linear version of Toupin's [9] variational principle for the classical equations of piezoelectricity. For the special case of equilibrium of one-dimensional fields in a centrosymmetric cubic or isotropic material, in the absence of extrinsic fields, the field equations were shown to reduce to [8]

$$\begin{aligned} c_{11}\partial^2 u + d_{11}\partial^2 P &= 0, \\ d_{11}\partial^2 u + b_{11}\partial^2 P - a_{11}P - \partial\varphi &= 0, \\ -\varepsilon_0\partial^2\varphi + \partial P &= 0, \end{aligned} \tag{1}$$

where $\partial = \mathrm{d}/\mathrm{d}x$, u and P are the x-components of the mechanical displacement and electronic polarization and φ is the potential of the Maxwell, electric self-field E:

$$E = -\partial\varphi. \tag{2}$$

Also, c_{11} is an elastic stiffness, ε_0 is the permittivity of a vacuum and $\varepsilon_0 a_{11}$ is the reciprocal dielectric susceptibility ($= \eta^{-1}$). The remaining two constants (b_{11} and d_{11}) are associated with terms, in the energy density, involving the polarization gradient: $\frac{1}{2}b_{11}$ is the coefficient of a quadratic term, $(\partial P)^2$, and d_{11} is the coefficient of a product of polarization gradient and strain: $\partial P \partial u$.

If the contribution of the polarization gradient is omitted, b_{11} and d_{11} are zero and (1) reduce to

$$\begin{aligned} c_{11}\partial^2 u &= 0, \\ a_{11}P + \partial\varphi &= 0, \\ -\varepsilon_0\partial^2\varphi + \partial P &= 0. \end{aligned} \tag{3}$$

This is the simplest illustration of the fact that, in the classical theory of piezoelectricity, there is no electromechanical coupling in centrosymmetric materials. For a dielectric layer with traction-free surfaces at $x = \pm h$, on which are impressed voltages $\pm V$, the boundary conditions, according to the classical theory, are

$$\begin{aligned} (\partial u)_{x=\pm h} &= 0, \\ (\varphi)_{x=\pm h} &= \pm V; \end{aligned} \tag{4}$$

and the solution of (3), subject to these boundary conditions, is, except for additive constants in u and φ,

$$\begin{aligned} u &= 0, \\ P &= -\varepsilon_0\eta V/h, \\ \varphi &= Vx/h. \end{aligned} \tag{5}$$

The capacitance (per unit area) is the ratio of the surface charge (per unit area) to the voltage drop across the layer:

$$C = \frac{(\varepsilon_0\partial\varphi - P)_{x=\pm h}}{2V} = \frac{\varepsilon_0(1+\eta)}{2h} = \frac{\varepsilon}{2h}, \tag{6}$$

where ε is the permittivity of the dielectric. These familar results of the classical theory are illustrated by the dashed lines in Fig. 2 and the straight line through the origin in Fig. 1.

If, again, the contribution of the polarization gradient is omitted, but the dielectric is noncentrosymmetric, a coupling term, say $f_{11}P\partial u$, where f_{11} is a piezoelectric constant, appears in the energy density and the one-dimensional differential equations of the classical theory are

$$\begin{aligned} c_{11}\partial^2 u + f_{11}\partial P &= 0, \\ f_{11}\partial u + a_{11}P + \partial\varphi &= 0, \\ -\varepsilon_0\partial^2\varphi + \partial P &= 0. \end{aligned} \tag{7}$$

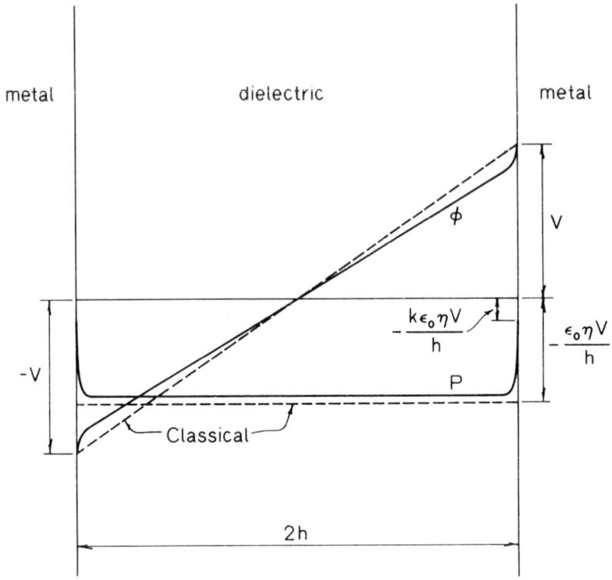

FIG. 2. Distributions of potential and polarization through dielectric layer.

The boundary conditions corresponding to (4) are now

$$\begin{aligned}(c_{11}\partial u+f_{11}P)_{x=\pm h} &= 0,\\ (\varphi)_{x=\pm h} &= \pm V;\end{aligned} \tag{8}$$

and the solution is, again except for additive constants in u and φ,

$$\begin{aligned}u &= -\varepsilon_0\eta' f_{11}Vx/c_{11}h,\\ P &= -\varepsilon_0\eta' V/h,\\ \varphi &= Vx/h,\\ C &= \varepsilon_0(1+\eta')/2h,\end{aligned} \tag{9}$$

where

$$\eta' = \eta/(1-f_{11}^2\varepsilon_0\eta/c_{11}). \tag{10}$$

Thus, the only alterations introduced by the classical piezoelectric effect are the introduction of a homogeneous strain and the replacement of the susceptibility η by η'. In view of this, the classical piezoelectric effect will be omitted in the sequel. It should be noted, however, that, for both centrosymmetric and noncentrosymmetric materials, the classical equations admit only two boundary conditions: one mechanical and one electrical.

On the other hand, solutions of (1), as shown previously [8], require an additional boundary condition—which may be taken as the specification of the surface polarization. Thus, the augmented equations permit the independent specification of the surface potential and polarization, whereas the classical equations of piezoelectricity do not. Now, the polarization at a boundary of the dielectric in a metal–dielectric–metal sandwich will depend on the physical properties of the adjacent electrode and metal–dielectric interface; and these properties are outside the compass of the theory of dielectrics. However, since the electronic polarization in the metal is zero, it is reasonable to suppose that the surface polarization in the dielectric will lie between the classical value, given by the second of (5), and zero. Thus, assuming that the two electrodes and interfaces are the same, their influence on the surface polarization may be introduced, phenomenologically, by setting the boundary conditions

$$(P)_{x=\pm h} = -k\varepsilon_0\eta V/h, \qquad 0 \le k \le 1. \tag{11a}$$

The classical condition is $k = 1$ while $k = 0$ describes continuity of polarization across the interfaces. We suppose, also, that such mechanical forces, as may be acting across the metal–dielectric interfaces, have no resultant. This boundary condition is, as shown previously [8],

$$(c_{11}\,\partial u + d_{11}\,\partial P)_{x=\pm h} = 0. \tag{11b}$$

Finally, we suppose that the voltages applied to the dielectric at the interfaces are

$$(\varphi)_{x=\pm h} = \pm V. \tag{11c}$$

We have now to solve (1) subject to the boundary conditions (11). Let

$$\begin{aligned} u &= B_1 \cosh(x/l), \\ P &= A_2 + B_2 \cosh(x/l), \\ \varphi &= A_3 x + B_3 \sinh(x/l), \end{aligned} \tag{12}$$

where l is a material constant, with the dimension of length, to be determined by the equations. Upon substituting (12) in (1), we find

$$A_2 = \varepsilon_0\eta A_3, \qquad B_2 = \varepsilon_0 B_3/l = -c_{11}B_1/d_{11}, \tag{13}$$

$$l = [\varepsilon_0(b_{11}c_{11} - d_{11}^2)/c_{11}(1+\eta^{-1})]^{\frac{1}{2}}. \tag{14}$$

The requirement of positive definiteness of the energy density makes l real.

As for boundary conditions, (11b) is satisfied identically while (11a) and (11c) become

$$\begin{aligned} A_2 + B_2 \cosh(h/l) &= -k\varepsilon_0\eta V/h, \\ A_3 h + B_3 \sinh(h/l) &= V, \end{aligned} \tag{15}$$

respectively. From (13) and (15),

$$\begin{aligned} A_3 &= (B_3/\eta l)\cosh(h/l) + kV/h, \\ B_3 &= (1-k)\eta V/[\eta \sinh(h/l) + (h/l)\cosh(h/l)]. \end{aligned} \tag{16}$$

The remaining constants, A_2, B_1 and B_2, are obtained easily from (16) and (13). The capacitance, ignoring any voltage drop that may occur in the electrodes, is

$$C = \frac{(\varepsilon_0\,\partial\varphi - P)_{x=\pm h}}{2V} = \frac{\varepsilon}{2h}\,\frac{1+(k\eta l/h)\tanh(h/l)}{1+(\eta l/h)\tanh(h/l)}. \tag{17}$$

In Fig. 3 is illustrated the relation between normalized inverse capacitance and normalized thickness, described by (17), for the case $k = 0{\cdot}1$, $\eta = 10$. Calculations from Mead's data, for small k, indicate that l is of the order of a few angstroms. Hence Mead's

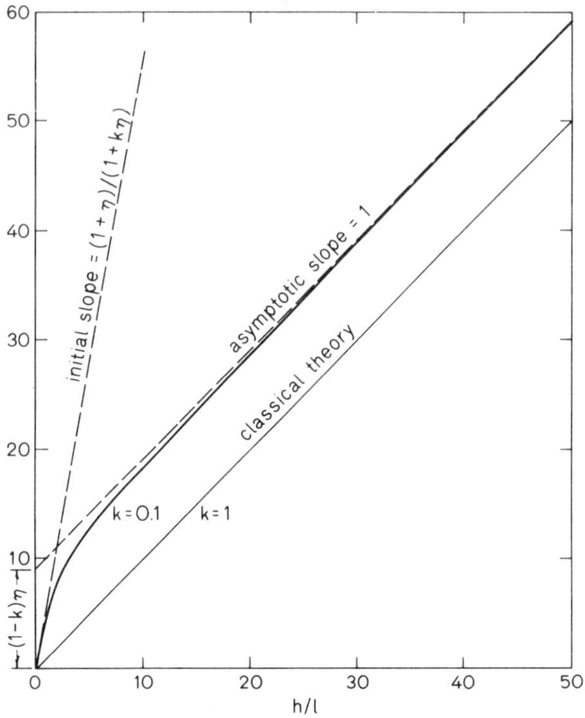

FIG. 3. Inverse capacitance vs. thickness according to augmented continuum theory; also according to lattice theory if l is replaced by λ in both abscissa and ordinate.

data, which do not extend below a thickness of 30 Å, would be well to the right of the knee of the curve and so, even if they did conform to (17), would give the appearance of a linear relation which, if extended to zero thickness, would have a non-zero intercept of inverse capacitance. This intercept, according to (17), is

$$C_0^{-1} = 2l(1-k)\eta/\varepsilon. \tag{18}$$

If $k = 1$, the intercept reduces to zero and, in fact, the whole solution reduces to the classical one. However, it seems unlikely that the presence of the metal would not influence the polarization of the dielectric at the metal–dielectric interface.

The variations of polarization and potential across the thickness of the dielectric are illustrated by the curves in Fig. 2. The absolute value of the polarization is almost uniform across the major portion of the thickness and slightly less than the uniform polarization of the classical theory; but then drops sharply, near the interfaces, to boundary values of k times the classical polarization, as specified. The potential has an almost uniform gradient, less than the uniform gradient of the classical theory, over most of the thickness, but then increases sharply on approaching the boundaries.

LATTICE THEORY

We consider a single line of atoms, in a three-dimensional lattice of shell-model atoms, as illustrated in Fig. 4. Only the intra-atomic interaction between core and shell of an atom

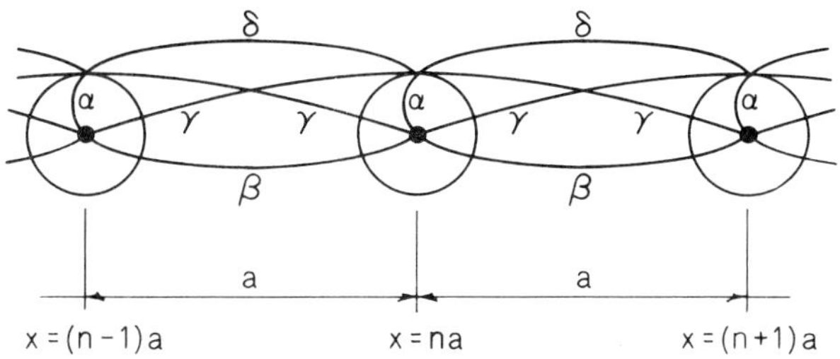

FIG. 4. Lattice of shell-model atoms.

and the interatomic interactions between nearest neighbors in the line are taken into account. The displacements of the core and shell of the atom at $x = na$ are designated by u_n and s_n respectively. The force constant of the intra-atomic core–shell interaction is designated by α and the force constants of the interatomic core–core, core–shell and shell–shell interactions are designated by β, γ and δ, respectively. All four force constants have the dimensions of force per unit length.

The equation of equilibrium of the nth interior atom is obtained by setting equal to zero the sum of the forces on its core and shell exerted by the cores and shells of its two nearest neighbor atoms:

$$\begin{aligned}\beta(u_{n+1}-u_n)+\gamma(s_{n+1}-u_n)+\gamma(u_{n+1}-s_n)+\delta(s_{n+1}-s_n)\\ -\beta(u_n-u_{n-1})-\gamma(u_n-s_{n-1})-\gamma(s_n-u_{n-1})-\delta(s_n-s_{n-1})=0.\end{aligned}\tag{19}$$

The equilibrium of the shell of the nth interior atom is expressed by

$$\begin{aligned}\alpha(u_n-s_n)+\gamma(u_{n+1}-s_n)+\delta(s_{n+1}-s_n)\\ -\gamma(s_n-u_{n-1})-\delta(s_n-s_{n-1})+qE_n=0,\end{aligned}\tag{20}$$

where E_n is the value, at $x = na$, of the Maxwell, electric self-field and q is a constant with the dimension of charge.

The polarization of the nth atom, per unit area of the three-dimensional lattice, is defined by

$$P_n = (s_n - u_n)q/a^3 \tag{21}$$

and we also adopt a symbol for the second central difference (divided by a^2):

$$\Delta^2 f_n = (f_{n+1}+f_{n-1}-2f_n)/a^2. \tag{22}$$

Then (19) and (20) may be rewritten as

$$\begin{aligned}(\beta+2\gamma+\delta)a^{-1}\Delta^2 u_n+(\gamma+\delta)a^2q^{-1}\Delta^2 P_n=0,\\ (\gamma+\delta)a^2q^{-1}\Delta^2u_n+\delta a^5q^{-2}\Delta^2P_n-(\alpha+2\gamma)a^3q^{-2}P_n+E_n=0.\end{aligned}\tag{23}$$

The lowest order continuum approximation to (23) is obtained by expanding the difference operator Δ^2 into an infinite series of derivatives and dropping all but the first term. Then

$$\Delta^2 \to \partial^2. \tag{24}$$

Also,

$$E_n \to E = -\partial\varphi. \tag{25}$$

The resulting differential equations have exactly the same form as the first two of (1). Accordingly, if we set

$$\begin{aligned} c_{11} &= (\beta+2\gamma+\delta)a^{-1}, & b_{11} &= \delta a^5 q^{-2}, \\ d_{11} &= (\gamma+\delta)a^2 q^{-1}, & a_{11} &= (\alpha+2\gamma)a^3 q^{-2} = \varepsilon_0^{-1}\eta^{-1}, \end{aligned} \tag{26}$$

the first two of (1) become the lowest order continuum approximations to (23).

As may be seen from (26), the polarization gradient terms (those with coefficients d_{11} and b_{11}) in (1) stem from shell–shell and interatomic core–shell interactions, identified by the force constants δ and γ, respectively. In fact, the form of (1) is preserved if the interatomic core–shell interaction is dropped but the shell–shell interaction is retained. If both of these interactions are omitted, the continuum approximation reduces to the classical form given by the first two of (3). Now, whereas the second of (23) is the equation of equilibrium of the shells in the lattice of shell-model atoms, the second of (3) stems, via a constitutive equation, from Toupin's [9] "equation of intramolecular force balance" based on his dumbbell model of a particle in an elastic dielectric; and the only material constant in the continuum equation is here identified as one proportional to the intra-atomic core–shell force constant, α, in the shell-model. Thus, in the case of the classical theory for centrosymmetric materials, it is shown that the shell model and the dumbbell model are identical and the equation of equilibrium of the shell is the counterpart of Toupin's equation of intramolecular force balance. The addition of the contribution of the polarization gradient to the stored energy, in the continuum theory, converts the second of (3) to the second of (1) and extends Toupin's force balance at a point to include force balance in the neighborhood of a point. That is, balance of both intramolecular and intermolecular forces is accommodated. In relation to the lattice, the second of (1) is still a continuum approximation to the equation of equilibrium of the shell, but with the shell–shell and interatomic core–shell interactions included in the latter.

We have yet to derive the lattice counterpart of the third of (1) and establish the boundary conditions. It is illuminating to reach these results from considerations of energy.

The ratio of the potential energy of one line of atoms, of the three-dimensional lattice, to the volume a^3 is

$$\begin{aligned} \mathscr{W} = \sum_n [&\tfrac{1}{2}\varepsilon_0(\partial_+\varphi_n)^2 + \tfrac{1}{2}\varepsilon_0^{-1}\eta^{-1}P_n^2 + \tfrac{1}{2}b_{11}(\Delta_+P_n)^2 \\ &+ c_{11}(\Delta_+u_n)^2 + d_{11}(\Delta_+P_n)(\Delta_+u_n)], \end{aligned} \tag{27}$$

where

$$\partial_+ f_n = \sum_{m=1}^{\infty} (-1)^{m-1} m^{-1} a^{m-1} \Delta_+^m f_n \tag{28}$$

and

$$\begin{aligned} \Delta_+ f_n &= (f_{n+1} - f_n)/a, \\ \Delta_+^2 f_n &= (f_{n+2} - 2f_{n+1} + f_n)/a^2, \\ \Delta_+^3 f_n &= (f_{n+3} - 3f_{n+2} + 3f_{n+1} - f_n)/a^3, \\ &\vdots \qquad \vdots \end{aligned} \tag{29}$$

i.e. ∂_+ is the Taylor series expansion of the derivative, ∂, in terms of forward differences. Thus, the first term in (27) is the contribution of the energy of the Maxwell self-field. The remaining terms constitute the contribution of the energy of the short range interactions between atoms.

As shown by Toupin [9], the equations of the classical, continuum theory of elastic dielectrics are derived from the electric enthalpy rather than the energy. An analogous procedure will be employed here. We define an electric enthalpy by

$$\begin{aligned}\mathscr{H} &= \mathscr{W} - \sum_n E_n(\varepsilon_0 E_n + P_n) \\ &= \sum_n [-\tfrac{1}{2}\varepsilon_0(\partial_+\varphi_n)^2 + P_n\partial_+\varphi_n + \tfrac{1}{2}\varepsilon_0^{-1}\eta^{-1}P_n^2 \\ &\quad + \tfrac{1}{2}b_{11}(\Delta_+P_n)^2 + \tfrac{1}{2}c_{11}(\Delta_+u_n)^2 + d_{11}(\Delta_+P_n)(\Delta_+u_n)].\end{aligned} \tag{30}$$

Then the equations of equilibrium and the equation of electrostatics are obtained from

$$\begin{aligned}-\partial\mathscr{H}/\partial u_n &= c_{11}\Delta^2 u_n + d_{11}\Delta^2 P_n = 0, \\ -\partial\mathscr{H}/\partial P_n &= d_{11}\Delta^2 u_n + b_{11}\Delta^2 P_n - \varepsilon_0^{-1}\eta^{-1}P_n - \partial_+\varphi_n = 0, \\ -\partial\mathscr{H}/\partial\varphi_n &= -\varepsilon_0\partial_-\partial_+\varphi_n + \partial_- P_n = 0,\end{aligned} \tag{31}$$

where

$$\partial_- f_n = \sum_{m=1}^{\infty} m^{-1}a^{m-1}\Delta_-^m f_n \tag{32}$$

and

$$\begin{aligned}\Delta_- &= (f_n - f_{n-1})/a, \\ \Delta_-^2 &= (f_n - 2f_{n-1} + f_{n-2})/a^2, \\ \Delta_-^3 &= (f_n - 3f_{n-1} + 3f_{n-2} - f_{n-3})/a^3, \\ &\ \vdots\end{aligned} \tag{33}$$

i.e. ∂_- is the Taylor series expansion of the derivative in terms of backward differences.

It may be seen that the first two of (31) are the same as (23), which have the first two of (1) as continuum approximations. and the third of (31) has the third of (1) as its continuum form.

If the lattice is of finite thickness spanning an odd number of atoms with the end ones at $n = \pm N$, admissible boundary conditions are the specification of one member of each of the three products

$$u_{\pm N}\frac{\partial\mathscr{H}}{\partial(\Delta_+u_{\pm N})}, \qquad P_{\pm N}\frac{\partial\mathscr{H}}{\partial(\Delta_+P_{\pm N})}, \qquad \varphi_{\pm N}\frac{\partial\mathscr{H}}{\partial(\partial_+\varphi_{\pm N})}, \tag{34}$$

where

$$\Delta_+ f_{\pm N} = (f_{\pm(N+1)} - f_N)/a, \qquad \partial_+\varphi_{\pm N} = (\partial_+\varphi_n)_{n=\pm N}. \tag{35}$$

We find

$$
\begin{aligned}
-\partial\mathscr{H}/\partial(\Delta_+ u_{\pm N}) &= c_{11}\Delta_+ u_{\pm N} + d_{11}\Delta_+ P_{\pm N},\\
-\partial\mathscr{H}/\partial(\Delta_+ P_{\pm N}) &= d_{11}\Delta_+ u_{\pm N} + b_{11}\Delta_+ P_{\pm N},\\
-\partial\mathscr{H}/\partial(\partial_+ \varphi_{\pm N}) &= \varepsilon_0 \partial_+ \varphi_{\pm N} - P_{\pm N}.
\end{aligned}
\tag{36}
$$

The problem analogous to the one solved in the preceding section is to find the solution of (31) satisfying the boundary conditions

$$
\begin{aligned}
c_{11}\Delta_+ u_{\pm N} + d_{11}\Delta_+ P_{\pm N} &= 0,\\
P_{\pm N} &= -k\varepsilon_0 \eta V/h,\\
\varphi_{\pm N} &= \pm V,
\end{aligned}
\tag{37}
$$

where $h = Na$. We take

$$
\begin{aligned}
u_n &= B_1' \cosh(na/\lambda),\\
P_n &= A_2' + B_2' \cosh(na/\lambda),\\
\varphi_n &= [A_3' x + B_3' \sinh(x/\lambda)]_{x=na},
\end{aligned}
\tag{38}
$$

and note that $\partial_+(na) = 1$ and

$$
\begin{aligned}
\Delta^2 \cosh(na/\lambda) &= 4a^{-2}\sinh^2(a/2\lambda)\cosh(na/\lambda),\\
\partial_+ \sinh(na/\lambda) &= \lambda^{-1}\cosh(na/\lambda),\\
\partial_- \cosh(na/\lambda) &= \lambda^{-1}\sinh(na/\lambda),\\
\partial_-\partial_+ \sinh(na/\lambda) &= \lambda^{-2}\sinh(na/\lambda).
\end{aligned}
\tag{39}
$$

Then, substituting (38) in (31), we find

$$
A_2' = \varepsilon_0 \eta A_3', \qquad B_2' = \varepsilon_0 B_3'/\lambda = -c_{11}B_1'/d_{11}, \tag{40}
$$

$$
\sinh(a/2\lambda) = a/2l, \tag{41}
$$

where l is the same as in (14).

Application of the boundary conditions (37) leads, by the same procedure as in the preceding section, to

$$
\begin{aligned}
A_3' &= (B_3'/\eta\lambda)\cosh(h/\lambda) + kV/h,\\
B_3' &= (1-k)\eta V/[\eta \sinh(h/\lambda) + (h/\lambda)\cosh(h/\lambda)]
\end{aligned}
\tag{42}
$$

and capacitance

$$
C = \frac{\varepsilon}{2h}\,\frac{1+(k\eta\lambda/h)\tanh(h/\lambda)}{1+(\eta\lambda/h)\tanh(h/\lambda)}. \tag{43}
$$

Thus, the entire solution is identical with the continuum one except that the displacement and polarization have significance only at the atom sites and l is replaced by λ. In particular, the curve in Fig. 3 is applicable to the lattice if l is replaced by λ in both abscissa and ordinate.

If the material constant λ is greater than the lattice constant a, λ and l, by (41), differ by less than 4% and the continuum equations are good approximations to the lattice equations. The condition $l \approx \lambda$ is all that is left, in the present case of equilibrium, of the requirements for the "long wave approximation" to be a good approximation. This statement stems from an inspection of the dispersion relations for wave solutions of (1) and (31) (after insertion of the appropriate inertia term, $\rho\ddot{u}$ or $\rho\ddot{u}_n$, in the first equation of each set). We find, for the continuum,

$$l^2\xi^4 + \left[1 - \frac{b_{11}\varepsilon_0\rho\omega^2}{c_{11}(1+\eta^{-1})}\right]\xi^2 - \frac{\rho\omega^2}{c_{11}} = 0 \tag{44}$$

and, for the lattice,

$$\frac{16l^2}{a^4}\sin^4\left(\frac{\xi a}{2}\right) + \left[1 - \frac{b_{11}\varepsilon_0\rho\omega^2}{c_{11}(1+\eta^{-1})}\right]\frac{4}{a^2}\sin^2\left(\frac{\xi a}{2}\right) - \frac{\rho\omega^2}{c_{11}} = 0, \tag{45}$$

where ω is the circular frequency and ξ is the wave number. In each case, the dispersion relation (ω vs.ξ) has two branches: one real (acoustic) and one pure imaginary, as shown in Fig. 5. The slopes of the real branches at long wave length approach the same limit $(c_{11}/\rho)^{\frac{1}{2}}$ so that, as far as propagating waves are concerned, the single restriction to long wave length is sufficient to make the continuum theory a good approximation. At long wave lengths, the real branches correspond to low frequencies; and at zero frequency the

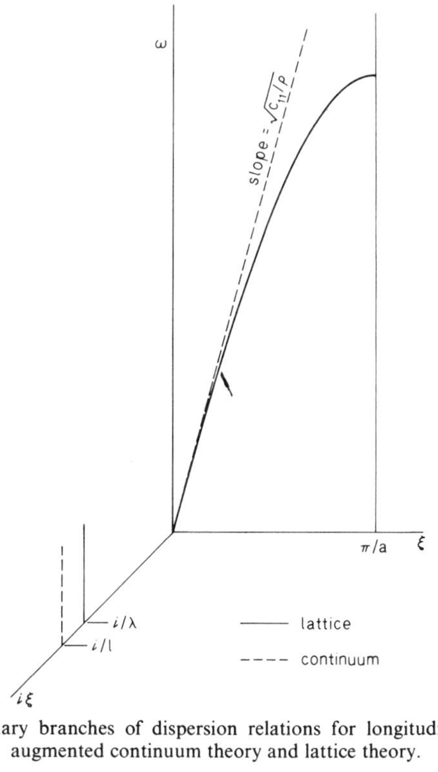

FIG. 5. Real and imaginary branches of dispersion relations for longitudinal waves according to augmented continuum theory and lattice theory.

imaginary branches have the same slope (infinite) but different intercepts: with magnitudes l^{-1} for the continuum and λ^{-1} for the lattice. If the body has a boundary at which the variables associated with the real and imaginary branches couple, amplitudes and spatial decay rates of the displacement, polarization and potential will be affected by a difference between l and λ. Thus, the quality of the approximation will depend not only on the wave length but also on a property of the material: λ. In the case of a problem of equilibrium, rather than motion, the real wave length is infinite, so that the sole remaining criterion for a good approximation is a condition on a material property.

The system of difference equations (31), with or without the inertia term, is sometimes approximated by retaining only a few early terms of the infinite series expansions, (28) and (32), for ∂_+ and ∂_-. However, in many cases, as in the present one, analytic solutions of the equations are much simpler if the full infinite series are used—preserving, at the same time, the long range electrostatic action in full. Often, the third of (31) does not appear at all: the dependent variable φ (or the Maxwell self-field) is eliminated between the second and third of (31) and only two equations (on the displacement and polarization) are employed. These are suitable for the calculation of dispersion relations for waves in an infinite lattice, which is the most frequent application; or for problems of bounded lattices with mixed ("cyclic") boundary conditions: which cannot be attained physically, but are suitable for calculations of bulk properties. For general boundary conditions, including the free boundary, the full set of three equations (seven, in three dimensions, since **u** and **P** are vectors) must be employed if surface effects are to be studied.

REFERENCES

[1] C. A. MEAD, *Phys. Rev. Lett.* **6**, 545 (1961).
[2] C. A. MEAD, *Phys. Rev.* **128**, 2088 (1962).
[3] M. MCCOLL and C. A. MEAD, *Trans. metall. Soc. A.I.M.E.* **233**, 502 (1965).
[4] C. A. MEAD, *Proc. Int. Symp. on Basic Problems in Thin Film Physics* p. 674. Vandenhoeck & Ruprecht (1966).
[5] H. Y. KU and F. G. ULLMAN, *J. appl. Phys.* **35**, 265 (1964).
[6] W. COCHRAN, *Proc. R. Soc.* **A253**, 260 (1959).
[7] B. G. DICK and A. W. OVERHAUSER, *Phys. Rev.* **112**, 90 (1958).
[8] R. D. MINDLIN, *Int. J. Solids Struct.* **4**, 637 (1968).
[9] R. A. TOUPIN, *J. ration. Mech. Analysis* **5**, 849 (1956).

(*Received* 24 *March* 1969)

Абстракт—В работе предполагается, что наблюдаемая аномалия Ц. А. Мздом в измерениях тонких диелектрических слоев, может быть вызвана злектромеханическим зффектом, неучитываемым в классической, сплошной теории пьезоэлектричества. Показано, что решение уравнений, заключающих добавочный зффект, согласуется с зкспериментальными данными. Показано также, что уравнения усилителя, по сравнению с классическими, правинельнее отображают сплошное приближение уравнений сетки электронной оболочки атомов. Аналогические решения равновесия уравнений сетки и уравнений сплошной среды являются почти идентическими.

Int. J. Solids Structures, 1970, Vol. 6, pp. 725 to 738. Pergamon Press. Printed in Great Britain

LATTICE THEORY OF SHEAR MODES OF VIBRATION AND TORSIONAL EQUILIBRIUM OF SIMPLE-CUBIC CRYSTAL PLATES AND BARS

R. D. MINDLIN

Department of Civil Engineering, Columbia University

Abstract—With a view toward helping to bridge the gap, from the discrete side, between discrete and continuum models of crystalline, elastic solids, analytic solutions, in closed form, are obtained of the Gazis–Herman–Wallis finite difference equations of a simple-cubic, crystal lattice for the cases of thickness-shear vibrations of a plate, face-shear and thickness-twist waves in a plate, axial shear vibrations of a rectangular bar and torsional equilibrium of a rectangular bar. The simple character of the solutions facilitates detailed studies of frequencies and deformations as the dimensions of the bodies and the wave lengths (or the dimensions alone, in the case of equilibrium) increase from interatomic distances to the sizes at which the classical continuum theory may be used.

INTRODUCTION

IT IS known that the description of the motions of elastic solids, by means of the classical theory of the elastic continuum, is limited to bodily dimensions and wave lengths large in comparison with a representative dimension of the structure of the material. Estimates of the magnitudes of the errors, as wave lengths and sizes diminish to structural dimensions, require data outside the compass of classical elasticity. One source of such data, applicable to crystalline solids, is the dynamical theory of crystal lattices [1]. The difference equations and boundary conditions of lattice theory have, as their long wave, low frequency limit, the differential equations and boundary conditions of classical elasticity. Analogous lattice and continuum solutions of vibration and wave propagation problems, or problems of equilibrium, for bodies with at least one finite dimension, can be used to calculate the errors in the continuum solutions as dimensions or wave lengths approach interatomic distances. Comparisons between the two types of solution are simplified if both are expressed in simple terms. In classical elasticity there are many such simple examples for bodies with one or two pairs of parallel, free boundaries; but analogous solutions of the difference equations of crystal lattice theory are rare.

The difference equations employed in this study are those formulated by Gazis, Herman and Wallis [2] for a simple-cubic lattice with nearest and next nearest neighbor central force interactions and angular interactions between three, successive, non-collinear atoms. Simple-cubic is the simplest of all lattice structures and the Gazis–Herman–Wallis equations are the simplest lattice equations that do not require a relation among the three elastic constants of cubic symmetry. Although no natural crystals with simple-cubic structure are known to exist, the idea is a suitable one for the present study.

Four problems are considered: Thickness-shear vibrations of a plate, face-shear and thickness-twist waves in a plate, axial shear vibrations of a rectangular bar and torsional equilibrium of a rectangular bar. In all cases, the faces of the plate or bar are free of traction.

The solutions are known, simple ones in classical elasticity and almost equally simple, analogous solutions of the lattice equations are derived, here, for comparison.

In the case of the vibration and wave propagation problems, explicit, closed, analytic solutions of the difference equations and boundary conditions are obtained straightforwardly. In the problem of torsional equilibrium of a rectangular bar, a complication is encountered, at an intermediate stage of solution, in the form of a system of simultaneous, linear, algebraic equations for the coefficients of a finite series representation of the warping function. The number of equations is equal to half the number of atoms (for an even number of atoms) or half the number of spaces between atoms (for an odd number of atoms) along one side of the cross section. However, an explicit solution is obtained for the typical unknown of the system of equations so that the final expression for the warping function is also an explicit, closed, analytic one.

In general, it is found that wave lengths or dimensions do not have to be greater than about half a dozen times the distance between nearest neighbor atoms for the continuum theory to be adequate. However, marked differences between analogous lattice and continuum solutions occur for smaller wave lengths or dimensions.

DIFFERENCE EQUATIONS OF MOTION AND BOUNDARY CONDITIONS

In rectangular coordinates x, y, z, the atoms of a simple-cubic, crystal lattice are taken to be at points $x = la$, $y = ma$, $z = na$, where l, m, n are positive or negative integers and a is the distance between nearest neighbor atoms. The force constants between nearest neighbors and between next nearest neighbors are designated by α and β, respectively, while γ is the angular force constant between three, successive, non-collinear atoms. Then, if $u_{l,m,n}$, $v_{l,m,n}$, $w_{l,m,n}$ are the rectangular components of displacement of the atom at point l, m, n, Gazis, Herman and Wallis [2] find three equations of motion of the type

$$\begin{aligned}
&\alpha(u_{l+1,m,n}+u_{l-1,m,n}-2u_{l,m,n})\\
&\quad+\beta(u_{l+1,m+1,n}+u_{l-1,m-1,n}+u_{l+1,m-1,n}+u_{l-1,m+1,n}+u_{l+1,m,n+1}+u_{l-1,m,n-1}\\
&\qquad\qquad+u_{l+1,m,n-1}+u_{l-1,m,n+1})\\
&\quad+(\beta+\gamma)(v_{l+1,m+1,n}+v_{l-1,m-1,n}-v_{l+1,m-1,n}-v_{l-1,m+1,n}+w_{l+1,m,n+1}+w_{l-1,m,n-1}\\
&\qquad\qquad-w_{l+1,m,n-1}-w_{l-1,m,n+1})\\
&\quad+4\gamma(u_{l,m+1,n}+u_{l,m-1,n}+u_{l,m,n+1}+u_{l,m,n-1}-4u_{l,m,n}) = \rho a^3\ddot{u}_{l,m,n},
\end{aligned}\tag{1}$$

where ρa^3 is the mass of an atom. The remaining two equations of motion are obtained by cyclical permutation of u, v, w and l, m, n.

At free boundaries $l = \pm L$, the conditions to be satisfied are

$$\begin{aligned}
&\pm\alpha(u_{\pm(L+1),m,n}-u_{\pm L,m,n})\\
&\quad\pm\beta(u_{\pm(L+1),m+1,n}+u_{\pm(L+1),m-1,n}+u_{\pm(L+1),m,n+1}+u_{\pm(L+1),m,n-1}-4u_{\pm L,m,n})\\
&\quad+\beta(v_{\pm(L+1),m+1,n}-v_{\pm(L+1),m-1,n}+w_{\pm(L+1),m,n+1}-w_{\pm(L+1),m,n-1})\\
&\quad\pm2\gamma(u_{\pm L,m+1,n}+u_{\pm L,m-1,n}+u_{\pm L,m,n+1}+u_{\pm L,m,n-1}-4u_{\pm L,m,n})\\
&\quad+\gamma(v_{\pm(L+1),m+1,n}-v_{\pm(L+1),m-1,n}-v_{\pm L,m+1,n}+v_{\pm L,m-1,n})\\
&\quad+\gamma(w_{\pm(L+1),m,n+1}-w_{\pm(L+1),m,n-1}-w_{\pm L,m,n+1}+w_{\pm L,m,n-1}) = 0,
\end{aligned}$$

$$\begin{aligned}
&\pm\beta(v_{\pm(L+1),m+1,n}+v_{\pm(L+1),m-1,n}-2v_{\pm L,m,n})\\
&+\beta(u_{\pm(L+1),m+1,n}-u_{\pm(L+1),m-1,n})\\
&\pm 4\gamma(v_{\pm(L+1),m,n}-v_{\pm L,m,n})\\
&+\gamma(u_{\pm(L+1),m+1,n}-u_{\pm(L+1),m-1,n}+u_{\pm L,m+1,n}-u_{\pm L,m-1,n})=0,
\end{aligned} \tag{2}$$

$$\begin{aligned}
&\pm\beta(w_{\pm(L+1),m,n+1}+w_{\pm(L+1),m,n-1}-2w_{\pm L,m,n})\\
&+\beta(u_{\pm(L+1),m,n+1}-u_{\pm(L+1),m,n-1})\\
&\pm 4\gamma(w_{\pm(L+1),m,n}-w_{\pm L,m,n})\\
&+\gamma(u_{\pm(L+1),m,n+1}-u_{\pm(L+1),m,n-1}+u_{\pm L,m,n+1}-u_{\pm L,m,n-1})=0.
\end{aligned}$$

Although solutions are given, in the sequel, only for odd numbers of atoms between boundaries, the solutions for even numbers can be obtained in a similar manner.

THICKNESS-SHEAR VIBRATIONS OF A PLATE

In a plate bounded by $l = \pm L$, consider displacements

$$\begin{aligned}
u_{l,m,n} &= v_{l,m,n} = 0,\\
w_{l,m,n} &= (A_1\cos\xi la + A_2\sin\xi la)\,e^{i\omega t}, \qquad 0\leqslant \xi a\leqslant \pi.
\end{aligned} \tag{3}$$

With these displacements, the first two equations of the type (1) are satisfied identically and the third is satisfied if

$$\rho a^2\omega^2 = 4\mu\sin^2\tfrac{1}{2}\xi a, \tag{4}$$

where

$$\mu = 2(\beta+2\gamma)/a. \tag{5}$$

Upon substituting the displacements (3) in the boundary conditions (2), we find that the first two conditions are satisfied identically and the third is satisfied if

$$\xi a = p\pi/(2L+1), \qquad p = 0,1,2,\ldots,<2L+1, \tag{6}$$

where even and odd p apply to symmetric and antisymmetric modes, respectively. Thus, the frequencies are

$$\omega = \frac{2}{a}\left(\frac{\mu}{\rho}\right)^{\frac{1}{2}}\sin\frac{p\pi}{2(2L+1)}, \qquad p = 0,1,2,\ldots,<2L+1, \tag{7}$$

and the displacements are

$$w_{l,m,n} = \left(A_1\cos\frac{p\pi l}{2L+1}+A_2\sin\frac{p\pi l}{2L+1}\right)e^{i\omega t}. \tag{8}$$

To compare the solution with the continuum one, we consider a continuum plate of thickness $2h$, where

$$2h = (2L+1)a, \tag{9}$$

i.e. each layer of atoms is replaced by a continuum layer of thickness a. The frequencies of the thickness-shear modes of such a plate are given by

$$\omega = (p\pi/2h)(\mu/\rho)^{\frac{1}{2}}, \qquad p = 0, 1, 2, \ldots, \tag{10}$$

and the displacements are

$$w = [A_1 \cos(p\pi x/2h) + A_2 \sin(p\pi x/2h)]\, e^{i\omega t}. \tag{11}$$

In the long wave, low frequency approximation ($la \to x, p \ll 2L+1$), it may be seen that (7) and (8) approach (10) and (11), respectively.

If the frequency for $p = 1$, in (10), i.e.

$$\omega_1 = \frac{\pi}{2h}\left(\frac{\mu}{\rho}\right)^{\frac{1}{2}} = \frac{\pi}{(2L+1)a}\left(\frac{\mu}{\rho}\right)^{\frac{1}{2}}, \tag{12}$$

is used as a reference frequency, the normalized frequencies for the lattice are

$$\Omega = \frac{\omega}{\omega_1} = \frac{2(2L+1)}{\pi} \sin\frac{p\pi}{2(2L+1)}, \qquad p = 0, 1, 2, \ldots, <2L+1, \tag{13}$$

and these are to be compared with the normalized frequencies, p, for the continuum.

The normalized frequencies of the lattice are given in Table 1 for plates $1, 3, 5, \ldots, 15$ atomic layers thick. It may be seen how, for a mode of any order $p > 0$, the normalized frequencies approach the normalized frequency, p, of the continuum as the number of layers increases. Also, as the number of layers increases, the number of lower modes that have frequencies nearly equal to those of the corresponding modes of the continuum increases. For a given number of layers, the discrepancy between the frequencies of the lattice and the continuum increases as the order of the mode increases. The corresponding

TABLE 1. NORMALIZED FREQUENCIES, Ω, OF THICKNESS-SHEAR MODES OF VIBRATION, OF ORDER p, IN PLATES $2L+1$ ATOMIC LAYERS THICK

	$L = 0$	$L = 1$	$L = 2$	$L = 3$	$L = 4$	$L = 5$	$L = 6$	$L = 7$
$p = 0$	0	0	0	0	0	0	0	0
$p = 1$		0·955	0·984	0·990	0·995	0·997	0·998	0·998
$p = 2$		1·654	1·871	1·934	1·960	1·972	1·978	1·985
$p = 3$			2·575	2·778	2·865	2·909	2·935	2·951
$p = 4$			3·027	3·484	3·683	3·786	3·845	3·884
$p = 5$				4·015	4·389	4·586	4·701	4·775
$p = 6$				4·344	4·962	5·292	5·488	5·613
$p = 7$					5·385	5·891	6·196	6·390
$p = 8$					5·643	6·370	6·811	7·097
$p = 9$						6·719	7·328	7·726
$p = 10$						6·932	7·738	8·270
$p = 11$							8·038	8·724
$p = 12$							8·216	9·082
$p = 13$								9·340
$p = 14$								9·497

departure of the lattice mode-shape from the sinusoidal form of the continuum mode is illustrated, for the case of fifteen layers, in Fig. 1. It may be seen that, for the first few modes, where the lattice mode-shapes are nearly sinusoidal, the normalized frequencies p and Ω are nearly the same; but, as the mode shape departs from sinusoidal for increasing orders, the frequencies separate—by almost 50 per cent for the highest mode, in this case. As the number of layers increases, the discrepancy between the frequencies of the highest lattice

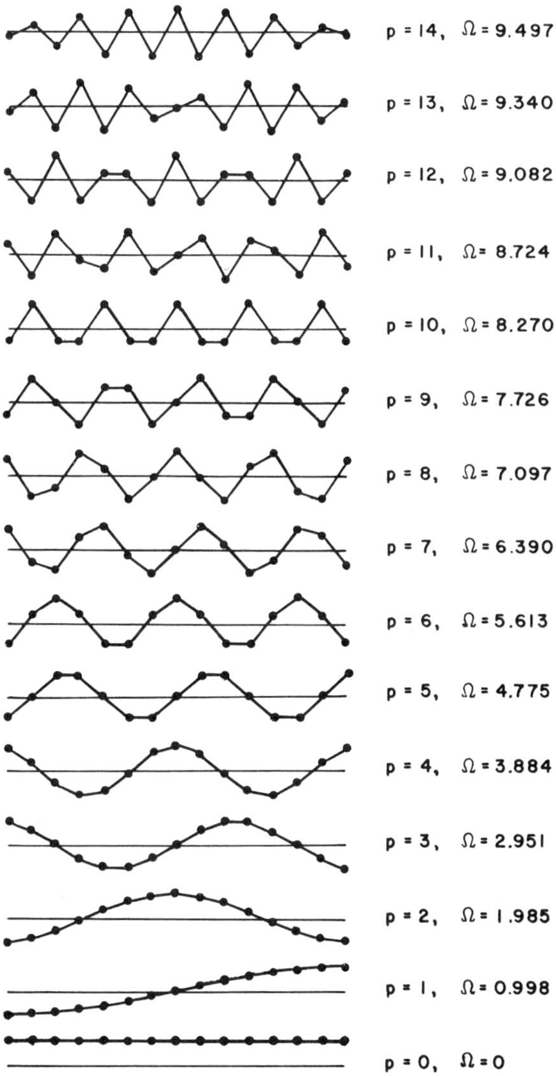

FIG. 1.

mode and the corresponding continuum mode approaches (50π–100) per cent. It should be noted that, whereas the number of modes in the continuum is unlimited, the number of modes in the lattice is equal to the number of layers. A wave described by N points can have no more than $N-1$ nodes.

FACE-SHEAR AND THICKNESS-TWIST WAVES IN A PLATE

Waves, in a plate, with displacement and wave normal at right angles to each other and parallel to the faces of the plate, are called thickness-twist waves except for the wave of zero order (in which the displacement does not vary through the thickness of the plate) which is called a face-shear wave. In the lattice plate considered in the preceding section, such waves are represented by displacements of the form

$$u_{l,m,n} = v_{l,m,n} = 0,$$
$$w_{l,m,n} = (A_1 \cos \xi la + A_2 \sin \xi la)\, e^{i(\eta ma - \omega t)} \begin{cases} 0 \leqslant \xi a \leqslant \pi, \\ 0 \leqslant \eta a \leqslant \pi. \end{cases} \tag{14}$$

This time, the equations of motion require

$$\rho a^2 \omega^2 = 4\mu(\sin^2 \tfrac{1}{2}\xi a + \sin^2 \tfrac{1}{2}\eta a) \tag{15}$$

and the boundary conditions again require (6). Thus, the normalized frequencies are

$$\Omega = \frac{\omega}{\omega_1} = \frac{2(2L+1)}{\pi}\left(\sin^2 \frac{p\pi}{2(2L+1)} + \sin^2 \tfrac{1}{2}\eta a\right)^{\frac{1}{2}}, \qquad p = 1, 2, \ldots, <2L+1, \tag{16}$$

and the displacements are

$$w_{l,m,n} = \left(A_1 \cos \frac{p\pi l}{2L+1} + A_2 \sin \frac{p\pi l}{2L+1}\right) e^{i(\eta ma - \omega t)}. \tag{17}$$

In the long wave, low frequency limit ($la \to x$, $ma \to y$, $p \ll 2L+1$), these become the known results from elasticity theory:

$$\Omega = [p^2 + (2\eta h/\pi)^2]^{\frac{1}{2}}, \qquad p = 0, 1, 2, \ldots, \tag{18}$$

$$w = [A_1 \cos(p\pi x/2h) + A_2 \sin(p\pi x/2h)]\, e^{i(\eta y - \omega t)}. \tag{19}$$

The real branches of the dispersion relation (16), for the lattice, are illustrated, in Fig. 2, for a plate fifteen layers thick ($L = 7$). At infinite wave length ($\eta = 0$) along the plate, the modes reduce to the thickness-shear modes illustrated in Fig. 1. These variations across the thickness of the plate are maintained for all wave lengths, $2\pi/\eta$, along the plate. The major differences between the dispersion relations for the lattice and the continuum are: Each of the finite number, $2L+1$, of branches of the lattice dispersion relation (16) has a high frequency cutoff, in addition to a low frequency cutoff (13), and all have the same upper cutoff, $\eta = \pi/a$, of wave number, i.e. wave length equal to $2a$; whereas all the infinity of real branches of the continuum dispersion relation (18) are hyperbolic curves extending from low frequency cutoffs, at $\eta = 0$, $\Omega = 1, 2, \ldots, \infty$, to infinite frequencies and wave numbers—asymptotic to

$$\Omega = (2L+1)\eta a/\pi, \tag{20}$$

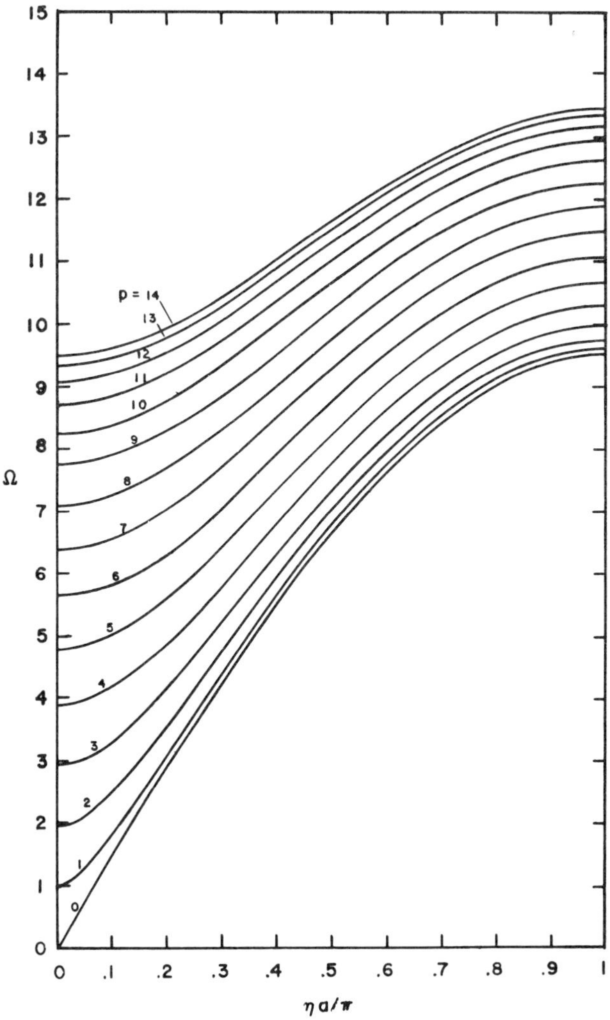

FIG. 2.

which is a straight line starting from the lower left corner and passing through the upper right corner of Fig. 2 and is, in fact, the face-shear branch of the continuum dispersion relation. Thus, the dispersion relation for the continuum is a good approximation to that for the lattice only in the lower left hand region of Fig. 2; i.e. for long wave lengths, in comparison with a, both along the plate and across its thickness. For any given order of mode ($p < 2L+1$), wave length ($2\pi/\eta$) and thickness ($2h = [2L+1]a$) or number of layers, the error in the continuum approximation may be found, readily, by comparing (18) and (19) with (16) and (17).

As in the case of the continuum, the lattice dispersion relation has branches for real frequency and imaginary wave number which are not shown in Fig. 2.

AXIAL SHEAR VIBRATIONS OF A RECTANGULAR BAR

We consider a bar bounded by $l = \pm L$, $m = \pm M$, with displacements

$$\begin{aligned} u_{l,m,n} &= v_{l,m,n} = 0, \\ w_{l,m,n} &= (A_1 \cos \xi la \sin \eta ma + A_2 \sin \xi la \sin \eta ma)\, e^{i\omega t} \\ &\quad + (A_3 \cos \xi la \cos \eta ma + A_4 \sin \xi la \cos \eta ma)\, e^{i\omega t}. \end{aligned} \tag{21}$$

Upon substituting these displacements in the equations of motion (1) and the boundary conditions (2), we again find (15) and (6). The boundary conditions at $m = \pm M$ are obtained from (2) by one cyclical permutation of u, v, w and l, m, n. Substitution of (21) in the result yields

$$\eta a = q\pi/(2M+1), \qquad q = 0, 1, 2, \ldots, <2M+1. \tag{22}$$

Hence, the normalized frequencies of axial shear vibrations of the rectangular bar are given by

$$\Omega = \frac{2(2L+1)}{\pi}\left(\sin^2 \frac{p\pi}{2(2L+1)} + \sin^2 \frac{q\pi}{2(2M+1)}\right)^{\frac{1}{2}} \begin{cases} p = 0, 1, 2, \ldots, <2L+1, \\ q = 0, 1, 2, \ldots, <2M+1, \end{cases} \tag{23}$$

i.e. simply by the replacement of ηa, in (16), by its values (22). For a bar with a cross section $2L+1$ atoms deep and $2M+1$ atoms wide, the frequency of any mode with p nodes across the depth and q nodes across the width is readily calculated from (23). The total number of modes is equal to the number of atoms in the cross section.

If $p/(2L+1)$ and $q/(2M+1)$ become small enough, the frequencies approach the continuum limit

$$\Omega = (p^2 + q^2 h_1^2/h_2^2)^{\frac{1}{2}}, \qquad p, q = 0, 1, 2, \ldots, \tag{24}$$

where h_1 and h_2 are the half-depth and half-width, respectively, of the cross section.

To see how the frequencies vary in a bar fifteen atoms deep and any number of atoms wide, enter Fig. 2 at abscissa $q/(2M+1)$ and note the frequencies for the branches $p = 0, \ldots, 14$ at that abscissa. As the ratio of the number, q, of nodes across the width to the number, $2M+1$, of atoms across the width diminishes, the abscissa at entry is displaced to the left and the frequencies become lower. If, simultaneously, the ratio of the number, p, of nodes across the depth to the number, $2L+1$, of atoms across the depth diminishes, the frequencies approach the continuum limit (24).

TORSIONAL EQUILIBRIUM OF A RECTANGULAR BAR

The bar is bounded by free faces at $l = \pm L$ and $m = \pm M$ and is in equilibrium under a twist about the axis of z with angle of twist τ per unit length.

By analogy with the St. Venant solution of the equations of classical elasticity for the analogous problem [3], it is assumed that

$$\begin{aligned} u_{l,m,n} &= u_{m,n} = -\tau a^2 mn, \\ v_{l,m,n} &= v_{l,n} = \tau a^2 ln, \\ w_{l,m,n} &= w_{l,m} = \tau a^2 (lm + A \sin l\theta \sinh m\varphi), \qquad 0 < \theta < \pi. \end{aligned} \tag{25}$$

With these displacements, the first two equations of the type (1), with the right hand side zero, are satisfied identically and the third equation is satisfied if

$$\cosh \varphi = 2-\cos \theta. \tag{26}$$

The conditions of the type (2), for free boundaries, reduce, in view of (25), to

$$\pm 2(w_{\pm(L+1),m}-w_{\pm L,m})+u_{m,n+1}-u_{m,n-1} = 0 \quad \text{on} \quad l = \pm L, \tag{27}$$

$$\pm 2(w_{l,\pm(M+1)}-w_{l,\pm M})+v_{l,n+1}-v_{l,n-1} = 0 \quad \text{on} \quad m = \pm M. \tag{28}$$

Upon substituting (25) in (27), we find

$$\theta = \theta_p = (2p-1)\pi/(2L+1), \qquad p = 1,2,\ldots, L. \tag{29}$$

Hence, the third of (25) may be written as a finite series:

$$w_{l,m} = \tau a^2\left(lm+\sum_{p=1}^{p=L} A_p \sin l\theta_p \sinh m\varphi_p\right). \tag{30}$$

Substitution of (30) in the boundary conditions (28) yields

$$\sum_{p=1}^{p=L} A_p \sin l\theta_p \sinh \tfrac{1}{2}\varphi_p \cosh(M+\tfrac{1}{2})\varphi_p = -l, \qquad l = 1,2,\ldots, L, \tag{31}$$

i.e. a set of L simultaneous, linear, algebraic equations on the coefficients A_p. The system of equations can be solved explicitly for the A_p by a method analogous to that for determining Fourier coefficients. Multiply both sides of (31) by $\sin l\theta_q$ and sum over l from $l = 1$ to $l = L$. Now [4],

$$\sum_{l=1}^{l=L} l \sin l\theta_p = \frac{\sin L\theta_p}{4\sin^2 \frac{1}{2}\theta_p}-\frac{L\cos(L+\frac{1}{2})\theta_p}{2\sin\frac{1}{2}\theta_p}. \tag{32}$$

Also, employing (29), we find

$$\sum_{l=1}^{l=L} \sin l\theta_p \sin l\theta_q = \begin{cases} 0, & q \neq p, \\ \frac{1}{4}(2L+1), & q = p. \end{cases} \tag{33}$$

Hence:

$$A_p = \frac{2L\sin\frac{1}{2}\theta_p \cos(L+\frac{1}{2})\theta_p-\sin L\theta_p}{(2L+1)\sinh^3\frac{1}{2}\varphi_p \cosh(M+\frac{1}{2})\varphi_p}. \tag{34}$$

Substitution of (34) in (30) completes the solution for the warping function $w_{l,m}$.

The approach of the warping from shapes peculiar to the lattice to the shapes, depicted in Figs. 3 and 4, found in the St. Venant solution for the continuum, is illustrated by a few

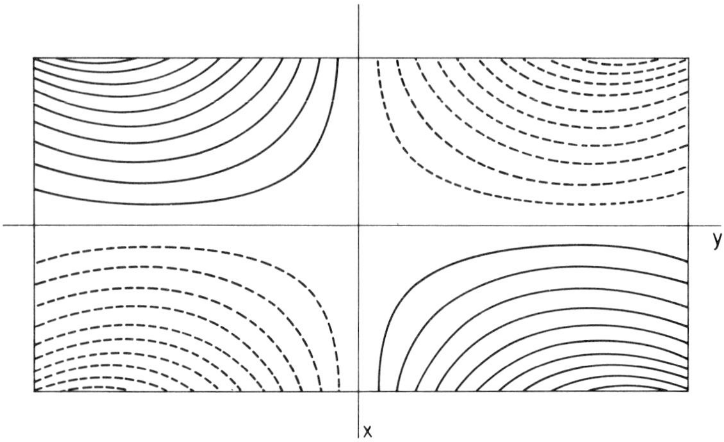

FIG. 3.

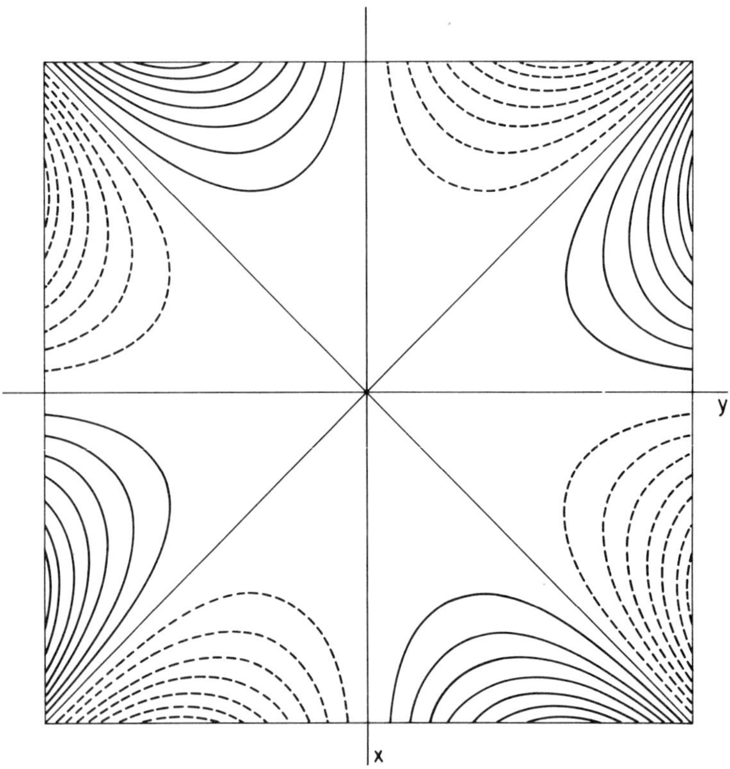

FIG. 4.

examples of $w_{l,m}$ for successively larger cross sections. First, for $L = 1$, $M = 1, 2, 3, 4$, we find

$$L = 1, M = 1: \quad w_{0,0} = w_{0,\pm 1} = w_{\pm 1,0} = 0$$
$$w_{1,1} = w_{-1,-1} = w_{1,-1} = w_{-1,1} = 0$$

$$L = 1, M = 2: \quad w_{0,0} = w_{0,\pm 1} = w_{0,\pm 2} = w_{\pm 1,0} = 0$$
$$w_{1,1} = w_{-1,-1} = -w_{1,-1} = -w_{-1,1} = 3\tau a^2/5$$
$$w_{1,2} = w_{-1,-2} = -w_{1,-2} = -w_{-1,2} = 4\tau a^2/5$$

$$L = 1, M = 3: \quad w_{0,0} = w_{0,\pm 1} = w_{0,\pm 2} = w_{0,\pm 3} = w_{\pm 1,0} = 0$$
$$w_{1,1} = w_{-1,-1} = -w_{1,-1} = -w_{-1,1} = 11\tau a^2/13$$
$$w_{1,2} = w_{-1,-2} = -w_{1,-2} = -w_{-1,2} = 20\tau a^2/13$$
$$w_{1,3} = w_{-1,-3} = -w_{1,-3} = -w_{-1,3} = 23\tau a^2/13$$

$$L = 1, M = 4: \quad w_{0,0} = w_{0,\pm 1} = w_{0,\pm 2} = w_{0,\pm 3} = w_{0,\pm 4} = w_{\pm 1,0} = 0$$
$$w_{1,1} = w_{-1,-1} = -w_{1,-1} = -w_{-1,1} = 16\tau a^2/17$$
$$w_{1,2} = w_{-1,-2} = -w_{1,-2} = -w_{-1,2} = 31\tau a^2/17$$
$$w_{1,3} = w_{-1,-3} = -w_{1,-3} = -w_{-1,3} = 43\tau a^2/17$$
$$w_{1,4} = w_{-1,-4} = -w_{1,-4} = -w_{-1,4} = 47\tau a^2/17.$$

These displacements are displayed in Fig. 5.

It may be seen that, with one dimension of the cross section restricted to three layers of atoms ($L = 1$), the warping bears little resemblance to that found by St. Venant; namely there is no warping of the square section ($L = 1$, $M = 1$) and, in each succeeding section, the displacements at a long side ($l = \pm 1, m = 0, \pm 1, \pm 2, \ldots$) do not have a maximum and a minimum between the center and the ends, as they have in the continuum solution illustrated in Figs. 3 and 4. In fact, it may be shown that, for $L = 1$ and $l = 1$,

$$\frac{w_{1,m}}{\tau a^2} = m - \frac{2^{M-m+2}[(3+\sqrt{5})^m - (3-\sqrt{5})^m]}{(1+\sqrt{5})(3+\sqrt{5})^M - (1-\sqrt{5})(3-\sqrt{5})^M}, \tag{35}$$

which is a monotonically increasing function of m, from zero to M, for any M.

However, for $L = 2$, $M = 2, 3, \ldots$, the dissimilarity disappears. We have:

$$L = 2, M = 2: \quad w_{0,0} = w_{0,\pm 1} = w_{0,\pm 2} = w_{\pm 1,0} = w_{\pm 2,0} = 0$$
$$w_{1,\pm 1} = w_{-1,\pm 1} = w_{2,\pm 2} = w_{-2,\pm 2} = 0$$
$$w_{2,1} = w_{-1,2} = w_{-2,-1} = w_{1,-2} = \tau a^2/3$$
$$w_{1,2} = w_{-2,1} = w_{-1,-2} = w_{2,-1} = -\tau a^2/3$$

$$L = 2, M = 3: \quad w_{0,0} = w_{0,\pm 1} = w_{0,\pm 2} = w_{0,\pm 3} = w_{\pm 1,0} = w_{\pm 2,0} = 0$$

$$w_{1,1} = w_{-1,-1} = -w_{1,-1} = -w_{-1,1} = 121\tau a^2/281$$

$$w_{1,2} = w_{-1,-2} = -w_{1,-2} = -w_{-1,2} = 184\tau a^2/281$$

$$w_{1,3} = w_{-1,-3} = -w_{1,-3} = -w_{-1,3} = 117\tau a^2/281$$

$$w_{2,1} = w_{-2,-1} = -w_{2,-1} = -w_{-2,1} = 300\tau a^2/281$$

$$w_{2,2} = w_{-2,-2} = -w_{2,-2} = -w_{-2,2} = 498\tau a^2/281$$

$$w_{2,3} = w_{-2,-3} = -w_{2,-3} = -w_{-2,3} = 448\tau a^2/281$$

$$L = 2, M = 4: \quad w_{0,0} = w_{0,\pm 1} = w_{0,\pm 2} = w_{0,\pm 3} = w_{0,\pm 4} = w_{\pm 1,0} = w_{\pm 2,0} = 0$$

$$w_{1,1} = w_{-1,-1} = -w_{1,-1} = -w_{-1,1} = 1539\tau a^2/2245$$

$$w_{1,2} = w_{-1,-2} = -w_{1,-2} = -w_{-1,2} = 2814\tau a^2/2245$$

$$w_{1,3} = w_{-1,-3} = -w_{1,-3} = -w_{-1,3} = 3475\tau a^2/2245$$

$$w_{1,4} = w_{-1,-4} = -w_{1,-4} = -w_{-1,4} = 3006\tau a^2/2245$$

$$w_{2,1} = w_{-2,-1} = -w_{2,-1} = -w_{-2,1} = 3342\tau a^2/2245$$

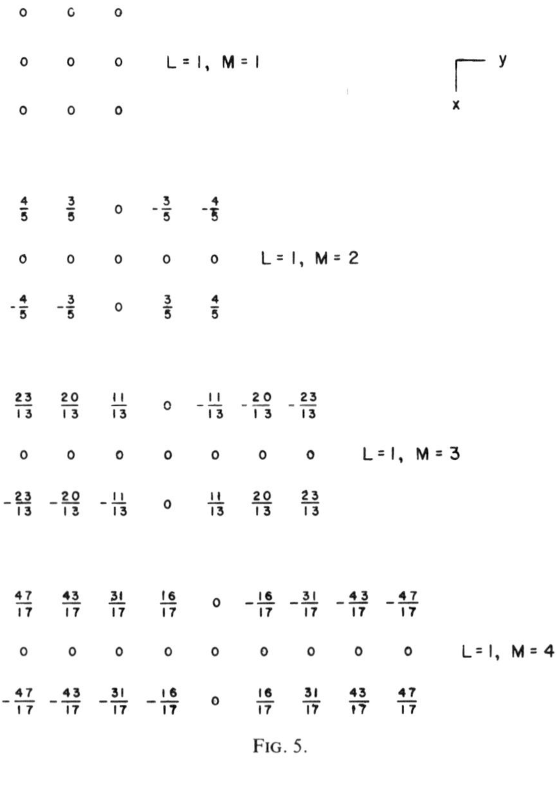

FIG. 5.

$$w_{2,2} = w_{-2,-2} = -w_{2,-2} = -w_{-2,2} = 6242\tau a^2/2245$$
$$w_{2,3} = w_{-2,-3} = -w_{2,-3} = -w_{-2,3} = 8080\tau a^2/2245$$
$$w_{2,4} = w_{-2,-4} = -w_{2,-4} = -w_{-2,4} = 7788\tau a^2/2245.$$

These displacements are displayed in Fig. 6.

L = 2, M = 2

0	$\frac{1}{3}$	0	$-\frac{1}{3}$	0
$-\frac{1}{3}$	0	0	0	$\frac{1}{3}$
0	0	0	0	0
$\frac{1}{3}$	0	0	0	$-\frac{1}{3}$
0	$-\frac{1}{3}$	0	$\frac{1}{3}$	0

y, x

L = 2, M = 3

$\frac{448}{281}$	$\frac{498}{281}$	$\frac{300}{281}$	0	$-\frac{300}{281}$	$-\frac{498}{281}$	$-\frac{448}{281}$
$\frac{117}{281}$	$\frac{184}{281}$	$\frac{121}{281}$	0	$-\frac{121}{281}$	$-\frac{184}{281}$	$-\frac{117}{281}$
0	0	0	0	0	0	0
$-\frac{117}{281}$	$-\frac{184}{281}$	$-\frac{121}{281}$	0	$\frac{121}{281}$	$\frac{184}{281}$	$\frac{117}{281}$
$-\frac{448}{281}$	$-\frac{498}{281}$	$-\frac{300}{281}$	0	$\frac{300}{281}$	$\frac{498}{281}$	$\frac{448}{281}$

L = 2, M = 4

$\frac{7788}{2245}$	$\frac{8080}{2245}$	$\frac{6242}{2245}$	$\frac{3342}{2245}$	0	$-\frac{3342}{2245}$	$-\frac{6242}{2245}$	$-\frac{8080}{2245}$	$-\frac{7788}{2245}$
$\frac{3006}{2245}$	$\frac{3475}{2245}$	$\frac{2814}{2245}$	$\frac{1539}{2245}$	0	$-\frac{1539}{2245}$	$-\frac{2814}{2245}$	$-\frac{3475}{2245}$	$-\frac{3006}{2245}$
0	0	0	0	0	0	0	0	0
$-\frac{3006}{2245}$	$-\frac{3475}{2245}$	$-\frac{2814}{2245}$	$-\frac{1539}{2245}$	0	$\frac{1539}{2245}$	$\frac{2814}{2245}$	$\frac{3475}{2245}$	$\frac{3006}{2245}$
$-\frac{7788}{2245}$	$-\frac{8080}{2245}$	$-\frac{6242}{2245}$	$-\frac{3342}{2245}$	0	$\frac{3342}{2245}$	$\frac{6242}{2245}$	$\frac{8080}{2245}$	$\frac{7788}{2245}$

FIG. 6.

For the square ($L = 2, M = 2$), it may be seen that the cross section is divided into eight sectors, with alternating signs of displacement, instead of the usual four sectors for a long, rectangular section. This is precisely the result found by St. Venant for the continuum, as illustrated in Fig. 4. For the rectangular sections $L = 2, M = 3$ and $L = 2$, $M = 4$, the displacements along a long side reach a maximum and a minimum, near the ends, while, along a short side, the displacements vary monotonically. Again, these results are the same as the corresponding ones in St. Venant's solution, as illustrated in Fig. 3. Thus, the cross section need have only as many as five atoms along the shorter side for the warping function to have all the qualitative properties found in the continuum solution.

It may be observed that, for both $L = 1$ and $L = 2$, the normalized axial displacements, $w_{l,m}/\tau a^2$, are expressed simply as ratios of integers. This is made possible by the fact that trigonometric functions of

$$\theta_p = (2p-1)\pi/(2L+1), \qquad p = 1, 2, \ldots, L, \tag{29}$$

can be expressed in closed, algebraic form if $L = 1$ or $L = 2$. Thus, for

$$L = 1: \quad \cos\theta_1 = \cos(\pi/3) = \tfrac{1}{2}; \tag{36}$$

$$\begin{aligned} L = 2: \quad \cos\theta_1 &= \cos(\pi/5) = \tfrac{1}{4}(\sqrt{(5)}+1),\\ \cos\theta_2 &= \cos(3\pi/5) = -\tfrac{1}{4}(\sqrt{(5)}-1). \end{aligned} \tag{37}$$

The next larger value of L for which such expressions are known is $L = 7$:

$$\begin{aligned} \cos\theta_1 &= \cos(\pi/15) = \tfrac{1}{8}(\sqrt{(5)}-1)+\tfrac{1}{8}\sqrt{[6\sqrt{(5)}(\sqrt{(5)}+1)]},\\ \cos\theta_2 &= \cos(3\pi/15) = \tfrac{1}{4}(\sqrt{(5)}+1),\\ \cos\theta_3 &= \cos(5\pi/15) = \tfrac{1}{2},\\ \cos\theta_4 &= \cos(7\pi/15) = -\tfrac{1}{8}[\sqrt{(5)}+1]+\tfrac{1}{8}\sqrt{[6\sqrt{(5)}(\sqrt{(5)}-1)]},\\ \cos\theta_5 &= \cos(9\pi/15) = -\tfrac{1}{4}[\sqrt{(5)}-1],\\ \cos\theta_6 &= \cos(11\pi/15) = \tfrac{1}{8}[\sqrt{(5)}-1]-\tfrac{1}{8}\sqrt{[6\sqrt{(5)}(\sqrt{(5)}+1)]},\\ \cos\theta_7 &= \cos(13\pi/15) = -\tfrac{1}{8}[\sqrt{(5)}+1]-\tfrac{1}{8}\sqrt{[6\sqrt{(5)}(\sqrt{(5)}-1)]}. \end{aligned}$$

Acknowledgements—These investigations were supported by the Office of Naval Research under a contract with Columbia University. The solution of the problem of torsional equilibrium of a rectangular bar was presented at the Twelfth International Congress of Applied Mechanics held at Stanford University, August, 1968. The solutions of the vibration and wave propagation problems were presented at the Symposium on Dynamics of Structured Solids held during the Winter Annual Meeting of the American Society of Mechanical Engineers in New York City, December, 1968.

REFERENCES

[1] M. BORN and K. HUANG, *Dynamical Theory of Crystal Lattices*. Oxford University Press (1956).
[2] D. C. GAZIS, R. HERMAN and R. F. WALLIS, Surface elastic waves in cubic crystals. *Phys. Rev.* **119**, 533 (1960).
[3] I. TODHUNTER and J. PEARSON, *A History of the Theory of Elasticity*, Vol. II, Part 1, p. 24. Cambridge University Press (1893).
[4] L. B. W. JOLLEY, *Summation of Series*, formula 427, p. 80, 2nd edition. Dover (1961).

(*Received* 2 *September* 1969)

Абстракт—Для устранения разниц, со стороны дискретных систем, между дискретной и сплошной моделью кристаллических, упругих тел, получаются аналитические решения, в замкнутом виде, для уравнения Газиса-Германа-Валлиса в конечных разностях, описывающие поведение простой кубической, кристаллической решетки. Этие решения касаются случая сдвиговых колебаний пластинки, поверхностных сдвиговых и крутильных волн в пластинке, осевых сдвиговых колебаний прямоугольного сечения и крутильного равновесия прямоугольного стержня. Простой характер решений, облегчает подробные исследования частот и деформаций, если размеры тел и длины волн /или только размеры, в случае равновесия/ увеличиваются от междуатомных размеров к размерам, при которых можно пользоваться классической теорией сплошной среды.

INTERNATIONAL CENTRE FOR MECHANICAL SCIENCES

RAYMOND D. MINDLIN
COLUMBIA UNIVERSITY, NEW YORK

POLARIZATION GRADIENT IN ELASTIC DIELECTRICS

COURSE HELD AT THE DEPARTMENT
FOR MECHANICS OF DEFORMABLE BODIES
JULY 1970

UDINE 1970

COURSES AND LECTURES - No. 24

P r e f a c e

During the past decade, there has been renewed interest in efforts to extend the classical theories of continua to accommodate the effects of the structure of materials and to bridge the gap between the classical continuum theories and the dynamical theories of crystal lattices. The Born- von Karman theory of monatomic lattices of mass points has the classical theory of elasticity as its long wave limit; so that the efforts there have been to extend the connection to shorter wave lengths and to compound lattices. The modern theories of lattices of electronically polarizable atoms do not, however, have the classical continuum theory of the elastic dielectric as their long wave limit. The discrepancy lies with the continuum theory and, for monatomic lattices, can be eliminated by adding, to the stored energy of deformation and polarization, a functional dependence on the polarization gradient.

I wish to thank the Rectors and the Secretary General of the International Centre for Mechanical Sciences for their kind invitation to me to present the following series of lectures in which the derivation of the augmented continuum theory of elastic dielectrics and some of its consequences are described.

July, 1970 *R.D.Mindlin*

1. Introduction.

The linear piezoelectric effect,which occurs in elastic dielectrics, is a physical phenomenon widely used in technology. The classical continuum theory [1] incorporating the effect is about seventy-five years old and is today universally employed in the analysis and design of crystal oscillators, filters and transducers [2,3]. In systematic derivations of the classical, phenomenological theory, piezoelectricity is usually expressed as an interaction between mechanical strain and one of the electrical variables: the electric field [3] the electric displacement [4] or the electric polarization [5]. There is, however, no fundamental reason for restricting attention to the interaction of the strain, a second rank tensor, with <u>only</u> a first rank electrical quantity (field, displacement, polarization). It is not illogical to examine the consequences of considering an additional, linear, electromechanical effect: an interaction between the strain and, say, the polarization gradient - a second rank tensor quantity. The added complexity is justified by the fact that the resulting mathematical theory [6] has interesting novel properties: (1) it accomodates the mathematical representation of a surface energy of deformation and polarization which is absent from the classical theory but which has been measured in the laboratory [7] and calculated from atomic considerations [8];

(2) it can account [9] for an apparent anomaly observed in measurements of the electrical capacitance of thin, dielectric films [10]; (3) the additional electromechanical effect is not confined to non-centrosymmetric materials, as is the classical piezoelectric effect; (4) the resulting equations, rather than the classical ones, are the correct, long wave limit [9,11] of the equations of the modern dynamical theory of crystal lattices of electronically polarizable atoms [12,13,14,15].

2. Classical Equations.

The classical, linear theory of elastic dielectrics is expressed in terms of 25 dependent variables [3,4]:

u_i	Mechanical displacement
S_{ij}	Strain
T_{ij}	Stress
E_i	Electric field
D_i	Electric displacement
P_i	Polarization
φ	Electric potential

These variables are linked by 25 equations:

(a) the stress-equations of motion

$$T_{ij,i} + f_j = \rho \ddot{u}_j , \qquad (2.1)$$

where ρ is the mass density and f_j is the body force;

(b) the equations of the electrostatic field

$$D_{i,i} = 0 , \qquad (2.2)$$

$$E_i = - \varphi_{,i} ; \qquad (2.3)$$

(c) the strain-displacement relations

$$S_{ij} = \frac{1}{2} \left(u_{j,i} + u_{i,j} \right) ; \qquad (2.4)$$

and (d) the constitutive relations

$$T_{ij} = c^E_{ijkl}\, S_{kl} - e_{kij}\, E_k\,, \tag{2.5}$$

$$D_j = e_{jkl}\, S_{kl} + \epsilon^S_{jk}\, E_k\,, \tag{2.6}$$

$$P_i = D_i - \epsilon_0\, E_i\,, \tag{2.7}$$

where

c^E_{ijkl} = elastic stiffness (E_i = constant),

e_{ijk} = piezoelectric stress-constant

ϵ^S_{ij} = permittivity (S_{ij} = constant),

ϵ_0 = permittivity of a vacuum.

The equations of motion may be reduced to four by elimination of all the variables except u_i and φ :

$$c^E_{ijkl}\, u_{k,li} + e_{kij}\, \varphi_{,ki} + f_j = \rho\, \ddot{u}_j \tag{2.8}$$

$$e_{kij}\, u_{i,jk} - \epsilon^S_{ij}\, \varphi_{,ij} = 0\,. \tag{2.9}$$

3. Toupin's Variational Principle for the Classical Theory.

The classical equations of equilibrium of elastic dielectrics may be derived in an illuminating form by means of a variational principle due to Toupin [5]. A linear version of the variational principle is described in this section.

First, separate the energy density, W, of the dielectric into the stored energy density of deformation and polarization, W^L, and a remainder which is the energy density of the Maxwell electric self-field:

$$W = W^L(S_{ij}, P_i) + \frac{1}{2}\epsilon_0 \varphi_{,i}\varphi_{,i}. \tag{3.1}$$

Then define an electric enthalpy

$$H = W - E_i D_i. \tag{3.2}$$

Upon substituting (3.1), (2.7) and (2.3) in (3.2) we find

$$H = W^L(S_{ij}, P_i) - \frac{1}{2}\epsilon_0 \varphi_{,i}\varphi_{,i} + \varphi_{,i}P_i. \tag{3.3}$$

Consider a body occupying a volume V bounded by a surface S separating V from an outer vacuum V'. For such a system, the version of Toupin's principle that leads to equations equivalent to those in the preceding section, for the case of equilibrium, is

$$(3.4)\qquad -\delta\int_{V^*} H\,dV + \int_V (f_i\,\delta u_i + E_i^0\,\delta P_i)\,dV + \int_S t_i\,\delta u_i\,dS = 0$$

for independent variations of u_i, P_i and φ. In (3.4), $V^* = V + V'$ and t_i is the surface traction across S and E_i^0 is the external electric field. Now,

$$(3.5)\qquad \delta H = \frac{\partial W^L}{\partial S_{ij}}\,\delta S_{ij} + \frac{\partial W^L}{\partial P_i}\,\delta P_i - \epsilon_0\,\varphi_{,i}\,\delta\varphi_{,i} + \varphi_{,i}\,\delta P_i + P_i\,\delta\varphi_{,i}.$$

Define a stress T_{ij} and an effective local electric force E_i^L by

$$(3.6)\qquad T_{ij} = \frac{\partial W^L}{\partial S_{ij}}\,,\quad E_i^L = -\frac{\partial W^L}{\partial P_i}\,.$$

Note that, since $T_{ij} = T_{ji}$ and

$$(3.7)\qquad S_{ij} = \frac{1}{2}(u_{j,i} + u_{i,j}),$$

$$T_{ij}\,\delta S_{ij} = T_{ij}\,\delta u_{j,i} = (T_{ij}\,\delta u_j)_{,i} - T_{ij,i}\,\delta u_j\,,$$

$$\varphi_{,i}\,\delta\varphi_{,i} = (\varphi_{,i}\,\delta\varphi)_{,i} - \varphi_{,ii}\,\delta\varphi\,,$$

$$P_i\,\delta\varphi_{,i} = (P_i\,\delta\varphi)_{,i} - P_{i,i}\,\delta\varphi\,,$$

by the chain rule of differentiation.

Then

$$\delta H = -T_{ij,i}\,\delta u_i - (E_i^L - \varphi_{,i})\delta P_i - (-\epsilon_0\varphi_{,ii} + P_{i,i})\delta\varphi$$

$$(3.8)\qquad + \left[T_{ij}\,\delta u_j + (-\epsilon_0\varphi_{,i} + P_i)\delta\varphi\right]_{,i}\,.$$

Noting that $V^* = V + V'$ and that u_i and P_i do not exist in V', we find, after applying the divergence theorem,

$$-\delta \int_V H\,dV = \int_V \left[T_{ij,i}\,\delta u_j + (E_i^L - \varphi_{,i})\,\delta P_i + (-\epsilon_0 \varphi_{,ii} + P_{i,i})\,\delta\varphi\right] dV$$

$$- \int_S n_i \left[T_{ij}\,\delta u_j + (-\epsilon_0 \varphi_{,i} + P_i)\,\delta\varphi\right] dS\,, \tag{3.9a}$$

$$- \int_{V'} H\,dV = - \int_{V'} \epsilon_0 \varphi_{,ii}\,\delta\varphi\,dV - \int_S \epsilon_0 n_i \varphi_{,i}\,\delta\varphi\,dS. \tag{3.9b}$$

Hence, (3.4) becomes

$$\int_V \left[(T_{ij,i} + f_j)\,\delta u_j + (E_i^L - \varphi_{,i} + E_i^0)\,\delta P_i + (-\epsilon_0 \varphi_{,ii} + P_{i,i})\,\delta\varphi\right] dV - \int_{V'} \epsilon_0 \varphi_{,ii}\,\delta\varphi\,dV$$

$$+ \int_S \left[(t_j - n_i T_{ij})\,\delta u_j + n_i \left(\epsilon_0 [\![\varphi_{,i}]\!] - P_i\right)\delta\varphi\right] dS = 0\,, \tag{3.10}$$

where $[\![\varphi_{,i}]\!]$ is the jump in $\varphi_{,i}$ across S. Then the Euler equations

$$\left.\begin{aligned} T_{ij,i} + f_j &= 0 \\ E_i^L - \varphi_{,i} + E_i^0 &= 0 \\ -\epsilon_0 \varphi_{,ii} + P_{i,i} &= 0 \end{aligned}\right\} \text{ in } V, \tag{3.11}$$

$$\varphi_{,ii} = 0 \quad \text{in } V' \tag{3.12}$$

and the natural boundary conditions

$$\left.\begin{aligned} n_i T_{ij} &= t_j\,, \\ n_i\left(-\epsilon_0 [\![\varphi_{,i}]\!] + P_i\right) &= 0, \end{aligned}\right\} \tag{3.13}$$

follow from (3.10). In (3.13), $n_i(-\epsilon_0[\![\varphi_{,i}]\!] + P_i)$ is the surface charge.

The energy density of deformation and polarization is taken to be

$$W^L = \frac{1}{2} a^s_{ij} P_i P_j + \frac{1}{2} c^P_{ijkl} S_{ij} S_{kl} + f_{kij} S_{ij} P_k \tag{3.14}$$

so that, from (3.6),

$$\begin{cases} -E^L_j = a^s_{jk} P_k + f_{jkl} S_{kl}, \\ T_{ij} = f_{kij} P_k + c^P_{ijkl} S_{kl}. \end{cases} \tag{3.15}$$

Equations (3.7), (3.11), (3.12) and (3.15), with boundary conditions (3.13), constitute a linear version of the equations of equilibrium of elastic dielectrics in the form given by Toupin.

The relations between the new constants a^s_{ij}, f_{ijk} and c^P_{ijkl} and more familiar (or, at least, standard [16] ones) are found as follows.

From (2.3) and the second of (3.11) (omitting E^0_i as in the usual formulation) $E^L_i = -E_i$. Hence the first of (3.15) becomes

$$E_j = a^s_{jk} P_k + f_{jkl} S_{kl}. \tag{3.16}$$

Now, the ratio of P to $\epsilon_0 E$ is the dielectric susceptibility [17]. Hence a^s_{ij} is proportional to the <u>reciprocal</u> susceptibility at constant strain $(\chi^{S}_{ij})^{*)}$:

$$a^{S}_{ij} = \epsilon_0^{-1} \chi^{S}_{ij}. \tag{3.17}$$

With (3.17), (3.16) becomes

$$E_j = \epsilon_0^{-1} \chi^{S}_{jk} P_k + f_{jkl} S_{kl}. \tag{3.18}$$

Define susceptibility at constant strain, η^{S}_{ij}, according to

$$\eta^{S}_{ij} \chi^{S}_{jk} = \delta_{ik} \tag{3.19}$$

and multiply both sides of (3.18) by $\epsilon_0 \eta_{mj}$ to get

$$\epsilon_0 \eta^{S}_{mj} E_j = P_m + \epsilon_0 \eta^{S}_{mj} f_{jkl} S_{kl}. \tag{3.20}$$

Then eliminate P between (3.20) and (2.7), with the result

$$D_i = -\epsilon_0 \eta^{S}_{ij} f_{jkl} S_{kl} + \epsilon_0 (\delta_{ij} + \eta^{S}_{ij}) E_j. \tag{3.21}$$

Accordingly, comparing (3.21) with (2.6),

$$e_{jkl} = -\epsilon_0 \eta^{S}_{ij} f_{ikl}, \quad \text{or} \quad f_{jkl} = -\epsilon_0^{-1} \chi^{S}_{ij} e_{ikl}; \tag{3.22}$$

$$e^{S}_{ij} = \epsilon_0 (\delta_{ij} + \eta^{S}_{ij}), \quad \text{or} \quad \eta^{S}_{ij} = \epsilon_0^{-1} \epsilon^{S}_{ij} - \delta_{ij}. \tag{3.23}$$

To find c^{P}_{ijkl} in terms of c^{E}_{ijkl}, first substitute the expression for P_i, given in terms of E_i and S_{ij} in (3.20),

---

*) Note that, according to the IRE Standards on Piezoelectric Crystals [16] reciprocal susceptibility is designated by χ whereas many text-books, e.g. [17] employ χ for susceptibility, In the IRE Standards, the susceptibility is η, following Voigt [1].

into the second of (3.15):

$$T_{ij} = f_{mij}\left(\epsilon_0 \eta^S_{mk} E_k - \epsilon_0 \eta^S_{mn} f_{nkl} S_{kl}\right) + c^P_{ijkl} S_{kl}$$

or, from (3.22),

$$T_{ij} = \left(c^P_{ijkl} - \epsilon_0 \eta^S_{mn} f_{mij} f_{nkl}\right) S_{kl} - e_{kij} E_k . \tag{3.24}$$

Comparing (3.24) with (2.5), we see that

$$c^E_{ijkl} = c^P_{ijkl} - \epsilon_0 \eta^S_{mn} f_{mij} f_{nkl} \tag{3.25}$$

or, from (3.22),

$$c^P_{ijkl} = c^E_{ijkl} + \epsilon_0^{-1} \chi^S_{mn} e_{mij} e_{nkl} . \tag{3.26}$$

With the aid of the foregoing expressions for the constants a^S_{ij} c^P_{ijkl} and f_{ijk}, we may readily reduce the three equations of equilibrium (3.11) and the two constitutive equations (3.15) to the classical form of equations on u_i and φ given at the end of the preceding section – by elimination of P_i . First, to solve the first of (3.15) for P_i, recall that the reciprocal of a^S_{ij} is $\epsilon_0 \eta^S_{ij}$. Then, upon multiplication, we have

$$P_i = -\epsilon_0 \eta^S_{ij}\left(E^L_j + f_{jkl} S_{kl}\right). \tag{3.27}$$

Substitute this for P_i in the second of (3.15) to obtain

$$T_{ij} = -f_{kij}\,\epsilon_0 \eta^S_{km}\left(E^L_m + f_{mpq} S_{pq}\right) + c^P_{ijkl} S_{kl} .$$

From the second equilibrium equation $E^L_i = \varphi_{,i}$, again omitting E^0_i.

Also $S_{ij} = \frac{1}{2}(u_{j,i} + u_{i,j})$. Hence

$$T_{ij,i} = \left(c^{P}_{ijk\ell} - \epsilon_0 \eta^{S}_{mn} f_{mij} f_{nk\ell}\right) u_{k,\ell i} - \epsilon_0 \eta^{S}_{km} f_{kij} \varphi_{,mi} .$$

Then, from the first of (3.22) and from (3.25),

$$T_{ij,i} + f_j = c^{E}_{ijk\ell} u_{k,\ell i} + e_{kij} \varphi_{,ki} + f_j$$

which is the left hand side of (2.8), as required.

Next, from (3.27),

$$\begin{aligned} P_{i,i} &= -\epsilon_0 \eta^{S}_{ij} \left(E^{L}_{j,i} + f_{jk\ell} S_{k\ell,i}\right) \\ &= -\epsilon_0 \eta^{S}_{ij} \varphi_{,ij} - \epsilon_0 \eta^{S}_{ij} f_{jk\ell} u_{k,\ell i} . \end{aligned}$$

Hence, the left hand side of the last of the equilibrium equations (3.11) becomes

$$-\epsilon_0 \varphi_{,ii} + P_{i,i} = -\epsilon_0 \eta^{S}_{ij} f_{jk\ell} u_{k,\ell i} - \epsilon_0 \left(\delta_{ij} + \eta^{S}_{ij}\right) \varphi_{,ij}$$

or, from the first of (3.22) and (3.23),

$$-\epsilon_0 \varphi_{,ii} + P_{i,i} = e_{ik\ell} u_{k,\ell i} - \epsilon^{S}_{ij} \varphi_{,ij}$$

which is the left hand side of (2.9), as required.

4. Polarization Gradient.

Toupin's form of the classical equations of elastic dielectrics reveals an omission in the classical theory. His equation

$$E_i^L - \varphi_{,i} + E_i^0 = 0$$

is the "equation of intramolecular force balance" [5] which he derived from fundamental considerations of the equilibrium of electrical forces, but which does not appear in the usual formulations. Granted the validity of the equation, it is significant that no boundary condition is associated with it. Whereas there is an equilibrium equation associated with each of the variables u_i, φ and P_i, only the variables u_i and φ are accompanied by boundary conditions. There is no coefficient of δP_i in the surface integral in (3.10) to complement that in the volume integral. This lack can be traced back to the absence of a functional dependence of W^L on the polarization gradient $P_{j,i}$. In fact, if we were to start by assuming dependence of W^L on the displacement and polarization and their gradients and truncate after the first gradient, $P_{j,i}$ would remain. Only u_i and the antisymmetric part of $u_{j,i}$ would have to be discarded - on the grounds of required translational and rotational invariance of W^L. The only justification for discarding $P_{j,i}$ would be its possible lack of importance judged on practical considerations. However, as we shall see in Sections 8 and 9, inclusion of $P_{j,i}$ in W^L results in the extension of the classical theory to accomodate interesting phenomena which are observed in nature but are not included in the classical theory. Furthermore, as shown in Section 10, the polarization gradient represents the continuum approximation to certain interatomic interactions which are known to contribute to fitting the predictions of crystal lattice theories to the results of neutron diffraction experiments.

To extend Toupin's variational principle to account for the contribution of the polarization gradient, it is only necessary to replace (3.3) with

$$H = W^L(S_{ij}, P_i, P_{j,i}) - \frac{1}{2}\epsilon_0 \varphi_{,i}\varphi_{,i} + \varphi_{,i}P_i . \qquad (4.1)$$

In addition, we shall include the kinetic energy so that (3.4) becomes

$$\delta\int_{t_0}^{t_1} dt \int_{V^*} \left(\frac{1}{2}\rho \dot{u}_i \dot{u}_i - H\right) dV + \int_{t_0}^{t_1} dt \left[\int_V (f_i \delta u_i + E_i^0 \delta P_i) dV + \int_S t_i \delta u_i dS\right] = 0 , \qquad (4.2)$$

for independent variations of u_i, φ and P_i between fixed limits at times t_0 and t_1.

The only additional operations involved are a new definition

$$E_{ij} = \frac{\partial W^L}{\partial P_{j,i}} \qquad (4.3)$$

and two integrations:

$$\frac{1}{2}\delta\int_{t_0}^{t_1}\dot{u}_i\dot{u}_i\,dt=\int_{t_0}^{t_1}\dot{u}_i\delta\dot{u}_i\,dt=\dot{u}_i\delta u_i\Big]_{t_0}^{t_1}-\int_{t_0}^{t_1}\ddot{u}_i\delta u_i\,dt=-\int_{t_0}^{t_1}\ddot{u}_i\delta u_i\,dt,$$

$$\int_V E_{ij}\delta P_{j,i}\,dV=\int_V\left[(E_{ij}\delta P_j)_{,i}-E_{ij,i}\delta P_j\right]dV=\int_S n_iE_{ij}\delta P_j\,dS-\int_V E_{ij,i}\delta P_j\,dV.$$

With these results and (3.9), (4.2) becomes

$$\int_{t_0}^{t_1}dt\int_V\Big[(T_{ij,i}+f_j-\rho\ddot{u}_j)\delta u_j+(E_{ij,i}+E_j^L-\varphi_{,j}+E_j^0)\delta P_i$$

$$+(-\epsilon_0\varphi_{,ii}+P_{i,i})\delta\varphi\Big]dV-\int_{t_0}^{t_1}dt\int_{V'}\epsilon_0\varphi_{,ii}\,\delta\varphi\,dV$$

$$(4.4)\qquad+\int_{t_0}^{t_1}dt\int_S\Big[(t_j-n_iT_{ij})\delta u_j-n_iE_{ij}\delta P_j+n_i(\epsilon_0[\![\varphi_{,i}]\!]-P_i)\delta\varphi\Big]dS=0.$$

Then the Euler equations are

$$(4.5)\qquad \text{in } V\quad\begin{cases} T_{ij,i}+f_j=\rho\ddot{u}_j,\\ E_{ij,i}+E_j^L-\varphi_{,i}+E_j^0=0,\\ -\epsilon_0\varphi_{,ii}+P_{i,i}=0;\end{cases}$$

$$(4.6)\qquad \varphi_{,ii}=0\quad\text{in } V',$$

with natural boundary conditions

$$(4.7)\qquad\begin{cases} n_iT_{ij}=t_j,\\ n_iE_{ij}=0,\\ n_i(-\epsilon_0[\![\varphi_{,i}]\!]+P_i)=0.\end{cases}$$

For W^L, we take

$$W^L = b^{\circ}_{ij} P_{j,i} + \frac{1}{2} a^{SG}_{ij} P_i P_j + \frac{1}{2} b^{PS}_{ijkl} P_{j,i} P_{l,k} + \frac{1}{2} c^{PG}_{ijkl} S_{ij} S_{kl}$$
$$+ d^{P}_{ijkl} P_{j,i} S_{kl} + f^{G}_{ijk} P_i S_{jk} + j^{S}_{ijk} P_i P_{k,j} . \tag{4.8}$$

The superscripts S, P and G, designating strain, polarization and polarization gradient, respectively, will be omitted in the sequel where no ambiguity results.

Substituting (4.8) in (3.6) and (4.3), we find

$$\left.\begin{aligned} -E^L_j &= a_{jk} P_k + j_{jkl} P_{l,k} + f_{jkl} S_{kl} , \\ E_{ij} &= j_{kij} P_k + b_{ijkl} P_{l,k} + d_{ijkl} S_{kl} + b^{\circ}_{ij} , \\ T_{ij} &= f_{kij} P_k + d_{klij} P_{l,k} + c_{ijkl} S_{kl} . \end{aligned}\right\} \tag{4.9}$$

The field equations (4.5) and (4.6), the constitutive equations (4.9) and the strain displacement relation $S_{ij} = \frac{1}{2}(u_{j,i} + u_{i,j})$ comprise the equations of the augmented theory [6]. It is apparent, from the form of the integrand of the surface integral in (4.4), what boundary conditions other than (4.7) are admissible. Thus, in place of the traction $n_i T_{ij}$, may be substituted the displacement u_i or a component of either and the resultant of the other in the plane at right angles. The same possibilities are open for $n_i E_{ij}$ and the polarization P_i. Finally either the surface charge $n_i(-\epsilon_0 [\![\varphi_{,i}]\!] + P_i)$ or the potential φ

may be specified. Of prime importance is the fact that both the potential φ and the polarization P_i may be specified. This is a latitude not permissible in the classical theory.

5. Centrosymmetric Materials: Cubic.

One of the novel properties of the augmented theory of elastic dielectrics is its accomodation of an electromechanical interaction even for materials with centrosymmetric physical properties. This may be seen by inspection of the energy density W^L as given in (4.8). For centrosymmetry, $f_{ijk} = j_{ijk} = 0$, since there are no centrosymmetric tensors of odd rank. However, this still leaves d_{ijkl} which is the coefficient of an electromechanical coupling term in the augmented theory but does not appear in the classical theory.

As a simple example of crystallographic centrosymmetry, consider the cubic point group $m3m$ (or O_h) the generators for which are [19]

$$\begin{pmatrix} -1 & 0 & 0 \\ 0 & -1 & 0 \\ 0 & 0 & -1 \end{pmatrix} \begin{pmatrix} 0 & 0 & 1 \\ 1 & 0 & 0 \\ 0 & 1 & 0 \end{pmatrix} \begin{pmatrix} 1 & 0 & 0 \\ 0 & 0 & 1 \\ 0 & -1 & 0 \end{pmatrix} \tag{5.1}$$

Applying these to the coefficients in (4.8), we find

$$\begin{aligned} & f_{ijk} = 0 , \quad j_{ijk} = 0 , \\ & a_{ij} = a_{11}\delta_{ij} , \; b^0_{ij} = b_0\delta_{ij}, \end{aligned} \tag{5.2a}$$

$$b_{ijkl} = b\delta_{ijkl} + b_{12}\delta_{ij}\delta_{kl} + b_{44}(\delta_{ik}\delta_{jl} + \delta_{il}\delta_{jk}) + b_{77}(\delta_{ik}\delta_{jl} - \delta_{il}\delta_{jk}),$$

(5.2b) $$c_{ijkl} = c\delta_{ijkl} + c_{12}\delta_{ij}\delta_{kl} + c_{44}(\delta_{ik}\delta_{jl} + \delta_{il}\delta_{jk}),$$

$$d_{ijkl} = d\delta_{ijkl} + d_{12}\delta_{ij}\delta_{kl} + d_{44}(\delta_{ik}\delta_{jl} + \delta_{il}\delta_{jk}),$$

where the coefficients $a_{11}, b_0, b_{pq}, c_{pq}$ and d_{pq} are constants and

$$b = b_{11} - b_{12} - 2b_{44},$$

(5.3) $$c = c_{11} - c_{12} - 2c_{44},$$

$$d = d_{11} - d_{12} - 2d_{44}.$$

In (5.2), δ_{ij} is the Kronecker delta and δ_{ijkl} is unity if all its indices are alike but zero otherwise.

Insertion of (5.2) in the constitutive equations (4.9) reduces the latter to

$$-E_i^L = a_{11} P_i,$$

$$E_{ij} = b\delta_{ijkl} P_{l,k} + b_{12}\delta_{ij} P_{k,k} + b_{44}(P_{j,i} + P_{i,j}) + b_{77}(P_{j,i} - P_{i,j})$$

(5.4) $$+ d\delta_{ijkl} S_{kl} + d_{12}\delta_{ij} S_{kk} + 2d_{44} S_{ij} + b_0\delta_{ij},$$

$$T_{ij} = d\delta_{ijkl} P_{l,k} + d_{12}\delta_{ij} P_{k,k} + d_{44}(P_{j,i} + P_{i,j})$$

$$+ c\delta_{ijkl} S_{kl} + c_{12}\delta_{ij} S_{kk} + 2c_{44} S_{ij}.$$

The "displacement" equations of motion are obtained by substituting (5.4) into (4.5) and employing $S_{ij}=\frac{1}{2}(u_{j,i}+u_{i,j})$ with the result

$$c\delta_{ijkl}u_{l,ki}+c_{12}u_{k,kj}+c_{44}(u_{j,ii}+u_{i,ji})+$$

$$+d\delta_{ijkl}P_{l,ki}+d_{12}P_{k,kj}+d_{44}(P_{j,ii}+P_{i,ji})+f_j=\rho\ddot{u}_j,$$

$$d\delta_{ijkl}u_{l,ki}+d_{12}u_{k,kj}+d_{44}(u_{j,ii}+u_{i,ji})-a_{11}P_j-\varphi_{,j}+ \tag{5.5}$$

$$+b\delta_{ijkl}P_{l,ki}+b_{12}P_{k,kj}+b_{44}(P_{j,ii}+P_{i,ji})+b_{77}(P_{j,ii}-P_{i,ji})+E_j^0=0,$$

$$-\epsilon_0\varphi_{,ii}+P_{i,i}=0.$$

6. Centrosymmetric Materials: Isotropic.

For centrosymmetric isotropic materials it is only necessary to set

$$b_{11} = b_{12} + 2b_{44}, \quad c_{11} = c_{12} + 2c_{44}, \quad d_{11} = d_{12} + 2d_{44}; \tag{6.1}$$

so that

$$b = c = d = 0 \tag{6.2}$$

in the equations for the centrosymmetric cubic case. The equations of motion (5.5) then become, in vector notation,

$$\begin{aligned} &c_{44}\nabla^2\mathbf{u} + (c_{12}+c_{44})\boldsymbol{\nabla}\boldsymbol{\nabla}\cdot\mathbf{u} + d_{44}\nabla^2\mathbf{P} + (d_{12}+d_{44})\boldsymbol{\nabla}\boldsymbol{\nabla}\cdot\mathbf{P} + \mathbf{f} = \rho\ddot{\mathbf{u}}, \\ &d_{44}\nabla^2\mathbf{u} + (d_{12}+d_{44})\boldsymbol{\nabla}\boldsymbol{\nabla}\cdot\mathbf{u} + (b_{44}+b_{77})\nabla^2\mathbf{P} + (b_{12}+b_{44}-b_{77})\boldsymbol{\nabla}\boldsymbol{\nabla}\cdot\mathbf{P} - a_{11}\mathbf{P} - \boldsymbol{\nabla}\varphi + \mathbf{E}^0 = 0, \\ &-\epsilon_0\nabla^2\varphi + \boldsymbol{\nabla}\cdot\mathbf{P} = 0. \end{aligned} \tag{6.3}$$

Schwartz [18] has given a complete solution of (6.3) in terms of two vector functions, a scalar function and the potential φ:

$$\begin{aligned} \mathbf{u} = {} & \mathbf{B} - \tfrac{1}{2}(1-k)\boldsymbol{\nabla}(\mathbf{r}\cdot\mathbf{B} + B_0) + a_{11}^{-1}c_{44}k_2(k_2-k_1)\boldsymbol{\nabla}\boldsymbol{\nabla}\cdot\mathbf{B} - k_2(\mathbf{K} - \ell_2^2\boldsymbol{\nabla}\boldsymbol{\nabla}\cdot\mathbf{K}) \\ & -\epsilon_0 k_1\boldsymbol{\nabla}\varphi + a_{11}^{-1}k_2(1+a_{11}\epsilon_0)(1-\ell_1^2\nabla^2)\boldsymbol{\nabla}\varphi, \end{aligned} \tag{6.4}$$

$$\mathbf{P} = -a_{11}^{-1}c_{44}(k_2-k_1)\boldsymbol{\nabla}\boldsymbol{\nabla}\cdot\mathbf{B} + \epsilon_0\boldsymbol{\nabla}\varphi - a_{11}^{-1}(1+a_{11}\epsilon_0)(1-\ell_1^2\nabla^2)\boldsymbol{\nabla}\varphi + \mathbf{K} - \ell_2^2\boldsymbol{\nabla}\boldsymbol{\nabla}\cdot\mathbf{K},$$

where $\underset{\sim}{B}, B_0, \underset{\sim}{K}$ and φ must satisfy the equations

$$
\begin{aligned}
c_{44}\nabla^2 \underset{\sim}{B} &= -\underset{\sim}{f}, \\
c_{44}\nabla^2 B_0 &= \underset{\sim}{r}\cdot\underset{\sim}{f}, \\
a_{11}(1-\ell_2^2\nabla^2)\underset{\sim}{K} &= \underset{\sim}{E}^0 - k_2\underset{\sim}{f}, \\
(1+a_{11}\epsilon_0)(1-\ell_1^2\nabla^2)\nabla^2\varphi &= \nabla\cdot\underset{\sim}{E}^0 - k_1\underset{\sim}{\nabla}\cdot\underset{\sim}{f},
\end{aligned}
\tag{6.5}
$$

in which $\underset{\sim}{r}$ is the position vector and

$$k = c_{44}/c_{11}, \quad k_1 = d_{11}/c_{11}, \quad k_2 = d_{44}/c_{44}, \tag{6.6}$$

$$\ell_1^2 = (b_{11}c_{11} - d_{11}^2)/c_{11}(a_{11}+\epsilon_0^{-1}), \quad \ell_2^2 = \left[(b_{44}+b_{77})c_{44} - d_{44}^2\right]/a_{11}c_{44}. \tag{6.7}$$

Among other solutions, Schwartz gives the fundamental one for the concentrated force $\underset{\sim}{F}$:

$$\underset{\sim}{B} = \frac{\underset{\sim}{F}}{4\pi c_{44} r}, \quad B_0 = 0, \tag{6.8}$$

$$\underset{\sim}{K} = -\frac{k_2 e^{-r/\ell_2}}{4\pi a_{11}\ell_2^2 r}\underset{\sim}{F}, \quad \varphi = \frac{k_1}{4\pi(1+\epsilon_0 a_{11})}\underset{\sim}{F}\cdot\underset{\sim}{\nabla}\left(\frac{1-e^{-r/\ell_1}}{r}\right). \tag{6.9}$$

7. Surface Energy of Deformation and Polarization.

It will be observed that the internal energy density of polarization and deformation, W^L, contains the linear term, $b^o_{ij} P_{j,i}$, in the polarization gradient. The analogous term linear in the strain, say $c^o_{ij} S_{ij}$, is omitted as it is of no consequence. It leads to a homogeneous stress which, in a bounded body with a free surface, can be removed by a homogeneous strain which, in turn, can be regarded as the reference configuration. The situation is otherwise for the term linear in the polarization gradient. Although this, too, produces a homogeneous field, it is a field of E_{ij} rather than T_{ij}. The removal of $n_i E_{ij}$ to free a surface results, as is illustrated in Section 8, in a polarization and strain which decay exponentially into the interior of the body.

To separate a body into two parts along a surface, a <u>bond</u> energy must be overcome. This is the energy that would be required to break the atomic bonds across the surface if the strain and polarization were <u>prevented</u> from developing, say by some hypothetical external field. The <u>release</u> of such a field would result in a deformation and polarization, localized near the surface, with which is associated a surface energy of deformation and polarization - always negative. Thus, the energy required to separate a body into two parts along a surface is the bond energy less the absolute value of the surface energy of deformation and polarization. The former is not represented in the present theory, but the latter is introduced via the linear term $b^0_{ij} P_{j,i}$ in W^L and a formula for that energy may be found as follows.

The total energy in the system in equilibrium is

$$\int_{V^*} W\,dV = \int_{V^*} (W^L + \tfrac{1}{2}\epsilon_0 \varphi_{,i}\varphi_{,i})dV = \int_V W^L dV + \int_{V^*} \tfrac{1}{2}\epsilon_0 \varphi_{,i}\varphi_{,i}\,dV. \tag{7.1}$$

By means of the constitutive equations (4.9), W^L may be expressed in the form

$$W^L = \tfrac{1}{2} T_{ij} S_{ij} + \tfrac{1}{2} E_{ij} P_{j,i} - \tfrac{1}{2} E^L_i P_i + \tfrac{1}{2} b^0_{ij} P_{j,i}. \tag{7.2}$$

Through the use of the chain rule of differentiation and the divergence theorem,

$$\int_V T_{ij} S_{ij}\,dV = \int_V T_{ij} u_{j,i}\,dV = \int_V [(T_{ij} u_j)_{,i} - T_{ij,i} u_j]\,dV =$$

$$= \int_S n_i T_{ij} u_j\,dS - \int_V T_{ij,i} u_j\,dV,$$

$$\int_V E_{ij} P_{j,i}\,dV = \int_V [(E_{ij} P_j)_{,i} - E_{ij,i} P_j]dV = \int_S n_i E_{ij} P_j\,dS - \int_V E_{ij,i} P_j\,dV,$$

$$\int_V b_{ij} P_{j,i}\,dV = \int_S n_i b^0_{ij} P_j\,dS,$$

$$\int_{V^*} \varphi_{,i}\varphi_{,i}\,dV = \int_{V^*} [(\varphi_{,i}\varphi)_{,i} - \varphi_{,ii}\varphi]\,dV =$$

$$= \int_S n_i [\![\varphi_{,i}]\!] \varphi\,dS - \int_V \varphi_{,ii}\varphi\,dV - \int_{V'} \varphi_{,ii}\varphi\,dV.$$

Upon assembling these results into (7.1), we have

$$\int_{V*} W\,dV = \frac{1}{2}\int_S n_i\left[b^0_{ij}P_j + T_{ij}u_j + E_{ij}P_j + \epsilon_0[\![\varphi_{,i}]\!]\varphi\right]dS -$$

$$(7.3)\quad -\frac{1}{2}\int_V (T_{ij,i}u_j + E_{ij,i}P_j + E^L_j P_j + \epsilon_0\varphi_{,ii}\varphi)dV - \frac{1}{2}\int_{V'}\epsilon_0\varphi_{,ii}\varphi\,dV.$$

Application of the equilibrium equations, (4.5) and (4.6) with $\rho\ddot{u}_j = 0$, to (7.3) reduces it to

$$(7.4)\quad \int_{V*} W\,dV = \frac{1}{2}\int_V (f_j u_j + E^0_j P_j)dV + \frac{1}{2}\int_S n_i\left[T_{ij}u_j + E_{ij}P_j - (-\epsilon_0[\![\varphi_{,i}]\!] + P_i) + b^0_{ij}P_j\right]dS.$$

Then, in the absence of external fields $(f_j = E_j = 0)$ and with the boundary free, i.e. (4.7) with $\mathbf{t} = 0$, (7.4) reduces to

$$(7.5)\qquad \int_{V*} W\,dV = \frac{1}{2}\int_S n_i b^0_{ij} P_j\,dS .$$

This is the energy of deformation and polarization which must be added to the bond energy to obtain the total energy required to separate the material into two parts along the surface S. As indicated by Schwartz [18], the additional energy is always negative as a consequence of the positive-definiteness of the quadratic part of the total energy which, in turn, is required for stability – i.e. because energy must be stored, rather than generated, during the application of external body and surface forces. Thus:

$$\int_{V*} W\,dV = \int_{V*}\left[\frac{1}{2}\varphi_{,i}\varphi_{,i} + (W^L - b^0_{ij}P_{j,i})\right]dV + \int_V b^0_{ij}P_{j,i}\,dV ,$$

$$(7.6)\qquad = \int_{V*}\left[\frac{1}{2}\varphi_{,i}\varphi_{,i} + (W^L - b^0_{ij}P_{j,i})\right]dV + \int_S n_i b^0_{ij}P_j\,dS .$$

Combining (7.5) and (7.6):

$$\int_{V^*} \left[\frac{1}{2} \varphi_{,i} \varphi_{,i} + \left(W^L - b^0_{ij} P_{j,i} \right) \right] dV + \frac{1}{2} \int_S n_i b_{ij} P_j \, dS = 0 . \qquad (7.7)$$

But the volume integral in (7.7) is greater than zero. Hence

$$\int_S n_i b^0_{ij} P_j \, dS < 0 . \qquad (7.8)$$

Finally, we may define

$$T = \frac{1}{2} \left[n_i b^0_{ij} P_j \right]_S \qquad (7.9)$$

as the surface energy of deformation and polarization per unit area, sometimes called the <u>surface tension</u>.

8. Energy at a Plane, Free Surface.

In this section the solution is given for the displacement, polarization, potential and surface energy of deformation and polarization in the half-space of a centrosymmetric cubic crystal bounded by a free (100) face. The same solution applies to the isotropic half-space if the conditions (6.1) are applied.

For this problem, the fields are one-dimensional and in equilibrium so that, for the half-space $x \geq 0$, the equations of motion (5.5) reduce to

$$
\begin{aligned}
& c_{11}\,\partial^2 u + d_{11}\,\partial^2 P = 0\,, \\
(8.1) \qquad & d_{11}\,\partial^2 u + b_{11}\,\partial^2 P - a_{11} P_1 - \partial\varphi = 0\,, \\
& -\epsilon_0\,\partial^2\varphi + \partial P = 0\,;
\end{aligned}
$$

and the boundary conditions (4.7), on $x = 0$, become

$$
\begin{aligned}
& c_{11}\,\partial u + d_{11}\,\partial P = 0\,, \\
(8.2) \qquad & d_{11}\,\partial u + b_{11}\,\partial P = -b_0\,, \\
& -\epsilon_0\,\partial\varphi + P = 0\,,
\end{aligned}
$$

where $\partial = d/dx$ and u, P are the x-components of u_i and P_i .

Consider

$$u = A_1 e^{-x/\ell} ,$$
$$P = A_2 e^{-x/\ell} , \tag{8.3}$$
$$\varphi = A_3 e^{-x/\ell} .$$

Upon substituting (8.3) into (8.1) we find

$$A_3 = -\ell A_2/\epsilon_0 = \ell c_{11} A_1/\epsilon_0 d_{11} , \tag{8.4}$$

$$\ell = \left(\frac{b_{11} c_{11} - d_{11}^2}{c_{11}(a_{11}+\epsilon_0^{-1})}\right)^{\frac{1}{2}} ; \tag{8.5}$$

i.e., ℓ is the same as ℓ_1 in (6.7). Positive-definiteness of W^L requires ℓ to be real.

With (8.4), the first and third of the boundary conditions (8.2) are satisfied identically and the second boundary condition yields

$$A_1 = -\frac{b_0 d_{11}}{c_{11}\ell(a_{11}+\epsilon_0^{-1})} . \tag{8.6}$$

Then, from (8.4),

$$A_2 = \frac{b_0}{\ell(a_{11}+\epsilon_0^{-1})} , \quad A_3 = -\frac{b_0}{1+a_{11}\epsilon_0^{-1}} . \tag{8.7}$$

It may be seen that the freeing of the boundary results in a field of displacement, polarization and potential localized at the surface, decaying into the interior with decay constant ℓ. Associated with the localized field is a surface energy of deformation and polarization, per unit area, given by (7.9):

$$T = -\frac{b_0^2}{2\ell a_{11}(1+a_{11}^{-1}\epsilon_0^{-1})}. \tag{8.8}$$

Note that, from (3.17), (3.19) and (3.23), $1+a_{11}^{-1}\epsilon_0^{-1} = \epsilon_{11}^{s}/\epsilon_0$, i.e. the dielectric constant in the [100] direction.

Askar, Lee and Cakmak [11] have calculated a_{11}, b_{11}, c_{11} and d_{11} for several alkali halide crystals by relating them to analogous constants in a Cochran [13] lattice of Dick--Overhauser [12] shell-model atoms in an NaCl structure. The correspondence is not complete because the NaCl lattice is diatomic whereas the continuum is a limiting form of a monatomic lattice. However, the calculation at least yields order of magnitude results. For NaCl, for example, they find

$$\ell = 1\cdot 3 \times 10^{-8}\,\text{cm}, \quad T = -59\ \text{erg}/\text{cm}^2. \tag{8.9}$$

The decay constant ℓ is about half the distance between nearest neighbor atoms in the NaCl crystal, indicating extremely rapid decay away from the surface. The surface energy of deformation and polarization is about thirty percent of the bond energy, so that it is far from negligible. Both of these results conform to experimental observations [24, 7].

Additional solutions for surface energies of deformation and polarization in isotropic materials, have been obtained by Schwartz [18], for internal and external spherical surfaces, and by Askar, Lee and Cakmak [25] for internal spherical and cylindrical surfaces and a crack. The latter have used average values of their previous calculations of constants for alkali halides [11] to obtain some quantitative results. They find that the absolute values of T are reduced by curvature of the surface and they also conclude that T is finite at the root of a crack.

9. Capacitance of Thin, Dieletric Films.

As a second example, consider the capacitance of a metal-dielectric-metal sandwich. Suppose the middle plane of the structure is the plane of y and z. If end effects are neglected, the fields are one-dimensional and the equations of equilibrium are given by (8.1). Also, assume that the conditions of isotropy (6.1) hold.

Let us first solve the problem within the framework of the classical theory. Then b_{11} and d_{11} are zero and the equations of equilibrium reduce to

$$
\begin{aligned}
& c_{11}\, \partial^2 u = 0 , \\
(9.1) \qquad & a_{11} P + \partial \varphi = 0 , \\
& -\epsilon_0 \partial^2 \varphi + \partial P = 0 .
\end{aligned}
$$

For a dielectric layer with traction-free surfaces at $x \pm h$, on which are impressed voltages $\pm V$, the boundary conditions, according to the classical theory, are

$$
\begin{aligned}
& (\partial u)_{x=\pm h} = 0 , \\
(9.2) \qquad & (\varphi)_{x=\pm h} = \pm V ;
\end{aligned}
$$

and the solution of (9.1), subject to these boundary conditions, is, except for additive constants in u and φ,

$$u = 0 ,$$
$$P = -\epsilon_0 \eta V h , \tag{9.3}$$
$$\varphi = V x / h ,$$

where $\eta = (\epsilon_0^{-1} a_{11}^{-1})$ is the dielectric susceptibility.

The capacitance (per unit area) is the ratio of the surface charge (per unit area) to the voltage drop across the layer:

$$C = \frac{(\epsilon_0 \partial\varphi - P)_{x=\pm h}}{2V} = \frac{\epsilon_0(1+\eta)}{2h} = \frac{\epsilon}{2h}, \tag{9.4}$$

where ϵ is the permittivity of the dielectric.

Thus, according to the classical theory, there is no strain, the polarization is uniform, the potential varies linearly through the thickness and the capacitance is inversely proportional to the thickness, so that a graph of inverse capacitance vs. thickness is a straight line through the origin. However, in a series of experiments with a variety of very thin dielectric films between metal electrodes, Mead [20,21,22] found a different relation between inverse capacitance and thickness. His experimental data fall on straight lines which, if extended to zero thickness, have positive intercepts of inverse capacitance, as illustrated in Fig.9.1. Initially [20], Mead suggested that the anomaly might be due to penetration of the electric field into the electrodes; but subsequently [21,22] he abandoned that view, although it has been supported by Ku and Ullman [23].

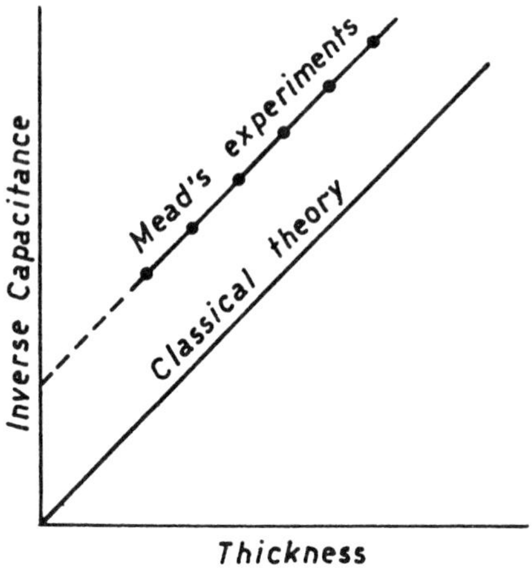

Fig.9.1

An alternative explanation is found in the augmented theory of elastic dielectrics.

The solution of the augmented equations of equilibrium (8.1), analogous to the one just given for the classical equations, requires a boundary condition in addition to the surface traction and potential. The additional condition can be the specification of the polarization at the surface of the dielectric. Now, the polarization at a boundary of the dielectric in a metal-dielectric-metal sandwich will depend on the physical properties of the adjacent electrode and metal-dielectric interface; and these properties are outside the compass of the theory of dielectrics. However, since the polarization in the metal is zero, it is reasonable to assume that the surface polarization in the dielectric, if not actually zero, will lie between zero and the classical value given by the second of (9.3). Thus, assuming that the two electrodes and interfaces are the same, their influence on the surface polarization may be introduced, phenomenologically, by setting the boundary condition

$$(P)_{x=\pm h} = -k\epsilon_0 \eta V/h, \qquad 0 \leq k \leq 1. \tag{9.5a}$$

The classical condition is $k=1$ while $k=0$ describes continuity of polarization across the interfaces. We suppose also, as is assumed in the classical theory, that the traction across the interfaces is zero. From (5.4) this condition is

$$(T_{11})_{x=\pm h} = (c_{11}\partial u + d_{11}\partial P)_{x=\pm h} = 0. \tag{9.5b}$$

Finally, we suppose that the voltages applied to the dielectric at the interfaces are again

$$(\varphi)_{x=\pm h} = \pm V. \tag{9.5c}$$

We have now to solve (8.1) subject to the boundary conditions (9.5a,b,c). Let

$$\begin{aligned} u &= B_1 \cosh(x/\ell), \\ P &= A_2 + B_2 \cosh(x/\ell), \\ \varphi &= A_3 x + B_3 \sinh(x/\ell). \end{aligned} \tag{9.6}$$

Upon substituting (9.6) in (8.1), we find

$$(9.7) \qquad A_2 = \epsilon_0 \eta A_3 , \quad B_3 = \ell B_2/\epsilon_0 = -\ell c_{11} B_1 / \epsilon_0 d_{11} ,$$

$$(9.8) \qquad \ell = \left(\frac{b_{11} c_{11} - d_{11}^2}{c_{11}(a_{11} + \epsilon_0^{-1})} \right)^{\frac{1}{2}} = \left(\frac{\epsilon_0 (b_{11} c_{11} - d_{11}^2)}{c_{11}(1+\eta^{-1})} \right)^{\frac{1}{2}}$$

i.e., ℓ is the same as in (8.5).

As for boundary conditions, (9.5b) is satisfied identically while (9.5a) and (9.5c) become

$$(9.9) \qquad \begin{aligned} &A_2 + B_2 \cosh(h/\ell) = -k\epsilon_0 \eta V/h, \\ &A_3 h + B_3 \sinh(h/\ell) = V, \end{aligned}$$

respectively. From (9.7) and (9.9),

$$(9.10) \qquad \begin{aligned} &A_3 = (B_3/\eta\ell) \cosh(h/\ell) + kVh, \\ &B_3 = (1-k)\eta V / \left[\eta \sinh(h/\ell) + (h/\ell)\cosh(h/\ell)\right]. \end{aligned}$$

The remaining constants are obtained easily from (9.10) and (9.7).

The capacitance, ignoring any voltage drop that may occur in the electrodes, is

$$(9.11) \qquad C = \frac{(\cdot\epsilon_0 \partial\varphi - P)_{x=\pm h}}{2V} = \frac{\epsilon}{2h} \; \frac{1 + (k\eta\ell/h)\tanh(h/\ell)}{1 + (\eta\ell/h)\tanh(h/\ell)} .$$

In Fig.9.2 the heavy, full line shows the relation between normalized inverse capacitance and normalized thickness, according to (9.11), for the case $k = 0.1$, $\eta = 10$. Calculations from Mead's data, for small k, indicate that the material property ℓ for all his dielectrics is of the order of magnitude of a few angstroms; and this is supported by Askar, Lee and Cakmak's calculations for alkali halides [11]. Hence, Mead's data, which do not extend below a thickness of 30 Å, would be well to the right of the knee of the curve, in Fig.9.2, and so would give the appearance of a linear relation which, if extended to zero thickness, would have a positive, non-zero intercept of inverse capacitance. This intercept, according to (9.11), is

$$C_0^{-1} = 2\ell(1-k)\eta/\epsilon. \tag{9.12}$$

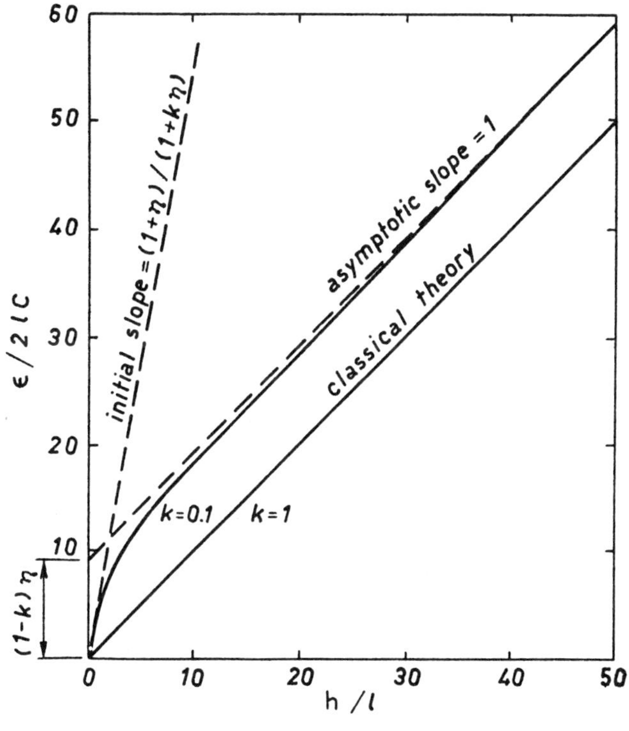

Fig.9.2

If $k=1$, the intercept reduces to zero and, in fact, the whole solution reduces to the classical one. However, it seems unlikely that the presence of the metal would not influence the polarization of the dielectric at the metal-dielectric interface.

The variations of polarization and potential across the thickness of the dielectric are illustrated by the curves in Fig.9.3. The absolute value of the polarization is almost uniform across the major portion of the thickness and slightly less than the uniform polarization of the classical theory; but then drops sharply, near the surfaces, to boundary values of k times the classical polarization, as specified. The potential has an almost uniform gradient, less than the uniform gradient of the classical theory, over most of the thickness, but then increases sharply on approaching the boundaries.

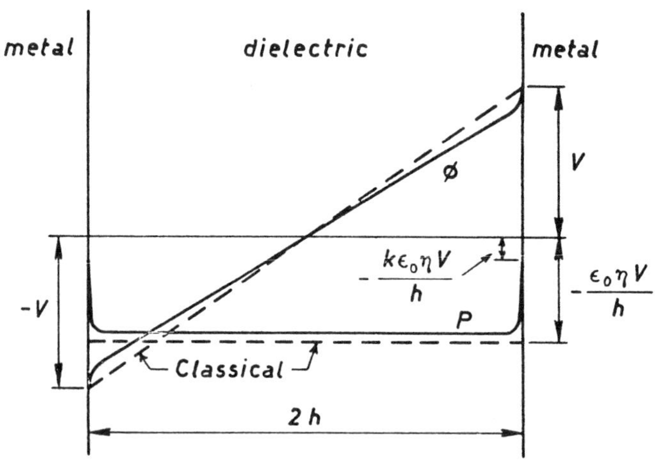

Fig.9.3

10. Lattice of Shell-Model Atoms.

In this section the equations of equilibrium are derived for a monatomic, one-dimensional lattice of the Cochran type [13] based on the Dick-Overhauser [12] shell-model of the atom: a core, comprising the nucleus and inner electrons, surrounded by a shell of outer electrons. The polarization is proportional to the relative displacement of the core and shell of the atom. In addition to this intra-atomic interaction, account is taken of interatomic interactions between core and core, core and shell and shell and shell of nearest neighbor atoms. It is shown that the equations of the lattice have, as their continuum limit, the equations of the augmented theory, including the contribution of the polarization gradient to the stored energy, rather than the classical theory of elastic dielectrics. The additional effects associated with the new constants b_{11} and d_{11} stem primarily from the shell-shell interaction. This interaction is known to be important in the matching of lattice dispersion relations to neutron diffraction dispersion data at short wave lengths [14,15].

We consider a single line of atoms extracted from a three-dimensional, monatomic lattice of shell-model atoms (Fig.10.1). The atoms are situated at $x = na$ where $\underline{n}$ is a positive or negative integer and $\underline{a}$ is the lattice constant, i.e. the distance between nearest neighbor atoms.

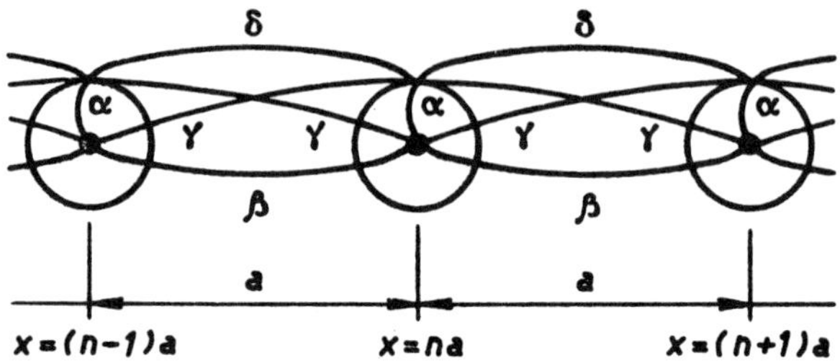

Fig.10.1

The displacements of the core and shell of the atom at $x = na$ are designated by u_n and s_n, respectively. The force constant, i.e. the force per unit relative displacement, of the intra-atomic core-shell interaction is designated by α and the force constants of the interatomic core-core, core-shell and shell-shell interactions are designated by β, γ and δ, respectively.

The equation of equilibrium of the nth interior <u>atom</u> (core and shell combined) is obtained by setting equal to zero the sum of the forces on its core and shell exerted by the cores and shells of its two nearest neighbor atoms:

$$\begin{aligned}&\beta(u_{n+1}-u_n)+\gamma(s_{n+1}-u_n)+\gamma(u_{n+1}-s_n)+\delta(s_{n+1}-s_n)\\ &-\beta(u_n-u_{n-1})-\gamma(u_n-s_{n-1})-\gamma(s_n-u_{n-1})-\delta(s_n-s_{n-1})=0\,.\end{aligned}\tag{10.1}$$

The <u>shell</u>, alone, of the nth atom is acted upon by its own core, the core and shell of each of the two neighboring atoms and, also, the Maxwell, electric self-field (which occupies all space). Hence,

$$\begin{aligned}\alpha(u_n - s_n) + \gamma(u_{n+1} - s_n) + \delta(s_{n+1} - s_n)\\ -\gamma(s_n - u_{n-1}) - \delta(s_n - s_{n-1}) + q E_n = 0,\end{aligned} \tag{10.2}$$

where q is the charge of the atom and E_n is the value, at $x = na$, of the Maxwell field. Now, $E_n = -[\partial\varphi]_{x=na}$ and this may be expressed in terms of φ_n, the value of the potential φ at $x = na$, by expanding the derivative, ∂, in an infinite series of forward differences [26]:

$$-E_n = \partial_+ \varphi_n = \sum_{m=1}^{\infty} (-1)^{m-1} m^{-1} a^{m-1} \Delta_+^m \varphi_n, \tag{10.3}$$

where

$$\begin{aligned}\Delta_+ \varphi_n &= (\varphi_{n+1} - \varphi_n)/a,\\ \Delta_+^2 \varphi_n &= (\varphi_{n+2} - 2\varphi_{n+1} + \varphi_n)/a^2,\\ \Delta_+^3 \varphi_n &= (\varphi_{n+3} - 3\varphi_{n+2} + 3\varphi_{n+1} - \varphi_n)/a^3,\\ &\vdots \qquad\qquad \vdots\end{aligned} \tag{10.4}$$

The polarization of the nth atom, per unit area of the three-dimensional lattice, is defined by

$$P_n = (s_n - u_n)q/a^3. \tag{10.5}$$

Further, let us adopt a symbol for the second central difference (divided by a^2):

$$\Delta^2 f_n = (f_{n+1} - 2f_n + f_{n-1})/a^2. \tag{10.6}$$

Then the equations of equilibrium (10.1) and (10.2) may be rearranged to the following forms:

$$(\beta + 2\gamma + \delta)a^{-1}\Delta^2 u_n + (\gamma+\delta)a^2 q^{-1}\Delta^2 P_n = 0, \tag{10.7}$$

$$(\gamma+\delta)a^2 q^{-1}\Delta^2 u_n + \delta a^5 q^{-2}\Delta^2 P_n - (\alpha + 2\gamma)a^3 q^{-2} P_n - \partial_+ \varphi_n = 0. \tag{10.8}$$

It will be observed that the difference equations (10.7) and (10.8) have exactly the same form as the first two of (8.1), which are the one-dimensional differential equations of equilibrium of the augmented theory of elastic dielectrics. Accordingly, if we set

$$\begin{aligned} a_{11} &= (\alpha + 2\gamma)a^3 q^{-2} = \epsilon_0^{-1}\eta^{-1}, \qquad & b_{11} &= \delta a^5 q^{-2}, \\ c_{11} &= (\beta + 2\gamma + \delta)a^{-1}, & d_{11} &= (\gamma+\delta)a^2 q^{-1}, \end{aligned} \tag{10.9}$$

and recall that [26]

$$\Delta^2 = \partial^2 + \frac{a^2\partial^4}{12} + \frac{a^4\partial^6}{360} + \ldots, \tag{10.10}$$

the first two of (8.1) are the lowest order continuum approximations to the equations of equilibrium of the atom and its shell, respectively.

As may be seen from (10.9), the polarization gradient terms (those with coefficients b_{11} and d_{11}) in (8.1) stem from the shell-shell and interatomic core-shell interactions, identified by the force constants δ and γ, respectively. In fact, the form of (8.1) is preserved if the interatomic core-shell interaction is omitted but the shell-shell interaction is retained. If both of these interactions are omitted, the continuum approximation reduces to (9.1), which is Toupin's form of the classical equations. The second of (9.1), Toupin's "equation of intramolecular force balance", does not appear in the traditional equations of elastic dielectrics; but we see that it is a fundamental equilibrium condition: the equilibrium of the shell under the action of the core of the same atom and the surrounding Maxwell field.The corresponding equation in the extended theory (the second of (8.1)) includes, in addition, the action of the adjacent atoms on the shell (in the continuum approximation).

We have yet to derive the lattice counterpart of the third of (8.1) and establish the boundary conditions. It is illuminating to reach these results from considerations of energy.

By analogy with (3.3) and (4.8), let us take, for the electric enthalpy of the one-dimensional lattice (per unit volume a^3)

$$\mathcal{H} = \sum_n \Big[b_0 \Delta_+ P_n + \frac{1}{2} \epsilon_0^{-1} \eta^{-1} P_n^2 + \frac{1}{2} b_{11} (\Delta_+ P_n)^2 + c_{11} (\Delta_+ u_n)^2 + d_{11} (\Delta_+ P_n)(\Delta_+ u_n) - \frac{1}{2} \epsilon_0 (\partial_+ \varphi_n)^2 + P_n (\partial_+ \varphi_n) \Big]. \quad (10.11)$$

We have to find the derivatives of $\mathcal{H}$ with respect to u_n, P_n and φ. The first two are conventional and straightforward, for example,

$$\frac{\partial(\Delta_+ u_m)^2}{\partial u_n} = \frac{1}{a^2}\,\frac{\partial(u_{m+1}-u_m)^2}{\partial u_n} = \frac{2}{a^2}\left(u_{m+1}-u_m\right)\,\frac{\partial(u_{m+1}-u_m)}{\partial u_n},$$

$$= \frac{2}{a^2}\left(u_{m+1}-u_m\right)\left(\delta_n^{m+1}-\delta_n^{m}\right),$$

$$= \frac{2}{a^2}\left(u_n - u_{n-1} - u_{n+1} + u_n\right) = 2\,\Delta^2 u_n\,, \tag{10.12}$$

where δ_q^p is the Kronecker delta.

To find the derivative of $\partial_+\varphi$ with respect to φ, note first that, from (10.3),

$$\partial_+\varphi_n = a^{-1}\Big[\varphi_{n+1}-\varphi_n-\frac{1}{2}\left(\varphi_{n+2}-2\varphi_{n+1}+\varphi_n\right)$$
$$+\frac{1}{3}\left(\varphi_{n+3}-3\varphi_{n+2}+3\varphi_{n+1}-\varphi_n\right)-\dots\Big].$$

Then

$$P_m\,\frac{\partial(\partial_+\varphi_m)}{\partial\varphi_n} = a^{-1}P_m\Big[\delta_n^{m+1}-\delta_n^{m}-\frac{1}{2}\left(\delta_n^{m+2}-2\delta_n^{m+1}+\delta_n^{m}\right)$$
$$+\frac{1}{3}\left(\delta_n^{m+3}-3\delta_n^{m+2}+3\delta_n^{m+1}-\delta_n^{m}\right)-\dots\Big],$$

$$= a^{-1}\Big[P_{n-1}-P_n-\frac{1}{2}\left(P_{n-2}-2P_{n-1}+P_n\right)$$
$$+\frac{1}{3}\left(P_{n-3}-3P_{n-2}+3P_{n-1}-P_n\right)-\dots\Big],$$

$$= -\,\partial_-P_n\,, \tag{10.13}$$

where

$$\partial_- P_n = \sum_{m=1}^{\infty} m^{-1} a^{m-1} \Delta_-^m P_n \tag{10.14}$$

and

$$\begin{aligned} \Delta_- P_n &= (P_n - P_{n-1})/a, \\ \Delta_-^2 P_n &= (P_n - 2P_{n-1} + P_{n-2})/a^2, \\ \Delta_-^3 P_n &= (P_n - 3P_{n-1} + 3P_{n-2} - P_{n-3})/a^3, \\ &\vdots \qquad\qquad \vdots \end{aligned} \tag{10.15}$$

i.e., $\partial_- P_n$ is the expansion of ∂P in an infinite series of backward differences [26]. Finally,

$$\frac{\partial(\partial_+ \varphi_m)^2}{\partial \varphi_n} = 2\,\partial_+ \varphi_m \frac{\partial(\partial_+ \varphi_m)}{\partial \varphi_n} = -2\,\partial_- \partial_+ \varphi_n. \tag{10.16}$$

We now find the equations of equilibrium:

$$\begin{aligned} -\frac{\partial \mathcal{H}}{\partial u_n} &= c_{11} \Delta^2 u_n + d_{11} \Delta^2 P_n = 0, \\ -\frac{\partial \mathcal{H}}{\partial P_n} &= d_{11} \Delta^2 u_n + b_{11} \Delta^2 P_n - \epsilon_0^{-1} \eta^{-1} P_n - \partial_+ \varphi_n = 0, \\ -\frac{\partial \mathcal{H}}{\partial \varphi_n} &= -\epsilon_0 \partial_- \partial_+ \varphi_n + \partial_- P_n = 0. \end{aligned} \tag{10.17}$$

The first two of (10.17) are the same as (10.7) and (10.8), with (10.9). This justifies the assumption (10.11) for the electric enthalpy. Also, we have found the third of (10.17), which has the third of (8.1) as its continuum form.

If the lattice is of finite thickness spanning an

odd number of atoms with the end ones at $n = \pm N$, the conditions for free boundaries are

$$
(10.18)\qquad
\begin{aligned}
-\frac{\partial \mathcal{H}}{\partial(\Delta_+ u_{\pm N})} &= c_{11}\Delta_+ u_{\pm N} + d_{11}\Delta_+ P_{\pm N} = 0\,,\\
-\frac{\partial \mathcal{H}}{\partial(\Delta_+ P_{\pm N})} &= d_{11}\Delta_+ u_{\pm N} + b_{11}\Delta_+ P_{\pm N} + b_0 = 0\,,\\
-\frac{\partial \mathcal{H}}{\partial(\partial_+ \varphi_{\pm N})} &= \epsilon_0 \partial_+ \varphi_{\pm N} - P_{\pm N} = 0\,,
\end{aligned}
$$

where

$$
(10.19)\qquad \Delta_+ f_{\pm N} = (f_{\pm(N+1)} - f_N)/a\,,\quad \partial_+\varphi_{\pm N} = (\partial_+\varphi_n)_{n=\pm N}\,.
$$

In general, admissible boundary conditions are the specification of one member of each of the three products $u_{\pm N}(c_{11}\Delta_+ u_{\pm N} + d_{11}\Delta_+ P_{\pm N})$, $P_{\pm N}(d_{11}\Delta_+ u_{\pm N} + b_{11}\Delta_+ P_{\pm N} + b_0)$, $\varphi_{\pm N}(\epsilon_0\partial_+\varphi_{\pm N} - P_{\pm N})$; i.e. eight combinations in all, for the one-dimensional case.

11. Solutions of the Lattice Equations.

As an example of solutions of the lattice equations, consider the one analogous to that in Section 9 for the capacitance of thin dielectric films. In place of the equations of equilibrium (8.1) we now have (10.17) and in place of the boundary conditions (9.5) we now have

$$\begin{aligned} c_{11}\Delta_+ u_{\pm N} + d_{11}\Delta_+ P_{\pm N} &= 0, \\ P_{\pm N} &= -k\epsilon_0 \eta V/h, \\ \varphi_{\pm N} &= \pm V, \end{aligned} \tag{11.1}$$

where $h = Na$. We take

$$\begin{aligned} u_n &= B_1' \cosh(na/\lambda), \\ P_n &= A_2' + B_2' \cosh(na/\lambda), \\ \varphi_n &= \left[A_3' x + B_3' \sinh(x/\lambda)\right]_{x=na}, \end{aligned} \tag{11.2}$$

and note that $\partial_\pm [f_n(na)] = [\partial f(x)]_{x=na}$ and

$$\begin{aligned} \Delta^2 \cosh\frac{na}{\lambda} &= a^{-2}\left[\cosh\frac{(n+1)a}{\lambda} + \cosh\frac{(n-1)a}{\lambda} - 2\cosh\frac{na}{\lambda}\right], \\ &= 4a^{-2}\sinh^2(a/2\lambda)\cosh(na/\lambda). \end{aligned} \tag{11.3}$$

Then, substituting (11.2) in (10.17), we find, analogous to (9.7),

(11.4) $$A_2' = \epsilon_0 \eta A_3', \quad B_3' = \lambda B_2' / \epsilon_0 = \lambda c_{11} B_1' / \epsilon_0 d_{11}$$

and, in place of (9.8),

(11.5) $$\sinh (a / 2\lambda) = a / 2\ell,$$

where ℓ is the same as in (9.8). Thus (11.4) is the same as (9.7) with ℓ replaced by λ; but ℓ and λ are not the same: being related in accordance with (11.5).

Application of the boundary conditions (11.1) leads, by the same procedure as in Section 8, to

(11.6) $$A_3' = (B_3' / \eta\lambda) \cosh (h/\lambda) + kV/h,$$

$$B_3' = (1-k)\eta V' / [\eta \sinh (h/\lambda) + (h/\lambda) \cosh (h/\lambda)],$$

analogous to (9.10). Finally, the capacitance is

(11.7) $$C = \frac{\epsilon_0 \partial_+ \varphi_{\pm N} - P_{\pm N}}{2V} = \frac{\epsilon}{2h} \frac{1 + (k\eta\lambda/h) \tanh (h/\lambda)}{1 + (\eta\lambda/h) \tanh (h/\lambda)}.$$

Thus, the entire solution is identical with the continuum one except that the displacement and polarization have significance only at the atom sites and ℓ is replaced by λ. In particular, the curve in Fig. 9.3 is applicable to the lattice if ℓ is replaced by λ in both abscissa and ordinate.

A somewhat similar result is obtained for the strain and polarization at a free surface, analogous to the problem solved in Section 8. The main difference is that, owing to the geometric asymmetry of the half-space, there is a shift of $a/2$ between the continuum and lattice solutions.

References.

[1] W. Voigt: Lehrbuch der Kristallphysik. B.G.Teubner, Leipzig, 2nd Edition, 1928; Edwards Brothers, Ann Arbor, 1946.

[2] W.P. Mason: Piezoelectric Crystals and their Application to Ultrasonics. D.Van Nostrand, New York, 1950.

[3] H.F. Tiersten: Linear Piezoelectric Plate Vibrations. Plenum Press, New York, 1969.

[4] R.D. Mindlin: Problems of Continuum Mechanics. J.R.M.Radok, Editor, Soc.Industrial and Appl.Math., Philadelphia, 1961, p.282.

[5] R.A. Toupin: J.Rational Mech.and Anal.Vol.5, 1956, pp.849--915.

[6] R.D. Mindlin: Int.J.Solids Structures, Vol.4, 1968, pp.637--642.

[7] M.P. Tosi: Solid State Physics Vol.16, Academic Press, New York, 1964, p.92.

[8] G.C. Benson and K.S. Yun: The Solid-Gas Interface. E.A.Flood, Editor, M.Dekker, Inc., New York, 1967, p.203.

[9] R.D. Mindlin: Int.J.Solids Structures, Vol.5, 1969, pp.1197--1208.

[10] C.A. Mead: Phys.Rev.Letters, Vol.6, 1961, p.545.

[11] A. Askar, P.C.Y. Lee and A.S. Cakmak: Phys.Rev.B, May 1970.

[12] B.G. Dick and A.W. Overhauser: Phys.Rev.Vol.112, 1958, p.90.

[13] W. Cochran: Proc.Roy.Soc., Vol.A 253, 1959, p.260.

[14] W. Cochran: Phonons in Perfect Lattices... R.W.H.Stevenson, Editor, Plenum Press, New York, 1966, p.53.

[15] W. Ludwig: Recent Developments in Lattice Theory. Springer Verlag, Berlin, 1967.

[16] Standards on Piezoelectric Crystals, 1949, Proc.Inst.Radio Engineers, Vol.37, 1949, pp.1378-1395.

[17] J.C. Slater and H. Frank: Electromagnetism. Mc Graw-Hill, N.Y., 1947.

[18] J. Schwartz: Int.J.Solids Structures, Vol.5, 1969, pp.1209--1220.

[19] J.S. Lomont: Applications of Finite Groups. Academic Press, N.Y.,1959.

[20] C.A. Mead: Phys.Rev.,Vol.128, 1962, p.2088.

[21] C.A. Mead and M.Mc Coll: Trans.Metall.Soc.A.I.M.E. Vol.233, 1965, p.502.

[22] C.A. Mead: Proc.Int.Symp.on Basic Problems in Thin Film Physics, Vandenhoeck and Ruprecht, Göttingen, 1966, p.674.

[23] H.Y. Ku and F.G. Ullmann: J.Appl.Phys., Vol.35, 1964, p.265.

[24] L.H. Germer, A.U. Mac Rae and C.D. Hartman: J.Appl.Phys., Vol.32, 1961, p.2432.

[25] A. Askar, P.C.Y. Lee and A.S. Cakmak: Int.J.Solids Structures, forthcoming.

[26] M.G. Salvadori and M.L. Baron: Numerical Methods in Engineering, Prentice-Hall, New Jersey, 2nd Edition, 1961.

Contents.

Int. J. Solids Structures, 1971, Vol. 7, pp. 1 to 4. Pergamon Press. Printed in Great Britain

THICKNESS-TWIST VIBRATIONS OF A QUARTZ STRIP

R. D. MINDLIN

Department of Civil Engineering, Columbia University, New York, N. Y.

Abstract—An exact solution of the three-dimensional equations of elasticity has been found for the thickness-twist modes of vibration of a rotated-Y-cut quartz plate with a pair of parallel, free edges. The solution has a simple form: made possible by relaxing the condition that the edge-planes be perpendicular to the main faces of the plate. For the AT-cut, the edges are off perpendicular by about five degrees.

INTRODUCTION

IN A previous paper [1], the solution was given for the thickness-twist modes of vibration of infinite, rotated-Y-cut, quartz crystal plates. In the present paper, the solution is recast in a different form [2] and extended to accommodate similar plates of finite width.

A rotated-Y-cut plate is one whose middle plane contains a digonal axis of elastic symmetry of the quartz crystal and whose normal makes an angle (35° 15′ for the widely used AT-cut) with the trigonal axis [3]. In an isotropic plate, the thickness-twist modes of vibration are those in which the displacement direction is parallel to the middle plane of the plate and the wave normal is perpendicular to the displacement. In infinite, rotated-Y-cut quartz plates, analogous modes are possible [1] if the displacement direction is parallel to the digonal axis in the plane of the plate. Again in the isotropic plate, simple reflections of thickness-twist waves occur at free, plane boundaries of the plate parallel to the displacement direction and perpendicular to the middle plane. The same is not true in the quartz plate: with the result that the thickness-twist vibrations of a quartz strip, with a rectangular cross-section perpendicular to the digonal axis, are not expressible in terms of a finite number of elementary functions. The mode shape, if it were found, would comprise an infinite number of elementary shapes; and those with the longer wave lengths would be designated as "unwanted modes" or "spurious modes" which cause difficulties in crystal filter applications in electric circuits.

In the present paper, it is shown that, if the free, plane boundaries of the strip are maintained parallel to the digonal axis but cut slightly off normal to the middle plane (about 5° off normal for the AT-cut) simple reflections occur, a simple, closed solution is obtained and modes with simple shapes and well spaced frequencies result.

SOLUTION

If the x-axis is a digonal axis and the y-axis is normal to the middle plane of the plate, the stress–strain relations for the rotated-Y-cuts of quartz are [3]

$$\begin{aligned} T_1 &= c_{11}S_1+c_{12}S_2+c_{13}S_3+c_{14}S_4, & T_2 &= c_{21}S_1+c_{22}S_2+c_{23}S_3+c_{24}S_4,\\ T_3 &= c_{31}S_1+c_{32}S_2+c_{33}S_3+c_{34}S_4, & T_4 &= c_{41}S_1+c_{42}S_2+c_{43}S_3+c_{44}S_4,\\ T_5 &= c_{55}S_5+c_{56}S_6, & T_6 &= c_{65}S_5+c_{66}S_6, \end{aligned} \tag{1}$$

where $c_{pq} = c_{qp}$ and the strains, S_p, are related to the displacements u, v, w, according to

$$S_1 = \frac{\partial u}{\partial x}, \qquad S_4 = \frac{\partial w}{\partial y}+\frac{\partial v}{\partial z},$$
$$S_2 = \frac{\partial v}{\partial y}, \qquad S_5 = \frac{\partial u}{\partial z}+\frac{\partial w}{\partial x}, \tag{2}$$
$$S_3 = \frac{\partial w}{\partial z}, \qquad S_6 = \frac{\partial v}{\partial x}+\frac{\partial u}{\partial y}.$$

The stresses must satisfy the equations of motion:

$$\frac{\partial T_1}{\partial x}+\frac{\partial T_6}{\partial y}+\frac{\partial T_5}{\partial z} = \rho\frac{\partial^2 u}{\partial t^2}, \qquad \frac{\partial T_6}{\partial x}+\frac{\partial T_2}{\partial y}+\frac{\partial T_4}{\partial z} = \rho\frac{\partial^2 v}{\partial t^2}, \qquad \frac{\partial T_5}{\partial x}+\frac{\partial T_4}{\partial y}+\frac{\partial T_3}{\partial z} = \rho\frac{\partial^2 w}{\partial t^2}. \tag{3}$$

We require a steady state solution, of these equations, of the form

$$u = U(y, z)\,e^{i\omega t}, \qquad v = w = 0, \tag{4}$$

for a strip of finite thickness in the y-direction and of finite width in the z-direction—with all four boundaries free of traction.

Substitution of (4) in (1)–(3) reduces the latter to

$$T_1 = T_2 = T_3 = T_4 = 0, \qquad T_5 = c_{55}\frac{\partial U}{\partial z}+c_{56}\frac{\partial U}{\partial y}, \qquad T_6 = c_{65}\frac{\partial U}{\partial z}+c_{66}\frac{\partial U}{\partial y}; \tag{5}$$

$$S_1 = S_2 = S_3 = S_4 = 0, \qquad S_5 = \frac{\partial U}{\partial z}, \qquad S_6 = \frac{\partial U}{\partial y}; \tag{6}$$

$$c_{66}\frac{\partial^2 U}{\partial y^2}+2c_{56}\frac{\partial^2 U}{\partial y \partial z}+c_{55}\frac{\partial^2 U}{\partial z^2} = -\rho\omega^2 U; \tag{7}$$

with the exponential factor omitted.

Now, the four functions [2]

$$U = A \sin \eta y \cos \zeta\left(\frac{c_{56}}{c_{66}}y-z\right)+B \sin \eta y \sin \zeta\left(\frac{c_{56}}{c_{66}}y-z\right)$$
$$+C \cos \eta y \cos \zeta\left(\frac{c_{56}}{c_{66}}y-z\right)+D \cos \eta y \sin \zeta\left(\frac{c_{56}}{c_{66}}y-z\right) \tag{8}$$

are solutions of (7) if

$$\rho\omega^2 = c_{66}\eta^2+\gamma_{55}\zeta^2, \tag{9}$$

where

$$\gamma_{55} = c_{55}-c_{56}^2/c_{66}. \tag{10}$$

For a plate of finite thickness, say $2h$, we require

$$T_2]_{y=\pm h} = T_4]_{y=\pm h} = T_6]_{y=\pm h} = 0; \tag{11}$$

and these boundary conditions are satisfied by (8) if

$$2\eta h = m\pi, \tag{12}$$

where m is an odd integer for solutions A and B and an even integer (including zero—which corresponds to a face-shear mode) for solutions C and D.

For traction-free boundaries at $z = \pm c$, we should require

$$T_3]_{z=\pm c} = T_4]_{z=\pm c} = T_5]_{z=\pm c} = 0. \tag{13}$$

The first two of these are satisfied identically, but the third cannot be satisfied by functions as simple as (8). An infinite series of such functions would be required. However, consider a rotation of coordinate axes, through an angle α about the x-axis, according to the scheme of direction cosines

$$\begin{array}{cccc} & x & y & z \\ x' & 1 & 0 & 0 \\ y' & 0 & \cos\alpha & \sin\alpha \\ z' & 0 & -\sin\alpha & \cos\alpha \end{array} \tag{14}$$

With (5), the components of stress, referred to the rotated axes, are given in terms of the original components by

$$T'_1 = T'_2 = T'_3 = T'_4 = 0, \qquad T'_5 = T_5 \cos\alpha - T_6 \sin\alpha, \qquad T'_6 = T_5 \sin\alpha + T_6 \cos\alpha; \tag{15}$$

and, for traction-free boundaries at $z' = \pm c \cos\alpha$, as shown in Fig. 1, we require

$$T'_3]_{z'=\pm c\cos\alpha} = T'_4]_{z'=\pm c\cos\alpha} = T'_5]_{z'=\pm c\cos\alpha} = 0. \tag{16}$$

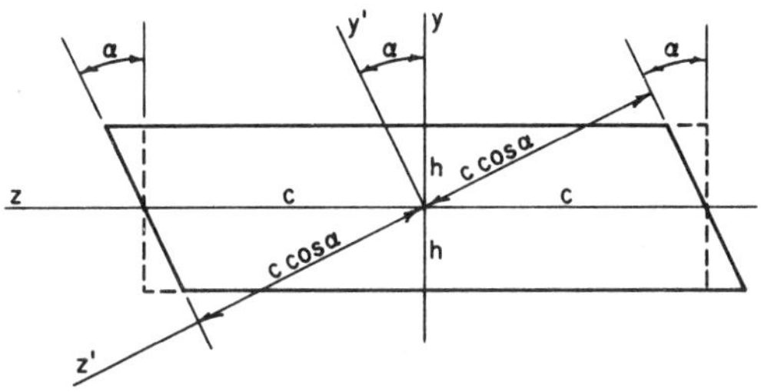

FIG. 1. Rotated axes and cross-section of quartz strip.

Now, T'_3 and T'_4 are identically zero whereas

$$\begin{aligned} T'_5 = {} & A\eta(c_{56}\cos\alpha - c_{66}\sin\alpha)\cos\eta y\cos\zeta\left(\frac{c_{56}}{c_{66}}y - z\right) + A\zeta\gamma_{55}\cos\alpha\sin\eta y\sin\zeta\left(\frac{c_{56}}{c_{66}}y - z\right) \\ & + B\eta(c_{56}\cos\alpha - c_{66}\sin\alpha)\cos\eta y\sin\zeta\left(\frac{c_{56}}{c_{66}}y - z\right) - B\zeta\gamma_{55}\cos\alpha\sin\eta y\cos\zeta\left(\frac{c_{56}}{c_{66}}y - z\right) \\ & - C\eta(c_{56}\cos\alpha - c_{66}\sin\alpha)\sin\eta y\cos\zeta\left(\frac{c_{56}}{c_{66}}y - z\right) + C\zeta\gamma_{55}\cos\alpha\cos\eta y\sin\zeta\left(\frac{c_{56}}{c_{66}}y - z\right) \\ & - D\eta(c_{56}\cos\alpha - c_{66}\sin\alpha)\sin\eta y\sin\zeta\left(\frac{c_{56}}{c_{66}}y - z\right) - D\zeta\gamma_{55}\cos\alpha\cos\eta y\cos\zeta\left(\frac{c_{56}}{c_{66}}y - z\right) \end{aligned} \tag{17}$$

In (17), set

$$\alpha = \arctan(c_{56}/c_{66}) \tag{18}$$

and note that $(c_{56}/c_{66})y - z = -z'/\cos\alpha$. Then (17) reduces to

$$\begin{aligned} T'_5 = {} & A\zeta\gamma_{55}\cos\alpha\sin\eta y\sin(\zeta z'/\cos\alpha) - B\zeta\gamma_{55}\cos\alpha\sin\eta y\cos(\zeta z'/\cos\alpha) \\ & + C\zeta\gamma_{55}\cos\alpha\cos\eta y\sin(\zeta z'/\cos\alpha) - D\zeta\gamma_{55}\cos\alpha\cos\eta y\cos(\zeta z'/\cos\alpha). \end{aligned} \tag{19}$$

Hence, it is apparent that $T'_5 = 0$ on $z' = \pm c\cos\alpha$ if

$$2\zeta c = n\pi, \tag{20}$$

where n is an even integer (including zero—which gives simple thickness-shear modes) for solutions A and C, and an odd integer for solutions B and D.

Upon substituting (12) and (20) in (8) and (9), we find the required simple solutions:

$$\begin{aligned} U = {} & A\sin\frac{m\pi y}{2h}\cos\frac{n\pi}{2c}\left(\frac{c_{56}}{c_{66}}y - z\right) + B\sin\frac{m\pi y}{2h}\sin\frac{n\pi}{2c}\left(\frac{c_{56}}{c_{66}}y - z\right) \\ & + C\cos\frac{m\pi y}{2h}\cos\frac{n\pi}{2c}\left(\frac{c_{56}}{c_{66}}y - z\right) + D\cos\frac{m\pi y}{2h}\sin\frac{n\pi}{2c}\left(\frac{c_{56}}{c_{66}}y - z\right). \end{aligned} \tag{21}$$

with circular frequencies

$$\omega = \frac{m\pi}{2h}\left(\frac{c_{66}}{\rho}\right)^{\frac{1}{2}}\left(1 + \frac{\gamma_{55}n^2h^2}{c_{66}m^2c^2}\right)^{\frac{1}{2}} \tag{22}$$

and m and n odd or even integers, for solutions A, B, C, D, according to

A: m odd, n even;

B: m odd, n odd;

C: m even, n even;

D: m even, n odd.

For the AT-cut of quartz $c_{56} = 2{\cdot}53$ and $c_{66} = 29{\cdot}01$, in units of 10^{10} dyn/cm^2, as calculated from Bechmann's values [4] of the principal constants of quartz. Hence, by (18), α is approximately 5° for the AT-cut. Thus, only a slight inclination of the edge-cuts is required to reduce the undesirable, complicated mode-shapes to simple ones.

REFERENCES

[1] R. D. MINDLIN, *Int. J. Solids Struct.* **1**, 141–145 (1965).
[2] R. D. MINDLIN, *J. acoust. Soc. Am.* **41**, 969–973 (1967).
[3] W. P. MASON, *Piezoelectric Crystals and their Application to Ultrasonics.* Van Nostrand (1950).
[4] R. BECHMANN, *Phys. Rev.* **110**, 1060–1061 (1958).

(*Received* 29 *June* 1970)

Абстракт—Приводится точное решение для трехмерных уравнений упругости для случая крутильных по толщине видов колебаний вращающейся пластинки из крарца Y–среза с парой параллельных, свободных краев.

Решение имеет простую форму и даёт возможность ослабить условия так, что краевые плоскости могут быть перпедикулярными к главным поверхностям пластинки. Для кварца AT–среза, края отдаляются от перпедикулятности около пяти степеней.

Int. J. Solids Structures, 1971, Vol. 7, pp. 1219 to 1227. Pergamon Press. Printed in Great Britain

ACOUSTICAL AND OPTICAL ACTIVITY IN ALPHA QUARTZ

R. D. MINDLIN

Department of Civil Engineering, Columbia University, New York, New York

and

R. A. TOUPIN

IBM Thomas J. Watson Research Center, Yorktown Heights, New York

Abstract—It is shown that a theory of elastic dielectrics, in which the stored electromechanical energy depends on the polarization gradient, accounts for both acoustical and optical activity. Formulas for the acoustical and optical rotatory powers of α-quartz are derived and the values of new material constants appearing in them are calculated from experimental data.

1. INTRODUCTION

ACOUSTICAL activity (rotation of the direction of mechanical displacement along the path of a transverse, elastic wave) has recently been observed by Pine [1] and by Joffrin and Levelut [2] in α-quartz. The possibility of the phenomenon appears first to have been mentioned by Silin [3]. It was accounted for by one of us [4] on the basis of the theory of elasticity in which the stored energy is a function of the gradient of the strain, in addition to the strain. Portigal and Burstein [5] found an equivalent result by assigning dependence of the elastic stiffness on the wave vector. The purpose of the present paper is to show that both acoustical and optical activity are accounted for in the theory of elastic dielectrics in which the stored electromechanical energy is a function of the polarization gradient [6] in addition to the usual strain and polarization. It has already been shown [7] that the differential equations of the resulting theory, rather than the equations of the classical theory of piezoelectricity, are the correct, long wave, low frequency limit of the finite difference equations of a lattice of shell model atoms if the shell–shell interaction between adjacent atoms is taken into account. It has also been shown that the polarization gradient can account [6] for the surface energy of deformation and polarization and it can also account [7] for an anomaly observed in measurements of electrical capacitance of thin, dielectric films. In the present paper, the field equations are exhibited for the coupled elastic–electric–magnetic system, the problem of shear waves along the trigonal axis of α-quartz is solved, formulas are obtained for the optical and acoustical rotatory powers and numerical values of the new material constants, in the formulas, are calculated from experimental data. Essentially, the theory has it that the appearance of optical activity depends on an interaction between the polarization and the polarization gradient; the appearance of acoustical activity depends on interactions of the strain with both the polarization and the polarization gradient, and is absent if either interaction is missing

—provided, of course, that the phenomenon does not depend on the strain gradient, as assumed, tentatively, in this paper.

2. COUPLED ELASTIC, ELECTRIC AND MAGNETIC FIELDS

The linear equations of an elastic, dielectric continuum, with the contribution of the polarization gradient taken into account, but without the coupling to the magnetic field, may be written as [6]

$$\begin{aligned} T_{ij,i} &= \rho \ddot{u}_j, \\ E_{ij,i}+E_j^L+E_j &= 0, \\ \varepsilon_{ijk}E_{k,j} &= 0, \\ \varepsilon_0 E_{i,i}+P_{i,i} &= 0, \end{aligned} \tag{1}$$

where

$$T_{ij} = \partial W^L/\partial S_{ij}, \qquad E_j^L = -\partial W^L/\partial P_j, \qquad E_{ij} = \partial W^L/\partial P_{j,i} \tag{2}$$

and

$$W^L = \tfrac{1}{2}a_{ij}P_iP_j+\tfrac{1}{2}b_{ijkl}P_{j,i}P_{l,k}+\tfrac{1}{2}c^P_{ijkl}S_{ij}S_{kl}+d_{ijkl}P_{j,i}S_{kl}+f_{ijk}P_iS_{jk}+j_{ijk}P_iP_{k,j}, \tag{3}$$

$$S_{ij} = \tfrac{1}{2}(u_{j,i}+u_{i,j}). \tag{4}$$

In (1), u_i is the mechanical displacement, P_i is the polarization density, E_i is the Maxwell electric self-field, ρ is the mass density, ε_0 is the permittivity of a vacuum and ε_{ijk} is the alternating tensor. In (3), W^L is the stored energy density of deformation and polarization in which b_{ijkl}, d_{ijkl} and j_{ijk} are constants associated with the polarization gradient, $P_{j,i}$, while a_{ij}, f_{ijk} and c^P_{ijkl} belong to the classical theory of piezoelectricity and are related to the I.R.E. standard [8] symbols for the reciprocal dielectric susceptibility, χ_{ij}, the piezoelectric stress constant, e_{ijk}, and the elastic stiffness at constant electric field, c^E_{ijkl}, according to [9]

$$a_{ij} = \varepsilon_0^{-1}\chi_{ij}, \qquad f_{ijk} = -\varepsilon_0^{-1}\chi_{il}e_{ljk}, \qquad c^P_{ijkl} = c^E_{ijkl}+\varepsilon_0^{-1}\chi_{mn}e_{mij}e_{nkl}. \tag{5}$$

To couple (1) to the equations of the magnetic field, it is only necessary to replace the third of (1) by

$$\varepsilon_{ijk}E_{k,j}+\dot{B}_i = 0 \tag{6}$$

and add the equations

$$\mu_0^{-1}\varepsilon_{ijk}B_{k,j}-\varepsilon_0\dot{E}_i-\dot{P}_i = 0, \tag{7}$$

$$B_{i,i} = 0, \tag{8}$$

where B_i is the magnetic flux density and μ_0 is the magnetic permeability, assumed to be that of a vacuum.

It is convenient to eliminate B_i by subtracting the curl of (6) from the time derivative of (7), with the result

$$E_{j,ii}-E_{i,ij} = \varepsilon_0\mu_0\ddot{E}_j+\mu_0\ddot{P}_j. \tag{9}$$

The last of (1) and (8) are not independent of (7) and (6), respectively, and may be disregarded for the present purpose.

Thus, (9), along with the first two of (1):

$$T_{ij,i} = \rho \ddot{u}_j,$$
$$E_{ij,i} + E_j^L + E_j = 0, \tag{10}$$
$$E_{j,ii} - E_{i,ji} = \varepsilon_0 \mu_0 \ddot{E}_j + \mu_0 \ddot{P}_j,$$

are the field equations governing mechanical and electromagnetic waves, coupled through the constitutive equations:

$$T_{ij} = c^P_{ijkl} S_{kl} + f_{kij} P_k + d_{klij} P_{l,k},$$
$$-E_j^L = f_{jkl} S_{kl} + a_{jk} P_k + j_{jkl} P_{l,k}, \tag{11}$$
$$E_{ij} = d_{ijkl} S_{kl} + j_{kij} P_k + b_{ijkl} P_{l,k},$$

which are obtained from (2) and (3).

For a crystal of class 32 (international) or D_3 (Schoenflies) [10], to which α-quartz belongs, the constitutive equations take the form shown in Fig. 1, in which x_3 has been taken as the trigonal axis and x_1 one of the digonal axes. In Fig. 1, an abridged notation is used for the subscripts attached to the material constants:

$$11 \to 1, \quad 22 \to 2, \quad 33 \to 3, \quad 23 \to 4, \quad 32 \to 7, \quad 31 \to 5, \quad 13 \to 8, \quad 12 \to 6, \quad 21 \to 9.$$

3. PLANE WAVES ALONG THE TRIGONAL AXIS

We consider plane, transverse waves propagating along the x_3-axis; i.e. u_i, P_i and E_i are functions of x_3 and t only, but u_3, P_3 and E_3 are zero. Then the field equations (10) reduce to

$$T_{13,3} = \rho \ddot{u}_1, \qquad T_{23,3} = \rho \ddot{u}_2,$$
$$E_{31,3} + E_1^L + E_1 = 0, \qquad E_{32,3} + E_2^L + E_2 = 0, \tag{12}$$
$$E_{1,33} = \varepsilon_0 \mu_0 \ddot{E}_1 + \mu_0 \ddot{P}_1, \qquad E_{2,33} = \varepsilon_0 \mu_0 \ddot{E}_2 + \mu_0 \ddot{P}_2$$

and the constitutive equations (11) reduce to

$$T_{32} = c^P_{44} u_{2,3} + f_{14} P_1 + d_{74} P_{2,3},$$
$$T_{31} = c^P_{44} u_{1,3} - f_{14} P_2 + d_{74} P_{1,3},$$
$$-E_1^L = f_{14} u_{2,3} + \varepsilon_0^{-1} \chi_{11} P_1 + j_{17} P_{2,3},$$
$$-E_2^L = -f_{14} u_{1,3} + \varepsilon_0^{-1} \chi_{11} P_2 - j_{17} P_{1,3}, \tag{13}$$
$$E_{32} = d_{74} u_{2,3} + j_{17} P_1 + b_{55} P_{2,3},$$
$$E_{31} = d_{74} u_{1,3} - j_{17} P_2 + b_{55} P_{1,3}.$$

	S_{11}	S_{22}	S_{33}	$2S_{23}$	$2S_{31}$	$2S_{12}$	P_1	P_2	P_3	$P_{1,1}$	$P_{2,2}$	$P_{3,3}$	$P_{3,2}$	$P_{2,3}$	$P_{1,3}$	$P_{3,1}$	$P_{2,1}$	$P_{1,2}$
T_{11}	c_{11}^P	c_{12}^P	c_{13}^P	c_{14}^P	0	0	f_{11}	0	0	d_{11}	d_{12}	d_{31}	d_{41}	d_{71}	0	0	0	0
T_{22}	c_{12}^P	c_{11}^P	c_{13}^P	$-c_{14}^P$	0	0	$-f_{11}$	0	0	d_{12}	d_{11}	d_{31}	$-d_{41}$	$-d_{71}$	0	0	0	0
T_{33}	c_{13}^P	c_{13}^P	c_{33}^P	0	0	0	0	0	0	d_{13}	d_{13}	d_{33}	0	0	0	0	0	0
T_{23}	c_{14}^P	$-c_{14}^P$	0	c_{44}^P	0	0	f_{14}	0	0	d_{14}	$-d_{14}$	0	d_{44}	d_{74}	0	0	0	0
T_{31}	0	0	0	0	c_{44}^P	c_{14}^P	0	$-f_{14}$	0	0	0	0	0	0	d_{74}	d_{44}	d_{14}	d_{14}
T_{12}	0	0	0	0	c_{14}^P	c_{66}^P	0	$-f_{11}$	0	0	0	0	0	0	d_{71}	d_{41}	d_{66}	d_{66}
$-E_1^L$	f_{11}	$-f_{11}$	0	f_{14}	0	0	a_{11}	0	0	j_{11}	$-j_{11}$	0	j_{14}	j_{17}	0	0	0	0
$-E_2^L$	0	0	0	0	$-f_{14}$	$-f_{11}$	0	a_{11}	0	0	0	0	0	0	$-j_{17}$	$-j_{14}$	$-j_{11}$	$-j_{11}$
$-E_3^L$	0	0	0	0	0	0	0	0	a_{33}	0	0	0	0	0	0	0	j_{36}	$-j_{36}$
E_{11}	d_{11}	d_{12}	d_{13}	d_{14}	0	0	j_{11}	0	0	b_{11}	b_{12}	b_{13}	b_{14}	b_{17}	0	0	0	0
E_{22}	d_{12}	d_{11}	d_{13}	$-d_{14}$	0	0	$-j_{11}$	0	0	b_{12}	b_{11}	b_{13}	$-b_{14}$	$-b_{17}$	0	0	0	0
E_{33}	d_{31}	d_{31}	d_{33}	0	0	0	0	0	0	b_{13}	b_{13}	b_{33}	0	0	0	0	0	0
E_{23}	d_{41}	$-d_{41}$	0	d_{44}	0	0	j_{14}	0	0	b_{14}	$-b_{14}$	0	b_{44}	b_{47}	0	0	0	0
E_{32}	d_{71}	$-d_{71}$	0	d_{74}	0	0	j_{17}	0	0	b_{17}	$-b_{17}$	0	b_{47}	b_{55}	0	0	0	0
E_{31}	0	0	0	0	d_{74}	d_{71}	0	$-j_{17}$	0	0	0	0	0	0	b_{55}	b_{47}	b_{17}	b_{17}
E_{13}	0	0	0	0	d_{44}	d_{41}	0	$-j_{14}$	0	0	0	0	0	0	b_{47}	b_{44}	b_{14}	b_{14}
E_{12}	0	0	0	0	d_{14}	d_{66}	0	$-j_{11}$	j_{36}	0	0	0	0	0	b_{17}	b_{14}	b_{66}	b_{69}
E_{21}	0	0	0	0	d_{14}	d_{66}	0	$-j_{11}$	$-j_{36}$	0	0	0	0	0	b_{17}	b_{14}	b_{69}	b_{66}

$c_{66}^P = \frac{1}{2}(c_{11}^P - c_{12}^P)$, $\quad d_{66} = \frac{1}{2}(d_{11} - d_{12})$, $\quad b_{66} + b_{69} = b_{11} - b_{12}$

FIG. 1. Constitutive relations for crystal class 32 (D_3).

Inserting (13) in (12), we have

$$\begin{aligned}
c_{44}^{P}u_{1,33}-f_{14}P_{2,3}+d_{74}P_{1,33} &= \rho\ddot{u}_1,\\
c_{44}^{P}u_{2,33}+f_{14}P_{1,3}+d_{74}P_{2,33} &= \rho\ddot{u}_2,\\
d_{74}u_{1,33}-2j_{17}P_{2,3}+b_{55}P_{1,33}-f_{14}u_{2,3}-\varepsilon_0^{-1}\chi_{11}P_1+E_1 &= 0,\\
d_{74}u_{2,33}+2j_{17}P_{1,3}+b_{55}P_{2,33}+f_{14}u_{1,3}-\varepsilon_0^{-1}\chi_{11}P_2+E_2 &= 0,\\
E_{1,33} &= \varepsilon_0\mu_0\ddot{E}_1+\mu_0\ddot{P}_1,\\
E_{2,33} &= \varepsilon_0\mu_0\ddot{E}_2+\mu_0\ddot{P}_2.
\end{aligned} \tag{14}$$

Now, take

$$\begin{aligned}
u_1 &= A_1\sin\zeta(x_3-vt), & u_2 &= A_2\cos\zeta(x_3-vt),\\
P_1 &= B_1\sin\zeta(x_3-vt), & P_2 &= B_2\cos\zeta(x_3-vt),\\
E_1 &= C_1\sin\zeta(x_3-vt), & E_2 &= C_2\cos\zeta(x_3-vt)
\end{aligned} \tag{15}$$

and substitute these functions in (14) to find

$$\begin{aligned}
(c_{44}^{P}-\rho v^2)\zeta A_1-f_{14}B_2+d_{74}\zeta B_1 &= 0,\\
(c_{44}^{P}-\rho v^2)\zeta A_2-f_{14}B_1+d_{74}\zeta B_2 &= 0,\\
d_{74}\zeta^2A_1-f_{14}\zeta A_2+(b_{55}\zeta^2+\varepsilon_0^{-1}\chi_{11})B_1-2j_{17}\zeta B_2-C_1 &= 0,\\
d_{74}\zeta^2A_2-f_{14}\zeta A_1+(b_{55}\zeta^2+\varepsilon_0^{-1}\chi_{11})B_2-2j_{17}\zeta B_1-C_2 &= 0,\\
\mu_0v^2B_1+(\varepsilon_0\mu_0v^2-1)C_1 &= 0,\\
\mu_0v^2B_2+(\varepsilon_0\mu_0v^2-1)C_2 &= 0.
\end{aligned} \tag{16}$$

Adding and subtracting these equations in pairs, we have

$$\begin{aligned}
(c_{44}^{P}-\rho v^2)(A_1\pm A_2)+(d_{74}\zeta\mp f_{14})(B_1\pm B_2) &= 0,\\
(d_{74}\zeta\mp f_{14})(A_1\pm A_2)+(b_{55}\zeta^2+\varepsilon_0^{-1}\chi_{11}\mp 2j_{17}\zeta)(B_1\pm B_2)-(C_1\pm C_2) &= 0,\\
\mu_0v^2(B_1\pm B_2)+(\varepsilon_0\mu_0v^2-1)(C_1\pm C_2) &= 0.
\end{aligned} \tag{17}$$

Thus, there are two solutions, each corresponding to circularly polarized waves [11, p. 222] since the amplitudes must satisfy

$$A_1 = \pm A_2, \qquad B_1 = \pm B_2, \qquad C_1 = \pm C_2 \tag{18}$$

with either all upper signs or all lower signs. Upon substituting (18) into (15), we see that the upper and lower signs give right and left circular polarization, respectively. The velocities are obtained by setting the determinant of the coefficients of the amplitudes in (17) equal to zero:

$$\begin{vmatrix}
c_{44}^{P}-\rho v^2 & d_{74}\zeta\mp f_{14} & 0\\
d_{74}\zeta\mp f_{14} & b_{55}\zeta^2+\varepsilon_0^{-1}\chi_{11}\mp 2j_{17}\zeta & 1\\
0 & 1 & \mu_0^{-1}v^{-2}-\varepsilon_0
\end{vmatrix} = 0. \tag{19}$$

This is a quadratic equation in v^2, so that there are two pairs of oppositely circularly polarized waves. Each pair of such waves combines to produce a linearly polarized wave with a rotating direction of polarization [11, p. 222]. Thus, we have two cases of rotary polarization. These may be identified as optical and acoustical by separating out first the electromagnetic part of (19) and then the electromechanical part. The two may, in fact, be considered separately owing to the large ratio of the frequencies (of the order of 10^5) at which the two effects are observed.

4. OPTICAL ACTIVITY

The electromagnetic part of the determinant in (19) is the minor of the upper left element. Thus, the pair of optical velocities is given by

$$\begin{vmatrix} b_{55}\zeta^2+\varepsilon_0^{-1}\chi_{11}\mp 2j_{17} & 1 \\ 1 & \mu^{-1}v^{-2}-\varepsilon_0 \end{vmatrix} = 0, \qquad (20)$$

which yields the dispersion formula (cf. [11, p. 426])

$$n_\pm^2-1 = (\chi_{11}\mp 2\varepsilon_0 j_{17}\zeta+\varepsilon_0 b_{55}\zeta^2)^{-1}, \qquad (21)$$

where $n_\pm$ are the indices of refraction:

$$n_\pm = c/v_\pm \qquad (22)$$

and c is the velocity of light *in vacuo*:

$$c = (\varepsilon_0\mu_0)^{-1}. \qquad (23)$$

From (21), we have

$$(n_-^2-1)^{-1}-(n_+^2-1)^{-1} = 4\varepsilon_0 j_{17}\zeta. \qquad (24)$$

Now, define $n = \frac{1}{2}(n_+ + n_-)$ and assume

$$|n_+ - n_-| \ll n_+ + n_-. \qquad (25)$$

Then (24) becomes, approximately,

$$n_+ - n_- = 2(n^2-1)^2\varepsilon_0 j_{17}\zeta_0, \qquad (26)$$

where $\zeta_0(=\zeta/n)$ is the wave number *in vacuo*.

The optical rotatory power, in radians per unit length, is given by [11, p. 222]

$$\theta_{OP} = \tfrac{1}{2}\zeta_0(n_- - n_+). \qquad (27)$$

Accordingly, from (26) and (27),

$$\theta_{OP} = -(n^2-1)^2\varepsilon_0 j_{17}\zeta_0^2 \qquad (28)$$

is the formula for the optical rotatory power in terms of the average index of refraction, n, along the optic axis, the wave length *in vacuo*, $\lambda_0(=2\pi/\zeta_0)$, the fundamental constant ε_0 and the material constant $j_{17}(=j_{132})$ which, as may be seen in (3), measures the interaction between the polarization and the polarization gradient.

5. ACOUSTICAL ACTIVITY

The electromechanical part of the determinant in (19) is the minor of the lower right element, so that we have

$$\begin{vmatrix} c_{44}^{P}-\rho v^2 & d_{74}\zeta \mp f_{14} \\ d_{74}\zeta \mp f_{14} & b_{55}\zeta^2+\varepsilon_0^{-1}\chi_{11}\mp 2j_{17}\zeta \end{vmatrix} = 0 \tag{29}$$

for the equation determining the velocities of the two acoustical waves, as influenced by the quasi-static polarization and polarization gradient. From (29),

$$\rho v_{\pm}^2 = c_{44}^{P}-(d_{74}\zeta \mp f_{14})^2/(b_{55}\zeta^2+\varepsilon_0^{-1}\chi_{11}\mp 2j_{17}\zeta). \tag{30}$$

In view of (18) and the inequality of v_+ and v_-, the superposition of the two waves results in rotary polarization (acoustical activity) with acoustical rotatory power

$$\theta_{AC} = \tfrac{1}{2}\omega(v_-^{-1}-v_+^{-1}), \tag{31}$$

where ω is the circular frequency.

Both waves are dispersive. At the zero frequency (long wave) limit, from (30) and (5),

$$\lim_{\zeta\to 0} \rho v_{\pm}^2 = c_{44}^{P}-\varepsilon_0 f_{14}^2/\chi_{11} = c_{44}^{E}, \tag{32}$$

which is the result (without acoustical activity, since $v_+ = v_-$) that would be obtained if the contribution of the polarization gradient were omitted, i.e. if d_{74}, b_{55} and j_{17} were assumed to be zero. As the frequency increases from zero, the absolute velocity difference, $|v_+ - v_-|$, at first increases, so that the acoustical activity appears and increases. With further increase of frequency, the velocity difference again approaches zero, since

$$\lim_{\zeta\to\infty} \rho v_{\pm}^2 = c_{44}^{P}-d_{74}^2/b_{55}, \tag{33}$$

so that the acoustical activity diminishes; but this is undoubtedly beyond the range of applicability of the continuum theory. Up to moderately large wave numbers, (30) and (5) give, to the first order in ζ,

$$v_0/v_{\pm} = 1\mp(d_{74}-j_{17}e_{14})e_{14}\zeta/c_{44}^{E} \tag{34}$$

where $v_0^2 = c_{44}^{E}/\rho$. In this range, the frequency is approximately proportional to the wave number: $\omega = v_0\zeta$; so that, from (31) and (34),

$$\theta_{AC} = (d_{74}-j_{17}e_{14})e_{14}\rho\omega^2/(c_{44}^{E})^2. \tag{35}$$

Thus, at frequencies up to, say, 10^{10} cps, the acoustical rotatory power is approximately proportional to the square of the frequency and depends on the constants ρ, e_{14} and c_{44}^{E}, which are commonly encountered in piezoelectricity theory, and also on the constants d_{74} and j_{17} which control the coupling of the polarization gradient with the strain and polarization, respectively.

6. APPLICATION TO QUARTZ

For α-quartz, all the quantities in the formula (28) for optical rotatory power are known except j_{17}. Thus, for left-handed quartz and the sodium D line,

$$\theta_{OP} = -379 \text{ rad./m} \qquad [11, \text{p. } 481];$$

$$\lambda_0 = 5893\times 10^{-10}\text{ m} \qquad [11, \text{p. } 481],$$

$$n = 1{\cdot}5533 \qquad [11, \text{p. } 481],$$

$$\varepsilon_0 = 8{\cdot}854\times 10^{-12}\text{ F/m} \qquad [10].$$

Hence,

$$j_{17} = -\theta_{OP}\lambda_0^2/4\pi^2\varepsilon_0(n^2-1)^2 = 0{\cdot}19\text{ m}^2/\text{F}. \tag{36}$$

With the value of j_{17} known, all quantities in the formula (35) for acoustical rotatory power are known except d_{74}. In particular, Pine [1] finds that the acoustical and optical activities have opposite signs and the acoustical rotary power along the trigonal axis is about 220 rad./m at one gigahertz. Thus, for left handed α-quartz,

$$\theta_{AC} = 220\text{ rad./m} \qquad [1],$$

$$\omega = 2\pi\times 10^9\text{ rad./sec} \qquad [1],$$

$$\rho = 2{\cdot}65\times 10^3\text{ kg/m}^3 \qquad [12],$$

$$c_{44}^E = 57{\cdot}94\times 10^9\text{ Newton/m}^2 \qquad [13],$$

$$e_{14} = -0{\cdot}0406\text{ C/m}^2 \qquad [13].$$

Hence,

$$d_{74} = \theta_{AC}(c_{44}^E)^2/e_{14}\rho\omega^2 + j_{17}e_{14} = -174 - 0{\cdot}0077\text{ V}. \tag{37}$$

The second term, $j_{17}e_{14}$, is negligible in comparison with the first, so that we may drop the dependence of acoustical rotatory power on j_{17} and replace (35) with

$$\theta_{AC} = d_{74}e_{14}\rho\omega^2/(c_{44}^E)^2. \tag{38}$$

Thus, according to this theory, the presence of acoustical activity in α-quartz depends on the existence of the piezoelectric stress constant $e_{14}(=e_{123})$ and the interaction constant $d_{74}(=d_{3223})$ between strain and polarization gradient, whereas the presence of optical activity depends only on the existence of the interaction constant $j_{17}(=j_{132})$ between polarization and polarization gradient.

REFERENCES

[1] A. S. PINE, Direct observation of acoustical activity in α-quartz. *Phys. Rev.* **B2**, 2049–2054 (1970).
[2] J. JOFFRIN and A. LEVELUT, Mise en évidence et mesure du pouvoir rotatoire acoustique naturel du quartz-α, Laboratoire d'Ultrasons, Faculté des Sciences de Paris, forthcoming.
[3] V. P. SILIN, Contribution to the theory of absorption of ultrasound in metals. *JETP* **38**, 977–983 (1960); *Am. Phys. Soc. Trans. Sov. Phys. JETP* **11**, 703–707 (1960).
[4] R. A. TOUPIN, Elastic materials with couple-stresses. *Arch. Rat. Mech. Anal.* **11**, 385–414 (1962).
[5] D. L. PORTIGAL and E. BURSTEIN, Acoustical activity and other first-order spatial dispersion effects in crystals. *Phys. Rev.* **170**, 673–678 (1968).

[6] R. D. Mindlin, Polarization gradient in elastic dielectrics. *Int. J. Solids Struct.* **4**, 637–642 (1968).
[7] R. D. Mindlin, Continuum and lattice theories of influence of electromechanical coupling on capacitance of thin, dielectric films. *Int. J. Solids Struct.* **5**, 1197–1208 (1969).
[8] Standards on piezoelectric crystals. *Proc. Inst. Radio Engrs* **37**, 1378–1385 (1949).
[9] R. D. Mindlin, Lectures on Polarization Gradient in Elastic Dielectrics, Int. Centre for Mechanical Sciences, Udine, Italy (1970).
[10] J. F. Nye, *Physical Properties of Crystals.* Oxford University Press (1959).
[11] R. A. Houston, *A Treatise on Light.* Longmans (1934).
[12] W. G. Cady, *Piezoelectricity.* McGraw-Hill (1946).
[13] R. Bechmann, Elastic and piezoelectric constants of alpha-quartz. *Phys. Rev.* **110**, 1060–1061 (1958).

(*Received* 26 *January* 1971)

Абстракт—Указывается, что теория упругих диэлектриков, в которых сохраняеиая электромеханическая энергия зависит от градиента поляризации, отвечает так за активность акустическую, как и оптическую. Определяются формулы для акустической мощности и оптической мощности вращения, для кварца α. Подсчитываются из экспериментальных данных значения новых постоянных материала, сущесмвующих в этих выажениях.

Reprinted from
PMM-Journal of Mechanics and Applied Mathematics
35 (1971) 404-408

ELECTROMECHANICAL VIBRATIONS OF CENTROSYMMETRIC CUBIC CRYSTAL PLATES

PMM Vol. 35, №3, 1971, pp. 446-450
R. D. MINDLIN
(Columbia University, New York, U. S. A.)
(This paper was copied from the original manuscript kindly supplied by the Author)

Introduction. According to the classical theory of piezoelectricity, there can be no piezoelectric effect in centrosymmetric crystals. Consequently the theory has it that vibrations of a centrosymmetric crystal plate cannot be excited, for example, by applying an alternating voltage drop between electrodes on the opposing faces of the plate. This conclusion is a direct result of the assumption, in the theory, that the stored energy of deformation and polarization is a function of the strain and polarization only [1]. Hence the only possible electromechanical interaction energy is the product of a second rank tensor (strain) and a first rank tensor (polarization) – with a third rank material coefficient (a piezoelectric constant). Since there are no third rank centrosymmetric tensors, there is no piezoelectric effect in centrosymmetric materials.

There is reason to believe, however, that the stored energy of deformation and polarization should depend not only on the strain and polarization, but also on the polarization gradient. The augmented theory [2] based on such an assumption can find justification on the grounds that: (1) it accommodates the mathematical representation of a surface energy of deformation and polarization [2] which is absent from the classical theory but which has been measured in the laboratory and calculated from atomic considerations; (2) it can account [3] for an apparent anomaly observed in measurements of the electrical capacitance of thin, dielectric films; (3) the resulting equations, rather than the classical ones, are the correct, low frequency limit [3, 5] of the modern dynamical theory of crystal lattices of electronically polarizable atoms.

If the polarization gradient is admitted as a variable in the stored energy function, an additional interaction energy, between the two second rank tensors of strain and polarization gradient, is possible. Since the accompanying material coefficient is fourth rank, it represents a possible electromechanical effect in centrosymmetric materials.

In the following paragraphs, it is shown that, at least in principle, it is possible to excite thickness vibrations in a centrosymmetric cubic crystal plate by applying an alternating voltage drop between electrodes on a pair of (100) faces of the plate. Approximate mode shapes and frequencies are calculated for the case of sodium chloride; for which it is found, also, that the electromechanical coupling coefficient can be as large as one four-hundredth of that of an X-cut quartz plate.

1. Field equations and boundary conditions. The linear equations of an elastic dielectric continuum, including the contribution of the polarization gradient to the stored energy, have been derived elsewhere [2] by means of a simple extension of Toupin's [1] variational principle for the classical equations of piezoelectricity. For the present case, we consider a plate of cubic crystal class $m3m$ (O_h) [4] bounded by a pair of (100) faces, at $x = \pm h$ on which are deposited electrode films to which are applied voltages $\pm Ve^{i\omega t}$. If end effects are neglected, the mechanical and electrical fields are one-dimensional and are governed by the equations [2]

$$c_{11}\partial^2 u + d_{11}\partial^2 P = -\rho\omega^2 u, \qquad d_{11}\partial^2 u + b_{11}\partial^2 P - a_{11}P - \partial\varphi = 0$$
$$\varepsilon_0\partial^2\varphi - \partial P = 0 \tag{1}$$

where $(\partial = d/dx)$, u and P are the x-components of the mechanical displacement and electronic polarization, respectively, and φ is the potential of the Maxwell, electric self-field. Also, ρ is the mass density, c_{11} is an elastic stiffness, ε_0 is the permittivity of a vacuum and $\varepsilon_0 a_{11}$ is the reciprocal dielectric susceptibility. The remaining two constants, b_{11} and d_{11}, are associated with terms, in the energy density, involving polarization gradient: $^1/_2 b_{11}$ is the coefficient of a quadratic term and d_{11} is the coefficient of a product of polarization gradient and strain: $\partial P \partial u$. Thus, d_{11} is the additional electromechanical constant without which, as may be seen from (1), the mechanical and electrical fields would not be coupled.

As for the boundary conditions, we suppose that the mass and stiffnes of the electrodes are negligible, in comparison with those of the plate, so that the tractions across $x =$ $= \pm h$ are zero.

This condition, as shown previously [2], is

$$(c_{11}\partial u + d_{11}\partial P)_{x=\pm h} = 0 \tag{2a}$$

Also, as shown previously [3], we may set

$$(P)_{x=\pm h} = -kV/a_{11}h \qquad (0 \leqslant k \leqslant 1) \tag{2b}$$

with zero the most likely value of k: corresponding to continuity of polarization across the crystal-electrode interfaces. Finally, the applied voltage is taken as

$$(\varphi)_{x=\pm h} = \pm V \tag{2c}$$

It may be noted that independent specification of boundary values of polarization and potential is permissible in the augmented theory, but not in the classical theory.

2. Solution. In view of the boundary conditions (2), a solution of (1), with appropriate symmetry and form, is

$$u = A\cos\xi x, \quad P = B_0 + B\cos\xi x, \quad \varphi = C_0 x + C\sin\xi x \tag{3}$$

Upon substituting (3) in (1), we find

$$B_0 = -a_{11}^{-1}C_0, \qquad B = \varepsilon_0\xi C \tag{4}$$

$$\frac{A}{C} = -\frac{\varepsilon_0 d_{11}\xi^3}{c_{11}\xi^2 - \rho\omega^2} = -\frac{1+\varepsilon_0 a_{11} + \varepsilon_0 b_{11}\xi^2}{d_{11}\xi} \tag{5}$$

The second of (5) is a quadratic in ξ^2:

$$(b_{11}c_{11} - d_{11}^2)\varepsilon_0\xi^4 - [\varepsilon_0 b_{11}\rho\omega^2 - (1+\varepsilon_0 a_{11})c_{11}]\xi^2 - (1+\varepsilon_0 a_{11})\rho\omega^2 = 0 \tag{6}$$

Now, positive-definiteness of the energy density requires

$$b_{11}c_{11} - d_{11}^2 > 0, \qquad a_{11} > 0 \tag{7}$$

Hence, one of the roots ξ^2, of (6), is positive real and the other is negative real; so that one ξ is real and the other is imaginary. Accordingly, from (4) to (7), we may rewrite the solution (3) as

$$\begin{aligned} u &= \alpha_1 C_1\cos\xi_1 x + \alpha_2 C_2\,ch\,\xi_2 x \\ P &= -a_{11}^{-1}C_0 + \varepsilon_0\xi_1 C_1\cos\xi_1 x + \varepsilon_0\xi_2 C_2\,ch\,\xi_2 x \\ \varphi &= C_0 x + C_1\sin\xi_1 x + C_2\,sh\,\xi_2 x \end{aligned} \tag{8}$$

where

$$\alpha_i = -\frac{\varepsilon_0 d_{11}\xi_i^3}{c_{11}\xi_i^2 + (-1)^i\rho\omega^2} = \frac{1+\varepsilon_0 a_{11} - (-1)^i\varepsilon_0 b_{11}\xi_i^2}{(-1)^i d_{11}\xi_i} \tag{9}$$

Upon substituting (8) in the boundary conditions (2), we find

$$\begin{aligned} (c_{11}\alpha_1 + \varepsilon_0 d_{11}\xi_1)\xi_1 C_1\sin\xi_1 h - (c_{11}\alpha_2 + \varepsilon_0 d_{11}\xi_2)\xi_2 C_2\,\mathrm{sh}\,\xi_2 h &= 0 \\ -C_0 + \varepsilon_0 a_{11}\xi_1 C_1\cos\xi_1 h + \varepsilon_0 a_{11}\xi_2 C_2\,\mathrm{ch}\,\xi_2 h &= kV/h \\ C_0 h + C_1\sin\xi_1 h + C_2\,\mathrm{sh}\,\xi_2 h &= V \end{aligned} \tag{10}$$

It may be seen, from (10), that the applied voltage V forces both the mechanical and electrical fields.

Resonance occurs when the determinant of the coefficients of C_0, C_1 and C_2 in (10), vanishes; or, when

$$\frac{1+\varepsilon_0 a_{11}\xi_1 h\,\mathrm{ctg}\,\xi_1 h}{1+\varepsilon_0 a_{11}\xi_2 h\,\mathrm{cth}\,\xi_2 h} = \frac{1+\rho\omega^2/c_{11}\xi_2^2}{1-\rho\omega^2/c_{11}\xi_1^2} \tag{11}$$

in which (9) has been employed.

3. Application of sodium chloride. For sodium chloride, the following are values of the material constants appearing in the solution:

$$\begin{aligned}
\varepsilon_0^{-1}a_{11}^{-1} &= 4.6 \text{ [1, pp. 68, 69]},\\
b_{11} &= 6.88\times 10^{3} \text{ dyn-cm}^4/\text{coul}^2 \text{ [5]},\\
c_{11} &= 4.83\times 10^{11} \text{ dyn/cm}^2 \text{ [6, p. 39]},\\
d_{11} &= 4.67\times 10^{7} \text{ dyn-cm/coul [5]},\\
\varepsilon_0^{-1} &= 36\pi\times 10^{9} \text{ dyn-cm}^2/\text{coul}^2 \text{ [4, p. 68]},\\
\rho &= 2.214 \text{ gm/cm}^3 \text{ [6. pp. 39, 88]}.
\end{aligned}$$

With these numerical values of the constants, simple approximate expressions for the resonance frequencies and mode shapes can be found.

At low frequencies, we find, from (6),

$$\lim_{\omega\to 0} \xi_2^{-1} = \left[\frac{b_{11} - d_{11}^2/c_{11}}{\varepsilon_0^{-1}(1+\varepsilon_0 a_{11})}\right]^{1/2} \approx 1.3\times 10^{-8} \text{ cm}. \tag{12}$$

Considering plate thicknesses of the order of millimeters, (12) indicates an extremely rapid decay, away from the surfaces of the plate, of the exponential parts of the fields, leaving

$$\begin{gathered}
u \approx \alpha_1 C_1 \cos \xi_1 x, \qquad P \approx -a_{11}^{-1}C_0 + \varepsilon_0\xi_1 C_1 \cos \xi_1 x\\
\varphi \approx C_0 x + C_1 \sin \xi_1 x
\end{gathered} \tag{13}$$

over most of the thickness of the plate. In fact, the decay modulus (12) is less than one-half the nearest neighbor distance,

$$a = 2.83\cdot 10^{-8} \text{ cm},$$

of the sodium chloride lattice. Also, with (12), the roots of (11) are very nearly roots of

$$\sin \xi_1 h = 0, \qquad \xi_1 h \neq 0, \tag{14}$$

or

$$\xi_1 h = n\pi, \qquad n = 1,\ 2,\ 3, \ldots \tag{15}$$

Finally, rewriting (6) in the form

$$\left(1 - \frac{\rho\omega^2}{c_{11}\xi^2}\right)\frac{1+\varepsilon_0 a_{11} + \varepsilon_0 b_{11}\xi^2}{\xi^2} - \frac{\varepsilon_0 d_{11}^2}{c_{11}} = 0 \tag{16}$$

and noting that

$$\frac{\varepsilon_0 d_{11}^2}{c_{11}} \approx 4\cdot 10^{-17}\text{cm}^2 \tag{17}$$

we see that the frequencies given by (16) are approximately

$$\omega = \xi_1 (c_{11}/\rho)^{1/2} \tag{18}$$

Hence, from (15),

$$\omega = \frac{n\pi}{h}\left(\frac{c_{11}}{\rho}\right)^{1/2} \tag{19}$$

or

$$f = \frac{\omega}{2\pi} \approx \frac{4.7\, n}{2h} \times 10^6 \text{ cyc/sec}, \tag{20}$$

where $2h$ is the thickness of the plate in millimeters. Thus, for a plate one millimeter thick, the lowest resonance ($n = 1$) has a frequency of about 4.7 megacycles per second. It may be noted that this mode has two nodes: the central half of the thickness of the plate and the outer quarters move in opposite directions. This is the mode most likely to be detected; but the response is very sharp in comparison with that of a piezoelectric crystal such as quartz, as may be seen from a calculation of an electromechanical coupling coefficient.

An electromechanical coupling coefficient may be defined as the square root of the ratio of the strain energy to the total energy – both at zero frequency.

At $\omega = 0$, the solution (8) reduces to [3]

$$u_0 = B_1 \operatorname{ch} \xi_2 x, \quad P_0 = A_2 + B_2 \operatorname{ch} \xi_2 x, \quad \varphi_0 = A_3 x + B_3 \operatorname{sh} \xi_2 x \tag{21}$$

where ξ_2 is given by (12) and

$$A_2 = A_3 / a_{11}, \quad B_2 = \varepsilon_0 \xi_2 B_3 = -c_{11} B_1 / d_{11}, \quad A_3 = B_3 \varepsilon_0 a_{11} \xi_2 \operatorname{ch} \xi_2 h$$
$$B_3 = (1-k) V (\operatorname{sh} \xi_2 h + \varepsilon_0 a_{11} \xi_2 h \operatorname{ch} \xi_2 h)^{-1} \tag{22}$$

The strain energy per unit area is

$$W_s = \frac{c_{11}}{2} \int_{-h}^{h} (\partial u_0)^2 \, dx = \frac{\varepsilon_0^2 \xi_2^2 d_{11}^2 (1-k)^2 V^2 (\operatorname{sh} 2\xi_2 h - 2\xi_2 h)}{4 c_{11} (\operatorname{sh} \xi_2 h + \varepsilon_0 a_{11} \xi_2 h \operatorname{ch} \xi_2 h)^2} \tag{23}$$

The total energy per unit area is V times the surface charge:

$$W = V(\varepsilon_0 \partial \varphi_0 - P_0)_{x=h} = \frac{\varepsilon_0 \xi_2 (1 + \varepsilon_0 a_{11})(1-k) V^2 \operatorname{ch} \xi_2 h}{\operatorname{sh} \xi_2 h + \varepsilon_0 a_{11} \xi_2 h \operatorname{ch} \xi_2 h} + \frac{k V^2 (1 + \varepsilon_0 a_{11})}{h a_{11}} \tag{24}$$

Taking into account the fact that $\xi_2 h$ is very large, the electromechanical coupling coefficient, for the case $k = 0$, is

$$\left(\frac{W_s}{W}\right)^{1/2} \approx \left(\frac{\xi_2 d_{11}^2}{2 \varepsilon_0 a_{11}^2 c_{11} (1 + \varepsilon_0^{-1} a_{11}^{-1}) h}\right)^{1/2} \approx 2.4 \cdot 10^{-4} \tag{25}$$

(for $h = 1$ mm)

This is to be compared with 9.5×10^{-2} for an X-cut quartz plate [7]. Thus, for $k = 0$, the coefficient for the sodium chloride plate is one four-hundredth that for the quartz plate, resulting in a resonance band width to be much narrower [7]. It is interesting to note that the coupling coefficient (25) is inversely proportional to the square root of the thickness of the plate, whereas the coefficient in the piezoelectric case is independent of the thickness.

BIBLIOGRAPHY

1. Toupin, R. A., "The Elastic Dielectric", Journal of Rational Mechanics and Analysis, Vol. 5, 1956, pp. 849-916.
2. Mindlin, R. D., "Polarization Gradient in Elastic Dielectrics". International Journal of Solids and Structures, Vol. 4, 1968, pp. 637-642.
3. Mindlin, R. D., "Continuum and Lattice Theories of Influence of Electromechanical Coupling on Capacitance of Thin Dielectric Films", International Journal of Solids and Structures, Vol. 5, 1969, pp. 1197-1208.
4. Nye, J. F., "Physical Properties of Crystals", Oxford University Press, London, 1957.
5. Askar, A, Lee, P. C. Y. and Cakmak, A. S., "A Lattice Dynamics Approach to the Theory of Elastic Dielectrics with Polarization Gradient", Physical Review, Series B, May 1970.
6. Hearmon, R. F. S., "An Introduction to Applied Anisotropic Elasticity, Oxford University Press, London, 1961.
7. Mason, W. P., "Piezoelectric Crystals and Their Application to Ultrasonics", D. Van Nostrand Company, New York, 1950, p. 96.

Int. J. Solids Structures, 1972, Vol. 8, pp. 369 to 383. Pergamon Press. Printed in Great Britain

A CONTINUUM THEORY OF A DIATOMIC, ELASTIC DIELECTRIC

R. D. MINDLIN

Department of Civil Engineering, Columbia University, New York, N.Y.

Abstract—The field equations and boundary conditions of the continuum theory of elastic dielectrics, including the contribution of the polarization gradient, are extended to apply to a compound material. A formula is obtained for the surface energy of deformation and polarization of a cubic crystal. The solution of a problem of plane waves leads to the identification of transverse and longitudinal optical, as well as acoustical, branches in the dispersion relation. A one-dimensional model of the NaCl-type crystal lattice of shell-model atoms is constructed and its finite difference equations of motion are shown to have the corresponding equations of the continuum theory as their long wave limit, without restriction to low frequency.

1. INTRODUCTION

THE differential equations of the classical theory of elasticity are known to be the long wave, low frequency limit of the finite difference equations of the theory of crystal lattices of mass-particles. If the lattice is a simple one of the Bravais type, with one particle per cell, there is no distinction between the long wave limit and the long wave, low frequency limit—as there are no optical branches in the dispersion relations for waves in Bravais lattices. However, in the case of a compound lattice, which has more than one particle per cell, there are high frequency optical branches as well as low frequency acoustical branches at the long wave limit. In that case, classical elasticity accounts for the long wave portions of only the acoustical branches. It has already been shown [1] how the classical equations of elasticity can be extended to include the long wave portions of the optical branches in the case of two particles per cell in a cubic crystal.

The applications now contemplated include those involving electromechanical interactions, so that it is appropriate to incorporate the electronic polarization, in the equations, in addition to the mechanical displacement. When this is done, the equations become the extension of the classical theory of elastic dielectrics to include the long wave, high frequency behavior of diatomic, polarizable crystals. Furthermore, even at low frequencies, wave lengths in the continuum approximation can be short enough and, at both low frequencies and in static equilibrium, dimensions can be small enough, for the effect of the polarization gradient to be significant. Consequently, the contribution of the polarization gradient is also included here. The resulting equations are then the extension, to diatomic materials, of a previous extension [2] of the classical theory of elastic dielectrics to account for the effects of the polarization gradient. At the same time, the equations are the long wave limit, not restricted to low frequencies, of the equations of a diatomic crystal lattice of polarizable atoms.

In Section 2, the potential and kinetic energy densities of the compound continuum are defined and a restriction to linearity is imposed. The field equations and boundary conditions are derived, in Section 3, by means of an extension of a linear version of Toupin's [3] variational principle for the equations of the classical theory of elastic dielectrics. This is followed, in Section 4, by a derivation of the constitutive equations for a special case of cubic symmetry. In Section 5, a formula is derived for the surface energy of deformation and polarization. A problem of plane waves in an infinite medium is solved in Section 6 and the acoustical and optical branches of the dispersion relations for transverse and longitudinal waves are identified. A limiting form of the equations, valid at long wave lengths, is equivalent to equations, given by Born and Huang [4], leading to a dispersion formula for the dielectric constant. In the final section, a one-dimensional Cochran-type [5] lattice of Dick–Overhauser [6] shell-model atoms in an NaCl-type structure is devised and the associated finite difference equations of motion are formulated. In the long wave limit, the difference equations of the lattice become the differential equations for longitudinal optical and acoustical waves in the diatomic continuum.

2. POTENTIAL AND KINETIC ENERGIES

In a continuum theory of a crystal lattice with two atoms per cell, each material point X_i, $i = 1, 2, 3$, is the site of two atoms, say 1 and 2, as illustrated in Fig. 1. Upon deformation, the two atoms move, with separate displacements $u_i^{(1)}$ and $u_i^{(2)}$ to a single spatial point x_i at which, if the atoms are polarizable, they take on electronic polarizations $P_i^{(1)}$ and $P_i^{(2)}$. A neighboring pair of atoms, initially at $X_i + \mathrm{d}X_i$, has displacements $u_i^{(\kappa)} + \mathrm{d}u_i^{(\kappa)}$, $\kappa = 1, 2$, and polarizations $P_i^{(\kappa)} + \mathrm{d}P_i^{(\kappa)}$. In a linear theory, to which the present treatment is confined,

$$\mathrm{d}u_j^{(\kappa)} \approx \mathrm{d}x_i u_{j,i}^{(\kappa)}, \qquad \mathrm{d}P_j^{(\kappa)} \approx \mathrm{d}x_i P_{j,i}^{(\kappa)}$$

where the summation convention for repeated indices and the comma notation for derivative (with respect to x_i) are employed.

Each point x_i may be regarded as being occupied by a point of each of two deformable and polarizable continua. The stored energy density of deformation and polarization, W^L, is assumed to be a quadratic function of the small strains, $S_{ij}^{(\kappa)}$, polarizations, $P_i^{(\kappa)}$, and

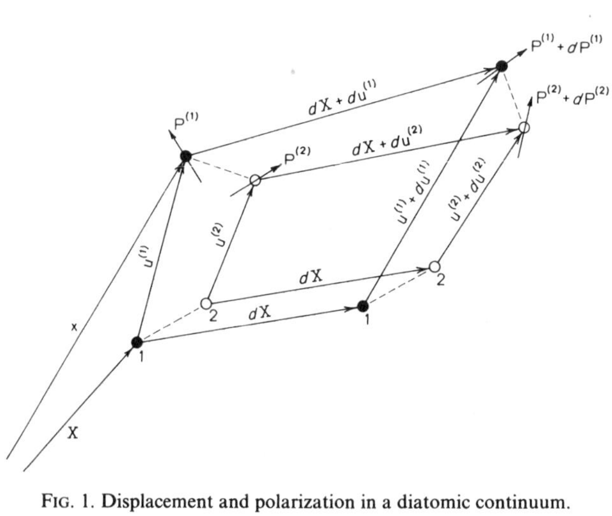

FIG. 1. Displacement and polarization in a diatomic continuum.

polarization gradients, $P_{j,i}^{(\kappa)}$, of the individual continua, and of the relative displacement, u_i^*, and relative rotation, ω_{ij}^*, of the two continua:

$$W^L = W^L(S_{ij}^{(1)}, S_{ij}^{(2)}, P_i^{(1)}, P_i^{(2)}, P_{j,i}^{(1)}, P_{j,i}^{(2)}, u_i^*, \omega_{ij}^*), \tag{1}$$

where

$$S_{ij}^{(\kappa)} = \tfrac{1}{2}(u_{j,i}^{(\kappa)} + u_{i,j}^{(\kappa)}),$$
$$u_i^* = u_i^{(2)} - u_i^{(1)},$$
$$\omega_{ij}^* = \tfrac{1}{2}(u_{j,i}^* - u_{i,j}^*).$$

With these variables, W^L is invariant in a rigid translation and rotation of the deformed and polarized body.

If the polarization and polarization gradient were omitted, W^L would be the energy density of a classical, linear, elastic body extended to accommodate a diatomic material [1]. If only the polarization gradient were omitted, W^L would be the energy density of deformation and polarization of a classical, linear, elastic dielectric extended to accommodate a diatomic material. As it stands in (1), W^L is the extension, to a diatomic material, of a previous energy function in which the inclusion of the polarization gradient has been shown to account for interesting, observed phenomena [2, 7, 8].

The total potential energy density, W, is the energy density of deformation and polarization augmented by the energy density of the Maxwell, electric self-field:

$$W = W^L + \tfrac{1}{2}\varepsilon_0\varphi_{,i}\varphi_{,i}, \tag{2}$$

where ε_0 is the permittivity of a vacuum and φ is the potential of the Maxwell self-field E_i^{MS}, i.e.

$$E_i^{MS} = -\varphi_{,i}. \tag{3}$$

The associated kinetic energy density is taken as

$$T = \sum_{\kappa} \tfrac{1}{2}\rho^{(\kappa)}u_{j,t}^{(\kappa)}u_{j,t}^{(\kappa)}, \qquad \kappa = 1, 2; \qquad j = 1, 2, 3, \tag{4}$$

where $\rho^{(1)}$ and $\rho^{(2)}$ are the mass densities of the two continua and the "comma t" notation designates the differentiation with respect to time.

3. FIELD EQUATIONS AND BOUNDARY CONDITIONS

The field equations and boundary conditions, corresponding to the potential and kinetic energy densities formulated in the preceding section, may be derived by means of an extension of a linear version of Toupin's [3] variational principle for the classical theory of elastic dielectrics. The extension accounts for the contributions of the polarization gradients, the two continua and the kinetic energy.

First we define an electric enthalpy density:

$$H = W - (\varepsilon_0 E_i^{MS} + P_i^{(1)} + P_i^{(2)} + q^* u_i^*)E_i^{MS} \tag{5}$$

i.e. the potential energy density diminished by the product of the Maxwell self-field and the electric displacement. The inclusion of the term $q^*u_i^*$, where q^* is a charge density, was suggested by P. C. Y. Lee to accommodate applications to ionic crystals; i.e. $q^*u_i^*$ is the ionic, or atomic, polarization as distinguished from the electronic polarizations $P_i^{(\kappa)}$. When the ionic polarization is included in what follows, the superscript $\kappa = 1$ identifies the positive ion and $\kappa = 2$ the negative ion.

Inserting (2) and (3) in (5), we find

$$H = W^L - \tfrac{1}{2}\varepsilon_0\varphi_{,i}\varphi_{,i} + \varphi_{,i}(P_i^{(1)} + P_i^{(2)} + q^*u_i^*). \tag{6}$$

In a body occupying a volume V bounded by a surface S separating V from a vacuum V', the extension of Toupin's variational principle takes the form

$$\delta\int_{t_0}^{t_1}\mathrm{d}t\int_{V''}(T-H)\,\mathrm{d}V + \sum_\kappa\int_{t_0}^{t_1}\mathrm{d}t\int_V (f_i^{(\kappa)}\delta u_i^{(\kappa)} + E_i^0\delta P_i^{(\kappa)} + E_i^0 q^*\delta u_i^*)\,\mathrm{d}V$$
$$+\sum_\kappa\int_{t_0}^{t_1}\mathrm{d}t\int_S t_i^{(\kappa)}\,\delta u_i^{(\kappa)}\,\mathrm{d}S = 0, \qquad \kappa = 1, 2, \tag{7}$$

for independent variations of $u_i^{(\kappa)}$, $P_i^{(\kappa)}$ and φ between fixed limits at times t_0 and t_1. In (7), $V'' = V + V'$, $f_i^{(\kappa)}$ and $t_i^{(\kappa)}$ are the external body forces and surface tractions on the two continua and E_i^0 is the external electric field. Now,

$$\delta W^L = \sum_\kappa (T_{ij}^{(\kappa)}\,\delta S_{ij}^{(\kappa)} - E_i^{(\kappa)}\,\delta P_i^{(\kappa)} + E_{ij}^{(\kappa)}\,\delta P_{j,i}^{(\kappa)}) + T_i^*\,\delta u_i^* + T_{ij}^*\delta\omega_{ij}^*, \tag{8}$$

where

$$T_{ij}^{(\kappa)} = \frac{\partial W^L}{\partial S_{ij}^{(\kappa)}},\quad E_i^{(\kappa)} = -\frac{\partial W^L}{\partial P_i^{(\kappa)}},\quad E_{ij}^{(\kappa)} = \frac{\partial W^L}{\partial P_{j,i}^{(\kappa)}},\quad T_i^* = \frac{\partial W^L}{\partial u_i^*},\quad T_{ij}^* = \frac{\partial W^L}{\partial \omega_{ij}^*}. \tag{9}$$

With the chain rule of differentiation, (8) becomes

$$\delta W^L = -\sum_\kappa [T_{ij,i}^{(\kappa)} + (-1)^\kappa(T_{ij,i}^* - T_j^*)]\,\delta u_j^{(\kappa)} - \sum_\kappa (E_j^{(\kappa)} + E_{ij,i}^{(\kappa)})\delta P_j^{(\kappa)}$$
$$+\sum_\kappa \{[T_{ij}^{(\kappa)} + (-1)^\kappa T_{ij}^*]\,\delta u_j^{(\kappa)}\}_{,i} + \sum_\kappa (E_{ij}^{(\kappa)}\,\delta P_j^{(\kappa)})_{,i}.$$

Treating the remaining part of H in (6) similarly, we have, in V:

$$\delta H = -\sum_\kappa [T_{ij,i}^{(\kappa)} + (-1)^\kappa(T_{ij,i}^* - T_j^* - q^*\varphi_{,i})]\,\delta u_j^{(\kappa)} - \sum_\kappa (E_j^{(\kappa)} + E_{ij,i}^{(\kappa)} - \varphi_{,i})\delta P_j^{(\kappa)}$$
$$+\sum_\kappa [\varepsilon_0\varphi_{,ii} - P_{i,i}^{(\kappa)} - (-1)^\kappa q^* u_{i,i}^{(\kappa)}]\delta\varphi + \sum_\kappa \{[T_{ij}^{(\kappa)} + (-1)^\kappa T_{ij}^*]\,\delta u_j^{(\kappa)}\}_{,i}$$
$$+\sum_\kappa (E_{ij}^{(\kappa)}\,\delta P_j^{(\kappa)})_{,i} - \sum_\kappa \{[\varepsilon_0\varphi_{,i} - P_i^{(\kappa)} - (-1)^\kappa q^* u_i^{(\kappa)}]\,\delta\varphi\}_{,i}; \tag{10}$$

and, in V':

$$\delta H = \varepsilon_0\varphi_{,ii}\,\delta\varphi - \varepsilon_0(\varphi_{,i}\,\delta\varphi)_{,i}. \tag{11}$$

Further, from (4),

$$\delta\int_{t_0}^{t_1} T\,\mathrm{d}t = -\sum_\kappa\int_{t_0}^{t_1}\rho^{(\kappa)}u_{i,tt}^{(\kappa)}\,\delta u_j^{(\kappa)}\,\mathrm{d}t. \tag{12}$$

Inserting (10)–(12) in (7) and applying the divergence theorem where appropriate, we find

$$\sum_{\kappa}\int_{t_0}^{t_1}\mathrm{d}t\int_V [T^{(\kappa)}_{ij,i}+(-1)^{\kappa}(T^{*}_{ij,i}-T^{*}_{j}-q^{*}\varphi_{,j})+(-1)^{\kappa}q^{*}E^{0}_{j}+f^{(\kappa)}_{j}-\rho^{(\kappa)}u^{(\kappa)}_{j,tt}]\,\delta u^{(\kappa)}_{j}\,\mathrm{d}V$$

$$+\sum_{\kappa}\int_{t_0}^{t_1}\mathrm{d}t\int_V \{(E^{(\kappa)}_{j}+E^{(\kappa)}_{ij,i}-\varphi_{,j}+E^{0}_{j})\delta P^{(\kappa)}_{j}+[-\varepsilon_0\varphi_{,ii}+P^{(\kappa)}_{i,i}+(-1)^{\kappa}q^{*}u^{(\kappa)}_{i,i}]\,\delta\varphi\}\,\mathrm{d}V$$

$$-\sum_{\kappa}\int_{t_0}^{t_1}\mathrm{d}t\int_S \{[n_i(T^{(\kappa)}_{ij}+(-1)^{\kappa}T^{*}_{ij})-t^{(\kappa)}]\,\delta u^{(\kappa)}_{j}+n_iE^{(\kappa)}_{ij}\,\delta P^{(\kappa)}_{j}+n_i[-\varepsilon_0[\![\varphi_{,i}]\!]$$

$$+P^{(\kappa)}_{i}+q^{*}(-1)^{\kappa}u^{(\kappa)}_{i}]\,\delta\varphi\}\,\mathrm{d}S-\int_{t_0}^{t_1}\mathrm{d}t\int_{V'}\varepsilon_0\varphi_{,ii}\,\mathrm{d}V=0, \tag{13}$$

where n_i is the outward normal to S and $[\![\varphi_{,i}]\!]$ is the jump in $\varphi_{,i}$ across S.

From (13), we have the Euler equations in V:

$$\begin{aligned} T^{(\kappa)}_{ij,i}+(-1)^{\kappa}(T^{*}_{ij,i}-T^{*}_{j}-q^{*}\varphi_{,i})+f^{(\kappa)}_{j}+(-1)^{\kappa}q^{*}E^{0}_{j} &= \rho^{(\kappa)}u^{(\kappa)}_{j,tt},\\ E^{(\kappa)}_{j}+E^{(\kappa)}_{ij,i}-\varphi_{,j}+E^{0}_{j} &= 0,\\ -\varepsilon_0\varphi_{,ii}+P^{(1)}_{i,i}+P^{(2)}_{i,i}+q^{*}u^{*}_{i,i} &= 0; \end{aligned} \tag{14}$$

and, in V':

$$\varphi_{,ii}=0; \tag{15}$$

and also the boundary conditions on S:

$$\begin{aligned} n_i[T^{(\kappa)}_{ij}+(-1)^{\kappa}T^{*}_{ij}] &= t^{(\kappa)}_{j},\\ n_iE^{(\kappa)}_{ij} &= 0,\\ n_i(-\varepsilon_0[\![\varphi_{,i}]\!]+P^{(1)}_{i}+P^{(2)}_{i}+q^{*}u^{*}_{i}) &= 0. \end{aligned} \tag{16}$$

As may be seen from (13), boundary conditions alternative to (16) are the specification of $u^{(\kappa)}_{i}$, $P^{(\kappa)}_{i}$ and φ, respectively.

4. CONSTITUTIVE EQUATIONS

We consider the constitutive equations for a crystal with NaCl-type lattice structure. Such a crystal belongs to cubic class *m*3*m* (International) or O_h (Schoenflies) [9] with generators [10]

$$\begin{pmatrix}-1&0&0\\0&-1&0\\0&0&-1\end{pmatrix}\begin{pmatrix}0&0&1\\1&0&0\\0&1&0\end{pmatrix}\begin{pmatrix}1&0&0\\0&0&1\\0&-1&0\end{pmatrix} \tag{17}$$

and, as illustrated in Fig. 2, comprises two, interpenetrating, f.c.c. lattices [11] each of which has *m*3*m* symmetry. Owing to the first generator in (17) (the centrosymmetry generator) there can be no coefficients of odd rank and no products of symmetric and antisymmetric variables in the quadratic function W^L. This reduces the interactions to those between variables connected by full lines in Fig. 3; whereas the dashed line segments denote excluded

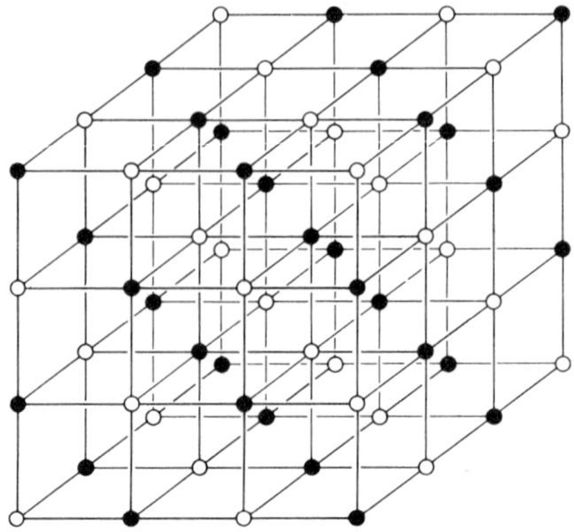

FIG. 2. NaCl-type lattice.

interactions. The corresponding energy function is

$$W^L = \tfrac{1}{2}\sum_{\kappa,\lambda}(a_{ij}^{\kappa\lambda}P_i^{(\kappa)}P_j^{(\lambda)} + b_{ijkl}^{\kappa\lambda}P_{j,i}^{(\kappa)}P_{l,k}^{(\lambda)} + c_{ijkl}^{\kappa\lambda}S_{ij}^{(\kappa)}S_{kl}^{(\lambda)} + 2d_{ijkl}^{\kappa\lambda}P_{(j,i)}^{(k)}S_{kl}^{(\lambda)})$$

$$+ \sum_{\kappa}(a^{*\kappa}u_i^*P_i^{(\kappa)} + d^{*\kappa}\omega_{ij}^*P_{[j,i]}^{(\kappa)}) + \tfrac{1}{2}a^{**}u_i^*u_i^* + c^{**}\omega_{ij}^*\omega_{ij}^*$$

$$+ \sum_{\kappa}(b^{\kappa 0}P_{i,i}^{(\kappa)} + c^{\kappa 0}S_{ii}^{(\kappa)}), \qquad (18)$$

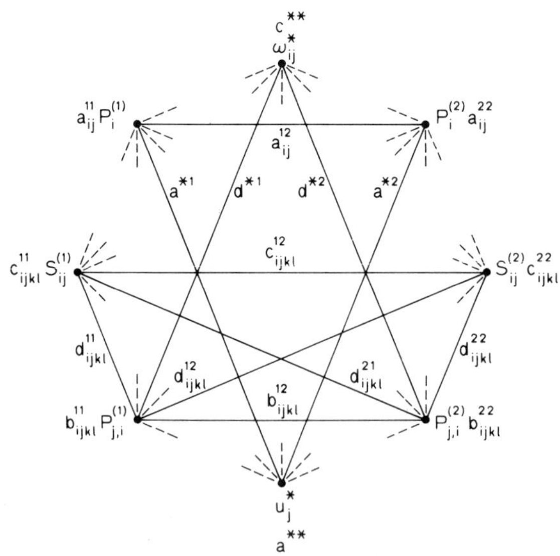

FIG. 3. Interactions among strains, polarizations, relative displacement and relative rotation in a diatomic, elastic, dielectric continuum with group $m3m$ (O_h) material symmetry.

where $P^{(\kappa)}_{(j,i)}$ and $P^{(\kappa)}_{[j,i]}$ denote the symmetric and antisymmetric parts of $P^{(\kappa)}_{j,i}$, respectively. We note that, in general,

$$a^{\kappa\lambda}_{ij} = a^{\lambda\kappa}_{ij}, \quad b^{\kappa\lambda}_{ijkl} = b^{\lambda\kappa}_{ijkl}, \quad c^{\kappa\lambda}_{ijkl} = c^{\lambda\kappa}_{ijkl} \tag{19}$$

and, from (17),

$$\begin{aligned}
a^{\kappa\lambda}_{ij} &= a^{\kappa\lambda}_{11}\delta_{ij}, \\
b^{\kappa\lambda}_{ijkl} &= b^{\kappa\lambda}\delta_{ijkl} + b^{\kappa\lambda}_{12}\delta_{ij}\delta_{kl} + b^{\kappa\lambda}_{44}(\delta_{ik}\delta_{jl} + \delta_{il}\delta_{jk}) + b^{\kappa\lambda}_{77}(\delta_{ik}\delta_{jl} - \delta_{il}\delta_{jk}), \\
c^{\kappa\lambda}_{ijkl} &= c^{\kappa\lambda}\delta_{ijkl} + c^{\kappa\lambda}_{12}\delta_{ij}\delta_{kl} + c^{\kappa\lambda}_{44}(\delta_{ik}\delta_{jl} + \delta_{il}\delta_{jk}), \\
d^{\kappa\lambda}_{ijkl} &= d^{\kappa\lambda}\delta_{ijkl} + d^{\kappa\lambda}_{12}\delta_{ij}\delta_{kl} + d_{44}(\delta_{ik}\delta_{jl} + \delta_{il}\delta_{jk}),
\end{aligned} \tag{20}$$

where δ_{ij} (or δ_{ijkl}) is unity when its indices are alike and zero otherwise;

$$b^{\kappa\lambda} = b^{\kappa\lambda}_{11} - b^{\kappa\lambda}_{12} - 2b^{\kappa\lambda}_{44}, \quad c^{\kappa\lambda} = c^{\kappa\lambda}_{11} - c^{\kappa\lambda}_{12} - 2c^{\kappa\lambda}_{44}, \quad d^{\kappa\lambda} = d^{\kappa\lambda}_{11} - d^{\kappa\lambda}_{12} - 2d^{\kappa\lambda}_{44}; \tag{21}$$

for $c^{\kappa\lambda}_{ijkl}$ and $d^{\kappa\lambda}_{ijkl}$ the abbreviated notation for pairs of indices, ij or kl has been used:

$$11 \to 1, \quad 22 \to 2, \quad 33 \to 3, \quad 23 \text{ or } 32 \to 4, \quad 31 \text{ or } 13 \to 5, \quad 12 \text{ or } 21 \to 6$$

and the same for $b^{\kappa\lambda}_{ijkl}$ except that

$$32 \to 7, \quad 13 \to 8, \quad 21 \to 9.$$

It may be observed that material constants with superscripts $\kappa\lambda = 11$ or 22 denote interactions within one of the two component continua whereas constants with superscripts $\kappa\lambda = 12$ or 21 denote interactions between the two component continua.

From (9) and (18)–(21), we find the constitutive equations

$$\begin{aligned}
T^{(\kappa)}_{ij} &= c^{\kappa 0}\delta_{ij} + \sum_\lambda (c^{\kappa\lambda}\delta_{ijkl}S^{(\lambda)}_{kl} + c^{\kappa\lambda}_{12}\delta_{ij}S^{(\lambda)}_{kk} + 2c^{\kappa\lambda}_{44}S^{(\lambda)}_{ij}) \\
&\quad + \sum_\lambda (d^{\lambda\kappa}\delta_{ijkl}P_{l,k} + d^{\lambda\kappa}_{12}\delta_{ij}P^{(\lambda)}_{k,k} + 2d^{\lambda\kappa}_{44}P^{(\lambda)}_{(j,i)}), \\
-E^{(\kappa)}_j &= \sum_\lambda a^{\kappa\lambda}_{11}P^{(\lambda)}_j + a^{*\kappa}u^*_j, \\
E^{(\kappa)}_{ij} &= b^{\kappa 0}\delta_{ij} + \sum_\lambda (b^{\kappa\lambda}\delta_{ijkl}P^{(\lambda)}_{l,k} + b^{\kappa\lambda}_{12}\delta_{ij}P^{(\lambda)}_{k,k} + 2b^{\kappa\lambda}_{44}P^{(\lambda)}_{(j,i)} + 2b^{\kappa\lambda}_{77}P^{(\lambda)}_{[j,i]}) \\
&\quad + 2d^{*\kappa}\omega^*_{ij} + \sum_\lambda (d^{\kappa\lambda}\delta_{ijkl}S^{(\lambda)}_{kl} + d^{\kappa\lambda}_{12}\delta_{ij}S^{(\lambda)}_{kk} + 2d^{\kappa\lambda}_{44}S^{(\lambda)}_{ij}), \\
T^*_j &= \sum_\lambda a^{*\lambda}P^{(\lambda)}_j + a^{**}u^*_j, \\
T^*_{ij} &= 2\sum_\lambda d^{*\lambda}P^{(\lambda)}_{[j,i]} + 2c^{**}\omega^*_{ij}.
\end{aligned} \tag{22}$$

We shall assume that, in the initial state, i.e. when

$$P^{(\kappa)}_i, P^{(\kappa)}_{j,i}, S^{(\kappa)}_{ij}, u^*_i, \omega^*_{ij}, \varphi_{,i}, f^{(\kappa)}_i, E^0_i = 0, \tag{23}$$

there is no resultant force across any surface—exterior or interior. That is,

$$t^{(1)}_j + t^{(2)}_j = 0.$$

Then, from the first of (16),

$$n_i T_{ij}^{(1)} + n_i T_{ij}^{(2)} = 0;$$

so that, from the first of (22),

$$c^{10} + c^{20} = 0. \tag{24}$$

Accordingly, from (23) and (24), in the initial state:

$$E_i^{(\kappa)}, T_i^*, T_{ij}^* = 0, \qquad T_{ij}^{(1)} + T_{ij}^{(2)} = 0,$$

but there is a self-equilibrated

$$E_{ij}^{(\kappa)} = b^{\kappa 0}\delta_{ij}$$

in each component continuum.

5. SURFACE ENERGY OF DEFORMATION AND POLARIZATION

In a state of equilibrium, the total energy in V'' is, from (2),

$$\mathscr{W} = \int_V W^L \, \mathrm{d}V + \tfrac{1}{2}\int_{V''} \varepsilon_0 \varphi_{,i}\varphi_{,i} \, \mathrm{d}V. \tag{25}$$

We can write W^L in the form

$$2W^L = \sum_\kappa (T_{ij}^{(\kappa)} S_{ij}^{(\kappa)} - E_i^{(\kappa)} P_i^{(\kappa)} + E_{ij}^{(\kappa)} P_{j,i}^{(\kappa)}) + T_i^* u_i^* + T_{ij}^* \omega_{ij}^* + \sum_\kappa b^{\kappa 0} P_{i,i}^{(\kappa)} + c^{20}(S_{ii}^{(2)} - S_{ii}^{(1)}).$$

Then, by the same procedure as that employed in arriving at (13),

$$\begin{aligned}\int_V W^L \, \mathrm{d}V = {} & -\tfrac{1}{2}\sum_\kappa \int_V \{[T_{ij,i}^{(\kappa)} + (-1)^\kappa (T_{ij,i}^* - T_j^*)]u_j^{(\kappa)} + (E_j^{(\kappa)} + E_{ij,i}^{(\kappa)})P_j^{(\kappa)}\} \, \mathrm{d}V \\ & + \tfrac{1}{2}\sum_\kappa \int_S n_i\{[T_{ij}^{(\kappa)} + (-1)^\kappa T_{ij}^*]u_j^{(\kappa)} + E_{ij}^{(\kappa)} P_j^{(\kappa)}\} \, \mathrm{d}S \\ & + \tfrac{1}{2}\sum_\kappa \int_S n_i(b^{\kappa 0} P_i^{(\kappa)} + c^{20} u_i^*) \, \mathrm{d}S. \end{aligned} \tag{26}$$

Also,

$$\int_{V''} \varphi_{,i}\varphi_{,i} \, \mathrm{d}V = -\int_{V''} \varphi_{,ii}\varphi \, \mathrm{d}V + \int_S n_i [\![\varphi_{,i}]\!] \varphi \, \mathrm{d}S. \tag{27}$$

Inserting (26) and (27) in (25) and applying the equations of equilibrium [i.e. (14) with $u_{j,tt}^{(\kappa)} = 0$], in the absence of external forces ($f_j^{(\kappa)}, E_j^0 = 0$), we find

$$\begin{aligned} 2\mathscr{W} = {} & -\sum_\kappa \int_V [(-1)^\kappa q^* \varphi_{,i} u_i^{(\kappa)} + \varphi_{,i} P_i^{(\kappa)} + (P_{i,i}^{(\kappa)} + q^* u_{i,i}^*)\varphi] \, \mathrm{d}V \\ & + \sum_\kappa \int_S n_i\{[T_{ij}^{(\kappa)} + (-1)^\kappa T_{ij}^*]u_j^{(\kappa)} + E_{ij}^{(\kappa)} P_j^{(\kappa)} + \varepsilon_0 [\![\varphi_{,i}]\!]\varphi\} \, \mathrm{d}S \\ & + \sum_\kappa \int_S n_i(b^{\kappa 0} P_i^{(\kappa)} + c^{20} u_i^*) \, \mathrm{d}S. \end{aligned}$$

Now,

$$\int_V P_{i,i}^{(\kappa)}\varphi \, dV = -\int_V \varphi_{,i} P_i^{(\kappa)} \, dV + \int_S n_i P_i^{(\kappa)}\varphi \, dS,$$

$$\int_V q^*(u_i^*\varphi_{,i} + u_{i,i}^*\varphi) \, dV = \int_S n_i q^* u_i^* \varphi \, dS.$$

Hence,

$$2\mathscr{W} = \sum_\kappa \int_S n_i\{[T_{ij}^{(\kappa)} + (-1)^\kappa T_{ij}^*]u_j^{(\kappa)} + E_{ij}^{(\kappa)}P_j^{(\kappa)} + (\varepsilon_0[\varphi_{,i}] - P_i^{(\kappa)} - q^*u_i^*)\varphi\} \, dS$$
$$+ \sum_\kappa \int_S n_i(b^{\kappa 0}P_i^{(\kappa)} + c^{20}u_i^*) \, dS. \tag{28}$$

From (16), the first integral in (28) vanishes if the boundary is entirely free. Then

$$2\mathscr{W} = \sum_\kappa \int_S n_i(b^{\kappa 0}P_i^{(\kappa)} + c^{20}u_i^*) \, dS.$$

Accordingly, the surface energy of deformation and polarization, per unit area, is

$$W^S = \tfrac{1}{2}n_i[b^{10}P_i^{(1)} + b^{20}P_i^{(2)} + c^{20}u_i^*]_S. \tag{29}$$

6. WAVE MOTION

The equations of motion in terms of $u_i^{(\kappa)}$, $P_i^{(\kappa)}$ and φ are obtained by substituting the constitutive equations (22) into the field equations (14), with the result, for each κ,

$$\begin{aligned}
&\sum_\lambda [c^{\kappa\lambda}\delta_{ijkl}u_{l,ki}^{(\lambda)} + c_{12}^{\kappa\lambda}u_{i,ij}^{(\lambda)} + c_{44}^{\kappa\lambda}(u_{j,ii}^{(\lambda)} + u_{i,ij}^{(\lambda)})] \\
&\quad + \sum_\lambda [d^{\lambda\kappa}\delta_{ijkl}P_{l,ki}^{(\lambda)} + d_{12}^{\lambda\kappa}P_{i,ij}^{(\lambda)} + d_{44}^{\lambda\kappa}(P_{j,ii}^{(\lambda)} + P_{i,ij}^{(\lambda)})] \\
&\quad + (-1)^\kappa \sum_\lambda [d^{*\lambda}(P_{j,ii}^{(\lambda)} - P_{i,ij}^{(\lambda)}) - (-1)^\lambda c^{**}(u_{j,ii}^{(\lambda)} - u_{i,ij}^{(\lambda)})] \\
&\quad - (-1)^\kappa[a^{*1}P_j^{(1)} + a^{*2}P_j^{(2)} + a^{**}(u_j^{(2)} - u_j^{(1)}) + q^*\varphi_{,j} - q^*E_j^0] + f_j^{(\kappa)} = \rho^{(\kappa)}u_{j,tt}^{(\kappa)}, \\
&\sum_\lambda [d^{\kappa\lambda}\delta_{ijkl}u_{l,ki}^{(\lambda)} + d_{12}^{\kappa\lambda}u_{i,ij}^{(\lambda)} + d_{44}^{\kappa\lambda}(u_{j,ii}^{(\lambda)} + u_{i,ij}^{(\lambda)}) + (-1)^\lambda d^{*\kappa}(u_{j,ii}^{(\lambda)} - u_{i,ij}^{(\lambda)})] \\
&\quad + \sum_\lambda [b^{\kappa\lambda}\delta_{ijkl}P_{l,ki}^{(\lambda)} + b_{12}^{\kappa\lambda}P_{i,ij}^{(\lambda)} + b_{44}^{\kappa\lambda}(P_{j,ii}^{(\lambda)} + P_{i,ij}^{(\lambda)}) + b_{77}^{\kappa\lambda}(P_{j,ii}^{(\lambda)} - P_{i,ij}^{(\lambda)})] \\
&\quad - a_{11}^{1\kappa}P_j^{(1)} - a_{11}^{2\kappa}P_j^{(2)} - a^{*\kappa}(u_j^{(2)} - u_j^{(1)}) - \varphi_{,j} + E_j^0 = 0, \\
&\qquad -\varepsilon_0\varphi_{,ii} + P_{i,i}^{(1)} + P_{i,i}^{(2)} + q^*(u_{i,i}^{(2)} - u_{i,i}^{(1)}) = 0.
\end{aligned} \tag{30}$$

The essential features of wave motion are revealed by an examination of plane waves in, say, the x_1, or (100), direction:

$$u_j^{(\kappa)} = A_j^{(\kappa)} e^{i(\xi x_1 - \omega t)}, \qquad P_j^{(\kappa)} = B_j^{(\kappa)} e^{i(\xi x_1 - \omega t)}, \qquad \varphi = C e^{i(\xi x_1 - \omega t)} \tag{31}$$

and $f_j^{(\kappa)} = E_j^0 = 0$. With these functions, the five equations (30) reduce to those for longitudinal waves if $j = 1$ and transverse waves if $j = 2$ or 3. Thus:

Longitudinal ($j = 1$)

$$\begin{aligned}
&c_{11}^{11}u_{1,11}^{(1)} + c_{11}^{12}u_{1,11}^{(2)} + d_{11}^{11}P_{1,11}^{(1)} + d_{11}^{21}P_{1,11}^{(2)} + a^{*1}P_1^{(1)} + a^{*2}P_1^{(2)} + a^{**}(u_1^{(2)} - u_1^{(1)}) + q^*\varphi_{,1} \\
&\qquad = \rho^{(1)}u_{1,tt}^{(1)}, \\
&c_{11}^{21}u_{1,11}^{(1)} + c_{11}^{22}u_{1,11}^{(2)} + d_{11}^{12}P_{1,11}^{(1)} + d_{11}^{22}P_{1,11}^{(2)} - a^{*1}P_1^{(1)} - a^{*2}P_1^{(2)} - a^{**}(u_1^{(2)} - u_1^{(1)}) - q^*\varphi_{,1} \\
&\qquad = \rho^{(2)}u_{1,tt}^{(2)}, \\
&d_{11}^{11}u_{1,11}^{(1)} + d_{11}^{12}u_{1,11}^{(2)} + b_{11}^{11}P_{1,11}^{(1)} + b_{11}^{12}P_{1,11}^{(2)} - a_{11}^{11}P_1^{(1)} - a_{11}^{12}P_1^{(2)} - a^{*1}(u_1^{(2)} - u_1^{(1)}) - \varphi_{,1} = 0, \\
&d_{11}^{21}u_{1,11}^{(1)} + d_{11}^{22}u_{1,11}^{(2)} + b_{11}^{21}P_{1,11}^{(1)} + b_{11}^{22}P_{1,11}^{(2)} - a_{11}^{21}P_1^{(1)} - a_{11}^{22}P_1^{(2)} - a^{*2}(u_1^{(2)} - u_1^{(1)}) - \varphi_{,1} = 0, \\
&P_{1,1}^{(1)} + P_{1,1}^{(2)} + q^*(u_{1,1}^{(2)} - u_{1,1}^{(1)}) - \varepsilon_0\varphi_{,11} = 0.
\end{aligned} \tag{32}$$

Transverse ($j = 2$)

$$\begin{aligned}
&(c_{44}^{11} - c^{**})u_{2,11}^{(1)} + (c_{44}^{12} + c^{**})u_{2,11}^{(2)} + (d_{44}^{11} - d^{*1})P_{2,11}^{(2)} + (d_{44}^{21} - d^{*2})P_{2,11}^{(2)} \\
&\qquad + a^{*1}P_2^{(1)} + a^{*2}P_2^{(2)} + a^{**}(u_2^{(2)} - u_2^{(1)}) = \rho^{(1)}u_{2,tt}^{(1)}, \\
&(c_{44}^{21} + c^{**})u_{2,11}^{(1)} + (c_{44}^{22} - c^{**})u_{2,11}^{(2)} + (d_{44}^{12} + d^{*1})P_{2,11}^{(1)} + (d_{44}^{22} + d^{*2})P_{2,11}^{(2)} \\
&\qquad - a^{*1}P_2^{(1)} - a^{*2}P_2^{(2)} - a^{**}(u_2^{(2)} - u_2^{(1)}) = \rho^{(2)}u_{2,tt}^{(2)}, \\
&(d_{44}^{11} - d^{*1})u_{2,11}^{(1)} + (d_{44}^{12} + d^{*1})u_{2,11}^{(2)} + (b_{44}^{11} + b_{77}^{11})P_{2,11}^{(1)} + (b_{44}^{12} + b_{77}^{12})P_{2,11}^{(2)} \\
&\qquad - a_{11}^{11}P_2^{(1)} - a_{11}^{12}P_2^{(2)} - a^{*1}(u_2^{(2)} - u_1^{(1)}) = 0, \\
&(d_{44}^{21} - d^{*2})u_{2,11}^{(1)} + (d_{44}^{22} + d^{*2})u_{2,11}^{(2)} + (b_{44}^{21} + b_{77}^{21})P_{2,11}^{(1)} + (b_{44}^{22} + b_{77}^{22})P_{2,11}^{(2)} \\
&\qquad - a_{11}^{21}P_2^{(1)} - a_{11}^{22}P_2^{(2)} - a^{*2}(u_2^{(2)} - u_2^{(1)}) = 0.
\end{aligned} \tag{33}$$

It may be noted that the Maxwell self-field is coupled to the mechanical displacement in the longitudinal waves, but not in the transverse waves.

Upon substituting (31) in (33) and eliminating $A_2^{(\kappa)}$ and $B_2^{(\kappa)}$, we find the dispersion relation for transverse waves:

$$\Delta_t = \begin{vmatrix}
\rho^{(1)}\omega^2 - a^{**} - (c_{44}^{11} - c^{**})\xi^2 & a^{**} - (c_{44}^{12} + c^{**})\xi^2 & a^{*1} - (d_{44}^{11} - d^{*1})\xi^2 & a^{*2} - (d_{44}^{21} - d^{*2})\xi^2 \\
a^{**} - (c_{44}^{21} + c^{**})\xi^2 & \rho^{(2)}\omega^2 - a^{**} - (c_{44}^{22} - c^{**})\xi^2 & -a^{*1} - (d_{44}^{12} + d^{*1})\xi^2 & -a^{*2} - (d_{44}^{22} + d^{*2})\xi^2 \\
a^{*1} - (d_{44}^{11} - d^{*1})\xi^2 & -a^{*1} - (d_{44}^{12} + d^{*1})\xi^2 & -a_{11}^{11} - (b_{44}^{11} + b_{77}^{11})\xi^2 & -a_{11}^{12} - (b_{44}^{12} + b_{77}^{12})\xi^2 \\
a^{*2} - (d_{44}^{21} - d^{*2})\xi^2 & -a^{*2} - (d_{44}^{22} + d^{*2})\xi^2 & -a_{11}^{21} - (b_{44}^{21} + b_{77}^{21})\xi^2 & -a_{11}^{22} - (b_{44}^{22} + b_{77}^{22})\xi^2
\end{vmatrix} = 0. \tag{34}$$

This is a quartic equation in ξ^2. Of the four branches, two are real: one acoustical and one optical, as may be verified by showing that one real branch passes through $\xi = 0$, $\omega = 0$ and one through $\xi = 0$, $\omega \neq 0$. We find, in fact,

$$\lim_{\omega,\xi \to 0} \Delta_t = \lim_{\omega,\xi \to 0} [(c_{44}^{11} + c_{44}^{22} + 2c_{44}^{12})\xi^2 - (\rho^{(1)} + \rho^{(2)})\omega^2]D,$$

where

$$D = \begin{vmatrix} a_{11}^{11} & a_{11}^{12} & a^{*1} \\ a_{11}^{21} & a_{11}^{22} & a^{*2} \\ a^{*1} & a^{*2} & a^{**} \end{vmatrix}.$$

Hence, there is a transverse acoustical branch with the limiting, low frequency behavior

$$\omega = \xi[(c_{44}^{11}+c_{44}^{22}+2c_{44}^{12})/(\rho^{(1)}+\rho^{(2)})]^{1/2}. \tag{35}$$

It may be seen, from (35), that the usual low frequency shear stiffness, c_{44}, is given by

$$c_{44} = c_{44}^{11}+c_{44}^{22}+2c_{44}^{12},$$

i.e. the sum of the shear stiffnesses of the two component continua augmented by twice the intercomponent shear stiffness. Further

$$\lim_{\xi\to 0} \Delta_t = \omega^2[\rho^{(1)}\rho^{(2)}\omega^2(a_{11}^{11}a_{11}^{22}-a_{11}^{12}a_{11}^{21})-(\rho^{(1)}+\rho^{(2)})D].$$

Thus, there is a transverse optical branch with limiting (infrared absorption) frequency

$$\omega = [D/\bar{\rho}(a_{11}^{11}a_{11}^{22}-a_{11}^{12}a_{11}^{21})]^{1/2}, \tag{36}$$

where $\bar{\rho}$ is the reduced density:

$$\bar{\rho} = \rho^{(1)}\rho^{(2)}/(\rho^{(1)}+\rho^{(2)}).$$

For the longitudinal waves, substitution of (31) in (32) and elimination of $A_1^{(\kappa)}$, $B_1^{(\kappa)}$ and C yields a dispersion relation the same as that for transverse waves except that the subscript 44 is changed to 11, the constants $d^{*\kappa}$, c^{**} and $b_{77}^{\kappa\lambda}$ are zero, the constants $a_{11}^{\kappa\lambda}$ are replaced by $a_{11}^{\kappa\lambda}+\varepsilon_0^{-1}$, the constants $a^{*\kappa}$ are replaced by $a^{*\kappa}+q^*\varepsilon_0^{-1}$ and a^{**} is replaced by $a^{**}+(q^*)^2\varepsilon_0^{-1}$. Subject to these changes, the low frequency behavior of the longitudinal acoustical branch and the long wave limit of the longitudinal optical branch are given by (35) and (36), respectively. Note that the constants $\varepsilon_0 a_{11}^{\kappa\lambda}$ are reciprocal dielectric susceptibilities, as may be seen from the expression for $E_j^{(\kappa)}$ in the constitutive relations (22). Note, also, that q^* is zero in a non-ionic crystal.

It should be observed that the real parts of the optical branches have no valid role in solutions, of the equations of motion, in which optical and acoustical modes are coupled, say through boundary conditions. The frequencies of the two types of mode must be the same, when they are coupled, but any frequency on a real segment of an optical branch is associated with a wave length, on an acoustical branch, that is too short to be within the range of validity of the continuum theory. In such coupled situations, there is an upper limit of frequency, set by the wave length limitation of the acoustic branches, to which the solution is restricted. Below such a frequency, the optical branches are imaginary or complex and correspond to displacements and polarizations confined essentially to the neighborhood of the boundary with exponential, or oscillating exponential, spatial decay into the interior.

The special field equations governing the behavior of the continuum at infinite wave length and high frequencies may be obtained from (30) by omitting all spatial derivative terms, after employing (3), with the result:

$$a^{*1}P_i^{(1)}+a^{*2}P_i^{(2)}+a^{**}(u_i^{(2)}-u_i^{(1)})-q^*E_i^{MS} = \rho^{(1)}u_{i,tt}^{(1)},$$
$$-a^{*1}P_i^{(1)}-a^{*2}P_i^{(2)}-a^{**}(u_i^{(2)}-u_i^{(1)})+q^*E_i^{MS} = \rho^{(2)}u_{i,tt}^{(2)},$$
$$a_{11}^{11}P_i^{(1)}+a_{11}^{12}P_i^{(2)}+a^{*1}(u_i^{(2)}-u_i^{(1)})-E_i^{MS} = 0,$$
$$a_{11}^{21}P_i^{(1)}+a_{11}^{22}P_i^{(2)}+a^{*2}(u_i^{(2)}-u_i^{(1)})-E_i^{MS} = 0.$$

These are equivalent to the equations, given by Born and Huang [4], from which a dispersion formula for the dielectric constant is obtained. Thus, the weighted difference between the first two equations gives

$$-\bar{\rho}u^*_{i,tt} = a^{*1}P_i^{(1)} + a^{*2}P_i^{(2)} + a^{**}u_i^* - q^*E_i^{MS}$$

and the second two equations yield

$$P_i^{(1)} = [(a_{11}^{22} - a_{11}^{21})E_i^{MS} - (a_{11}^{22}a^{*1} - a_{11}^{21}a^{*2})u_i^*]/(a_{11}^{11}a_{11}^{22} - a_{11}^{12}a_{11}^{21}),$$

$$P_i^{(2)} = [(a_{11}^{11} - a_{11}^{12})E_i^{MS} - (a_{11}^{11}a^{*2} - a_{11}^{12}a^{*1})u_i^*]/(a_{11}^{11}a_{11}^{22} - a_{11}^{12}a_{11}^{21}).$$

Substituting the latter two in the former and defining the total polarization density as

$$P_i = P_i^{(1)} + P_i^{(2)} + q^*u_i^*,$$

we find equations of the same form as (7.1) and (7.2) in Born and Huang [4]:

$$\bar{\rho}u^*_{i,tt} = \beta_{11}u_i^* + \beta_{12}E_i^{MS},$$

$$P_i = \beta_{21}u_i^* + \beta_{22}E_i^{MS},$$

where

$$\beta_{11} = -D/(a_{11}^{11}a_{11}^{22} - a_{11}^{12}a_{11}^{21}), \qquad \beta_{22} = (a_{11}^{11} + a_{11}^{22} - 2a_{11}^{12})/(a_{11}^{11}a_{11}^{22} - a_{11}^{12}a_{11}^{21}),$$

$$\beta_{12} = \beta_{21} = -[a_{11}^{11}a^{*2} + a_{11}^{22}a^{*1} - (a^{*1} + a^{*2})a_{11}^{12} - q^*(a_{11}^{11}a_{11}^{22} - a_{11}^{22}a_{11}^{21})]/(a_{11}^{11}a_{11}^{22} - a_{11}^{12}a_{11}^{21}).$$

7. ONE-DIMENSIONAL LATTICE

The purpose of this section is to show, for the one-dimensional, longitudinal case, that the equations of motion (32), exhibited in the preceding section, are the long wave limit of the equations of motion of an NaCl-type lattice of the kind devised by Cochran [5] based on the Dick–Overhauser [6] shell model of the atom: a core, comprising the nucleus and inner electrons, surrounded by a shell of outer electrons—with the electronic polarization proportional to the relative displacement of the shell with respect to the core. In addition to the intra-atomic interaction producing the electronic polarization, certain interatomic core–core, shell–shell and core–shell interactions are taken into account. The one-dimensional NaCl-type lattice is most conveniently represented by two lines of alternating atoms (positive and negative ions, in the case of an ionic crystal) with one atom of each type at each lattice site, as shown in Fig. 4, where the two types of atom are identified by the digits 1 and 2. Nearest neighbor interactions between unlike (adjacent) atoms in each line and next nearest neighbor interactions between like atoms in the two lines are those taken into account, but no interactions between the unlike atoms at the same site are considered. With each of p and q equal to 1 or 2, the force constants of the interactions are denoted by α_p for the intra-atomic core–shell interactions and $\beta_{pq}, \gamma_{pq}, \delta_{pq}$ for the core–core, core–shell, and shell–shell interatomic interactions, respectively, between like atoms for $p = q$ and unlike atoms for $p \neq q$. We note that $\beta_{pq} = \beta_{qp}$ and $\delta_{pq} = \delta_{qp}$ but $\gamma_{pq} \neq \gamma_{qp}$. The two lines of atoms are taken to be parallel to the x_1 axis with the atom sites at $x_1 = na$, where n is a positive or negative integer. The displacements of the cores and shells of the atoms at $x_1 = na$ are denoted by $u_n^{(\kappa)}$ and $s_n^{(\kappa)}$, where $\kappa = 1, 2$ to designate the atoms of types 1 and 2 (1 for the positive ion and 2 for the negative ion, in the case of an ionic crystal).

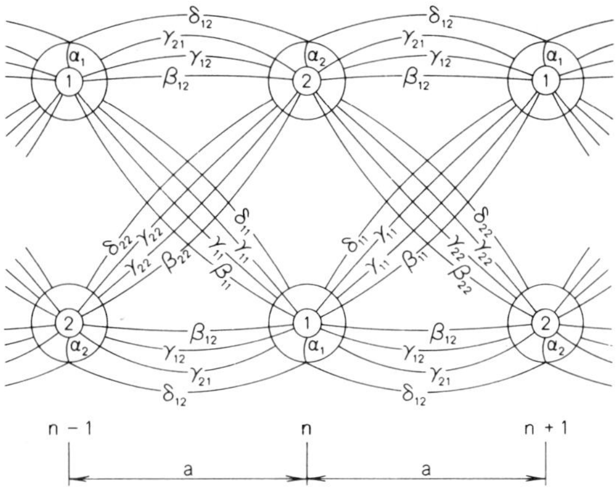

FIG. 4. One-dimensional NaCl-type lattice of shell-model atoms.

The equation of motion of the nth atom of type 1 is obtained by equating its inertia force to the sum of the forces on its core and shell exerted by the cores and shells of the two nearest neighbor (unlike) atoms, the two next nearest neighbor (like) atoms and the Maxwell, electric self-field at $x_1 = na$:

$$\begin{aligned}
&\beta_{11}(u^{(1)}_{n+1}-u^{(1)}_n)+\gamma_{11}(s^{(1)}_{n+1}-s^{(1)}_n+u^{(1)}_{n+1}-u^{(1)}_n)+\delta_{11}(s^{(1)}_{n+1}-s^{(1)}_n)\\
&\quad -\beta_{11}(u^{(1)}_n-u^{(1)}_{n-1})-\gamma_{11}(u^{(1)}_n-u^{(1)}_{n-1}+s^{(1)}_n-s^{(1)}_{n-1})-\delta_{11}(s^{(1)}_n-s^{(1)}_{n-1})\\
&\quad +\beta_{12}(u^{(2)}_{n+1}-u^{(1)}_n)+\gamma_{12}(s^{(2)}_{n+1}-u^{(1)}_n)+\gamma_{21}(u^{(2)}_{n+1}-s^{(1)}_n)+\delta_{12}(s^{(2)}_{n+1}-s^{(1)}_n)\\
&\quad -\beta_{12}(u^{(1)}_n-u^{(2)}_{n-1})-\gamma_{12}(u^{(1)}_n-s^{(2)}_{n-1})-\gamma_{21}(s^{(1)}_n-u^{(2)}_{n-1})-\delta_{12}(s^{(1)}_n-s^{(2)}_{n-1})-q^*E^{MS}_n\\
&\quad = m_1 u^{(1)}_{n,tt}.
\end{aligned} \tag{37}$$

Similarly, for the shell, alone, of the type 1 atom:

$$\begin{aligned}
&\alpha_1(u^{(1)}_n-s^{(1)}_n)+\gamma_{11}(u^{(1)}_{n+1}-s^{(1)}_n)+\delta_{11}(s^{(1)}_{n+1}-s^{(1)}_n)-\gamma_{11}(s^{(1)}_n-u^{(1)}_{n-1})-\delta_{11}(s^{(1)}_n-s^{(1)}_{n-1})\\
&\quad +\gamma_{21}(u^{(2)}_{n+1}-s^{(1)}_n)+\delta_{12}(s^{(2)}_{n+1}-s^{(1)}_n)-\gamma_{21}(s^{(1)}_n-u^{(2)}_{n-1})-\delta_{12}(s^{(1)}_n-s^{(2)}_{n-1})+qE^{MS}_n = 0,
\end{aligned} \tag{38}$$

where q is the electronic charge and the mass of the shell is neglected.

We adopt the following definitions of the second central difference operation, the electronic polarization densities and the mass densities:

$$\begin{aligned}
\Delta^2 f_n &= (f_{n+1}+f_{n-1}-2f_n)/a^2,\\
P^{(\kappa)}_n &= (s^{(\kappa)}_n-u^{(\kappa)}_n)q/a^3,\\
\rho^{(\kappa)} &= m_\kappa/a^3, \qquad \kappa = 1, 2,
\end{aligned} \tag{39}$$

and also make the following identifications:

$$\begin{aligned} a^{11}_{11} &= (\alpha_1+2\gamma_{11}+2\gamma_{21}+2\delta_{12})a^3q^{-2}, & b^{12}_{11} &= a^{12}_{11}a^2/2 = \delta_{12}a^5q^{-2}, \\ b^{11}_{11} &= \delta_{11}a^5q^{-2}, & c^{12}_{11} &= a^{**}a^2/2 = (\beta_{12}+\gamma_{12}+\gamma_{21}+\delta_{12})a^{-1}, \\ c^{11}_{11} &= (\beta_{11}+2\gamma_{11}+\delta_{11})a^{-1}, & d^{21}_{11} &= a^{*2}a^2/2 = (\gamma_{12}+\delta_{12})a^2q^{-1}, \\ d^{11}_{11} &= (\gamma_{11}+\delta_{11})a^2q^{-1}, & d^{12}_{11} &= -a^{*1}a^2/2 = (\gamma_{21}+\delta_{12})a^2q^{-1}. \end{aligned} \tag{40}$$

Then (37) and (38) can be written as

$$\begin{aligned} &c^{11}_{11}\Delta^2u^{(1)}_n+c^{12}_{11}\Delta^2u^{(2)}_n+d^{11}_{11}\Delta^2P^{(1)}_n+d^{21}_{11}\Delta^2P^{(2)}_n+a^{*1}P^{(1)}_n \\ &\quad +a^{*2}P^{(2)}_n+a^{**}(u^{(2)}_n-u^{(1)}_n)-q^*E^{MS}_n = \rho^{(1)}u^{(1)}_{n,tt}, \end{aligned} \tag{41}$$

$$\begin{aligned} &d^{11}_{11}\Delta^2u^{(1)}_n+d^{12}_{11}\Delta^2u^{(2)}_n+b^{11}_{11}\Delta^2P^{(1)}_n+b^{12}_{11}\Delta^2P^{(2)}_n-a^{11}_{11}P^{(1)}_n \\ &\quad -a^{12}_{11}P^{(2)}_n-a^{*1}(u^{(2)}_n-u^{(1)}_n)+E^{MS}_n = 0. \end{aligned} \tag{42}$$

In the same way, for the type 2 atom:

$$\begin{aligned} &c^{21}_{11}\Delta^2u^{(1)}_n+c^{22}_{11}\Delta^2u^{(2)}_n+d^{12}_{11}\Delta^2P^{(1)}_n+d^{22}_{11}\Delta^2P^{(2)}_n-a^{*1}P^{(1)}_n \\ &\quad -a^{*2}P^{(2)}_n-a^{**}(u^{(2)}_n-u^{(1)}_n)+q^*E^{MS}_n = \rho^{(2)}u^{(2)}_{n,tt}, \end{aligned} \tag{43}$$

$$\begin{aligned} &d^{21}_{11}\Delta^2u^{(1)}_n+d^{22}_{11}\Delta^2u^{(2)}_n+b^{21}_{11}\Delta^2P^{(1)}_n+b^{22}_{11}\Delta^2P^{(2)}_n-a^{21}_{11}P^{(1)}_n \\ &\quad -a^{22}_{11}P^{(2)}_n-a^{*2}(u^{(2)}_n-u^{(1)}_n)+E^{MS}_n = 0, \end{aligned} \tag{44}$$

where, in addition to (39) and (40), we identify:

$$\begin{aligned} a^{22}_{11} &= (\alpha_2+2\gamma_{22}+2\gamma_{12}+2\delta_{12})a^3q^{-2}, & b^{22}_{11} &= \delta_{22}a^5q^{-2}, \\ c^{22}_{11} &= (\beta_{22}+2\gamma_{22}+\delta_{22})a^{-1}, & d^{22}_{11} &= (\gamma_{22}+\delta_{22})a^2q^{-1}. \end{aligned} \tag{45}$$

Also, as in [7], the charge equation may be written as

$$-\varepsilon_0\partial_-\partial_+\varphi_n+\partial_-[P^{(1)}_n+P^{(2)}_n+q^*(u^{(2)}_n-u^{(1)}_n)] = 0, \tag{46}$$

where ∂_+ and ∂_- are the Taylor series expansions of the derivative $\partial/\partial x_1$ in terms of forward and backward differences, respectively.

In the long wave limit,

$$\begin{aligned} u^{(\kappa)}_n \to u^{(\kappa)}_1(x_1), \qquad P^{(\kappa)}_n \to P^{(\kappa)}_1(x_1), \qquad E^{MS}_n \to E^{MS}_1(x_1) = -\partial\varphi/\partial x_1, \\ \Delta^2 f_n \to \partial^2 f(x_1)/\partial x_1^2, \qquad \partial_\pm \to \partial/\partial x_1. \end{aligned} \tag{47}$$

Then (41), (43), (42), (44) and (46), in that order, reduce to the five equations (32) in the order given.

Alternatively, in the long wave, low frequency limit, we have, in addition to (47),

$$u^{(1)}_1 = u^{(2)}_1, \qquad P^{(1)}_1 = P^{(2)}_1.$$

Then, with the notations

$$\begin{aligned} u_1 &= u^{(1)}_1 = u^{(2)}_1, \qquad \tfrac{1}{2}P_1 = P^{(1)}_1 = P^{(2)}_1, \qquad \rho = \rho^{(1)}+\rho^{(2)}, \\ a_{11} &= \tfrac{1}{4}(a^{11}_{11}+a^{22}_{11}+2a^{12}_{11}), \qquad b_{11} = \tfrac{1}{4}(b^{11}_{11}+b^{22}_{11}+2b^{12}_{11}), \\ c_{11} &= c^{11}_{11}+c^{22}_{11}+2c^{12}_{11}, \qquad d_{11} = \tfrac{1}{2}(d^{11}_{11}+d^{22}_{11}+d^{12}_{11}+d^{21}_{11}), \end{aligned}$$

the sum of (41) and (43), the sum of (42) and (44) and, finally, (46) become, respectively,

$$
\begin{aligned}
c_{11}u_{1,11}+d_{11}P_{1,11} &= \rho u_{1,tt},\\
d_{11}u_{1,11}+b_{11}P_{1,11}-a_{11}P_1-\varphi_{,1} &= 0,\\
-\varepsilon_0\varphi_{,11}+P_{1,1} &= 0.
\end{aligned}
\tag{48}
$$

These are the equations found in [2] for longitudinal waves in a simple (as opposed to compound) dielectric continuum, with symmetry *m3m*, when the polarization gradient is taken into account. Thus, the long wave limit of the equations of the one-dimensional, diatomic lattice yields the equations for longitudinal waves in the compound continuum when no restriction is placed on the frequency; whereas the long wave, low frequency limit yields the equations of the simple continuum. In a three-dimensional treatment, Askar *et al.* [12] showed the connection between the equations of the NaCl-type lattice of polarizable atoms and the equations in [2]. A similar investigation, not restricted to low frequencies, would, presumably, reveal the connection with the three-dimensional continuum equations of the present paper.

REFERENCES

[1] R. D. Mindlin, *IUTAM Symposium on Mechanics of Generalized Continua*, Freudenstadt-Stuttgart, 1967, edited by E. Kröner, p. 312. Springer-Verlag (1968).
[2] R. D. Mindlin, *Int. J. Solids Struct.* **4**, 637 (1968).
[3] R. A. Toupin, *J. rat. Mech. Analysis* **5**, 849 (1956).
[4] M. Born and K. Huang, *Dynamical Theory of Crystal Lattices*, p. 82. Oxford University Press (1956).
[5] W. Cochran, *Proc. R. Soc.* **A253**, 260 (1959).
[6] B. G. Dick and A. W. Overhauser, *Phys. Rev.* **112**, 90 (1958).
[7] R. D. Mindlin, *Int. J. Solids Struct.* **5**, 1197 (1969).
[8] R. D. Mindlin and R. A. Toupin, *Int. J. Solids Struct.* **7**, 1219 (1971).
[9] I. F. Nye, *Physical Properties of Crystals*. Oxford University Press (1960).
[10] J. S. Lomont, *Applications of Finite Groups*. Academic Press (1959).
[11] C. Kittel, *Introduction to Solid State Physics*. John Wiley (1966).
[12] A. Askar, P. C. Y. Lee and A. S. Cakmak, *Phys. Rev.* **B1**, 3525 (1970).

(*Received* 8 *July* 1971)

Абстракт – Обобщаются уравнения поля и граничнве условия континуальной теории упругих диэлектриков, учитывающей влияние градиента поляризации, с целью применения ее к сложному материалу. Получается формула для поверхностной энергии деформации и поляризации кубического кристалла. Решение задачи плоских волн приводит к отождествению оптических продольных и поперечных ветвей, а также акустических, для зависимости дисперсии. Строится одномерная модель кристаллической сети типа NaCl оболочечной модели атомов. Уравнения движения этой модели, выраженные в конечных разностях, соответствуют уравнениям сплошной теории, в смысле ограничения длины волны, без ограничений низкой частоты.

Int. J. Solids Structures, 1972, Vol. 8, pp. 401–408. Pergamon Press. Printed in Great Britain

COUPLED ELASTIC AND ELECTROMAGNETIC FIELDS IN A DIATOMIC, DIELECTRIC CONTINUUM

R. D. MINDLIN

Department of Civil Engineering, Columbia University, New York

Abstract—Huang's equations for coupled mechanical and electromagnetic fields in diatomic, ionic crystals are extended to accommodate shorter wave lengths and to include the acoustic branches of the dispersion relations for plane waves.

1. INTRODUCTION

EQUATIONS governing coupled mechanical and electromagnetic fields in diatomic, ionic, optically isotropic crystals were given by Huang [1, 2] in terms of a polarization variable, the relative displacement of the ions and the usual variables of the electromagnetic field. In the present paper, the equations are extended to take into account the separate electronic polarizations and the separate displacements of the two ions, the ionic polarization, the two electronic polarization gradients and the two displacement gradients. Aside from a more detailed representation of the atomic and electronic interactions, there are two main effects of the additional considerations: first, the equations are applicable to shorter wave lengths, owing to the inclusion of the displacement and polarization gradients; second, the acoustic branches are included in the dispersion relations for plane waves, in addition to the optical branches and the electromagnetic branch.

In Section 2, the differential equations of motion are given for a diatomic, ionic crystal with NaCl structure. This is done simply by adding the equations involving the magnetic field to a previous set of equations [3] which was restricted to a quasi-static electric field.

Dispersion relations for longitudinal and transverse waves in the [100] direction are derived in Section 3 and an examination of the long wave limits reveals the presence of the acoustical branches as well as the optical branches and electromagnetic branch included in Huang's results. As previously found by Huang, the longitudinal and transverse optical branches have the same long wave limit when the electromagnetic field is taken into account. It is also found that the long wave limits of the optical and electromagnetic branches are not influenced by the polarization and displacement gradients and the long wave limits of the acoustic branches are independent of electrical and magnetic properties.

In the last section, the identity of the static dielectric constant and the low frequency index of refraction of the electromagnetic wave is verified.

2. EQUATIONS OF MOTION

As in a previous paper restricted to the quasi-static electric field, each of the two interpenetrating continua representing the diatomic ionic crystal has its own displacement $u_i^{(\kappa)}, \kappa = 1, 2$ and electronic polarization (per unit area) $P_i^{(\kappa)}$. The superscripts 1 and 2 identify the continuum representations of the lattices of positive and negative ions, respectively.

The stored energy density of deformation and polarization, W^L, is assumed to be a function of the individual small strains

$$S_{ij}^{(\kappa)} = \tfrac{1}{2}(u_{j,i}^{(\kappa)} + u_{i,j}^{(\kappa)}), \qquad \kappa = 1, 2, \tag{1}$$

the individual polarizations, $P_i^{(\kappa)}$, the individual polarization gradients, $P_{j,i}^{(\kappa)}$, the relative displacement

$$u_i^* = u_i^{(2)} - u_i^{(1)} \tag{2}$$

and the relative rotation

$$\omega_{ij}^* = \tfrac{1}{2}(u_{j,i}^* - u_{i,j}^*). \tag{3}$$

Thus,

$$W^L = W^L(S_{ij}^{(1)}, S_{ij}^{(2)}, P_i^{(1)}, P_i^{(2)}, P_{j,i}^{(1)}, P_{j,i}^{(2)}, u_i^*, \omega_{ij}^*). \tag{4}$$

The total polarization per unit area, P_i, is the sum of the electronic polarizations and the ionic polarization

$$P_i = P_i^{(1)} + P_i^{(2)} + q_* u_i^*, \tag{5}$$

where q_* is the ionic charge per unit volume.

Of the variables in (4), only P_i and u_i^* are accounted for in Huang's [1, 2] equations of motion.

When the Maxwell electric self-field, E_i^{MS}, is quasi-static, it satisfies

$$\varepsilon_{ijk} E_{k,j}^{MS} = 0, \tag{6}$$

where ε_{ijk} is the unit alternating tensor. In that case, the total potential energy density, W, is the sum of W^L and the energy density associated with E_i^{MS}:

$$W = W^L + \tfrac{1}{2}\varepsilon_0 E_i^{MS} E_i^{MS}, \tag{7}$$

where ε_0 is the permittivity of a vacuum. The accompanying kinetic energy density is

$$T = \sum_{\kappa} \tfrac{1}{2}\rho^{(\kappa)} \dot{u}_i^{(\kappa)} \dot{u}_i^{(\kappa)}, \qquad \kappa = 1, 2, \tag{8}$$

where $\rho^{(1)}$ and $\rho^{(2)}$ are the mass densities of the individual continua and the $\dot{u}_i^{(\kappa)}$ are the velocities.

Again for the case of the quasi-static electric field, it was shown previously [3], by means of an extension of Toupin's [4] variational principle for elastic dielectrics, that the field equations, in addition to (6), are

$$T_{ij,i}^{(\kappa)} + (-1)^\kappa (T_{ij,i}^* - T_j^* + q_* E_j^{MS}) + f_j^{(\kappa)} + (-1)^\kappa q_* E_j^0 = \rho^{(\kappa)} \ddot{u}_j^{(\kappa)}, \tag{9}$$

$$E_{ij,i}^{(\kappa)} + E_j^{(\kappa)} + E_j^{MS} + E_j^0 = 0, \tag{10}$$

$$\varepsilon_0 E_{i,i}^{MS} + P_{i,i}^{(1)} + P_{i,i}^{(2)} + q_* u_{i,i}^* = 0, \tag{11}$$

where

$$T_{ij}^{(\kappa)} = \frac{\partial W^L}{\partial S_{ij}^{(\kappa)}}, \qquad E_i^{(\kappa)} = -\frac{\partial W^L}{\partial P_i^{(\kappa)}}, \qquad E_{ij}^{(\kappa)} = \frac{\partial W^L}{\partial P_{j,i}^{(\kappa)}}, \qquad T_i^* = \frac{\partial W^L}{\partial u_i^*}, \qquad T_{ij}^* = \frac{\partial W^L}{\partial \omega_{ij}^*} \tag{12}$$

and $f_i^{(\kappa)}$ and E_i^0 are the body forces and external electric field, respectively.

The magnetic field is incorporated into these equations simply by changing (6) to

$$\varepsilon_{ijk}E^{MS}_{k,j}+\dot{B}_i=0 \tag{13}$$

and adding the equations

$$\mu_0^{-1}\varepsilon_{ijk}B_{k,j}=\varepsilon_0\dot{E}^{MS}_i+\dot{P}^{(1)}_i+\dot{P}^{(2)}_i+q_*\dot{u}^*_i, \tag{14}$$

$$B_{i,i}=0, \tag{15}$$

where B_i is the magnetic flux density and μ_0 is the magnetic permeability of a vacuum (the magnetic susceptibility being neglected). The potential energy density becomes

$$W=W^L+\tfrac{1}{2}\varepsilon_0E^{MS}_iE^{MS}_i+\tfrac{1}{2}\mu_0^{-1}B_iB_i, \tag{16}$$

while the energy density of deformation and polarization (7), the kinetic energy density (8), the field equations (9)–(11) and the constitutive equations (12) remain unchanged.

It was shown previously [3] that, for the NaCl-type structure, the energy density of deformation and polarization is

$$\begin{aligned}W^L=&\tfrac{1}{2}\sum_{\kappa,\lambda}(a^{\kappa\lambda}_{ij}P^{(\kappa)}_iP^{(\lambda)}_j+b^{\kappa\lambda}_{ijkl}P^{(\kappa)}_{j,i}P^{(\lambda)}_{l,k}+c^{\kappa\lambda}_{ijkl}S^{(\kappa)}_{ij}S^{(\lambda)}_{kl}+2d^{\kappa\lambda}_{ijkl}P^{(\kappa)}_{(j,i)}S^{(\lambda)}_{kl}\\&+\sum_{\kappa}(a^{*\kappa}u^*_iP^{(\kappa)}_i+d^{*\kappa}\omega^*_{ij}P^{(\kappa)}_{[j,i]}+a^{**}u^*_i\,u^*_i+c^{**}\omega^*_{ij}\omega^*_{ij})\\&+\sum_{\kappa}(b^{\kappa 0}P^{(\kappa)}_{i,i}+c^{\kappa 0}S^{(\kappa)}_{ii}),\end{aligned} \tag{17}$$

where $P^{(\kappa)}_{(j,i)}$ and $P^{(\kappa)}_{[j,i]}$ denote the symmetric and antisymmetric parts of $P^{(\kappa)}_{j,i}$, respectively. Also,

$$\begin{aligned}a^{\kappa\lambda}_{ij}&=a^{\lambda\kappa}_{ij}=a^{\kappa\lambda}_{11}\delta_{ij},\\b^{\kappa\lambda}_{ijkl}&=b^{\lambda\kappa}_{ijkl}=b^{\kappa\lambda}\delta_{ijkl}+b^{\kappa\lambda}_{12}\delta_{ij}\delta_{kl}+b^{\kappa\lambda}_{44}(\delta_{ik}\delta_{jl}+\delta_{il}\delta_{jk})+b^{\kappa\lambda}_{77}(\delta_{ik}\delta_{jl}-\delta_{il}\delta_{jk}),\\c^{\kappa\lambda}_{ijkl}&=c^{\lambda\kappa}_{ijkl}=c^{\kappa\lambda}\delta_{ijkl}+c^{\kappa\lambda}_{12}\delta_{ij}\delta_{kl}+c^{\kappa\lambda}_{44}(\delta_{ik}\delta_{jl}+\delta_{il}\delta_{jk}),\\d^{\kappa\lambda}_{ijkl}&=d^{\kappa\lambda}\delta_{ijkl}+d^{\kappa\lambda}_{12}\delta_{ij}\delta_{kl}+d^{\kappa\lambda}_{44}(\delta_{ik}\delta_{jl}+\delta_{il}\delta_{jk}),\end{aligned} \tag{18}$$

where δ_{ij} (or δ_{ijkl}) is unity when its indices are alike and zero otherwise and

$$b^{\kappa\lambda}=b^{\kappa\lambda}_{11}-b^{\kappa\lambda}_{12}-2b^{\kappa\lambda}_{44},\qquad c^{\kappa\lambda}=c^{\kappa\lambda}_{11}-c^{\kappa\lambda}_{12}-2c^{\kappa\lambda}_{44},\qquad d^{\kappa\lambda}=d^{\kappa\lambda}_{11}-d^{\kappa\lambda}_{12}-2d^{\kappa\lambda}_{44}. \tag{19}$$

It may be observed that material constants with superscripts $\kappa\lambda=11$ or 22 denote interactions within one of the two component continua whereas constants with superscripts $\kappa\lambda=12$ or 21 denote interactions between the two continua. Also, the asterisks identify material constants associated with the relative displacement u^*_i.

Upon substituting (17) in (12) and the resulting expressions in the field equations, we find

$$\begin{aligned}&\sum_{\lambda}[c^{\kappa\lambda}\delta_{ijkl}u^{(\lambda)}_{l,ki}+c^{\kappa\lambda}_{12}u^{(\lambda)}_{i,ij}+c^{\kappa\lambda}_{44}(u^{(\lambda)}_{j,ii}+u^{(\lambda)}_{i,ij})]\\&\quad+\sum_{\lambda}[d^{\lambda\kappa}\delta_{ijkl}P^{(\lambda)}_{l,ki}+d^{\lambda\kappa}_{12}P^{(\lambda)}_{i,ij}+d^{\lambda\kappa}_{44}(P^{(\lambda)}_{j,ii}+P^{(\lambda)}_{i,ij})]\\&\quad+(-1)^{\kappa}\sum_{\lambda}[d^{*\lambda}(P^{(\lambda)}_{j,ii}-P^{(\lambda)}_{i,ij})-(-1)^{\lambda}c^{**}(u^{(\lambda)}_{j,ii}-u^{(\lambda)}_{i,ij})]\\&\quad-(-1)^{\kappa}[a^{*1}P^{(1)}_j+a^{*2}P^{(2)}_j+a^{**}(u^{(2)}_j-u^{(1)}_j)-q_*E^{MS}_j-q_*E^0_j]+f^{(\kappa)}_j=\rho^{(\kappa)}\ddot{u}^{(\kappa)}_j,\end{aligned} \tag{20}$$

$$\sum_{\lambda}[d^{\kappa\lambda}\delta_{ijkl}u^{(\lambda)}_{l,ki}+d^{\kappa\lambda}_{12}u^{(\lambda)}_{i,ij}+d^{\kappa\lambda}_{44}(u^{(\lambda)}_{j,ii}+u^{(\lambda)}_{i,ij})+(-1)^{\lambda}d^{*\kappa}(u^{(\lambda)}_{j,ii}-u^{(\lambda)}_{i,ij})]$$

$$+\sum_{\lambda}[b^{\kappa\lambda}\delta_{ijkl}P^{(\lambda)}_{l,ki}+b^{\kappa\lambda}_{12}P^{(\lambda)}_{i,ij}+b^{\kappa\lambda}_{44}(P^{(\lambda)}_{j,ii}+P^{(\lambda)}_{i,ij})+b^{\kappa\lambda}_{77}(P_{j,ii}-P_{i,ij})]$$

$$-a^{1\kappa}_{11}P^{(1)}_j-a^{2\kappa}_{11}P^{(2)}_j-a^{**}(u^{(2)}_j-u^{(1)}_j)+E^{MS}_j+E^0_j = 0, \tag{21}$$

$$\varepsilon_0E^{MS}_{i,i}+P^{(1)}_{i,i}+P^{(2)}_{i,i}+q_*(u^{(2)}_{i,i}-u^{(1)}_{i,i}) = 0, \tag{22}$$

$$\varepsilon_{ijk}E^{MS}_{k,j}+\dot{B}_i = 0, \tag{23}$$

$$\mu_0^{-1}\varepsilon_{ijk}B_{k,j}-\varepsilon_0\dot{E}^{MS}_i-\dot{P}^{(1)}_i-\dot{P}^{(2)}_i-q_*\dot{u}^*_i = 0, \tag{24}$$

$$B_{i,i} = 0. \tag{25}$$

These are the equations of motion for the coupled elastic and electromagnetic fields in the continuum representation of a diatomic ionic crystal with NaCl structure.

3. WAVES IN THE [100] DIRECTION

We consider plane waves in the [100], or x_1, direction:

$$u^{(\kappa)}_i = K^{(\kappa)}\,e^{i\psi}, \qquad P^{(\kappa)}_i = L^{(\kappa)}_i\,e^{i\psi}, \qquad E^{MS}_i = M_i\,e^{i\psi}, \qquad B_i = N_i\,e^{i\psi}, \tag{26}$$

where $\psi = \xi x_1-\omega t$ and $K^{(\kappa)}_i$, $L^{(\kappa)}_i$, M_i and N_i are constants.

(a) *Longitudinal waves*

In the case of longitudinal waves,

$$u^{(\kappa)}_2 = u^{(\kappa)}_3 = 0, \qquad P^{(\kappa)}_2 = P^{(\kappa)}_3 = 0, \qquad E^{MS}_2 = E^{MS}_3 = 0, \tag{27}$$

$$u^{(\kappa)}_i = K^{(\kappa)}_1\,e^{i\psi}, \qquad P^{(\kappa)}_1 = L^{(\kappa)}_1\,e^{i\psi}, \qquad E^{MS}_1 = M_1\,e^{i\psi}. \tag{28}$$

Then, from (24), $B_i = 0$; i.e. there is no coupling with the magnetic field, but the Maxwell electric self-field is coupled with the displacement and polarization fields. The solution is, therefore, the same as that found previously [3] for the case of the quasi-static electric field, with dispersion relation

$$\Delta_L = 0, \tag{29}$$

where Δ_L is the determinant with elements

$$\begin{aligned}
\Delta_{11} &= \rho^{(1)}\omega^2-a^{**}-q_*\varepsilon_0^{-1}-c^{11}_{11}\xi^2, & \Delta_{31} &= \Delta_{13},\\
\Delta_{12} &= a^{**}+q_*^2\varepsilon_0^{-1}-c^{12}_{11}\xi^2, & \Delta_{32} &= \Delta_{23},\\
\Delta_{13} &= a^{*1}+q_*\varepsilon_0^{-1}-d^{11}_{11}\xi^2, & \Delta_{33} &= -a^{11}_{11}-\varepsilon_0^{-1}-b^{11}_{11}\xi^2,\\
\Delta_{14} &= a^{*2}+q_*\varepsilon_0^{-1}-d^{21}_{11}\xi^2, & \Delta_{34} &= -a^{21}_{11}-\varepsilon_0^{-1}-b^{21}_{11}\xi^2,\\
\Delta_{21} &= \Delta_{12}, & \Delta_{41} &= \Delta_{4},\\
\Delta_{22} &= \rho^{(2)}\omega^2-a^{**}-q_4^2\varepsilon_0^{-1}-c^{11}_{11}\xi^2, & \Delta_{42} &= \Delta_{24},\\
\Delta_{23} &= -a^{*1}-q_*\varepsilon_0^{-1}-d^{12}_{1}\xi^2, & \Delta_{43} &= \Delta_{34},\\
\Delta_{24} &= -a^{*2}-q_*\varepsilon_0^{-1}-d^{22}_{11}\xi^2, & \Delta_{44} &= -a^{22}_{11}-\varepsilon_0^{-1}-b^{22}_{11}\xi^2.
\end{aligned} \tag{30}$$

The long wave limit of the longitudinal optical branch is obtained from

$$\lim_{\xi\to 0} \Delta_L = \omega^2\{\rho^{(1)}\rho^{(2)}\omega^2[(a_{11}^{11}+\varepsilon_0^{-1})(a_{11}^{22}+\varepsilon_0^{-1})-(a_{11}^{12}+\varepsilon_0^{-1})^2]-(\rho^{(1)}+\rho^{(2)})D_L\} \quad (31)$$

where

$$D_L = \begin{vmatrix} a_{11}^{11}+\varepsilon_0^{-1} & a_{11}^{12}+\varepsilon_0^{-1} & a^{*1}+q_*\varepsilon_0^{-1} \\ a_{11}^{21}+\varepsilon_0^{-1} & a_{11}^{22}+\varepsilon_0^{-1} & a^{*2}+q_*\varepsilon_0^{-1} \\ a^{*1}+q_*\varepsilon_0^{-1} & a^{*2}+q_*\varepsilon_0^{-1} & a^{**}+q_*^2\varepsilon_0^{-1} \end{vmatrix}. \quad (32)$$

Hence, the limiting frequency of the longitudinal optical branch is

$$\omega_{LO} = \{D_L/\bar{\rho}[(a_{11}^{11}+\varepsilon_0^{-1})(a_{11}^{22}+\varepsilon_0^{-1})-(a_{11}^{12}+\varepsilon_0^{-1})^2]\}^{\frac{1}{2}}, \quad (33)$$

where $\bar{\rho} = \rho^{(1)}\rho^{(2)}/(\rho^{(1)}+\rho^{(2)})$, i.e. $\bar{\rho}$ is the reduced mass density.

The long wave behavior of the longitudinal acoustic branch of the dispersion relation is obtained from

$$\lim_{\omega,\xi\to 0} \Delta_L = \lim_{\omega,\xi\to 0} [(c_{11}^{11}+c_{11}^{22}+2c_{11}^{12})\xi^2-(\rho^{(1)}+\rho^{(2)})\omega^2]D_L. \quad (34)$$

Hence, at long wave lengths, the frequency of the longitudinal acoustic branch is given by

$$\omega_{LA} = \xi[(c_{11}^{11}+c_{11}^{22}+2c_{11}^{12})/(\rho^{(1)}+\rho^{(2)})]^{\frac{1}{2}}. \quad (35)$$

(b) *Transverse waves*

Of the two similar transverse waves, with displacements in the directions of x_2 and x_3, respectively, we choose the former for examination:

$$u_1^{(\kappa)} = u_3^{(\kappa)} = 0, \qquad P_1^{(\kappa)} = P_3^{(\kappa)} = 0, \qquad E_1^{MS} = E_3^{MS} = 0, \qquad B_1 = B_2 = 0, \quad (36)$$

$$u_2^{(\kappa)} = K_2^{(\kappa)}\,\mathrm{e}^{i\psi}, \qquad P_2^{(\kappa)} = L_2^{(\kappa)}\,\mathrm{e}^{i\psi}, \qquad E_2^{MS} = M_2\,\mathrm{e}^{i\psi}, \qquad B_3 = N_3\,\mathrm{e}^{i\psi}. \quad (37)$$

In the case of the quasi-static electric field the absence of (24) permitted E_2^{MS}, as well as B_3, to be zero; but both must be non-zero when the full electromagnetic equations are imposed.

To find a form of the dispersion determinant similar to (30), it is useful first to express M_2 and N_3 in terms of $K_2^{(\kappa)}$ and $L_2^{(\kappa)}$ through the use of (23) and (24):

$$M_2 = -i\omega N_3/\xi = -\varepsilon_\xi^{-1}[L_2^{(1)}+L_2^{(2)}+q_*(K_2^{(2)}-K_2^{(1)})], \quad (38)$$

where

$$\varepsilon_\xi = \varepsilon_0-\xi^2/\mu_0\omega^2. \quad (39)$$

When the results (38), along with (36) and (37), are inserted in the remaining equations of motion, the latter are satisfied if

$$\Delta_T^{EM} = 0, \quad (40)$$

where Δ_T^{EM} is the determinant with elements

$$\begin{aligned}
\Delta'_{11} &= \rho^{(1)}\omega^2 - a^{**} - q_*\varepsilon_\xi^{-1} - (c_{44}^{11} - c^{**})\xi^2, & \Delta'_{31} &= \Delta'_{13},\\
\Delta'_{12} &= a^{**} + q_*^2\varepsilon_\xi^{-1} - (c_{44}^{21} + c^{**})\xi^2, & \Delta'_{32} &= \Delta'_{23},\\
\Delta'_{13} &= a^{*1} + q_*\varepsilon_\xi^{-1} - (d_{44}^{11} - d^{*1})\xi^2, & \Delta'_{33} &= -a_{11}^{11} - \varepsilon_\xi^{-1} - (b_{44}^{11} + b_{77}^{11})\xi^2,\\
\Delta'_{14} &= a^{*2} + q_*\varepsilon_\xi^{-1} - (d_{44}^{21} - d^{*2})\xi^2, & \Delta'_{34} &= -a_{11}^{21} - \varepsilon_\xi^{-1} - (b_{44}^{21} + b_{77}^{21})\xi^2,\\
\Delta'_{21} &= \Delta'_{12}, & \Delta'_{41} &= \Delta'_{14},\\
\Delta'_{22} &= \rho^{(2)}\omega^2 - a^{**} - q_*\varepsilon_\xi^{-1} - (c_{44}^{22} - c^{**})\xi^2, & \Delta'_{42} &= \Delta'_{24},\\
\Delta'_{23} &= -a^{*1} - q_*\varepsilon_\xi^{-1} - (d_{44}^{12} + d^{*1})\xi^2, & \Delta'_{43} &= \Delta'_{34},\\
\Delta'_{24} &= -a^{*2} - q_*\varepsilon_\xi^{-1} - (d_{44}^{22} + d^{*2})\xi^2, & \Delta'_{44} &= -q_{11}^{22} - \varepsilon_\xi^{-1} - (b_{44}^{22} + b_{77}^{22})\xi^2.
\end{aligned} \tag{41}$$

The long wave limit of the transverse optical branch is obtained from

$$\lim_{\xi\to 0} \Delta_T^{EM} = \omega^4\{\rho^{(1)}\rho^{(2)}\omega^{(2)}[(a_{11}^{11}+\varepsilon_0^{-1})(a_{11}^{22}+\varepsilon_0^{-1})-(a_{11}^{12}+\varepsilon_0^{-1})^2]-(\rho^{(1)}+\rho^{(2)}D_L\}, \tag{42}$$

so that the limiting frequency is

$$\omega_{TO}^{EM} = \{D_L/\bar{\rho}[(a_{11}^{11}+\varepsilon_0^{-1})(a_{11}^{22}+\varepsilon_0^{-1})-(a_{11}^{12}+\varepsilon_0^{-1})^2]\}^{\frac{1}{2}}, \tag{43}$$

i.e. as found by Huang [1, 2], the same as the limiting frequency of the longitudinal optical branch.

The long wave, low frequency behavior is given by

$$\lim_{\omega,\xi\to 0} \Delta_T^{EM} = \lim_{\omega,\xi\to 0} [(c_{44}^{11}+c_{44}^{22}+2c_{44}^{12})\xi^2-(\rho^{(1)}+\rho^{(2)})\omega^2]D_\xi, \tag{44}$$

where

$$D_\xi = (\varepsilon_0\mu_0 D_L\omega^2 - \xi^2 D)/(\varepsilon_0\mu_0\omega^2-\xi^2) \tag{45}$$

and

$$D = \begin{vmatrix} a_{11}^{11} & a_{11}^{12} & a^{*1} \\ a_{11}^{21} & a_{11}^{22} & a^{*2} \\ a^{*1} & a^{*2} & a^{**} \end{vmatrix}. \tag{46}$$

Thus, there are two branches, with limiting behaviors

$$\omega_{EM} = \xi(D/\varepsilon_0\mu_0 D_L)^{\frac{1}{2}}, \tag{47}$$

$$\omega_{TA}^{EM} = \xi[(c_{44}^{11}+c_{44}^{22}+2c_{44}^{12})/(\rho^{(1)}+\rho^{(2)})]^{\frac{1}{2}}. \tag{48}$$

The first of these is the long wave end of the electromagnetic branch, as found by Huang [1, 2]; the second is the long wave end of the transverse acoustic branch: identical with that obtained for the case of the quasi-static electric field [3].

It will be observed that the long wave limits of the longitudinal and transverse optical branches, (33) and (43), and the electromagnetic branch (47) are independent of the material constants b, c and d, i.e. independent of the polarization and displacement gradients. Also the long wave limits of the acoustic branches, (35) and (48), depend only on the mass density and the elastic stiffness $c_{11}^{\kappa\lambda}$ and $c_{44}^{\kappa\lambda}$. Hence, as far as the limiting behavior at the long wave end is concerned, the results, here, conform with Huang's, with the addition of the acoustic branches. However, as may be seen from the determinants (30) and (41), when the wave length diminishes from infinity, the elastic part couples with the remainder and the results diverge from Huang's.

4. INDEX OF REFRACTION AND DIELECTRIC CONSTANT

Recalling that the velocity of electromagnetic waves in a vacuum is

$$c = (\varepsilon_0 \mu_0)^{-\frac{1}{2}}, \tag{49}$$

we see, from (47), that the low frequency index of refraction of electromagnetic waves in the dielectric is

$$n = c\xi/\omega = (D_L/D)^{\frac{1}{2}}. \tag{50}$$

That the square of the low frequency index of refraction, D_L/D, is equal to the static dielectric constant, defined as

$$K_0 = 1 + (P/\varepsilon_0 E^{MS})_{\omega=0}, \tag{51}$$

may be shown, conveniently, by deducing Huang's formula for the dielectric constant from (20) and (21). As shown in [3], Huang's equations of motion may be found from (20) and (21) by discarding all spatial derivative terms (thereby restricting the applicability of the equations to infinite or, at least, very long wave lengths) and expressing the residue in terms of $w_i(= \bar{\rho}^{\frac{1}{2}} u_i^*)$, P_i and E_i^{MS}, with the result (omitting $f_i^{(\kappa)}$ and E_i^0):

$$\ddot{w}_i = b_{11} w_i + b_{12} E_i^{MS}, \tag{52}$$

$$P_i = b_{21} w_i + b_{22} E_i^{MS}, \tag{53}$$

where Huang's constants b_{11}, b_{22}, b_{12}, b_{21} are expressed in terms of constants appearing in the present paper by

$$b_{11} = -D/\bar{\rho}(a_{11}^{11} a_{11}^{22} - a_{11}^{12} a_{11}^{21}), \qquad b_{22} = (a_{11}^{11} + a_{11}^{22} - 2a_{11}^{12})/(a_{11}^{11} a_{11}^{22} - a_{11}^{12} a_{11}^{21}), \tag{54}$$

$$b_{12} = b_{21} = [(a_{11}^{11} - a_{11}^{12})a^{*2} + (a_{11}^{22} - a_{11}^{12})a^{*1} - q_*(a_{11}^{11} a_{11}^{22} - a_{11}^{12} a_{11}^{21})/(a_{11}^{11} a_{11}^{22} - a_{11}^{12} a_{11}^{21})\bar{\rho}^{\frac{1}{2}}. \tag{55}$$

Introducing a factor $e^{i\omega t}$ and eliminating w_i between (52) and (53), we have

$$P_i/E_i^{MS} = [b_{22}(b_{11} + \omega^2) - b_{12}^2]/(b_{11} + \omega^2). \tag{56}$$

Inserting this in (51), we find Huang's formula for the static dielectric constant

$$K_0 = 1 + (b_{11} b_{22} - b_{12}^2)/\varepsilon_0 b_{11}. \tag{57}$$

Upon substituting the formulas for b_{11}, b_{22} and b_{12}, given in (54) and (55), into (57), we find, after some algebraic manipulations,

$$K_0 = D_L/D = n^2, \tag{58}$$

as required.

REFERENCES

[1] K. HUANG, On the interaction between the radiation field and ionic crystals. *Proc. R. Soc.* **A208**, 352–365 (1951).
[2] M. BORN and K. HUANG, *Dynamical Theory of Crystal Lattices*, Chapter II, Section 8. Oxford University Press (1956).
[3] R. D. MINDLIN, A continuum theory of a diatomic, elastic dielectric. *Int. J. Solids Struct.* **8**, 369–383 (1972).
[4] R. A. TOUPIN, The elastic dielectric. *J. rat. Mech. Analysis* **5**, 849–915 (1956).

(*Received* 12 *October* 1971)

Абстракт—Обобщаются уравнения Хуанга, касающиеся сопряженных механических и электромагнитных полей, в двухатомных, ионных кристаллах, для приспособления более коротких волн и учёта акустических ветвей для зависимостей дисперсии плоских волн.

Int. J. Solids Structures, 1972, Vol. 8, pp. 895 to 906. Pergamon Press. Printed in Great Britain

HIGH FREQUENCY VIBRATIONS OF PIEZOELECTRIC CRYSTAL PLATES

R. D. MINDLIN

Department of Civil Engineering, Columbia University, New York, N.Y.

Abstract—Two-dimensional equations of motion of piezoelectric crystal plates are obtained by retaining early terms of power series expansions of the mechanical displacement and electric potential in a variational principle for the three-dimensional equations of piezoelectricity.

INTRODUCTION

THERE have been two previous formulations of two-dimensional equations of motion of piezoelectric crystal plates applicable to frequencies as high as those of the fundamental thickness-shear modes. In the first [1], the reduction of the classical equations of piezoelectricity from three to two dimensions was based on an approximation involving the early terms of series expansions of the mechanical displacement and electric potential in powers of the thickness coordinate of the plate. Only flexure, thickness–shear and thickness–twist modes were taken into account. The equations were extended, subsequently [2], to include the face–extension and face–shear modes (the "contour" modes) in a treatment based on power series expansions of the mechanical displacement and electric displacement. In the present paper, equations of somewhat simpler form, including the flexure, thickness–shear, thickness–twist and contour modes, are obtained by means of power series expansions of the mechanical displacement and electric potential. The derivation employed is a revision (to introduce a variational principle) and an extension (to include the electric field and the thickness modes) of one devised by Cauchy [3] to deduce the classical equations of low frequency vibrations of anisotropic plates (flexure and contour modes only) from the three-dimensional equations of elasticity. The two-dimensional equations are derived in condensed form for the triclinic crystal, but they are displayed in detail for the case of the rotated-Y-cuts of quartz—the crystal cuts most widely used for resonators at the present time. Shear correction factors, of the type employed by Bresse [4] and Timoshenko [5] in the theory of beams, are introduced and their values are determined by equating the thickness–shear frequencies obtained from the two- and three-dimensional equations [1].

THREE-DIMENSIONAL EQUATIONS

The three-dimensional equations of piezoelectricity, from which the two-dimensional equations are to be deduced, follow.

The field equations are

$$T_{ij,i} = \rho \ddot{u}_j, \qquad D_{i,i} = 0, \tag{1}$$

where the T_{ij}, u_j and D_i are the components of stress, mechanical displacement and electric displacement, respectively, and ρ is the mass density.

The constitutive equations are

$$T_{ij} = c_{ijkl}S_{kl} - e_{kij}E_k, \qquad D_i = e_{ijk}S_{jk} + \varepsilon_{ij}E_j, \tag{2}$$

where the c_{ijkl}, e_{kij} and ε_{ij} are the components of elastic stiffness, piezoelectric strain constant and dielectric permittivity, respectively, and the S_{ij} and E_i are the components of strain and electric field—expressed in terms of the u_i and the electric potential φ by

$$S_{ij} = \tfrac{1}{2}(u_{j,i}+u_{i,j}), \qquad E_i = -\varphi_{,i}. \tag{3}$$

Combining (1)–(3), we have the equations governing the u_i and φ:

$$c_{ijkl}u_{k,li} + e_{kij}\varphi_{,ki} = \rho\ddot{u}_j, \qquad e_{kij}u_{i,jk} - \varepsilon_{ij}\varphi_{,ij} = 0. \tag{4}$$

These equations may be derived from the following variational principle: in a region V bounded by a surface S, with outward normal $\mathbf{n}$,

$$\delta\int_{t_0}^{t_1} \mathrm{d}t \int_V (K-H)\,\mathrm{d}V + \int_{t_0}^{t_1} \mathrm{d}t \int_S (t_j\,\delta u_j + \sigma\,\delta\varphi)\,\mathrm{d}S = 0 \tag{5}$$

for independent variations of the u_j and φ between fixed limits at times t_0 and t_1. In (5), K is the kinetic energy density:

$$K = \tfrac{1}{2}\rho\dot{u}_i\dot{u}_i$$

and H is the electric enthalpy density (the energy density less E_iD_i):

$$H = \tfrac{1}{2}c_{ijkl}S_{ij}S_{kl} - \tfrac{1}{2}\varepsilon_{ij}E_iE_j - e_{ijk}E_iS_{jk}$$

so that

$$T_{ij} = \partial H/\partial S_{ij}, \qquad D_i = -\partial H/\partial E_i.$$

Also, in (5) t_j is the surface traction and σ is the surface charge. We have

$$\delta\int_{t_0}^{t_1} K\,\mathrm{d}t = \int_{t_0}^{t_1} \rho\dot{u}_j\,\delta\dot{u}_j\,\mathrm{d}t = [\rho\dot{u}_j\,\delta u_j]_{t_0}^{t_1} - \int_{t_0}^{t_1} \rho\ddot{u}_j\,\delta u_j\,\mathrm{d}t = -\int_{t_0}^{t_1} \rho\ddot{u}_j\,\delta u_j\,\mathrm{d}t,$$

$$\begin{aligned}
\delta\int_V H\,\mathrm{d}V &= \int_V [(\partial H/\partial S_{ij})\,\delta S_{ij} + (\partial H/\partial E_i)\,\delta E_i]\,\mathrm{d}V = \int_V (T_{ij}\,\delta u_{j,i} + D_i\,\delta\varphi_{,i})\,\mathrm{d}V \\
&= \int_V [(T_{ij}\,\delta u_j)_{,i} - T_{ij,i}\,\delta u_j + (D_i\,\delta\varphi)_{,i} - D_{i,i}\,\delta\varphi]\,\mathrm{d}V \\
&= \int_S n_i(T_{ij}\,\delta u_j + D_i\,\delta\varphi)\,\mathrm{d}S - \int_V (T_{ij,i}\,\delta u_j + D_{i,i}\,\delta\varphi)\,\mathrm{d}V.
\end{aligned}$$

Hence, (5) becomes

$$\int_{t_0}^{t_1} \mathrm{d}t \int_V [(T_{ij,i} - \rho\ddot{u}_j)\,\delta u_j + D_{i,i}\,\delta\varphi]\,\mathrm{d}V + \int_{t_0}^{t_1} \mathrm{d}t \int_S [(t_j - n_iT_{ij})\,\delta u_j + (\sigma - n_iD_i)\,\delta\varphi]\,\mathrm{d}S = 0,$$

from which follow the natural boundary conditions

$$n_iT_{ij} = t_j, \qquad n_iD_i = \sigma \quad \text{on } S \tag{6}$$

or, alternatively, specification of surface displacement $\bar{\mathbf{u}}$ and surface potential $\bar{\varphi}$:

$$u_i = \bar{u}_i, \qquad \varphi = \bar{\varphi} \quad \text{on } S \tag{7}$$

and the field equations (1) as Euler equations which, with (2) and (3), produce the equations (4) on the u_i and φ.

SERIES OF TWO-DIMENSIONAL EQUATIONS

The plate is referred to rectangular coordinates x_i with the faces, of area A, at $x_2 = \pm b$ and with x_1 and x_3 the coordinates of the middle plane which intersects the right cylindrical or prismatic boundary of the plate in a curve C. We assume that the mechanical displacement and electric potential can be approximated by power series in x_2:

$$u_i = \sum_n x_2^n u_i^{(n)}, \qquad \varphi = \sum_n x_2^n \varphi^{(n)}, \tag{8}$$

where $u_i^{(n)}$ and $\varphi^{(n)}$, $n = 1, 2 \ldots k$, are functions of x_1, x_3 and t only. Then, from (8) and (3),

$$S_{ij} = \sum_n x_2^n S_{ij}^{(n)}, \qquad E_i = \sum_n x_2^n E_i^{(n)}, \tag{9}$$

where

$$S_{ij}^{(n)} = \tfrac{1}{2}[u_{j,i}^{(n)} + u_{i,j}^{(n)} + (n+1)(\delta_{i2} u_j^{(n+1)} + \delta_{2j} u_i^{(n+1)})], \qquad E_i^{(n)} = -\varphi_{,i}^{(n)} - (n+1)\delta_{2i}\varphi^{(n+1)}. \tag{10}$$

We have now to reduce the variational principle (5) to two dimensions. Consider, first, the surface integral and separate it into integrals over the faces and the cylindrical boundary:

$$\begin{aligned}\int_S (t_j\,\delta u_j + \sigma\,\delta\varphi)\,\mathrm{d}S &= \int_S n_i(T_{ij}\,\delta u_j + D_i\,\delta\varphi)\,\mathrm{d}S \\ &= \sum_n \int_A [x_2^n(T_{2j}\,\delta u_j^{(n)} + D_2\,\delta\varphi^{(n)}]_{-b}^{b}\,\mathrm{d}A \\ &\quad + \sum_n \oint_C \int_{-b}^{b} x_2^n n_a(T_{aj}\,\delta u_j^{(n)} + D_a\,\delta\varphi^{(n)})\,\mathrm{d}x_2\,\mathrm{d}s,\end{aligned}$$

where the index a ranges over 1 and 3 only and s is the coordinate measured along the curve C. Define face-tractions $T_j^{(n)}$, face-charges $D^{(n)}$, edge-tractions $t^{(n)}$ and edge-charges $d^{(n)}$ according to

$$T_j^{(n)} = B_n^{-1}[x_2^n T_{2j}]_{-b}^{b}, \qquad D^{(n)} = B_n^{-1}[x_2^n D_2]_{-b}^{b},$$

$$t_j^{(n)} = B_n^{-1}\int_{-b}^{b} x_2^n(n_a T_{aj})_C\,\mathrm{d}x_2, \qquad d^{(n)} = B_n^{-1}\int_{-b}^{b} x_2^n(n_a D_a)_C\,\mathrm{d}x_2,$$

where

$$B_n = 2b^{2n+1}/(2n+1).$$

Also, define

$$\bar{K} = \int_{-b}^{b} K\,\mathrm{d}x_2, \qquad \bar{H} = \int_{-b}^{b} H\,\mathrm{d}x_2.$$

Then the variational principle (5) becomes the two-dimensional one: for independent variations $\delta u_j^{(n)}$ and $\delta\varphi^{(n)}$ between fixed limits at t_0 and t_1:

$$\delta \int_{t_0}^{t_1} \mathrm{d}t \int_A (\bar{K}-\bar{H})\,\mathrm{d}A + \int_{t_0}^{t_1} \mathrm{d}t \int_A \sum_n B_n(T_j^{(n)}\,\delta u_j^{(n)} + D^{(n)}\,\delta\varphi^{(n)})\,\mathrm{d}A$$

$$+ \int_{t_0}^{t_1} \mathrm{d}t \oint_C \sum_n B_n(t_j^{(n)}\,\delta u_j^{(n)} + d^{(n)}\,\delta\varphi^{(n)})\,\mathrm{d}s = 0. \tag{11}$$

For the kinetic energy density $\bar{K}$, we have

$$\bar{K} = \tfrac{1}{2}\rho \int_{-b}^{b} \dot{u}_j\dot{u}_j\,\mathrm{d}x_2 = \tfrac{1}{2}\rho \int_{-b}^{b} \left(\sum_m x_2^m \dot{u}_j^{(m)}\right)\left(\sum_n x_2^n \dot{u}_j^{(n)}\right)\mathrm{d}x_2 = \tfrac{1}{2}\rho \sum_m\sum_n B_{mn}\dot{u}_j^{(m)}\dot{u}_j^{(n)},$$

where

$$B_{mn} = 2b^{m+n+1}/(m+n+1), \qquad m+n \text{ even}$$

and $B_{mn} = 0$ for $m+n$ odd. Further,

$$\int_{t_0}^{t_1} \delta\bar{K}\,\mathrm{d}t = \rho \sum_m\sum_n B_{mn}[\dot{u}_j^{(m)}\,\delta u_j^{(n)}]_{t_0}^{t_1} - \rho\int_{t_0}^{t_1} \sum_m\sum_n B_{mn}\ddot{u}_j^{(m)}\,\delta u_j^{(n)}\,\mathrm{d}t$$

or

$$\int_{t_0}^{t_1} \delta\bar{K}\,\mathrm{d}t = -\rho\int_{t_0}^{t_1} \sum_m\sum_n B_{mn}\ddot{u}_j^{(m)}\,\delta u_j^{(n)}\,\mathrm{d}t. \tag{12}$$

As for the electric enthalpy density $\bar{H}$, we have

$$\delta\bar{H} = \int_{-b}^{b} \delta H\,\mathrm{d}x_2 = \int_{-b}^{b} (T_{ij}\,\delta S_{ij} - D_i\,\delta E_i)\,\mathrm{d}x_2$$

or, with (9),

$$\delta\bar{H} = \int_{-b}^{b} \sum_n x_2^n(T_{ij}\,\delta S_{ij}^{(n)} - D_i\,\delta E_i^{(n)})\,\mathrm{d}x_2 = \sum_n (T_{ij}^{(n)}\,\delta S_{ij}^{(n)} - D_i^{(n)}\,\delta E_i^{(n)}), \tag{13}$$

where $T_{ij}^{(n)}$ and $D_i^{(n)}$ are the components of stress and electric displacement of order n:

$$T_{ij}^{(n)} = \int_{-b}^{b} x_2^n T_{ij}\,\mathrm{d}x_2, \qquad D_i^{(n)} = \int_{-b}^{b} x_2^n D_i\,\mathrm{d}x_2.$$

Upon substituting (10) into (13), we have

$$\delta\bar{H} = \sum_n \{T_{ij}^{(n)}[\delta u_{j,i}^{(n)} + (n+1)\delta_{i2}\,\delta u_j^{(n+1)}] + D_i^{(n)}[\delta\varphi_{,i}^{(n)} + (n+1)\,\delta\varphi^{(n+1)}]\}$$

$$= \sum_n T_{ij}^{(n)}\,\delta u_j^{(n)} + D_i^{(n)}\,\delta\varphi^{(n)})_{,i} - \sum_n [(T_{ij,i}^{(n)} - nT_{2j}^{(n-1)})\,\delta u_j^{(n)} + (D_{i,i}^{(n)} - nD_2^{(n-1)})\,\delta\varphi^{(n)}],$$

whence

$$\int_A \delta\bar{H}\,\mathrm{d}A = \oint_C \sum_n n_a(T_{aj}^{(n)}\,\delta u_j^{(n)} + D_a^{(n)}\,\delta\varphi^{(n)})\,\mathrm{d}s$$

$$- \int_A \sum_n [(T_{ij,i}^{(n)} - nT_{2j}^{(n-1)})\,\delta u_j^{(n)} + (D_{i,i}^{(n)} - nD_2^{(n-1)})\,\delta\varphi^{(n)}]\,\mathrm{d}A. \qquad (14)$$

Inserting (12) and (14) in (11), we arrive at the series form of the two-dimensional variational equation:

$$\sum_n \int_{t_0}^{t_1} \mathrm{d}t \int_A \left(T_{ij,i}^{(n)} - nT_{2j}^{(n-1)} + B_n T_j^{(n)} - \rho \sum_m B_{mn}\ddot{u}_j^{(m)}\right) \delta u_j^{(n)}\,\mathrm{d}A$$

$$+ \sum_n \int_{t_0}^{t_1} \mathrm{d}t \int_A (D_{i,i}^{(n)} - nD_2^{(n-1)} + B_n D^{(n)})\,\delta\varphi^{(n)}\,\mathrm{d}A$$

$$+ \sum_n \int_{t_0}^{t_1} \mathrm{d}t \oint_C [(B_n t_j^{(n)} - n_a T_{aj}^{(n)})\,\delta u_j^{(n)} + (B_n d^{(n)} - n_a D_a^{(n)})\,\delta\varphi^{(n)}]\,\mathrm{d}s = 0,$$

from which follow the field equations, in A,

$$T_{ij,i}^{(n)} - nT_{2j}^{(n-1)} + B_n T_j^{(n)} = \rho \sum_m B_{mn}\ddot{u}_j^{(m)}, \qquad D_{i,i}^{(n)} - nD_2^{(n-1)} + B_n D^{(n)} = 0$$

and boundary conditions

$$n_a T_{aj}^{(n)} = B_n t_j^{(n)}, \qquad n_a D_a^{(n)} = B_n d^{(n)} \quad \text{on } C.$$

Alternative boundary conditions are prescribed displacements $\bar{u}_j^{(n)}$ and prescribed potentials $\bar{\varphi}^{(n)}$ on C:

$$u_j^{(n)} = \bar{u}_j^{(n)}, \qquad \varphi^{(n)} = \bar{\varphi}^{(n)} \quad \text{on } C.$$

The constitutive equations of order n are obtained as follows:

$$T_{ij}^{(n)} = \int_{-b}^{b} x_2^n T_{ij}\,\mathrm{d}x_2 = \int_{-b}^{b} x_2^n(c_{ijkl}S_{kl} - e_{kij}E_k)\,\mathrm{d}x_2 = \int_{-b}^{b} \sum_m x_2^m x_2^n(c_{ijkl}S_{kl}^{(m)} - e_{kij}E_k^{(m)})\,\mathrm{d}x_2,$$

$$D_i^{(n)} = \int_{-b}^{b} x_2^n D_i\,\mathrm{d}x_2 = \int_{-b}^{b} x_2^n(e_{ijk}S_{jk} + \varepsilon_{ij}E_j)\,\mathrm{d}x_2 = \int_{-b}^{b} \sum_m x_2^m x_2^n(e_{ijk}S_{jk}^{(m)} + \varepsilon_{ij}E_j^{(m)})\,\mathrm{d}x_2,$$

whence,

$$T_{ij}^{(n)} = \sum_m B_{mn}(c_{ijkl}S_{kl}^{(m)} - e_{kij}E_k^{(m)}), \qquad D_i^{(n)} = \sum_m B_{mn}(e_{ijk}S_{jk}^{(m)} + \varepsilon_{ij}E_j^{(m)}). \qquad (15)$$

TRUNCATION OF SERIES AND ADJUSTMENT

The process of truncation of the series and adjustment of the remaining terms begins with the discard of the strains and electric fields of order higher than the first, leaving $S_{ij}^{(0)}$, $S_{ij}^{(1)}$, $E_i^{(0)}$, $E_i^{(1)}$ which contain the zero, first and second order mechanical displacements, $u_j^{(0)}$, $u_j^{(1)}$, $u_j^{(2)}$ and electric potentials, $\varphi^{(0)}$, $\varphi^{(1)}$, $\varphi^{(2)}$, some of which will be eliminated subsequently. However, at this stage the constitutive equations (15) reduce to

$$T_{ij}^{(0)} = 2b(c_{ijkl}S_{kl}^{(0)} - e_{kij}E_k^{(0)}), \qquad D_i^{(0)} = 2b(e_{ijk}S_{jk}^{(0)} + \varepsilon_{ij}E_j^{(0)}), \tag{16}$$

$$T_{ij}^{(1)} = \tfrac{2}{3}b^3(c_{ijkl}S_{kl}^{(1)} - e_{kij}E_k^{(1)}), \qquad D_i^{(1)} = \tfrac{2}{3}b^3(e_{ijk}S_{jk}^{(1)} + \varepsilon_{ij}E_j^{(1)}) \tag{17}$$

and these are derivable from the electric enthalpy density

$$\bar{H} = b(c_{ijkl}S_{ij}^{(0)}S_{kl}^{(0)} - \varepsilon_{ij}E_i^{(0)}E_j^{(0)} - 2e_{ijk}E_i^{(0)}S_{jk}^{(0)}) + \tfrac{1}{3}b^3(c_{ijkl}S_{ij}^{(1)}S_{kl}^{(1)} - \varepsilon_{ij}E_i^{(1)}E_j^{(1)} - 2e_{ijk}E_i^{(1)}S_{jk}^{(1)})$$

according to

$$T_{ij}^{(n)} = \partial\bar{H}/\partial S_{ij}^{(n)}, \qquad D_i^{(n)} = -\partial\bar{H}/\partial E_i^{(n)}, \qquad n = 0, 1.$$

Next, following Cauchy [3], we neglect the velocity $\dot{u}_2^{(1)}$ in the kinetic energy density, and provide for free development of the strain $S_{22}^{(0)}(=u_2^{(1)})$ by setting $T_{22}^{(0)} = 0$ in (16). Thus,

$$T_{22}^{(0)} = 2b(c_{22kl}S_{kl}^{(0)} - e_{k22}E_k^{(0)}) = 0. \tag{18}$$

Add $c_{2222}S_{22}^{(0)}$ to each side of (18) and divide by c_{2222}:

$$S_{22}^{(0)} = -c_{22kl}S_{kl}^{(0)}/c_{2222} + S_{22}^{(0)} + e_{k22}E_k^{(0)}/c_{2222}. \tag{19}$$

Now, add and subtract $c_{ij22}S_{22}^{(0)}$ to the expression for $T_{ij}^{(0)}$ in (16):

$$(2b)^{-1}T_{ij}^{(0)} = (c_{ijkl}S_{kl}^{(0)} - c_{ij22}S_{22}^{(0)}) + c_{ij22}S_{22}^{(0)} - e_{kij}E_k^{(0)}. \tag{20}$$

Finally, substitute the expression for $S_{22}^{(0)}$, given in (19), for the $S_{22}^{(0)}$ outside the parentheses in (20), to obtain

$$T_{ij}^{(0)} = 2b(\bar{c}_{ijkl}S_{kl}^{(0)} - \bar{e}_{kij}E_k^{(0)}), \tag{21}$$

where

$$\bar{c}_{ijkl} = c_{ijkl} - c_{ij22}c_{22kl}/c_{2222}, \qquad \bar{e}_{kij} = e_{kij} - e_{k22}c_{ij22}/c_{2222}.$$

Note that, in (21), $T_{22}^{(0)}$ is now zero and $S_{22}^{(0)}$ $(=u_2^{(1)})$ is no longer present.

The first order terms are treated similarly except that all three velocities $\dot{u}_j^{(2)}$ are neglected in the kinetic energy density and free development of the three strains $S_{2j}^{(1)}$ is accommodated by setting $T_{2j}^{(1)} = 0$ in (17). When so many components of stress are to be set equal to zero, it is simpler to start with expressions for the strains in terms of the stresses. To do this, first define elastic compliances, s_{ijkl}, and piezoelectric strain constants, d_{ijk}, according to [6]

$$s_{ijmn}c_{ijkl} = I_{mnkl}, \qquad d_{kmn} = s_{ijmn}e_{kij},$$

where I_{mnkl} is the fourth rank unit tensor. Then multiply the formula for $T_{ij}^{(1)}$, in (17), by s_{ijmn} to obtain

$$s_{ijmn}T_{ij}^{(1)} = \tfrac{2}{3}b^3(S_{mn}^{(1)} - d_{kmn}E_k^{(1)}). \tag{22}$$

In (22), set $T_{2j}^{(1)} = 0$, $\varphi^{(2)} = 0$, so that we may write

$$s_{abcd}T_{cd}^{(1)} = \tfrac{2}{3}b^3(S_{ab}^{(1)} - d_{cab}E_c^{(1)}), \tag{23}$$

where the subscripts a, b, c, d range over 1 and 3 only. Now, solve the three equations (23) for the three independent $T_{cd}^{(1)}$:

$$T_{ab}^{(1)} = \tfrac{2}{3}b^3A_{abcd}(S_{cd}^{(1)} - d_{ecd}E_e^{(11)})/|s_{abcd}|, \tag{24}$$

where the determinant $|s_{abcd}|$ is given by

$$|s_{abcd}| = \begin{vmatrix} s_{1111} & s_{3311} & s_{1311} \\ s_{1133} & s_{3333} & s_{1333} \\ s_{1113} & s_{3313} & s_{1313} \end{vmatrix}$$

and A_{abcd} is the cofactor of s_{abcd} in $|s_{abcd}|$. The expression (24) may be written in the form

$$T_{ab}^{(1)} = \tfrac{2}{3}b^3(c_{abcd}^{(1)}\, S_{cd}^{(1)} - e_{cab}^{(1)}E_c^{(1)}), \tag{25}$$

where

$$c_{abcd}^{(1)} = A_{abcd}/|s_{abcd}|, \qquad e_{cab}^{(1)} = d_{cde}c_{abde}^{(1)}.$$

The constants $c_{abcd}^{(1)}$ are Voigt's [7] constants γ_{pq}, $p, q = 1, 3, 5$.

At this stage, an electric enthalpy density that produces the stresses $T_{ij}^{(0)}$ in (21) and $T_{ab}^{(1)}$ in (24) is

$$\begin{aligned} \bar{H} = {} & b(\bar{c}_{ijkl}S_{ij}^{(0)}S_{kl}^{(0)} - \varepsilon_{ij}E_i^{(0)}E_j^{(0)} - 2\bar{e}_{kij}E_k^{(0)}S_{ij}^{(0)}) \\ & + \tfrac{1}{3}b^3(c_{abcd}^{(1)}S_{ab}^{(1)}S_{cd}^{(1)} - \varepsilon_{ab}E_a^{(1)}E_b^{(1)} - 2e_{abc}^{(1)}E_a^{(1)}S_{bc}^{(1)}) \end{aligned}$$

and this form also fixes $D_i^{(0)}$ and $D_a^{(1)}$ through

$$D_i^{(0)} = -\partial\bar{H}/\partial E_i^{(0)}, \qquad D_a^{(1)} = -\partial\bar{H}/\partial E_a^{(1)}.$$

The final adjustment is made by introducing shear correction factors analogous to those employed in beam theory by Bresse [4] and Timoshenko [5]. The thickness-shear strains $S_{21}^{(0)}$ and $S_{23}^{(0)}$ are replaced by $\kappa_1 S_{21}^{(0)}$ and $\kappa_3 S_{23}^{(0)}$ in the electric enthalpy density, where κ_1 and κ_3 are correction factors whose values may be chosen in such a way [1] that the important thickness-shear frequencies have the correct values, thus compensating, in part, for the omission of terms of higher order in the series expansions. The final form of the electric enthalpy density is, thus,

$$\begin{aligned} \bar{H} = {} & b(c_{ijkl}^{(0)}S_{ij}^{(0)}S_{kl}^{(0)} - \varepsilon_{ij}E_i^{(0)}E_j^{(0)} - 2e_{kij}^{(0)}E_k^{(0)}S_{ij}^{(0)}) \\ & + \tfrac{1}{3}b^3(c_{abcd}^{(1)}S_{ab}^{(1)}S_{cd}^{(1)} - \varepsilon_{ab}E_a^{(1)}E_b^{(1)} - 2e_{abc}^{(1)}E_a^{(1)}S_{bc}^{(1)}) \end{aligned}$$

where

$$c_{ijkl}^{(0)} = \kappa_{i+j-2}^{\mu}\kappa_{k+l-2}^{\nu}\bar{c}_{ijkl}, \qquad e_{kij}^{(0)} = \kappa_{i+j-2}^{\mu}\bar{e}_{kij} \quad \text{(not summed)}$$

and μ and ν are the powers

$$\mu = \cos^2(ij\pi/2), \qquad \nu = \cos^2(kl\pi/2).$$

Thus, κ_{i+j-2}^{μ} (or κ_{k+l-2}^{ν}) is equal to κ_1, κ_3 or unity according as $i+j$ (or $k+l$) is 3, 5 or neither, respectively.

RECAPITULATION

The equations and variables remaining, after truncation of the series and adjustment of the terms retained, are

Kinetic energy density

$$\bar{K} = \rho b(\dot{u}_j^{(0)}\dot{u}_j^{(0)} + \tfrac{1}{3}b^2\dot{u}_a^{(1)}\dot{u}_a^{(1)}).$$

Electric enthalpy density

$$\bar{H} = b(c^{(0)}_{ijkl}S^{(0)}_{ij}S^{(0)}_{kl} - \varepsilon_{ij}E^{(0)}_i E^{(0)}_j - 2e^{(0)}_{kij}E^{(0)}_k S^{(0)}_{ij}) + \tfrac{1}{3}b^3(c^{(1)}_{abcd}S^{(1)}_{ab}S^{(1)}_{cd} - \varepsilon_{ab}E^{(1)}_a E^{(1)}_b - 2e^{(1)}_{abc}E^{(1)}_a S^{(1)}_{bc}).$$

Strain–displacement relations

$$S^{(0)}_{ij} = \tfrac{1}{2}(u^{(0)}_{j,i} + u^{(0)}_{i,j} + \delta_{i2}u^{(1)}_j + \delta_{2j}u^{(1)}_i), \qquad S^{(1)}_{ab} = \tfrac{1}{2}(u^{(1)}_{b,a} + u^{(1)}_{a,b}).$$

Electric field–potential relations

$$E^{(0)}_i = -\varphi^{(0)}_{,i} - \delta_{i2}\varphi^{(1)}, \qquad E^{(1)}_a = -\varphi^{(0)}_{,a}.$$

Constitutive equations

$$T^{(0)}_{ij} = \partial\bar{H}/\partial S^{(0)}_{ij} = 2b(c^{(0)}_{ijkl}S^{(0)}_{kl} - e^{(0)}_{kij}E^{(0)}_k), \qquad D^{(0)}_i = -\partial\bar{H}/\partial E^{(0)}_i = 2b(e^{(0)}_{ijk}S^{(0)}_{jk} + \varepsilon_{ij}E^{(0)}_j)$$

$$T^{(1)}_{ab} = \partial\bar{H}/\partial S^{(1)}_{ab} = \tfrac{2}{3}b^3(c^{(1)}_{abcd}S^{(1)}_{cd} - e^{(1)}_{cab}E^{(1)}_c), \qquad D^{(1)}_a = -\partial\bar{H}/\partial E^{(1)}_a = \tfrac{2}{3}b^3(e^{(1)}_{abc}S^{(1)}_{bc} + \varepsilon_{ab}E^{(1)}_{ab}).$$

Variational principle

$$\int_{t_0}^{t_1} \mathrm{d}t \int_A (\bar{K} - \bar{H})\,\mathrm{d}A + \int_{t_0}^{t_1} \mathrm{d}t \int_A [2b(T^{(0)}_j\,\delta u^{(0)}_j + D^{(0)}\,\delta\varphi^{(0)}) + \tfrac{2}{3}b^3(T^{(1)}_a\,\delta u^{(1)}_a + D^{(1)}\,\delta\varphi^{(1)})]\,\mathrm{d}A$$
$$+ \int_{t_0}^{t_1} \mathrm{d}t \oint_C [(2bt^{(0)}_j - n_a T^{(0)}_{aj})\,\delta u^{(0)}_j + (2bd^{(0)} - n_a D^{(0)}_a)\,\delta\varphi^{(0)}]\,\mathrm{d}s$$
$$+ \int_{t_0}^{t_1} \mathrm{d}t \oint_C [(\tfrac{2}{3}b^3 t^{(1)}_b - n_a T^{(1)}_{ab})\,\delta u^{(1)}_b + (\tfrac{2}{3}b^3 d^{(1)} - n_a D^{(1)}_a)\delta\varphi^{(1)}]\,\mathrm{d}s = 0.$$

Field equations

$$T^{(0)}_{ij,i} + 2bT^{(0)}_j = 2b\rho\ddot{u}^{(0)}_j$$
$$D^{(0)}_{i,i} + 2bD^{(0)} = 0$$
$$T^{(1)}_{ab,a} - T^{(0)}_{2b} + \tfrac{2}{3}b^3 T^{(1)}_b = \tfrac{2}{3}b^3\rho\ddot{u}^{(1)}_b$$
$$D^{(1)}_{a,a} - D^{(0)}_2 + \tfrac{2}{3}b^3 D^{(1)} = 0.$$

Edge conditions

$$n_a T^{(0)}_{aj} = 2bt^{(0)}_j \quad \text{or} \quad u^{(0)}_j = \bar{u}^{(0)}_j$$
$$n_a D^{(0)}_a = 2bd^{(0)} \quad \text{or} \quad \varphi^{(0)} = \bar{\varphi}^{(0)}$$
$$n_a T^{(1)}_{ab} = \tfrac{2}{3}b^3 t^{(1)}_b \quad \text{or} \quad u^{(1)}_b = \bar{u}^{(1)}_b$$
$$n_a D^{(1)}_a = \tfrac{2}{3}b^3 d^{(1)} \quad \text{or} \quad \varphi^{(1)} = \bar{\varphi}^{(1)}.$$

Equations on $u^{(0)}_j$, $u^{(1)}_a$, $\varphi^{(0)}$, $\varphi^{(1)}$

$$c^{(0)}_{ijkl}(u^{(0)}_{k,li} + \delta_{2l}u^{(1)}_{k,i}) + e^{(0)}_{kij}\varphi^{(0)}_{,ki} + T^{(0)}_j = \rho\ddot{u}^{(0)}_j$$
$$e^{(0)}_{kij}(u^{(0)}_{j,ki} + \delta_{2i}u^{(1}_{j,k}) - \varepsilon_{ij}\varphi^{((0)}_{,ij} + D^{(0)} = 0$$
$$c^{(1)}_{abcd}u^{(1)}_{d,ca} + e^{(1)}_{cab}\varphi^{(1)}_{,ca} - 3b^{-2}[c^{(0)}_{2bkl}(u^{(0)}_{l,k} + \delta_{2k}u^{(1)}_l) + e^{(0)}_{a2b}\varphi^{(0)}_{,a} + e^{(0)}_{22b}\varphi^{(1)}] + T^{(1)}_b = \rho\ddot{u}^{(1)}_b$$
$$e^{(1)}_{cab}u^{(1)}_{b,ca} - \varepsilon_{bc}\varphi^{(1)}_{,bc} - 3b^{-2}[e^{(2)}_{2ij}(u^{(0)}_{j,i} + \delta_{2i}u^{(1)}_j) - \varepsilon_{a2}\varphi^{(0)}_{,a} - \varepsilon_{22}\varphi^{(1)}] + D^{(1)} = 0.$$

APPLICATION TO ROTATED-Y-CUTS OF QUARTZ

After performing the summations over repeated indices in the preceding equations, it is convenient to employ the abbreviated notation in which a pair of indices, ij or kl, is replaced by a single index, p or q, ranging over 1–6 according to [6]

$$\begin{array}{lcccccc} ij \text{ or } kl = & 11 & 22 & 33 & 23,32 & 31,13 & 12,21 \\ p \text{ or } q = & 1 & 2 & 3 & 4 & 5 & 6 \end{array}$$

Then

$$c_{ijkl} \to c_{pq}, \qquad c_{ijkl}^{(1)} \to \gamma_{pq}, \qquad e_{kij} \to e_{kp},$$

$$s_{ijkl} \to \begin{cases} s_{pq} \text{ if } i = j,\, k = l, \\ \tfrac{1}{2}s_{pq} \text{ if } i = j,\, k \neq l \quad \text{or} \quad i \neq j,\, k = l, \\ \tfrac{1}{4}s_{pq} \text{ if } i \neq j,\, k \neq l, \end{cases}$$

$$d_{kij} \to \begin{cases} d_{kp} \text{ if } i = j, \\ \tfrac{1}{2}d_{kp} \text{ if } i \neq j. \end{cases}$$

For the rotated-Y-cuts of quartz, many of the constants are zero. If x_1 is the digonal axis in the plane of the plate [8]:

$$c_{15} = c_{16} = c_{25} = c_{26} = c_{35} = c_{36} = c_{45} = c_{46} = 0,$$

$$s_{15} = s_{16} = s_{25} = s_{26} = s_{35} = s_{36} = s_{45} = s_{46} = 0,$$

$$e_{21} = e_{31} = e_{22} = e_{32} = e_{23} = e_{33} = e_{24} = e_{34} = e_{15} = e_{16} = 0,$$

$$d_{21} = d_{31} = d_{22} = d_{32} = d_{23} = d_{33} = d_{24} = d_{34} = d_{15} = d_{16} = 0,$$

$$\varepsilon_{12} = \varepsilon_{13} = 0.$$

The equations on $u_j^{(0)}$, $u_a^{(1)}$, $\varphi^{(0)}$, $\varphi^{(1)}$ then reduce to

$$\begin{aligned} &\bar{c}_{11}u_{1,11}^{(0)} + c_{55}u_{1,33}^{(0)} + (\kappa_1 c_{56} + \kappa_3\bar{c}_{14})u_{2,13}^{(0)} + (\bar{c}_{13} + c_{55})u_{3,13}^{(0)} \\ &\quad + \kappa_1 c_{56}u_{1,3}^{(1)} + \kappa_3\bar{c}_{14}u_{3,1}^{(1)} + \bar{e}_{11}\varphi_{,11}^{(0)} + e_{35}\varphi_{,33}^{(0)} + T_1^{(0)} = \rho\ddot{u}_1^{(0)}, \end{aligned} \tag{26}$$

$$\begin{aligned} &(\kappa_1 c_{56} + \kappa_3\bar{c}_{14})u_{1,13}^{(0)} + \kappa_1^2 c_{66}u_{2,11}^{(0)} + \kappa_3^2\bar{c}_{44}u_{2,33}^{(0)} + \kappa_3\bar{c}_{34}u_{3,33}^{(0)} \\ &\quad + \kappa_1^2 c_{66}u_{1,1}^{(1)} + \kappa_3^2\bar{c}_{44}u_{3,3}^{(1)} + (\kappa_1 e_{36} + \kappa_3\bar{e}_{14})\varphi_{,13}^{(0)} + T_2^{(0)} = \rho\ddot{u}_2^{(0)}, \end{aligned} \tag{27}$$

$$\begin{aligned} &(\bar{c}_{13} + c_{55})u_{1,13}^{(0)} + \kappa_1 c_{56}u_{2,11}^{(0)} + \kappa_3\bar{c}_{34}u_{2,33}^{(0)} + c_{55}u_{3,11}^{(0)} + \bar{c}_{33}u_{3,33}^{(0)} \\ &\quad + \kappa_1 c_{56}u_{1,1}^{(1)} + \kappa_3\bar{c}_{34}u_{3,3}^{(1)} + (\bar{e}_{13} + e_{35})\varphi_{,13}^{(0)} + T_3^{(0)} = \rho\ddot{u}_3^{(0)}, \end{aligned} \tag{28}$$

$$\begin{aligned} &\bar{e}_{11}u_{1,11}^{(0)} + e_{35}u_{1,35}^{(0)} + (\kappa_1 e_{36} + \kappa_3\bar{e}_{14})u_{2,13}^{(0)} + (\bar{e}_{13} + e_{35})u_{3,13}^{(0)} \\ &\quad + \kappa_1 e_{36}u_{1,3}^{(1)} + \kappa_3\bar{e}_{14}u_{3,1}^{(1)} - \varepsilon_{11}\varphi_{,11}^{(0)} - \varepsilon_{33}\varphi_{,33}^{(0)} + D^{(0)} = 0, \end{aligned} \tag{29}$$

$$\begin{aligned} &\gamma_{11}u_{1,11}^{(1)} + \gamma_{55}u_{1,33}^{(1)} + (\gamma_{13} + \gamma_{55})u_{3,13}^{(1)} + e_{11}^{(1)}\varphi_{,11}^{(1)} + e_{35}^{(1)}\varphi_{,33}^{(1)} \\ &\quad - 3b^{-2}[\kappa_1^2 c_{66}(u_{2,1}^{(0)} + u_1^{(1)}) + \kappa_1 c_{56}(u_{3,1}^{(0)} + u_{1,3}^{(0)}) + \kappa_1 e_{36}\varphi_{,3}^{(0)} + \kappa_1 e_{26}\varphi^{(1)}] + T_1^{(1)} = \rho\ddot{u}_1^{(1)}, \end{aligned} \tag{30}$$

$$\begin{aligned} &(\gamma_{13} + \gamma_{55})u_{1,13}^{(1)} + \gamma_{55}u_{3,11}^{(1)} + \gamma_{33}u_{3,33}^{(1)} + (e_{13}^{(1)} + e_{35}^{(1)})\varphi_{,13}^{(1)} \\ &\quad - 3b^{-2}[\kappa_3^2\bar{c}_{44}(u_{2,3}^{(0)} + u_3^{(1)}) + \kappa_3\bar{c}_{14}u_{1,1}^{(0)} + \kappa_3\bar{c}_{34}u_{3,3}^{(0)} + \kappa_3\bar{e}_{14}\varphi_{,1}^{(0)}] + T_3^{(1)} = \rho\ddot{u}_3^{(1)}, \end{aligned} \tag{31}$$

$$e^{(1)}_{11}u^{(1)}_{1,11}+e^{(1)}_{35}u^{(1)}_{1,33}+(e^{(1)}_{13}+e^{(1)}_{35})u^{(1)}_{3,13}-\varepsilon_{11}\varphi^{(1)}_{,11}-\varepsilon_{33}\varphi^{(1)}_{,33}$$
$$-3b^{-2}[\kappa_1 e_{26}(u^{(0)}_{2,1}+u^{(1)}_1)+e_{25}(u^{(0)}_{3,1}+u^{(0)}_{1,3})-\varepsilon_{23}\varphi^{(0)}_{,3}-\varepsilon_{22}\varphi^{(1)}]+D^{(1)}=0. \quad (32)$$

In these equations,

$$\bar{c}_{11}=c_{11}-c^2_{12}/c_{22} \qquad \bar{c}_{13}=c_{13}-c_{12}c_{23}/c_{22}$$
$$\bar{c}_{33}=c_{33}-c^2_{23}/c_{22} \qquad \bar{c}_{14}=c_{14}-c_{12}c_{24}/c_{22}$$
$$\bar{c}_{44}=c_{44}-c^2_{24}/c_{22} \qquad \bar{c}_{34}=c_{34}-c_{23}c_{24}/c_{22}$$

$$\gamma_{11}=s_{33}/(s_{11}s_{33}-s^2_{13}) \qquad \gamma_{13}=-s_{13}/(s_{11}s_{33}-s^2_{13})$$
$$\gamma_{33}=s_{11}/(s_{11}s_{33}-s^2_{13}) \qquad \gamma_{55}=1/s_{55}$$

$$\bar{e}_{11}=e_{11}-e_{12}c_{12}/c_{22}, \qquad e^{(1)}_{11}=d_{11}\gamma_{11}+d_{13}\gamma_{13}$$
$$\bar{e}_{13}=e_{13}-e_{12}c_{23}/c_{22} \qquad e^{(1)}_{13}=d_{11}\gamma_{13}+d_{13}\gamma_{33}$$
$$\bar{e}_{14}=e_{14}-e_{12}c_{24}/c_{22} \qquad e^{(1)}_{35}=d_{35}\gamma_{55}.$$

Note that the primes employed by Sykes [8] are omitted here.

It remains only to fix the values of the shear coefficients κ_1 and κ_3. This is done [1] by equating the thickness-shear frequencies obtained from the two-dimensional equations (26)–(32) to the corresponding frequencies obtained from the three-dimensional equations.

In (26)–(32), set all spatial differential coefficients equal to zero. The last three equations then reduce to

$$-3b^{-2}(\kappa_1^2 c_{66}u^{(1)}_1+\kappa_1 e_{26}\varphi^{(1)})+T^{(1)}_1=\rho\ddot{u}^{(1)}_1, \quad (33)$$

$$3b^{-2}\kappa_3^2\bar{c}_{44}u^{(1)}_3+T^{(1)}_3=\rho\ddot{u}^{(1)}_3, \quad (34)$$

$$-3b^{-2}(\kappa_1 e_{26}u^{(1)}_1-\varepsilon_{22}\varphi^{(1)})+D^{(1)}=0. \quad (35)$$

It may be seen that the thickness-shear displacement $u^{(1)}_1$ is coupled to the electric field, but $u^{(1)}_3$ is not. In the former case, a thickness-shear vibration may be driven by an alternating voltage $Ve^{i\omega t}$ applied to electrode films deposited on the faces of the plate: Then, if the two films are alike, the surface conditions are

$$T_{21}]_{\pm b}=\mp 2\rho' b'\ddot{u}_1]_{\pm b}, \qquad \varphi]_{\pm b}=\pm V\,e^{i\omega t},$$

where $2\rho' b'$ is the mass per unit area of each electrode film. Accordingly, in (33) and (35),

$$T^{(1)}_1=-3R\rho\ddot{u}^{(1)}_1, \qquad \varphi^{(1)}=V\,e^{i\omega t}/b,$$

where $R(=2\rho' b'/\rho b)$ is the ratio of the mass per unit area of both electrode films to the mass per unit area of the plate. If we set

$$u^{(1)}_1=A\,e^{i\omega t},$$

we find, from (33),

$$A=3\kappa_1 e_{26}V/b[b^2(1+3R)\rho\omega^2-3\kappa_1^2 c_{66}]. \quad (36)$$

Hence, resonance occurs when

$$\omega^2 = 3\kappa_1^2 c_{66}/\rho b^2(1+3R), \qquad (\kappa_1 \text{ with electrodes}) \tag{37}$$

and the surface charge, off resonance, is obtained from $D^{(1)}$ in (35):

$$D_2]_{-b}^{b} = \tfrac{2}{3}b^2 D^{(1)} = 2b^{-1}(\kappa_1 e_{26}A - \varepsilon_{22}V)\,e^{i\omega t},$$

where A is given in (36).

In the absence of electrode films, the traction and charge on the faces are zero, i.e.

$$T^{(1)} = 0, \qquad D^{(1)} = 0$$

in (33) and (35). Then with

$$u_1^{(1)} = A\,e^{i\omega t}, \qquad \varphi^{(1)} = B\,e^{i\omega t},$$

(33) and (35) become

$$(\kappa_1^2 c_{66} - \tfrac{1}{3}\rho b^2\omega^2)A + \kappa_1 e_{26}B = 0,$$

$$\kappa_1 e_{26}A - \varepsilon_{22}B = 0,$$

from which, by elimination of A and B,

$$\omega^2 = 3\kappa_1^2 \hat{c}_{66}/\rho b^2 \quad (\kappa_1 \text{ without electrodes}) \tag{38}$$

where

$$\hat{c}_{66} = c_{66} + e_{26}^2/\varepsilon_{22}.$$

In the case of thickness-shear in the x_3-direction, (34) applies. When there are electrode films, we set

$$T_3^{(1)} = -3R\rho\ddot{u}_3^{(1)}, \qquad u_3^{(1)} = A\,e^{i\omega t}$$

and find

$$\omega^2 = 3\kappa_3^2 \bar{c}_{44}/\rho b^2(1+3R) \quad (\kappa_3 \text{ with electrodes}). \tag{39}$$

In the absence of electrodes, $T_3^{(1)} = 0$ and (34) yields

$$\omega^2 = 3\kappa_3^2 \bar{c}_{44}/\rho b^2 \quad (\kappa_3 \text{ without electrodes}). \tag{40}$$

The four frequencies given in (37)–(40) are to be equated to the corresponding frequencies obtained from the three-dimensional equations (4) and boundary conditions (6) or (7).

For thickness-shear in the x_1-direction, the pertinent solutions of the three-dimensional equations are given by Bleustein and Tiersten [9]. For

$$R = 2\rho' b'/\rho b \ll 1, \qquad \kappa_{26}^2 = e_{26}^2/\varepsilon_{22}\hat{c}_{66} \ll 1,$$

they find, for the plate with electrodes,

$$\omega^2 = \pi^2 \hat{c}_{66}(1 - R - 4k_{26}^2/\pi^2)^2 4\rho b^2 \tag{41}$$

and, without electrodes,

$$\omega^2 = \pi^2 \hat{c}_{66}/4\rho b^2. \tag{42}$$

Hence, equating (41) and (42) to (37) and (38), respectively, we have, for $R \ll 1$, $k_{26}^2 \ll 1$,

$$\kappa_1^2 = (\pi^2/12)(1+R-8k_{26}^2/\pi^2) \quad \text{(with electrodes)}$$

$$\kappa_1^2 = \pi^2/12 \quad \text{(without electrodes).}$$

For thickness-shear in the x_3-direction, the solution of the three-dimensional equations is described in [10]. For $R \ll 1$, the frequency is found to be, after correcting an error in equation (43) of [10];

$$\omega^2 = \pi^2 c_3(1-R)^2/4\rho b^2, \tag{43}$$

where

$$c_3 = \tfrac{1}{2}\{c_{22}+c_{44}-[(c_{22}-c_{44})^2+4c_{24}^2]^{\frac{1}{2}}\}.$$

Without electrodes, the corresponding frequency is [11]

$$\omega^2 = \pi^2 c_3/4\rho b^2. \tag{44}$$

Hence, for $R \ll 1$, we have, upon equating (43) and (44) to (39) and (40), respectively,

$$\kappa_3^2 = (\pi^2/12)(1+R)c_3/\bar{c}_{44} \quad \text{(with electrodes)}$$

$$\kappa_3^2 = (\pi^2/12)c_3/\bar{c}_{44} \quad \text{(without electrodes).}$$

REFERENCES

[1] R. D. MINDLIN, Forced thickness-shear and flexural vibrations of piezoelectric crystal plates. *J. appl. Phys.* **23**, 83–88 (1952).
[2] H. F. TIERSTEN and R. D. MINDLIN, Forced vibrations of piezoelectric crystal plates. *Q. appl. Math.* **22**, 107–119 (1962).
[3] A. L. CAUCHY, Sur l'équilibre et le mouvement d'une plaque élastique dont l'élasticité n'est pas la même dans tous les sens. *Exercices Math.* **4**, 1–14 (1829); *Oeuvres* Sér. 2, **19**, 9–22 (1891).
[4] M. BRESSE, *Cours de Mécanique Appliqué*, p. 126. Mallet-Bachelier (1859).
[5] S. P. TIMOSHENKO, On the correction for shear of the differential equation for transverse vibrations of prismatic bars. *Phil. Mag.* **41**, 744–746 (1921); On the transverse vibrations of bars of uniform section. *Phil. Mag.* **43**, 125–131 (1922).
[6] I. R. E. Standards on Piezoelectric Crystals. *Proc. I. R. E.* **37**, 1377–1395 (1949).
[7] W. VOIGT, *Lehrbuch der Kristallphysik*, p. 681. Teubner (1928).
[8] R. A. SYKES, Modes of Motion of Quartz Crystals, the Effects of Coupling and Methods of Design, *Quartz Crystals for Electrical Circuits*, edited by R. A. HEISING, chapter VI, p. 247. Nostrand (1946).
[9] J. L. BLEUSTEIN and H. F. TIERSTEN, Forced thickness-shear vibrations of discontinuously plated piezoelectric crystal plates. *J. acoust. Soc. Am.* **43**, 1311–1318 (1968).
[10] R. D. MINDLIN, High Frequency Vibrations of Plated, Crystal Plates, *Progress in Applied Mechanics*, the Prager Anniversary Volume, pp. 73–84. Macmillan (1963).
[11] R. D. MINDLIN, High frequency vibrations of crystal plates, *Q. appl. Math.* **19**, 51–61 (1961).

(*Received* 6 *December* 1971)

Абстракт—Приводятся двухмерные уравнения движения пьезоэлектрических кристаллических пластинок, путём сохранения первых выражений в разложениях в степенный ряд, для механического перемещения и электрического потенциала, в вариационном принципе для трехмерных уравнений пьезоэлектричности.

Int. J. Solids Structures, 1972, Vol. 8, pp. 1407 to 1411. Pergamon Press. Printed in Great Britain

GALERKIN STRESS FUNCTIONS FOR NON-LOCAL THEORIES OF ELASTICITY†

R. D. MINDLIN

Department of Civil Engineering, Columbia University, New York

INTRODUCTION

IN THE classical theory of elasticity, the potential energy density is a function of the strain whereas in the nonlocal theories the energy is a function of the strain and gradients of the strain, or parts thereof such as the rotation gradient which leads to couple stresses. For the classical theory, Galerkin [1] exhibited a complete solution of the displacement equation of equilibrium in terms of a single vector function satisfying an equation of higher order but of simpler and more tractable form. In the present paper, after a review of the solutions for the classical and rotation gradient cases, the analogous solutions are exhibited for the non-local equations of higher and higher orders up to Cauchy's [2] equation of infinite order for his asymptotic theory of an elastic medium with long range interactions between molecules in a periodic structure.

CLASSICAL THEORY

The displacement equation of equilibrium in the classical, linear theory of elasticity, for the isotropic case, is

$$k\nabla\nabla\cdot\mathbf{u}-\nabla\times\nabla\times\mathbf{u}+\mathbf{f} = 0. \tag{1}$$

In (1) $\mathbf{u}$ is the displacement, $k = (\lambda+2\mu)/\mu$ where λ is Lamé's constant and μ is the shear modulus, $\mathbf{f}$ is the ratio of the body force density to μ and ∇, $\nabla\cdot$, $\nabla\times$ are the gradient, divergence and curl operators, respectively.

Galerkin's solution of (1) may be obtained as follows. In a region V, let x, y, z be the coordinates of a field point, P, and ξ, η, ζ be the coordinates of a source point Q. Let

$$r = (x^2+y^2+z^2)^{1/2},\ r' = (\xi^2+\eta^2+\zeta^2)^{1/2},\ r_1 = [(x-\xi)^2+(y-\eta)^2+(z-\zeta)^2]^{1/2}.$$

Suppose $\mathbf{u}$ is *any* (sufficiently smooth) vector function and define a vector function $\mathbf{H}$ and a scalar function φ by

$$4\pi\mathbf{H} = -\int_V r_1^{-1}\mathbf{u}_Q\,\mathrm{d}V_Q,\qquad 4\pi\varphi = \int_V r_1^{-1}\mathbf{r}'\cdot\nabla_Q^2\mathbf{H}_Q\,\mathrm{d}V_Q.$$

Then

$$\nabla^2\mathbf{H} = \mathbf{u},\qquad \nabla^2\varphi = -\mathbf{r}\cdot\nabla^2\mathbf{H}, \tag{2}$$

where ∇^2 is Laplace's operator.

† Paper prepared for the Galerkin Centennial 1871–1971, U.S.S.R. Academy of Sciences.

Now, define the Galerkin function by

$$\mathbf{G} = \mathbf{H}+\tfrac{1}{2}(k-1)\nabla(\mathbf{r}\cdot\mathbf{H}+\varphi). \tag{3}$$

Then

$$\nabla\nabla\cdot\mathbf{G} = \nabla\nabla\cdot\mathbf{H}+\tfrac{1}{2}(k-1)\nabla(\mathbf{r}\cdot\nabla^2\mathbf{H}+2\nabla\cdot\mathbf{H}+\nabla^2\varphi)$$

so that, from the second of (2),

$$\nabla\nabla\cdot\mathbf{G} = k\nabla\nabla\cdot\mathbf{H}. \tag{4}$$

Also, from (3),

$$\nabla\times\nabla\times\mathbf{G} = \nabla\times\nabla\times\mathbf{H} = \nabla\nabla\cdot\mathbf{H}-\nabla^2\mathbf{H}.$$

This becomes, from (4) and the first of (2),

$$\nabla\times\nabla\times\mathbf{G} = k^{-1}\nabla\nabla\cdot\mathbf{G}-\mathbf{u}.$$

Thus, *any* sufficiently smooth vector function $\mathbf{u}$ may be represented by

$$\mathbf{u} = k^{-1}\nabla\nabla\cdot\mathbf{G}-\nabla\times\nabla\times\mathbf{G}. \tag{5}$$

Suppose, now, that $\mathbf{u}$ is a solution of (1). To find the resulting equation governing $\mathbf{G}$, first note that, from (5),

$$\nabla\nabla\cdot\mathbf{u} = k^{-1}\nabla^2\nabla\nabla\cdot\mathbf{G},$$

$$\nabla\times\nabla\times\mathbf{u} = -\nabla\times\nabla\times\nabla\times\nabla\times\mathbf{G} = \nabla\times\nabla\times\nabla^2\mathbf{G} = \nabla^2\nabla\nabla\cdot\mathbf{G}-\nabla^4\mathbf{G},$$

where $\nabla^4 = \nabla^2\nabla^2$. Then upon substituting (5) in (1), we find

$$\nabla^4\mathbf{G} = -\mathbf{f}. \tag{6}$$

Accordingly, Galerkin's complete solution of (1) is given by (5), provided the Galerkin function $\mathbf{G}$ satisfies (6).

ROTATION GRADIENT

The theory of elasticity in which the potential energy density is a function of the strain and the rotation gradient was given by Aero and Kuvshinskii [3], Grioli [4], Rajagopal [5] and Truesdell and Toupin [6]. The displacement equation of equilibrium for the centrosymmetric, isotropic case is

$$k\nabla\nabla\cdot\mathbf{u}-(1-l_2^2\nabla^2)\nabla\times\nabla\times\mathbf{u}+\mathbf{f} = 0, \tag{7}$$

where l_2 is a material constant.

To find the Galerkin function for (7), first define $\mathbf{H}$ and φ, in this case, by

$$4\pi\mathbf{H} = -\int_V r_1^{-1}(1-e^{-r_1/l_2})\mathbf{u}_Q\,dV_Q, \qquad 4\pi\varphi = \int_V r_1^{-1}\mathbf{r}'\cdot(1-l_2^2\nabla_Q^2)\nabla_Q^2\mathbf{H}_Q\,dV_Q.$$

Then [7, equation (10.12)],

$$(1-l_2^2\nabla^2)\nabla^2\mathbf{H} = \mathbf{u}, \qquad \nabla^2\varphi = -\mathbf{r}\cdot(1-l_2^2\nabla^2)\nabla^2\mathbf{H}. \tag{8}$$

Now, define the Galerkin function, in this case, by

$$\mathbf{G} = (1-l_2^2\nabla^2)\mathbf{H}+\tfrac{1}{2}(k-1)\nabla[\mathbf{r}\cdot(1-l^2\nabla^2)\mathbf{H}+\varphi],$$

where

$$l^2 = l_2^2/(1-k).$$

Then, using the second of (8), we find

$$\nabla\nabla\cdot\mathbf{G} = k\nabla\nabla\cdot\mathbf{H}. \tag{9}$$

Also, taking into account (9) and the first of (8), we find

$$\mathbf{u} = k^{-1}(1-l_2^2\nabla^2)\nabla\nabla\cdot\mathbf{G}-\nabla\times\nabla\times\mathbf{G}. \tag{10}$$

To find the equation governing $\mathbf{G}$, substitute (10) in (7). The result is

$$(1-l_2^2\nabla^2)\nabla^4\mathbf{G} = -\mathbf{f}. \tag{11}$$

The solution (10) and (11) was given by Mindlin and Tiersten [7, equations (11.24) and (11.25)]. An alternative proof of completeness was given by Doyle [8].

FIRST STRAIN GRADIENT

The theory of elasticity in which the potential energy density is a function of the strain and the first gradient of the strain was given by Toupin [9]. The displacement equation of equilibrium for the centrosymmetric, isotropic case is [10, equation (13.1)]

$$k(1-l_1^2\nabla^2)\nabla\nabla\cdot\mathbf{u}-(1-l_2^2\nabla^2)\nabla\times\nabla\times\mathbf{u}+\mathbf{f} = 0, \tag{12}$$

where l_1 and l_2 are constants. Define $\mathbf{H}$ and φ, now, as

$$4\pi\mathbf{H} = -\int_V r_1^{-1}[1+(l_2^2\,e^{-r_1/l_2}-l_1^2\,e^{-r_1/l_1})/(l_1^2-l_2^2)]\mathbf{u}_Q\,dV_Q,$$

$$4\pi\varphi = \int_V r_1^{-1}\mathbf{r}'\cdot(1-l^2\nabla_Q^2)\nabla_Q^2\mathbf{H}_Q\,dV_Q,$$

where

$$l^2 = (kl_1^2-l_2^2)/(k-1). \tag{13}$$

Then [11, equation (67)]

$$(1-l_1^2\nabla^2)(1-l_2^2\nabla^2)\nabla^2\mathbf{H} = \mathbf{u}, \qquad \nabla^2\varphi = -\mathbf{r}\cdot(1-l^2\nabla^2)\nabla^2\mathbf{H}. \tag{14}$$

Next, define the Galerkin function for this case by

$$\mathbf{G} = (1-l_2^2\nabla^2)\mathbf{H}+\tfrac{1}{2}(k-1)\nabla[\mathbf{r}\cdot(1-l^2\nabla^2)\mathbf{H}+\varphi].$$

Then, employing (13) and the second of (14), we find

$$\nabla\nabla\cdot\mathbf{G} = k(1-l_1^2\nabla^2)\nabla\nabla\cdot\mathbf{H}. \tag{15}$$

Also,

$$\nabla\times\nabla\times\mathbf{G} = (1-l_2^2\nabla^2)\nabla\times\nabla\times\mathbf{H} = (1-l_2^2\nabla^2)(\nabla\nabla\cdot\mathbf{H}-\nabla^2\mathbf{H}). \tag{16}$$

From (15) and (16), and employing the first of (14), we find

$$\mathbf{u} = k^{-1}(1-l_2^2)\nabla\nabla\cdot\mathbf{G}-(1-l_1^2\nabla^2)\nabla\times\nabla\times\mathbf{G}.$$

Finally, upon substituting this representation in the equation of equilibrium (12), we find the equation governing the Galerkin function for the strain gradient theory:

$$(1-l_1^2\nabla^2)(1-l_2^2\nabla^2)\nabla^4\mathbf{G} = -\mathbf{f}.$$

SECOND AND HIGHER GRADIENTS OF STRAIN

When the potential energy density is a function of the strain and the first and second gradients of the strain, the displacement equation of equilibrium for the centrosymmetric isotropic case is [11, equation (25)]

$$k(1-l_{11}^2\nabla^2)(1-l_{12}^2\nabla^2)\nabla\nabla\cdot\mathbf{u}-(1-l_{21}^2\nabla^2)(1-l_{22}^2\nabla^2)\nabla\times\nabla\times\mathbf{u}+\mathbf{f} = 0, \tag{17}$$

where l_{ij} are constants. The pattern of representation having been established in the preceding sections, it may be inferred that any solution of (17) can be expressed in terms of a Galerkin function, $\mathbf{G}$, according to

$$\mathbf{u} = k(1-l_{21}^2\nabla^2)(1-l_{22}^2\nabla^2)\nabla\nabla\cdot\mathbf{G}-(1-l_{11}^2\nabla^2)(1-l_{12}^2\nabla^2)\nabla\times\nabla\times\mathbf{G}. \tag{18}$$

Upon substituting (18) into (17) we find the following equation to be satisfied by the Galerkin stress function for the second strain gradient theory:

$$(1-l_{11}^2\nabla^2)(1-l_{12}^2\nabla^2)(1-l_{21}^2\nabla^2)(1-l_{22}^2\nabla^2)\nabla^4\mathbf{G} = -\mathbf{f}.$$

Similarly, if the potential energy density is a function of the strain and all gradients of strain up to and including the nth, it may be inferred that the equation of equilibrium, for the centrosymmetric, isotropic case, would be

$$\begin{aligned} &k[(1-l_{11}^2\nabla^2)(1-l_{12}^2\nabla^2)\dots(1-l_{1n}^2\nabla^2)]\nabla\nabla\cdot\mathbf{u} \\ &\quad -[(1-l_{21}^2\nabla^2)(1-l_{22}^2\nabla^2)\dots(1-l_{2n}^2\nabla^2)]\nabla\times\nabla\times\mathbf{u}+\mathbf{f} = 0 \end{aligned} \tag{19}$$

and the complete representation of the displacement in terms of the Galerkin function would be

$$\begin{aligned} \mathbf{u} = k^{-1}&[(1-l_{21}^2\nabla^2)(1-l_{22}^2\nabla^2)\dots(1-l_{2n}^2\nabla^2)]\nabla\nabla\cdot\mathbf{G} \\ &-[(1-l_{11}^2\nabla^2)(1-l_{12}^2\nabla^2)\dots(1-l_{1n}^2\nabla^2)]\nabla\times\nabla\times\mathbf{G}, \end{aligned}$$

where $\mathbf{G}$ satisfies the equation

$$[(1-l_{21}^2\nabla^2)(1-l_{22}^2\nabla^2)\dots(1-l_{2n}^2\nabla^2)][(1-l_{11}^2\nabla^2)(1-l_{12}^2\nabla^2)\dots(1-l_{1n}^2\nabla^2)]\nabla^4\mathbf{G} = -\mathbf{f}.$$

The equation of equilibrium (19) may be written in the form

$$(E_1\nabla^2+E_2\nabla^4+\cdots E_{n+1}\nabla^{2(n+1)})\mathbf{u}+(F_1+F_2\nabla^2+\cdots F_n\nabla^{2n})\nabla\nabla\cdot\mathbf{u}+\mathbf{f} = 0,$$

where $E_1, E_2\dots$ and $F_1, F_2\dots$ are constants. This equation is to be compared with Cauchy's asymptotic equation for an elastic medium with long range interactions between molecules in a periodic structure. In the case of equilibrium of a centrosymmetric, isotropic material, Cauchy's equation is [2]

$$E\mathbf{u}+F\nabla\nabla\cdot\mathbf{u}+\mathbf{f} = 0,$$

where E and F are entire functions of ∇^2. Upon expanding E and F in series of powers of ∇^2, we see that the nth gradient theory comprises the first n terms of Cauchy's theory.

Finally, it may be observed that the Galerkin functions for the gradient theories of various orders are all governed by equations of the form

$$[(1-a_1\nabla^2)(1-a_2\nabla^2)\dots(1-a_n\nabla^2)]\nabla^4\mathbf{G} = -\mathbf{f}.$$

Since the differential operator, of order $2n+4$, is a product of the double Laplacian and n second order linear operators, the Galerkin function may be resolved into the sum of a particular integral, a biharmonic function, a harmonic function and n functions satisfying homogeneous second order equations:

$$\mathbf{G} = \mathbf{G}_p+\mathbf{G}''+\mathbf{G}'+\sum_{i=1}^{i=n}\mathbf{G}_i,$$

where

$$\nabla^4\mathbf{G}'' = 0, \qquad \nabla^2\mathbf{G}' = 0, \qquad (1-a_i\nabla^2)\mathbf{G}_i = 0, \qquad i = 1, 2\dots n,$$

and G_p is the particular integral.

REFERENCES

[1] B. G. Galerkin, *C. r. hebd. Séanc. Acad. Sci., Paris* **190**, 1047 (1930).
[2] A. L. Cauchy, *C. r. hebd. Séanc. Acad. Sci., Paris* **32**, 323 (1851).
[3] E. L. Aero and E. V. Kuvshinskii, *Fiz. tverd. Tela* **2**, 1399 (1960).
[4] G. Grioli, *Annali. Mat. pura appl.* **50**, 389 (1960).
[5] E. S. Rajagopal, *Ann. Phys.* **6**, 192 (1960).
[6] C. A. Truesdell and R. A. Toupin, *Encyclopedia of Physics*, Vol. III/1. Springer (1960).
[7] R. D. Mindlin and H. F. Tiersten, *Arch. ration. Mech. Analysis* **11**, 415 (1962).
[8] J. M. Doyle, *J. appl. Mech.* **31**, 728 (1964).
[9] R. A. Toupin, *Arch. ration. Mech. Analysis* **11**, 385 (1962).
[10] R. D. Mindlin, *Arch. ration. Mech. Analysis* **16**, 51 (1964).
[11] R. D. Mindlin, *Int. J. Solids Struct.* **1**, 917 (1965).

(*Received* 17 *April* 1972)

Journal of Elasticity, vol. 2, no 4, December 1972
Wolters-Noordhoff Publishing Groningen
Printed in the Netherlands

Elasticity, piezoelectricity and crystal lattice dynamics

R. D. MINDLIN
Department of Civil Engineering, Columbia University, New York

ABSTRACT

A review of some recent developments in the area between the dynamical theory of crystal lattices, in the harmonic approximation, and the classical, linear theories of elasticity and piezoelectricity.

RÉSUMÉ

Un examen de quelques évolutions récentes dans le champ entre la théorie dynamique des réseaux cristallins (approximation harmonique) et les théories classiques, linéaires, de l'élasticité et de la piézoélectricité.

1. Introduction

The differential equations of the classical theory of the anisotropic, elastic continuum are the long wave, low frequency limit of the finite difference equations of the dynamical theory of crystal lattices of mass particles. Consequently, solutions of the difference equations converge to the corresponding solutions of the differential equations as periods and wave lengths increase. Conversely, the two diverge as periods and wave lengths diminish or, in the case of equilibrium, as dimensions diminish. The discrepancy is even greater between classical piezoelectricity and modern theories of the crystal lattice of polarizable atoms because these two do not even converge at the long wave, low frequency limit. In both cases, the defects are due to insufficient consideration, in the continuum theories, of the structure of crystals and the interactions between atoms or molecules.

During the past dozen years, there has been a revival of activity aimed at extending the range of the classical theories of elasticity and piezoelectricity to account for more aspects of structure and interatomic interactions. Developments have progressed along numerous paths, including: (1) rediscoveries of the early terms and the whole of Cauchy's [1] infinite series representation of an elastic solid with a periodic structure – equivalent to taking into account all of the gradients of strain in the potential energy density; (2) formulation of continuum theories of deformation and polarization of crystals with compound (non-primitive) lattice structure; (3) revival and extensions of the Cosserat [2] continuum representation of molecular crystals; (4) augmentation of the classical (Voigt [3]) theory of piezoelectricity to include the contribution of the gradient of the electronic polarization.

Not all of the new developments contribute to reduction of the gap between continuum and lattice theories if they are judged on the grounds that the extended range of the continuum theory must conform to the lattice theory in that range and that the augmented equations must accommodate observed or observable physical phenomena not accounted for by the classical continuum theory. However, the following favorable conclusions can be reached:

(1) The Cauchy-type theories are acceptable if at least the first and second gradients of strain are included. Then the additional terms in the dispersion relation for plane waves can be matched to the early terms of the lattice dispersion relation without sacrificing positive definiteness of

the potential energy density; and the phenomenon of surface energy of deformation makes its appearance for centrosymmetric as well as noncentrosymmetric crystals (the latter requiring only the strain and its first gradient).

(2) The continuum equations developed recently for diatomic crystals are the correct, long wave limits of the difference equations of the NaCl-type lattice of both mass particles and polarizable atoms and they give the correct behavior of the optical branches at long wave lengths, accommodate surface energy of deformation (and polarization) and, with the inclusion of the magnetic field, yield a dispersion relation exhibiting the long wave portions of the coupled acoustic, optical and electromagnetic branches.

(3) The equations of the Cosserat continuum are the long wave limit of the difference equations of motion of monomolecular crystals and contribute the long wave behavior of the soft optical mode. The recent extension of the Cosserat theory to include a micro-structure that is deformable, as well as rotatable, may prove to describe the higher modes.

(4) The addition of the first gradient of the electronic polarization to the energy density in classical piezoelectricity supplies the missing terms required to match the long wave limit of the difference equations of the lattice of polarizable atoms and serves to accommodate several observed phenomena otherwise not included: surface energy of deformation and polarization, anomalous capacitance of thin dielectric films, acoustical activity and, with the inclusion of the magnetic field, optical acivity.

In the following pages, typical examples of these developments and results are described within the framework of linear continuum theories (linear differential equations) and lattice theories in the harmonic approximation (linear difference equations). The point of view is from the side of continuum theories and they are treated in more general terms than are the lattice theories. For the latter, reference may be made to: "Theory of Lattice Dynamics in the Harmonic Approximation" by A. A. Maradudin, E. W. Montroll, G. H. Weiss and I. P. Ipatova (Academic Press, (1971); "Recent Developments in Lattice Theory" by W. Ludwig (Springer-Verlag, 1967); "Phonons in Perfect Lattices and in Lattices with Point Imperfections" edited by R. W H. Stevenson(Plenum Press, 1966).

2. Simple-Cubic Lattice of Mass Points and Classical Elasticity

The fact that the equations of classical elasticity are the long wave, low frequency limit of the difference equations of a lattice of mass particles, indicates that classical elasticity is limited to wave lengths and bodily dimensions large in comparison with the distance between nearest neighbor atoms (the "lattice parameter"). For the actual magnitudes of the errors, it is necessary to compare detailed solutions of analogous problems in the two types of theory. For this purpose, it is convenient to employ, for the lattice equations, those formulated by Gazis, Herman and Wallis [4] for a simple-cubic lattice with nearest and next nearest neighbor central force interactions and their novel angular interactions between three, successive, noncollinear atoms. Simple-cubic is the simplest of all lattice structures and the Gazis-Herman-Wallis equations are the simplest, valid, cubic-lattice equations that do not require a relation among the three elastic constants of cubic symmetry. Although no natural crystals with simple-cubic structure are known to exist, the equations are suitable for a study of the major effects of wave length and size.

i. Difference Equations.

In rectangular coordinates x_i, $i = 1, 2, 3$, the atoms of the simple-cubic lattice are taken to be at the points $x_1 = la$, $x_2 = ma$, $x_3 = na$, where l, m, n are positive or negative integers and a is the lattice parameter. The central force constants (i.e., force per unit relative displacement) between nearest neighbor atoms and between next nearest neighbor atoms are designated by α and β, respectively, while γ is the angular force constant between three, successive, non-collinear atoms. The rôles of the three force constants are depicted schematically by dashed lines in Fig. 1.

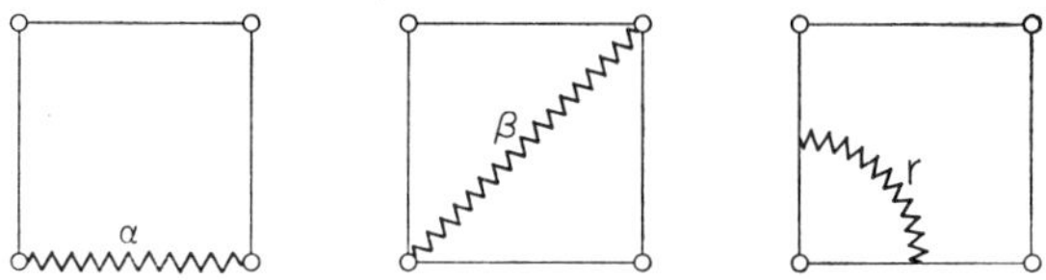

Figure 1

With $u_i^{l,m,n}$, $i = 1, 2, 3$, the rectangular components of displacement of the atom at l, m, n, Gazis, Herman and Wallis [4] find three equations of motion of the type

$$\begin{aligned} M\ddot{u}_1^{l,m,n} = {} & \alpha(u_1^{l+1,m,n}+u_1^{l-1,m,n}-2u_1^{l,m,n})+\beta(u_1^{l+1,m+1,n}+u_1^{l-1,m-1,n}+u_1^{l+1,m-1,n} \\ & +u_1^{l-1,m+1,n}+u_1^{l+1,m,n+1}+u_1^{l-1,m,n-1}+u_1^{l+1,m,n-1}+u_1^{l-1,m,n+1}-8u_1^{l,m,n}) \\ & +(\beta+\gamma)(u_2^{l+1,m+1,n}+u_2^{l-1,m-1,n}-u_2^{l+1,m-1,n}-u_2^{l-1,m+1,n}+u_3^{l+1,m,n+1} \\ & +u_3^{l-1,m,n-1}-u_3^{l+1,m,n-1}-u_3^{l-1,m,n+1})+4\gamma(u_1^{l,m+1,n}+u_1^{l,m-1,n} \\ & +u_1^{l,m,n+1}+u_1^{l,m,n-1}-4u_1^{l,m,n}), \end{aligned} \tag{2.1}$$

where M is the mass of the atom. The remaining two equations of motion are obtained by cyclical permutation of subscripts and superscripts. Born and von Kármán, in their classical paper [5], would have found the same equations if they had taken their α, β, γ, δ, κ to be the present α, 4γ, β, 0, $\beta+\gamma$, respectively.

At free boundaries $l = \pm L$, the conditions to be satisfied are

$$\begin{aligned} & \pm\alpha(u_1^{\pm(L+1),m,n}-u_1^{\pm L,m,n}) \\ & \pm\beta(u_1^{\pm(L+1),m+1,n}+u_1^{\pm(L+1),m-1,n}+u_1^{\pm(L+1),m,n+1}+u_1^{\pm(L+1),m,n-1}-4u_1^{\pm L,m,n}) \\ & +\beta(u_2^{\pm(L+1),m+1,n}-u_2^{\pm(L+1),m-1,n}+u_3^{\pm(L+1),m,n+1}-u_3^{\pm(L+1),m,n-1}) \\ & \pm 2\gamma(u_1^{\pm L,m+1,n}+u_1^{\pm L,m-1,n}+u_1^{\pm L,m,n+1}+u_1^{\pm L,m,n-1}-4u_1^{\pm L,m,n}) \\ & +\gamma(u_2^{\pm(L+1),m+1,n}-u_2^{\pm(L+1),m-1,n}-u_2^{\pm L,m+1,n}+u_2^{\pm L,m-1,n}) \\ & +\gamma(u_3^{\pm(L+1),m,n+1}-u_3^{\pm(L+1),m,n-1}-u_3^{\pm L,m,n+1}+u_3^{\pm L,m,n-1}) = 0, \\ & \pm\beta(u_2^{\pm(L+1),m+1,n}+u_2^{\pm(L+1),m-1,n}-2u_2^{\pm L,m,n}) \\ & +\beta(u_1^{\pm(L+1),m+1,n}-u_1^{\pm(L+1),m-1,n}) \\ & \pm 4\gamma(u_2^{\pm(L+1),m,n}-u_2^{\pm L,m,n}) \\ & +\gamma(u_1^{\pm(L+1),m+1,n}-u_1^{\pm(L+1),m-1,n}+u_1^{\pm L,m+1,n}-u_1^{\pm L,m-1,n}) = 0, \\ & \pm\beta(u_3^{\pm(L+1),m,n+1}+u_3^{\pm(L+1),m,n-1}-2u_3^{\pm L,m,n}) \\ & +\beta(u_1^{\pm(L+1),m,n+1}-u_1^{\pm(L+1),m,n-1}) \\ & \pm 4\gamma(u_3^{\pm(L+1),m,n}-u_3^{\pm L,m,n}) \\ & +\gamma(u_1^{\pm(L+1),m,n+1}-u_1^{\pm(L+1),m,n-1}+u_1^{\pm L,m,n+1}-u_1^{\pm L,m,n-1}) = 0. \end{aligned} \tag{2.2}$$

ii. Reduction to Equations of Classical Elasticity.

To show how (2.1) reduces to the corresponding equation in classical elasticity, it is convenient to put it in a different form [6]. Define difference operators as follows:

$$\Delta_1^+ u_i^{l,m,n} = (u_i^{l+1,m,n} - u_i^{l,m,n})/a, \; \Delta_1^- u_i^{l,m,n} = (u_i^{l,m,n} - u_i^{l-1,m,n})/a,$$
$$\Delta_1^2 u_i^{l,m,n} = \Delta_1^+ \Delta_1^- u_i^{l,m,n} = (u_i^{l+1,m,n} - 2u_i^{l,m,n} + u_i^{l-1,m,n})/a^2, \tag{2.3}$$
$$\Delta_1 \Delta_2 u_i^{l,m,n} = \tfrac{1}{4}(\Delta_1^+ \Delta_2^+ + \Delta_1^- \Delta_2^- + \Delta_1^+ \Delta_2^- + \Delta_1^- \Delta_2^+) u_i^{l,m,n}$$

and analogous definitions, e.g. $\Delta_2^+ u_i^{l,m,n} = (u_i^{l,m+1,n} - u_i^{l,m,n})/a$, for forward, backward, second central and cross differences in the remaining coordinate directions. Then, noting that

$$u_i^{l+1,m+1,n} + u_i^{l-1,m-1,n} + u_i^{l+1,m-1,n} + u_i^{l-1,m+1,n} - 4u_i^{l,m,n} = a^2(a^2 \Delta_1^2 \Delta_2^2 + 2\Delta_1^2 + 2\Delta_2^2) u_i^{l,m,n},$$
$$u_i^{l+1,m+1,n} + u_i^{l-1,m-1,n} - u_i^{l+1,m-1,n} - u_i^{l-1,m+1,n} = 4a^2 \Delta_1 \Delta_2 u_i^{l,m,n}, \tag{2.4}$$

we can write (2.1) in the form

$$\rho \ddot{u}_1^{l,m,n} = [c_{11} \Delta_1^2 + c_{44}(\Delta_2^2 + \Delta_3^2) + \tfrac{1}{2} a^2 c_{12} \Delta_1^2 (\Delta_2^2 + \Delta_3^2)] u_1^{l,m,n} + (c_{12} + c_{44})(\Delta_1 \Delta_2 u_2^{l,m,n} + \Delta_1 \Delta_3 u_3^{l,m,n}), \tag{2.5}$$

where

$$\rho = M/a^3, \; ac_{11} = \alpha + 4\beta, \; ac_{12} = 2\beta, \; ac_{44} = 2(\beta + 2\gamma). \tag{2.6}$$

Now, expand the difference operators, on the $u^{l,m,n}$, in Taylor series of partial derivatives ∂_i, $= \partial/\partial x_i$, of continuous displacement functions $u_i(x_i)$[7]:

$$\Delta_i^2 u_j^{l,m,n} = (1 + \tfrac{1}{12} a^2 \partial_i^2 + \ldots) \partial_i^2 u_j,$$
$$\Delta_i \Delta_j u_k^{l,m,n} = (1 + \tfrac{1}{6} a^2 \partial_i^2 + \tfrac{1}{6} a^2 \partial_j^2 + \ldots) \partial_i \partial_j u_k, \tag{2.7}$$
$$\Delta_i^2 \Delta_j^2 u_k^{l,m,n} = (1 + \tfrac{1}{12} a^2 \partial_i^2 + \tfrac{1}{12} a^2 \partial_j^2 + \ldots) \partial_i^2 \partial_j^2 u_k.$$

If only second derivatives are retained, (2.5) reduces to

$$\rho \ddot{u}_1 = [c_{11} \partial_1^2 + c_{44}(\partial_2^2 + \partial_3^2)] u_1 + (c_{12} + c_{44})(\partial_1 \partial_2 u_2 + \partial_1 \partial_3 u_3), \tag{2.8}$$

which is the typical equation of motion of classical elasticity for materials with the constants c_{11}, c_{12}, c_{44} of cubic symmetry.

iii. Dispersion Relations.

The designation "long wave, low frequency limit", for (2.8), stems from a comparison of dispersion relations (frequency vs. wave number) for plane waves from (2.5) and (2.8). Consider, for example, longitudinal waves in the x-direction. For (2.5), we set

$$u_2^{l,m,n} = u_3^{l,m,n} = 0, \; u_1^{l,m,n} = A\, e^{i(\xi l a - \omega t)}, \; 0 \leqslant \xi a \leqslant \pi. \tag{2.9}$$

Then (2.5) is satisfied if

$$\omega = \frac{2}{a} \sqrt{\frac{c_{11}}{\rho}} \sin \frac{\xi a}{2}. \tag{2.10}$$

For (2.8), we set

$$u_2 = u_3 = 0, \; u_1 = A\, e^{i(\xi x_1 - \omega t)} \tag{2.11}$$

and find

$$\omega = \xi \sqrt{\frac{c_{11}}{\rho}}. \tag{2.12}$$

The two dispersion relations, (2.10) and (2.12), are illustrated in Fig. 2. It may be seen that the lattice dispersion relation (2.10) approaches the continuum relation (2.12) as the wave length

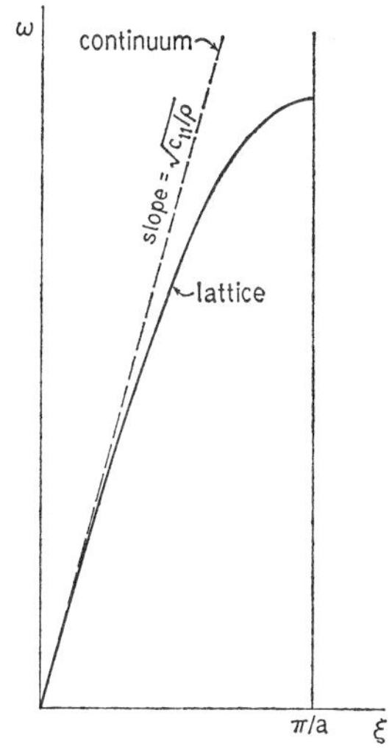

Figure 2

approaches infinity, i.e. as the wave number, ζ, approaches zero. Simultaneously, the frequency, in a monatomic Bravais lattice, also approaches zero.

For more detailed examinations of the differences between solutions of lattice and continuum equations, it is necessary to consider problems for bodies with at least one finite dimension with which to compare the lattice parameter a.

iv. Thickness-Shear Vibrations of a Plate [8].

In a plate bounded by $l = \pm L$, consider the displacements

$$u_1^{l,m,n} = u_2^{l,m,n} = 0,\ u_3^{l,m,n} = (A_1 \cos \xi la + A_2 \sin \xi la)\, e^{i\omega t},\ 0 \leqslant \xi a \leqslant \pi. \tag{2.13}$$

With these displacements, the first two equations of the type (2.5) are satisfied identically and the third is satisfied if

$$\rho a^2 \omega^2 = 4c_{44} \sin^2 \tfrac{1}{2}\xi a. \tag{2.14}$$

Upon substituting the displacements (2.13) in the boundary conditions (2.2), we find that the first two conditions are satisfied identically and the third is satisfied if

$$\xi a = p\pi/(2L+1),\ p = 0, 1, 2, \ldots < 2L+1, \tag{2.15}$$

where even and odd p apply to symmetric and antisymmetric modes, respectively. From (2.14) and (2.15), the frequencies are

$$\omega = \frac{2}{a}\left(\frac{c_{44}}{\rho}\right)^{\frac{1}{2}} \sin \frac{p\pi}{2(2L+1)},\ p = 0, 1, 2, \ldots < 2L+1 \tag{2.16}$$

and the displacements, from (2.13) and (2.15), are

$$u_3^{l,m,n} = \left(A_1 \cos \frac{p\pi l}{2L+1} + A_2 \sin \frac{p\pi l}{2L+1}\right) e^{i\omega t}. \tag{2.17}$$

To compare the lattice solution with the continuum one, we consider a continuum plate of thickness $2h$, where

$$2h = (2L+1)a, \tag{2.18}$$

i.e. each layer of atoms is replaced by a thickness a of the continuum. The frequencies of the thickness-shear modes of such a plate are given by

$$\omega = \frac{p\pi}{2h}\left(\frac{c_{44}}{\rho}\right)^{\frac{1}{2}}, \quad p = 0, 1, 2 \ldots \tag{2.19}$$

and the displacements are

$$u_3 = \left(A_1 \cos\frac{p\pi x_1}{2h} + A_2 \sin\frac{p\pi x_1}{2h}\right) e^{i\omega t}. \tag{2.20}$$

In the long wave, low frequency approximation,

$$la \to x_1, \; p \ll 2L+1, \tag{2.21}$$

in which case (2.16) and (2.17) approach (2.19) and (2.20), respectively.

If the frequency of the lowest, antisymmetric, thickness-shear mode in the continuum plate, i.e. $p = 1$ in (2.19), is used as a reference frequency ω_1:

$$\omega_1 = \frac{\pi}{2h}\left(\frac{c_{44}}{\rho}\right)^{\frac{1}{2}} = \frac{\pi}{(2L+1)a}\left(\frac{c_{44}}{\rho}\right)^{\frac{1}{2}}, \tag{2.22}$$

the normalized frequencies for the lattice are

$$\Omega = \frac{\omega}{\omega_1} = \frac{2(2L+1)}{\pi}\sin\frac{p\pi}{2(2L+1)}, \quad p = 0, 1, 2, \ldots < 2L+1, \tag{2.23}$$

and these are to be compared with the normalized frequencies, $p = 0, 1, 2\ldots$ for the continuum.

The discrepancy between the frequencies p and Ω and the corresponding departure of the lattice mode shape from the sinusoidal form characteristic of the continuum, as the order of the mode increases, are illustrated in Fig. 3 for the case of fifteen layers. It may be seen that, for the first few modes, where the lattice mode shapes are nearly sinusoidal, the normalized frequencies p and Ω are nearly the same. Inspection of the case $p = 3$ shows that the half wave length can be as small as five lattice parameters with a good representation of the mode shape by a sinusoid and an error in frequency of less than 2 %. However, as the mode shape departs from the sinusoidal for increasing orders, the frequencies p and Ω separate – by almost 50 % for the highest mode, in the case of fifteen layers. As the number of layers increases, the percent discrepancy between the frequencies of the highest lattice mode and the corresponding continuum mode approaches $50\pi - 100 = 57$ %.

It should be noted that, whereas the number of modes in the continuum plate is unlimited, the number of modes in the lattice plate is equal to the number of layers. This is becase a wave described by N points can have no more than $N-1$ nodes.

v. Face-shear and Thickness-Twist waves in a plate [8].

Waves, in a plate, with displacement and wave normal at right angles to each other and parallel to the faces of the plate, are called thickness-twist waves except for the wave of zero order (in which the displacement does not vary through the thickness of the plate) which is called a face-shear wave. In the continuum, these waves are special cases of Love's [9] transverse waves in a superficial layer.

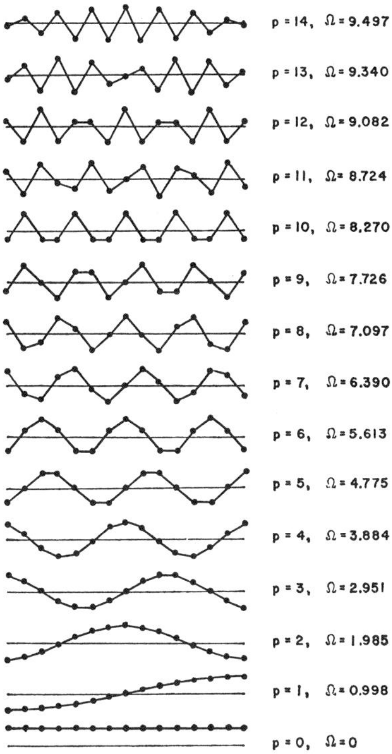

Figure 3

In the lattice plate, the waves are represented by displacements of the form

$$u_1^{l,m,n} = u_2^{l,m,n} = 0,\ u_3^{l,m,n} = (A_1 \cos \xi la + A_2 \sin \xi la)\, e^{i(\eta ma - \omega t)}, \tag{2.24}$$

where $0 \leqslant \xi a \leqslant \pi$, $0 \leqslant \eta a \leqslant \pi$. This time, the equations of motion of the type (2.5) require

$$\rho a^2\omega^2 = 4c_{44}(\sin^2 \tfrac{1}{2}\xi a + \sin^2 \tfrac{1}{2}\eta a) \tag{2.25}$$

and the boundary conditions (2.2) again require (2.15). Thus, the normalized frequencies are

$$\Omega = \frac{\omega}{\omega_1} = \frac{2(2L+1)}{\pi}\left(\sin^2 \frac{p\pi}{2(2L+1)} + \sin^2 \tfrac{1}{2}\eta a\right)^{\frac{1}{2}},\ p = 0, 1, 2 \ldots < 2L+1, \tag{2.26}$$

and the displacements are

$$u_3^{l,m,n} = \left(A_1 \cos \frac{p\pi l}{2L+1} + A_2 \sin \frac{p\pi l}{2L+1}\right) e^{i(\eta ma - \omega t)}. \tag{2.27}$$

In the long wave, low frequency limit ($la \to x_1$, $ma \to x_2$, $p \ll 2L+1$) these become the known results for the special case of Love waves [9]:

$$\Omega = [p^2 + (2\eta h/\pi)]^{\frac{1}{2}},\ p = 0, 1, 2 \ldots \tag{2.28}$$

$$u_3 = [A_1 \cos (p\pi x_1/2h) + A_2 \sin (p\pi x_1/2h)]\, e^{i(\eta x_2 - \omega t)}. \tag{2.29}$$

The real branches of the dispersion relation (2.26) are illustrated in Fig. 4 for a plate fifteen layers thick ($L = 7$). At infinite wave length ($\eta = 0$) along the plate, the modes reduce to the

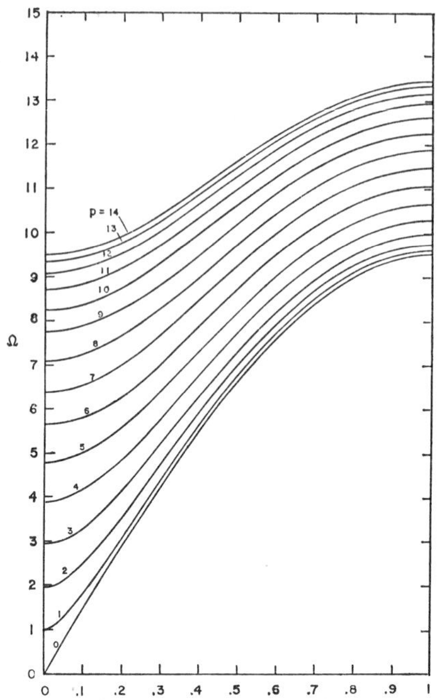

Figure 4

thickness-shear modes illustrated in Fig. 3. These variations of displacement across the thickness of the plate are maintained for all wavelengths, $2\pi/\eta$, along the plate. The major differences between the dispersion relations for the lattice and the continuum are: each of the finite number, $2L+1$, of branches of the lattice dispersion relation (2.26) has a high frequency cut-off, in addition to a low frequency cut-off (2.23), and all have the same upper cut-off of wave number $\eta = \pi/a$, i.e. wave length equal to $2a$; whereas all the infinity of real branches of the continuum dispersion relation (2.28) are hyperbolic curves extending from low frequency cut-offs $\Omega = 1, 2, \ldots, \infty$, at $\eta = 0$, to infinite frequencies and wave numbers – asymptotic to

$$\Omega = (2L+1)\eta a/\pi, \tag{2.30}$$

which is the straight line starting from the lower left corner and passing through the upper right corner of Fig. 4 and is, in fact, the face-shear branch of the continuum dispersion relation. Thus, the dispersion relation for the continuum is a good approximation to that of the lattice only in the lower left region of Fig. 4; i.e. for long wave lengths, in comparison with a, both along the plate and across its thickness. Two or three times a is a sufficient for accuracy within 2 %.

As in the case of the continuum, the lattice dispersion relation has branches for real frequency and imaginary wave number which are not shown in Fig. 4.

The problem of face-shear and thickness-shear waves in a plate has also been solved for the face-centered cubic lattice (Brady [10]) and for the body centered cubic lattice (Gong [11]).

vi. Torsional Equilibrium of a Rectangular Bar.

The bar is bounded by free faces at $l = \pm L$ and $m = \pm M$ and is in equilibrium under a twist about the axis of x_3 with angle of twist τ per unit length.

By analogy with the St. Venant solution of the equations of classical elasticity for the analogous problem [12], it is assumed that

$$\begin{aligned} u_1^{l,m,n} &= -\tau a^2 mn, \\ u_2^{l,m,n} &= \tau a^2 ln, \\ u_3^{l,m,n} &= \tau a^2(lm + A \sin l\theta \sinh m\varphi), \quad 0 < \theta < \pi. \end{aligned} \tag{2.31}$$

With these displacements, the first two equations of the type (2.5), with the left hand side zero, are satisfied identically and the third equation is satisfied if

$$\cosh \varphi = 2 - \cos \theta. \tag{2.32}$$

The conditions of the type (2.2), for free boundaries, reduce, in view of (2.31), to

$$\pm 2(u_3^{\pm(L+1),m,n} - u_3^{\pm L,m,n}) + u_1^{l,m,n+1} - u_1^{l,m,n-1} = 0 \text{ on } l = \pm L, \tag{2.33}$$

$$\pm 2(u_3^{l,\pm(M+1),n} - u_3^{l,\pm M,n}) + u_2^{l,m,n+1} - u_2^{l,m,n-1} = 0 \text{ on } m = \pm M. \tag{2.34}$$

Upon substituting (2.31) in (2.33), we find

$$\theta = \theta_p = (2p-1)\pi/(2L+1), \; p = 1, 2, \ldots, L. \tag{2.35}$$

Hence, the third of (2.31) may be written as a finite series:

$$u_3^{l,m,n} = \tau a^2(lm + \sum_{p=1}^{p=L} A_p \sin l\theta_p \sinh m\varphi_p). \tag{2.36}$$

Substitution of (2.36) in the boundary conditions (2.34) yields

$$\sum_{p=1}^{p=L} A_p \sin l\theta_p \sinh \tfrac{1}{2}\varphi_p \cosh (M+\tfrac{1}{2})\varphi_p = -l, \quad l = 1, 2, \ldots, L, \tag{2.37}$$

i.e. a set of L simultaneous, linear, algebraic equations on the coefficients A_p. The system of equations can be solved explicitly for the A_p by a method analogous to that for determining Fourier coefficients. Multiply both sides of (2.37) by $\sin l\theta_q$ and sum over l from $l = 1$ to $l = L$. Now [13],

$$\sum_{l=1}^{l=L} l \sin l\theta_p = \frac{\sin L\theta_p}{4 \sin^2 \frac{1}{2}\theta_p} - \frac{L \cos (L+\frac{1}{2})\theta_p}{2 \sin \frac{1}{2}\theta_p}. \tag{2.38}$$

Also, employing (2.35), we find

$$\sum_{l=1}^{l=L} \sin l\theta_p \sin l\theta_q = \begin{cases} 0, & q \neq p \\ \frac{1}{4}(2L+1), & q = p. \end{cases} \tag{2.39}$$

Hence

$$A_p = \frac{2L \sin \frac{1}{2}\theta_p \cos (L+\frac{1}{2})\theta_p - \sin L\theta_p}{(2L+1) \sinh^3 \frac{1}{2}\varphi_p \cosh (M+\frac{1}{2})\varphi_p}. \tag{2.40}$$

Substitution of (2.40) in (2.36) completes the solution for the warping function $u_3^{l,m,n}$. The results are shown in Fig. 5 in which the ratios of integers at the lattice points give $u_3^{l,m,n}/\tau a^2$ at those points. Fig. 5 is to be compared with Fig. 6 which depicts typical contours of the warping functions for square and long rectangular sections from the St. Venant solution. It may be seen that, with one side of the cross section restricted to three atoms ($L = 1$), the warping bears little resemblance to that found by St. Venant; namely, there is no warping of the square sec-

0	0	0
0	0	0
0	0	0

L = 1, M = 1

$\frac{4}{5}$	$\frac{3}{5}$	0	$-\frac{3}{5}$	$-\frac{4}{5}$
0	0	0	0	0
$-\frac{4}{5}$	$-\frac{3}{5}$	0	$\frac{3}{5}$	$\frac{4}{5}$

L = 1, M = 2

$\frac{23}{13}$	$\frac{20}{13}$	$\frac{11}{13}$	0	$-\frac{11}{13}$	$-\frac{20}{13}$	$-\frac{23}{13}$
0	0	0	0	0	0	0
$-\frac{23}{13}$	$-\frac{20}{13}$	$-\frac{11}{13}$	0	$\frac{11}{13}$	$\frac{20}{13}$	$\frac{23}{13}$

L = 1, M = 3

$\frac{47}{17}$	$\frac{43}{17}$	$\frac{31}{17}$	$\frac{16}{17}$	0	$-\frac{16}{17}$	$-\frac{31}{17}$	$-\frac{43}{17}$	$-\frac{47}{17}$
0	0	0	0	0	0	0	0	0
$-\frac{47}{17}$	$-\frac{43}{17}$	$-\frac{31}{17}$	$-\frac{16}{17}$	0	$\frac{16}{17}$	$\frac{31}{17}$	$\frac{43}{17}$	$\frac{47}{17}$

L = 1, M = 4

Figure 5a

0	$\frac{1}{3}$	0	$-\frac{1}{3}$	0
$-\frac{1}{3}$	0	0	0	$\frac{1}{3}$
0	0	0	0	0
$\frac{1}{3}$	0	0	0	$-\frac{1}{3}$
0	$-\frac{1}{3}$	0	$\frac{1}{3}$	0

L = 2, M = 2

$\frac{448}{281}$	$\frac{498}{281}$	$\frac{300}{281}$	0	$-\frac{300}{281}$	$-\frac{498}{281}$	$-\frac{448}{281}$
$\frac{117}{281}$	$\frac{184}{281}$	$\frac{121}{281}$	0	$-\frac{121}{281}$	$-\frac{184}{281}$	$-\frac{117}{281}$
0	0	0	0	0	0	0
$-\frac{117}{281}$	$-\frac{184}{281}$	$-\frac{121}{281}$	0	$\frac{121}{281}$	$\frac{184}{281}$	$\frac{117}{281}$
$-\frac{448}{281}$	$-\frac{498}{281}$	$-\frac{300}{281}$	0	$\frac{300}{281}$	$\frac{498}{281}$	$\frac{448}{281}$

L = 2, M = 3

$\frac{7788}{2245}$	$\frac{8080}{2245}$	$\frac{6242}{2245}$	$\frac{3342}{2245}$	0	$-\frac{3342}{2245}$	$-\frac{6242}{2245}$	$-\frac{8080}{2245}$	$-\frac{7788}{2245}$
$\frac{3006}{2245}$	$\frac{3475}{2245}$	$\frac{2814}{2245}$	$\frac{1539}{2245}$	0	$-\frac{1539}{2245}$	$-\frac{2814}{2245}$	$-\frac{3475}{2245}$	$-\frac{3006}{2245}$
0	0	0	0	0	0	0	0	0
$-\frac{3006}{2245}$	$-\frac{3475}{2245}$	$-\frac{2814}{2245}$	$-\frac{1539}{2245}$	0	$\frac{1539}{2245}$	$\frac{2814}{2245}$	$\frac{3475}{2245}$	$\frac{3006}{2245}$
$-\frac{7788}{2245}$	$-\frac{8080}{2245}$	$-\frac{6242}{2245}$	$-\frac{3342}{2245}$	0	$\frac{3342}{2245}$	$\frac{6242}{2245}$	$\frac{8080}{2245}$	$\frac{7788}{2245}$

L = 2, M = 4

Figure 5b

tion ($L = 1$, $M = 1$) and, in each succeeding section ($L = 1$, $M = 2, 3, 4$), the displacements at a long side ($l = \pm 1, m = 0, \pm 1, \pm 2, \ldots$) do not have a maximum and a minimum between the center and the ends, as they have in the continuum solution. However, if the smaller dimension is increased to five atoms ($L = 2$, $M = 2, 3, \ldots$) the dissimilarity disappears. For the square ($L = 2$, $M = 2$), it may be seen that the cross section is divided into eight sectors, with alternating signs of displacements, instead of the usual four sectors for long rectangular sections. This is precisely the result found by St. Venant for the continuum, as illustrated in

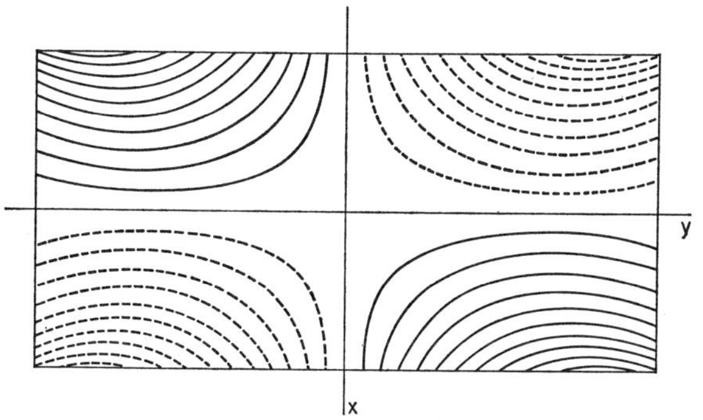

Figure 6a

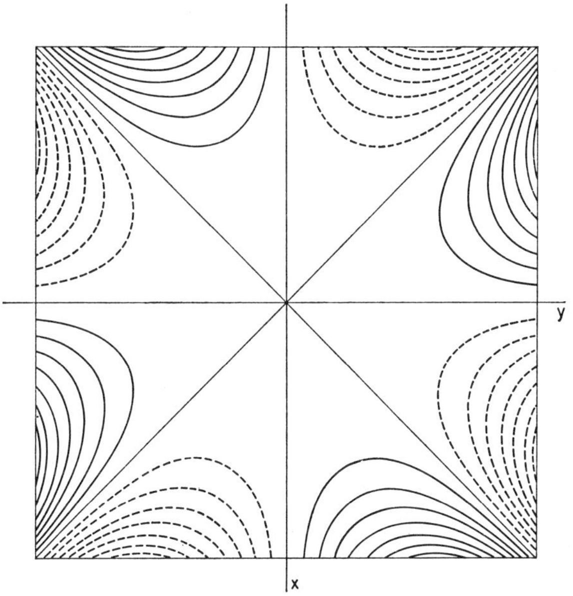

Figure 6b

Fig. 6. For the rectangular sections $L = 2$, $M = 3$ and $L = 2$, $M = 4$, the displacements along a long side reach a maximum and a minimum, near the ends, while, along a short side, the displacements vary monotonically. Again, these results are the same as the corresponding ones in St. Venant's solution, as illustrated in Fig. 6. Thus, the cross section need have only as many as five atoms along the shorter side for the displacements to have the major qualitative properties found in the continuum solution.

3. Strain Gradient Theories

Extensions of classical elasticity to account for crystal structure began in 1851 with Cauchy's [1] infinite series representation of an isotropic material with a periodic structure. Although

Cauchy did not carry his idea very far, it is now known that his theory corresponds to an infinite series expansion of difference operators, in a lattice theory, in terms of differential operators; or, equivalently, augmentation of classical elasticity through the inclusion of all the gradients of strain in the potential energy density. Little attention was paid to the work for many years – possibly owing to the complexity introduced by Cauchy's consideration of non-centrosymmetric isotropy; possibly owing to an error of omission in the constitutive equation for the symmetric part of the stress; possibly owing to Todhunter's assessment: "This consists merely of generalities, and is apparently of no importance" [12, Vol. 1, p. 374].

i. Rotation Gradient.

Interest in the gradient type of theory was stimulated, in 1960, by Aero and Kuvshinskii [14], Grioli [15], Rajagopal [16] and Truesdell and Toupin [17] who took into account the first gradient of the rotation:

$$\kappa_{ij} = \tfrac{1}{2}\delta_{jkl}\partial_i\partial_k u_l, \tag{3.1}$$

where δ_{ijk} is the unit alternating tensor:

$$\delta_{ijk} = \begin{cases} +1, & ijk = 123,\ 231,\ 312, \\ -1, & ijk = 321,\ 213,\ 132, \\ 0, & \text{otherwise.} \end{cases} \tag{3.2}$$

A connection with crystal lattice theory was noted by Krumhansl [18], in 1963, who showed, by expansion of the differences in the expression for the potential energy of a general Bravais lattice, that the second order terms, in the resulting series of derivatives, contain the rotation gradient.

The novel features accompanying the rotation gradient are the appearances of an antisymmetric part of the stress (indeterminate), a couple stress and a material constant with the dimension of length; but, except for a tenuous connection with molecular crystals (see Article 5), no relation to crystal lattice theory appears to have been established beyond Krumhansl's brief remark.

ii First Strain Gradient.

The rotation gradient has eight independent components which are eight of the eighteen components of the first gradient of the strain. The additional ten components are

$$\tfrac{1}{3}(\partial_i\partial_j u_k + \partial_j\partial_k u_i + \partial_k\partial_i u_j). \tag{3.3}$$

Alternatively, in terms of the strain,

$$S_{ij} = \tfrac{1}{2}(\partial_i u_j + \partial_j u_i), \tag{3.4}$$

all eighteen components may be expressed as $\partial_i S_{jk}$; but perhaps the simplest form is the second gradient of the displacement:

$$S_{ijk} = \partial_i\partial_j u_k. \tag{3.5}$$

The fundamental equations for the elastic material in which account is taken of the full first gradient of the strain were given by Toupin [19] in 1962. In the linear case, the potential energy density may be expressed in the form

$$W = c_{ijk}S_{ijk} + \tfrac{1}{2}c_{ijkl}S_{ij}S_{kl} + \tfrac{1}{2}c_{ijklmn}S_{ijk}S_{lmn} + c_{ijklm}S_{ij}S_{klm}, \tag{3.6}$$

where the $c_{ij}\ldots$ are material constants. The most interesting feature of (3.6) is the linear term $c_{ijk}S_{ijk}$ which corresponds to a self-equilibrating initial stress and gives rise to a surface energy of deformation: an observed physical phenomenon not contained in classical elasticity.

The surface energy of deformation is a part of the energy associated with fracture along a surface. Part of the separation energy is the bond energy — that required to break the atomic bonds across the surface while the relative positions of the atoms in each of the two resulting portions are held fixed — say, by fictitious forces. The release of these forces is accompanied by a deformation, localized near the surfaces, and an associated (negative) energy: the surface energy of deformation — estimates of the magnitude of which range as high as 40 % of the bond energy [20]. The extreme localization of deformation at the surface was first observed in 1961 by Germer, MacRae and Hartman [21] by means of low energy electron diffraction measurements at nickel surfaces. They found that the displacement of the superficial layer of atoms toward the interior was five times as large as that of the next layer. Toupin and Gazis [22], in 1963, found mathematical solutions for a similar surface deformation within the framework of Toupin's strain gradient equations and also exhibited the correspondence with a one-dimensional, monatomic lattice with nearest and next nearest neighbor interactions.

The first strain gradient theory, however, suffers from two deficiencies. First of all, it contains a surface energy of deformation only for non-centrosymmetric materials. This is evidenced by the third rank tensor coefficient of the linear term in (3.6); whereas the elasticity of centrosymmetric materials can be characterized only be tensors of even rank. Secondly, the dispersion relation for plane waves can match that for a lattice beyond the linear range only if the requirement of a positive strain energy density is abandoned — as shown in the following section. Both of these defects disappear if the second gradient of the strain is included.

iii. Second Strain Gradient [23].

If the strain and its first two gradients are taken into account, the potential energy density for a centrosymmetric material has the form

$$W = \tfrac{1}{2}c_{ijkl}S_{ij}S_{kl}+\tfrac{1}{2}\alpha_{ijklmn}S_{ijk}S_{lmn}+\tfrac{1}{2}\beta_{ijklmnpq}S_{ijkl}S_{mnpq}+\gamma_{ijklmn}S_{ij}S_{klmn}+\beta^{0}_{ijkl}S_{ijkl}, \tag{3.7}$$

where, in addition to (3.4) and (3.5),

$$S_{ijkl} = \partial_i\partial_j\partial_k u_l. \tag{3.8}$$

This energy and the usual kinetic energy density $\tfrac{1}{2}\rho\dot{u}_i\dot{u}_i$ lead to the stress equations of motion

$$\partial_i T_{il}-\partial_i\partial_j T_{ijl}+\partial_i\partial_j\partial_k T_{ijkl} = \rho\ddot{u}_l, \tag{3.9}$$

where the stresses T_{ij}, T_{ijk} and T_{ijkl} are given by

$$\begin{aligned} T_{ij} &= \partial W/\partial S_{ij} = c_{ijkl}S_{kl}+\gamma_{ijklmn}S_{klmn}, \\ T_{ijk} &= \partial W/\partial S_{ijk} = \alpha_{ijklmn}S_{lmn}, \\ T_{ijkl} &= \partial W/\partial S_{ijkl} = \gamma_{ijklmn}S_{mn}+\beta_{ijklmnpq}S_{mnpq}+\beta^{0}_{ijkl}, \end{aligned} \tag{3.10}$$

and to three, vector boundary tractions:

$$\begin{aligned} t_l^{(1)} &= n_i(T_{il}-\partial_j T_{ijl}+\partial_j\partial_k T_{ijkl})+L_i[n_j(T_{ijl}-\partial_k T_{ijkl})]+L_iL_j(n_kT_{ijkl}) \\ &\quad -L_i[(\delta_{ip}-n_in_p)(\partial_p n_q)n_jn_kT_{qjkl}], \end{aligned} \tag{3.11}$$

$$t_l^{(2)} = n_in_j(T_{ijl}-\partial_k T_{ijkl})+n_iL_j(n_kT_{ijkl})+L_k(n_in_jT_{ijkl}),$$

$$t_l^{(3)} = n_in_jn_kT_{ijkl},$$

where

$$L_i = n_i(\delta_{jk}-n_j n_k)\delta_k n_j-(\partial_{ij}-n_i n_j)\partial_j \tag{3.12}$$

and n_i is the unit outward normal to the boundary.

It will be observed that the strain energy density (3.7) has no term linear in S_{ij}. This is because, just as in classical elasticity, the material configuration, to which W is referred, can be chosen so that such a term does not appear. Also, there is no term in (3.7) linear in S_{ijk} as its material coefficient would be a tensor of odd rank, which cannot exist for centrosymmetric materials. This leaves the only linear term to be the one in S_{ijkl}, which can be shown to produce a surface energy of deformation, as follows.

From an unstrained body under no external forces, remove a portion, of volume V, bounded by a surface S. The energy in such a portion, in equilibrium under no external forces, is

$$\mathscr{W} = \int_V W\, dV, \tag{3.13}$$

or, from (3.7) and (3.10),

$$\mathscr{W} = \tfrac{1}{2}\int_V (T_{ij} S_{ij}+T_{ijk} S_{ijk}+T_{ijkl} S_{ijkl})\, dV+\tfrac{1}{2}\int_V \beta^0_{ijkl} S_{ijkl}\, dV. \tag{3.14}$$

The second integral in (3.14) enters because, as may be seen in (3.10), it corresponds to a constant term β^0_{ijkl} in T_{ijkl} whereas the remaining parts of T_{ij}, T_{ijk} and T_{ijkl} vary in proportion to the strain and its gradients. The first integral in (3.14) can be transformed to

$$\tfrac{1}{2}\int_S [t_i^{(1)}u_i+t_i^{(2)}(n_j\partial_j)u_i+t_i^{(3)}(n_j\partial_j)^2 u_i]\, dS, \tag{3.15}$$

which vanishes since the surface S is free of traction. This leaves

$$\mathscr{W} = \tfrac{1}{2}\int_V \beta^0_{ijkl} S_{ijkl}\, dV = \tfrac{1}{2}\int_S n_i \beta^0_{ijkl}\partial_j\partial_k u_l\, dS \tag{3.16}$$

as an energy that remains in V although the body is under no external forces. The energy per unit area of the surface is, from (3.16),

$$W^S = \tfrac{1}{2}[n_i \beta^0_{ijkl}\partial_j\partial_k u_l]_S. \tag{3.17}$$

To show that this energy is localized near the surface, it is sufficient to consider the case of the half-space $x_1 \geqq 0$ with $x_1 = 0$ free of traction. Assume

$$u_1 = u(x_1), \quad u_2 = u_3 = 0. \tag{3.18}$$

Then the stress equations of equilibrium, from (3.9), reduce to

$$\partial_1 T_{11}-\partial_1^2 T_{111}+\partial_1^3 T_{1111} = 0, \tag{3.19}$$

while the boundary conditions, from (3.11), become

$$T_{11}-\partial_1 T_{111}+\partial_1^2 T_{1111} = 0, \quad T_{111}-\partial_1 T_{1111} = 0, \quad T_{1111} = 0 \text{ on } x_1 = 0. \tag{3.20}$$

Also, with the assumed form of displacements (3.18), the constitutive equations (3.10) take the form

$$T_{11} = c\partial_1 u_1+\gamma\partial_1^3 u_1, \quad T_{111} = \alpha\partial_1^2 u_1, \quad T_{1111} = \beta_0+\gamma\partial_1 u_1+\beta\partial_1^3 u_1 \tag{3.21}$$

and the surface energy per unit area (3.17) becomes, after noting that $n_1 = -1$ and $\beta^0_{ijkl} \to \beta_0$,

$$W^S = -\tfrac{1}{2}[\beta_0 \partial_1^2 u_1]_{x_1=0}. \tag{3.22}$$

Substitution of (3.21) in (3.19) gives the displacement equation of equilibrium

$$[c-(\alpha-2\gamma)\partial_1^2+\beta\partial_1^4]\partial_1^2 u_1 = 0, \tag{3.23}$$

or

$$(1-l_1^2\partial_1^2)(1-l_2^2\partial_1^2)\partial_1^2 u_1 = 0, \tag{3.24}$$

where

$$2cl_i^2 = \alpha-2\gamma\pm[(\alpha-2\gamma)^2-4\beta c]^{\frac{1}{2}},\ i = 1, 2. \tag{3.25}$$

Regarding boundary conditions, it may be seen that the first of (3.20) is automatically satisfied if the stress equation of equilibrium (3.19) is satisfied. The remaining two boundary conditions, from (3.20) and (3.21), are

$$[(\alpha-\gamma)\partial_1^2 u_1-\beta\partial_1^4 u_1]_{x_1=0} = 0, \quad (\gamma\partial_1 u_1+\beta\partial_1^3 u_1)_{x_1=0} = -\beta_0. \tag{3.26}$$

The solution of (3.24), vanishing at infinity, is

$$u_1 = A_1 e^{-x_1/l_1}+A_2 e^{-x_1/l_2}. \tag{3.27}$$

The values of A_1 and A_2 may be determined from the boundary conditions (3.26) and then the surface energy of deformation is, from (3.22),

$$W^S = -\tfrac{1}{2}\beta_0(A_1 l_1^{-2}+A_2 l_2^{-2}). \tag{3.28}$$

The associated strain diminishes exponentially into the interior with decay rate controlled by the magnitudes of l_1 and l_2. Thè appearance of material constants l_i, with the dimension of length, is typical of the gradient type theories and yields a length scale characteristic of the structure of the material. Such a scale is absent in classical elasticity. An estimate of the order of magnitude of the length scale may be obtained from the electron diffraction measurements at nickel surfaces by Germer, MacRae and Hartman [21]. Their finding that the displacement of the superficial layer of atoms is five times as large as that of the next layer would lead, on the on the assumption of a simple exponential decay $e^{-x/l}$, to an l of about five eights of the distance between adjacent layers of atoms. Thus, the deformation is significant only at the first few layers.

Upon departure into the interior, the displacement need not maintain the same sign. The conditions for positive strain energy density do not include relations between α and β or between α and γ. Hence, from (3.25), the l_i may be complex, corresponding to an oscillating decay of displacement into the interior.

To obtain the corresponding solution for a lattice, consider a single row of particles distributed along the x_1-axis, at unit distance apart, with interactions as far as third neighbors and including self-equilibrating initial forces. Now, separate the lattice between the particles at $n = 0$ and $n = -1$ and consider the semi-infinite lattice $n \geqq 0$. The separation is effected by the addition of forces P_0, P_1, P_2 on particles at 0, 1, 2, respectively, equal and opposite to the resultants of the initial forces exerted on those particles by the particles at -1, -2, -3. Since the initial forces are self-equilibrating,

$$P\ +P_1+P_2 = 0. \tag{3.29}$$

With force constants α_1, α_2, α_3 for the interactions between first, second and third neighbors, respectively, and with u_n the displacement of the nth particle, the equilibrium of that particle is expressed by

$$\sum_{i=1}^{i=3} \alpha_i(u_{n+i}-2u_n+u_{n-i}) = 0, \quad n \geqslant 3 \tag{3.30}$$

$$\begin{aligned} &\sum_{i=1}^{i=2} \alpha_i(u_{2+i}-2u_2+u_{2-i})+\alpha_3(u_5-u_2) = P_2, \quad n = 2, \\ &\alpha_1(u_2-2u_1+u_0)+\alpha_2(u_3-u_1)+\alpha_3(u_4-u_1) = P_1, \quad n = 1, \\ &\alpha_1(u_1-u_0)+\alpha_2(u_2-u_0)+\alpha_3(u_3-u_0) = P_0, \quad n = 0. \end{aligned} \tag{3.31}$$

We adopt the notations $\Delta u_n = u_{n+1}-u_n$,

$$\Delta^2 u_n = u_{n+1}-2u_n+u_{n-1}, \; \Delta^4 u_n = u_{n+2}-4u_{n+1}+6u_n-4u_{n-1}+u_{n-2}. \tag{3.32}$$

Then, with

$$\beta_1 = \alpha_1+4\alpha_2+9\alpha_3, \; \beta_2 = -\alpha_2-6\alpha_3, \; \beta_3 = \alpha_3, \tag{3.33}$$

the general equation of equilibrium (3.30) becomes

$$(\beta_1-\beta_2\Delta^2+\beta_3\Delta^4)\Delta^2 u_n = 0, \tag{3.34}$$

or

$$(1-\lambda_1^2\Delta^2)(1-\lambda_2^2\Delta^2)\Delta^2 u_n = 0, \tag{3.35}$$

where

$$2\beta_1\lambda_i^2 = \beta_2 \pm (\beta_2^2-4\beta_1\beta_3)^{\frac{1}{2}}, \; i = 1, 2. \tag{3.36}$$

The solution of (3.35), vanishing at infinity, is

$$u_n = A_1 e^{-n\theta_1}+A_2 e^{-n\theta_2} \tag{3.37}$$

provided that

$$\theta_i = \operatorname{arccosh}(1+\tfrac{1}{2}\lambda_i^{-2}), \quad i = 1, 2. \tag{3.38}$$

As for the boundary conditions, there are three for the two constants A_1 and A_2, just as in the case of the continuum solution. Following Gazis and Wallis [24]; we sum the three equations (3.31) to obtain

$$(1-\lambda_1^2\Delta^2)(1-\lambda_2^2\Delta^2)\Delta u_2 = 0, \tag{3.39}$$

which is satisfied identically by the solution (3.37) with (3.38). Thus, as in the case of the continuum, the general condition of equilibrium disposes of one of the three boundary conditions. The remaining two conditions serve to determine A_1 and A_2.

The similarity in form between the differential equations, boundary conditions and solutions for the continuum and the lattice is to be remarked and the correspondence of the continuum expressions to the long wave wave behavior in the case of the lattice may be verified by noting that

$$\begin{aligned} \Delta^2 u_n &= \partial_1^2 u_1+\tfrac{1}{12}\partial_1^4 u_1+\tfrac{1}{360}\partial_1^6 u_1+\dots \\ \Delta^4 u_n &= \partial_1^4 u_1+\tfrac{1}{6}\partial_1^6 u_1+\dots \\ \Delta^6 u_n &= \partial_1^6 u_1+\dots \end{aligned} \tag{3.40}$$

and, in (3.38),

$$\theta_i \to 1/\lambda_i. \tag{3.41}$$

It is also necessary to compare the long wave behavior of the dispersion relations for the continuum and the lattice. The one-dimensional equations of motion, for the two cases, are: for the continuum, from (3.23),

$$[c-(\alpha-2\gamma)\partial_1^2+\beta\partial_1^4]\partial_1^2 u_1 = \rho\ddot{u}_1 \tag{3.42}$$

and, for the lattice, from (3.34),

$$(\beta_1-\beta_2\Delta^2+\beta_3\Delta^4)\Delta^2 u_n = M\ddot{u}_n. \tag{3.43}$$

Inserting the wave forms

$$u_1 = A\,e^{i(\xi x_1-\omega t)}, \quad u_n = A\,e^{i(n\theta-\omega t)}, \tag{3.44}$$

respectively, we find, to the fourth power of the wave number,

$$\rho\omega^2 = c\xi^2+(\alpha-2\gamma)\xi^4+\ldots, \tag{3.45}$$

$$M\omega^2 = \beta_1\theta^2-(\tfrac{1}{2}\beta_1+\beta_2)\theta^4+\ldots, \tag{3.46}$$

If only the strain and its first gradient were considered, (3.45) would be

$$\rho\omega^2 = c\xi^2+\alpha\xi^4$$

and α is required to be positive by the condition of positive definitiness of the energy density. Hence, in the first gradient theory, the group velocity, $d\omega/d\xi$, must increase as the wave number increases from zero; but, in the lattice the opposite is generally true, as illustrated in Fig. 7. In the second gradient theory, α is replaced by $\alpha-2\gamma$ as the coefficient of the second term in the dispersion relation; and the conditions of positive definiteness place no restrictions on the relation between α and γ, so that the group velocities may either increase or decrease with increasing wave number – and, accordingly, can match the dispersion relation of the lattice beyond the linear portion.

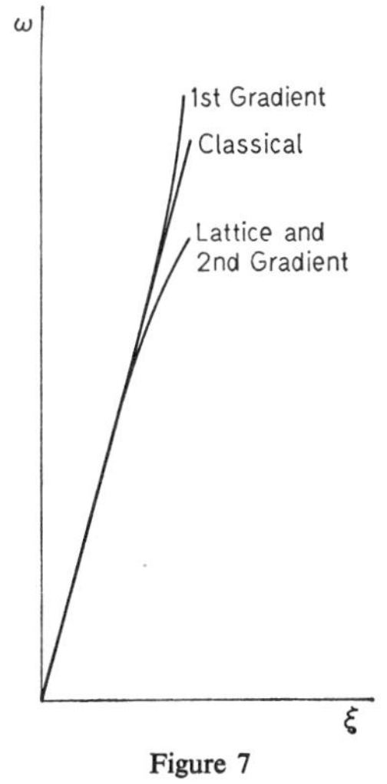

Figure 7

iv. Higher gradients and exact representations.

Shorter and shorter wave lengths can be accommodated as higher and higher gradients of strain are introduced into the potential energy density. From (3.7), (3.9) and (3.10), it can be seen that, if the potential energy density is expressed as a quadratic function of strain gradients to an arbitrary order:

$$W = W(S_{ij}, \partial_k S_{ij}, \partial_k \partial_l S_{ij}, \ldots), \tag{3.47}$$

the equation of motion becomes

$$\partial_i \tau_{ij} = \rho \ddot{u}_j, \tag{3.48}$$

where τ_{ij} has the form

$$\tau_{ij} = (c_{ijkl} + C^{(1)}_{ijklm}\partial_m + C^{(2)}_{idklmn}\partial_m\partial_n + C^{(3)}_{ijklmnp}\partial_m\partial_n\partial_p + \ldots)S_{kl}. \tag{3.49}$$

With an infinity of gradients, these equations would be a generalization, to anisotropy, of Cauchy's [1] continuum equations for a periodic, non-centrosymmetric, isotropic structure. Equivalent equations were deduced directly from lattice equations by Krumhansl [18] who pointed out that the resulting continuum field contains the same infromation as that of the lattice.

Kunin, Krumhansl and Kroner (see their review articles in "Mechanics of Generalized Continua", E. Kröner, Editor, Springer-Verlag, 1968) have devised integro-differential equations equivalent to lattice equations by deducing

$$\tau_{ij} = c_{ijkl} S_{kl}(x_i) + \int_V C_{ijkl}(x_i - x_i') S_{kl}(x_i') \mathrm{d}V', \tag{3 50}$$

from the lattice equations, in place of (3.49). The equations have been developed extensively and applied to a variety of phenomena on the atomic scale by Kunin and his group. As all three authors have pointed out, a Taylor series expansion of $S_{kl}(x_i)$ about x_i would bring (3.50) back to (3.49) and a suitable truncation of the series would produce strain gradient equations of any order.

Recently, Eringen and Edelen [71] have included nonlinear and thermodynamic effects in a derivation from the continuum side and Eringen [72] has given an exact, plane wave solution of a one-dimensional, linearized version illustrating the capacity of the equations to yield the correct form of interatomic interactions and the dispersion relation over the entire Brillouin zone.

In the case of waves and vibrations for which the spatial part of the displacement has the exponential form

$$e^{i(\theta_1 x_1 + \theta_2 x_2 + \theta_3 x_3)/a}, \tag{3.51}$$

simple differential equations completely equivalent to the difference equations of a lattice may be deduced in the following manner [6]. Consider the series expansions of the forward and backward differences (2.3) in terms of derivatives:

$$\Delta_i^{\pm} u_j^{l,m,n} = \left(1 \pm \frac{a\partial_i}{2!} + \frac{a^2\partial_i^2}{3!} \pm \frac{a^3\partial_i^3}{4!} + \ldots\right)\partial_i u_j. \tag{3.52}$$

If the spatial part of the displacement has the form (3.51), the series in (3.52) can be summed:

$$\Delta_i^{\pm} u_j^{l,m,n} = \kappa_i^{\pm}\partial_i u_j, \tag{3.53}$$

where

$$\kappa_i^{\pm} = \pm i\theta_i^{-1}(1-e^{\pm i\theta_i}) \tag{3.54}$$

and $\kappa_i^{\pm}\partial_i$, in (3.53), is not summed over the repeated index i. We also adopt the notations

$$\kappa_i = \tfrac{1}{2}(\kappa_i^{+}+\kappa_i^{-}) = 2\theta_i^{-1}\sin\theta_i, \quad \text{(not summed)} \tag{3.55}$$

$$\kappa_i^2 = \kappa_i^{+}\kappa_i^{-} = (\sin\tfrac{1}{2}\theta/\tfrac{1}{2}\theta)^2, \qquad \text{(not summed)}. \tag{3.56}$$

Then, for exponential functions,

$$\Delta_i^2 = \kappa_i^2\partial_i^2, \; \Delta_i\Delta_j = \kappa_i\kappa_j\partial_i\partial_j \; \text{(not summed)}. \tag{3.57}$$

As an application, take the Gazis-Herman-Wallis equations for a simple cubic lattice. Upon inserting (3.57) into the three difference equations of the type (2.5), we obtain three differential equations of the type

$$\begin{aligned}\rho\ddot{u}_1^{l,m,n} = {} & [c_{11}\kappa_1^2\partial_1^2+c_{44}(\kappa_2^2\partial_2^2+\kappa_3^2\partial_3^2) \\ & +\tfrac{1}{2}a^2c_{12}\kappa_1^2\partial_1^2(\kappa_2^2\partial_2^2+\kappa_3^2\partial_3^2)]u_1^{l,m,n} \\ & +(c_{12}+c_{44})(\kappa_1\kappa_2\partial_1\partial_2u_2^{l,m,n}+\kappa_1\kappa_3\partial_1\partial_3u_3^{l,m,n}).\end{aligned} \tag{3.58}$$

Whether we substitute the lattice displacements

$$u_i^{l,m,n} = A_i e^{i(l\theta_1+m\theta_2+n\theta_3)}e^{i\omega t} \tag{3.59}$$

in the three difference equations of the type (2.5) or substitute the continuous displacements

$$u_i = A_i e^{i(\theta_1x_1+\theta_2x_2+\theta_3x_3)/a}e^{i\omega t} \tag{3.60}$$

in the three differential equations of the type (3.58), we obtain the same dispersion relation:

$$|d_{ij}| = 0, \; i, j = 1, 2, 3 \tag{3.61}$$

where

$$\begin{aligned}d_{ij} = {} & \rho a^2\omega^2-2(c_{11}-2c_{12})(1-\cos\theta_i)-2c_{12}(2-\cos\theta_i\sum_{k\neq i}\cos\theta_k) \\ & -2(c_{44}-c_{12})(2-\sum_{k\neq i}\cos\theta_k) \quad (i = j),\end{aligned}$$

$$d_{ij} = -\tfrac{1}{4}(c_{12}+c_{44})\cos\theta_i\cos\theta_j \; (i \neq j).$$

Thus, the difference equations, with constant material coefficients, have been converted to completely equivalent differential equations with material coefficients that are functions of the wave numbers, e.g. $c_{11} \to \kappa_1^2 c_{11}$.

4. Diatomic Crystals and Compound Continua

If the lattice structure of a crystal is a simple one of the Bravais type, with one atom per cell there is no distinction between the long wave limit and the long wave, low frequency limit of the finite difference equations of motion or the dispersion relation. However, if the lattice has a "basis", i.e. a group of two or more atoms per cell, there are high frequency "optical" branches as well as low frequency acoustic branches at the long wave limit, as illustrated in Fig. 8. In that case, classical elasticity accounts for the long wave portions of only the acoustic branches. The appropriate extension [6] of classical elasticity to accommodate the long wave behavior of such polyatomic crystals is suggested by the fact that a lattice with a basis can be resolved into two or more interpenetrating Bravais lattices.

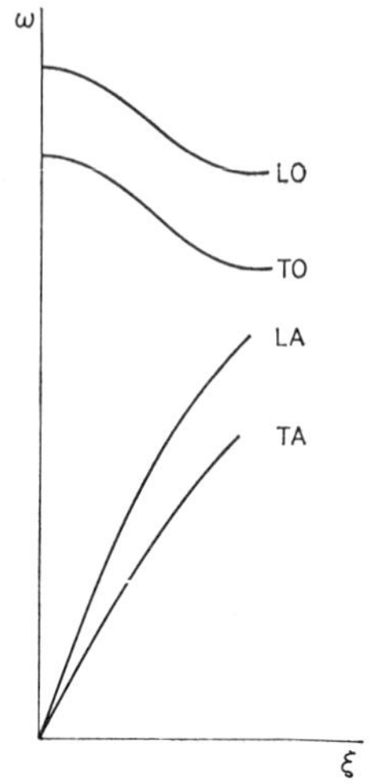

Figure 8

The simplest lattice with a basis is the NaCl-type which has particles at the points (*la*, *ma*, *na*) of a simple-cubic lattice; but the particles at $l+m+n$ even have mass M^1, say, and those at $l+m+n$ odd have mass M^2, as represented in Fig. 9 by circles and dots, respectively. It may be

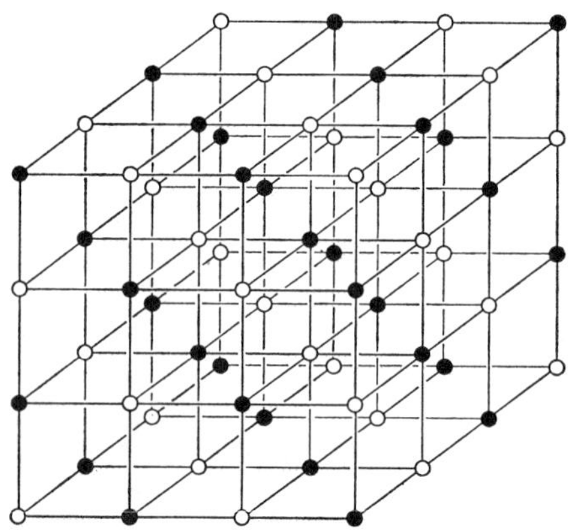

Figure 9

seen that the lattice can be resolved into two, interpenetrating, face-centered, cubic lattices: one with particles of mass M^1 and the other with particles of mass M^2. The interactions between particles may be taken to be similar to those for the Gazis-Herman-Wallis lattice described in Article 2.1. The force constant between nearest neighbor (unlike) particles is designated by α. The force constants between next nearest neighbor (like) particles are assumed to be different: β^1 and β^2 for masses M^1 and M^2, respectively. The angular force constant, γ, is taken to be the same whether a particle of mass M^1 or M^2 is at the apex. Then Gazis and Wallis [25] find, for the particles at $l+m+n$ even, three equations of the type (2.1) with M and β replaced by M^1 and β^1; and, for the particles at $l+m+n$ odd, three equations of the type (2.1) with M and β replaced by M^2 and β^2. When the displacements

$$\overset{\nu}{u}{}_j^{l,m,n} = \overset{\nu}{A}_j e^{i(l\theta_1+m\theta_2+n\theta_3-\omega t)} \tag{4.1}$$

with $\nu = 1$ and 2 for $l+m+n$ even and odd, respectively, are substituted in the six difference equations of motion, there results the dispersion relation

$$\begin{vmatrix} \overset{1}{d}_{11} & \overset{1}{d}_{12} & \overset{1}{d}_{13} & d_{11} & 0 & 0 \\ \overset{1}{d}_{21} & \overset{1}{d}_{22} & \overset{1}{d}_{23} & 0 & d_{22} & 0 \\ \overset{1}{d}_{31} & \overset{1}{d}_{32} & \overset{1}{d}_{33} & 0 & 0 & d_{33} \\ d_{11} & 0 & 0 & \overset{2}{d}_{11} & \overset{2}{d}_{12} & \overset{2}{d}_{13} \\ 0 & d_{22} & 0 & \overset{2}{d}_{21} & \overset{2}{d}_{22} & \overset{2}{d}_{23} \\ 0 & 0 & d_{33} & d_{31} & d_{32} & d_{33} \end{vmatrix} = 0, \tag{4.2}$$

where

$$\begin{aligned} \overset{\nu}{d}_{ij} &= \overset{\nu}{M}\omega^2 - 2(\alpha+8\gamma) - 4\overset{\nu}{\beta}(2-\cos\theta_i \sum_{k\neq i} \cos\theta_k), \quad i = j, \\ \overset{\nu}{d}_{ij} &= -4(\overset{\nu}{\beta}+\gamma)\sin\theta_i \sin\theta_j, \quad i \neq j, \\ d_{ii} &= 2\alpha\cos\theta_i + 8\gamma \sum_{k\neq i} \cos\theta_k, \end{aligned} \tag{4.3}$$

in which i, j and k range over 1, 2, 3 and ν over 1, 2.

If we designate u_i^1 and u_i^2 as the displacements of particles with mass M^1 and M^2, respectively and write

$$\overset{\nu}{u}{}_j^{l,m,n} = \overset{\nu}{A}_j e^{i(l\theta_1+m\theta_2+n\theta_3-\omega t)} \tag{4.4}$$

for all l, m, n, and adopt the notation (2.3) for finite difference operators, the same dispersion relation as (4.2) results from three equations of the type

$$\begin{aligned} \overset{1}{M}\overset{1}{u}{}_{1,tt}^{l,m,n} &= 2(\alpha+8\gamma)(\overset{2}{u}{}_1^{l,m,n} - \overset{1}{u}{}_1^{l,m,n}) + 2a^2\overset{1}{\beta}(2\Delta_1^2+\Delta_2^2+\Delta_3^2)\overset{1}{u}{}_1^{l,m,n} \\ &\quad + 4(\overset{1}{\beta}+\gamma)a^2(\Delta_1\Delta_2\overset{1}{u}{}_2^{l,m,n} + \Delta_1\Delta_3\overset{1}{u}{}_3^{l,m,n}) \\ &\quad + \overset{1}{\beta}a^4(\Delta_1^2\Delta_2^2+\Delta_1^2\Delta_3^2)\overset{1}{u}{}_1^{l,m,n} \\ &\quad + \alpha a^2\Delta_1^2\overset{2}{u}{}_1^{l,m,n} + 4\gamma a^2(\Delta_2^2+\Delta_3^2)\overset{2}{u}{}_1^{l,m,n} \end{aligned} \tag{4.5}$$

and three more obtained by interchange of superscripts 1 and 2. This form [6] is more convenient for passing to continuum limits since the necessity for distinguishing between $l+m+n$ even and odd is dispensed with.

At long wave length and low frequency: $u_i^1 = u_i^2 = u_i$, say. Then employing the expansions (2.7), discarding derivatives higher than the second and adding corresponding members of the two sets of three equations of motion, we recover the equations of classical elasticity of the type (2.8) with

$$\rho = (\overset{1}{M}+\overset{2}{M})/2a^3 \tag{4.6}$$

and stiffness constants given by (2.6) with

$$\beta = (\overset{1}{\beta}+\overset{2}{\beta})/2. \tag{4.7}$$

At long wave lengths but not necessarily low frequency, (4.5) becomes, with $M^1 = 2a^3\rho^1$,

$$\begin{aligned} a\overset{1}{\rho}\overset{1}{u}_{1,tt} = {} & a^{-2}(\alpha+8\gamma)(\overset{2}{u}_1-\overset{1}{u}_1)+\overset{1}{\beta}(2\partial_1^2+\partial_2^2+\partial_3^2)\overset{1}{u}_1 \\ & +2(\overset{1}{\beta}+\gamma)(\partial_1\partial_2\overset{1}{u}_2+\partial_1\partial_3\overset{1}{u}_3) \\ & +\tfrac{1}{2}\alpha\partial_1^2\overset{2}{u}_1+2\gamma(\partial_1^2+\partial_2^2)\overset{2}{u}_1 . \end{aligned} \tag{4.8}$$

Thus, for the long wave approximation, there are three equations of the type (4.8) and three more obtained by interchange of superscripts 1 and 2. The six equations yield the long wave limits of the dispersion relation (4.2) for both the acoustic and optical branches.

A continuum theory that produces equations of the same form as (4.8) may be constructed as follows [6]. We consider two interpenetrating continua, identified by superscripts 1 and 2, representing the two face-centered cubic sub-lattices of the NaCl structure. The potential energy density is taken as a quadratic function of the strains of the two continua and their relative displacement and rotation. For the crystal class *m*3*m*, to which the face-centered cubic and NaCl lattices belong, this is, with $\kappa = 1, 2$ and $\lambda = 1, 2$,

$$W = c_0^*(\overset{2}{S}_{ii}-\overset{1}{S}_{ii})+\tfrac{1}{2}a^{**}u_i^*u_i^*+c^{**}\omega_{ij}^*\omega_{ij}^*+\tfrac{1}{2}\sum_{\kappa,\lambda} c_{ijkl}^{\kappa\lambda}\overset{\kappa}{S}_{ij}\overset{\lambda}{S}_{kl}, \tag{4.9}$$

where

$$\begin{aligned} & u_i^* = (\overset{2}{u}_i-\overset{1}{u}_i),\ \overset{\kappa}{S}_{ij} = \tfrac{1}{2}(\partial_i\overset{\kappa}{u}_j+\partial_j\overset{\kappa}{u}_i),\ \omega_j^* = \tfrac{1}{2}(\partial_i u_j^*-\partial_j u_i^*), \\ & c_{ijkl}^{\kappa\lambda} = c_{ijkl}^{\lambda\kappa} = (c_{11}^{\kappa\lambda}-c_{12}^{\kappa\lambda}-2c_{44}^{\kappa\lambda})\delta_{ijkl}+c_{12}^{\kappa\lambda}\delta_{ij}\delta_{kl}+c_{44}^{\kappa\lambda}(\delta_{ik}\delta_{jl}+\delta_{il}\delta_{jk}) \end{aligned} \tag{4.10}$$

and the $\delta_{ij\ldots}$ are unity if all indices are alike and zero otherwise.

With kinetic energy density

$$T = \tfrac{1}{2}\sum_{\kappa} \overset{\kappa}{\rho}\overset{\kappa}{u}_{i,t}\overset{\kappa}{u}_{i,t} \tag{4.11}$$

and the potential energy density (4.9), Hamilton's variational principle, for independent variations δu_i^1 and δu_i^2, yields the six Euler equations

$$\begin{aligned} \overset{\kappa}{\rho}\overset{\kappa}{u}_{j,tt} = {} & -(-1)^\kappa[a^{**}(\overset{2}{u}_j-\overset{1}{u}_j)+c^{**}\sum_{\lambda}\partial_i(\partial_i\overset{\lambda}{u}_j-\partial_j\overset{\lambda}{u}_i)] \\ & +\sum_{\lambda}[(c_{11}^{\kappa\lambda}-c_{12}^{\kappa\lambda}-2c_{44}^{\kappa\lambda})\delta_{ijkl}\partial_i\partial_k\overset{\lambda}{u}_l] \\ & +\sum_{\lambda}[c_{12}^{\kappa\lambda}\partial_i\partial_j\overset{\lambda}{u}_i+c_{44}^{\kappa\lambda}\partial_i(\partial_i\overset{\lambda}{u}_j+\partial_j\overset{\lambda}{u}_i)]. \end{aligned} \tag{4.12}$$

These become the six equations of the type (4.8) if we set

$$\begin{aligned} & ac_{11}^{\kappa\kappa} = 2ac_{44}^{\kappa\kappa} = 2\overset{\kappa}{\beta}, && ac_{12}^{\kappa\kappa} = \beta+2\gamma, && c^{**} = 0, \\ & ac_{44}^{12} = -ac_{12}^{12} = 2\gamma, && ac_{11}^{12} = \tfrac{1}{2}\alpha, && a^{**} = (\alpha+8\gamma)/a^3. \end{aligned} \tag{4.13}$$

Thus, the theory of a compound continuum, represented by the potential and kinetic energy densities (4.9) and (4.11), produces equations of motion which are the long wave limit of the Gazis-Wallis finite difference equations of an NaCl-type lattice of mass particles.

It may be noted that the linear term in (4.9) is the energy density of a self-equilibrated initial stress which produces a surface energy of deformation analogous to that in the second strain gradient equations described in Article 3iii, but with the continuum representation of only nearest and next nearest neighbor interactions.

5. Molecular Crystals and Cosserat Continua

Crystals with identical groups of closely packed atoms, forming molecules, situated at the points of a lattice are termed molecular crystals if the intramolecular forces are large in comparison with the intermolecular forces [26]. In a monomolecular crystal, there is one molecule at each point of a Bravais lattice. The displacements of the mass centers of the molecules correspond to the displacements of the atoms of a monatomic lattice; and the associated equations of motion reduce to those of classical elasticity in the long wave approximation. However, the rotation and strain of the molecules themselves may also be taken into account. In the long wave approximation, these motions, which may be termed the micro-rotation and the micro-strain of a micro-structure (molecule), can be represented by an asymmetric tensor of the second rank, say ψ_{ij}. The antisymmetric part of the tensor:

$$\psi_{[ij]} = \tfrac{1}{2}(\psi_{ij}-\psi_{ji}) \tag{5.1}$$

describes the micro-rotation and the symmetric part:

$$\psi_{(ij)} = \tfrac{1}{2}(\psi_{ij}+\psi_{ji}) \tag{5.2}$$

describes the strain – both of which may be different from the rotation and strain calculated from the displacements of the mass centers.

The micro-rotation was taken into account in a continuum theory by E. and F. Cosserat [27] in 1909. Their equations were revived and applied to two-dimensional problems by Schaefer [28] in 1962. Subsequently, the Cosserat theory was termed "micropolar elasticity" by Eringen [29]. Comparisons of various forms of the theory and general solutions of the equations by Eringen [29], Aero and Kuvshinkii [30, 31], Neuber [32] and Mindlin [33] were given by Cowin [34, 35] along with a more concise general solution. As is shown below, the Cosserat theory is the long wave approximation to a lattice theory of monomolecular crystals. A direct correspondence between the Cosserat theory and the molecular crystal KNO_3 has recently been exhibited by Askar [36].

The long wave, low frequency limit of both the molecular lattice and Cosserat theories is the rotation gradient theory mentioned in Article 3i. This limiting form is the same as that obtained from the Cosserat theory by constraining the micro-rotation to be the same as the rotation calculated from the displacements. In fact, Toupin [37] refers to the rotation gradient theory as the "Cosserat theory with constrained rotations".

Extensions of the Cosserat theory to take into account the strain of the micro-structure were made in 1964 by Eringen [38] ("micromorphic" theory) Green and Rivlin [39] ("multipolar" theory) and Mindlin [40] ("micro-structure" theory). In their simplest form, these theories are characterized by the potential energy density [40]

$$W = \tfrac{1}{2}c_{ijkl}S_{ij}S_{kl}+\tfrac{1}{2}b_{ijkl}\gamma_{ij}\gamma_{kl}+\tfrac{1}{2}a_{ijklmn}\kappa_{ijk}\kappa_{lmn} + d_{ijklm}\gamma_{ij}\kappa_{klm}+f_{ijklm}\kappa_{ijk}S_{lm}+g_{ijkl}\gamma_{ij}S_{kl}, \tag{5.3}$$

where S_{ij} is the ordinary strain (macro-strain):

$$S_{ij} = \tfrac{1}{2}(\delta_i u_j+\partial_j u_i), \tag{5.4}$$

γ_{ij} is the relative deformation (the difference between the displacement gradient and the micro-deformation):

$$\gamma_{ij} = \partial_i u_j - \psi_{ij} \tag{5.5}$$

and κ_{ijk} is the gradient of the micro-deformation:

$$\kappa_{ijk} = \partial_i \psi_{jk}. \tag{5.6}$$

In the centrosymmetric, isotropic case [40], d_{ijklm} and f_{ijklm}, in (5.3), are zero and

$$\begin{aligned} c_{ijkl} &= \lambda\delta_{ij}\delta_{kl} + \mu(\delta_{ik}\delta_{jl} + \delta_{il}\delta_{jk}), \\ g_{ijkl} &= g_1\delta_{ij}\delta_{kl} + g_2(\delta_{ik}\delta_{jl} + \delta_{il}\delta_{jk}), \\ b_{ijkl} &= b_1\delta_{ij}\delta_{kl} + b_2\delta_{ik}\delta_{jl} + b_3\delta_{il}\delta_{jk}, \\ a_{ijklmn} &= a_1(\delta_{ij}\delta_{lk}\delta_{mn} + \delta_{kj}\delta_{in}\delta_{lm}) + a_2(\delta_{ij}\delta_{km}\delta_{nl} + \delta_{ki}\delta_{jn}\delta_{lm}) \\ &\quad + a_3\delta_{ij}\delta_{kn}\delta_{lm} + a_4\delta_{jk}\delta_{il}\delta_{mn} + a_5(\delta_{jk}\delta_{im}\delta_{nl} + \delta_{ki}\delta_{jl}\delta_{mn}) \\ &\quad + a_8\delta_{ki}\delta_{jm}\delta_{nl} + a_{10}\delta_{il}\delta_{jm}\delta_{kn} + a_{11}(\delta_{jl}\delta_{km}\delta_{in} + \delta_{kl}\delta_{im}\delta_{jn}) \\ &\quad + a_{13}\delta_{il}\delta_{jn}\delta_{km} + a_{14}\delta_{jl}\delta_{kn}\delta_{im} + a_{15}\delta_{kl}\delta_{in}\delta_{jm}. \end{aligned} \tag{5.7}$$

The associated kinetic energy density is

$$T = \tfrac{1}{2}\rho\dot{u}_j\dot{u}_j + \tfrac{1}{6}\rho d^2\dot{\psi}_{ij}\dot{\psi}_{ij}, \tag{5.8}$$

where ρ is the mass density and d is a characteristic length dimension of the molecule (half the edge length in the case of a cube).

The complete set of dispersion relations for plane waves, in such a material with a deformable microstructure, is given in [40] and exhibits twelve branches: the usual three acoustic branches and, in addition, nine optical branches, i.e. branches with non-zero frequency at infinite wave length. At low frequencies and long wave lengths, the differential equations of motion and the dispersion relations reduce to those of the first strain gradient theory.

In molecular crystals, the most interesting of the optical branches is the relatively low frequency "librational" or "soft optical" branch (labelled TRO, for "transverse rotational optical" in [40]) which is coupled to a transverse acoustic branch. The equations of motion for this pair of coupled modes are obtained by setting $\psi_{(12)} = 0$ (thereby eliminating the shear micro-strain) in the first and third of Eqs. (8.5) in [40], with the result

$$\begin{aligned} \bar{k}_{11}\partial_1^2 u_2 - \bar{k}_{13}\partial_1\psi_{[12]} &= \rho\ddot{u}_2, \\ \bar{k}_{13}\partial_1 u_2 + \bar{k}_{33}\partial_1^2\psi_{[12]} - 2\bar{k}_{13}\psi_{[12]} &= \tfrac{2}{3}\rho d^2\,\ddot{\psi}_{[12]}, \end{aligned} \tag{5.9}$$

where

$$\begin{aligned} \bar{k}_{11} &= \mu + 2g_2 + b_2, \quad \bar{k}_{13} = b_2 - b_3, \\ \bar{k}_{33} &= -2a_2 + a_3 + a_8 + 2a_{10} - 2a_{13} + a_{14} + a_{15}. \end{aligned} \tag{5.10}$$

Since $\psi_{(ij)} = 0$ in (5.9), they are equations of a Cosserat continuum. Setting

$$u_2 = A e^{i(\xi x_1 - \omega t)}, \quad \psi_{[12]} = B e^{i(\xi x_1 - \omega t)} \tag{5.11}$$

and eliminating A and B, we obtain the dispersion relation

$$\begin{vmatrix} \bar{k}_{11}\xi^2 - \rho\omega^2 & \bar{k}_{13}\xi \\ \bar{k}_{13}\xi & \bar{k}_{33}\xi^2 + 2\bar{k}_{13} - \tfrac{2}{3}\rho d^2\omega^2 \end{vmatrix} = 0. \tag{5.12}$$

The two branches of (5.12) have ordinates and slopes at $\xi = 0$ as follows:

$$\text{acoustic branch: } \omega]_{\xi=0} = 0, \quad \frac{d\omega}{d\xi}\bigg]_{\xi=0} = \left(\frac{2\bar{k}_{11}-\bar{k}_{13}}{2\rho}\right)^{\frac{1}{2}};$$
$$\text{soft optical branch: } \omega]_{\xi=0} = \left(\frac{3\bar{k}_{13}}{\rho d^2}\right)^{\frac{1}{2}}, \quad \frac{d\omega}{d\xi}\bigg]_{\xi=0} = 0. \tag{5.13}$$

It may be verified that (5.9) are the long wave limits of the finite difference equations of motion of a linear lattice of dumbbell molecules [26, Article 10]. We consider a line of molecules of length $2d$, spacing a, mass m and moment of inertia I with transverse displacement u_n and rota-

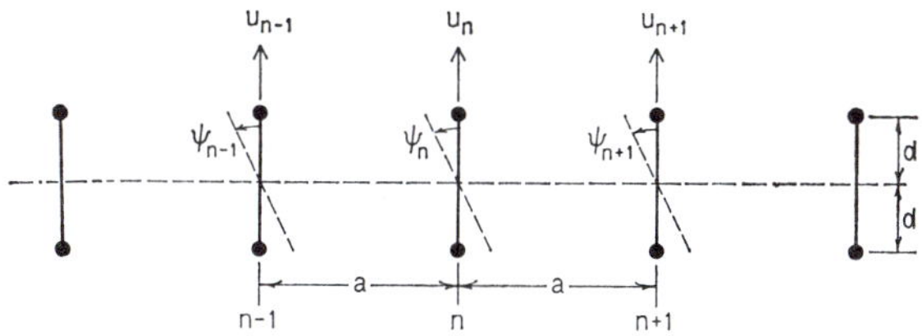

Figure 10

tion ψ_n at lattice point n, as illustrated in Fig. 10. The potential energy of the lattice may be taken as

$$\varphi = a^{-2} \sum \left[\tfrac{1}{2}\alpha_1(u_{n+1}-u_n)^2+\tfrac{1}{2}\alpha_2(u_{n+1}-u_n-d\psi_n)^2 + \tfrac{1}{2}\alpha_3 d^2(\psi_{n+1}-\psi_n)^2+\alpha_4(u_{n+1}-u_n)(u_{n+1}-u_n-d\psi_n)\right] \tag{5.14}$$

where $\alpha_1 \ldots \alpha_4$ are force constants. Comparing with (5.3) we see that (5.14) is the one-dimensional, finite difference analogue for the centrosymmetric case, i.e.

$$d_{ijklm} = 0, \quad f_{ijklm} = 0$$
$$S_{ij} = \tfrac{1}{2}(\partial_i u_j+\partial_j u_i) \to (u_{n+1}-u_n)/a, \quad \gamma_{ij} = \partial_i u_j - \psi_{ij} \to (u_{n+1}-u_n-d\psi_n)/a, \tag{5.15}$$
$$\kappa_{ijk} = \partial_i \psi_{jk} \to (\psi_{n+1}-\psi_n)/a.$$

With finite difference operators Δ^+, Δ^- and Δ^2 again defined as in (2.3), the equations of motion are given by

$$-\frac{\partial\varphi}{\partial u_n} = (\alpha_1+\alpha_2+2\alpha_4)\Delta^2 u_n - \frac{d}{a}(\alpha_2+\alpha_4)\Delta^-\psi_n = m\ddot{u}_n a^{-2}$$
$$-\frac{\partial\varphi}{\partial\psi_n} = \frac{d}{a}(\alpha_2+\alpha_4)\Delta^+ u_n + \alpha_3 d^2\Delta^2\psi_n - \frac{d^2}{a^2}\alpha_2\psi_n = I\ddot{\psi}_n a^{-2} \tag{5.16}$$

Insert

$$u_n = A\,e^{i(\xi na-\omega t)}, \quad \psi_n = B\,e^{i(\xi na-\omega t)} \tag{5.17}$$

in (5.16) and find the dispersion relation

$$\begin{vmatrix} 2(\alpha_1+\alpha_2+2\alpha_4)(\cos\xi a-1)-m\omega^2 & d(\alpha_2+\alpha_4)(\cos\xi a-1-\sin\xi a) \\ d(\alpha_2+\alpha_4)(\cos\xi a-1+\sin\xi a) & 2\alpha_3 d^2(\cos\xi a-1)-\alpha_2 d^2+I\omega^2 \end{vmatrix} = 0 \tag{5.18}$$

which is illustrated in Fig. 11. The ordinates and slopes of the two branches at $\xi = 0$ are:

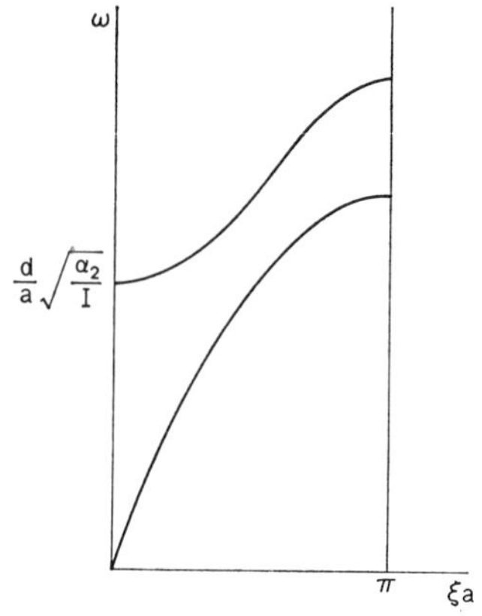

Figure 11

$$
\begin{aligned}
&\text{acoustic branch: } \omega]_{\xi=0} = 0, \quad \frac{d\omega}{d\xi}\bigg]_{\xi=0} = \left[\frac{\alpha_2(\alpha_1+2\alpha_2)+\alpha_4(4\alpha_2+\alpha_4)}{m\alpha_2 a^{-2}}\right]^{\frac{1}{2}} \\
&\text{soft optical branch: } \omega]_{\xi=0} = d\sqrt{\frac{\alpha_2}{I}}, \quad \frac{d\omega}{d\xi}\bigg]_{\xi=0} = 0.
\end{aligned}
\tag{5.19}
$$

If, in the finite difference equations of motion (5.16), we set

$$
a\bar{k}_{11} = \alpha_1+\alpha_2+2\alpha_4, \quad a\bar{k}_{13} = (\alpha_2+\alpha_4)d/a = \alpha_2 d^2/2a^2, \quad a\bar{k}_{33} = \alpha_3 d^2, \tag{5.20}
$$

$$
a^3\rho \to m, \quad I \to \tfrac{2}{3}\rho d^2 a^3
$$

and retain only the first term in the expansion of each of the difference operators in series of differential operators, we find that (5.16) reduce to the differential equations of motion (5.9) of the Cosserat continuum. Similarly, if we insert (5.20) in the expressions (5.9) for the long wave limits of the two branches of the lattice dispersion relation, we find the corresponding expressions (5.13) for the Cosserat continuum. Thus, it is verified that the Cosserat continuum is the long wave limit of the dumbbell model of a molecular crystal – including the acoustic and moderately high frequency soft optical branches.

An alternative formulation of the lattice equations and an application to KNO_3 are given by Askar [36].

6. Classical Piezoelectricity

The classical, linear theory of elastic dielectrics (piezoelectricity) is expressed in terms of twenty-five variables [3, 41]:

u_i Mechanical displacement
S_{ij} Strain
T_{ij} Stress
E_i Electric field
D_i Electric displacement
P_i Polarization
φ Electric potential

These variables are linked by twenty-five equations:

(a) stress-equations of motion

$$T_{ij,i}+f_j = \rho\ddot{u}_j; \tag{6.1}$$

(b) equations of the electrostatic field

$$D_{i,i} = 0, \tag{6.2}$$

$$E_i = -\varphi_{,i}; \tag{6.3}$$

(c) strain-displacement relations

$$S_{ij} = \tfrac{1}{2}(u_{j,i}+u_{i,j}) \tag{6.4}$$

(d) constitutive relations

$$T_{ij} = c^E_{ijkl}S_{kl}-e_{kij}E_k, \tag{6.5}$$

$$D_j = e_{jkl}S_{kl}+\varepsilon^S_{jk}E_k, \tag{6.6}$$

$$P_i = D_i-\varepsilon_0 E_i, \tag{6.7}$$

where

c^E_{ijkl} = elastic stiffness (E_i = constant)
e_{ijk} = piezoelectric stress-constant
ε_{ij} = permittivity (S_{ij} = constant)
ε_0 = permittivity of a vacuum

The equations of motion may be reduced to four by elimination of all the variables except u_i and φ:

$$c_{ijkl}u_{k,li}+e_{kij}\varphi_{,ki}+f_j = \rho\ddot{u}_j, \tag{6.8}$$

$$e_{kij}u_{i,jk}-\varepsilon^S_{ij}\varphi_{,ij} = 0. \tag{6.9}$$

A linear version of a variational principle due to Toupin [42] may be employed to produce equivalent equations of equilibrium, as follows.

First, separate the energy density, W, of the dielectric into the stored energy of deformation and polarization, W^L, and a remainder which is the energy density of the Maxwell electric self-field:

$$W = W^L(S_{ij}, P_i)+\tfrac{1}{2}\varepsilon_0\varphi_{,i}\varphi_{,i}. \tag{6.10}$$

Then define an electric enthalpy [41]

$$H = W-E_iD_i. \tag{6.11}$$

Upon substituting (6.10), (6.7) and (6.3) in (6.11) we find

$$H = W^L(S_{ij}, P_i)-\tfrac{1}{2}\varepsilon_0\varphi_{,i}\varphi_{,i}\varphi_{,i}P_i. \tag{6.12}$$

Consider a body occupying a volume V bounded by a surface S separating V from an outer vacuum V'. For such a system, the version of Toupin's principle that leads to equations equivalent to those preceding, for the case of equilibrium, is

$$-\delta\int_{V^*}H\,dV+\int_V(f_i\delta u_i+E^0_i\delta P_i)\,dV+\int_S t_i\delta u_i\,dS = 0, \tag{6.13}$$

for independent variations of u_i, P_i and φ. In (6.13), $V^* = V+V'$, t_i is the surface traction across S and E^0_i is the external electric field. Now,

$$\delta H = \frac{\partial W^L}{\partial S_{ij}}\,\delta S_{ij} + \frac{\partial W^L}{\partial P_i}\,\delta P_i - \varepsilon_0\,\varphi_{,i}\,\delta\varphi_{,i} + \varphi_{,i}\,\delta P_i + P_i\,\delta\varphi_{,i}. \tag{6.14}$$

Define a stress T_{ij} and an effective local electric force E_i^L by

$$T_{ij} = \frac{\partial W^L}{\partial S_{ij}}, \quad E_i^L = -\frac{\partial W^L}{\partial P_i}. \tag{6.15}$$

Note that, since $T_{ij} = T_{ji}$ and

$$\begin{aligned} &S_{ij} = \tfrac{1}{2}(u_{j,i} + u_{i,j}), \\ &T_{ij}\,\delta S_{ij} = T_{ij}\,\delta u_{j,i} = (T_{ij}\,\delta u_j)_{,i} - T_{ij,i}\,\delta u_j, \\ &\varphi_{,i}\,\delta\varphi_{,i} = (\varphi_{,i}\,\delta\varphi)_{,i} - \varphi_{,ii}\,\delta\varphi, \\ &P_i\,\delta\varphi_{,i} = (P_i\,\delta\varphi)_{,i} - P_{i,i}\,\delta\varphi, \end{aligned} \tag{6.16}$$

by the chain rule of differentiation. Then

$$\delta H = -T_{ij,i}\,\delta u_j - (E_i^L - \varphi_{,i})\delta P_i - (-\varepsilon_0\varphi_{,ii} + P_{i,i})\delta\varphi + [T_{ij}\,\delta u_j + (-\varepsilon_0\varphi_{,i} + P_i)\delta\varphi]_{,i}. \tag{6.17}$$

Noting that $V^* = V + V'$ and that u_i and P_i do not exist in V', we find, after applying the divergence theorem,

$$-\delta\int_V H\,dV = \int_V [T_{ij,i}\,\delta u_j + (E_i^L - \varphi_{,i})\delta P_i + (-\varepsilon_0\,\varphi_{,ii} + P_{i,i})\delta\varphi]\,dV - \int_S n_i[T_{ij}\,\delta u_j + (-\varepsilon_0\,\varphi_{,i} + P_i)\delta\varphi]\,dS, \tag{6.18a}$$

$$-\delta\int_{V'} H\,dV = -\int_{V'} \varepsilon_0\,\varphi_{,ii}\,\delta\varphi\,dV - \int_S \varepsilon_0\,n_i\,\varphi_{,i}\,\partial\varphi\,dS. \tag{6.18b}$$

Hence, (6.13) becomes

$$\int_V [(T_{ij,i} + f_j)\delta u_j + (E_i^L - \varphi_{,i} + E_i^0)\delta P_i + (-\varepsilon_0\varphi_{,ii} + P_{i,i})\delta\varphi]\,dV - \int_{V'} \varepsilon_0\,\varphi_{,ii}\,\delta\varphi\,dV + \int_S [(t_j - n_i\,T_{ij})\delta u_j + n_i(\varepsilon_0[\varphi_{,i}] - P_i)\delta\varphi]\,dS = 0, \tag{6.19}$$

where $[\varphi_{,i}]$ is the jump in $\varphi_{,i}$ across S. Then the Euler equations

$$\left.\begin{aligned} T_{ij,i} + f_j &= 0, \\ E_i^L - \varphi_{,i} + E_i^0 &= 0, \\ -\varepsilon_0\,\varphi_{,ii} + P_{i,i} &= 0, \end{aligned}\right\}\ \text{in } V, \tag{6.20}$$

$$\varphi_{,ii} = 0 \text{ in } V' \tag{6.21}$$

and the natural boundary conditions

$$\left.\begin{aligned} n_i\,T_{ij} &= t_j \\ n_i(-\varepsilon_0\,[\varphi_{,i}] + P_i) &= 0, \end{aligned}\right\}\ \text{on } S, \tag{6.22}$$

follow from (6.19). In (6.22), $n_i(-\varepsilon_0[\varphi_{,i}] + P_i)$ is the surface charge.

The energy density of deformation and polarization is taken to be

$$W^L = \tfrac{1}{2}a_{ij}^S\,P_i\,P_j + \tfrac{1}{2}c_{ijkl}^P\,S_{ij}\,S_{kl} + f_{kij}\,S_{ij}\,P_k \tag{6.23}$$

so that, from (6.15),

$$-E_j^L = a_{jk}^S P_k + f_{jkl} S_{kl},$$
$$T_{ij} = f_{kij} P_k + c_{ijkl}^P S_{kl}. \tag{6.24}$$

Equations (6.16), (6.20), (6.21) and (6.24), with boundary conditions (6.22), constitute a linear version of the equations of equilibrium of elastic dielectrics in the form given by Toupin [42].

The relations between the new constants a_{ij}^S, f_{ijk} and c_{ijkl}^P and the IRE Standard [43] ones are found as follows.

From (6.3) and the second of (6.20) (omitting E_i^0 as in the usual formulation) $E_i^L = -E_i$. Hence, the first of (6.24) becomes

$$E_j = a_{jk}^S P_k + f_{jkl} S_{kl}. \tag{6.25}$$

Now the ratio of P to $\varepsilon_0 E$ is the dielectric susceptibility. Hence a_{ij}^S is proportional to the reciprocal susceptibility at constant strain (χ_{ij}^s):

$$a_{ij}^S = \varepsilon_0^{-1} \chi_{ij}^S. \tag{6.26}$$

With (6.26), (6.25) becomes

$$E_j = \varepsilon_0^{-1} \chi_{jk} P_k + f_{jkl} S_{kl}. \tag{6.27}$$

Define susceptibility at constant strain, η_{ij}^S, according to

$$\eta_{ij}^S \chi_{jk}^S = \delta_{ik} \tag{6.28}$$

and multiply both sides of (6.27) by $\varepsilon_0 \eta_{mj}$ to get

$$\varepsilon_0 \eta_{mj}^S E_j = P_m + \varepsilon_0 \eta_{mj}^S f_{jkl} S_{kl}. \tag{6.29}$$

Then eliminate P between (6.29) and (6.7), with the result

$$D_i = -\varepsilon_0 \eta_{ij}^S f_{jkl} S_{kl} + \varepsilon_0 (\delta_{ij} + \eta_{ij}^S) E_j. \tag{6.30}$$

Accordingly, comparing (6.30) with (6.6), we find

$$e_{jkl} = -\varepsilon_0 \eta_{ij}^S f_{ikl}, \text{ or } f_{jkl} = -\varepsilon_0^{-1} \chi_{ij}^S e_{ikl}; \tag{6.31}$$

$$\varepsilon_{ij}^S = \varepsilon_0 (\delta_{ij} + \eta_{ij}^S), \quad \text{or } \eta_{ij}^S = \varepsilon_0^{-1} \varepsilon_{ij}^S - \delta_{ij}. \tag{6.32}$$

To find c_{ijkl}^P in terms of c_{ijkl}^E, first substitute the expression for P_i, given in terms of E_i and S_{ij} in (6.29), into the second of (6.24):

$$T_{ij} = f_{mij}(\varepsilon_0 \eta_{mk}^S E_k - \varepsilon_0 \eta_{mn}^S f_{nkl} S_{kl}) + c_{ijkl}^P S_{kl}$$

or, from (6.31),

$$T_{ij} = (c_{ijkl}^P - \varepsilon_0 \eta_{mn}^S f_{mij} f_{nkl}) S_{kl} - e_{kij} E_k. \tag{6.33}$$

Comparing (6.33), with (6.5) we see that

$$c_{ijkl}^E = c_{ijkl}^P - \varepsilon_0 \eta_{mn}^S f_{mij} f_{nkl} \tag{6.34}$$

or, from (6.31),

$$c_{ijkl}^P = c_{ijkl}^E + \varepsilon_0^{-1} \chi_{mn}^S e_{mij} e_{nkl}. \tag{6.35}$$

With the aid of the foregoing expressions for the constants a_{ij}^S, c_{ijkl}^P and f_{ijk}, we may reduce the equations of equilibrium (6.20) and the constitutive equations (6.24) to the classical form of equations on u_i and φ, given in (6.8) and (6.9), by elimination of P_i. First, to solve the first of (6.24) for P_i, recall that the reciprocal of a_{ij}^S is $\varepsilon_0 \eta_{ij}^S$. Then, upon multiplication, we have

$$P_i = -\varepsilon_0 \eta_{ij}^S (E_j^L + f_{jkl} S_{kl}). \tag{6.36}$$

Substitute this for P_i in the second of (6.24) to obtain

$$T_{ij} = -f_{kij} \varepsilon_0 \eta_{km}^S (E_m^L + f_{mpq} S_{pq}) + c_{ijkl}^P S_{kl}.$$

From the second equilibrium equation, $E_i^0 = \varphi_{,i}$, again omitting E_i^L. Also, $S_{ij} = \frac{1}{2}(u_{j,i} + u_{i,j})$. Hence

$$T_{ij,i} = (c_{ijkl}^P - \varepsilon_0 \eta_{mn}^S f_{mij} f_{nkl}) u_{k,li} - \varepsilon_0 \eta_{km}^S f_{kij} \varphi_{,mi}$$

Then, from the first of (6.31) and from (6.34),

$$T_{ij,i} + f_j = c_{ijkl}^E u_{k,li} + e_{kij} \varphi_{,ki} + f_j$$

which is the left hand side of (6.8), as required.

Next, from (6.36),

$$P_{i,i} = -\varepsilon_0 \eta_{ij}^S (E_{j,i}^L + f_{jkl} S_{kl,i}) = -\varepsilon_0 \eta_{ij}^S \varphi_{,ij} - \varepsilon_0 \eta_{ij}^S f_{jkl} u_{k,li}.$$

Hence, the left hand side of the last of the equilibrium equations (6.20) becomes

$$-\varepsilon_0 \varphi_{,ii} + P_{i,i} = -\varepsilon_0 \eta_{ij}^S f_{jkl} u_{k,li} - \varepsilon_0 (\delta_{ij} + \eta_{ij}^S) \varphi_{,ij}$$

or, from the first of (6.31) and (6.32),

$$-\varepsilon_0 \varphi_{,ii} + P_{i,i} = e_{ikl} u_{k,li} - \varepsilon_{ij}^S \varphi_{,ij}$$

which is the left hand side of (6.9), as required.

7. Polarization Gradient

Toupin's form of the classical equations of elastic dielectrics reveals an omission in the classical theory. His equation

$$E_i^L - \varphi_{,i} + E_i^0 = 0$$

is the "equation of intramolecular force balance" [42] which he derived from fundamental considerations of the equilibrium of electrical forces, but which does not appear in the usual formulations. Granted the validity of the equation, it is significant that no boundary condition is associated with it. Whereas there is an equilibrium equation associated with each of the variables u_i, φ and P_i, only u_i and φ are accompanied by boundary conditions. There is no coefficient of δP_i in the surface integral in (6.19) to complement that in the volume integral. This lack can be traced back to the absence of a functional dependence of W^L on the polarization gradient $P_{j,i}$. In fact, if we were to start by assuming dependence of W^L on the displacement and polarization and their gradients and truncate after the first gradient, $P_{j,i}$ would remain along with S_{ij} and P_i. Only u_i and the antisymmetric part of $u_{j,i}$ would have to be discarded – on the grounds of required translational and rotational invariance of W^L. The only justification for discarding $P_{j,i}$ would be its possible lack of importance judged on practical considerations. However, as will be shown, $P_{j,i}$ supplies terms found in the long wave limits of finite difference equations of lattice theories of crystals and also extends the continuum theory to accommodate observed physical phenomena.

To extend Toupin's variational principle to account for the contribution of the polarization gradient, it is only necessary to replace (6.12) with

$$H = W^L(S_{ij}, P_i, P_{j,i}) - \tfrac{1}{2}\varepsilon_0 \varphi_{,i}\varphi_{,i} + \varphi_{,i}P_i. \tag{7.1}$$

In addition, we shall include the kinetic energy so that (6.13) becomes

$$\delta\int_{t_0}^{t_1} dt \int_{V^*} (\tfrac{1}{2}\rho \dot{u}_i \dot{u}_i - H)\, dV + \int_{t_0}^{t_1} dt \left[\int_V (f_i \delta u_i + E_i^0 \delta P_i)\, dV + \int_S t_i \delta u_i \, dS\right] = 0, \tag{7.2}$$

for independent variations of u_i, φ and P_i between fixed limits at times t_0 and t_1.

The only additional operations involved are a new definition,

$$E_{ij} = \frac{\partial W^L}{\partial P_{j,i}}, \tag{7.3}$$

and two integrations:

$$\tfrac{1}{2}\delta\int_{t_0}^{t_1} \dot{u}_i \dot{u}_i \, dt = \left[\int_{t_0}^{t_1} \dot{u}_i \delta u_i \, dt = \dot{u}_i \delta u_i\right]_{t_0}^{t_1} - \int_{t_0}^{t_1} \ddot{u}_i \delta u_i \, dt = -\int_{t_0}^{t_1} \ddot{u}_i \delta u_i \, dt,$$

$$\int_V E_{ij}\delta P_{j,i}\, dV = \int_V [(E_{ij}\delta P_j)_{,i} - E_{ij,i}\delta P_j]\, dV = \int_S n_i E_{ij}\delta P_j \, dS - \int_V E_{ij,i}\delta P_j \, dV.$$

With these results and (6.18), (7.2) becomes

$$\begin{aligned}&\int_{t_0}^{t_1} \mathrm{d}t \int_V [(T_{ij,i} + f_j - \rho\ddot{u}_j)\delta u_j + (E_{ij,i} + E_j^L - \varphi_{,j} + E_j^0)\delta P_i + (-\varepsilon_0 \varphi_{,ii} + P_{i,i})\delta\varphi]\, \mathrm{d}V \\ &- \int_{t_0}^{t_1} dt \int_{V'} \varepsilon_0 \varphi_{,ii}\delta\varphi \, dV + \int_{t_0}^{t_1} dt \int_S [(t_j - n_i T_{ij})\delta u_j - n_i E_{ij}\delta P_j + n_i(\varepsilon_0 [\varphi_{,i}] - P_i)\delta\varphi]\, dS = 0.\end{aligned} \tag{7.4}$$

Then the Euler equations are

$$\left.\begin{aligned} T_{ij,i} + f_j &= \rho\ddot{u}_j, \\ E_{ij,i} + E_j^L - \varphi_{,j} + E_j^0 &= 0, \\ -\varepsilon_0 \varphi_{,ii} + P_{i,i} &= 0; \end{aligned}\right\} \text{ in } V \tag{7.5}$$

$$\varphi_{,ii} = 0 \text{ in } V', \tag{7.6}$$

with natural boundary conditions

$$\begin{aligned} n_i T_{ij} &= t_j, \\ n_i E_{ij} &= 0, \\ n_i(-\varepsilon_0[\varphi_{,i}] + P_i) &= 0. \end{aligned} \tag{7.7}$$

For W^L, we take

$$\begin{aligned} W^L = b_{ij}^0 P_{j,i} + \tfrac{1}{2}a_{ij}^{SG} P_i P_j + \tfrac{1}{2}b_{ijkl}^{PS} P_{j,i} P_{l,k} + \tfrac{1}{2}c_{ijkl}^{PG} S_{ij} S_{kl} \\ + d_{ijkl}^P P_{j,i} S_{kl} + f_{ijk}^G P_i S_{jk} + j_{ijk}^S P_i P_{k,j}. \end{aligned} \tag{7.8}$$

The superscripts S, P, G, designating fixed strain, polarization and polarization gradient, respectively, will be omitted in the sequel where no ambiguity results.

Substituting (7.8) in (6.15) and (7.3), we find

$$
\begin{aligned}
-E_j^L &= a_{jk}P_k + j_{jkl}P_{l,k} + f_{jkl}S_{kl},\\
E_{ij} &= j_{kij}P_k + b_{ijkl}P_{l,k} + d_{ijkl}S_{kl} + b_{ij}^0,\\
T_{ij} &= f_{kij}P_k + d_{klij}P_{l,k} + c_{ijkl}S_{kl}.
\end{aligned} \tag{7.9}
$$

The field equations (7.5) and (7.6), the constitutive equations (7.9) and the strain-displacement relation $S_{ij} = \frac{1}{2}(u_{j,i}+u_{i,j})$ comprise the equations of the augmented theory [44]. It is apparent, from the form of the integrand of the surface integral in (7.4), what boundary conditions other than (7.7) are admissible. Thus, in place of the traction n_iT_{ij}, may be substituted the displacement u_i or a component of either and the resultant of the other in the plane at right angles. The same possibilities are open for n_iE_{ij} and P_i. Finally, either the surface charge $n_i(-\varepsilon_0[\varphi_{,i}]+P_i)$ or the potential φ may be specified. Of prime importance is the fact that *both* the potential φ and the polarization P_i may be specified. This is a latitude not permissible in the classical theory.

One of the novel properties of the augmented theory of elastic dielectrics is its accommodation of an electromechanical interaction even for materials with centrosymmetric physical properties. This may be seen by inspection of the energy density W^L as given in (7.8). For centrosymmetry, $f_{ijk} = j_{ijk} = 0$, since there are no centrosymmetric tensors of odd rank. However, this still leaves d_{ijkl} which is the coefficient of an electromechanical coupling term in the augmented theory, but does not appear in the classical theory.

As an example of crystallographic centrosymmetry, consider the cubic point group *m*3*m* (International), O_h(Schoenflies), the generators for which are [45]

$$
\begin{vmatrix} -1 & 0 & 0\\ 0 & -1 & 0\\ 0 & 0 & -1 \end{vmatrix}
\quad
\begin{vmatrix} 0 & 0 & 1\\ 1 & 0 & 0\\ 0 & 1 & 0 \end{vmatrix}
\quad
\begin{vmatrix} 1 & 0 & 0\\ 0 & 0 & 1\\ 0 & -1 & 0 \end{vmatrix} \tag{7.10}
$$

Applying these to the coefficients in (7.8), we find

$$
\begin{aligned}
f_{ijk} &= 0, \quad j_{ijk} = 0,\\
a_{ij} &= a_{11}\delta_{ij}, \quad b_{ij}^0 = b_0\delta_{ij},\\
b_{ijkl} &= b\delta_{ijkl} + b_{12}\delta_{ij}\delta_{kl} + b_{44}(\delta_{ik}\delta_{jl}+\delta_{il}\delta_{jk}) + b_{77}(\delta_{ik}\delta_{jl}-\delta_{il}\delta_{jk}),\\
c_{ijkl} &= c\delta_{ijkl} + c_{12}\delta_{ij}\delta_{kl} + c_{44}(\delta_{ik}\delta_{jl}+\delta_{il}\delta_{jk}),\\
d_{ijkl} &= d\delta_{ijkl} + d_{12}\delta_{ij}\delta_{kl} + d_{44}(\delta_{ik}\delta_{jl}+\delta_{il}\delta_{jk}),
\end{aligned} \tag{7.11}
$$

where

$$
b = b_{11} - b_{12} - 2b_{44}, \quad c = c_{11} - c_{12} - 2c_{44}, \quad d = d_{11} - d_{12} - 2c_{44}. \tag{7.12}
$$

Insertion of (7.11) in the constitutive equations (7.9) reduces the latter to

$$
\begin{aligned}
-E_i^L &= a_{11}P_i,\\
E_{ij} &= b\delta_{ijkl}P_{l,k} + b_{12}\delta_{ij}P_{k,k} + b_{44}(P_{j,i}+P_{i,j}) + b_{77}(P_{j,i}-P_{i,j}) + b_0\delta_{ij}\\
&\quad + d\delta_{ijkl}S_{kl} + d_{12}\delta_{ij}S_{kk} + 2d_{44}S_{ij},\\
T_{ij} &= d\delta_{ijkl}P_{l,k} + d_{12}\delta_{ij}P_{k,k} + d_{44}(P_{j,i}+P_{i,j}) + c\delta_{ijkl}S_{kl} + c_{12}\delta_{ij}S_{kk} + 2c_{44}S_{ij}.
\end{aligned} \tag{7.13}
$$

The equations of motion on u_i, P_i and φ are obtained by substituting (7.13) in (7.5) and employing the strain-displacement relation, with the result:

$$
\begin{aligned}
&c\delta_{ijkl}u_{l,ki} + c_{12}u_{k,kj} + c_{44}(u_{j,ii}+u_{i,ji}) + d\delta_{ijkl}P_{l,ki} + d_{12}P_{k,kj} + d_{44}(P_{j,ii}+P_{i,ji}) + f_j = \rho\ddot{u}_j,\\
&d\delta_{ijkl}u_{l,ki} + d_{12}u_{k,kj} + d_{44}(u_{j,ii}+u_{i,ji}) + b\delta_{ijkl}P_{l,ki} + b_{12}P_{k,kj} + b_{44}(P_{j,ii}+P_{i,ji})\\
&\qquad + b_{77}(P_{j,ii}-P_{i,ji}) - a_{11}P_j - \varphi_{,j} + E_j^0 = 0, \quad -\varepsilon_0\varphi_{,ii} + P_{i,i} = 0.
\end{aligned} \tag{7.14}
$$

In the one dimensional case,

$$u_1 = u_1(x_1, t),\ u_2 = u_3 = 0,\ P_1 = P_1(x_1, t),\ P_2 = P_3 = 0,\ \varphi = \varphi(x_1, t), \tag{7.15}$$

and in the absence of extrinsic fields f_i and E_i^0, (7.14) reduce to

$$\begin{aligned} c_{11}\partial_1^2 u_1 + d_{11}\partial_1^2 P_1 &= \rho \ddot{u}_1, \\ d_{11}\partial_1^2 u_1 + b_{11}\partial_1^2 P_1 - a_{11} P_1 - \partial_1 \varphi &= 0, \\ -\varepsilon_0 \partial_1^2 \varphi + \partial_1 P_1 &= 0. \end{aligned} \tag{7.16}$$

To obtain the dispersion relation for (7.16), let

$$u_1 = A e^{i(\xi x_1 - \omega t)}, \quad P_1 = B e^{i(\xi x_1 - \omega t)}, \quad \varphi = C e^{i(\xi x_1 - \omega t)}. \tag{7.17}$$

Upon substituting (7.17) into (7.16) and eliminating A, B and C, we obtain

$$\begin{vmatrix} c_{11}\xi^2 - \rho\omega^2 & d_{11}\xi^2 \\ d_{11}\xi^2 & b_{11}\xi^2 + a_{11} + \varepsilon_0^{-1} \end{vmatrix} = 0. \tag{7.18}$$

This is a quadratic equation in ξ^2, so that there are two branches. Positive definiteness of the energy density requires

$$a_{11} > 0, \quad b_{11} c_{11} - d_{11}^2 > 0. \tag{7.19}$$

Consequently one of the roots ξ^2 is positive real and the other is negative real. Thus, one of the branches of the dispersion relation is real and the other is imaginary. The behavior of the real branch at low frequency is given by

$$\lim_{\substack{\omega \to 0 \\ \xi \to 0}} \frac{\omega}{\xi} = \sqrt{\frac{c_{11}}{\rho}}. \tag{7.20}$$

The limit of the imaginary branch at zero frequency is

$$\lim_{\substack{\omega \to 0 \\ \xi \neq 0}} \xi = i/l, \tag{7.21}$$

where

$$l = \left[\frac{(b_{11} c_{11} - d_{11}^2)\varepsilon_0}{(1 + \varepsilon_0 a_{11}) c_{11}}\right]^{\frac{1}{2}}. \tag{7.22}$$

If the contribution of the polarization gradient is omitted, b_{11} and d_{11} are zero and (7.16) reduces to

$$\begin{aligned} c_{11}\partial_1^2 u_1 &= \rho \ddot{u}_1, \\ a_{11} P_1 + \partial_1 \varphi &= 0, \\ -\varepsilon_0 \partial_1^2 \varphi + \partial_1 P_1 &= 0, \end{aligned} \tag{7.23}$$

which is Toupin's form of the classical equations for the one-dimensional, linear, centrosymmetric case. In (7.23), there is no electromechanical coupling; the wave motion is dispersionless with frequency given by (7.20) for all ξ.

If, again, the contribution of the polarization gradient is omitted but the material is non-centrosymmetric, (7.23) becomes

$$\begin{aligned} c_{11}\partial_1^2 u_1 + f_{11}\partial_1 P_1 &= \rho \ddot{u}_1, \\ f_{11}\partial_1 u_1 + a_{11}P_1 + \partial_1\varphi &= 0, \\ -\varepsilon_0\partial_1^2\varphi + \partial_1 P_1 &= 0, \end{aligned} \tag{7.24}$$

so that electromechanical coupling is regained. Inserting (7.17) in (7.24) and eliminating A, B and C, we find

$$\rho\omega^2 = [c_{11} - f_{11}^2/(a_{11} + \varepsilon_0^{-1})]\xi^2. \tag{7.25}$$

Thus, even though non-centrosymmetry induces electromechanical coupling, there is still no imaginary branch and the real branch is dispersionless – in the absence of the contribution of the polarization gradient.

8. Monatomic Lattice of Shell-Model Atoms

In this Article, the difference equations of motion are derived for a one-dimensional, monatomic lattice of the Cochran type [46] based on the Dick-Overhauser [47] "shell model" of the atom: a core, comprising the nucleus and inner electrons, surrounded by a shell of outer electrons. The electronic polarization is proportional to the relative displacement of the core and shell of the atom. In addition to this intra-atomic interaction, account is taken of interatomic interactions between core and core, core and shell and shell and shell of nearest neighbor atoms. It is shown that the equations of the lattice have, as their continuum limit, the augmented equations described in the preceding Article rather than the equations of the classical theory of elastic dielectrics. The additional effects associated with the new material constants b_{ijkl} and d_{ijkl} stem primarily from the shell-shell interaction. This interaction is known to be important in the matching of lattice dispersion relations to dispersion data from neutron diffraction measurements at short wave lengths [26, 48].

We consider a single line of atoms (Fig. 12) extracted from a three-dimensional, monatomic

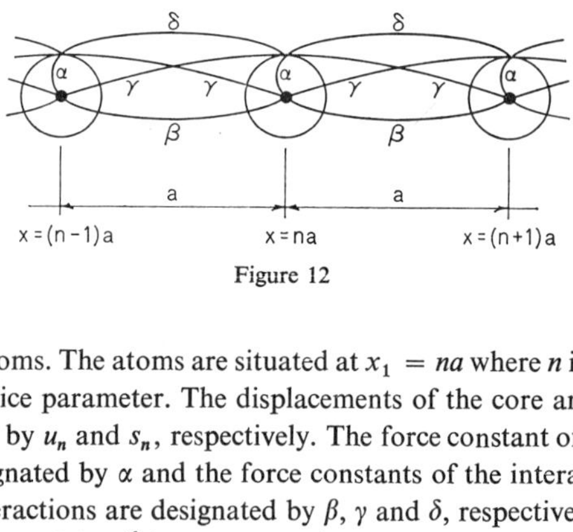

Figure 12

lattice of shell-model atoms. The atoms are situated at $x_1 = na$ where n is a positive or negative integer and a is the lattice parameter. The displacements of the core and shell of the atom at $x_1 = na$ are designated by u_n and s_n, respectively. The force constant of the intra-atomic core-shell interaction is designated by α and the force constants of the interatomic core-core, core-shell and shell-shell interactions are designated by β, γ and δ, respectively.

The equation of motion of the nth interior atom (core and shell combined) is obtained by setting the inertia of the atom equal to the sum of the forces on its core and shell by the cores and shells of its two nearest neighbor atoms:

$$\begin{aligned} &\beta(u_{n+1} - u_n) + \gamma(s_{n+1} - u_n) + \gamma(u_{n+1} - s_n) + \delta(s_{n+1} - s_n) \\ &-\beta(u_n - u_{n-1}) - \gamma(u_n - s_{n-1}) - \gamma(s_n - u_{n-1}) - \delta(s_n - s_{n-1}) = m\ddot{u}_n, \end{aligned} \tag{8.1}$$

where m is the mass of the atom. The shell, alone, of the nth atom is acted upon by its own core, the core and shell of each of the two neighboring atoms and, also, the Maxwell, electric self-field (which occupies all space). Hence

$$\alpha(u_n - s_n) + \gamma(u_{n+1} - s_n) + \delta(s_{n+1} - s_n)$$
$$-\gamma(s_n - u_{n-1}) - \delta(s_n - s_{n-1}) + qE_n = 0, \tag{8.2}$$

in which the inertia of the shell is neglected, q is the charge of the atom and E_n is the value, at $x_1 = na$, of the Maxwell field. Now, $E_n' = -[\partial_1 \varphi]_{x_1 = na}$ and this may be expressed in terms of φ_n, the value of the potential φ at $x_1 = na$, by expanding the derivative in an infinite series of forward differences [7]:

$$E_n = \partial^+ \varphi_n = \sum_{m=1}^{\infty} (-1)^{m-1} m^{-1} a^{m-1} \Delta_+^m \varphi_n, \tag{8.3}$$

where

$$\begin{aligned} \Delta_+ \varphi_n &= (\varphi_{n+1} - \varphi_n)/a, \\ \Delta_+^2 \varphi_n &= (\varphi_{n+2} - 2\varphi_{n+1} + \varphi_n)/a^2 \\ \Delta_+^3 \varphi_n &= (\varphi_{n+3} - 3\varphi_{n+2} + 3\varphi_{n+1} - \varphi_n)/a^3 \\ &\vdots \end{aligned} \tag{8.4}$$

The polarization of the nth atom, per unit area of the three-dimensional lattice, is defined by

$$P_n = (s_n - u_n)q/a^3. \tag{8.5}$$

Then the equations of motion (8.1) and (8.2) may be rearranged to the following forms:

$$\begin{aligned} &(\beta + 2\gamma + \delta)a^{-1}\Delta_1^2 u_n + (\gamma + \delta)a^2 q^{-1} \Delta_1^2 P_n = \rho \ddot{u}_n, \\ &(\gamma + \delta)a^2 q^{-1} \Delta_1^2 u_n + \delta a^5 q^{-2} \Delta_1^2 P_n - (\alpha + 2\gamma)a^3 q^{-2} P_n - \partial^+ \varphi_n = 0, \end{aligned} \tag{8.6}$$

where $\rho = m/a^3$ and Δ_1^2 is the second central difference operator (divided by a^2) as defined in (2.3).

It will be observed that the difference equations (8.6) have the same form as the first two of (7.16), which are the one-dimensional differential equations of motion of the augmented theory of elastic dielectrics. Accordingly, if we set

$$\begin{aligned} &a_{11} = (\alpha + 2\gamma)a^3 q^{-2} = \varepsilon_0^{-1}\eta^{-1}, \quad b_{11} = \delta a^5 q^{-2}, \\ &c_{11} = (\beta + 2\gamma + \delta)a^{-1}, \quad d_{11} = (\gamma + \delta)a^2 q^{-1}, \end{aligned} \tag{8.7}$$

and employ the expansion (2.7) of the second central difference in series of derivatives, we find that the first two of (7.16) are the lowest order continuum approximation to the equations of motion of the atom and its shell, respectively.

As may be seen from (8.7) the polarization gradient terms (those with coefficients b_{11} and d_{11}) in (7.16) stem from the shell-shell and interatomic core-shell interactions – identified by the force constants δ and γ, respectively. In fact, the form of (7.16) is preserved if the interatomic core-shell interaction is omitted but the shell-shell interaction is retained. If both of these interactions are omitted, the continuum approximation reduces to (7.17), which is Toupin's form of the classical equations (in the one-dimensional, linear case). The second of (7.17), Toupin's "equation of intramolecular force balance", does not appear in the traditional equations of elastic dielectrics; but we see that it is a fundamental equilibrium condition: the equilibrium of the shell (i.e. the outer electrons) under the action of the core of the same atom and the surrounding Maxwell field. The corresponding equation in the extended theory (the second of (7.16)) includes, in addition, the action of the adjacent atoms on the shell (in the continuum approximation).

We have yet to derive the lattice counterpart of the third of (7.16) (the "charge equation") and establish the boundary conditions. It is illuminating to reach these results from considerations of energy.

By analogy with (6.12) and (7.8), let us take, for the electric enthalpy of the one-dimensional lattice (per unit volume a^3)

$$\mathscr{H} = \sum_n [b_0 \Delta_+ P_n + \tfrac{1}{2}\varepsilon_0^{-1}\eta^{-1}P_n^2 + \tfrac{1}{2}b_{11}(\Delta_+ P_n)^2 + \tfrac{1}{2}c_{11}(\Delta_+ u_n)^2 + d_{11}(\Delta_+ P_n)(\Delta_+ u_n) - \tfrac{1}{2}\varepsilon_0(\partial^+\varphi_n)^2 + P_n(\partial^+\varphi_n)]. \quad (8.8)$$

We have to find the derivatives of $\mathscr{H}$ with respect to u_n, P_n and φ_n. The first two are conventional and straightforward. For example,

$$\begin{aligned}\frac{\partial(\Delta_+ u_m)^2}{\partial u_n} &= \frac{1}{a^2}\frac{\partial(u_{m+1}-u_m)^2}{\partial u_n} = \frac{2}{a^2}(u_{m+1}-u_m)\frac{\partial(u_{m+1}-u_m)}{\partial u_n},\\ &= \frac{2}{a^2}(u_{m+1}-u_m)(\delta_n^{m+1}-\delta_n^m),\\ &= \frac{2}{a^2}(u_n-u_{n-1}-u_{n+1}+u_n) = 2\Delta_1^2 u_n,\end{aligned} \quad (8.9)$$

where δ_g^p is the Kronecker delta. To find the derivative of $\partial^+\varphi$ with respect to φ_n, note first that, from (8.3),

$$\partial^+\varphi_n = a^{-1}[\varphi_{n+1}-\varphi_n-\tfrac{1}{2}(\varphi_{n+2}-2\varphi_{n+1}+\varphi_n)+\tfrac{1}{3}(\varphi_{n+3}-3\varphi_{n+2}+3\varphi_{n+1}-\varphi_n)-\ldots].$$

Then

$$\begin{aligned}P_m\frac{\partial(\partial^+\varphi_m)}{\partial\varphi_n} &= a^{-1}P_m[\delta_n^{m+1}-\delta_n^m-\tfrac{1}{2}(\delta_n^{m+2}-2\delta_n^{m+1}+\delta_n^m)\\ &\quad+\tfrac{1}{3}(\delta_n^{m+2}-3\delta_n^{m+2}+3\delta_n^{m+1}-\delta_n^m)-\ldots],\\ &= a^{-1}[P_{n-1}-P_n-\tfrac{1}{2}(P_{n-2}-2P_{n-1}+P_n)+\tfrac{1}{3}(P_{n-3}-3P_{n-2}+3P_{n-1}-P_n)-\ldots],\\ &= -\partial^-P_n,\end{aligned} \quad (8.10)$$

where

$$\partial^-P_n = \sum_{m=1}^{\infty} m^{-1}a^{m-1}\Delta_-^m P_n \quad (8.11)$$

and

$$\begin{aligned}\Delta_- P_n &= (P_n-P_{n-1})/a,\\ \Delta_-^2 P_n &= (P_n-2P_{n-1}+P_{n-2})/a^2,\\ \Delta_-^3 P_n &= (P_n-3P_{n-1}+3P_{n-2}-P_{n-3})/a^3,\\ &\vdots\end{aligned} \quad (8.12)$$

i.e. ∂^-P_n is the expansion of $\partial_1 P$ in an infinite series of backward differences [7]. Finally

$$\frac{\partial(\partial^+\varphi_m)^2}{\partial\varphi_n} = 2\partial^+\varphi_m\frac{\partial(\partial^+\varphi_m)}{\partial\varphi_n} = -2\partial^-\partial^+\varphi_n. \quad (8.13)$$

We now find the equations of motion:

$$-\frac{\partial \mathscr{H}}{\partial u_n} = c_{11}\Delta_1^2 u_n + d_{11}\Delta_1^2 P_n = \rho \ddot{u}_n,$$
$$-\frac{\partial \mathscr{H}}{\partial P_n} = d_{11}\Delta_1^2 u_n + b_{11}\Delta_1^2 P_n - \varepsilon_0{}^{-1}\eta^{-1}P_n - \partial^+\varphi_n = 0, \tag{8.14}$$
$$-\frac{\partial \mathscr{H}}{\partial \varphi_n} = -\varepsilon_0\partial^-\partial^+\varphi_n + \partial^- P_n = 0.$$

The first two of (8.14) are the same as (8.6), with (8.7). This justifies the assumption (8.8) for the electric enthalpy. Also, we have found the third of (8.14), which has the third of (7.16) as its continuum form.

If the lattice is of finite thickness spanning an odd number of atoms with the end ones at $n = \pm N$, the conditions for free boundaries are

$$-\frac{\partial \mathscr{H}}{\partial(\Delta_+ u_{\pm N})} = c_{11}\Delta_+ u_{\pm N} + d_{11}\Delta_+ P_{\pm N} = 0,$$
$$-\frac{\partial \mathscr{H}}{\partial(\Delta_+ P_{\pm N})} = d_{11}\Delta_+ u_{\pm N} + b_{11}\Delta_+ P_{\pm N} + b_0 = 0, \tag{8.15}$$
$$-\frac{\partial \mathscr{H}}{\partial(\partial^+\varphi_{\pm N})} = \varepsilon_0\partial^+\varphi_{\pm N} - P_{\pm N} = 0,$$

where

$$\Delta_+ f_{\pm N} = (f_{\pm(N+1)} - f_N)/a, \quad \partial^+\varphi_{\pm N} = (\partial^+\varphi_n)_{n=\pm N}. \tag{8.16}$$

In general, admissible boundary conditions are: the specification of one member of each of the three products, at each end,

$$u_{\pm N}(c_{11}\Delta_+ u_{\pm N} + d_{11}\Delta_+ P_{\pm N}),\ P_{\pm N}(d_{11}\Delta_+ u_{\pm N} + b_{11}\Delta_+ P_{\pm N} + b_0),\ \varphi_{\pm N}(\varepsilon_0\partial^+\varphi_{\pm N} - P_{\pm N});$$

i.e. eight possible combinations at each end, for the one-dimensional case.

The dispersion relation for (8.14) is obtained by inserting the functions

$$u_n = A e^{i(\xi na - \omega t)}, \quad P_n = B e^{i(\xi na - \omega t)}, \quad \varphi_n = C e^{i(\xi na - \omega t)} \tag{8.17}$$

in the equations and eliminating A, B and C, with the result

$$\begin{vmatrix} 2a^{-2}c_{11}(1-\cos\xi a) - \rho\omega^2 & 2a^{-2}d_{11}(1-\cos\xi a) \\ 2a^{-2}d_{11}(1-\cos\xi a) & 2a^{-2}b_{11}(1-\cos\xi a) + a_{11} + \varepsilon_0^{-1} \end{vmatrix} = 0. \tag{8.18}$$

This two-branch dispersion relation is to be compared with (7.18) – the dispersion relation for the long wave approximation. For $\xi a \ll 1$ and real,

$$2a^{-2}(1-\cos\xi a) \to \xi^2,$$

so that (8.18) reduces to (7.18), as expected. Thus, the long wave behavior of the real branch of (8.18) is again given by (7.20). As for the imaginary branch of (8.18), the wave number at zero frequency is given by

$$\sin\frac{\xi a}{2} = \frac{ia}{2l}; \text{ or } \xi a = \frac{ia}{\lambda}, \text{ where } \sinh\frac{a}{2\lambda} = \frac{a}{2l}, \tag{8.19}$$

instead of $\xi a = ia/l$. That is, although the low frequency ends of the branches for the long wave approximation (7.18) are qualitatively the same as the low frequency limit of the branches for the lattice, only the real branches are quantitatively the same there (Fig. 13). The

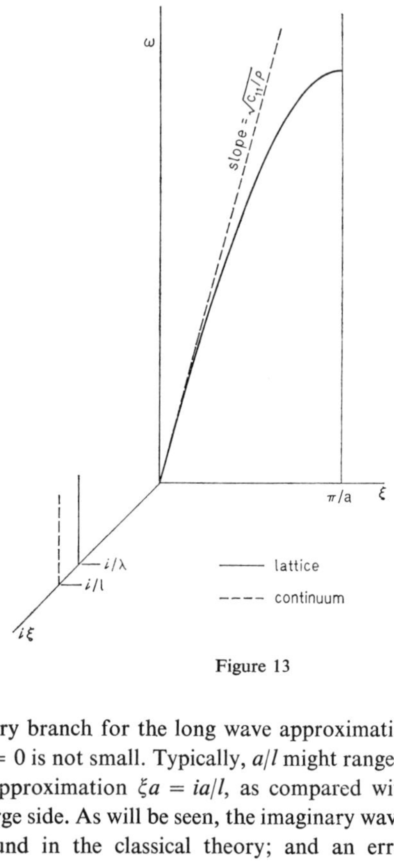

Figure 13

limit of the imaginary branch for the long wave approximation is somewhat in error, as the wave number at $\omega = 0$ is not small. Typically, a/l might range between 1 and 2; then the error in the continuum approximation $\xi a = ia/l$, as compared with ia/λ, would range from 4 % to 12 % – on the large side. As will be seen, the imaginary wave number is associated with surface effects not found in the classical theory; and an error in the wave number affects the amplitude of these effects and their spatial rate of decay into the interior.

The range of a/l from 1 to 2 stems from calculations, by Askar, Lee and Cakmak [49], of the values of the new constants b_{11} and d_{11} (and, consequently, l) for several alkali halide crystals. They found the relations of these constants to the constants in a three-dimensional Cochran [46] lattice of Dick-Overhauser [47] shell-model atoms in an NaCl structure. The correspondence is not complete because the NaCl lattice is diatomic whereas the continuum in Article 7 is a simple (i.e. not compound) one. However the long wave, *low frequency* limit of the diatomic lattice equations has the same form as the continuum equations in Article 7, as is shown in Article 12.

9. Capacitance of Thin, Dielectric Films

The example of the electrical capacitance of thin, dielectric films supplies a vehicle suitable for comparisons of experimental data with the predictions of the classical theory of elastic dielectrics, polarization gradient theory and lattice theory.

Consider a metal-dielectric-metal sandwich whose middle plane, $x_2 - x_3$, is a plane of geometrical and material symmetry. If end effects are neglected, the fields are one-dimensional and the equations of equilibrium are given by (7.16) with $\ddot{u}_1 = 0$.

Let us first solve the problem within the framework of the classical theory. Then b_{11} and d_{11} are zero and the equations of equilibrium reduce to (7.23) with $\ddot{u}_1 = 0$:

$$c_{11}\partial_1^2 u_1 = 0,\ a_{11}P_1 + \partial_1\varphi = 0,\ -\varepsilon_0\partial_1^2\varphi + \partial_1 P_1 = 0. \tag{9.1}$$

For a dielectric layer with traction-free surfaces at $x_1 = \pm h$, on which are impressed voltages $\pm V$, the boundary conditions, according to the classical theory, are

$$[\partial_1 u_1]_{x_1=\pm h} = 0,\quad [\varphi]_{x_1=\pm h} = \pm V; \tag{9.2}$$

and the solution of (9.1), subject to these boundary conditions, is, except for additive constants in u_1 and φ,

$$u_1 = 0,\ P_1 = -\varepsilon_0\eta Vh,\ \varphi = Vx_1/h, \tag{9.3}$$

where η, $= \varepsilon_1^{-1}a_{11}^{-1}$, is the dielectric susceptibility.

The capacitance (per unit area) is the ratio of the surface charge (per unit area) to the voltage drop across the layer:

$$C = \frac{[\varepsilon_0\partial\varphi - P_1]_{x_1=\pm h}}{2V} = \frac{\varepsilon_0(1+\eta)}{2h} = \frac{\varepsilon}{2h}, \tag{9.4}$$

where ε is the permittivity of the dielectric.

Thus, according to the classical theory, there is no strain, the polarization is uniform, the potential varies linearly through the thickness and the capacitance is inversely proportional to the thickness, so that a graph of inverse capacitance vs. thickness is a straight line through the origin. However, in a series of experiments with a variety of very thin dielectric films between metal electrodes, Mead [50, 51] found a different relation between inverse capacitance and thickness. His experimental data fall on straight lines which, if extended to zero thickness, have positive intercepts of inverse capacitance, as illustrated in Fig. 14. Initially [50], Mead sug-

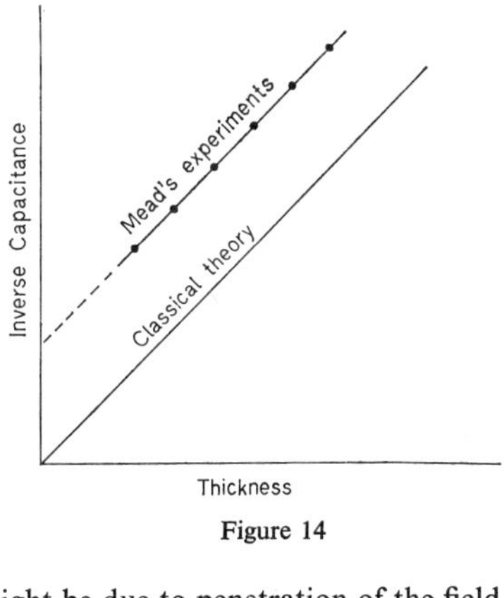

Figure 14

gested that the anomaly might be due to penetration of the field into the electrodes; but subsequently [51] he abandoned that view, although, meanwhile, it had been supported by Ku and Ullman [52]. An alternative explanation is found in the augmented theory of elastic dielectrics [53].

The solution of the augmented equations of equilibrium, (7.16) with $\ddot{u}_1 = 0$, analogous to the one just given for the classical equations, requires a boundary condition in addition to the surface traction and potential. The additional condition can be the specification of the polarization at the surface of the dielectric. Now, the polarization at a boundary of the dielectric in a metal-dielectric-metal sandwich will depend on the physical properties of the adjacent electrode and metal-dielectric interface; and these properties are outside the compass of the theory of dielectrics. However, since the polarization in the metal is zero, it is reasonable to assume that the surface polarization in the dielectric, if not actually zero, will lie between zero and the classical value given by the second of (9.3). Thus, assuming that the two electrodes and interfaces are the same, their influence on the surface polarization may be introduced, phenomenologically, by setting the boundary condition

$$[P_1]_{x_1=\pm h} = -k\varepsilon_0 \eta V/h, \quad 0 \leqslant k \leqslant 1. \tag{9.5a}$$

The classical condition is $k = 1$ while $k = 0$ describes continuity of polarization across the interfaces. We suppose also, as is assumed in the classical theory, that the traction across the interfaces is zero. From (7.13), this condition is

$$[T_{11}]_{x_1=\pm h} = [c_{11}\partial_1 u_1 + d_{11}\partial_1 P_1]_{x_1=\pm h} = 0. \tag{9.5b}$$

Finally, we suppose that the voltages applied to the dielectric at the interfaces are again

$$[\varphi]_{x_1=\pm h} = \pm V. \tag{9.5c}$$

We have now to solve (7.16), with $\ddot{u}_1 = 0$, subject to the boundary conditions (9.5a, b, c). Let

$$u_1 = B_1 \cosh\frac{x_1}{l}, \quad P_1 = A_2 + B_2\cosh\frac{x_1}{l}, \quad \varphi = A_3 x_1 + B_3 \sinh\frac{x_1}{l}. \tag{9.6}$$

Upon substituting (9.6) in (7.16), we find

$$A_2 = \varepsilon_0 \eta A_3, \quad B_3 = lB_2/\varepsilon_0 = -lc_{11}B_1/\varepsilon_0 d_{11}, \tag{9.7}$$

where

$$l = \left(\frac{b_{11}c_{11} - d_{11}^2}{c_{11}(a_{11}+\varepsilon_0^{-1})}\right)^{\frac{1}{2}} = \left(\frac{\varepsilon_0(b_{11}c_{11}-d_{11}^2)}{c_{11}(1+\eta^{-1})}\right)^{\frac{1}{2}} \tag{9.8}$$

i.e. l is the same as in (7.22).

As for boundary conditions, (9.5b) is satisfied identically while (9.5a) and (9.5c) become

$$\begin{aligned} A_2 + B_2\cosh(h/l) &= -k\varepsilon_0\eta V/h, \\ A_3 h + B_3 \sinh(h/l) &= V, \end{aligned} \tag{9.9}$$

respectively. From (9.7) and (9.9),

$$\begin{aligned} A_3 &= (B_3/\eta l)\cosh(h/l) + kV/h, \\ B_3 &= (1-k)\eta V/[\eta\sinh(h/l) + (h/l)\cosh(h/l)]. \end{aligned}$$

The remaining constants are obtained from (9.10) and (9.7).

The capacitance is

$$C = \frac{[\varepsilon_0\partial\varphi - P_1]_{x_1=\pm h}}{2V} = \frac{\varepsilon}{2h}\,\frac{1+(k\eta l/h)\tanh(h/l)}{1+(\eta l/h)\tanh(h/l)}, \tag{9.11}$$

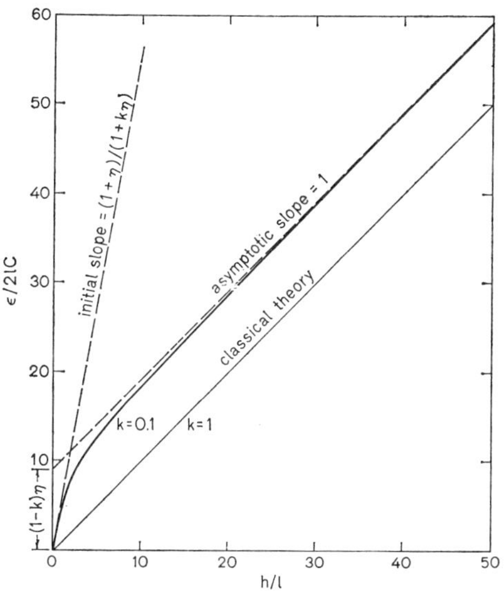

Figure 15

if we ignore any voltage drop that may occur in the electrodes.

In Fig. 15, the curve marked $k = 0.1$ shows the relation between normalized inverse capacitance and normalized thickness, according to (9.11), for the case $\eta = 10$, $k = 0.1$. Calculations from Mead's data, for small k, indicate that the material property l for all his dielectrics is of the order of magnitude of a few angstroms; and this is supported by Askar, Lee and Cakmak's calculations for alkali halides [49]. Hence, Mead's data, which do not extend below a thickness of 30A, would be well to the right of the knee of the curve $k = 0.1$, in Fig. 15, and so would give the appearance of a linear relation which, if extended to zero thickness, would have a non-zero, positive, intercept of inverse capacitance. This intercept, according to (9.11), is

$$C_0^{-1} = 2l(1-k)\eta/\varepsilon. \tag{9.12}$$

If $k = 1$, the intercept reduces to zero and, in fact, the whole solution reduces to the classical one. However, it seems unlikely that the presence of the metal would not influence the polarization of the dielectric at the metal-dielectric interface.

The polarization and potential vary across the thickness of the dielectric as shown in Fig. 16. The absolute value of the polarization is almost uniform across the major portion of the thickness and slightly less than the uniform polarization of the classical theory; but then drops sharply, near the surfaces, to boundary values k times the classical polarization, as specified. The potential has an almost uniform gradient, less than the uniform gradient of the classical theory, over most of the thickness, but then increases sharply on approaching the boundaries. The extremely localized surface effects are accounted for by the contribution of the polarization gradient to the stored energy and are associated, mathematically, with the imaginary branch of the dispersion relation — a branch not present in the classical theory.

Now consider the analogous solution of the lattice equations (8.14), with $\ddot{u}_n = 0$, subject to boundary conditions analogous to (9.5):

$$P_{\pm N} = -k\varepsilon_0\eta V/h, \; c_{11}\Delta_+ u_{\pm N} + d_{11}\Delta_+ P_{\pm N} = 0, \; \varphi_{\pm N} = \pm V, \tag{9.13}$$

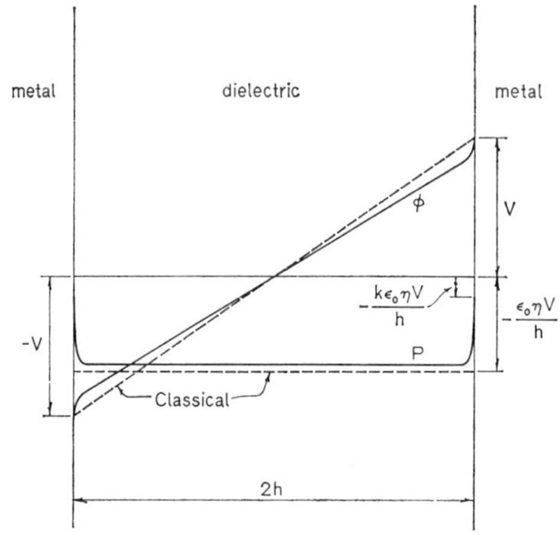

Figure 16

where $h = Na$. We take

$$u_n = B_1' \cosh\frac{na}{\lambda}, \; P_n = A_2' + B_2' \cosh\frac{na}{\lambda}, \; \varphi_n = A_3' na + B_3' \sinh\frac{na}{\lambda} \tag{9.14}$$

and note that $\partial^+[f_n(na)] = [\partial_1 f(x_1)]_{x_1 = na}$ and

$$\Delta^2 \cosh\frac{na}{\lambda} = a^{-2}\left[\cosh\frac{(n+1)a}{\lambda} + \cosh\frac{(n-1)a}{\lambda} - 2\cosh\frac{na}{\lambda}\right] = 4a^{-2}\sinh^2\frac{a}{2\lambda}\cosh\frac{na}{\lambda}. \tag{9.15}$$

Then, substituting (9.14) in (8.14), we find, analogous to (9.7),

$$A_2' = \varepsilon_0 \eta A_3', \; B_3' = \lambda B_2'/\varepsilon_0 = -\lambda c_{11} B_1'/\varepsilon_0 d_{11} \tag{9.16}$$

and, in place of (9.8),

$$\sinh(a/2\lambda) = a/2l, \tag{9.17}$$

where l is the same as in (9.8). Thus, (9.16) is the same as (9.7) with l replaced by λ; but l and λ are not quite the same: being related in accordance with (9.17) and differing by some 4 to 12 percent, as remarked following (8.19).

Application of the boundary conditions (9.13) leads, by the same procedure as for the continuum, to

$$A_3' = \frac{B_3'}{\eta\lambda}\cosh\frac{h}{\lambda} + \frac{kV}{h}, \; B_3' = \frac{(1-k)\eta V}{\eta \sinh\frac{h}{\lambda} + \frac{h}{\lambda}\cosh\frac{h}{\lambda}}, \tag{9.18}$$

analogous to (9.10). Finally, the capacitance is

$$C = \frac{\varepsilon_0 \partial^+ \varphi_{\pm N} - P_{\pm N}}{2V} = \frac{\varepsilon}{2h}\,\frac{1 + (k\eta\lambda/h)\tanh(h/\lambda)}{1 + (\eta\lambda/h)\tanh(h/\lambda)}. \tag{9.19}$$

Thus, the entire solution is identical with the continuum one except that the displacement and polarization have significance only at the atom sites and l is replaced by λ. In particular, the curve in Fig. 15 is applicable to the lattice if l is replaced by λ in both abscissa and ordinate.

10. Surface Energy of Deformation and Polarization

It will be observed that the internal energy density of deformation and polarization, (7.8), contains the term $b_{ij}^0 P_{j,i}$ linear in the polarization gradient. As remarked in Article 3, the analogous term linear in the strain, say $c_{ij}^0 S_{ij}$, is omitted as it is of no consequence. It leads to a homogeneous stress which, in a bounded body with a free surface, can be removed by a homogeneous strain which, in turn, can be regarded as the reference configuration. The situation is otherwise for the term linear in the polarization gradient. Although this, too, produces a homogeneous field, it is a field of E_{ij} rather than T_{ij}, as may be seen by referring to the constitutive equations (7.9). The removal of $n_i E_{ij}$ to free a surface (see (7.7)) results, as is illustrated below, in a polarization and strain which decay exponentially into the interior of the body.

As is described in Article 3, the energy required to separate a body into two parts comprises a bond energy and a remainder. The former is what would be required to break the atomic bonds across the surface if the strain, and now the polarization, were prevented from developing, say by hypothetical mechanical and electrical external fields. The release of the hypothetical fields would result in a deformation and polarization, localized near the surface, with which is associated a surface energy of deformation and polarization – always negative. Thus, the energy required to separate a body into two parts is the bond energy less the absolute value of the surface energy of deformation and polarization. The latter is introduced by the linear term, $b_{ij}^0 P_{j,i}$, in W^L and a formula for that part of the surface energy may be found as follows.

The total energy in the system, in equilibrium, is

$$\int_{V^*} W \,\mathrm{d}V = \int_{V^*} (W^L + \tfrac{1}{2}\varepsilon_0 \varphi_{,i}\varphi_{,i}) \mathrm{d}V = \int_V W^L \mathrm{d}V + \int_{V^*} \tfrac{1}{2}\varepsilon_0 \varphi_{,i}\varphi_{,i} \mathrm{d}V, \tag{10.1}$$

where V is the volume occupied by the dielectric solid bounded by a surface S separating V from an outer vacuum V'; and $V^* = V + V'$.

By means of the consititutive equations (7.9), the form of W^L may be converted from (7.8) to

$$2W^L = T_{ij} S_{ij} + E_{ij} P_{j,i} - E_i^L P_i + b_{ij}^0 P_{j,i}. \tag{10.2}$$

Through the use of the chain rule of differentiation and the divergence theorem,

$$\begin{aligned}
\int_V T_{ij} S_{ij} dV &= \int_V T_{ij} u_{j,i} dV = \int_V [(T_{ij} u_j)_{,i} - T_{ij,i} u_j] \, dV \\
&= \int_S n_i T_{ij} u_j dS - \int_V T_{ij,i} u_j dV,
\end{aligned}$$

$$\int_V E_{ij} P_{j,i} dV = \int_V [(E_{ij} P_j)_{,i} - E_{ij,i} P_j] dV = \int_S n_i E_{ij} P_j dS - \int_V E_{ij,i} P_j dV,$$

$$\int_V b_{ij}^0 P_{j,i} dV = \int_S n_i b_{ij}^0 P_j dS,$$

$$\begin{aligned}
\int_{V^*} \varphi_{,i}\varphi_{,i} dV &= \int_{V^*} [(\varphi_{,i}\varphi)_{,i} - \varphi_{,ii}\varphi] \, dV \\
&= \int_S n_i [\varphi_{,i}] \varphi \, dS - \int_V \varphi_{,ii}\varphi \, dV - \int_{V'} \varphi_{,ii}\varphi \, dV,
\end{aligned}$$

where $[\varphi_{,i}]$ is the jump in $\varphi_{,i}$ across S. Upon inserting these results in (10.1), we have

$$\int_{V^*} W\,dV = \tfrac{1}{2}\int_S n_i[b^0_{ij}P_j + T_{ij}u_j + E_{ij}P_j + \varepsilon_0[\varphi_{,i}]\varphi]\,dS$$
$$-\tfrac{1}{2}\int_V (T_{ij,i}u_j + E_{ij,i}P_j + E^L_j P_j + \varepsilon_0\varphi_{,ii}\varphi)\,dV - \int_{V'}\varepsilon_0\varphi_{,ii}\varphi\,dV. \quad (10.3)$$

Application of the equilibrium equations, i.e. (7.5) and (7.6) with $\ddot{u}_j = 0$, to (10.3) reduces it to

$$\int_{V^*} W\,dV = \tfrac{1}{2}\int_V (f_j u_j + E^0_j P_j)\,dV + \tfrac{1}{2}\int_S n_i[T_{ij}u_j + E_{ij}P_j - (-\varepsilon_0[\varphi_{,i}] + P_i) + b^0_{ij}P_j]\,dS. \quad (10.4)$$

Then, in the obsence of external fields ($f_j = E^0_j = 0$) and with the boundary free, i.e. (7.7) with $t_j = 0$, (10.4) reduces to

$$\int_{V^*} W\,dV = \tfrac{1}{2}\int_S n_i b^0_{ij} P_j\,dS. \quad (10.5)$$

This is the energy of deformation and polarization which must be added to the bond energy to obtain the total energy required to separate the material along the surface S. As indicated by Schwartz [54], the additional energy is always negative as a consequence of the positive definiteness of the quadratic part of the energy density which, in turn, is required for stability – i.e. because energy must be stored, rather than generated, during the application of external body or surface forces. Thus,

$$\int_{V^*} W\,dV_i = \int_{V^*} [\tfrac{1}{2}\varphi_{,i}\varphi_{,i} + (W^L - b^0_{ij}P_{j,i})]\,dV + \int_V b^0_{ij}P_{j,i}\,dV$$
$$= \int_{V^*} [\tfrac{1}{2}\varphi_{,i}\varphi_{,i} + (W^L - b^0_{ij}P_{j,i})]\,dV + \int_S n_i b^0_{ij}P_j\,dS. \quad (10.6)$$

Subtracting (10.5) from (10.6), we have

$$\int_{V^*} [\tfrac{1}{2}\varphi_{,i}\varphi_{,i} + (W^L - b^0_{ij}P_{j,i})]\,dV + \tfrac{1}{2}\int_S n_i b^0_{ij}P_j\,dS = 0. \quad (10.7)$$

But the volume integral in (10.8) is positive. Hence

$$\int_S n_i b^0_{ij}P_j\,dS < 0. \quad (10.8)$$

Finally, we may define

$$W^S = \tfrac{1}{2}[n_i b^0_{ij} P_j]_S \quad (10.9)$$

as the surface energy of deformation and polarization per unit area, sometimes called the surface tension.

There follows the solution for the displacement, polarization, potential and surface energy of deformation and polarization in the half-space of a centrosymmetric cubic crystal bounded by a free (100) face. For this problem, the fields are one-dimensional and in equilibrium so that, for the half-space $x_1 = 0$, the equations of motion (7.14) or (7.16) reduce to

$$\begin{aligned} c_{11}\partial_1^2 u_1 + d_{11}\partial_1^2 P_1 &= 0,\\ d_{11}\partial_1^2 u_1 + b_{11}\partial_1^2 P_1 - a_{11}P_1 - \partial_1\varphi &= 0,\\ -\varepsilon_0\partial_1^2\varphi + \partial_1 P_1 &= 0; \end{aligned} \quad (10.10)$$

and the boundary conditions (7.7) become, on $x_1 = 0$,

$$\begin{aligned} c_{11}\partial_1 u_1 + d_{11}\partial_1 P_1 &= 0, \\ d_{11}\partial_1 u_1 + b_{11}\partial_1 P_1 &= -b_0, \\ -\varepsilon_0\partial_1\varphi + P_1 &= 0. \end{aligned} \tag{10.11}$$

Consider

$$u_1 = A_1 e^{-x_1/l}, \; P_1 = A_2 e^{-x_1/l}, \; \varphi = A_3 e^{-x_1/l}. \tag{10.12}$$

Upon substituting (10.12) into (10.10), we find

$$A_3 = -lA_2/\varepsilon_0 = lc_{11}A_1/\varepsilon_0 d_{11} \tag{10.13}$$

where l is again given by (7.22). With (10.13), the first and third of the boundary conditions (10.11) are satisfied identically and the second boundary condition yields

$$A_1 = -\frac{b_0 d_{11}}{c_{11}\, l(a_{11}+\varepsilon_0^{-1})}. \tag{10.14}$$

Then, from (10.13),

$$A_2 = \frac{b_0}{l(a_{11}+\varepsilon_0^{-1})}, \; A_3 = -\frac{b_0}{1+a_{11}\varepsilon_0}. \tag{10.15}$$

It may be seen that the freeing of the boundary results in a field of displacement, polarization and potential localized at the surface, decaying into the interior with decay constant l. Associated with the localized field is a surface energy of deformation and polarization, per unit area, given by (10.9):

$$W^S = -\frac{b_0^2}{2la_{11}(1+a_{11}^{-1}\varepsilon_0^{-1})}. \tag{10.16}$$

Note that, from (6.26), (6.28) and (6.32),

$$1+a_{11}^{-1}\varepsilon_0^{-1} = \varepsilon_{11}^S/\varepsilon_0 = \kappa_{11} \tag{10.17}$$

i.e. the dielectric constant in the [100] direction.

As noted at the end of Article 8, Askar, Lee and Cakmak [49] have made calculations of the new material constants for several alkali halide crystals. From these they find, in the case of NaCl, for example,

$$l = 1.3\times 10^{-8}\text{ cm}, \; W^S = -59 \text{ erg/cm}^2. \tag{10.18}$$

The decay constant l is about half the distance between nearest neighbor atoms in the NaCl crystal, indicating extremely rapid decay from the surface into the interior. The surface energy of deformation and polarization is about 30 % of the bond energy, so that it is far from negligible. Both of these results conform to experimental observations [20, 21].

Additional solutions for surface energies of deformation and polarization (in isotropic materials) have been obtained by Schwartz [54] for internal and external spherical surfaces and by Askar, Lee and Cakmak [55] for internal spherical and cylindrical surfaces and a crack. The latter have used average values of their previous calculations of constants for alkali halides [49] to obtain some quantitative results. They find that the absolute values of W^S are reduced by curvature of internal surfaces and they also conclude that W^S is finite at the root of a crack.

11. Acoustical and Optical Activity in Alpha Quartz

Acoustical activity (rotation of the direction of mechanical displacement along the path of a transverse, elastic wave) has recently been observed by Pine [56] in α-quartz. The possibility of the phenomenon appears first to have been mentioned by Silin [57]. It was accounted for by Toupin [19] on the basis of the first strain gradient theory (Article 3ii). Portigal and Burstein [58] found an equivalent result by assigning dependence of the elastic stiffness on the wave vector. In the present Article, it is shown [59] that both acoustical and optical activities are accounted for in the polarization gradient theory (Article 7). In what follows, the field equations are exhibited for the coupled elastic-electric-magnetic system, the problem of shear waves along the trigonal axis of α-quartz is solved, formulas are obtained for the optical and acoustical rotatory powers and numerical values of the new material constants, in the formulas, are calculated from experimental data. Essentially, the theory has it that the appearance of optical activity depends on an interaction between the polarization and the polarization gradient; the appearance of acoustical activity depends on interactions of the strain with both the polarization and the polarization gradient, and is absent if either interaction is missing.

When the magnetic field is taken into account, the field equations of Article 7 become

$$T_{ij,i}+f_j = \rho\ddot{u}_j, \tag{11.1}$$

$$E_{ij,i}+E_j^L+E_j+E_j^0 = 0, \tag{11.2}$$

$$\delta_{ijk}E_{k,j}+\dot{B}_i = 0, \tag{11.3}$$

$$\mu_0^{-1}\delta_{ijk}B_{k,j}-\varepsilon_0\dot{E}_i-\dot{P}_i = 0, \tag{11.4}$$

$$\varepsilon_0 E_{i,i}+P_{i,i} = 0, \tag{11.5}$$

$$B_{i,i} = 0, \tag{11.6}$$

where B_i is the magnetic flux density, μ_0 is the magnetic permeability of a vacuum and δ_{ijk} is the unit alternating tensor. Also, as before,

$$T_{ij} = \frac{\partial W^L}{\partial S_{ij}}, \quad E_i^L = -\frac{\partial W^L}{\partial P_i}, \quad E_{ij} = \frac{\partial W^L}{\partial P_{j,i}}, \tag{11.7}$$

where

$$W^L = \tfrac{1}{2}a_{ij}P_iP_j+\tfrac{1}{2}b_{ijkl}P_{j,i}P_{l,k}+\tfrac{1}{2}c^P_{klij}S_{ij}S_{kl}+d_{ijkl}P_{j,i}S_{kl}+f_{ijk}P_iS_{jk}+j_{ijk}P_iP_{k,j}, \tag{11.8}$$

$$S_{ij} = \tfrac{1}{2}(u_{j,i}+u_{i,j}). \tag{11.9}$$

It is convenient to eliminate B_i at the start – by subtracting the curl of (11.3) from the time derivative of (11.4), with the result:

$$E_{j,ii}-E_{i,ij} = \varepsilon_0\mu_0\ddot{E}_j+\mu_0\ddot{P}_j. \tag{11.10}$$

Equations (11.5) and (11.6) are not independent of (11.4) and (11.3), respectively, and may be disregarded for the present purpose. Thus, (11.10), along with (11.1) and (11.2):

$$\begin{aligned} &T_{ij,i}+f_j = \rho\ddot{u}_j,\\ &E_{ij,i}+E_j^L+E_j+E_j^0 = 0,\\ &E_{j,ii}-E_{i,ji} = \varepsilon_0\mu_0\ddot{E}_j+\mu_0\ddot{P}_j, \end{aligned} \tag{11.11}$$

are the field equations governing mechanical and electromagnetic waves – coupled through the constitutive equations:

$$\begin{aligned} T_{ij} &= c^P_{ijkl}S_{kl}+f_{kij}P_k+d_{klij}P_{l,k},\\ -E^L_j &= f_{jkl}S_{kl}+a_{jk}P_k+j_{jkl}P_{l,k},\\ E_{ij} &= d_{ijkl}S_{kl}+j_{kij}P_k+b_{ijkl}P_{l,k}, \end{aligned} \tag{11.12}$$

which are obtained from (11.7) and (11.8).

We consider plane, transverse waves propagating along the x_3-axis; i.e. u_3. P_3 and E_3 are zero and the remaining u_i, P_i, and E_i are functions of x_3 and t only. Then the field equations (11.11), with f_j and E^0_j zero, reduce to

$$\begin{aligned} &T_{13,3} = \rho\ddot{u}_1, \quad T_{23,3} = \rho\ddot{u}_2,\\ &E_{31,3}+E^L_1+E_1 = 0, \quad E_{32,3}+E^L_2+E_2 = 0,\\ &E_{1,33} = \varepsilon_0\mu_0\ddot{E}_1+\mu_0 E_1, \quad E_{2,33} = \varepsilon_0\mu_0\ddot{E}_2+\mu_0 P_2, \end{aligned} \tag{11.13}$$

and the constitutive equations (11.12) reduce to [59]

$$\begin{aligned} T_{31} &= c^P_{44}u_{1,3}-f_{14}P_2+d_{74}P_{1,3}, & T_{32} &= c^P_{44}u_{2,3}+f_{14}P_1+d_{74}P_{2,3},\\ -E^L_1 &= f_{14}u_{2,3}+\varepsilon_0^{-1}\chi_{11}P_1+j_{17}P_{2,3}, & -E^L_2 &= -f_{14}u_{1,3}+\varepsilon_0^{-1}\chi_{11}P_2-j_{17}P_{1,3},\\ E_{31} &= d_{74}u_{1,3}-j_{17}P_2+b_{55}P_{1,3}, & E_{32} &= d_{74}u_{2,3}+j_{17}P_1+b_{55}P_{2,3}. \end{aligned} \tag{11.14}$$

for the crystal class 32 (International) or D_3 (Schoenflies) [60] to which α-quartz belongs. In (11.14), the abridged notation

$$11 \to 1, 22 \to 2, 33 \to 3, 23 \to 4, 32 \to 7, 31 \to 5, 13 \to 8, 12 \to 6, 21 \to 9$$

is used for the subscripts attached to the material constants.

Inserting (11.14) in (11.13), we have

$$\begin{aligned} &c^P_{44}u_{1,33}-f_{14}P_{2,3}+d_{74}P_{1,33} = \rho\ddot{u}_1,\\ &c^P_{44}u_{2,33}+f_{14}P_{1,3}+d_{74}P_{2,33} = \rho\ddot{u}_2,\\ &d_{74}u_{1,33}-2j_{17}P_{2,3}+b_{55}P_{1,33}-f_{14}u_{2,3}-\varepsilon_0^{-1}\chi_{11}P_1+E_1 = 0,\\ &d_{74}u_{2,33}+2j_{17}P_{1,3}+b_{55}P_{2,33}+f_{14}u_{1,3}-\varepsilon_0^{-1}\chi_{11}P_2+E_2 = 0,\\ &E_{1,33} = \varepsilon_0\mu_0\ddot{E}_1+\mu_0\ddot{P}_1,\\ &E_{2,33} = \varepsilon_0\mu_0\ddot{E}_2+\mu_0\ddot{P}_2. \end{aligned} \tag{11.15}$$

Now, take

$$\begin{aligned} u_1 &= A_1\sin\zeta(x_3-vt), & u_2 &= A_2\cos\zeta(x_3-vt),\\ P_1 &= B_1\sin\zeta(x_3-vt), & P_2 &= B_2\cos\zeta(x_3-vt),\\ E_1 &= C_1\sin\zeta(x_3-vt), & E_2 &= C_2\cos\zeta(x_3-vt) \end{aligned} \tag{11.16}$$

and substitute these functions in (11.15) to find

$$\begin{aligned} &(c^P_{44}-\rho v^2)\zeta A_1-f_{14}B_2+d_{74}\zeta B_1 = 0,\\ &(c^P_{44}-\rho v^2)\zeta A_2-f_{14}B_1+d_{74}\zeta B_2 = 0,\\ &d_{74}\zeta^2A_1-f_{14}\zeta A_2+(b_{55}\zeta^2+\varepsilon_0^{-1}\chi_{11})B_1-2j_{17}\zeta B_2-C_1 = 0,\\ &d_{74}\zeta^2A_2-f_{14}\zeta A_1+(b_{55}\zeta^2+\varepsilon_0^{-1}\chi_{11})B_2-2j_{17}\zeta B_1-C_2 = 0,\\ &\mu_0v^2B_1+(\varepsilon_0\mu_0v^2-1)C_1 = 0,\\ &\mu_0v^2B_2+(\varepsilon_0\mu_0v^2-1)C_2 = 0. \end{aligned} \tag{11.17}$$

Adding and subtracting these equations in pairs, we have

$$\begin{aligned}&(c_{44}^{P}-\rho v^2)(A_1\pm A_2)+(d_{74}\zeta\mp f_{14})(B_1\pm B_2) = 0,\\&(d_{74}\zeta\mp f_{14})(A_1\pm A_2)+(b_{55}\zeta^2+\varepsilon_0^{-1}\chi_{11}\mp 2j_{17}\zeta)(B_1\pm B_2)-(C_1\pm C_2) = 0,\\&\mu_0 v^2(B_1\pm B_2)+(\varepsilon_0\mu_0 v^2-1)(C_1\pm C_2) = 0.\end{aligned}\tag{11.18}$$

Thus, there are two solutions: each corresponding to circularly polarized waves [61, p. 222] since the amplitudes must satisfy

$$A_1 = \pm A_2,\ B_1 = \pm B_2,\ C_1 = \pm C_2 \tag{11.19}$$

with either all upper signs or all lower signs. Upon substituting (11.19) into (11.16), we see that the upper and lower signs give right and left circular polarization, respectively. The equation for the velocities is obtained by setting the determinant of the coefficients of the amplitudes in (11.18) equal to zero:

$$\begin{vmatrix} c_{44}^{P}-\rho v^2 & d_{74}\zeta\mp f_{14} & 0\\ d_{74}\zeta\mp f_{14} & b_{55}\zeta^2+\varepsilon_0^{-1}\chi_{11}\mp 2j_{17}\zeta & 1\\ 0 & 1 & \mu_0^{-1}v^{-2}-\varepsilon_0\end{vmatrix} = 0. \tag{11.20}$$

This is a quadratic equation in v^2, so that there are two pairs of oppositely circularly polarized waves. Each pair of such waves combines to produce a linearly polarized wave with a rotating direction of polarization [61, p. 222]. Thus, we have two cases of rotary polarization. These may be identified as optical and acoustical by separating out first the electromagnetic part of (11.20) and then the electromechanical part. The two may, in fact, be considered separately owing to the large ratio of frequencies (of the order of 10^5) at which the two effects are observed.

The electromagnetic part of the determinant in (11.20) is the minor of the upper left element. Thus, the pair of optical velocities is given by

$$\begin{vmatrix} b_{55}\zeta^2+\varepsilon_0^{-1}\chi_{11}\mp 2j_{17} & 1\\ 1 & \mu_0^{-1}v^{-2}-\varepsilon_0\end{vmatrix} = 0, \tag{11.21}$$

which yields the dispersion formula [61, p. 426]

$$n_{\pm}^2-1 = (\chi_{11}\mp 2\varepsilon_0 j_{17}\zeta+\varepsilon_0 b_{55}\zeta^2)^{-1}, \tag{11.22}$$

where $n_{\pm}$ are the indexes of refraction:

$$n_{\pm} = c/v_{\pm} \tag{11.23}$$

and c is the velocity of light in vacuo:

$$c = (\varepsilon_0\mu_0)^{-1}. \tag{11.24}$$

From (11.22), we have

$$(n_-^2-1)^{-1}-(n_+^2-1)^{-1} = 4\varepsilon_0 j_{17}\zeta. \tag{11.25}$$

Now, define $n = \frac{1}{2}(n_+ + n_-)$ and assume

$$|n_+ - n_-| \ll n_+ + n_-. \tag{11.26}$$

Then (11.25) becomes, approximately,

$$n_+ - n_- = 2(n^2-1)^2\varepsilon_0 j_{17}\zeta_0, \tag{11.27}$$

where ζ_0, $= \zeta/n$, is the wave number in vacuo.

The optical rotatory power, in radians per unit length, is given by [61, p. 222]

$$\theta_{OP} = \tfrac{1}{2}\zeta_0(n_- - n_+). \tag{11.28}$$

Accordingly, from (11.27) and (11.28),

$$\theta_{OP} = -(n^2-1)^2\varepsilon_0 j_{17}\zeta_0^2 \tag{11.29}$$

is the formula for the optical rotatory power in terms of the average index of refraction, n, along the optic axis, the wave length in vacuo, λ_0, $= 2\pi/\zeta_0$, the fundamental constant ε_0 and the material constant j_{17}, $= j_{132}$, which, as may be seen in (11.8), measures the interaction between the polarization and the polarization gradient.

The electromechanical part of the determinant in (11.20) is the minor of the lower right element, so that we have

$$\begin{vmatrix} c_{44}^P - \rho v^2 & d_{74}\zeta \mp f_{14} \\ d_{74}\zeta \mp f_{14} & b_{55}\zeta^2 + \varepsilon_0^{-1}\chi_{11} \mp 2j_{17}\zeta \end{vmatrix} = 0 \tag{11.30}$$

for the equation determining the velocities of the two acoustical waves, as influenced by the quasi-static polarization and polarization gradient. From (11.30),

$$\rho v_\pm^2 = c_{44}^P - (d_{74}\zeta \mp f_{14})^2/(b_{55}\zeta^2 + \varepsilon_0^{-1}\chi_{11} \mp 2j_{17}\zeta). \tag{11.31}$$

In view of (11.19) and the inequality of v_+ and v_-, the superposition of the two waves results in rotary polarization (acoustical activity) with acoustical rotatory power

$$\theta_{AC} = \tfrac{1}{2}\omega(v_-^{-1} - v_+^{-1}) \tag{11.32}$$

where ω is the circular frequency. Both waves are dispersive. At the zero frequency (long wave) limit, from (11.31) and (6.34), with (6.28),

$$\lim_{\zeta\to 0} \rho v_\pm^2 = c_{44}^P - \varepsilon_0 f_{14}^2/\chi_{11} = c_{44}^E, \tag{11.33}$$

which is the result (without acoustical acitivity, since $v_+ = v_-$) that would be obtained if the contribution of the polarization gradient were omitted, i.e. if d_{74}, b_{55} and j_{17} were assumed to be zero. As the frequency increases from zero, the absolute velocity difference, $|v_+ - v_-|$, at first increases, so that the acoustical activity appears and increases. With further increase of frequency, the velocity difference again approaches zero, since

$$\lim_{\xi\to\infty} \rho v_\pm^2 = c_{44}^P - d_{74}^2/b_{55}, \tag{11.34}$$

so that the acoustical activity diminishes; but this is undoubtedly beyond the range of applicability of the continuum theory. Up to moderately large wave numbers, (11.31), with (6.28) and (6.34), gives, to the first order in ζ,

$$v_0/v_\pm = 1 \mp (d_{74} - j_{17}e_{14})e_{14}\zeta/c_{44}^E, \tag{11.35}$$

where $v_0^2 = c_{44}^E/\rho$. In this range, the frequency is approximately proportional to the wave number: $\omega = v_0\zeta$; so that, from (11.32) and (11.35),

$$\theta_{AC} = (d_{74} - j_{17}e_{14})e_{14}\rho\omega^2/(c_{44}^E)^2. \tag{11.36}$$

Thus, at frequencies up to, say, 10^{10} cps, the acoustical rotatory power is approximately proportional to the square of the frequency and depends on the constants ρ, e_{14} and c_{44}^E, which are commonly encountered in piezoelectricity theory, and also on the constants d_{74} and j_{17}, which control the coupling of the polarization gradient with the strain and polarization, respectively.

For α-quartz, all the quantities in the formula (11.29) for optical rotatory power are known except j_{17}. Thus, for left-handed quartz and the sodium D line,

$$\begin{aligned}
\theta_{OP} &= -379 \text{ radian/meter [61, p. 481]},\\
\lambda_0 &= 5893 \times 10^{-10} \text{ meter [61, p. 481]},\\
n &= 1.5533 \text{ [61, p. 481]},\\
\varepsilon_0 &= 8.854 \times 10^{-12} \text{ farad/meter [60]}.
\end{aligned}$$

Hence,

$$j_{17} = -\theta_{OP}\lambda_0^2/4\pi^2\varepsilon_0(n^2-1)^2 = 0.19 \text{ meter}^2/\text{farad}. \tag{11.37}$$

With the value of j_{17} known, all quantities in the formula (11.36) for acoustical rotatory power are known except d_{74}. In particular, Pine [56] finds that the acoustical and optical activities have opposite signs and the acoustical rotatory power along the trigonal axis is about 220 radians/meter at one gigahertz. Thus, for left-handed α-quartz,

$$\begin{aligned}
\theta_{AC} &= 220 \text{ radian/meter [56]},\\
\omega &= 2\pi \times 10^9 \text{ radian/second [56]},\\
\rho &= 2.65 \times 10^3 \text{ kilogram/meter}^3 \text{ [62]},\\
c_{44}^E &= 57.94 \times 10^9 \text{ newton/mreter}^2 \text{ [63]}.\\
e_{14} &= -0.0406 \text{ coulomb/meter}^2 \text{ [63]}.
\end{aligned}$$

Hence

$$d_{74} = \theta_{AC}(c_{44}^E)^2/e_{14}\rho\omega^2 + j_{17}e_{14} = -174 - 0.0077 \text{ volt}. \tag{11.38}$$

The second term, $j_{17}\, e_{14}$, is negligible in comparison with the first, so that we may drop the dependence of acoustical rotatory power on j_{17}, i.e. on the interaction between polarization and polarization gradient, and replace (11.36) with

$$\theta_{AC} = d_{74}e_{14}\rho\omega^2/(c_{44}^E)^2. \tag{11.39}$$

Thus, according to this theory, the presence of acoustical activity in α-quartz depends on the existence of the piezoelectric stress constant e_{14}, $= e_{123}$, and the interaction constant d_{74}, $= d_{3223}$, between strain and polarization gradient; whereas the presence of optical activity depends only on the existence of the interaction constant j_{17}, $= j_{132}$, between polarization and polarization gradient.

12. Diatomic, Elastic, Dielectric Continuum and Lattice

i. Potential and Kinetic Energies.

A continuum theory of a crystal with two polarizable atoms per cell may be produced by combining the procedures employed in Articles 4 and 7 [64]. Atoms in the two component continua are again identified by 1 and 2 with displacements now designated by $u_i^{(\kappa)}$, $\kappa = 1, 2$, and

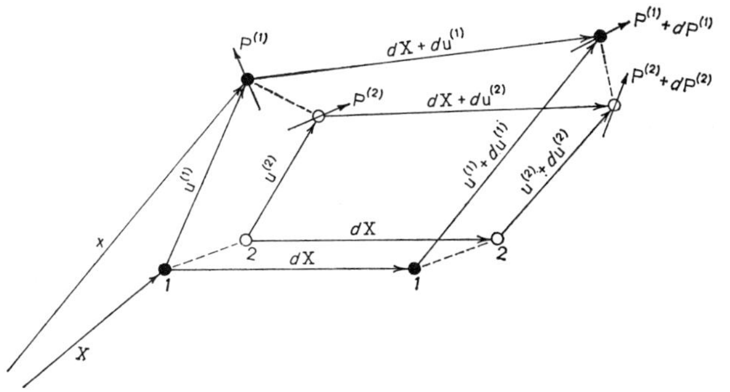

Figure 17

polarizations by $P_i^{(\kappa)}$, as illustrated in Fig. 17; while polarization gradients, strains, relative displacement and relative rotation are denoted by

$$P_{j,i},\ S_{ij}^{(\kappa)} = \tfrac{1}{2}(u_{j,i}^{(\kappa)}+u_{i,j}^{(\kappa)}),\ u_i^* = u_i^{(2)}-u_i^{(1)},\ \omega_{ij}^* = \tfrac{1}{2}(u_{j,i}^*-u_{i,j}^*), \tag{12.1}$$

respectively. The potential energy density of deformation and polarization is assumed to be a quadratic function of $S_{ij}^{(\kappa)}$, $P_i^{(\kappa)}$, $P_{j,i}^{(\kappa)}$, u_i^* and ω_{ij}^*:

$$W^L = W^L(S_{ij}^{(1)}, S_{ij}^{(2)}, P_i^{(1)}, P_i^{(2)}, P_{j,i}^{(1)}, P_{j,i}^{(2)}, u_i^*, \omega_{ij}^*). \tag{12.2}$$

The total potential energy density, W, is again the energy density of deformation and polarization augmented by the energy density of the Maxwell, electric self-field $E_i = -\varphi_{,i}$:

$$W = W^L+\tfrac{1}{2}\varepsilon_0\varphi_{,i}\varphi_{,i}. \tag{1.23}$$

The associated kinetic energy density is, again

$$T = \sum_{\kappa} \tfrac{1}{2}\rho^{(\kappa)}\dot{u}_i^{(\kappa)}\dot{u}_i^{(\kappa)},\ \kappa = 1, 2;\ i = 1, 2, 3, \tag{12.4}$$

where $\rho^{(1)}$ and $\rho^{(2)}$ are the mass densities of the two continua and the "over-dot" notation designates differentiation with respect to time.

ii. Field Equations and Boundary Conditions.

The field equations and boundary conditions, corresponding to the potential and kinetic energy densities (12.3) and (12.4), may be derived, as in Article 7, by means of an extension of a linear version of Toupin's [42] variational principle for the classical theory of elastic dielectrics. The extension accounts for the contributions of the polarization gradient, the two continua and the kinetic energy.

First, we define an electric enthalpy density — the energy density diminished by the product of the Maxwell self-field, E_i, and the electric displacement D_i:

$$H = W-E_iD_i, \tag{12.5}$$

where

$$D_i = \varepsilon_0 E_i+P_i, \tag{12.6}$$

$$E_i = -\varphi_{,i}, \tag{12.7}$$

$$P_i = P_i^{(1)}+P_i^{(2)}+q_* u_i^*. \tag{12.8}$$

The inclusion of the term $q_* u_i^*$, in the polarization density P_i, was suggested by P. C. Y. Lee to accommodate applications to ionic crystals. The product of the charge density q_* and the relative displacement u_i^* represents the ionic, or atomic, polarization as distinguished from the electronic polarizations $P_i^{(\kappa)}$. In what follows, if q_* is not set equal to zero, the superscript $\kappa = 1$ identifies the positive ion and $\kappa = 2$ the negative ion.

Inserting (12.3) and (12.6)-(12.8) in (12.5), we find

$$H = W^L - \tfrac{1}{2}\varepsilon_0 \varphi_{,i}\varphi_{,i} + \varphi_{,i}(P_i^{(1)} + P_i^{(2)} + q_* u_i^*). \tag{12.9}$$

In a body occupying a volume V bounded by a surface S separating V from a vacuum V', the extension of Toupin's variational principle takes the form

$$\delta\int_{t_0}^{t_1} dt \int_{V^*} (T-H)\,dV + \sum_\kappa \int_{t_0}^{t_1} dt \int_V (f_i^{(\kappa)}\delta u_i^{(\kappa)} + E_i^0 \delta P_i^{(\kappa)} + E_i^{(0)} q_* \delta u_i^*)\,dV + \sum_\kappa \int_{t_0}^{t_1} dt \int_S t_i^{(\kappa)} \delta u_i^{(\kappa)} dS = 0, \quad \kappa = 1, 2, \tag{12.10}$$

for independent variations of $u_i^{(\kappa)}$, $P_i^{(\kappa)}$ and φ between fixed limits at times t_0 and t_1. In (12.10), $V^* = V + V'$, $f_i^{(\kappa)}$ and $t_i^{(\kappa)}$ are the external body forces and surface tractions on the two continua and E_i^0 is the external electric field. Now,

$$\delta W^L = \sum_\kappa (T_{ij}^{(\kappa)} \delta S_{ij}^{(\kappa)} - E_i^{(\kappa)} \delta P_i^{(\kappa)} + E_{ij}^{(\kappa)} \delta P_{j,i}^{(\kappa)}) + T_i^* \delta u_i^* + T_{ij}^* \delta \omega_{ij}^*, \tag{12.11}$$

where

$$T_{ij}^{(\kappa)} = \frac{\partial W^L}{\delta S_{ij}^{(\kappa)}}, \quad E_i^{(\kappa)} = -\frac{\partial W^L}{\partial P_i^{(\kappa)}}, \quad E_{ij}^{(\kappa)} = \frac{\partial W^L}{\partial P_{j,i}^{(\kappa)}}, \quad T_i^* = \frac{\partial W^L}{\delta u_i^*}, \quad T_{ij}^* = \frac{\partial W^L}{\partial \omega_{ij}^*}. \tag{12.12}$$

Note that the $E_i^{(\kappa)}$ are the effective local fields of the two continua: as distinguished from the Maxwell self-field E_i.

With the chain rule of differentiation, (12.11) becomes

$$\delta W^L = -\sum_\kappa [T_{ij,i}^{(\kappa)} + (-1)^\kappa (T_{ij,i}^* - T_j^*)]\delta u_j^{(\kappa)} - \sum_\kappa (E_j^{(\kappa)} + E_{ij,i}^{(\kappa)})\delta P_j^{(\kappa)} + \sum_\kappa \{[T_{ij}^{(\kappa)} + (-1)^\kappa T_{ij}^*]\delta u_j^{(\kappa)}\}_{,i} + \sum_\kappa (E_{ij}^{(\kappa)} \delta P_j^{(\kappa)})_{,i}.$$

Treating the remaining part of H in (12.9) similarly, we have, in V:

$$\begin{aligned} \delta H = &-\sum_\kappa [T_{ij,i}^{(\kappa)} + (-1)^\kappa (T_{ij,i}^* - T_j^* - q_* \varphi_{,i})]\delta u_j^{(\kappa)} - \sum_\kappa (E_j^{(\kappa)} + E_{ij,i}^{(\kappa)} - \varphi_{,i})\delta P_j^{(\kappa)} \\ &+ \sum_\kappa [\varepsilon_0 \varphi_{,ii} - P_{i,i}^{(\kappa)} - (-1)^\kappa q_* u_{i,i}^{(\kappa)}]\delta\varphi + \sum_\kappa \{[T_{ij}^{(\kappa)} + (-1)^\kappa T_{ij}^*]\delta u_j^{(\kappa)}\}_{,i} \\ &+ \sum_\kappa (E_{ij}^{(\kappa)} \delta P_j^{(\kappa)})_{,i} - \sum_\kappa \{[\varepsilon_0 \varphi_{,i} - P_i^{(\kappa)} - (-1)^\kappa q_* u_i^{(\kappa)}]\delta\varphi\}_{,i}; \end{aligned} \tag{12.13}$$

and, in V':

$$\delta H = \varepsilon_0 \varphi_{,ii} \delta\varphi - \varepsilon_0 (\varphi_{,i} \delta\varphi)_{,i}. \tag{12.14}$$

Further, from (12.4),

$$\delta \int_{t_0}^{t_1} T\,dt = -\sum_\kappa \int_{t_0}^{t_1} \rho^{(\kappa)} \ddot{u}_i^{(\kappa)} \delta u_i^{(\kappa)} dt. \tag{12.15}$$

Inserting (12.13)-(12.15) in (12.10) and applying the divergence theorem where appropriate, we find

$$\sum_{\kappa}\int_{t_0}^{t_1} dt\int_V [T^{(\kappa)}_{ij,i}+(-1)^{\kappa}(T^*_{ij,i}-T^*_j-q_*\varphi_{,j})+(-1)^{\kappa}q_*E^0_j+f^{(\kappa)}_j-\rho^{(\kappa)}\ddot{u}^{(\kappa)}_j]\delta u^{(\kappa)}_j dV$$
$$+\sum_{\kappa}\int_{t_0}^{t_1} dt\int_V \{(E^{(\kappa)}_j+E^{(\kappa)}_{ij,i}-\varphi_{,j}+E^0_j)\delta P^{(\kappa)}_j+[-\varepsilon_0\varphi_{,ii}+P^{(\kappa)}_{i,i}+(-1)^{\kappa}q_*u^{(\kappa)}_{i,i}]\delta\varphi\}\,dV$$
$$-\sum_{\kappa}\int_{t_0}^{t_0} dt\int_S \{n_i[T^{(\kappa)}_{ij}+(-1)^{\kappa}T^*_{ij}]-t^{(\kappa)}_j\}\delta u^{(\kappa)}_j dS$$
$$-\sum_{\kappa}\int_{t_0}^{t_1} dt\int_S \{n_i E^{(\kappa)}_{ij}\delta P^{(\kappa)}_j+n_i[-\varepsilon_0[\varphi_{,i}]+P_i^{(\kappa)}+q_*(-1)^{\kappa}u^{(\kappa)}_i]\delta\varphi\, dS$$
$$-\int_{t_0}^{t_1} dt\int_{V'} \varepsilon_0\varphi_{,ii}\delta\varphi\, dV = 0. \tag{12.16}$$

where n_i is the unit outward normal to S and $[\varphi_{,i}]$ is the jump in $\varphi_{,i}$ across S.

From (12.16), we have the Euler equations in V:

$$\begin{aligned} &T^{(\kappa)}_{ij,i}+(-1)^{\kappa}(T^*_{ij,i}-T^*_j-q_*\varphi_{,i})+f^{(\kappa)}_j+(-1)^{\kappa}q_*E^0_j = \rho^{(\kappa)}\ddot{u}^{(\kappa)}_j,\\ &E^{(\kappa)}_j+E^{(\kappa)}_{ij,i}-\varphi_{,j}+E^0_j = 0,\\ &-\varepsilon_0\varphi_{,ii}+P^{(1)}_{i,i}+P^{(2)}_{i,i}+q_*u^*_{i,i} = 0; \end{aligned} \tag{12.17}$$

and, in V':

$$\varphi_{,ii} = 0; \tag{12.18}$$

and also the boundary conditions on S:

$$\begin{aligned} &n_i[T^{(\kappa)}_{ij}+(-1)^{\kappa}T^*_{ij}] = t^{(\kappa)}_j,\\ &n_iE^{(\kappa)}_{ij} = 0,\\ &n_i(-\varepsilon_0[\varphi_{,i}]+P^{(1)}_i+P^{(2)}_i+q_*u^*_i) = 0. \end{aligned} \tag{12.19}$$

As may be seen from (12.16), boundary conditions alternative to each of (12.19) are the specification of $u^{(\kappa)}_i$, $P^{(\kappa)}_i$ and φ, respectively.

iii Constitutive Equations.

We consider the constitutive equations for a crystal with NaCl-type lattice structure, which falls under cubic point group $m3m$, the generators for which are given in (7.10). Owing to the first generator in (7.10) (the centrosymmetry generator) there can be no coefficients of odd rank and no products of symmetric and antisymmetric variables in the quadratic function W^L. This reduces the interactions to those between variables connected by full lines in Fig. 18; whereas the dashed line segments denote excluded interactions. The corresponding energy function is

$$\begin{aligned} W^L = {}&\tfrac{1}{2}\sum_{\kappa,\lambda}(a^{\kappa\lambda}_{ij}P^{(\kappa)}_iP^{(\lambda)}_j+b^{\kappa\lambda}_{ijkl}P^{(\kappa)}_{j,i}P^{(\lambda)}_{l,k}+c^{\kappa\lambda}_{ijkl}S^{(\kappa)}_{ij}S^{(\lambda)}_{kl}+2d^{\kappa\lambda}_{ijkl}P^{(\kappa)}_{(j,i)}S^{(\lambda)}_{kl})\\ &+\sum_{\kappa}(a^{*\kappa}u^*_iP^{(\kappa)}_i+2d^{*\kappa}\omega^*_{ij}P^{(\kappa)}_{[j,i]})+\tfrac{1}{2}a^{**}u^*_iu^*_i+c^{**}\omega^*_{ij}\omega^*_{ij}\\ &+\sum_{\kappa}(b^{\kappa 0}P^{(\kappa)}_{i,i}+c^{\kappa 0}S^{(\kappa)}_{ii}), \end{aligned} \tag{12.20}$$

where $P^{(\kappa)}_{(j,i)}$ and $P^{(\kappa)}_{[j,i]}$ denote the symmetric and antisymmetric parts of $P^{(\kappa)}_{j,i}$, respectively. We note that, in general,

$$a^{\kappa\lambda}_{ij} = a^{\lambda\kappa}_{ij},\ b^{\kappa\lambda}_{ijkl} = b^{\lambda\kappa}_{ijkl},\ c^{\kappa\lambda}_{ijkl} = c^{\lambda\kappa}_{ijkl},\ d^{\kappa\lambda}_{ijkl} \neq d^{\lambda\kappa}_{ijkl} \tag{12.21}$$

and, from (7.10),

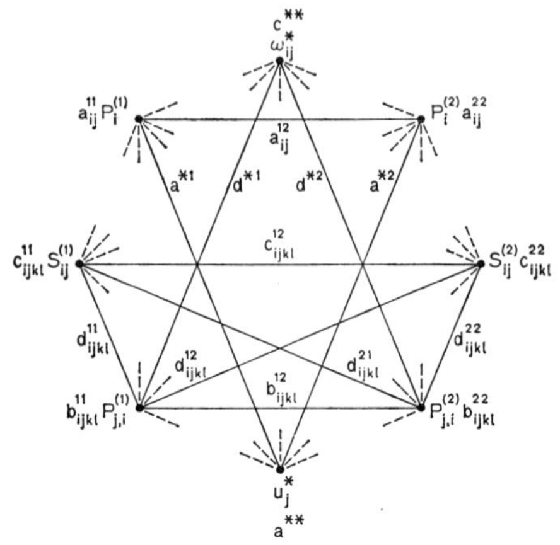

Figure 18

$$
\begin{aligned}
a_{ij}^{\kappa\lambda} &= a_{11}^{\kappa\lambda}\delta_{ij}, \\
b_{ijkl}^{\kappa\lambda} &= b^{\kappa\lambda}\delta_{ijkl}+b_{12}^{\kappa\lambda}\delta_{ij}\delta_{kl}+b_{44}^{\kappa\lambda}(\delta_{ik}\delta_{jl}+\delta_{il}\delta_{jk})+b_{77}^{\kappa\lambda}(\delta_{ik}\delta_{jl}-\delta_{il}\delta_{jk}), \\
c_{ijkl}^{\kappa\lambda} &= c^{\kappa\lambda}\delta_{ijkl}+c_{12}^{\kappa\lambda}\delta_{ij}\delta_{kl}+c_{44}^{\kappa\lambda}(\delta_{ik}\delta_{jl}+\delta_{il}\delta_{jk}), \\
d_{ijkl}^{\kappa\lambda} &= d^{\kappa\lambda}\delta_{ijkl}+d_{12}^{\kappa\lambda}\delta_{ij}\delta_{kl}+d_{44}^{\kappa\lambda}(\delta_{ik}\delta_{jl}+\delta_{il}\delta_{jk}),
\end{aligned}
\tag{12.22}
$$

where δ_{ij} (or δ_{ijkl}) is unity when its indices are alike and zero otherwise;

$$
b^{\kappa\lambda} = b_{11}^{\kappa\lambda}-b_{12}^{\kappa\lambda}-2b_{44}^{\kappa\lambda},\ c^{\kappa\lambda} = c_{11}^{\kappa\lambda}-c_{12}^{\kappa\lambda}-2c_{44}^{\kappa\lambda},\ d^{\kappa\lambda} = d_{11}^{\kappa\lambda}-d_{12}^{\kappa\lambda}-2d_{44}^{\kappa\lambda};
\tag{12.23}
$$

for $b_{ijkl}^{\kappa\lambda}$, $c_{ijkl}^{\kappa\lambda}$ and $d_{ijkl}^{\kappa\lambda}$, the abbreviated notation for pairs of indices, ij or kl, has been used – as in Article 11. It may be observed that material constants with superscripts 11 or 22 denote interactions within one of the two component continua whereas constants with superscripts 12 or 21 denote interactions between the two component continua.

From (12.12) and (12.20)-(12.22), we find the constitutive equations

$$
\begin{aligned}
T_{ij}^{(\kappa)} &= c^{\kappa 0}\delta_{ij}+\sum_{\lambda}(c^{\kappa\lambda}\delta_{ijkl}S_{kl}^{(\lambda)}+c_{12}^{\kappa\lambda}\delta_{ij}S_{kk}^{(\lambda)}+2c_{44}S_{ij}^{(\lambda)}) \\
&\qquad +\sum_{\lambda}(d^{\lambda\kappa}\delta_{ijkl}P_{l,k}^{(\lambda)}+d_{12}^{\lambda\kappa}\delta_{ij}P_{k,k}^{(\lambda)}+2d_{44}^{\lambda\kappa}P_{(j,i)}^{(\lambda)}), \\
-E_j^{(\kappa)} &= \sum_{\lambda} a_{11}^{\kappa\lambda}P_j^{(\lambda)}+a^{**}u_j^{*}, \\
E_{ij}^{(\kappa)} &= b^{\kappa 0}\delta_{ij}+\sum_{\lambda}(b^{\kappa\lambda}\delta_{ijkl}P_{l,k}^{(\lambda)}+b_{12}^{\kappa\lambda}\delta_{ij}P_{k,k}^{(\lambda)}+2b_{44}^{\kappa\lambda}P_{(j,i)}^{(\lambda)}+2b_{77}^{\kappa\lambda}P_{[j,i]}^{(\lambda)}) \\
&\qquad +2d^{*\kappa}\omega_{ij}^{*}+\sum_{\lambda}(d^{\kappa\lambda}\delta_{ijkl}S_{kl}^{(\lambda)}+d_{12}^{\kappa\lambda}\delta_{ij}S_{kk}^{(\lambda)}+2d_{44}^{\kappa\lambda}S_{ij}^{(\lambda)}), \\
T_j^{*} &= \sum_{\lambda} a^{*\lambda}P_j^{(\lambda)}+a^{**}u_j^{*}, \\
T_{ij}^{*} &= 2\sum_{\lambda} d^{*\lambda}P_{[j,i]}^{(\lambda)}+2c^{**}\omega_{ij}^{*}.
\end{aligned}
\tag{12.24}
$$

We shall assume that, in the initial state, i.e. when

$$
P_i^{(\kappa)},\ P_{j,i}^{(\kappa)},\ S_{ij}^{(\kappa)},\ u_i^{*},\ \omega_{ij}^{*},\ \varphi_{,i},\ f_i^{(\kappa)},\ E_i^{0} = 0,
$$

there is no resultant force across any surface – exterior or interior. That is $t_j^{(1)}+t_j^{(2)} = 0$. Then, from the first of (12.19),

$$
n_i T_{ij}^{(1)}+n_i T_{ij}^{(2)} = 0;
$$

so that, from the first of (12.24),

$$c^{10}+c^{20} = 0. \tag{12.25}$$

Accordingly, from (12.22) and (12.25), in the initial state:

$$E_i^{(\kappa)},\ T_i^*,\ T_{ij}^* = 0,\ \ T_{ij}^{(1)}+T_{ij}^{(2)} = 0,$$

but there is a self-equilibrated

$$E_{ij}^{(\kappa)} = b^{\kappa 0}\delta_{ij}$$

in each component continuum.

iv. Surface Energy of Deformation and Polarization.

In a state of equilibrium, the total energy in V^* is, from (12.3),

$$\mathscr{W} = \int_V W^L dV + \tfrac{1}{2}\int_{V^*} \varepsilon_0\, \varphi_{,i}\varphi_{,i}\, dV. \tag{12.26}$$

We can write W^L in the form

$$2W^L = \sum_\kappa (T_{ij}^{(\kappa)}S_{ij}^{(\kappa)} - E_i^{(\kappa)}P_i^{(\kappa)} + E_{ij}^{(\kappa)}P_{j,i}^{(\kappa)}) + T_i^* u_i^* + T_{ij}^*\omega_{ij}^* + \sum_\kappa b^{\kappa 0}P_{i,i}^{(\kappa)} + c^{20}(S_{ii}^{(2)} - S_{ii}^{(1)}).$$

Then, by the same procedure as that employed in arriving at (12.16),

$$\begin{aligned}\int_V W^L dV = {} & -\tfrac{1}{2}\sum_\kappa \int_V \{[T_{ij,i}^{(\kappa)} + (-1)^\kappa (T_{ij,i}^* - T_j^*)]u_j^{(\kappa)} + (E_j^{(\kappa)} + E_{ij,i}^{(\kappa)})P_j^{(\kappa)}\}\, dV \\ & + \tfrac{1}{2}\sum_\kappa \int_S n_i\{[T_{ij}^{(\kappa)} + (-1)^\kappa T_{ij}^*]u_j^{(\kappa)} + E_{ij}^{(\kappa)}P_j^{(\kappa)}\}\, dS + \tfrac{1}{2}\sum_\kappa \int_S n_i(b^{\kappa 0}P_i^{(\kappa)} + c^{20}u_i^*)\, dS. \end{aligned} \tag{12.27}$$

Also,

$$\int_{V^*} \varphi_{,i}\varphi_{,i}\, dV = -\int_{V^*} \varphi_{,ii}\,\varphi\, dV + \int_S n_i[\varphi_{,i}]\varphi\, dS. \tag{12.28}$$

Inserting (12.27) and (12.28) in (12.26) and applying the equations of equilibrium, i.e. (12.17) with $\ddot{u}_j^\kappa = 0$, in the absence of external forces (f_j^κ, $E_j^0 = 0$) we find

$$\begin{aligned} 2\mathscr{W} = {} & -\sum_\kappa \int_V [(-1)^\kappa q_*\, \varphi_{,i} u_i^{(\kappa)} + \varphi_{,i} P_i^{(\kappa)} + (P_{i,i}^{(\kappa)} + q_* u_{i,i}^*)\varphi]\, dV \\ & + \sum_\kappa \int_S n_i\{[T_{ij}^{(\kappa)} + (-1)^\kappa T_{ij}^*]u_j^{(\kappa)} + E_{ij}^{(\kappa)}P_j^{(\kappa)} + \varepsilon_0[\varphi_{,i}]\varphi\}\, dS \\ & + \sum_\kappa \int_S n_i(b^{\kappa 0}P_i^{(\kappa)} + c^{20}u_i^*)\, dS. \end{aligned}$$

Now

$$\int_V P_{i,i}^{(\kappa)}\varphi\, dV = -\int_V \varphi_{,i} P_i^{(\kappa)} dV + \int_S n_i P_i^{(\kappa)}\varphi\, dS,$$

$$\int_V q_*(u_i^*\varphi_{,i} + u_{i,i}^*\varphi)\, dV = \int_S n_i\, q_*\, u_i^*\varphi\, dS.$$

Hence

$$2\mathscr{W} = \sum_{\kappa}\int_S n_i\{[T_{ij}^{(\kappa)}+(-1)^{\kappa}T_{ij}^{*}]u_j^{(\kappa)}+E_{ij}^{(\kappa)}P_j^{(\kappa)}+(\varepsilon_0\,[\varphi_{,i}]-P_i^{(\kappa)}-q_*u_i^*)\varphi\}\,dS$$
$$+\sum_{\kappa}\int_S n_i(b^{\kappa 0}P_i^{(\kappa)}+c^{20}u_i^*)\,dS. \tag{12.29}$$

From (12.19) the first integral in (12.29) vanishes if the boundary is entirely free. In that case,

$$2\mathscr{W} = \sum_{\kappa}\int_S n_i(b^{\kappa 0}P_i^{(\kappa)}+c^{20}u_i^*)\,dS.$$

Accordingly, the surface energy of deformation and polarization, per unit area, is

$$W^S = \tfrac{1}{2}n_i[b^{10}P_i^{(1)}+b^{20}P_i^{(2)}+c^{20}u_i^*]_S. \tag{12.30}$$

This result is to be compared with (10.9).

v. Wave Motion.

The equations of motion in terms of $u_i^{(\kappa)}$, $P_i^{(\kappa)}$ and φ are obtained by substituting the constitutive equations (12.24) into the field equations (12.17) with the result, for each κ,

$$\begin{aligned}
&\sum_{\lambda}[c^{\kappa\lambda}\delta_{ijkl}u_{l,ki}^{(\lambda)}+c_{12}^{\kappa\lambda}u_{i,ij}^{(\lambda)}+c_{44}^{\kappa\lambda}(u_{j,ii}^{(\lambda)}+u_{i,ij}^{(\lambda)})]\\
&\quad+\sum_{\lambda}[d^{\lambda\kappa}\delta_{ijkl}P_{l,ki}^{(\lambda)}+d_{12}^{\lambda\kappa}P_{i,ij}^{(\lambda)}+d_{44}^{\lambda\kappa}(P_{j,ii}^{(\lambda)}+P_{i,ij}^{(\lambda)})]\\
&\quad+(-1)^{\kappa}\sum_{\lambda}[d^{*\lambda}(P_{j,ii}^{(\lambda)}-P_{i,ij}^{(\lambda)})+(-1)^{\lambda}c^{**}(u_{j,ii}^{(\lambda)}-u_{i,ij}^{(\lambda)})]\\
&\quad-(-1)^{\kappa}[a^{*1}P_j^{(1)}+a^{*2}P_j^{(2)}+a^{**}(u_j^{(2)}-u_j^{(1)})+q_*\varphi_{,j}-q_*E_j^0]+f_j^{(\kappa)} = \rho^{(\kappa)}\ddot{u}_j^{(\kappa)},\\
&\sum_{\lambda}[d^{\kappa\lambda}\delta_{ijkl}u_{l,ki}^{(\lambda)}+d_{12}^{\kappa\lambda}u_{i,ij}^{(\lambda)}+d_{44}^{\kappa\lambda}(u_{j,ii}^{(\lambda)}+u_{i,ij}^{(\lambda)})+(-1)^{\lambda}d^{**}(u_{j,ii}^{(\lambda)}-u_{i,ij}^{(\lambda)})]\\
&\quad+\sum_{\lambda}[b^{\kappa\lambda}\delta_{ijkl}P_{l,ki}^{(\lambda)}+b_{12}^{\kappa\lambda}P_{i,ij}^{(\lambda)}+b_{44}^{\kappa\lambda}(P_{j,ii}^{(\lambda)}+P_{i,ij}^{(\lambda)})+b_{77}^{\kappa\lambda}(P_{j,ii}^{(\lambda)}-P_{i,ij}^{(\lambda)})]\\
&\quad-a_{11}^{1\kappa}P_j^{(1)}-a_{11}^{2\kappa}P_j^{(2)}-a^{*\kappa}(u_j^{(2)}-u_j^{(1)})-\varphi_{,j}+E_j^0 = 0,\\
&\quad-\varepsilon_0\varphi_{,ii}+P_{i,i}^{(1)}+P_{i,i}^{(2)}+q_*(u_{i,i}^{(2)}-u_{i,i}^{(1)}) = 0.
\end{aligned} \tag{12.31}$$

The essential features of wave motion are revealed by an examination of plane waves in, say, the x_1, or (100), direction:

$$u_j^{(\kappa)} = A_j^{(\kappa)}e^{i(\xi x_1-\omega t)},\quad P_j^{(\kappa)} = B_j^{(\kappa)}e^{i(\xi x_1-\omega t)},\quad \varphi = Ce^{i(\xi x_1-\omega t)} \tag{12.32}$$

and $f_j^{(\kappa)}$, $E_j^0 = 0$. With these functions, the five equations (12.31) reduce to those for longitudinal waves if $j = 1$ and transverse waves if $j = 2$ or 3. Thus:

Longitudinal $(j = 1)$

$$\begin{aligned}
&c_{11}^{11}u_{1,11}^{(1)}+c_{11}^{12}u_{1,11}^{(2)}+d_{11}^{11}P_{1,11}^{(1)}+d_{11}^{21}P_{1,11}^{(2)}+a^{*1}P_1^{(1)}+a^{*2}P_1^{(2)}+a^{**}(u_1^{(2)}-u_1^{(1)})+q_*\varphi_{,1}\\
&\qquad = \rho^{(1)}\ddot{u}_1^{(1)},\\
&c_{11}^{21}u_{1,11}^{(1)}+c_{11}^{22}u_{1,11}^{(2)}+d_{11}^{12}P_{1,11}^{(1)}+d_{11}^{22}P_{1,11}^{(2)}-a^{*1}P_1^{(1)}-a^{*2}P_1^{(2)}-a^{**}(u_1^{(2)}-u_1^{(1)})-q_*\varphi_{,1}\\
&\qquad = \rho^{(2)}\ddot{u}_1^{(2)},\\
&d_{11}^{11}u_{1,11}^{(1)}+d_{11}^{12}u_{1,11}^{(2)}+b_{11}^{11}P_{1,11}^{(1)}+b_{11}^{12}P_{1,11}^{(2)}-a_{11}^{11}P_1^{(1)}-a_{11}^{12}P_1^{(2)}-a^{*1}(u_1^{(2)}-u_1^{(1)})-\varphi_{,1} = 0,\\
&d_{11}^{21}u_{1,11}^{(1)}+d_{11}^{22}u_{1,11}^{(2)}+b_{11}^{21}P_{1,11}^{(1)}+b_{11}^{22}P_{1,11}^{(2)}-a_{11}^{21}P_1^{(1)}-a_{11}^{22}P_1^{(2)}-a^{*2}(u_1^{(2)}-u_1^{(1)})-\varphi_{,1} = 0,\\
&P_{1,1}^{(1)}+P_{1,1}^{(2)}+q_*(u_{1,1}^{(2)}-u_{1,1}^{(1)})-\varepsilon_0\varphi_{,11} = 0.
\end{aligned} \tag{12.33}$$

Transverse $(j = 2)$

$$(c_{44}^{11}+c^{**})u_{2,11}^{(1)}+(c_{44}^{12}-c^{**})u_{2,11}^{(2)}+(d_{44}^{11}-d^{*1})P_{2,11}^{(1)}+(d_{44}^{21}-d^{*2})P_{2,11}^{(2)} +a^{*1}P_2^{(1)}+a^{*2}P_2^{(2)}+a^{**}(u_2^{(2)}-u_2^{(1)}) = \rho^{(1)}\ddot{u}_2^{(1)},$$

$$(c_{44}^{21}-c^{**})u_{2,11}^{(1)}+(c_{44}^{22}+c^{**})u_{2,11}^{(2)}+(d_{44}^{12}+d^{*1})P_{2,11}^{(1)}+(d_{44}^{22}+d^{*2})P_{2,11}^{(2)} -a^{*1}P_2^{(1)}-a^{*2}P_2^{(2)}-a^{**}(u_2^{(2)}-u_2^{(1)}) = \rho^{(2)}\ddot{u}_2^{(2)}, \tag{12.34}$$

$$(d_{44}^{11}-d^{*1})u_{2,11}^{(1)}+(d_{44}^{12}+d^{*1})u_{2,11}^{(2)}+(b_{44}^{11}+b_{77}^{11})P_{2,11}^{(1)}+(b_{44}^{12}+b_{77}^{12})P_{2,11}^{(2)} -a_{11}^{11}P_2^{(1)}-a_{11}^{12}P_2^{(2)}-a^{*1}(u_2^{(2)}-u_2^{(1)}) = 0,$$

$$(d_{44}^{21}-d^{*2})u_{2,11}^{(1)}+(d_{44}^{22}+d^{*2})u_{2,11}^{(2)}+(b_{44}^{21}+b_{77}^{21})P_{2,11}^{(1)}+(b_{44}^{22}+b_{77}^{22})P_{2,11}^{(2)} -a_{11}^{21}P_2^{(1)}-a_{11}^{22}P_2^{(2)}-a^{*2}(u_2^{(2)}-u_2^{(1)}) = 0.$$

It may be noted that the Maxwell self-field is coupled to the mechanical displacement in the longitudinal waves, but not in the transverse waves. As is shown in the following Article, if the magnetic field is taken into account, the longitudinal wave is unchanged but there is full coupling of the elastic and electromagnetic fields in the transverse wave.

Upon substituting (12.32) in (12.33) and eliminating $A_1^{(\kappa)}$, $B_1^{(\kappa)}$ and C, we find the dispersion relation for longitudinal waves:

$$\Delta_L = 0, \tag{12.35}$$

where Δ^L is the determinant with elements

$$\begin{aligned}
\Delta_{11} &= \rho^{(1)}\omega^2-a^{**}-q_*^2\varepsilon_0^{-1}-c_{11}^{11}\xi^2, & \Delta_{23} &= -a^{*1}-q_*\varepsilon_0^{-1}-d_{11}^{12}\xi^2 = \Delta_{32},\\
\Delta_{12} &= a^{**}+q_*^2\varepsilon_0^{-1}-c_{11}^{12}\xi^2 = \Delta_{21}, & \Delta_{24} &= -a^{*2}-q_*\varepsilon_0^{-1}-d_{11}^{22}\xi^2 = \Delta_{42},\\
\Delta_{13} &= a^{*1}+q_*\varepsilon_0^{-1}-d_{11}^{11}\xi^2 = \Delta_{31}, & \Delta_{33} &= -a_{11}^{11}-\varepsilon_0^{-1}-b_{11}^{11}\xi^2,\\
\Delta_{14} &= a^{*2}+q_*\varepsilon_0^{-1}-d_{11}^{21}\xi^2 = \Delta_{41}, & \Delta_{34} &= -a_{11}^{21}-\varepsilon_0^{-1}-b_{11}^{21}\xi^2 = \Delta_{43},\\
\Delta_{22} &= \rho^{(2)}\omega^2-a^{**}-q_*^2\varepsilon_0^{-1}-c_{11}^{11}\xi^2, & \Delta_{44} &= -a_{11}^{22}-\varepsilon_0^{-1}-b_{11}^{22}\xi^2.
\end{aligned} \tag{12.36}$$

This is a quartic equation in ξ^2. Of the four branches, two are real: one acoustical and one optical, as may be verified by showing that one real branch passes through $\xi = 0, \omega = 0$ and one through $\xi = 0, \omega \neq 0$. We find, in fact,

$$\lim_{\omega,\xi\to 0} \Delta_L = \lim_{\omega,\xi\to 0} [(c_{11}^{11}+c_{11}^{22}+2c_{11}^{12})\xi^2-(\rho^{(1)}+\rho^{(2)})\omega^2]D_L, \tag{12.37}$$

where

$$D_L = \begin{vmatrix} a_{11}^{11}+\varepsilon_0^{-1} & a_{11}^{12}+\varepsilon_0^{-1} & a^{*1}+q_*\varepsilon_0^{-1}\\ a_{11}^{21}+\varepsilon_0^{-1} & a_{11}^{22}+\varepsilon_0^{-1} & a^{*2}+q_*\varepsilon_0^{-1}\\ a^{*1}+q_*\varepsilon_0^{-1} & a^{*2}+q_*\varepsilon_0^{-1} & a^{**}+q_*^2\varepsilon_0^{-1}\end{vmatrix}. \tag{12.38}$$

Hence, at long wave lengths, the frequency of the longitudinal acoustic wave is

$$\omega_{LA} = \xi[(c_{11}^{11}+c_{11}^{22}+2c_{11}^{12})/(\rho^{(1)}+\rho^{(2)})]^{\frac{1}{2}}. \tag{12.39}$$

It may be seen, from (12.39), that the usual low frequency extensional stiffness, c_{11}, is given by

$$c_{11} = c_{11}^{11}+c_{11}^{22}+2c_{11}^{12}, \tag{12.40}$$

i.e. the sum of the extensional stiffnesses, c_{11}^{11} and c_{11}^{22}, of the two component continua augmented by the intercomponent extensional stiffness $2c_{11}^{12}$.

The long wave limit of the longitudinal optical branch is obtained from

$$\lim_{\xi\to 0} \Delta_L = \omega^2\{\rho^{(1)}\rho^{(2)}\omega^2[(a_{11}^{11}+\varepsilon_0^{-1})(a_{11}^{22}+\varepsilon_0^{-1})-(a_{11}^{12}+\varepsilon_0^{-1})^2]-(\rho^{(1)}+\rho^{(2)})D_L\}, \tag{12.41}$$

Hence, the limiting frequency of the longitudinal optical branch is

$$\omega_{LO} = \{D_L\varepsilon_0^2/\bar{\rho}[(1+a_{11}^{11}\varepsilon_0)(1+a_{11}^{22}\varepsilon_0)-(1+a_{11}^{12}\varepsilon_0)^2\}^{\frac{1}{2}}, \tag{12.42}$$

where

$$\bar{\rho} = \rho^{(1)}\rho^{(2)}/(\rho^{(1)}+\rho^{(2)}), \tag{12.43}$$

i.e. $\bar{\rho}$ is the reduced density.

In a similar way, we find, from (12.34), the dispersion relation for transverse waves in the (100) direction:

$$\Delta_T = 0, \tag{12.44}$$

where Δ_T is the determinant with elements

$$\begin{aligned}
\Delta_{11} &= \rho^{(1)}\omega^2-a^{**}-(c_{44}^{11}+c^{**})\xi^2, & \Delta_{23} &= -a^{*1}-(d_{44}^{12}+d^{*1})\xi^2 = \Delta_{32},\\
\Delta_{12} &= a^{**}-(c_{44}^{12}-c^{**})\xi^2 = \Delta_{21}, & \Delta_{24} &= -a^{*2}-(d_{44}^{22}+d^{*2})\xi^2 = \Delta_{42},\\
\Delta_{13} &= a^{*1}-(d_{44}^{11}-d^{*1})\xi^2 = \Delta_{31}, & \Delta_{33} &= -a_{11}^{11}-(b_{44}^{11}+b_{77}^{11})\xi^2,\\
\Delta_{14} &= a^{*2}-(d_{44}^{21}-d^{*2})\xi^2 = \Delta_{41}, & \Delta_{34} &= -a_{11}^{12}-(b_{44}^{12}+b_{77}^{12})\xi^2, = \Delta_{43}\\
\Delta_{22} &= \rho^{(2)}\omega^2-a^{**}-(c_{44}^{22}+c^{**})\xi^2, & \Delta_{44} &= -a_{11}^{22}-(b_{44}^{22}+b_{77}^{22})\xi^2.
\end{aligned} \tag{12.45}$$

Again, this is a quartic in ξ^2 with two real branches. For the long wave behavior of the transverse acoustic branch:

$$\lim_{\omega,\,\xi\to 0} \Delta_T = \lim_{\omega,\,\xi\to 0} [(c_{44}^{11}+c_{44}^{22}+2c_{44}^{12})\xi^2-(\rho^{(1)}+\rho^{(2)})\omega^2]D, \tag{12.46}$$

where

$$D = \begin{vmatrix} a_{11}^{11} & a_{11}^{12} & a^{*1} \\ a_{11}^{21} & a_{11}^{22} & a^{*2} \\ a^{*1} & a^{*2} & a^{**} \end{vmatrix}. \tag{12.47}$$

Hence, there is a transverse acoustic branch with limiting, low frequency behavior

$$\omega_{TA} = \xi[(c_{44}^{11}+c_{44}^{22}+2c_{44}^{12})/(\rho^{(1)}+\rho^{(2)})]^{\frac{1}{2}}. \tag{12.48}$$

The long wave limit of the transverse optical branch is found from

$$\lim_{\xi\to 0} \Delta_T = \omega^2[\rho^{(1)}\rho^{(2)}\omega^2(a_{11}^{11}a_{11}^{22}-a_{11}^{12}a_{11}^{21})-(\rho^{(1)}+\rho^{(2)})D], \tag{12.49}$$

from which the limiting frequency of the transverse optical branch is

$$\omega_{TO} = [D/\bar{\rho}(a_{11}^{11}a_{11}^{22}+a_{11}^{12}a_{11}^{21})]^{\frac{1}{2}}. \tag{12.50}$$

vi. One-dimensional NaCl-type Lattice of Shell-model Atoms.

The purpose of this section is to show that the one-dimensional equations of longitudinal motion (12.33) are the long wave limit of the difference equations of motion of a one-dimensional NaCl-type lattice of shell model atoms. Such a lattice is conveniently represented by two lines of alternating atoms (positive and negative ions, in the case of an ionic crystal) with one atom of each type at each lattice site, as shown in Fig. 19 – where the two types of atom are

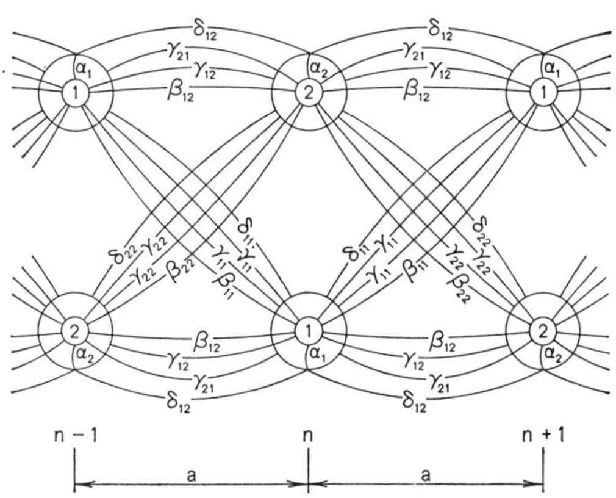

Figure 19

identified by the digits 1 and 2. Nearest neighbor interactions between unlike (adjacent) atoms in each line and next nearest neighbor interactions between like atoms in the two lines are those taken into account, but no interactions between unlike atoms at the same site nor between like atoms in the same line are considered. With each of p and q taking on the identifications 1 or 2, the force constants of the interactions are denoted by α_p for the intra-atomic core-shell interactions and $\beta_{pq}, \gamma_{pq}, \delta_{pq}$ for the core-core, core-shell and shell-shell interatomic interactions respectively, between like atoms for $p = q$ and unlike atoms for $p \neq q$. We note that $\beta_{pq} = \beta_{qp}$ and $\delta_{qp} = \delta_{pq}$ but $\gamma_{pq} \neq \gamma_{qp}$. The two lines of atoms are taken to be parallel to the x_1 axis with the atom sites at $x_1 = na$, where n is a positive or negative integer. The displacements of the cores and shells of the atoms at $x_1 = na$ are denoted by $u_n^{(\kappa)}$ and $s_n^{(\kappa)}$, respectively, where $\kappa = 1, 2$ to designate the atoms of types 1 and 2 (1 for the positive ion and 2 for the negative ion, in the case of an ionic crystal).

The equation of motion of the nth atom of type 1 is obtained by equating its inertia force to the sum of the forces on its core and shell exerted by the cores and shells of the two nearest neighbor (unlike) atoms, the two next nearest neighbor (like) atoms and the Maxwell, electric self-field at $x_1 = na$:

$$\begin{aligned}&\beta_{11}(u_{n+1}^{(1)}-u_n^{(1)})+\gamma_{11}(s_{n+1}^{(1)}-s_n^{(1)}+u_{n+1}^{(1)}-u_n^{(1)})+\delta_{11}(s_{n+1}^{(1)}-s_n^{(1)})\\&\quad-\beta_{11}(u_n^{(1)}-u_{n-1}^{(1)})-\gamma_{11}(u_n^{(1)}-u_{n-1}^{(1)}+s_n^{(1)}-s_{n-1}^{(1)})-\delta_{11}(s_n^{(1)}-s_{n-1}^{(1)})\\&\quad+\beta_{12}(u_{n+1}^{(2)}-u_n^{(1)})+\gamma_{12}(s_{n+1}^{(2)}-u_n^{(1)})+\gamma_{21}(u_{n+1}^{(2)}-s_n^{(1)})+\delta_{12}(s_{n+1}^{(2)}-s_n^{(1)})\\&\quad-\beta_{12}(u_n^{(1)}-u_{n-1}^{(2)})-\gamma_{12}(u_n^{(1)}-s_{n-1}^{(2)})-\gamma_{21}(s_n^{(1)}-u_{n-1}^{(2)})-\delta_{12}(s_n^{(1)}-s_{n-1}^{(2)})-q_*E_n = m_1\ddot{u}_n^{(1)}.\end{aligned} \tag{12.51}$$

Similarly, for the shell, alone, of the type 1 atom:

$$\begin{aligned}&\alpha_1(u_n^{(1)}-s_n^{(1)})+\gamma_{11}(u_{n+1}^{(1)}-s_n^{(1)})+\delta_{11}(s_{n+1}^{(1)}-s_n^{(1)})-\gamma_{11}(s_n^{(1)}-u_{n-1}^{(1)})-\delta_{11}(s_n^{(1)}-s_{n-1}^{(1)})\\&+\gamma_{21}(u_{n+1}^{(2)}-s_n^{(1)})+\delta_{12}(s_{n+1}^{(2)}-s_n^{(1)})-\gamma_{21}(s_n^{(1)}-u_{n-1}^{(2)})-\delta_{12}(s_n^{(1)}-s_{n-1}^{(2)})+qE_n = 0,\end{aligned} \tag{12.52}$$

where q is the electronic charge and the mass of the shell is neglected.

We adopt the following definitions of electronic polarization densities and mass densities:

$$P_n^{(\kappa)} = (s_n^{(\kappa)}-u_n^{(\kappa)})q/a^3,\ \rho^{(\kappa)} = m_\kappa/a^3,\ \kappa = 1, 2 \tag{12.53}$$

and also make the following identifications:

$$
\begin{aligned}
a_{11}^{11} &= (\alpha_1+2\gamma_{11}+2\gamma_{21}+2\delta_{12})a^3q^{-2}, & b_{11}^{12} &= a_{11}^{12}a^2/2 = \delta_{12}a^5q^{-2},\\
b_{11}^{11} &= \delta_{11}a^5q^{-2}, & c_{11}^{12} &= a^{**}a^2/2 = (\beta_{12}+\gamma_{12}+\gamma_{21}+\delta_{12})a^{-1},\\
c_{11}^{11} &= (\beta_{11}+2\gamma_{11}+\delta_{11})a^{-1}, & d_{11}^{21} &= a^{*2}a^2/2 = (\gamma_{12}+\delta_{12})a^2q^{-1},\\
d_{11}^{11} &= (\gamma_{11}+\delta_{11})a^2q^{-1}, & d_{11}^{12} &= -a^{*1}a^2/2 = (\gamma_{21}+\delta_{12})a^2q^{-1}.
\end{aligned} \tag{12.54}
$$

Then (12.51) and (12.52) can be written as

$$
\begin{aligned}
&c_{11}^{11}\Delta^2u_n^{(1)}+c_{11}^{12}\Delta^2u_n^{(2)}+d_{11}^{11}\Delta^2P_n^{(1)}+d_{11}^{21}\Delta^2P_n^{(2)}\\
&\qquad +a^{*1}P_n^{(1)}+a^{*2}P_n^{(2)}+a^{**}(u_n^{(2)}-u_n^{(1)})-q_*E_n = \rho^{(1)}\ddot{u}_n^{(1)},
\end{aligned} \tag{12.55}
$$

$$
\begin{aligned}
&d_{11}^{11}\Delta^2u_n^{(1)}+d_{11}^{12}\Delta^2u_n^{(2)}+b_{11}^{11}\Delta^2P_n^{(1)}+b_{11}^{12}\Delta^2P_n^{(2)}\\
&\qquad -a_{11}^{11}P_n^{(1)}-a_{11}^{12}P_n^{(2)}-a^{*1}(u_n^{(2)}-u_n^{(1)})+E_n = 0,
\end{aligned} \tag{12.56}
$$

where Δ^2 is, again, the second central difference divided by a^2.

In the same way, for the type 2 atom,

$$
\begin{aligned}
&c_{11}^{21}\Delta^2u_n^{(1)}+c_{11}^{22}\Delta^2u_n^{(2)}+d_{11}^{12}\Delta^2P_n^{(1)}+d_{11}^{22}\Delta^2P_n^{(2)}\\
&\qquad -a^{*1}P_n^{(1)}-a^{*2}P_n^{(2)}-a^{**}(u_n^{(2)}-u_n^{(1)})+q_*E_n = \rho^{(2)}\ddot{u}_n^{(2)},
\end{aligned} \tag{12.57}
$$

$$
\begin{aligned}
&d_{11}^{21}\Delta^2u_n^{(1)}+d_{11}^{22}\Delta^2u_n^{(2)}+b_{11}^{21}\Delta^2P_n^{(1)}+b_{11}^{22}\Delta^2P_n^{(2)}\\
&\qquad -a_{11}^{21}P_n^{(1)}-a_{11}^{22}P_n^{(2)}-a^{*2}(u_n^{(2)}-u_n^{(1)})+E_n = 0,
\end{aligned} \tag{12.58}
$$

where, in addition to (12.54), we identify

$$
\begin{aligned}
a_{11}^{22} &= (\alpha_2+2\gamma_{22}+2\gamma_{12}+2\delta_{12})a^3q^{-2}, & b_{11}^{22} &= \delta_{22}a^5q^{-2},\\
c_{11}^{22} &= (\beta_{22}+2\gamma_{22}+\delta_{22})a^{-1}, & d_{11}^{22} &= (\gamma_{22}+\delta_{22})a^2q^{-1}.
\end{aligned} \tag{12.59}
$$

Also, as in Article 8, the charge equation may be written as

$$
-\varepsilon_0\partial^-\partial^+\varphi_n+\partial^-[P_n^{(1)}+P_n^{(2)}+q_*(u_n^{(2)}-u_n^{(1)})] = 0, \tag{12.60}
$$

where ∂^+ and ∂^- are the Taylor series expansions of the derivative $\partial/\partial x_1$ in terms of forward and backward differences, respectively.

In the long wave limit,

$$
\begin{aligned}
&u_n^{(\kappa)} \to u_1^{(\kappa)}(x_1),\ P_n^{(\kappa)} \to P_1^{(\kappa)}(x_1),\ E_n \to E_1(x_1) = -\partial\varphi/\partial x_1\\
&\Delta^2 f_n \to \partial^2 f(x_1)/\partial x_1^2,\ \partial^{\pm} \to \partial/\partial x_1.
\end{aligned} \tag{12.61}
$$

Then (12.55), (12.57), (12.56), (12.58) and (12.60), in that order, reduce to the five equations (12.33) in the order given. Since the continuum equations are the long wave limits of the lattice equations, the two dispersion relations are *ipso facto* the same in the long wave limit.

It should be observed that, strictly speaking, the real segments of the optical branches, in the continuum approximation, have no valid role in cases where the optical and acoustic modes are coupled — as, for example in solutions of the equations for free boundary conditions. The frequencies of the two types of mode must be the same, when they are coupled; but, any frequency on a real segment of an optical branch is so high that the wave length of the acoustic branch, at that frequency, is too short to be within the range of validity of the continuum approxmiation. In such situations there is, in general, an upper limit of frequency, set by the wave length limitation of the acoustic branches, to which the solution is restricted.

However, there is an important exception. At frequencies close to ω_{TO} or ω_{LO}, the wave length of the optical branch is extremely long in comparison with that of the acoustic branch. In that case, the amplitude of the contribution of the acoustic part, to the coupled mode, is small in comparison with the amplitude contribution of the optical part; so that the error introduced by the inaccuracy of the acoustic branch is small. In fact, the acoustic branch may be discarded, then, without seriously affecting what remains – thus simplifying the equations of motion. The situation is analogous to that in the theory of thickness-shear and flexural vibrations of plates [65]. The thickness shear branch corresponds to the optical branch and the flexural branch corresponds to the acoustic branch. At frequencies near the cut-off frequency of the thickness-shear branch, the amplitude ratio of the flexural to the thickness-shear modes is small [65]. In that case, the flexural branch can be discarded entirely and a simpler equation of motion for the thickness-shear mode, alone, may be written [66].

In the long wave, low frequency limit, we have, in addition to (12.61),

$$u_1^{(1)} = u_1^{(2)}, \; P_1^{(1)} = P_1^{(2)}.$$

Then, with the notations

$$u_1 = u_1^{(1)} = u_1^{(2)}, \; \tfrac{1}{2}P_1 = P_1^{(1)} = P_1^{(2)}, \; \rho = \rho^{(1)}+\rho^{(2)},$$

$$a_{11} = \tfrac{1}{4}(a_{11}^{11}+a_{11}^{22}+2a_{11}^{12}), \qquad b_{11} = \tfrac{1}{4}(b_{11}^{11}+b_{11}^{22}+2b_{11}^{12}),$$
$$c_{11} = c_{11}^{11}+c_{11}^{22}+2c_{11}^{12}, \qquad d_{11} = \tfrac{1}{2}(d_{11}^{11}+d_{11}^{22}+d_{11}^{12}+d_{11}^{21}),$$

the sum of (12.55) and (12.57), the sum of (12.56) and (12.58) and, finally, (12.60) become, respectively,

$$c_{11}u_{1,11}+d_{11}P_{1,11} = \rho\ddot{u}_1,$$
$$d_{11}u_{1,11}+b_{11}P_{1,11}-a_{11}P_1-\varphi_{,1} = 0,$$
$$-\varepsilon_0\varphi_{,11}+P_{1,1} = 0.$$

These are the equations (7.16) exhibited in Article 7 for longitudinal waves in a simple, as opposed to a compound, dielectric continuum, with symmetry $m3m$, when the polarization gradient is taken into account. Thus, the long wave limit of the equations of the one-dimensional, diatomic lattice yields the equations for longitudinal waves in the compound continuum when no restriction is placed on the frequency; whereas the long wave, low frequency limit yields the equations of the simple continuum.

At very low frequencies, the acoustic branch is nearly exact and, at zero frequency, it is absent. The quality of the approximation then depends on the low or zero frequency values of the imaginary or complex branches of the dispersion relation. These affect behavior at or near surfaces. A typical example is given by the difference between il and $i\lambda$, the imaginary wave lengths at zero frequency of the imaginary branches of the continuum and lattice dispersion relations, in Article 8, and the role of this difference in the solutions of the continuum and lattice equations for capacitance of thin films, in Article 9.

13. Coupled Elastic and Electromagnetic Fields in a Diatomic, Dielectric Continuum

Equations governing coupled mechanical and electromagnetic fields in diatomic, ionic, optically isotropic crystals were given by Huang [67, 68] in terms of a polarization variable, the relative displacement of the ions and the usual variables of the electromagnetic field. These equations have been extended [69] taking into account the separate electronic polarizations and the separate displacements of the two ions, the ionic polarization, the two electronic polarization gradients and the two displacement gradients. Aside from a more detailed representation of the atomic and electronic interactions, there are two main effects of the additional considerations: first, the equations are applicable to shorter wave lengths, owing to the inclusion of the displacement and polarization gradients; second, the acoustic branches are included in the dispersion relations for plane waves, in addition to the optical branches and the electromagnetic branch.

As in Article 12, each of the two interpenetrating continua representing the diatomic crystal has its own displacement $u_i^{(\kappa)}$, $\kappa = 1, 2$, and electronic polarization $P_i^{(\kappa)}$. The stored energy of deformation and polarization is the same as before: a function of the individual small strains, $S_{ij}^{(\kappa)}$, the individual polarizations and polarization gradients, $P_i^{(\kappa)}$ and $P_{j,i}^{(\kappa)}$, and the relative displacement u_i^* and rotation, ω_{ij}^*:

$$W^L = W^L(S_{ij}^{(1)}, S_{ij}^{(2)}, P_i^{(1)}, P_i^{(2)}, P_{j,i}^{(1)}, P_{j,i}^{(2)}, u_i^*, \omega_{ij}^*). \tag{13.1}$$

Also, as before, the total polarization, per unit area, P_i, is the sum of the electronic polarizations and the ionic polarization:

$$P_i = P_i^{(1)} + P_i^{(2)} + q_* u_i^*. \tag{13.2}$$

Of the variables in (13.1), only P_i and u_i^* are accounted for in Huang's [67, 68] equations of motion.

When the Maxwell, electric self-field, E_i, is quasi-static, it satisfies

$$\delta_{ijk} E_{k,j} = 0, \tag{13.3}$$

where, as in (3.2), δ_{ijk} is the unit alternating tensor. In the electromagnetic case, (13.3) is replaced by

$$\delta_{ijk} E_{k,j} + \dot{B}_i = 0, \tag{13.4}$$

where, as in Article 11, the magnetic flux density, B_i, satisfies

$$\mu_0^{-1} \delta_{ijk} B_{k,j} = \varepsilon_0 \dot{E}_i + \dot{P}_i, \quad B_{i,i} = 0. \tag{13.5}$$

In addition to (13.4) and (13.5) the field equations (12.31) also hold, but with $\varphi_{,i}$ replaced by $-$ $-E_i$.

Plane waves in the [100] direction are expressed by

$$u_i^{(\kappa)} = K_i^{(\kappa)} e^{i\psi}, P_i^{(\kappa)} = L_i^{(\kappa)} e^{i\psi}, E_i = M_i e^{i\psi}, B_i = N_i e^{i\psi}, \tag{13.6}$$

where $\psi = \xi x_1 - \omega t$ and $K_i^{(\kappa)}$, $L_i^{(\kappa)}$, M_i and N_i are constants.

In the case of longitudinal waves,

$$u_2^{(\kappa)} = u_3^{(\kappa)} = 0, P_2^{(\kappa)} = P_3^{(\kappa)} = 0, E_2 = E_3 = 0, \tag{13.7}$$

$$u_1^{(\kappa)} = K_1^{(\kappa)} e^{i\psi}, P_1^{(\kappa)} = L_1^{(\kappa)} e^{i\psi}, E_1 = M_1 e^{i\psi}. \tag{13.8}$$

Then, from (13.4), $B_i = 0$; i.e. there is no coupling with the magnetic field. However the Maxwell, electric self-field is coupled with the displacement and polarization fields. The solution, therefore, is the same as that found in Article 12 for the case of the quasi-static electric field, with dispersion relation (12.35) as the result.

Of the two similar transverse waves, with displacements in the directions of x_2 and x_3, we choose the former for examination:

$$u_1^{(\kappa)} = u_3^{(\kappa)} = 0, P_1^{(\kappa)} = P_3^{(\kappa)} = 0, E_1 = E_3 = 0, B_1 = B_2 = 0, \tag{13.9}$$

$$u_2^{(\kappa)} = K_2^{(\kappa)} e^{i\psi}, P_2^{(\kappa)} = L_2^{(\kappa)} e^{i\psi}, E_2 = M_2 e^{i\psi}, B_3 = N_3 e^{i\psi}. \tag{13.10}$$

In the case of the quasi-static electric field, the absence of (13.5) permitted E_2, as well as B_3, to be zero; but both must be non-zero when the full electromagnetic equations are imposed.

To find a form of the dispersion determinant similar to (12.44), it is useful first to express M_2 and N_3 in terms of $K_2^{(\kappa)}$ and $L_2^{(\kappa)}$ through the use of (13.4) and (13.5):

$$M_2 = -i\omega N_3 \xi^{-1} = -\varepsilon_\xi^{-1}[L_2^{(1)}+L_2^{(2)}+q_*(K_2^{(2)}-K_2^{(1)})], \tag{13.11}$$

where

$$\varepsilon_\xi = \varepsilon_0 - \xi^2/\mu_0\omega^2. \tag{13.12}$$

When the results (13.11), along with (13.9) and (13.10), are inserted in the remaining equations of motion, the latter are satisfied if

$$\Delta_T^{EM} = 0, \tag{13.13}$$

where Δ_T^{EM} is the determinant with elements

$$\Delta_{11} = \rho^{(1)}\omega^2 - a^{**} - q_*^2\varepsilon_\xi^{-1} - (c_{44}^{11}+c^{**})\xi^2,$$
$$\Delta_{12} = a^{**} + q_*^2\varepsilon_\xi^{-1} - (c_{44}^{21}-c^{**})\xi^2 = \Delta_{21},$$
$$\Delta_{13} = a^{*1} + q_*\varepsilon_\xi^{-1} - (d_{44}^{11}-d^{*1})\xi^2 = \Delta_{31},$$
$$\Delta_{14} = a^{*2} + q_*\varepsilon_\xi^{-1} - (d_{44}^{21}-d^{*2})\xi^2 = \Delta_{41},$$
$$\Delta_{22} = \rho^{(2)}\omega^2 - a^{**} - q_*^2\varepsilon_\xi^{-1} - (c_{44}^{22}+c^{**})\xi^2,$$
$$\Delta_{23} = -a^{*1} - q_*\varepsilon_\xi^{-1} - (d_{44}^{12}+d^{*1})\xi^2 = \Delta_{32},$$
$$\Delta_{24} = -a^{*2} - q_*\varepsilon_\xi^{-1} - (d_{44}^{22}+d^{*2})\xi^2 = \Delta_{42},$$
$$\Delta_{33} = -a_{11}^{11} - \varepsilon_\xi^{-1} - (b_{44}^{11}+b_{77}^{11})\xi^2,$$
$$\Delta_{34} = -a_{11}^{21} - \varepsilon_\xi^{-1} - (b_{44}^{21}+b_{77}^{21})\xi^2 = \Delta_{43},$$
$$\Delta_{44} = -a_{11}^{22} - \varepsilon_\xi^{-1} - (b_{44}^{22}+b_{77}^{22})\xi^2.$$

The long wave, high frequency behavior is obtained from

$$\lim_{\xi\to 0} \Delta_T^{EM} = \omega^4\{\rho^{(1)}\rho^{(2)}\omega^2[(a_{11}^{11}+\varepsilon_0^{-1})(a_{11}^{22}+\varepsilon_0^{-1})-(a_{11}^{12}+\varepsilon_0^{-1})^2]-(\rho^{(1)}+\rho^{(2)})D_L\}, \tag{13.14}$$

where D_L is again given by (12.38). Thus, the liniting frequency is

$$\omega_{TO}^{EM} = \{D_L/\rho[(a_{11}^{11}+\varepsilon_0^{-1})(a_{11}^{22}+\varepsilon_0^{-1})-(a_{11}^{12}+\varepsilon_0^{-1})^2]\}^{\frac{1}{2}}, \tag{13.15}$$

i.e., as found by Huang [67, 68], the same as the limiting frequency of the longitudinal optical branch.

The long wave, low frequency behavior is given by

$$\lim_{\omega,\xi\to 0} \Delta_T^{EM} = \lim_{\omega,\xi\to 0} [(c_{44}^{11}+c_{44}^{22}+2c_{44}^{12})\xi^2-(\rho^{(1)}+\rho^{(2)})\omega^2]D_\xi, \tag{13.16}$$

where

$$D_\xi = (\varepsilon_0\mu_0 D_L\omega^2 - \xi^2 D)/(\varepsilon_0\mu_0\omega^2-\xi^2) \tag{13.17}$$

and D is again given by (12.47). Thus, there are two branches, with limiting, low frequency, long wave behaviors

$$\omega_{EM} = \xi(D/\varepsilon_0\mu_0 D_L)^{\frac{1}{2}}, \tag{13.18}$$

$$\omega_{TA}^{EM} = \xi[(c_{44}^{11}+c_{44}^{22}+2c_{44}^{12})/(\rho^{(1)}+\rho^{(2)})]^{\frac{1}{2}}. \tag{13.19}$$

The first of these is the long wave end of the electromagnetic branch, with slope equal to the quotient of the velocity of electromagnetic waves in a vacuum $(\varepsilon_0\mu_0)^{-\frac{1}{2}}$ and the index of refraction $(D_L/D)^{\frac{1}{2}}$ in the dielectric. The second is the long wave end of the transverse acoustic branch: identical with that for the quasi-static electric field (Article 12) and the purely elastic field (Article 4). The long wave portions of the real branches of the dispersion relations for both the longitudinal and transverse waves are illustrated in Fig. 20. The vertical scale of the acoustic branches is expanded: the ratio of their slopes to that of the electromagnetic branch is of the order of 10^{-5}.

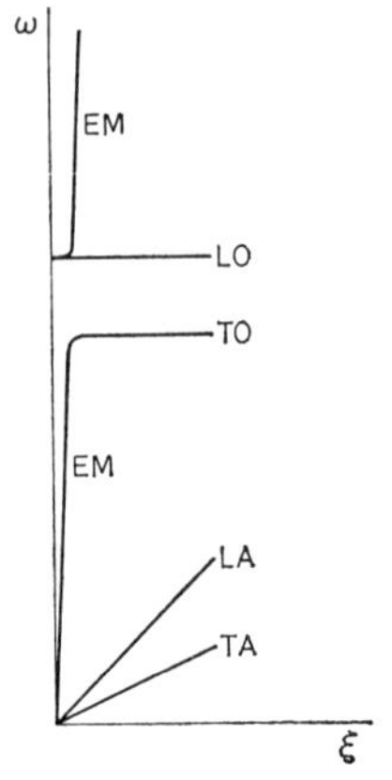

Figure 20

The long wave limits of the EM and LO branches are independent of the constants b_{ijkl}, c_{ijkl} and d_{ijkl}, i.e. independent of the polarization and displacement gradients. Also, the long wave limits of the acoustic branches depend only on the mass densities and the elastic stiffnesses. Hence, as far as the limiting behavior at the long wave end is concerned, the results here, conform with Huang's, with the addition of the acoustic branches. However, as may be seen from the determinants (12.35) and (13.13), when the wave length diminishes from infinity, the displacement and polarization gradients affect the dispersion relation, as does the coupling with the elastic field. Then the results diverge from Huang's.

A very general treatment, including crystals of arbitrary symmetry, an arbitrary number of ions and electrons per unit cell and non-linear, as well as linear, interactions, is given by Lax and Nelson [70].

REFERENCES

[1] Cauchy, A. L., "Note sur l'équilibre et les mouvements vibratoires des corps solides", *Comptes Rendus Acad. Sci*, Paris, *32*, pp. 323–326 (1851).

[2] Cosserat, E. et F., Théorie des corps déformables, A. Hermann et Fils, Paris (1909).

[3] Voigt, W., Lehrbuch der Kristallphysik, B. G. Teubner, Leipzig (1910).

[4] Gazis, D. C., Herman, R. and Wallis, R. F., "Surface elastic waves in cubic crystals," *Phys. Rev. 119*, pp. 533–544 (1960).

[5] Born, M. und Kármán, Th. v., "Über Schwingungen in Raumgittern", *Physikal. Zeit. 13*, pp. 297–309 (1912).

[6] Mindlin, R. D., "Theories of elastic continua and crystal lattice theories", in *IUTAM* Symposium Freudenstad-Stuttgart 1967, *Mechanics of Generalized Continua*, E . Kröner, editor, Springer-Verlag, Berlin pp. 312–320, (1968).

[7] Salvadori, M. G. and Baron, M. L., Numerical Methods in Engineering, Prentice-Hall, New York, 2nd Edition, 1961.

[8] Mindlin, R. D., "Lattice theory of shear modes of vibration and torsional equilibrium of simple-cubic crystal plates and bars", *Int. J. Solids Struct.*, *6*, pp. 725–738 (1970).

[9] Love, A. E. H., Some Problems of Geodynamics, *Cambridge Univ. Press*, London, p. 160, (1926).

[10] Brady, K. J., "Lattice theory of face-shear and thickness-twist waves in face-centered cubic crystal plates", *Int. J. Solids Struct.*, *7*, pp. 941–964 (1971).

[11] Chung Gong, "Lattice theory of face-shear and thickness-twist waves in body-centered cubic crystal plates", *Int. J. Solids Struct.*, *7*, pp. 751–787 (1971).

[12] Todhunter, I, and Pearson, J., A History of the Theory of Elasticity, Vol. 2, Part 1, p. 24, *Cambridge Univ. Press*, London, 1893.
[13] Jolley, L. B. W., Summation of Series, Dover, New York, 2nd Ed. (1968), p. 80, No. 427.
[14] Aero, E. L. and Kuvshinskii, E. V., "Fundamental equations of the theory of elastic media with rotationally interacting particles", *Fizika Tverdogo Tela, 2*, pp. 1399–1409 (1960); Translation: *Soviet Physics Solid State, 2*, 1272–1281 (1961).
[15] Grioli, G., "Elasticità asimmetrica", *Ann. di Mat. pura ed appl.*, Ser. IV, *50*, pp. 1399–1409 (1960).
[16] Rajagopal, E. S., "The existence of interfacial couples in infinitesimal elasticity", *Ann. der Physik, 6*, pp. 192–201 (1960).
[17] Truesdell, C. A. and Toupin, R. A., Encyclopedia of Physics, Vol. III/1, Springer-Verlag, Berlin (1960).
[18] Krumhansl, J. A., "Generalized continuum field representations for lattice vibrations," in Lattice Dynamics, R. F. Wallis, *Editor, Pergamon Press*, Oxford, 1964, pp. 627–634.
[19] Toupin, R. A., "Elastic materials with couple-stresses", *Arch. Rat. Mech. Anal., 11.* pp. 385–414(1962).
[20] Tosi, M. P., "IV. Surface energy", in Solid State Physics, Vol. 16, *Academic Press*, New York (1964) pp. 92–107.
[21] Germer, L. H., MacRae, A. U. and Hartman, C. D., "(110) Nickel surface", *J. Appl. Phys., 32*, pp. 2432–2439 (1961).
[22] Toupin, R. A. and Gazis, D. C., "Surface effects and initial stress in continuum and lattice models of elastic crystals", in Lattice Dynamics, R. F. Wallis, *Editor, Pergamon Press*, Oxford, 1964, pp. 597–605.
[23] Mindlin, R. D., "Second gradient of strain and surface-tension in linear elasticity", *Int. J. Solids Struct., 1*, pp. 417–438 (1965).
[24] Gazis, D. C. and Wallis, R. F., "Surface tension and surface modes in semi-infinite lattices", *Surface Science, 3*, pp. 19–32 (1964).
[25] Gazis, D. C. and Wallis, R. F., *Private communication* (1966).
[26] Ludwig, W., Recent Developments in Lattice Theory, Springer-Verlag, Berlin (1967).
[27] Cosserat, E. et F., Th*é*orie des corps d*é*formable, A. Hermann et Fils, Paris (1909).
[28] Schaefer, H., "Versuch einer Elastizitatstheorie des Zweidimensionalen ebenen Cosserat-Kontinuums", *Miszellaneen der Angewandten Mechanik*, pp. 277–292, Akademie-Verlag, Berlin, (1962).
[29] Eringen, A. C., "Linear theory of micropolar elasticity", *J. Math. Mech., 15*, pp. 909-924 (1966).
[30] Kuvshinskii, E. V. and Aero, E. L., "Continuum theory of asymmetric elasticity", *Fizika tverdogo tela, 5*, pp. 2591–2598 (1963); Translation: *Soviet Physics Solid State, 5*, pp. 1892–1897 (1964).
[31] Aero, E. L. and Kurshinkskii, E. V., "Continuum theory of asymmetric elasticity. Equilibrium of an isotropic body", *Fizika tverdogo tela, 6*, pp. 2689–2699 (1964); Translation: *Soviet Physics Solid State, 6*, pp. 2141–2148 (1965).
[32] Neuber, H., "On the general solution of linear-elastic problems in isotropic and anisotropic Cosserat continua", Proc. 11th *Int. Cong. Appl. Mech.* pp. 153–158, Springer-Verlag, Berlin (1965).
[33] Mindlin, R. D., "Stress functions for a Cosserat continuum", *Int. J. Solids Struct., 1*, pp. 265–271 (1965).
[34] Cowin, S. C., "Stress functions for Cosserat elasticity", *Int. J. Solids Struct., 6*, pp. 389–398 (1970).
[35] Cowin, S. C., "An incorrect inequality in micropolar elasticity theory", *Zeit. f. Ang. Math. u. Physik, 21*, pp. 494–497 (1970).
[36] Askar, A., "Molecular crystals and the polar theories of the continua. Experimental values of material coefficients for KNO_3", *Int. J. Enging. Sci., 10*, pp. 293–300 (1972).
[37] Toupin, R. A., "Theories of elasticity with couple-stress", *Arch. Rat. Mech. Anal., 17*, pp. 85–112 (1964).
[38] Eringen, A. C., "Mechanics of micromorphic materials", Proc. 11th *Int. Cong. Appl. Mech.*, Springer-Verlag, Berlin, 1966, pp. 131–138.
[39] Green, A. E. and Rivlin, R. S., "Multipolar continuum mechanics", *Arch. Rat. Mech. Anal., 17*, pp. 113–147 (1964).
[40] Mindlin, R. D., "Micro-structure in linear elasticity", *Arch. Rat. Mech. Anal., 16*, pp. 51–78 (1964).
[41] Mason, W. P., Piezoelectric Crystals and their Application to Ultrasonics, *van Nostrand*, New York, 1950.
[42] Toupin, R. A., "The elastic dielectric", *J. Rat. Mech. Anal., 5*, pp. 849–915 (1956).
[43] I. R. E., Standards on Piezoelectric Crystals, 1949, *Proc. Inst. Radio Engineers, 37*, pp. 1378–1395 (1949).
[44] Mindlin, R. D., "Polarization gradient in elastic dielectrics", *Int. J. Solids Struct., 4*, pp. 637–642 (1968).
[45] Lomont, J. S., Applications of Finite Groups, *Academic Press*, New York, 1959.
[46] Cochran, W., "Theory of lattice vibrations of germanium", *Proc. Roy. Soc., A 253*, pp. 260–276 (1959).
[47] Dick, Jr. B. G. and Overhauser, A. W., "Theory of the dielectric constants of alkali halide crystals", *Phys. Rev., 112*, pp. 90–103 (1958).

[48] Cochran, W., "Theory of phonon dispersion curves", in Phonons in Perfect Lattice and in Lattices with Point Imperfections, R. W. H. Stevenson, Editor, *Plenum Press*, New York, 1966, pp. 53–72.

[49] Askar, A., Lee, P. C. Y. and Cakmak, A. S., "Lattice-dynamics approach to the theory of elastic dielectrics with polarization gradient", *Phys. Rev.*, *B1*, pp. 3525–3537 (1970).

[50] Mead, C. A., "Electron transport mechanisms in thin insulating films", *Phys. Rev.*, *128*, pp. 2088–2093 (1962).

[51] Mead, C. A., "Electron transport in thin insulating films", *Proc. Int. Symp. on Basic Problems in Thin Film Physics*, Vandenhoeck & Ruprecht, Göttingen, 1966, pp. 674–678; M.McColl and C. A. Mead, "Electron current through thin mica films", Trans. *Metall. Soc. AIME*, *233*, pp. 502–511 (1965).

[52] Ku, H Y. and Ullman, F. G., "Capacitance of thin dielectric structures", *J. Appl. Phys.*, *35*, pp. 265–270 (1964).

[53] Mindlin, R. D., "Continuum and lattice theories of influence of electromechanical coupling on capacitance of thin dielectric films", *Int. J. Solids Struct.*, *5*, pp. 1197–1208 (1969).

[54] Schwartz, J., "Solutions of the equations of equilibrium of elastic dielectrics: stress functions, concentrated force, surface energy", *Int. J. Solids Struct.*, *5*, pp. 1209–1220 (1969).

[55] Askar, A., Lee, P. C. Y. and Cakmak, A. S., "The effect of surface curvature and discontinuity on the surface energy density and other induced fields in elastic dielectrics with polarization gradient", *Int. J. Solids Struct.*, *7*, pp. 523–537 (1971).

[56] Pine, A. S., "Direct observation of acoustical activity in α-quartz", *Phys. Rev. B2*, pp. 2049–2054 (1970).

[57] Silin, V. P., "Contribution to the theory of absorption of ultrasound in metals," *JETP 38*, pp. 977–983 (1960); Translation: *Sov. Phys. JETP 11*, pp. 703–707 (1960).

[58] Portigal, D. L. and Burstein, E., "Acoustical activity and other first order spatial dispersion effects in crystals", *Phys. Rev.*, *170*, 673–678 (1968).

[59] Mindlin, R. D. and Toupin, R. A., "Acoustical and optical activity in alpha quartz", *Int. J. Solids Struct.* *7*, pp. 1219–1227 (1971).

[60] Nye, J. F., Physical Properties of Crystals, *Oxford University Press* (1959).

[61] Houston, R. A., A Treatise on Light, *Longmans*, *Green*, London (1934).

[62] Cady, W. G., Piezoelectricity, *Mc Graw-Hill*, New York (1946).

[63] Bechmann, R., "Elastic and piezoelectric constants of alpha-quartz". *Phys. Rev. 110*, pp. 1060–1061 (1958).

[64] Mindlin, R. D., "A continuum theory of a diatomic, elastic dielectric", *Int. J. Solids Struct.*, *8*, pp. 369–383 (1972).

[65] Mindlin, R. D., "Thickness-shear and flexural vibrations of crystal plates", *J. Appl. Phys.*, *22*, pp. 316–323 (1951).

[66] Mindlin, R. D. and Forray, M., "Thickness-shear and flexural vibrations of contoured crystal plates", *J. Appl. Phys.*, *25*, pp. 12–20 (1954).

[67] Born, M. and Huang, K., Dynamical Theory of Crystal Lattices, *Oxford University Press* (1956) Chapter II.

[68] Huang, K., "On the interaction between the radiation field and ionic crystals", *Proc. Roy. Soc.*, *A208*, pp. 352–365 (1951).

[69] Mindlin, R. D., "Coupled elastic and electromagnetic fields in a diatomic, dielectric continuum", *Int. J. Solids Struct.*, *8*, pp. 401–408 (1972).

[70] Lax, M. and Nelson, D. F., "Linear and nonlinear electrodynamics in elastic anisotropic dielectrics", *Phys. Rev.*, *B4*, pp. 3694–3731 (1971).

[71] Eringen, A. C. and Edelen, D. G. B., On nonlocal elasticity, *Int. J. Engng. Sci.*, *10*, pp. 233-248 (1972).

[72] Eringen, A. C., Linear theory of nonlocal elasticity and dispersion of plane waves, *Int. J. Engng. Sci.* *10*, pp. 425-435 (1972).

Int. J. Solids Structures, 1973, Vol. 9, pp. 233 to 235. Pergamon Press. Printed in Great Britain

ON THE ELECTROSTATIC POTENTIAL OF A POINT CHARGE IN A DIELECTRIC SOLID

R. D. MINDLIN

Department of Civil Engineering, Columbia University, New York

THE electrostatic potential of a point charge, q, in an unbounded vacuum, is

$$\varphi = q/4\pi\varepsilon_0 r, \tag{1}$$

where ε_0 is the permittivity of the vacuum and r is the distance from the point charge. If the vacuum is replaced by an isotropic dielectric, the point charge polarizes the atoms of of the dielectric and, according to the classical theory of electrostatics, the potential is modified to

$$\varphi = q/4\pi\varepsilon_0(1+\eta)r = q/4\pi\varepsilon r, \tag{2}$$

where η is the dielectric susceptibility and ε is the permittivity of the dielectric. The result (2) is based on the assumption, fundamental to the classical theory, that the stored energy of polarization depends only on the local polarization. In terms of the shell model of the atom (a core, comprising the nucleus and inner electrons, surrounded by a shell of outer electrons) this corresponds to a contribution to the stored energy solely from the interaction between the shell of an atom and its own core. It is interesting to calculate the effect of the interaction between the shells of adjacent atoms, which is known to be important in the calculation of phonon dispersion relations in dielectric crystal lattices. In a continuum, the shell–shell interaction is represented by the contribution of the polarization gradient [1] to the stored energy of polarization. The appropriate field equations may be derived by means of a slight modification of Toupin's [2] variational principle for the classical theory of elastic dielectrics. Essentially, it is only necessary to replace the strain by the polarization gradient.

For a dielectric material occupying a volume V, bounded by a surface S separating V from an outer vacuum V', the variational principle, for the linear case, is

$$\delta\int_{V+V'} H\,\mathrm{d}V+\int_V \rho\delta\varphi\,\mathrm{d}V = 0, \tag{3}$$

where ρ is the charge density and H is the electric enthalpy density:

$$H = W^L(P_i,P_{j,i})-\tfrac{1}{2}\varepsilon_0\varphi_{,i}\varphi_{,i}+\varphi_{,i}P_i. \tag{4}$$

In (4), $W^L(P_i, P_{j,i})$ is the stored energy of polarization: a function of both the polarization and the polarization gradient. The polarization and the charge density are zero outside V whereas φ exists in V and V'.

Now,

$$\delta H = -E_i^L\delta P_i + E_{ij}\delta P_{j,i} - \varepsilon_0\varphi_{,i}\delta\varphi_{,i} + \varphi_{,i}\delta P_i + P_i\delta\varphi_{,i} \tag{5}$$

where

$$E_i^L \equiv -\partial W^L/\partial P_i, \qquad E_{ij} \equiv \partial W^L/\partial P_{j,i}. \tag{6}$$

By the chain rule of differentiation,

$$\delta H = -(E_{ij,i} + E_j^L - \varphi_{,j})\delta P_j - (-\varepsilon_0\varphi_{,ii} + P_{i,i})\delta\varphi + (E_{ij}\delta P_j)_{,i} + [(-\varepsilon_0\varphi_{,i} + P_i)\delta\varphi]_{,i}. \tag{7}$$

Inserting this result in (3) and applying the divergence theorem, we have

$$-\int_V [(E_{ij,i} + E_j^L - \varphi_{,i})\delta P_i + (-\varepsilon_0\varphi_{,ii} + P_{i,i} - \rho)\delta\varphi]\,\mathrm{d}V + \int_{V'} \varepsilon_0\varphi_{,ii}\delta\varphi\,\mathrm{d}V$$

$$+\int_S [n_iE_{ij}\delta P_j + n_i(\varepsilon_0[\varphi_{,i}] + P_i)\delta\varphi]\,\mathrm{d}S = 0, \tag{8}$$

where $[\varphi_{,i}]$ is the jump in $\varphi_{,i}$ across S. Thus, the field equations in V, which are all we need for the present application, are

$$E_{ij,i} + E_i^L - \varphi_{,i} = 0, \qquad \varepsilon_0\varphi_{,ii} - P_{i,i} + \rho = 0. \tag{9}$$

The stored energy of polarization is taken to be

$$W^L = \tfrac{1}{2}\alpha_{ij}P_iP_j + \tfrac{1}{2}\beta_{ijkl}P_{j,i}P_{l,k} + \gamma_{ijk}P_iP_{k,j}. \tag{10}$$

We shall consider only centrosymmetric, isotropic materials—in which case

$$\begin{aligned} \alpha_{ij} &= \varepsilon_0^{-1}\eta^{-1}\delta_{ij}, \\ \beta_{ijkl} &= \beta_1\delta_{ij}\delta_{kl} + \beta_2(\delta_{ik}\delta_{jl} + \delta_{il}\delta_{jk}) + \beta_3(\delta_{ik}\delta_{jl} - \delta_{il}\delta_{jk}), \\ \gamma_{ijk} &= 0, \end{aligned} \tag{11}$$

where δ_{ij} is the Kronecker delta and β_1, β_2, β_3 are constants associated with the polarization gradient.

From (6), (10) and (11), we have

$$\begin{aligned} E_i^L &= -\varepsilon_0^{-1}\eta^{-1}P_i, \\ E_{ij} &= \beta_1\delta_{ij}P_{k,k} + \beta_2(P_{j,i} + P_{i,j}) + \beta_3(P_{j,i} - P_{i,j}). \end{aligned} \tag{12}$$

Substituting these constitutive equations into (9) and converting to vector notation, we find the field equations for the centrosymmetric, isotropic dielectric:

$$(\beta_2+\beta_3)\nabla^2\mathbf{P} + (\beta_1+\beta_2-\beta_3)\nabla\nabla\,.\,\mathbf{P} - \varepsilon_0^{-1}\eta^{-1}\mathbf{P} - \nabla\varphi = 0, \qquad -\varepsilon_0\nabla^2\varphi + \nabla\,.\,\mathbf{P} = \rho. \tag{13}$$

To eliminate **P** from (13), operate on the first of the two equations with the divergence, to get

$$(\beta_1+2\beta_2)\nabla^2\nabla\,.\,\mathbf{P} - \varepsilon_0^{-1}\eta^{-1}\nabla\,.\,\mathbf{P} - \nabla^2\varphi = 0$$

and then use the second of (13) to eliminate $\nabla\,.\,\mathbf{P}$ with the result, after some manipulations,

$$\varepsilon(1 - l^2\nabla^2)\nabla^2\varphi = -[1 + (1+\eta)l^2\nabla^2]\rho, \tag{15}$$

where

$$l = [\varepsilon_0 \eta(\beta_1 + 2\beta_2)/(1+\eta)]^{\frac{1}{2}}. \tag{16}$$

The constant l has the dimension of length and is estimated to be of the order of the distance between adjacent atoms or molecules.

A particular integral of (15) is

$$4\pi\varepsilon\varphi = \int_V \psi[1+(1+\eta)l^2\nabla^2]\rho \, dV, \tag{17}$$

where

$$\psi = (1-e^{-r_1/l})/r_1 \tag{18}$$

and r_1 is the distance between a field point and a source point.

By Green's second identity,

$$\int_V \psi\nabla^2\rho \, dV = \int_S \mathbf{n} \, . \, (\psi\nabla\rho - \rho\nabla\psi) \, dS + \int_V \rho\nabla^2\psi \, dV. \tag{19}$$

Also,

$$l^2\nabla^2\psi = e^{-r_1/l}/r_1. \tag{20}$$

Hence

$$4\pi\varepsilon\varphi = \varepsilon \int_S \mathbf{n} \, . \, (\psi\nabla\rho - \rho\nabla\psi) \, dS + \int_V \rho r_1^{-1}(1+\eta \, e^{-r_1/l}) \, dV. \tag{21}$$

Accordingly, for a point charge, q, in an infinite dielectric,

$$\varphi = q(1+\eta \, e^{-r/l})/4\pi\varepsilon r, \tag{22}$$

where r is the distance from the point charge.

For materials with high dielectric susceptibility, the exponential correction term, in (22), can be significant even at distances of several times l from the point charge, where the continuum approximation is probably valid. A comparable result has been obtained from atomic considerations [3].

REFERENCES

[1] R. D. Mindlin, *Int. J. Solids Struct.* **4**, 637 (1968); **5**, 1197 (1969).
[2] R. A. Toupin, *J. ration. Mech. Analysis* **5**, 849 (1956).
[3] G. D. Mahan and R. M. Mazo, *Phys. Rev.* **175**, 1191 (1968).

(*Received* 7 *May* 1972)

Int. J. Solids Structures, 1973, Vol. 9, pp. 697 to 702. Pergamon Press. Printed in Great Britain

ELECTROMAGNETIC RADIATION FROM A VIBRATING QUARTZ PLATE

R. D. MINDLIN

Department of Civil Engineering, Columbia University, New York

Abstract—The linear equations of piezoelectromagnetism are solved for the case of thickness-shear vibrations of a quartz plate. For an AT-cut plate vibrating near resonance, the radiation from each face is about 25 μW/cm^2 for a shear strain of 10^{-5}.

1. INTRODUCTION

THE problem of electromagnetic radiation from a quartz plate vibrating in a predominantly thickness-shear mode was studied by Tiersten [1] who concluded that there is no radiation in the case of a "trapped energy" [2] mode. In the present paper, an exact solution of the linear equations of piezoelectromagnetism is obtained for a mode of vibration favorable to radiation: thickness-shear, independent of the coordinates in the plane of the plate. A simple formula is obtained for the power radiated, per unit of area, from each face of the plate:

$$P = \tfrac{1}{2}\varkappa_{33}\hat{c}_{66}^2 S_0^2/\rho c,$$

where $\varkappa_{33}$ is a dielectric constant, $\hat{c}_{66}$ is a shear stiffness, S_0 is the maximum shear strain, ρ is the mass density and c is the velocity of light *in vacuo*. For the AT-cut of quartz, the radiation is about 25μW/cm^2 for a shear strain of 10^{-5}.

2. EQUATIONS AND BOUNDARY CONDITIONS

The field equations of piezoelectromagnetism in a dielectric comprise the stress equations of motion

$$T_{ij,i} = \rho\ddot{u}_j \tag{1}$$

and the equations of the electromagnetic field:

$$\varepsilon_{ijk}E_{k,j} = -\dot{B}_i, \qquad \varepsilon_{ijk}B_{k,j} = \mu_0\dot{D}_i, \qquad B_{i,i} = 0, \qquad D_{i,i} = 0. \tag{2}$$

In these equations, **u** is the mechanical displacement, **E** is the Maxwell electric self-field, **B** is the magnetic flux density, ρ is the mass density, μ_0 is the magnetic permeability of a vacuum and ε_{ijk} is the unit alternating tensor. For the rotated-Y-cuts of quartz [3], the components of stress, T_{ij}, and the components of electric displacement, D_i, are expressed, in the linear theory, in terms of the components of strain,

$$S_{ij} = \tfrac{1}{2}(u_{j,i}+u_{i,j}), \tag{3}$$

and the components of the electric field, E_i, according to

$$\begin{aligned}
T_{11} &= c_{11}S_{11}+c_{12}S_{22}+c_{13}S_{33}+2c_{14}S_{23}-e_{11}E_1,\\
T_{22} &= c_{21}S_{11}+c_{22}S_{22}+c_{23}S_{33}+2c_{24}S_{23}-e_{12}E_1,\\
T_{33} &= c_{31}S_{11}+c_{32}S_{22}+c_{33}S_{33}+2c_{34}S_{23}-e_{13}E_1,\\
T_{23} &= c_{41}S_{11}+c_{42}S_{22}+c_{43}S_{33}+2c_{44}S_{23}-e_{14}E_1,\\
T_{31} &= 2c_{55}S_{31}+2c_{56}S_{12}-e_{25}E_2-e_{35}E_3,\\
T_{12} &= 2c_{56}S_{31}+2c_{66}S_{12}-e_{26}E_2-e_{36}E_3,\\
D_1 &= e_{11}S_{11}+e_{12}S_{22}+e_{13}S_{33}+2e_{14}S_{14}+\varepsilon_{11}E_1,\\
D_2 &= 2e_{25}S_{31}+2e_{26}S_{12}+\varepsilon_{22}E_2+\varepsilon_{23}E_3,\\
D_3 &= 2e_{35}S_{31}+2e_{36}S_{12}+\varepsilon_{23}E_2+\varepsilon_{33}E_3.
\end{aligned} \tag{4}$$

In (4), the c_{pq} are the elastic stiffnesses at constant electric field, the ε_{ij} are the dielectric permittivities and the e_{ip} are the piezoelectric stress constants.

We shall consider simple x_1-thickness-shear modes in an infinite plate bounded by planes $x_2 = \pm b$ which separate the plate from a vacuum. In that case, in $|x_2| \leq b$,

$$u_1 = u_1(x_2, t), \qquad u_2 = u_3 = 0; \qquad E_1 = 0, \qquad E_2 = E_2(x_2, t), \qquad E_3 = E_3(x_2, t), \tag{5}$$

and consequently equations (1)–(4) reduce to

$$T_{12,2} = \rho\ddot{u}_1, \tag{6}$$

$$E_{3,2} = -\dot{B}_1, \qquad B_{1,2} = -\mu_0\dot{D}_3, \qquad \dot{D}_2 = 0, \tag{7}$$

$$S_{12} = \tfrac{1}{2}u_{1,2}, \tag{8}$$

$$\begin{aligned}
T_{12} &= 2c_{66}S_{12}-e_{26}E_2-e_{36}E_3,\\
D_2 &= 2e_{26}S_{12}+\varepsilon_{22}E_2+\varepsilon_{23}E_3,\\
D_3 &= 2e_{36}S_{12}+\varepsilon_{23}E_2+\varepsilon_{33}E_3.
\end{aligned} \tag{9}$$

The remaining equations are either satisfied identically or are not coupled with (6)–(9). The latter may be condensed, finally, to the three equations

$$\begin{aligned}
c_{66}u_{1,2}-e_{26}E_{2,2}-e_{36}E_{3,2} &= \rho\ddot{u}_1,\\
e_{26}u_{1,2}+\varepsilon_{22}E_2+\varepsilon_{23}E_3 &= 0,\\
e_{36}\ddot{u}_{1,2}+\varepsilon_{23}\ddot{E}_2+\varepsilon_{33}\ddot{E}_3 &= \mu_0^{-1}E_{3,22}.
\end{aligned} \tag{10}$$

Outside the plate, $u_1 = E_2 = 0$, $E_3 \to E_3^0$, $e_{36} = \varepsilon_{23} = 0$ and $\varepsilon_{33} \to \varepsilon_0$, the permittivity of a vacuum. Equations (10) then reduce to

$$E_{3,22}^0 = \varepsilon_0\mu_0\ddot{E}_3^0. \tag{11}$$

For the mechanical boundary conditions on $x_2 = \pm b$, we shall specify oscillating tractions $n_i T_{ij}$, where $\mathbf{n}$ is the unit outward normal. The electromagnetic boundary conditions require continuity of

$$n_i D_i,\ n_j \varepsilon_{ijk} E_k,\ n_i B_i,\ n_j \varepsilon_{ijk} B_k$$

across $x_2 = \pm b$. In view of (5), the boundary conditions reduce to

$$T_{12} = T\cos\omega t, \qquad E_3 = E_3^0, \qquad E_{3,2} = E_{3,2}^{(0)} \tag{12}$$

on $x_2 = \pm b$, where T is the amplitude of the prescribed surface traction.

3. SOLUTION

Anticipating, from the forms of equations (10–12), that there will be two wave numbers and two phase components, we take

$$\begin{aligned} u_1 &= \sum_r A_r \sin \eta_r x_2 \cos(\omega t - \varepsilon_r), && r = 1, 2, \\ E_2 &= \sum_r B_r \cos \eta_r x_2 \cos(\omega t - \varepsilon_r), && r = 1,2, \\ E_3 &= \sum_r C_r \cos \eta_r x_2 \cos(\omega t - \varepsilon_r), && r = 1, 2 \end{aligned} \tag{13}$$

for $|x_2| \leq b$ and, for $|x_2| \geq b$,

$$E_3^0 = C_0 \cos \eta_0 (x_2 \mp b \mp ct \mp \delta), \tag{14}$$

where $c(= \varepsilon_0^{-\frac{1}{2}} \mu_0^{-\frac{1}{2}})$ is the velocity of electromagnetic waves in a vacuum, $\eta_0 = \omega/c$ and the upper and lower signs are for $x_2 \geq b$ and $x_2 \leq b$, respectively.

Upon substitution of (13) in (10), there results

$$\begin{aligned} (\rho\omega^2 - \eta_r^2 c_{66}) A_r + \eta_r e_{26} B_r + \eta_r e_{26} C_r &= 0, \\ \eta_r e_{26} A_r + \varepsilon_{22} B_r + \varepsilon_{23} C_r &= 0, \\ \eta_r e_{36} A_r + \varepsilon_{23} B_r + (\varepsilon_{33} - \eta_r^2 \mu_0^{-1} \omega^{-2}) C_r &= 0. \end{aligned} \tag{15}$$

From these equations follow the amplitude ratios

$$\begin{aligned} \beta_r &= \frac{B_r}{A_r} = \frac{(c_{66} + e_{26} e_{36}/\varepsilon_{23})\eta_r^2 - \rho\omega^2}{\eta_r (e_{26} - e_{36}\varepsilon_{22}/\varepsilon_{23})}, \\ \gamma_r &= \frac{C_r}{A_r} = \frac{\varepsilon_{22}[\rho\omega^2 - (c_{66} + e_{26}^2/\varepsilon_{22})\eta_r^2]}{\eta_r \varepsilon_{23}(e_{26} - e_{36}\varepsilon_{22}/\varepsilon_{23})}, \end{aligned} \tag{16}$$

in which the η_r are the two positive roots of the equation obtained by setting the determinant of the coefficients of A_r, B_r and C_r, in (15), equal to zero:

$$\eta^4 - \alpha_2 \eta^2 + \alpha_3 = 0, \tag{17}$$

where

$$\alpha_2 = \left\{1+\frac{\hat{v}^2}{\hat{c}^2}\left[1+\frac{e_{26}^2}{\varepsilon_{22}c_{66}}+\frac{e_{36}^2}{\varepsilon_{33}c_{66}}-\frac{\varepsilon_{23}}{\varepsilon_{22}}\left(\frac{\varepsilon_{23}}{\varepsilon_{22}}+\frac{2e_{26}e_{36}}{\varepsilon_{22}c_{66}}\right)\right]\right\}\frac{\omega^2}{\hat{v}^2},$$
$$\alpha_3 = \left(1-\frac{\varepsilon_{23}^2}{\varepsilon_{22}\varepsilon_{33}}\right)\frac{\omega^4}{\hat{v}^2\hat{c}^2} \tag{18}$$

and

$$\hat{v}^2 = (c_{66}+e_{26}^2/\varepsilon_{22})/\rho = \hat{c}_{66}/\rho, \qquad \hat{c}^2 = c^2\varepsilon_0/\varepsilon_{33}, \tag{19}$$

i.e. $\hat{v}$ is the velocity of the x_1-thickness-shear wave including the electric, but not the magnetic, influence and c is the velocity of an electromagnetic wave, in the quartz plate, with wave normal and electric vector in the x_2 and x_3 directions, respectively.

The functions (13) may now be written as

$$u_1 = \sum_r A_r \sin\eta_r x_2 \cos(\omega t-\varepsilon_r), \qquad r = 1, 2,$$
$$E_2 = \sum_r \beta_r A_r \cos\eta_r x_2 \cos(\omega t-\varepsilon_r), \qquad r = 1, 2, \tag{20}$$
$$E_3 = \sum_r \gamma_r A_r \cos\eta_r x_2 \cos(\omega t-\varepsilon_r), \qquad r = 1, 2,$$

where β_r and γ_r are given by (16) and η_r by the two positive roots of (17).

Substitution of (20) in the three boundary conditions (12), yields the following six equations on the six constants A_r', A_r'', C_0', C_0'':

$$a_1A_1'+a_2A_2' = T, \qquad a_1A_1''+a_2A_2'' = 0,$$
$$b_1A_1'+b_2A_2'-C_0' = 0, \qquad b_2A_1''+b_2A_2''+C_0'' = 0, \tag{21}$$
$$c_1A_1'+c_2A_2'+\eta_0C_0'' = 0, \qquad c_1A_1''+c_2A_2''+\eta_0C_0' = 0,$$

where

$$A_r' = A_r\cos\varepsilon_r, \qquad A_r'' = A_r\sin\varepsilon_r, \qquad r = 1, 2,$$
$$C_0' = C_0\cos\delta, \qquad C_0' = C_0\sin\delta, \tag{22}$$

$$a_r = (\eta_r c_{66}-\beta_r e_{26}-\gamma_r e_{36})\cos\eta_r b, \qquad r = 1, 2,$$
$$b_r = \gamma_r\cos\eta_r b, \qquad c_r = \eta_r\gamma_r\sin\eta_r b, \qquad r = 1, 2. \tag{23}$$

The solution of the six equations (21) is

$$A_1' = (b_2\eta_0^2\Delta_{ab}-c_2\Delta_{ca})T/\Delta, \qquad A_1'' = a_2\eta_0\Delta_{bc}T/\Delta,$$
$$A_2' = -(b_1\eta_0^2\Delta_{ab}-c_1\Delta_{ca})T/\Delta, \qquad A_2'' = -a_1\eta_0\Delta_{bc}T/\Delta, \tag{24}$$
$$C_0' = -\Delta_{bc}\Delta_{ca}T/\Delta, \qquad C_0'' = \eta_0\Delta_{ab}\Delta_{bc}T/\Delta,$$

where

$$\Delta_{ab} = a_1b_2-a_2b_1, \qquad \Delta_{bc} = b_1c_2-b_2c_1,$$
$$\Delta_{ca} = c_1a_2-c_2a_1, \qquad \Delta = \eta_0^2\Delta_{ab}^2+\Delta_{ca}^2. \tag{25}$$

This completes the formal solution of the problem.

4. POWER RADIATED

Since the only power loss from the vibrating plate is through electromagnetic radiation, the radiated power must be equal to the power input of the surface traction. Per unit area of each face of the plate, this is $T_{12}\dot{u}_1$, on $x_2 = \pm b$. The radiated power, per unit area of each face, averaged over the period $\tau(=2\pi/\omega)$, is

$$P = \tau^{-1}\int_0^{\tau}(T_{11}\dot{u}_1)_{x_2=b}\,\mathrm{d}t = \tfrac{1}{2}\omega T\sum_r A_r'' \sin\eta_r b. \tag{26}$$

A simple formula for P may be found after taking into account the magnitudes of the terms involved in η_r and A_r''.

For the AT-cut of quartz [3], the constants c, e and ε, appearing in (16), as calculated by Syke's formulae [4] from Bechmann's [5] values of the principal constants, are

$$\begin{aligned} c_{66} &= 29{\cdot}013\times10^{10}\ \text{dyn/cm}^2\\ e_{26} &= -9{\cdot}490\times10^{-6}\ \text{C/cm}^2\\ e_{36} &= 6{\cdot}707\times10^{-6}\ \text{C/cm}^2\\ \varepsilon_{22} &= 39{\cdot}816\times10^{-21}\ \text{C}^2/\text{dyn cm}^2\\ \varepsilon_{33} &= 40{\cdot}424\times10^{-21}\ \text{C}^2/\text{dyn cm}^2\\ \varepsilon_{23} &= 0{\cdot}8678\times10^{-21}\ \text{C}^2/\text{dyn cm}^2. \end{aligned}$$

Also,

$$\begin{aligned} \rho &= 2{\cdot}6485\ \text{g/cm}^3\\ \hat{v} &= 3{\cdot}322\times10^{5}\ \text{cm/sec}\\ \varepsilon_0 &= 8{\cdot}854\times10^{-21}\ \text{C}^2/\text{dyn cm}^2\\ c &= 2{\cdot}998\times10^{10}\ \text{cm/sec}. \end{aligned}$$

Accordingly, to a close approximation,

$$\alpha_2 = \omega^2/\hat{v}^2, \qquad \alpha_3 = \omega^4/\hat{v}^2\hat{c}^2 \tag{27}$$

and

$$\eta_1^2 = (\omega^2/\hat{v}^2)(1-\hat{v}^2/\hat{c}^2), \qquad \eta_2^2 = \omega^2/\hat{c}^2. \tag{28}$$

Thus, to the second degree in $\hat{v}/\hat{c}$, i.e. to 10^{-10}, the wave number of the essentially mechanical wave is slightly less than if the electromagnetic effect is neglected and the wave number of the electromagnetic wave is not influenced by the elasticity. From (28), it is apparent that the percent increase in resonance frequencies due to the electromagnetic effect is only about 6×10^{-7}.

We shall take the forcing frequency, ω, to be a resonance frequency as calculated in the absence of the electromagnetic field:

$$\omega = n\pi\hat{v}/2b, \qquad n \text{ odd}. \tag{29}$$

Then,

$$\eta_0 = n\pi\hat{v}/2bc, \qquad \eta_1 \approx n\pi(1-\tfrac{1}{2}\hat{v}^2/\hat{c}^2)/2b, \qquad \eta_2 \approx n\pi\hat{v}/2b\hat{c} \tag{30}$$

and

$$\cos \eta_1 b \approx -(-1)^{2n-1} n\pi \hat{v}^2/4\hat{c}^2, \qquad \cos \eta_2 b \approx 1,$$
$$\sin \eta_1 b \approx -(1)^{2n-1}, \qquad \sin \eta_2 b \approx n\pi\hat{v}/2\hat{c}. \tag{31}$$

The amplitude of the strain in the middle plane of the plate is

$$S_0 = |u_{1,2}|_{x_2=0} \approx A_1'' \eta_1 \approx a_2 \eta_0 \eta_1 \Delta_{bc} T/\Delta \tag{32}$$

and P is, approximately,

$$P \approx \tfrac{1}{2}\omega A_1'' T \approx \tfrac{1}{2}\omega \Delta S_0^2/a_2 \eta_0 \eta_1^2 \Delta_{bc}. \tag{33}$$

Now,

$$\Delta \approx \Delta_{ca}^2 \approx \rho^4 \omega^8 (\hat{v}\varepsilon_{22}/\hat{c}\varepsilon_{23})^2 (2b/n\pi)^2/e_{26}^2 (1 - e_{36}\varepsilon_{22}/e_{26}\varepsilon_{23})^2.$$
$$\Delta_{bc} \approx \rho^2 \omega^4 (\hat{v}/\hat{c})(\varepsilon_{22}/\varepsilon_{23})^2 (2b/n\pi)/e_{26}^2 (1 - e_{36}\varepsilon_{22}/e_{26}\varepsilon_{23})^2,$$
$$a_2 \approx \rho\omega^2/\eta_2, \qquad \rho\omega^2 \approx \hat{c}_{66}(n\pi/2b)^2.$$

Consequently we have, finally,

$$P \approx \tfrac{1}{2}\varkappa_{33} \hat{c}_{66}^2 S_0^2/\rho c, \tag{34}$$

where $\varkappa_{33} = \varepsilon_{33}/\varepsilon_0$, as the simple formula for the power radiated, per unit area, from each face of the plate. Employing the numerical values of the constants, we have

$$P \approx 25 \times 10^4 \, S_0^2 \text{ W/cm}^2. \tag{35}$$

Hence, for a strain of 10^{-5}, the power radiated is about 25 μW/cm^2.

REFERENCES

[1] H. F. TIERSTEN, The radiation and confinement of electromagnetic energy accompanying the oscillation of piezoelectric crystal plates, *Recent Advances in Engineering Science*, Vol. 5, pp. 63–90. Gordon and Breach (1970).

[2] W. S. MORTLEY, Energy trapping, *Marconi Rev.* **30** (165), 53 (1967).

[3] W. P. MASON, *Peizoelectric Crystals and their Application to Ultrasonics*. D. Van Nostrand (1950).

[4] R. A. SYKES, Modes of motion in quartz crystals, the effects of coupling and methods of design, in *Quartz Crystals for Electrical Circuits*, edited by R. A. HEISING. D. Van Nostrand (1946).

[5] R. BECHMANN, Elastic and piezoelectric constants of alpha quartz, *Phys. Rev.* **110**, 1060 (1958).

(*Received* 16 *June* 1972)

Абстракт—Решаются линейные уравнения пьезоэлектромагнетизма для случая сдвиговых колебаний по толщине для пластинки из кварца. Для пластинки среза *AT*, колебающейся близи резонанса, излучение из каждой стороны является приблизительно 25 мквт/см² для деформации сдвига 10^{-5}.

Int. J. Solids Structures, 1974, Vol. 10, pp. 453–459. Pergamon Press. Printed in Gt. Britain.

COUPLED PIEZOELECTRIC VIBRATIONS OF QUARTZ PLATES

R. D. MINDLIN
Department of Civil Engineering, Columbia University,
New York 10027, U.S.A.

(*Received* 4 *September* 1973)

Abstract—The contributions of coupling with the electric field and mass of electrode coatings are taken into account in solutions of equations governing coupled thickness-shear, flexure and face-shear vibrational modes in rotated-*Y*-cut quartz plates.

INTRODUCTION

In a recent paper [1], a derivation was given of a set of two-dimensional equations of motion of piezoelectric crystal plates—applicable up to the frequencies of the fundamental thickness-shear modes. Solutions of the equations were exhibited for the simple thickness-shear modes (i.e. thickness-shear motions independent of the coordinates in the plane of the plate) in infinite, rotated-*Y*-cut quartz plates executing free vibrations without electrode coatings on the plate-faces and forced vibrations induced by an alternating voltage applied to electrode coatings on the faces. In the present paper, solutions are given for free and forced vibrations in which there is variation of displacement along one direction in the plane of the plate owing to the presence of a pair of free edges in the orthogonal direction.

Quartz is a trigonal crystal: with one axis of trigonal symmetry and, in the plane at right angles, three digonal axes 120° apart. A rotated-*Y*-cut is a plate with one digonal axis (designated, here, by the coordinate x_1) in the plane of the plate and with an arbitrary angle ($+35°\ 15'$ for the AT-cut, $-49°\ 12'$ for the BT-cut) between the normal to the plate and the trigonal axis. In the solutions to be considered, the dependent variables are functions of x_1 alone, the free edges of the plate are at $x_1 = \pm a$ and the motion comprises coupled thickness-shear, flexure and face-shear modes. The corresponding solution without consideration of the electric field and the mass of the electrodes is the one designated as Group A in a previous paper [2].

In the case of free vibrations of the plate without electrodes, an additional branch appears in the dispersion relation for straight-crested waves when coupling with the electric field is taken into account. However, the additional wave is a non-propagating one so that no additional resonances are found when the edge conditions are applied. Furthermore, the three remaining branches are not altered very much, either qualitatively or quantitatively, because the additional (electrical) terms, for quartz, are very small in comparisons with the mechanical terms.

In the case of forced vibrations, the uniformity of the voltage applied to the electrode coatings on the faces of the plate results in the total suppression of the additional branch in the dispersion relation for straight-crested waves. Also, the electric field affects the three surviving branches even less than in the case of free vibrations without electrodes. The effect of the mass of the electrode films on those branches is somewhat larger, but not of order greater than 10^{-2} for the usual thicknesses of electrodes; so that, again, the frequency spectrum for the bounded plate is not changed very much from that calculated in [2]. The solution does, though, yield a formula for the surface charge.

FREE VIBRATIONS

From (27), (28), (30) and (32) of [1], the equations on the dependent variables $u_2^{(0)}$, $u_3^{(0)}$, $u_1^{(1)}$ and $\phi^{(1)}$, to be solved for the case of free vibrations without electrodes, are

$$\begin{aligned} \kappa_1 c_{56} u_{3,11}^{(0)} + \kappa_1^2 c_{66}(u_{2,11}^{(0)} + u_{1,1}^{(1)}) + \kappa_1 e_{26}\phi_{,1}^{(1)} &= \rho \ddot{u}_2^{(0)}, \\ c_{55} u_{3,11}^{(0)} + \kappa_1 c_{56}(u_{2,11}^{(0)} + u_{1,1}^{(1)}) + e_{25}\phi_{,1}^{(1)} &= \rho \ddot{u}_3^{(0)} \\ \gamma_{11}u_{1,11}^{(1)} - 3b^{-2}\kappa_1[c_{56}u_{3,1}^{(0)} + \kappa_1 c_{66}(u_{2,1}^{(0)} + u_1^{(1)}) + e_{26}\phi^{(1)}] + e_{11}^{(1)}\phi_{,11}^{(1)} &= \rho \ddot{u}_1^{(1)}, \\ e_{11}^{(1)}u_{1,11}^{(1)} - 3b^{-2}[e_{25}u_{3,1}^{(0)} + \kappa_1 e_{26}(u_{2,1}^{(0)} + u_1^{(1)}) - \varepsilon_{22}\phi^{(1)}] - \varepsilon_{11}\phi_{,11}^{(1)} &= 0, \end{aligned} \tag{1}$$

with edge conditions

$$T_{12}^{(0)} = T_{13}^{(0)} = T_{11}^{(1)} = D_1^{(1)} = 0 \quad \text{on} \quad x_1 = \pm a, \tag{2}$$

where

$$\begin{aligned} T_{12}^{(0)} &= 2b\kappa_1[c_{56}u_{3,1}^{(0)} + \kappa_1 c_{66}(u_{2,1}^{(0)} + u_1^{(1)}) + e_{26}\phi^{(1)}], \\ T_{13}^{(0)} &= 2b[c_{55}u_{3,1}^{(0)} + \kappa_1 c_{56}(u_{2,1}^{(0)} + u_1^{(1)}) + e_{25}\phi^{(1)}], \\ T_{11}^{(1)} &= \tfrac{2}{3}b^3(\gamma_{11}u_{1,1}^{(1)} + e_{11}^{(1)}\phi_{,1}^{(1)}), \\ D_1^{(1)} &= \tfrac{2}{3}b^3(e_{11}^{(1)}u_{1,1}^{(1)} - \varepsilon_{11}\phi_{,1}^{(1)}), \end{aligned} \tag{3}$$

and $\kappa_1^2 = \pi^2/12$. Note that the face-charge $D^{(1)}$ has been set equal to zero, in the fourth of (1), as the faces of the plate are not coated with electrodes. Note, also, that the terms in $u_3^{(0)}$ and $\phi^{(1)}$ in the first of (1) and the term in $\phi^{(1)}$ in the second of (1) were inadvertently omitted in (27) and (28), respectively, of [1].

For the solution of (1), we take, omitting a factor $e^{i\omega t}$,

$$\begin{aligned} u_2^{(0)} &= A_2 b \sin \xi x_1, & u_1^{(1)} &= A_4 \cos \xi x_1, \\ u_3^{(0)} &= A_3 b \sin \xi x_1, & \phi^{(1)} &= A_5 \cos \xi x_1. \end{aligned} \tag{4}$$

Then, upon substituting (4) in (1), we find

$$\begin{aligned} (\bar{\xi}^2 - 3\Omega^2)A_2 + \hat{c}_{56}\bar{\xi}^2 A_3 + \bar{\xi}A_4 + \hat{e}_{26}\bar{\xi}A_5 &= 0, \\ c_{56}\bar{\xi}^2 A_2 + (\hat{c}_{55}\bar{\xi}^2 - 3\Omega^2)A_3 + \hat{c}_{56}\bar{\xi}A_4 + \hat{e}_{25}\bar{\xi}A_5 &= 0, \\ \bar{\xi}A_2 + \hat{c}_{56}\bar{\xi}A_3 + (\hat{\gamma}_{11}\bar{\xi}^2 + 1 - \Omega^2)A_4 + (\hat{e}_{26} + \hat{e}_{11}^{(1)}\bar{\xi}^2)A_5 &= 0, \\ \hat{e}_{26}\bar{\xi}A_2 + \hat{e}_{25}\bar{\xi}A_3 + (\hat{e}_{26} + \hat{e}_{11}^{(1)}\bar{\xi}^2)A_4 - (\hat{\varepsilon}_{22} + \hat{\varepsilon}_{11}^{(1)}\bar{\xi}^2)A_5 &= 0, \end{aligned} \tag{5}$$

where

$$\Omega^2 = \rho b^2\omega^2/3\kappa_1^2 c_{66}, \qquad \bar{\xi} = \xi b,$$

$$\begin{aligned} \hat{c}_{55} &= c_{55}/\kappa_1^2 c_{66}, & \hat{c}_{56} &= c_{56}/\kappa_1 c_{66}, & \hat{\gamma}_{11} &= \gamma_{11}/3\kappa_1^2 c_{66}, \\ \hat{e}_{25} &= e_{25}/\kappa_1^2 c_{66}, & \hat{e}_{26} &= e_{26}/\kappa_1 c_{66}, & \hat{e}_{11}^{(1)} &= e_{11}^{(1)}\, 3\kappa_1^2 c_{66}, \\ \hat{\varepsilon}_{22} &= \varepsilon_{22}/\kappa_1^2 c_{66}, & & & \hat{\varepsilon}_{11} &= \varepsilon_{11}/3\kappa_1^2 c_{66}. \end{aligned}$$

If the electric terms are omitted ($e = \varepsilon = 0$), the four equations (5) reduce to the three equations (11) in [2]—after noting that the coefficient of A_2 in the third of (11) in [2] should be $\bar{\xi}$ instead of $\bar{\xi}^2$.

The dispersion relation for straight-crested waves, obtained by setting the determinant of the coefficients in (5) equal to zero, is a quartic equation in $\bar{\xi}^2$. Thus, there are four branches instead of the three (thickness-shear, flexure and face-shear) found in (11) of [2]. The fourth (electric) branch is an imaginary one for all frequencies, representing a non-propagating mode, and hence can contribute no additional resonances to the frequency spectrum of a plate of finite length. To see what effect the electric branch has on the remaining three (mechanical) branches, it is advantageous to eliminate A_5 from (5) so as to arrive at the quartic equation in the form

$$\begin{vmatrix} B_{22}\bar{\xi}^2 - 3\Omega^2 & B_{23}\bar{\xi}^2 & B_{24}\bar{\xi} \\ B_{23}\bar{\xi}^2 & B_{33}\bar{\xi}^2 - 3\Omega^2 & B_{34}\bar{\xi} \\ B_{24}\bar{\xi} & B_{34}\bar{\xi} & \hat{\gamma}_{11}\bar{\xi}^2 + B_{44} - \Omega^2 \end{vmatrix} = 0 \tag{6}$$

where

$$B_{22} = 1 + \hat{e}_{26}^2\varepsilon_\xi^{-1}, \qquad B_{23} = \hat{c}_{56} + \hat{e}_{25}\hat{c}_{26}\varepsilon_\xi^{-1},$$
$$B_{33} = \hat{c}_{55} + \hat{e}_{25}^2\varepsilon_\xi^{-1}, \qquad B_{34} = \hat{c}_{56} + \hat{e}_{25}(\hat{e}_{26} + \hat{e}_{11}^{(1)}\bar{\xi}^2)\varepsilon_\xi^{-1},$$
$$B_{44} = 1 + (\hat{e}_{26} + \hat{e}_{11}^{(1)}\bar{\xi}^2)^2\varepsilon_\xi^{-1}, \qquad B_{24} = 1 + \hat{e}_{26}(\hat{e}_{26} + \hat{e}_{11}^{(1)}\bar{\xi}^2)\varepsilon_\xi^{-1},$$

and $\varepsilon_\xi = \hat{\varepsilon}_{11}\bar{\xi}^2 + \hat{\varepsilon}_{22}$.

The second terms in the B_{pq} comprise the entire contribution of the electric field; and, when they are eliminated, as may be done by setting the piezoelectric constants e equal to zero, the quartic equation reduces to the cubic $(11)_1$ of [2]. Now, for the small $\bar{\xi}$ to which the equations of motion are restricted, the second terms in the B_{pq} are all of the order of $e^2/\varepsilon c$. For example, when $\bar{\xi} = 0$,

$$B_{22} = 1 + e_{26}^2/\varepsilon_{22}c_{66}.$$

For the AT-cut of quartz,

$$e_{26} = -9{\cdot}490 \times 10^{-6}\ \mathrm{C/cm^2},$$
$$\varepsilon_{22} = 39{\cdot}816 \times 10^{-21}\ \mathrm{C^2/dyn\ cm^2},$$
$$c_{66} = 29{\cdot}013 \times 10^{10}\ \mathrm{dyn/cm^2},$$

as calculated from Bechmann's constants for α-quartz [3] by Sykes's formulas [4] for rotated-Y-cuts. Hence, for the AT-cut,

$$B_{22} = 1 + 0{\cdot}0078;$$

with similar results for the other B_{pq}. Accordingly, the presence of the electric terms has little effect on the mechanical branches of the dispersion relation for straight-crested waves and, subsequently, on the frequency spectrum of a plate of finite length.

To satisfy the four edge-conditions (2), all four solutions of the type (4), corresponding to the four roots $\bar{\xi}^2$ of (6), are required. For each $\bar{\xi}$, (5) has a set of three amplitude ratios $A_2:A_3:A_4:A_5$. Let $\bar{A}_i$, $i = 1, 2, 3, 5$, be the value of A_4 for the ith root $\bar{\xi}_i$; and let

$$A_2/A_4 = \alpha_{2i}, \qquad A_3/A_4 = \alpha_{3i}, \qquad A_5/A_4 = \alpha_{5i}, \qquad i = 1, 2, 3, 5$$

for each root $\bar{\xi}_i$. Then, for the complete solution of the type (4), again omitting a factor $e^{i\omega t}$,

$$
\begin{aligned}
u_2^{(0)} &= \sum_{i=1}^{3} b\bar{A}_i\alpha_{2i} \sin \xi_i x_1 + b\bar{A}_5\alpha_{25} \sinh \xi_5 x_1, \\
u_3^{(0)} &= \sum_{i=1}^{3} b\bar{A}_i\alpha_{31} \sin \xi_i x_1 + b\bar{A}_5\alpha_{35} \sinh \xi_5 x_1, \\
u_1^{(1)} &= \sum_{i=1}^{3} \bar{A}_i \cos \xi_i x_1 + \bar{A}_5 \cosh \xi_5 x_1, \\
\phi^{(1)} &= \sum_{i=1}^{3} \bar{A}_i\alpha_{5i} \cos \xi_i x_1 + \bar{A}_5\alpha_{55} \cosh \xi_5 x_1.
\end{aligned} \tag{6}
$$

Upon substituting (6) in (3) and the result in the edge conditions (2), we obtain

$$
\begin{aligned}
&\sum_{i=1}^{3} \bar{A}_i\bar{\alpha}_{1i} \cos \xi_i \bar{a} + \bar{A}_5\bar{\alpha}_{15} \cosh \bar{\xi}_5 \bar{a} = 0, \\
&\sum_{i=1}^{3} \bar{A}_i\bar{\alpha}_{2i} \cos \xi_i \bar{a} + \bar{A}_5\bar{\alpha}_{25} \cosh \bar{\xi}_5 \bar{a} = 0, \\
&\sum_{i=1}^{3} \bar{A}_i\bar{\alpha}_{3i} \sin \bar{\xi}_i \bar{a} - \bar{A}_5\bar{\alpha}_{35} \sinh \bar{\xi}_5 \bar{a} = 0, \\
&\sum_{i=1}^{3} \bar{A}_i\bar{\alpha}_{5i} \sin \bar{\xi}_i \bar{a} - \bar{A}_5\bar{\alpha}_{55} \sinh \bar{\xi}_5 \bar{a} = 0,
\end{aligned} \tag{7}
$$

where $\bar{a} = a/b$ (i.e. the ratio of the half-length to the half-thickness of the plate) and

$$
\begin{aligned}
\bar{\alpha}_{1i} &= c_{55}\alpha_{3i}\bar{\xi}_i + \kappa_1 c_{56}(\alpha_{2i}\bar{\xi}_i + 1) + e_{25}\alpha_{5i}, \\
\bar{\alpha}_{2i} &= c_{56}\alpha_{3i}\bar{\xi}_i + \kappa_1 c_{66}(\alpha_{2i}\bar{\xi}_i + 1) + e_{26}\alpha_{5i}, \\
\bar{\alpha}_{3i} &= (1 + \gamma_{11}^{-1} e_{11}^{(1)}\alpha_{5i})\bar{\xi}_i, \\
\bar{\alpha}_{5i} &= (e_{11}^{(1)} - \varepsilon_{11}\alpha_{5i})\bar{\xi}_i.
\end{aligned} \tag{8}
$$

The frequency equation is obtained by setting the determinant of the coefficients of the $\bar{A}_i$ in (7) equal to zero. For convenience of comparison with the result in [2], the equation may be expressed in the form

$$
\tilde{A}_1 \tan \xi_1\bar{a} + \tilde{A}_2 \tan \xi_2\bar{a} + \tilde{A}_3 \tan \xi_3\bar{a} = 0, \tag{9}
$$

where

$$
\begin{aligned}
\tilde{A}_1 &= \beta_{31}(\beta_{12}\beta_{23} - \beta_{22}\beta_{13}), \\
\tilde{A}_2 &= \beta_{32}(\beta_{13}\beta_{21} - \beta_{23}\beta_{11}), \\
\tilde{A}_3 &= \beta_{33}(\beta_{11}\beta_{22} - \beta_{21}\beta_{12}),
\end{aligned} \tag{10}
$$

and

$$
\begin{aligned}
\beta_{ij} &= \bar{\alpha}_{ij} + \bar{\alpha}_{i5}\bar{\alpha}_{5j}\bar{\alpha}_{55}^{-1} \coth \xi_5\bar{a} \tan \bar{\xi}_j\bar{a}, \qquad i = 1, 2; j = 1, 2, 3 \\
\beta_{3j} &= \bar{\alpha}_{3j} - \bar{\alpha}_{35}\bar{\alpha}_{5j}, \qquad j = 1, 2, 3.
\end{aligned}
$$

When the electric terms are omitted, (6–10) reduce to (16–20) of [2]. Owing to the non-propagating character of the electric branch, resulting in no additional resonances, and its small effect on the mechanical branches, the frequencies obtained from (9) differ little from those obtained from (19) in [2]; so that the curves marked TS_1, F_1, and FS_1 (for thickness-shear, flexure and face-shear waves travelling parallel to the x_1 direction) in Fig. 3 of [2] would be only slightly changed and, hence, would still fit the Koga–Fukuyo data [5] shown there.

FORCED VIBRATIONS

We now consider a plate with like electrode films deposited on its two faces. An alternating voltage, with drop 2V across the thickness of the plate, is applied to the electrodes. The first order electric potential is therefore fixed throughout the plate:

$$\phi^{(1)} = b^{-1}Ve^{i\omega t}. \tag{11}$$

The equation of motion (27), (28), (30) and (32) of [1] are, then,

$$\begin{aligned} \kappa_1 c_{56} u_{3,11}^{(0)} + \kappa_1^2 c_{66}(u_{2,11}^{(0)} + u_{1,1}^{(1)}) &= \rho(1+R)\ddot{u}_2^{(0)}, \\ c_{55} u_{3,11}^{(0)} + \kappa_1 c_{56}(u_{2,11}^{(0)} + u_{1,1}^{(1)}) &= \rho(1+R)\ddot{u}_3^{(0)}, \\ \gamma_{11}u_{1,11}^{(1)} - 3b^{-2}\kappa_1[c_{56}u_{3,1}^{(0)} + \kappa_1 c_{66}(u_{2,1}^{(0)} + u_1^{(1)}) + e_{26}\phi^{(1)}] &= \rho(1+3R)\ddot{u}_1^{(1)}, \\ e_{11}^{(1)}u_{1,11}^{(1)} - 3b^{-2}[e_{25}u_{3,1}^{(0)} + \kappa_1 e_{26}(u_{2,1}^{(0)} + u_1^{(1)}) - \varepsilon_{22}\phi^{(1)}] + D^{(1)} &= 0, \end{aligned} \tag{12}$$

where R is the ratio of the mass per unit area of both films to the mass per unit area of the quartz plate alone; the factors $(1 + R)$ and $(1 + 3R)$ are obtained from [6]. The surface charge on each face is $\frac{1}{3}b^2D^{(1)}$ and κ_1 has the value given by Bleustein and Tiersten [7]:

$$\kappa_1^2 = \frac{\pi^2}{12}\left[1 + R - \frac{8}{\pi^2(1 + \varepsilon_{22}c_{66}/e_{26}^2)}\right]. \tag{13}$$

The conditions for free edges are again those in (2), but we note, from (3), that the last two conditions are identical owing to the fixed value (11) of $\phi^{(1)}$. The reduction of the number of boundary conditions from four to three is consistent with the reduction of the number of dependent variables from four ($u_2^{(0)}$, $u_3^{(0)}$, $u_1^{(1)}$ and $\phi^{(1)}$) to three: $u_2^{(0)}$, $u_3^{(0)}$, and $u_1^{(1)}$. The latter are to be obtained as solutions of the first three of (12); the fourth equation serves simply as a formula for the surface charge.

We first find a particular solution of (12) for the constant forcing term $\phi^{(1)}$:

$$u_2^{(0)} = u_3^{(0)} = 0, \qquad u_1^{(1)} = A_0, \qquad \phi^{(1)} = b^{-1}V, \qquad D^{(1)} = D_0 \tag{14}$$

(omitting a factor $e^{i\omega t}$). Substitution of (14) in (12) yields, as before [1],

$$A_0 = \frac{3\kappa_1 e_{26}V}{\rho b^3(1 + 3R)(\omega^2 - \bar{\omega}^2)}, \tag{15}$$

$$D_0 = \frac{3\varepsilon_{22}V(1 + e_{26}^2/\varepsilon_{22}c_{66})(1 - \omega^2/\bar{\omega}_a^2)}{b^3(\omega^2/\bar{\omega}^2 - 1)}, \tag{16}$$

where

$$\bar{\omega}^2 = \frac{3\kappa_1^2 c_{66}}{\rho b^2(1 + 3R)}, \tag{17}$$

$$\bar{\omega}_a^2 = \frac{3\kappa_1^2 c_{66}(1 + e_{26}^2/\varepsilon_{22}c_{66})}{\rho b^2(1 + 3R)}. \tag{18}$$

It may be seen, from (16), that $\bar{\omega}$ and $\bar{\omega}_a$ are the resonance and antiresonance frequencies, respectively, of the infinite, plated, piezoelectric plate vibrating in the thickness-shear mode.

For the complementary solution, we again take (4), but with $\phi^{(1)} = 0$. Upon substituting in the first three of (12), we find

$$
\begin{aligned}
(\xi^2 - 3R'\bar{\Omega}^2)A_2 + \hat{c}_{56}\bar{\xi}^2 A_3 + \bar{\xi}A_4 &= 0,\\
\hat{c}_{56}\bar{\xi}^2 A_2 + (\hat{c}_{55}\bar{\xi}^2 - 3R'\bar{\Omega}^2)A_3 + \hat{c}_{56}\bar{\xi}A_4 &= 0,\\
\bar{\xi}A_2 + \hat{c}_{56}\bar{\xi}A_3 + (\hat{\gamma}_{11}\bar{\xi}^2 + 1 - \bar{\Omega}^2)A_4 &= 0,
\end{aligned} \tag{19}
$$

where

$$
\bar{\Omega} = \omega/\bar{\omega}, \qquad R' = (1 + R)/(1 + 3R). \tag{20}
$$

Thus, the dispersion relation for straight-crested waves, obtained by setting the determinant of the coefficients of A_2, A_3, A_4, in (19), equal to zero, is

$$
\begin{vmatrix}
\xi^2 - 3R'\bar{\Omega}^2 & \hat{c}_{56}\bar{\xi}^2 & \bar{\xi}\\
\hat{c}_{56}\bar{\xi}^2 & \hat{c}_{55}\bar{\xi}^2 - 3R'\bar{\Omega}^2 & \hat{c}_{56}\bar{\xi}\\
\bar{\xi} & \hat{c}_{56}\bar{\xi} & \hat{\gamma}_{11}\bar{\xi}^2 + 1 - \bar{\Omega}^2
\end{vmatrix} = 0. \tag{21}
$$

The three roots of (21) give the usual thickness-shear, flexure and face-shear branches found in [2], but slightly modified by the influence of the electrode mass, incorporated in R' and $\bar{\Omega}$ in (21), and by both the electrode mass and the piezoelectric effect as represented in the formulas (13) and (17) for κ_1 and $\bar{\omega}$. However, contrary to the case of free vibrations of the piezoelectric plate, there is no electric branch: it is suppressed by the uniform electric potential applied over the electrodes.

For each root $\bar{\xi}_i$, $i = 1, 2, 3$, of (21), (19) has a pair of amplitude ratios $A_2 : A_3 : A_4$. Let $\bar{A}_i$, $i = 1, 2, 3$ be the value of A_4 for the ith root $\bar{\xi}_i$; and let

$$
A_2/A_4 = \alpha_{2i}, \qquad A_3/A_4 = \alpha_{3i}
$$

for each $\bar{\xi}_i$. Then, the complete solution may be written as

$$
\begin{aligned}
u_2^{(0)} &= \sum_{i=1}^{3} b\bar{A}_i\alpha_{2i}\sin\bar{\xi}_i x_1,\\
u_3^{(0)} &= \sum_{i=1}^{3} b\bar{A}_i\alpha_{3i}\sin\bar{\xi}_i x_1,\\
u_1^{(1)} &= A_0 + \sum_{i=1}^{3} \bar{A}_i\cos\bar{\xi}_i x_1,\\
\phi^{(1)} &= b^{-1}V.
\end{aligned} \tag{22}
$$

The edge conditions (2) now produce only three equations:

$$
\begin{aligned}
\sum_{i=1}^{3} \bar{A}_i\bar{\alpha}_{1i}\cos\bar{\xi}_i\bar{a} &= -\kappa_1 c_{56}A_0 - e_{25}b^{-1}V,\\
\sum_{i=1}^{3} \bar{A}_i\bar{\alpha}_{2i}\cos\bar{\xi}_i\bar{a} &= -\kappa_1 c_{66}A_0 - e_{26}b^{-1}V,\\
\sum_{i=1}^{3} \bar{A}_i\gamma_{11}\bar{\xi}_i\sin\bar{\xi}_i\bar{a} &= 0,
\end{aligned} \tag{23}
$$

where

$$
\begin{aligned}
\bar{\alpha}_{1i} &= c_{55}\alpha_{3i}\bar{\xi}_i + \kappa_1 c_{56}(\alpha_{2i}\bar{\xi}_i + 1),\\
\bar{\alpha}_{2i} &= c_{56}\alpha_{3i}\bar{\xi}_i + \kappa_1 c_{66}(\alpha_{2i}\bar{\xi}_i + 1),
\end{aligned} \tag{24}
$$

and A_0 is given by (15).

For each root ξ_i of (21), for a given frequency ω, the two amplitude ratios α_{2i} and α_{3i} are obtained from (19). Then $\bar{\alpha}_{1i}$ and $\bar{\alpha}_{2i}$ are calculated from (24) and inserted in (23), which may be solved for the three $\bar{A}_i$. Thus, $u_2^{(0)}$, $u_3^{(0)}$ and $u_1^{(1)}$, in (22), are fully determined. Finally, $D^{(1)}$, and hence the surface charge, is obtained from the fourth of (12).

The frequency equation is found by setting the determinant of the coefficients of the $\bar{A}_i$, in (23), equal to zero:

$$\tilde{A}_1 \tan \bar{\xi}_1 \bar{a} + \tilde{A}_2 \tan \bar{\xi}_2 \bar{a} + \tilde{A}_3 \tan \bar{\xi}_3 \bar{a} = 0, \tag{25}$$

where

$$\begin{aligned} \tilde{A}_1 &= \bar{\xi}_1(\bar{\alpha}_{12}\bar{\alpha}_{23} - \bar{\alpha}_{22}\bar{\alpha}_{13}), \\ \tilde{A}_2 &= \bar{\xi}_2(\bar{\alpha}_{13}\bar{\alpha}_{21} - \bar{\alpha}_{23}\bar{\alpha}_{11}), \\ \tilde{A}_3 &= \bar{\xi}_3(\bar{\alpha}_{11}\bar{\alpha}_{22} - \bar{\alpha}_{21}\bar{\alpha}_{12}). \end{aligned} \tag{26}$$

The frequency spectrum calculated from (25) differs little from that obtained from (19) of [2]. The influence of the electric field is contained entirely in the electrical term in the expression (13) for κ_1 which, for the AT-cut of quartz, for example, is only

$$8/\pi^2(1 + \varepsilon_{22} c_{66}/e_{26}^2) = 0{\cdot}006. \tag{27}$$

The influence of the mass of the electrodes makes its appearance in the dispersion relation (21) in the terms R' and $\bar{\Omega}$ as given, in terms of the mass ratio R, by (20), (17) and (13). As R is usually of the order of 10^{-2} or less, the influence of the mass of electrodes on the frequency spectrum is small qualitatively and, quantitatively, appears mostly as an alteration of the vertical scale ratio in the usual depiction of the frequency spectrum with frequency ratio as ordinate and length to thickness-ratio as abscissa, as in Fig. 3 of [2].

For the case of free vibrations of the plate with electrode coatings on its faces, it is only necessary to set $V = 0$.

REFERENCES

1. R. D. Mindlin, High frequency vibrations of piezoelectric crystal plates, *Int. J. Solids Struct.* **8**, 895–906 (1972).
2. R. D. Mindlin and D. C. Gazis, Strong resonances of rectangular AT-cut quartz plates, *Proc. 4th U.S. Natl. Cong. Appl. Mech. Am. Soc. Mech. Eng.* pp. 305–310, June (1962).
3. R. Bechmann, Elastic and piezoelectric constants of alpha quartz, *Phys. Rev.* **110**, 1060–1061 (1958).
4. R. A. Sykes, Modes of motion in quartz crystals, *Quartz Crystals for Electrical Circuits*, edited by R. A. Heising. Chap. VI, pp. 205–248. Van Nostrand (1946).
5. I. Koga and H. Fukuyo, Vibration of thin piezoelectric quartz crystal plates, especially an R_1-cut (AT-cut) rectangular plate, *J. Inst. Elec. Comm. Engrs. Japan*, **36**, 59–67 (1953).
6. R. D. Mindlin, High frequency vibrations of plated crystal plates, *Progress in Applied Mechanics*, pp. 73–84 (the Prager Anniversary Volume). Macmillan (1963).
7. J. L. Bleustein and H. F. Tiersten, Forced thickness-shear vibrations of discontinuously plated piezoelectric crystal plates, *J. Acoust. Soc. Am.* **43**, 1311–1318 (1968).

Резюме — При решении уравнений, определяющих сопряженную волну сдвигового типа, изгиб и виды колебаний поверхностного сдвига во вращаемых вилкообразно вырезанных кварцевых листах берутся во внимание способствующие факторы взаимодействия с напряженностью электрического поля и электродными покрытиями массы.

Int. J. Solids Structures, 1974, Vol. 10, pp. 625–637. Pergamon Press. Printed in Gt. Britain.

EQUATIONS OF HIGH FREQUENCY VIBRATIONS OF THERMOPIEZOELECTRIC CRYSTAL PLATES

R. D. MINDLIN

Department of Civil Engineering, Columbia University, New York 10027, U.S.A.

(*Received* 1 *November* 1973)

Abstract—A system of two-dimensional equations is derived for high frequency motions of crystal plates accounting for coupling of mechanical, electrical and thermal fields.

1. INTRODUCTION

In a recent paper [1], a system of two-dimensional equations, somewhat simpler in form than previous ones, was derived for problems of vibrations of crystal plates in which there is coupling between elastic and electric fields at frequencies as high as those of the fundamental thickness–shear modes. In the present paper, the equations are extended to accommodate coupling with a thermal field. There are already available the plate-equations derived by Tasi and Herrmann [2], which include coupling of elastic and thermal fields, but a different method for treating the thermal field is employed here in order to deal with all three fields in the same way as in [1]. The mechanical displacement, electric potential and temperature change are all expanded in power series of the thickness-coordinate of the plate. The expansions are inserted in an integral energy-equation including a dissipation function instead of in a variational principle with an adjoint field to receive the dissipated energy. The energy-equation is analogous to Biot's [3] variational principle for the equations of thermoelasticity.

Following the substitution of the series expressions in the integral equation of energy balance, the integration with respect to the thickness coordinate is performed and two-dimensional equations of order n, analogous to the three-dimensional equations, are extracted. This is followed by discard of variables and equations of the second and higher orders and the usual adjustments of the remaining members. Finally, a uniqueness theorem, analogous to Weiner's [4] for the thermoelastic case, is devised to establish various face- and edge-conditions sufficient to assure unique solutions of the two-dimensional equations.

2. THREE-DIMENSIONAL, LINEAR THERMOPIEZOELECTRICITY

The equations of the classical, linear theory of thermopiezoelectricity may be grouped as divergence, gradient and constitutive equations for the mechanical, electrical and thermal fields:

Divergence equations

$$T_{ij,i} = \rho \ddot{u}_j, \qquad D_{i,i} = 0, \qquad h_{i,i} = -\Theta_0 \dot{\eta} \tag{1}$$

where T_{ij}, D_i, h_i, u_j, η, ρ and Θ_0 are, respectively, the stress, electric displacement, heat flux, mechanical displacement, entropy density, mass density and reference temperature.

Gradient equations

$$S_{ij} = \tfrac{1}{2}(u_{j,i} + u_{i,j}), \qquad E_i = -\varphi_{,i}, \qquad h_i = -\kappa_{ij}\theta_{,j} \tag{2}$$

where S_{ij}, E_i, φ, κ_{ij} and θ are, respectively, the strain, electric field, heat conduction coefficient and small temperature change: $|\theta| \ll \Theta_0$.

Constitutive equations

$$T_{ij} = \partial G/\partial S_{ij}, \qquad D_i = -\partial G/\partial E_i, \qquad \eta = -\partial G/\partial \theta, \tag{3}$$

where G is the "electric Gibbs function" ([5], p. 34).

Equations (1), (2) and (3) comprise the twenty-seven equations of linear thermopiezoelectricity governing the twenty-seven dependent variables u_i, T_{ij}, S_{ij}, D_i, E_i, φ, h_i, η, θ. The thermodynamic potential G is related to the potential energy density U by

$$G = U - E_i D_i - \eta\Theta$$

in which Θ, $=\Theta_0 + \theta$, is the absolute temperature. Explicitly, in terms of the variables S_{ij}, E_i, θ and Mason's [5] symbols for the material coefficients,

$$G = \tfrac{1}{2}c^{E\theta}_{ijkl} S_{ij} S_{kl} - \varepsilon^{S\theta}_{ij} E_i E_j - \rho C^E_v \Theta_0^{-1} \theta^2 - e^{\theta}_{ijk} E_i S_{jk} - p^S_i \theta E_i - \lambda^E_{ij} S_{ij} \theta. \tag{4}$$

From (3) and (4), omitting, in the sequel, the superscripts E, S, θ on the symbols for the material coefficients, we find:

$$\begin{aligned} T_{ij} &= c_{ijkl} S_{kl} - e_{kij} E_k - \lambda_{ij}\theta, \\ D_j &= e_{jkl} S_{kl} + \varepsilon_{jk} E_k + p_j \theta, \\ \eta &= \lambda_{kl} S_{kl} + p_k E_k + \alpha\theta \end{aligned} \tag{5}$$

where $\alpha = \rho C^E_v \Theta_0^{-1}$.

By successive substitution, the twenty-seven equations may be reduced to five on u_i, φ and θ:

$$\begin{aligned} c_{ijkl} u_{k,li} + e_{kij}\varphi_{,ki} - \lambda_{ij}\theta_{,i} &= \rho \ddot{u}_j, \\ e_{kij} u_{i,jk} - \varepsilon_{ij}\varphi_{,ij} + p_i \theta_{,i} &= 0, \\ \lambda_{ij}\dot{u}_{i,j} - p_i \dot{\varphi}_{,i} + \alpha\dot{\theta} &= \Theta_0^{-1}\kappa_{ij}\theta_{,ij}. \end{aligned}$$

3. ENERGY EQUATION

In a body occupying a volume V, bounded by a surface S with outward normal $\mathbf{n}$, the principle of conservation of energy, appropriate to (1), (2) and (3), can be expressed [6] in a form analogous to Biot's [3] variational principle for the thermoelastic case:

$$\int_V (\dot{K} + \dot{B} + 2F)\,\mathrm{d}V = \int_S (t_j \dot{u}_j - \dot{\sigma}\varphi - \nu\theta\Theta_0^{-1})\,\mathrm{d}S, \tag{6}$$

where

$$t_j = n_i T_{ij}, \qquad \sigma = n_i D_i, \qquad \nu = n_i h_i, \quad \text{on } S, \tag{7}$$

i.e. t_j is the surface traction, σ is the surface charge and ν is the normal component of the heat flux (across the surface). In the volume integral in (6), K is the kinetic energy density:

$$K = \tfrac{1}{2}\rho \dot{u}_i \dot{u}_i,$$

F is the dissipation function:

$$F = \tfrac{1}{2}\kappa_{ij}\theta_{,i}\theta_{,j}\Theta_0^{-1} = -\tfrac{1}{2}h_i\theta_{,i}\Theta_0^{-1},$$

and B is Biot's generalized free energy density [3]:

$$B = B(S_{ij}, D_i, \eta) = U - \Theta_0\eta = G + E_i D_i + \eta\theta, \tag{8}$$

with

$$T_{ij} = \partial B/\partial S_{ij}, \qquad E_i = \partial B/\partial D_i, \qquad \theta = \partial B/\partial\eta.$$

Now,

$$\begin{aligned}\dot{B} &= (\partial B/\partial S_{ij})\dot{S}_{ij} + (\partial B/\partial D_i)\dot{D}_i + (\partial B/\partial\eta)\dot{\eta},\\ &= T_{ij}\dot{u}_{j,i} - \varphi_{,i}\dot{D}_i + \theta\dot{\eta},\\ &= (T_{ij}\dot{u}_j)_{,i} - T_{ij,i}\dot{u}_j - (\varphi\dot{D}_i)_{,i} + \dot{D}_{i,i}\varphi + \dot{\eta}\theta.\end{aligned}$$

Hence,

$$\int_V \dot{B}\,\mathrm{d}V = \int_S n_i(T_{ij}\dot{u}_j - \varphi\dot{D}_i)\,\mathrm{d}S - \int_V (T_{ij}\dot{u}_j - \dot{D}_{i,i}\varphi - \dot{\eta}\theta)\,\mathrm{d}V. \tag{9}$$

Also,

$$2F = -h_i\theta_{,i}\Theta_0^{-1} = -[(h_i\theta)_{,i} - h_{i,i}\theta]\Theta_0^{-1},$$

so that

$$\int_V 2F\,\mathrm{d}V = -\int_S n_i h_i\theta\Theta_0^{-1}\,\mathrm{d}S + \int_V h_{i,i}\theta\Theta_0^{-1}\,\mathrm{d}V. \tag{10}$$

Finally,

$$\dot{K} = \rho\ddot{u}_i\dot{u}_i. \tag{11}$$

Upon substituting (9), (10) and (11) in (6), we find

$$\int_V [(T_{ij,i} - \rho\ddot{u}_j)\dot{u}_j - \dot{D}_{i,i}\varphi - (\dot{\eta} + h_{i,i}\Theta_0^{-1})\theta]\,\mathrm{d}V$$

$$- \int_S [(n_i T_{ij} - t_j)\dot{u}_j - (n_i\dot{D}_i - \dot{\sigma})\varphi - (n_i h_i - v)\theta\Theta_0^{-1}]\,\mathrm{d}S = 0. \tag{12}$$

It will be observed that the coefficients of $\dot{u}_j$, φ and θ, set equal to zero, give the divergence equations (1) and the boundary values (7). This property will be employed in deriving analogous plate-equations after expanding u_j, φ and θ in power series of the thickness coordinate x_2 and integrating with respect to x_2.

4. UNIQUENESS OF SOLUTIONS

Paralleling Weiner's [4] uniqueness theorem for the thermoelastic case, the reverse of the procedure leading from (6) to (12) can be employed to establish boundary conditions sufficient to assure unique solutions of the twenty-seven equations (1–3).

Consider two sets of the twenty-seven dependent variables u_i, T_{ij}, S_{ij}, D_i, E_i, φ, h_i, η, θ (initially zero) distinguished by prime and double prime, each set comprising a solution of the twenty-seven equations. Let

$$u_i^* = u_i' - u_i'', \qquad T_{ij}^* = T_{ij}' - T_{ij}'', \quad \text{etc.}$$

Since the twenty-seven equations are linear, the difference variables, u_i^* etc. are also solutions. We write the equation

$$\int_V [(T_{ij,i}^* - \rho \ddot{u}_j^*)\dot{u}_j^* - \dot{D}_{i,i}^* \varphi^* - (\dot{\eta}^* + h_{i,i}^* \Theta_0^{-1})\theta^*]\,\mathrm{d}V = 0 \tag{13}$$

which is satisfied in view of (1). By the chain rule of differentiation and the divergence theorem (omitting the stars, temporarily):

$$\int_V T_{ij,i}\dot{u}_j\,\mathrm{d}V = \int_V [(T_{ij}\dot{u}_j)_{,i} - T_{ij}\dot{u}_{j,i}]\,\mathrm{d}V = \int_S n_i T_{ij}\dot{u}_j\,\mathrm{d}S - \int_V T_{ij}\dot{u}_{j,i}\,\mathrm{d}V,$$

$$\int_V \dot{D}_{i,i}\varphi\,\mathrm{d}V = \int_V [(\dot{D}_i\varphi)_{,i} - \dot{D}_i\varphi_{,i}]\,\mathrm{d}V = \int_S n_i\dot{D}_i\varphi\,\mathrm{d}S - \int_V \dot{D}_i\varphi_{,i}\,\mathrm{d}V,$$

$$\int_V h_{i,i}\theta\,\mathrm{d}V = \int_V [(h_i\theta)_{,i} - h_i\theta_{,i}]\,\mathrm{d}V = \int_S n_i h_i\theta\,\mathrm{d}S - \int_V h_i\theta_{,i}\,\mathrm{d}V.$$

Employing (2), we have

$$\int_V T_{ij}\dot{u}_{j,i}\,\mathrm{d}V = \int_V T_{ij}\dot{S}_{ij}\,\mathrm{d}V,$$

$$\int_V \dot{D}_i\varphi_{,i}\,\mathrm{d}V = -\int_V \dot{D}_i E_i\,\mathrm{d}V,$$

$$\int_V h_i\theta_{,i}\,\mathrm{d}V = -\int_V \kappa_{ij}\theta_{,i}\theta_{,j}\,\mathrm{d}V = -\Theta_0\int_V 2F\,\mathrm{d}V.$$

Now,

$$T_{ij}\dot{S}_{ij} + \dot{D}_i E_i + \dot{\eta}\theta = T_{ij}\dot{S}_{ij} - D_i\dot{E}_i - \eta\dot{\theta} + \frac{\mathrm{d}}{\mathrm{d}t}(E_i D_i + \eta\theta).$$

This becomes, with (3) (the last of the twenty-seven equations)

$$\begin{aligned} T_{ij}\dot{S}_{ij} + \dot{D}_i E_i + \dot{\eta}\theta &= \frac{\partial G}{\partial S_{ij}}\dot{S}_{ij} + \frac{\partial G}{\partial E_i}\dot{E}_i + \frac{\partial G}{\partial \theta}\dot{\theta} + \frac{\mathrm{d}}{\mathrm{d}t}(E_i D_i + \eta\theta) \\ &= \frac{\mathrm{d}}{\mathrm{d}t}(G + E_i D_i + \eta\theta) = \frac{\mathrm{d}B}{\mathrm{d}t}. \end{aligned}$$

Finally,

$$\rho\ddot{u}_i\dot{u}_i = \mathrm{d}K/\mathrm{d}t.$$

Assembling these results and inserting them in (18), we have, with the stars reinstated,

$$\frac{\mathrm{d}}{\mathrm{d}t}\int_V (K^* + B^*)\,\mathrm{d}V = \int_S n_i(T_{ij}^*\dot{u}_j^* - \dot{D}_i^*\varphi^* - h_i^*\theta^*\Theta_0^{-1})\,\mathrm{d}S - \int_V 2F^*\,\mathrm{d}V. \tag{14}$$

Now, the individual K, B and F are positive definite, by definition, and initially zero; so that K^*, B^* and F^*, calculated from the difference variables, have the same properties.

Hence, if the integrand of the surface integral in (14) vanishes, K^* and B^* must be zero and that is possible only if the two solutions are identical. Conditions sufficient to make the integrand of the surface integral vanish (and, hence, conditions sufficient for a unique solution) are the specification, at each point of S, of: one member of each of the three products of traction and displacement components (referred to orthogonal directions α, β, γ at the point) $T_{\alpha\alpha} u_\alpha$, $T_{\alpha\beta} u_\beta$, $T_{\alpha\gamma} u_\gamma$; *and* either the surface charge, $n_i D_i$, or the surface potential, φ; *and* either the heat flux across the surface, $n_i h_i$, or the surface temperature change, θ.

Various conditions arising from interactions with the exterior of V may also be specified. Of special interest, here, is the radiation condition:

$$n_i h_i - k\theta = 0, \quad \text{on } S, \tag{15}$$

where k is a positive constant ranging from zero for an adiabatic boundary to infinity (i.e. $\theta = 0$) for an isothermal boundary. Suppose suitable boundary conditions have been specified for the mechanical and electrical fields so that (14) reduces to

$$\frac{\mathrm{d}}{\mathrm{d}t}\int_V (K^* + B^*)\,\mathrm{d}V = -\Theta_0^{-1}\int_S n_i h_i^* \theta^*\,\mathrm{d}S - \int_V 2F^*\,\mathrm{d}V. \tag{16}$$

Add $\Theta_0^{-1}\int_S k\theta^*\theta^*\,\mathrm{d}S$ to each side of (16):

$$\frac{\mathrm{d}}{\mathrm{d}t}\int_V (K^* + B^*)\,\mathrm{d}V + \Theta_0^{-1}\int_S K\theta^*\theta^*\,\mathrm{d}S = -\Theta_0^{-1}\int_S (n_i h_i^* - k\theta^*)\theta^*\,\mathrm{d}S - \int_V 2F^*\,\mathrm{d}V.$$

Then, since $k\theta^*\theta^*\Theta_0^{-1}$ is positive, the same argument as before yields (15) as an acceptable boundary condition for the thermal field.

5. SERIES OF TWO-DIMENSIONAL EQUATIONS

The plate is referred to rectangular coordinates x_i with the faces, of area A, at $x_2 = \pm b$ and with x_1 and x_3 the coordinates in the middle plane which intersects the right cylindrical or prismatic boundary of the plate in a curve or polygon C. We assume that u_i, φ and θ can be approximated by a power series in x_2:

$$u_i = \sum_n x_2^n u_i^{(n)}, \qquad \varphi = \sum_n x_2^n \varphi^{(n)}, \qquad \theta = \sum_n x_2^n \theta^{(n)}, \tag{17}$$

where $u_i^{(n)}$, $\varphi^{(n)}$ and $\theta^{(n)}$, $n = 0, 1, 2 \ldots$ are functions of x_1, x_3 and t only. These expansions are to be substituted in the energy–balance equation (12) and the integrations with respect to x_2 performed. By the same procedure as in [1], the volume integral in (12) becomes

$$\begin{aligned}\sum_n \int_A \Big(T_{ij,i}^{(n)} - nT_{2j}^{(n-1)} + [x_2^n T_{2j}]_{-b}^{b} - \rho \sum_m B_{mn}\ddot{u}_j^{(m)}\Big)\dot{u}_j^{(n)}\,\mathrm{d}A \\ - \sum_n \int_A (\dot{D}_{i,i}^{(n)} - n\dot{D}_2^{(n-1)} + [x_2^n \dot{D}_2]_{-b}^{b})\varphi^{(n)}\,\mathrm{d}A \\ - \sum_n \int_A (H_{i,i}^{(n)} - nH_2^{(n-1)} + [x_2^n h_2]_{-b}^{b} + \Theta_0 \dot{\eta}^{(n)})\Theta_0^{-1}\theta^{(n)}\,\mathrm{d}A,\end{aligned} \tag{18}$$

where

$$(T_{ij}^{(n)}, D_i^{(n)}, \eta^{(n)}, H_i^{(n)}) = \int_{-b}^{b} x_2^n (T_{ij}, D_i, \eta, h_i)\,\mathrm{d}x_2, \tag{19}$$

$$B_{mn} = 2b^{m+n+1}/m + n + 1, \quad m + n \text{ even}$$

and $B_{mn} = 0$ for $m + n$ odd. Also, as in [1], the surface integral in (12) becomes the sum of area integrals over the two faces of the plate and a line integral around the boundary C:

$$-\sum_n \int_A [(T_{2j} - t_j)x_2^n \dot{u}_j^{(n)} - (\dot{D}_2 - \dot{\sigma})x_2^n \varphi^{(n)} - (h_2 - v)x_2^n \theta^{(n)}\Theta_0^{-1}]_{-b}^{b}\, dA$$

$$-\sum_n \oint_C \int_{-b}^{b} [(n_a T_{aj} - t_j)x_2^n \dot{u}_j^{(n)} - (n_a \dot{D}_a - \dot{\sigma})x_2^n \varphi^{(n)} - (n_a h_a - v)x_2^n \theta^{(n)}\Theta_0^{-1}]_C\, dx_2\, ds, \quad (20)$$

where the index a is planar. Define

$$(F_j^{(n)}, S^{(n)}, N^{(n)}) = [x_2^n(t_j, \sigma, v)]_{-b}^{b}$$

$$(t_j^{(n)}, \sigma^{(n)}, v^{(n)}) = \int_{-b}^{b} x_2^n(t_j, \sigma, v)\, dx_2$$

and insert these in (20). Then set the sum of (18) and (20) equal to zero to obtain the two-dimensional version of (12):

$$\sum_n \int_A (T_{ij,i}^{(n)} - nT_{2j}^{(n-1)} + F_j^{(n)} - \rho \sum_m B_{mn} \ddot{u}_j^{(m)})\dot{u}_j^{(n)}\, dA$$

$$-\sum_n \int_A (\dot{D}_{i,i}^{(n)} - n\dot{D}_2^{(n-1)} + S^{(n)})\varphi^{(n)}\, dA$$

$$-\sum_n \int_A (H_{i,i}^{(n)} - nH_2^{(n-1)} + N^{(n)} + \Theta_0 \dot{\eta}^{(n)})\theta^{(n)}\Theta_0^{-1}\, dA$$

$$-\sum_n \oint_C [(n_a T_{aj} - t_j^{(n)})\dot{u}_j^{(n)} - (n_a \dot{D}_a^{(n)} - \dot{\sigma}^{(n)})\varphi^{(n)} - (n_a H_a^{(n)} - v^{(n)})\theta^{(n)}\Theta_0^{-1}]\, ds = 0. \quad (21)$$

Following the argument at the end of Article 3, we extract, from (21), the divergence equations of order n:

$$T_{ij,i}^{(n)} - nT_{2j}^{(n-1)} + F_j^{(n)} = \rho \sum_m B_{mn} \ddot{u}_j^{(m)},$$

$$D_{i,i}^{(n)} - nD_2^{(n-1)} + S^{(n)} = 0,$$

$$H_{i,i}^{(n)} - nH_2^{(n-1)} + N^{(n)} = -\Theta_0 \dot{\eta}^{(n)}$$

and the edge-values of order n:

$$n_a T_{aj}^{(n)} = t_j^{(n)}, \qquad n_a D_a^{(n)} = \sigma^{(n)}, \qquad n_a H_a^{(n)} = v^{(n)} \quad \text{on } C.$$

Proceeding, now, to the gradient equations, we substitute (17) in (2) to obtain

$$S_{ij} = \sum_n x_2^n S_{ij}^{(n)}, \qquad E_i = \sum_n x_2^n E_i^{(n)}, \qquad h_i = \sum_n x_2^n h_i^{(n)}, \quad (22)$$

where $S_{ij}^{(n)}$, $E_i^{(n)}$ and $h_i^{(n)}$ are given by the gradient equations of order n:

$$\begin{aligned} S_{ij}^{(n)} &= \tfrac{1}{2}[u_{j,i}^{(n)} + u_{i,j}^{(n)} + (n+1)(\delta_{i2} u_j^{(n+1)} + \delta_{2j} u_i^{(n+1)})], \\ E_i^{(n)} &= -\varphi_{,i}^{(n)} - (n+1)\delta_{2i}\varphi^{(n+1)}, \\ h_i^{(n)} &= -\kappa_{ij}[\theta_{,j}^{(n)} + (n+1)\delta_{2j}\theta^{(n+1)}]. \end{aligned} \quad (23)$$

For the constitutive equations of order n we find, from (19) and (5):

$$T_{ij}^{(n)} = \sum_m B_{mn}(c_{ijkl} S_{kl}^{(m)} - e_{kij} E_k^{(m)} - \lambda_{ij} \theta^{(m)}),$$

$$D_j^{(n)} = \sum_m B_{mn}(e_{jkl} S_{kl}^{(m)} + \varepsilon_{kj} E_k^{(m)} + p_j \theta^{(m)}), \tag{24}$$

$$\eta^{(n)} = \sum_m B_{mn}(\lambda_{kl} S_{kl}^{(m)} + p_k E_k^{(m)} + \alpha\theta^{(m)}),$$

Finally, from (19) and the third of (22),

$$H_i^{(n)} = -\sum_m B_{mn} \kappa_{ij}[\theta_{,j}^{(m)} + (m+1)\delta_{2j}\theta^{(m+1)}], \tag{25}$$

which, when restricted to orders zero and one, is an alternative to the third of (23): the thermal gradient equation of order n.

This completes the establishment of the two-dimensional, thermopiezoelectric equations of order n.

6. TRUNCATION OF SERIES AND ADJUSTMENT

For frequencies up to those of the fundamental thickness–shear modes, the process of truncation of the series of two-dimensional equations and adjustment of the remaining terms begins with the discard of all the second and higher order terms. Thus, the first of (24) becomes

$$\begin{aligned} T_{ij}^{(0)} &= 2b(c_{ijkl} S_{kl}^{(0)} - e_{kij} E_k^{(0)} - \lambda_{ij} \theta^{(0)}), \\ T_{ij}^{(1)} &= \tfrac{2}{3} b^3 (c_{ijkl} S_{kl}^{(1)} - e_{kij} E_k^{(1)} - \lambda_{ij} \theta^{(1)}) \end{aligned} \tag{26}$$

and (25) becomes

$$H_i^{(0)} = -2b\kappa_{ij}(\theta_{,j}^{(0)} + \delta_{2j}\theta^{(1)}), \qquad H_i = -\tfrac{2}{3} b^3 \kappa_{ij} \theta_{,j}^{(1)}, \tag{27}$$

but these are only provisional as adjustments of the type employed by Cauchy [7], Poisson [8], Bresse [9] and Timoshenko [10] are to be made in order to compensate, in part, for the omission of the higher order terms.

First of all, free development of the thickness-stretch strain $S_{22}^{(0)}$, $=u_2^{(1)}$, is allowed by setting $T_{22}^{(0)} = 0$ in (26). The resulting equation is solved for $S_{22}^{(0)}$ and that expression is used to eliminate $S_{22}^{(0)}$ from the remaining $T_{ij}^{(0)}$, yielding

$$T_{ij}^{(0)} = 2b(\bar{c}_{ijkl} S_{kl}^{(0)} - \bar{e}_{kij} E_k^{(0)} - \bar{\lambda}_{ij} \theta^{(0)}),$$

where

$$\bar{c}_{ijkl} = c_{ijkl} - c_{ij22} c_{22kl}/c_{2222}, \qquad \bar{e}_{kij} = e_{kij} - c_{ij22} e_{k22}/c_{2222},$$

$$\bar{\lambda}_{ij} = \lambda_{ij} - c_{ij22} \lambda_{22}/c_{2222}.$$

Then the Bresse–Timoshenko shear-correction factors, κ_1 and κ_3, are introduced by the replacement of $\bar{c}$, $\bar{e}$ and $\bar{\lambda}$ by

$$c_{ijkl}^{(0)} = \kappa_{i+j-2}^{\mu} \kappa_{k+l-2}^{\nu} \bar{c}_{ijkl}, \qquad e_{kij}^{(0)} = \kappa_{i+j-2}^{\mu} \bar{e}_{kij}, \qquad \lambda_{ij}^{(0)} = \kappa_{i+j-2}^{\mu} \bar{\lambda}_{ij} \quad \text{(not summed)},$$

where μ and ν are the powers $\cos^2(ij\pi/2)$ and $\cos^2(kl\pi/2)$, respectively. Thus the final expression for $T_{ij}^{(0)}$ is

$$T_{ij}^{(0)} = 2b(c_{ijkl}^{(0)} S_{kl}^{(0)} - e_{kij}^{(0)} E_k^{(0)} - \lambda_{ij}^{(0)} \theta^{(0)}). \tag{28}$$

In the case of the first order strains, free development of all three $S_{2j}^{(1)}$ is required and effected by setting $T_{2j}^{(1)} = 0$ in (26) and using the resulting three equations to eliminate the $S_{2j}^{(1)}$ from the remaining $T_{ij}^{(1)}$ with the result:

$$T_{ab}^{(1)} = \tfrac{2}{3}b^3(c_{abcd}^{(1)}S_{cd}^{(1)} - e_{cab}^{(1)}E_c^{(1)} - \lambda_{ab}^{(1)}\theta^{(1)}), \tag{29}$$

in which the indices $a, b, c \dots$ range over 1 and 3 only; the $c_{abcd}^{(1)}$ are the Voigt constants γ [11] and

$$e_{cab}^{(1)} = e_{cij}s_{ijde}c_{abde}^{(1)}, \qquad \lambda_{ab}^{(1)} = \lambda_{ij}s_{ijcd}c_{abcd}^{(1)},$$

where the s_{ijkl} are the elastic compliances.

It will be observed that $u_2^{(1)}$ no longer appears in (28) and (29) and that displacement component is eliminated entirely by dropping it from the kinetic energy density.

Further, following Tasi and Herrmann [2], thermal correction factors κ_1^T and κ_3^T may be introduced by replacing the heat conduction coefficients κ_{ab}, in the expression for $H_a^{(1)}$, from (27), by

$$\kappa_{ab} \to \kappa_{ab}\kappa_a^T\kappa_b^T \quad \text{(not summed)}.$$

An "electric Gibbs function" $\bar{G}$ that produces (28 and 29) through

$$T_{ij}^{(0)} = \partial\bar{G}/\partial S_{ij}^{(0)}, \qquad T_{ab}^{(1)} = \partial\bar{G}/\partial S_{ab}^{(1)} \tag{30}$$

is

$$\begin{aligned}\bar{G} = b(c_{ijkl}^{(0)}S_{ij}^{(0)}S_{kl}^{(0)} - \varepsilon_{ij}E_i^{(0)}E_j^{(0)} - \alpha\theta^{(0)}\theta^{(0)} - 2e_{ijk}^{(0)}E_i^{(0)}S_{jk}^{(0)} - 2p_i\theta^{(0)}E_i^{(0)} - 2\lambda_{ij}^{(0)}S_{ij}^{(0)}\theta^{(0)})\\ + \tfrac{1}{3}b^3(c_{abcd}^{(1)}S_{ab}^{(1)}S_{cd}^{(1)} - \varepsilon_{ab}E_a^{(1)}E_b^{(1)} - \alpha\theta^{(1)}\theta^{(1)} - 2e_{abc}^{(1)}E_a^{(1)}S_{bc}^{(1)}\\ - 2p_a\theta^{(1)}E_a^{(1)} - 2\lambda_{ab}^{(1)}S_{ab}^{(1)}\theta^{(1)}).\end{aligned} \tag{31}$$

Finally, we define $D_i^{(0)}$, $D_a^{(1)}$, $\eta^{(0)}$ and $\eta^{(1)}$ by

$$D_i^{(0)} = -\partial\bar{G}/\partial E_i^{(0)}, \quad D_a^{(1)} = -\partial\bar{G}/\partial E_a^{(1)}, \quad \eta^{(0)} = -\partial\bar{G}/\partial\theta^{(0)}, \quad \eta^{(1)} = -\partial\bar{G}/\partial\theta^{(1)} \tag{32}$$

to serve as adjusted expressions to replace those obtained from (24).

7. RECAPITULATION

Divergence equations

$$\begin{aligned} &T_{aj,a}^{(0)} + F_j^{(0)} = 2b\rho\ddot{u}_j^{(0)}, && T_{ab,a}^{(1)} - T_{2b}^{(0)} + F_b^{(1)} = \tfrac{2}{3}b^3\rho\ddot{u}_b^{(1)},\\ &D_{a,a}^{(0)} + S^{(0)} = 0, && D_{a,a}^{(1)} - D_2^{(0)} + S^{(1)} = 0,\\ &H_{a,a}^{(0)} + N^{(0)} = -\Theta_0\dot{\eta}^{(0)}, && H_{a,a}^{(1)} - H_2^{(0)} + N^{(1)} = -\Theta_0\dot{\eta}^{(1)};\end{aligned} \tag{33}$$

Gradient equations

$$\begin{aligned} &S_{ij}^{(0)} = \tfrac{1}{2}[u_{j,i}^{(0)} + u_{i,j}^{(0)} + \delta_{i2}u_j^{(1)} + \delta_{2j}u_i^{(1)}], && S_{ab}^{(1)} = \tfrac{1}{2}(u_{b,a}^{(1)} + u_{a,b}^{(1)}),\\ &E_i^{(0)} = -(\varphi_{,i}^{(0)} + \delta_{i2}\varphi^{(1)}), && E_a^{(1)} = -\varphi_{,a}^{(1)},\\ &H_i^{(0)} = -2b\kappa_{ij}(\theta_{,j}^{(0)} + \delta_{2j}\theta^{(1)}), && H_a^{(1)} = -\tfrac{2}{3}b^3\kappa_{ab}^{(1)}\theta_{,b}^{(1)};\end{aligned} \tag{34}$$

Constitutive equations

$$
\begin{aligned}
T_{ij}^{(0)} &= 2b(c_{ijkl}^{(0)} S_{kl}^{(0)} - e_{kij}^{(0)} E_k^{(0)} - \lambda_{ij}^{(0)} \theta^{(0)}),\\
T_{ab}^{(1)} &= \tfrac{2}{3} b^3 (c_{abcd}^{(1)} S_{cd}^{(1)} - e_{cab}^{(1)} E_c^{(1)} - \lambda_{ab}^{(1)} \theta^{(1)}),\\
D_j^{(0)} &= 2b(e_{jkl}^{(0)} S_{kl}^{(0)} + \varepsilon_{jk} E_k^{(0)} + p_j \theta^{(0)}),\\
D_b^{(1)} &= \tfrac{2}{3} b^3 (e_{bcd}^{(1)} S_{cd}^{(1)} + \varepsilon_{bc} E_c^{(1)} + p_b \theta^{(1)}),\\
\eta^{(0)} &= 2b(\lambda_{kl}^{(0)} S_{kl} + p_k E_k^{(0)} + \alpha \theta^{(0)}),\\
\eta^{(1)} &= \tfrac{2}{3} b^3 (\lambda_{cd}^{(1)} S_{cd}^{(1)} + p_c E_c^{(1)} + \alpha \theta^{(1)}),
\end{aligned}
\tag{35a}
$$

or

$$
T_{ij}^{(0)} = \frac{\partial \bar{G}}{\partial S_{ij}^{(0)}}, \qquad T_{ab}^{(1)} = \frac{\partial \bar{G}}{\partial S_{ab}^{(1)}}, \qquad D_i^{(0)} = -\frac{\partial \bar{G}}{\partial E_i^{(0)}}, \qquad D_a^{(1)} = -\frac{\partial \bar{G}}{\partial E_a^{(1)}},
$$

$$
\eta^{(0)} = -\frac{\partial \bar{G}}{\partial \theta^{(0)}}, \qquad \eta^{(1)} = -\frac{\partial \bar{G}}{\partial \theta^{(1)}}, \tag{35b}
$$

where $\bar{G}$ is given by (31);

Equations† *on* $u_i^{(0)}$, $u_a^{(1)}$, $\varphi^{(0)}$, $\varphi^{(1)}$, $\theta^{(0)}$, $\theta^{(1)}$

$$
\begin{aligned}
&c_{ijkl}^{(0)}(u_{k,li}^{(0)} + \delta_{2l} u_{k,i}^{(1)}) + e_{kij}^{(0)}(\varphi_{,ki}^{(0)} + \delta_{2k} \varphi_{,i}^{(1)}) - \lambda_{ij}^{(0)} \theta_{,i}^{(0)} + \tfrac{1}{2} b^{-1} F_j^{(0)} = \rho \ddot{u}_j^{(0)},\\
&\quad e_{kij}^{(0)}(u_{j,ki}^{(0)} + \delta_{2l} u_{j,k}^{(1)}) - \varepsilon_{ij}(\varphi_{,ij}^{(0)} + \delta_{i2} \varphi_{,i}^{(1)}) + p_i \theta_{,i}^{(0)} + \tfrac{1}{2} b^{-1} S^{(0)} = 0,\\
&\lambda_{ij}^{(0)}(\dot{u}_{j,i}^{(0)} + \delta_{2i} \dot{u}_j^{(1)}) - p_i(\dot{\phi}_{,i}^{(0)} + \delta_{i2} \dot{\phi}^{(1)}) + \alpha \dot{\theta}^{(0)} + \tfrac{1}{2} b^{-1} \Theta_0^{-1} N^{(0)} = \Theta_0^{-1} \kappa_{ij}(\theta_{,ij}^{(0)} + \delta_{2j} \theta_{,i}^{(1)})
\end{aligned}
\tag{36}
$$

$$
\begin{aligned}
&c_{abcd}^{(1)} u_{d,ca}^{(1)} + e_{cab}^{(1)} \varphi_{,ca}^{(1)} - \lambda_{ab}^{(1)} \theta_{,a}^{(1)} - 3b^{-2}[c_{2bkl}^{(0)}(u_{l,k}^{(0)} + \delta_{2k} u_l^{(1)})\\
&\qquad + e_{i2b}^{(0)}(\varphi_{,i}^{(0)} + \delta_{i2} \varphi^{(1)}) - \lambda_{2b}^{(0)} \theta^{(0)}] + \tfrac{3}{2} b^{-3} F_b^{(1)} = \rho \ddot{u}_b^{(1)},\\
&e_{cab}^{(1)} u_{b,ca}^{(1)} - \varepsilon_{bc} \varphi_{,bc}^{(1)} + p_a \theta_{,a}^{(1)} - 3b^{-2}[e_{2ij}^{(0)}(u_{j,i}^{(0)} + \delta_{2i} u_j^{(1)})\\
&\qquad - \varepsilon_{i2}(\varphi_{,i}^{(0)} + \delta_{i2} \varphi^{(1)}) + p_2 \theta^{(0)}] + \tfrac{3}{2} b^{-3} S^{(1)} = 0,\\
&\lambda_{ab}^{(1)} \dot{u}_{b,a}^{(1)} - p_a \dot{\phi}_{,a}^{(1)} + \alpha \dot{\theta}_a^{(1)} - 3b^{-2} \kappa_{2j}(\theta_{,j}^{(0)} + \delta_{2j} \theta^{(1)}) + \tfrac{3}{2} b^{-3} \Theta_0^{-1} N^{(1)} = \Theta_0^{-1} \kappa_{ab}^{(1)} \theta_{,ab}^{(1)}.
\end{aligned}
\tag{37}
$$

In addition, it is convenient to define a kinetic energy density and a dissipation function:

$$
K = \rho b \dot{u}_j^{(0)} \dot{u}_j^{(0)} + \tfrac{1}{3} \rho b^3 \dot{u}_b^{(1)} \dot{u}_b^{(1)}, \tag{38}
$$

$$
F = b \Theta_0^{-1} \kappa_{ij}(\theta_{,i}^{(0)} + \delta_{i2} \theta^{(1)})(\theta_{,j}^{(0)} + \delta_{2j} \theta^{(1)}) + \tfrac{1}{3} b^3 \Theta_0^{-1} \kappa_{ab}^{(1)} \theta_{,a}^{(1)} \theta_{,b}^{(1)}. \tag{39}
$$

8. UNIQUENESS AND FACE- AND EDGE-CONDITIONS

The plate equations comprise the forty-two equations (33–35) (a or b) on the forty-two dependent variables $u_i^{(0)}, u_a^{(1)}, T_{ij}^{(0)}, T_{ab}^{(1)}, S_{ij}^{(0)}, S_{ab}^{(1)}, D_i^{(0)}, D_a^{(1)}, E_i^{(0)}, E_a^{(1)}, \varphi^{(0)}, \varphi^{(1)}, H_i^{(0)}, H_a^{(1)}$, $\eta^{(0)}, \eta^{(1)}, \theta^{(0)}, \theta^{(1)}$. As before, we consider two sets of solutions, initially zero, identified by prime and double prime, and their differences

$$
u_i^{(0)*} = u_i^{(0)\prime} - u_i^{(0)\prime\prime}, \quad \text{etc.}
$$

† Note that the first two of (36) contain terms in $\varphi^{(1)}$ which were inadvertently omitted in the corresponding equations in [1].

which are also solutions owing to the linearity of the equations. We form the equation

$$\int_A [(T^{(0)*}_{aj,a} + F^{(0)*}_j - 2b\rho\ddot{u}^{(0)*}_j)\dot{u}^{(0)*}_j + (T^{(1)*}_{ab,a} - T^{(0)*}_{2b} + F^{(1)*}_b - \tfrac{2}{3}b^3\rho\ddot{u}^{(1)*}_b)\dot{u}^{(1)*}_b]\,\mathrm{d}A$$

$$-\int_A [(\dot{D}^{(0)*}_{a,a} + \dot{S}^{(0)*})\varphi^{(0)*} + (\dot{D}^{(1)*}_{a,a} - \dot{D}^{(0)*}_2 + \dot{S}^{(1)*})\varphi^{(1)*}]\,\mathrm{d}A$$

$$-\int_A \Theta_0^{-1}[(H^{(0)*}_{a,a} + N^{(0)*} + \Theta_0\dot{\eta}^{(0)*})\theta^{(0)*}$$

$$+ (H^{(1)*}_{a,a} - H^{(0)*}_2 + N^{(1)*} + \Theta_0\dot{\eta}^{(1)*})\theta^{(1)*}]\,\mathrm{d}A = 0, \quad (40)$$

which is satisfied in view of (33). Omitting stars, temporarily, we have

$$\int_A [T^{(0)}_{aj,a}\dot{u}^{(0)}_j + (T^{(1)}_{ab,a} - T^{(0)}_{2b})\dot{u}^{(1)}_b]\,\mathrm{d}A = \oint_C n_a(T^{(0)}_{aj}\dot{u}^{(0)}_j + T^{(1)}_{ab}\dot{u}^{(1)}_b)\,\mathrm{d}s$$

$$-\int_A [T^{(0)}_{ij}(\dot{u}^{(0)}_{j,i} + \delta_{i2}\dot{u}^{(1)}_j) + T^{(1)}_{ab}\dot{u}^{(1)}_{b,a}]\,\mathrm{d}A,$$

$$\int_A [\dot{D}^{(0)}_{a,a}\varphi^{(0)} + (\dot{D}^{(1)}_{a,a} - \dot{D}^{(0)}_2)\varphi^{(1)}]\,\mathrm{d}A = \oint_C n_a(\dot{D}^{(0)}_a\varphi^{(0)} + \dot{D}^{(1)}_a\varphi^{(1)})\,\mathrm{d}s$$

$$-\int_A [\dot{D}^{(0)}_i(\varphi^{(0)}_{,i} + \delta_{i2}\varphi^{(1)}) + \dot{D}^{(1)}_a\varphi^{(1)}_{,a}]\,\mathrm{d}A,$$

$$\int_A [H^{(0)}_{a,a}\theta^{(0)} + (H^{(1)}_{a,a} - H^{(0)}_2)\theta^{(1)}]\,\mathrm{d}A = \oint_C n_a(H^{(0)}_a\theta^{(0)} + H^{(1)}_a\theta^{(1)})\,\mathrm{d}s$$

$$-\int_A [H^{(0)}_i(\theta^{(0)}_{,i} + \delta_{i2}\theta^{(1)}) + H^{(1)}_a\theta^{(1)}_{,a}]\,\mathrm{d}A.$$

Employing (34) we find

$$\int_A [T^{(0)}_{ij}(\dot{u}^{(0)}_{j,i} + \delta_{i2}\dot{u}^{(1)}_j) + T^{(1)}_{ab}u^{(1)}_{b,a}]\,\mathrm{d}A = \int_A (T^{(0)}_{ij}\dot{S}^{(0)}_{ij} + T^{(1)}_{ab}\dot{S}^{(1)}_{ab})\,\mathrm{d}A,$$

$$\int_A [\dot{D}^{(0)}_i(\varphi^{(0)}_{,i} + \delta_{i2}\varphi^{(1)}) + \dot{D}^{(1)}_a\varphi^{(1)}_{,a}]\,\mathrm{d}A = -\int_A (\dot{D}^{(0)}_i E^{(0)}_i + \dot{D}^{(1)}_a E^{(1)}_a)\,\mathrm{d}A,$$

$$\int_A [H^{(0)}_i(\theta^{(0)}_{,i} + \delta_{i2}\theta^{(1)}) + H^{(1)}_a\theta^{(1)}_{,a}]\,\mathrm{d}A = -\int_A [2b\kappa_{ij}(\theta^{(0)}_{,i} + \delta_{i2}\theta^{(1)})(\theta^{(0)}_{,j} + \delta_{2j}\theta^{(1)})$$

$$+ \tfrac{2}{3}b^3\kappa^{(1)}_{ab}\theta^{(1)}_{,a}\theta^{(1)}_{,b}]\,\mathrm{d}A$$

$$= -\Theta_0\int_A 2\bar{F}\,\mathrm{d}A, \quad \text{from (39).}$$

Now,

$$T^{(0)}_{ij}\dot{S}^{(0)}_{ij} + T^{(1)}_{ab}\dot{S}^{(1)}_{ab} + \dot{D}^{(0)}_i E^{(0)}_i + \dot{D}^{(1)}_a E^{(1)}_a + \dot{\eta}^{(0)}\theta^{(0)} + \dot{\eta}^{(1)}\theta^{(1)}$$

$$= T^{(0)}_{ij}\dot{S}^{(0)}_{ij} + T^{(1)}_{ab}\dot{S}^{(1)}_{ab} - D^{(0)}_i\dot{E}^{(0)}_i - D^{(1)}_a\dot{E}^{(1)}_a - \eta^{(0)}\dot{\theta}^{(0)} - \eta^{(1)}\dot{\theta}^{(1)}$$

$$+ \frac{\mathrm{d}}{\mathrm{d}t}(E^{(0)}_i D^{(0)}_i + E^{(1)}_a D^{(1)}_a + \eta^{(0)}\theta^{(0)} + \eta^{(1)}\theta^{(1)})$$

$$= \frac{d\bar{G}}{dt} + \frac{d}{dt}(E_i^{(0)}D_i^{(0)} + E_a^{(1)}D_a^{(1)} + \eta^{(0)}\theta^{(0)} + \eta^{(1)}\theta^{(1)}) \quad \text{from (35b)},$$

$$= d\bar{B}/dt,$$

where

$$\bar{B} = \bar{G} + E_i^{(0)}D_i^{(0)} + E_a^{(1)}D_a^{(1)} + \eta^{(0)}\theta^{(0)} + \eta^{(1)}\theta^{(1)}$$

i.e. by comparison with (8), $\bar{B}$ is the plate analogue of Biot's generalized free energy. Finally, from (38)

$$2b\rho\ddot{u}_j^{(0)}\dot{u}_j^{(0)} + \tfrac{2}{3}b^3\rho\ddot{u}_b^{(1)}\dot{u}_b^{(1)} = d\bar{K}/dt.$$

Assembling all these results and inserting them in (40), the latter becomes

$$\frac{d}{dt}\int_A (\bar{K}^* + \bar{B}^*)\, dA = \int_A A^*\, dA + \oint_C C^*\, ds - \int_A 2\bar{F}^*\, dA, \tag{41}$$

where

$$A^* = F_j^{(0)*}\dot{u}_j^{(0)*} + F_a^{(1)*}\dot{u}_a^{(1)*} - \dot{S}^{(0)*}\varphi^{(0)*} - \dot{S}^{(1)*}\varphi^{(1)*} - \Theta_0^{-1}N^{(0)*}\theta^{(0)*} - \Theta_0^{-1}N^{(1)*}\theta^{(1)*},$$

$$C^* = n_a(T_{aj}^{(0)*}\dot{u}_j^{(0)*} + T_{ab}^{(1)*}\dot{u}_b^{(1)*} - \dot{D}_a^{(0)*}\varphi^{(0)*} - \dot{D}_a^{(1)*}\varphi^{(1)*} - \Theta_0^{-1}H_a^{(0)*}\theta^{(0)*} - \Theta_0^{-1}H_a^{(1)*}\theta^{(1)*}).$$

As in the three-dimensional case, the individual $\bar{K}$, $\bar{B}$ and $\bar{F}$ are positive definite by definition and initially zero, so that $\bar{K}^*$, $\bar{B}^*$ and $\bar{F}^*$, calculated from the difference-variables, have the same properties. Hence, if A^* and C^* are zero, $\bar{K}^*$ and $\bar{B}^*$ must be zero and that is possible only if the two solutions are identical. Conditions sufficient to make A^* zero are: at each point of the interior of the plate, specification of one member of each of the five products of displacement and face–traction components (referred to orthogonal directions α, β, x_2):

$$F_\alpha^{(0)}u_\alpha^{(0)},\ F_\beta^{(0)}u_\beta^{(0)},\ F_2^{(0)}u_2^{(0)},\ F_\alpha^{(1)}u_\alpha^{(1)},\ F_\beta^{(1)}u_\beta^{(1)};$$

and one member of each of the two products of electric potential and face-charge:

$$\varphi^{(0)}S^{(0)}, \qquad \varphi^{(1)}S^{(1)};$$

and one member of each of the two products of temperature change and thermal flux:

$$\theta^{(0)}N^{(0)}, \qquad \theta^{(1)}N^{(1)}.$$

Conditions sufficient to make C^* zero are: at each point of the edge of the plate, specification of one member of each of the five products of edge–displacement and edge–traction components (referred to orthogonal directions, n, s, x_2, with n the outward normal to C):

$$T_{nn}^{(0)}u_n^{(0)},\ T_{ns}^{(0)}u_s^{(0)},\ T_{n2}^{(0)}u_2^{(0)},\ T_{nn}^{(1)}u_n^{(1)},\ T_{ns}^{(1)}u_s^{(1)};$$

and one member of each of the two products of edge–potential and edge–charge:

$$\varphi^{(0)}D_n^{(0)}, \qquad \varphi^{(1)}D_n^{(1)};$$

and one member of each of the two products of edge–temperature change and thermal edge–flux:

$$\theta^{(0)}H_n^{(0)}, \qquad \theta^{(1)}H_n^{(1)}.$$

To find thermal radiation conditions for the faces and edges of the plate, add

$$k\Theta_0^{-1}\int_A (\theta^{(0)*}+x_2\theta^{(1)*})^2_{x_2=b}\,dA + k\Theta_0^{-1}\int_A (\theta^{(0)*}+x_2\theta^{(1)*})^2_{x_2=-b}\,dA$$

$$+k\Theta_0^{-1}\oint_C\int_{-b}^{b}(\theta^{(0)*}+x_2\theta^{(1)*})^2\,dx_2\,ds$$

$$=2k\Theta_0^{-1}\int_A(\theta^{(0)*}\theta^{(0)*}+b^2\theta^{(1)*}\theta^{(1)*})\,dA+2bk\Theta_0^{-1}\oint_C(\theta^{(0)*}\theta^{(0)*}+\tfrac{1}{3}b^2\theta^{(1)*}\theta^{(1)*})\,ds$$

to each side of (41) and note that the additional integrals are positive. The thermal terms on the right-hand side of (41) are then

$$-\Theta_0^{-1}\int_A[(N^{(0)*}-2k\theta^{(0)*})\theta^{(0)*}+(N^{(1)*}-2kb^2\theta^{(1)*})\theta^{(1)*}]\,dA$$

$$-\Theta_0^{-1}\oint_C[(n_aH_a^{(0)*}-2bk\theta^{(0)*})\theta^{(0)*}+(n_aH_a^{(1)*}-\tfrac{2}{3}b^3\theta^{(1)*})\theta^{(1)*}]\,ds.$$

Hence, by the usual argument, the thermal radiation conditions are

$$N^{(0)}-2k\theta^{(0)}=0,\qquad N^{(1)}-2kb^2\theta^{(1)}=0$$

for the faces of the plate and

$$n_aH_a^{(0)}-2bk\theta^{(0)}=0,\qquad n_aH_a^{(1)}-\tfrac{2}{3}b^3k\theta^{(1)}=0$$

for the edge of the plate. The constant k can be different for the interior and edge of the plate as well as different for the two faces of the plate. In the latter case, the fluxes across the two faces have to be distinguished:

$$N^{(0)}=N_+^{(0)}-N_-^{(0)},\qquad N^{(1)}=N_+^{(1)}-N_-^{(1)}$$

where

$$N_\pm^{(0)}=[h_2]_{x_2=\pm b},\qquad N_\pm^{(1)}=[x_2h_2]_{x_2=\pm b}.$$

Then, for the faces of the plate, the radiation conditions are

$$N_+^{(0)}-k_+(\theta^{(0)}+b\theta^{(1)})=0,\qquad N_+^{(1)}-k_+b(\theta^{(0)}+b\theta^{(1)})=0,$$

$$N_-^{(0)}+k_-(\theta^{(0)}-b\theta^{(1)})=0,\qquad N_-^{(1)}-k_-b(\theta^{(0)}-b\theta^{(1)})=0,$$

where k_+ and k_- are the values of k for the faces $x_2=b$ and $x_2=-b$, respectively.

REFERENCES

1. R. D. Mindlin, High frequency vibrations of piezoelectric crystal plates. *Int. J. Solids Struct.* **8**, 895–906 (1972).
2. J. Tasi and G. Herrmann, Thermoelastic dissipation in high-frequency vibrations of crystal plates. *J. Acoust. Soc. Am.* **36**, 100–110 (1964).
3. M. A. Biot, Thermoelasticity and irreversible thermodynamics. *J. appl. Phys.* **27**, 240–253 (1956).
4. J. H. Weiner, A uniqueness theorem for the coupled thermoelastic problem. *Q. Appl. Math.* **15**, 102–105 (1957).
5. W. P. Mason, *Piezoelectric Crystals and their Application to Ultrasonics.* Van Nostrand, New York (1950).
6. R. D. Mindlin, On the equations of motion of piezoelectric crystals. In *Problems of Continuum Mechanics* (edited by J. Radok), pp. 282–290. Soc. Ind. Appl. Math., Philadelphia (1961).

7. A. L. Cauchy, Sur l'équilibre et le mouvement d'une plaque élastique dont l'élasticité n'est pas la même dans tous les sens. *Exercices Math.* **4**, 1–14 (1829).
8. S. D. Poisson, Mémoire sur l'équilibre et le mouvement des corps élastiques. *Mém. Acad. Sci.*, Paris, Ser. 2, **8**, 357–570 (1829).
9. J. A. C. Bresse, *Cours de Mécanique Appliquée*, p. 126. Mallet-Bachelier, Paris (1859).
10. S. Timoshenko, On the correction for shear of the differential equation for transverse vibrations of prismatic bars. *Phil. Mag.* **41**, 744–746 (1921).
11. W. Voigt, *Lehrbuch der Kristallphysik*, B. G. Teubner, Leipzig, 2nd ed. (1928); Reprint, Edwards Bros., Ann Arbor pp. 681, 698 (1946).

Абстракт—Разработана двухмерная система уравнений для высокочастотных движений кристаллических пластинок, принимающая во внимание взаимодействие механических, электростатических и термических полей.

MECH. RES. COMM. Vol.1, 57-59, 1974. Pergamon Press. Printed in USA.

NOTE ON TORSIONAL RIGIDITY OF ELASTIC BARS

R. D. Mindlin
Department of Civil Engineering, Columbia University,
New York, New York, U. S. A.

(Received 10 April 1974; accepted as ready for print 23 April 1974)

Introduction

In this note it is shown how the Bleustein-Stanley [1] equations of torsion of bars (obtained by expansion of the displacements in power series of the transverse coordinates, followed by truncation) can be extended to produce torsional rigidities and low frequency dispersion relations either exact, or as close to exact as desired, without the use of correction factors.

Exposition

Employing the notation of Bleustein and Stanley for displacement,

$$u_i = \sum_{n=0}^{\infty} \sum_{m=0}^{\infty} x_1^n x_2^m u_i^{(n,m)}(x_3, t), \tag{1}$$

and stress,

$$T_{ij}^{(n,m)} = \int_A x_1^n x_2^m T_{ij} \, dA, \tag{2}$$

and considering only equilibrium, for the sake of brevity, their equation (7) yields the equation of axial equilibrium in uniform torsion, for each $u_3^{(n,m)}$

$$nT_{13}^{(n-1,m)} + mT_{23}^{(n,m-1)} = 0, \tag{3}$$

and their equations (27) and (28) give the torsional rigidity:

Scientific Communication

$$C = \left(T_{23}^{(1,0)} - T_{13}^{(0,1)}\right) / \left(\theta_o/\ell\right), \tag{4}$$

where θ_o/ℓ is the angle of twist per unit length.

Bleustein and Stanley restricted u_3 to $u_3 = x_1x_2u_3^{(1,1)}$. This is the appropriate form for the elliptic section; for then (3) becomes $T_{13}^{(0,1)} + T_{23}^{(1,0)} = 0$ and (4) yields the St. Venant value $C = \mu\pi a^3b^3/(a^2 + b^2)$, in which a and b are the semi-principal axes of the ellipse and μ is the shear modulus.

The only departure from Bleustein's and Stanley's procedure that is suggested is the retention of all the terms in the series expansion of u_3 when truncating u_1 and u_2. Subsequently, for each cross-section to be considered, only a few of the terms $u_3^{(n,m)}$, judiciously chosen, need be used, along with the corresponding equations from (3). Thus, for the equilateral triangle, the expression for u_3 in (1) should be taken as

$$u_3 = x_1^2\, x_2u_3^{(2,1)} + x_2^3\, u_3^{(0,3)} \tag{5}$$

so that, from (3),

$$2T_{13}^{(1,1)} + T_{23}^{(2,0)} = 0, \quad T_{23}^{(0,2)} = 0. \tag{6}$$

With <u>a</u> the side-length, (4) becomes $C = \mu\sqrt{3}\, a^4/80$, which is the St. Venant value.

It may be seen, from the examples of the elliptic and triangular sections, that, if the axial displacement in the St. Venant solution is a polynomial in the transverse coordinates, the torsional rigidity from the approximate equations (as well as the dispersion relation for torsional waves at low enough frequencies) can be made exact. Otherwise, as close an approximation as is desired can be obtained by discriminating selection of the terms in u_3 to be retained. For example, for the rectangular section, with sides 2a and 2b, if u_3 and (3) are taken as

$$u_3 = x_1x_2u_3^{(1,1)} + x_1\, x_2^3u_3^{(1,3)} + x_1^3\, x_2u_3^{(3,1)} +$$

$$+ x_1^3 x_2^3 u_3^{(3,3)} \tag{7}$$

$$\left.\begin{aligned} T_{13}^{(0,1)} + T_{23}^{(1,0)} = 0, \quad T_{13}^{(0,3)} + 3T_{23}^{(1,2)} = 0, \\ 3T_{13}^{(2,1)} + T_{23}^{(3,0)} = 0, \quad T_{13}^{(2,3)} + T_{23}^{(3,2)} = 0, \end{aligned}\right\} \tag{8}$$

we find that (4) yields

$$C = k_1 \mu (2a)^3 (2b)/12, \tag{9}$$

where

$$k_1 = 1 + k^2 - \frac{(1-k^2)^2(7 + 212k^2 + 7k^4)}{(1+k^2)(7 + 121k^2 + 7k^4)} + \\ + \frac{7k^2 (13 - 25k^2 - 25k^4 + 13k^6)}{(1 + k^2)^2 (7 + 121k^2 + 7k^4)} \tag{10}$$

and $k = b/a$. The error in C increases from a value of about 0.1% at $k = 1$ to about 1.0% at $k = 3$ and 3.0% around $k = 10$ and then diminishes to zero as k approaches infinity. If smaller errors are desired, more terms in u_3 can be retained. It results that, when $k = 1$, the equations (8) require $u_3^{(1,1)}$ and $u_3^{(3,3)}$ to be zero. This accounts for a difficulty encountered by Bleustein and Stanley in applying their equations to the square cross-section.

Applications to vibrations of isotropic bars and to coupled longitudinal, flexural and torsional vibrations of anisotropic and piezoelectric bars will appear elsewhere [2].

References

1. J. L. Bleustein and R. M. Stanley, Int. J. Solids Struct., 6, 569 (1970)
2. R. D. Mindlin, Int. J. Solids Struct. (forthcoming)

Int. J. Solids Structures, 1974, Vol. 10, pp. 1307–1314. Pergamon Press. Printed in Gt. Britain.

ELECTROMAGNETIC RADIATION FROM A VIBRATING, ELASTIC SPHERE

R. D. MINDLIN
Department of Civil Engineering, Columbia University, New York, N. Y. 10027, U.S.A.

(*Received* 6 *March* 1974)

Abstract—When the contribution of the polarization gradient to the stored energy is taken into account, in the theory of elastic dielectrics, there is a small coupling between electromagnetic and elastic energies even in isotropic materials. It is found that the electromagnetic radiation from an isotropic, elastic, dielectric sphere vibrating in its fundamental rotatory mode, with a maximum shear strain of 10^{-3}, is of the order of 10^{-38} watts: independent of the radius of the sphere and proportional to the square of the strain.

INTRODUCTION

According to the classical theory of the elastic dielectric continuum, there is no coupling between the mechanical displacement and the electronic polarization in centrosymmetric materials. However, if the contribution of the polarization gradient to the stored energy of deformation and polarization is taken into account, in addition to the usual strain and polarization, such a coupling does exist. Even in the material of highest symmetry (centrosymmetric isotropic) an elastic shear wave induces a transverse polarization wave which, in turn, excites an electromagnetic wave. Thus, with suitable boundary conditions, an electromagnetic radiation may be expected to emanate from any dielectric solid vibrating in a mode involving shear. The simplest example, for a finite body, is that of an isotropic, elastic sphere in rotatory vibration—a mode of vibration in which every spherical surface concentric with the boundary rotates back and forth through a small angle about a single axis. The radiated energy can be expected, of course, to be extremely small; but there is some interest in discovering how small. In the case examined here (a sphere with the isotropized material constants of sodium chloride) the radiation rate accompanying the fundamental mode of rotatory vibration, with a maximum shear strain of 10^{-3}, is of the order of 10^{-38} watts: independent of the radius of the sphere and proportional to the square of the strain.

DIFFERENTIAL EQUATIONS AND BOUNDARY CONDITIONS

For the centrosymmetric isotropic case, the energy density, W^L, of strain, polarization and polarization gradient is [1]

$$2W^L = a_{11}\mathbf{P}\cdot\mathbf{P} + b_{12}\,\nabla\cdot\mathbf{P}\nabla\cdot\mathbf{P} + b_{44}(\nabla\mathbf{P}:\nabla\mathbf{P} + \nabla\mathbf{P}:\mathbf{P}\nabla) + b_{77}(\nabla\mathbf{P}:\nabla\mathbf{P} - \nabla\mathbf{P}:\mathbf{P}\nabla)$$
$$+ c_{12}\,\nabla\cdot\mathbf{u}\nabla\cdot\mathbf{u} + c_{44}(\nabla\mathbf{u}:\nabla\mathbf{u} + \nabla\mathbf{u}:\mathbf{u}\nabla) + 2d_{12}\,\nabla\cdot\mathbf{P}\,\nabla\cdot\mathbf{u} + d_{44}(\nabla\mathbf{P}:\nabla\mathbf{u} + \nabla\mathbf{P}:\mathbf{u}\nabla), \tag{1}$$

where $\mathbf{u}$ is the mechanical displacement, $\mathbf{P}$ is the electronic polarization, ∇ is the gradient operator and $\nabla\cdot$ is the divergence operator. Corresponding to (1), the mechanical and electrical equations of motion are [1]

$$c_{44}\,\nabla^2\mathbf{u} + (c_{12}+c_{44})\,\nabla\nabla\cdot\mathbf{u} + d_{44}\,\nabla^2\mathbf{P} + (d_{12}+d_{44})\,\nabla\nabla\cdot\mathbf{P} = \rho\ddot{\mathbf{u}} \tag{2}$$

$$d_{44}\,\nabla^2\mathbf{u} + (d_{12}+d_{44})\,\nabla\nabla\cdot\mathbf{u} + (b_{44}+b_{77})\,\nabla^2\mathbf{P} + (b_{12}+b_{44}-b_{77})\,\nabla\nabla\cdot\mathbf{P} - a_{11}\mathbf{P} + \mathbf{E} = 0 \tag{3}$$

where $\mathbf{E}$ is the Maxwell electric self-field, ρ is the mass density and ∇^2 is Laplace's operator To these equations are adjoined the equations of the electromagnetic field [2, p. 75]

$$c^2\nabla^2\mathbf{A} + \varepsilon_0^{-1}\dot{\mathbf{P}} = \ddot{\mathbf{A}}, \qquad c^2\nabla^2\varphi - \varepsilon_0^{-1}c^2\nabla\cdot\mathbf{P} = \ddot{\varphi}, \qquad c^2\nabla\cdot\mathbf{A} = -\dot{\varphi}, \tag{4}$$

where c and ε_0 are the velocity of electromagnetic waves and the dielectric permittivity in a vacuum and $\mathbf{A}$ and φ are the vector and scalar potentials defined in terms of $\mathbf{E}$ and the magnetic flux density, $\mathbf{B}$, through

$$\mathbf{B} = \nabla\times\mathbf{A}, \qquad \mathbf{E} = -\nabla\varphi - \dot{\mathbf{A}}. \tag{5}$$

We shall be concerned only with transverse waves, in which case φ and all divergences vanish. Then (2)–(5) reduce to

$$c_{44}\nabla^2\mathbf{u} + d_{44}\nabla^2\mathbf{P} = \rho\ddot{\mathbf{u}}, \tag{6}$$

$$d_{44}\,\nabla^2\mathbf{u} + (b_{44}+b_{77})\,\nabla^2\mathbf{P} - a_{11}\mathbf{P} - \dot{\mathbf{A}} = 0, \tag{7}$$

$$c^2\,\nabla^2\mathbf{A} + \varepsilon_0^{-1}\dot{\mathbf{P}} = \ddot{\mathbf{A}} \tag{8}$$

in the elastic dielectric and to

$$c^2\,\nabla^2\mathbf{A} = \ddot{\mathbf{A}} \tag{9}$$

in a vacuum.

If the surface S, with outward normal $\mathbf{n}$, separating the dielectric from a vacuum, supports a surface traction $\mathbf{t}$ but is otherwise free, the boundary conditions on S, when $\nabla\cdot\mathbf{u}$ and $\nabla\cdot\mathbf{P}$ are zero, are [1]

$$\mathbf{n}\cdot\partial W^L/\partial\tfrac{1}{2}(\nabla\mathbf{u}+\mathbf{u}\nabla) = \mathbf{n}\cdot[c_{44}(\nabla\mathbf{u}+\mathbf{u}\nabla) + d_{44}(\nabla\mathbf{P}+\mathbf{P}\nabla)] = \mathbf{t}, \tag{10}$$

$$\mathbf{n}\cdot\partial W^L/\partial\nabla\mathbf{P} = \mathbf{n}\cdot[d_{44}(\nabla\mathbf{u}+\mathbf{u}\nabla) + b_{44}(\nabla\mathbf{P}+\mathbf{P}\nabla) + b_{77}(\nabla\mathbf{P}-\mathbf{P}\nabla)] = 0, \tag{11}$$

along with the usual electromagnetic conditions of continuity of

$$\mathbf{n}\times\mathbf{B}, \ \mathbf{n}\cdot\mathbf{B}, \ \mathbf{n}\times\mathbf{E}, \ \mathbf{n}\cdot(\varepsilon_0\mathbf{E}+\mathbf{P}) \tag{12}$$

across S [2, p. 3].

SOLUTION

We consider the rotatory vibrations of an isotropic, elastic dielectric sphere, of radius a, surrounded by a vacuum. Guided by the solution for rotatory vibrations of a purely elastic sphere ([3] p. 285), we take, in spherical coordinates, r, θ, φ, for $r \le a$,

$$u_r = 0, \ u_\theta = 0, \ u_\varphi = C_1\psi(\alpha r)\sin\theta\cos(\omega t+\varepsilon), \tag{13}$$

$$P_r = 0, \ P_\theta = 0, \ P_\varphi = C_2\,\psi(\alpha r)\sin\theta\cos(\omega t+\varepsilon), \tag{14}$$

$$A_r = 0, \ A_\theta = 0, \ A_\varphi = C_3\,\psi(\alpha r)\sin\theta\sin(\omega t+\varepsilon), \tag{15}$$

where

$$\psi(\alpha r) = (\alpha r)^{-1}\cos\alpha r - (\alpha r)^{-2}\sin\alpha r. \tag{16}$$

Then, for example,

$$\nabla^2\mathbf{u} = \mathbf{e}_\varphi r^{-2}[(r^2\psi')' - 2\psi]C_1 \sin\theta\cos(\omega t + \varepsilon) = -\ \mathbf{e}_\varphi \alpha^2 u_\varphi, \tag{17}$$

where prime designates differentiation with respect to r and $\mathbf{e}_\varphi$ is a unit vector in the direction of φ increasing. Similar results hold for the Laplacians of **P** and **A**. Accordingly, upon substitution of (13), (14) and (15) in (6), (7) and (8), the latter become

$$(c_{44}\alpha^2 - \rho\omega^2)C_1 + d_{44}\alpha^2 C_2 = 0, \tag{18}$$

$$d_{44}\alpha^2 C_1 + [(b_{44} + b_{77})\alpha^2 + a_{11}]C_2 + \omega C_3 = 0, \tag{19}$$

$$\omega C_2 + \varepsilon_0(c^2\alpha^2 - \omega^2)C_3 = 0, \tag{20}$$

which have a nonzero solution if

$$\Delta \equiv \begin{vmatrix} c_{44}\alpha^2 - \rho\omega^2 & d_{44}\alpha^2 & 0 \\ d_{44}\alpha^2 & (b_{44} + b_{77})\alpha^2 + a_{11} & \omega \\ 0 & \omega & \varepsilon_0(c^2\alpha^2 - \omega^2) \end{vmatrix} = 0 \tag{21}$$

and this is a cubic equation in α^2 with three real roots: two positive and one negative. Thus, the dispersion relation (21) has two real branches and one imaginary branch. The characters of the branches may be identified by examining their behaviors at low frequencies. For the real branches,

$$\lim_{\omega,\alpha\to 0} \Delta = \lim_{\omega,\alpha\to 0} (c_{44}\alpha^2 - \rho\omega^2)(c^2\alpha^2 - K\omega^2)\varepsilon_0 a_{11} = 0, \tag{22}$$

where K is the dielectric constant:

$$K = 1 + (\varepsilon_0 a_{11})^{-1}, \tag{23}$$

since $\varepsilon_0 a_{11}$ is the reciprocal dielectric susceptibility. Hence, at low frequencies, the dispersion relations of the two real branches are

$$\alpha_1^2 = \rho\omega^2/c_{44} \text{ (acoustic branch)} \tag{24}$$

$$\alpha_2^2 = K\omega^2/c^2 \text{ (electromagnetic branch).} \tag{25}$$

For the third branch, we have

$$\lim_{\omega\to 0} \Delta = \varepsilon_0 c^2\alpha^4\{[c_{44}(b_{44} + b_{77}) - d_{44}^2]\alpha^2 + a_{11}c_{44}\} = 0 \tag{26}$$

or

$$\alpha_3^2 = -a_{11}c_{44}/[c_{44}(b_{44} + b_{77}) - d_{44}^2] \quad \text{(surface branch).} \tag{27}$$

Since positive definiteness of W^L requires $c_{44}(b_{44} + b_{77}) - d_{44}^2$, c_{44} and a_{11} to be positive, α_3 is imaginary.

Taking into account the three branches and separating the two phases of the functions u_φ, P_φ and A_φ, we have, as their forms for $r \le a$,

$$u_\varphi = \sum_{j=1}^{3} (C_{1j}\cos\omega t + C'_{1j}\sin\omega t)\psi(\alpha_j r)\sin\theta, \tag{28}$$

$$P_\varphi = \sum_{j=1}^{3} (C_{2j}\cos\omega t + C'_{2j}\sin\omega t)\psi(\alpha_j r)\sin\theta, \tag{29}$$

$$A_\varphi = \sum_{j=1}^{3} (C_{3j}\sin\omega t + C'_{3j}\cos\omega t\})\psi(\alpha_j r)\sin\theta, \tag{30}$$

where

$$\psi(\alpha_j r) = (\alpha_j r)^{-1} \cos \alpha_j r - (\alpha_j r)^{-2} \sin \alpha_j r, \; j = 1, 2, \tag{31}$$

$$\psi(\alpha_j r) = (\alpha_j r)^{-1} \cosh \alpha_j r - (\alpha_j r)^{-2} \sinh \alpha_j r, \; j = 3. \tag{32}$$

Note that, in (32), the sign of α_3 has been reversed from that in (27) so that, in the sequel, α_3 is real.

The eighteen constants C_{ij} and C'_{ij} are subject to twelve relations through equations (18), (19) and (20), from which we may define

$$\gamma_{2j} = \frac{C_{2j}}{C_{1j}} = \frac{C'_{2j}}{C'_{1j}} = \frac{\mp d_{44} \alpha_j^2 \varepsilon_0(\omega^2 \mp c^2\alpha_j^2)}{\varepsilon_0(\omega^2 \mp c^2\alpha_j^2)[a_{11} \pm (b_{44} + b_{77})\alpha_j^2] + \omega^2}, \tag{33}$$

$$\gamma_{3j} = \frac{C_{3j}}{C_{1j}} = \frac{C'_{3j}}{C'_{1j}} = \frac{\mp d_{44} \omega \alpha_j^2}{\varepsilon_0(\omega^2 \mp c^2\alpha_j^2)[a_{11} \pm (b_{44} + b_{77})\alpha_j^2] + \omega^2}, \tag{34}$$

where the upper signs are for $j = 1, 2$ and the lower signs are for $j = 3$. Then (28)–(30) may be written in terms of the six remaining constants C_{1j} and C'_{1j}:

$$u_\varphi = \sum_{j=1}^{3} (C_{1j} \cos \omega t + C'_{1j} \sin \omega t)\psi(\alpha_j r) \sin \theta, \tag{35}$$

$$P_\varphi = \sum_{j=1}^{3} (C_{1j} \cos \omega t + C'_{1j} \sin \omega t)\gamma_{2j}\psi(\alpha_j r) \sin \theta, \tag{36}$$

$$A_\varphi = \sum_{j=1}^{3} (C_{1j} \sin \omega t + C'_{1j} \cos \omega t)\gamma_{3j}\psi(\alpha_j r) \sin \theta. \tag{37}$$

In $r \geq a$, we take $\mathbf{u}$, $\mathbf{P}$, A_r, $A_\theta = 0$ and $A_\varphi = A_\varphi^0$, where

$$\begin{aligned} A_\varphi^0 &= C_0[(\alpha_0 r)^{-1} \cos \alpha_0(r - a - ct) - (\alpha_0 r)^{-2} \sin \alpha_0(r - a - ct)]\sin \theta \\ &\quad + C'_0[(\alpha_0 r)^{-1} \sin \alpha_0(r - a - ct) + (\alpha_0 r)^{-2} \cos \alpha_0(r - a - ct)]\sin \theta \end{aligned} \tag{38}$$

and

$$\alpha_0 = \omega/c. \tag{39}$$

To maintain the electromagnetic radiation expressed by (38), an external action on the sphere is required. This we take to be a surface traction on $r = a$:

$$\mathbf{t} = \mathbf{e}_\varphi T \sin \theta \cos \omega t. \tag{40}$$

Noting that, in the present case

$$\nabla \mathbf{u} + \mathbf{u}\nabla = (\mathbf{e}_r \mathbf{e}_\varphi + \mathbf{e}_\varphi \mathbf{e}_r)\left(\frac{\partial u_\varphi}{\partial r} - \frac{u_\varphi}{r}\right) + (\mathbf{e}_\theta \mathbf{e}_\varphi + \mathbf{e}_\varphi \mathbf{e}_\theta)\left(\frac{1}{r}\frac{\partial u_\varphi}{\partial \theta} - \frac{u_\varphi \cot \theta}{r}\right), \tag{41}$$

with a similar expression for $\nabla \mathbf{P} + \mathbf{P}\nabla$; and also

$$\nabla \mathbf{P} - \mathbf{P}\nabla = (\mathbf{e}_r \mathbf{e}_\varphi + \mathbf{e}_\varphi \mathbf{e}_r)\left(\frac{\partial P_\varphi}{\partial r} + \frac{P_\varphi}{r}\right) + (\mathbf{e}_\theta \mathbf{e}_\varphi + \mathbf{e}_\varphi \mathbf{e}_\theta)\left(\frac{1}{r}\frac{\partial P_\varphi}{\partial \theta} + \frac{P_\varphi \cot \theta}{r}\right) \tag{42}$$

and $\mathbf{n} = \mathbf{e}_r$, we find that the boundary conditions (10) and (11) become

$$\left[c_{44}\left(\frac{\partial u_\varphi}{\partial r} - \frac{u_\varphi}{r}\right) + d_{44}\left(\frac{\partial P_\varphi}{\partial r} - \frac{P_\varphi}{r}\right)\right]_{r=a} = T \sin\theta \cos\omega t, \tag{43}$$

$$\left[d_{44}\left(\frac{\partial u_\varphi}{\partial r} - \frac{u_\varphi}{\partial r}\right) + b_{44}\left(\frac{\partial P_\varphi}{\partial r} - \frac{P_\varphi}{r}\right) + b_{77}\left(\frac{\partial P_\varphi}{\partial r} + \frac{P_\varphi}{r}\right)\right]_{r=a} = 0. \tag{44}$$

As for the continuity conditions (12), we note that, in the present case,

$$\mathbf{B} = \nabla \times \mathbf{A} = \frac{\mathbf{e}_r}{r \sin\theta} \frac{\partial}{\partial\theta}(A_\varphi \sin\theta) - \frac{\mathbf{e}_\theta}{r} \frac{\partial}{\partial r}(rA_\varphi), \tag{45}$$

$$\mathbf{E} = -\,\mathbf{e}_\varphi \dot{\mathbf{A}}_\varphi, \tag{46}$$

$$\varepsilon_0 \mathbf{E} + \mathbf{P} = \mathbf{e}_\varphi(-\varepsilon_0 \dot{\mathbf{A}}_\varphi + P_\varphi), \tag{47}$$

whence:

$$\mathbf{n} \times \mathbf{B} = \mathbf{e}_r \times \mathbf{B} = -\mathbf{e}_\varphi(\partial A_\varphi/\partial r + A_\varphi/r), \tag{48}$$

$$\mathbf{n} \cdot \mathbf{B} = \mathbf{e}_r \cdot \mathbf{B} = \frac{1}{r \sin\theta} \frac{\partial}{\partial\theta}(A_\varphi \sin\theta), \tag{49}$$

$$\mathbf{n} \times \mathbf{E} = -\mathbf{e}_r \times \dot{\mathbf{A}} = \mathbf{e}_\theta \dot{\mathbf{A}}_\varphi, \tag{50}$$

$$\mathbf{n} \cdot (\varepsilon_0 \mathbf{E} + \mathbf{P}) = \mathbf{e}_r \cdot \mathbf{e}_\varphi(-\varepsilon_0 \dot{\mathbf{A}}_\varphi + P_\varphi) = 0. \tag{51}$$

Hence, the four continuity conditions (12) reduce to

$$[A_\varphi]_{r=a} = [A_\varphi^0]_{r=a}, \tag{52}$$

$$[\partial A_\varphi/\partial r]_{r=a} = [\partial A_\varphi^0/\partial r]_{r=a}. \tag{53}$$

Upon substituting (35)–(38) in the four conditions (43), (44), (52) and (53) and equating coefficients of $\cos\omega t$ and $\sin\omega t$ separately, we find the following eight equations on the eight constants C_{1j}, C'_{1j}, C_0 and C'_0:

$$\sum_{j=1}^{3} C_{1j}\beta_{1j} = T, \qquad \sum_{j=1}^{3} C'_{1j}\beta_{1j} = 0, \tag{54}$$

$$\sum_{j=1}^{3} C_{1j}\beta_{2j} = 0, \qquad \sum_{j=1}^{3} C'_{1j}\beta_{2j} = 0, \tag{55}$$

$$\sum_{j=1}^{3} C_{1j}\beta_{3j} + C_0[2(\alpha_0 a)^{-3} - (\alpha_0 a)^{-1}] - 2C'_0(\alpha_0 a)^{-2} = 0, \tag{56}$$

$$\sum_{j=1}^{3} C'_{1j}\beta_{3j} + 2C_0(\alpha_0 a)^{-2} + C'_0[2(\alpha_0 a)^{-3} - (\alpha_0 a)^{-1}] = 0, \tag{57}$$

$$\sum_{j=1}^{3} C_{1j}\gamma_{3j}\psi_j - C_0(\alpha_0 a)^{-2} + C'_0(\alpha_0 a)^{-1} = 0, \tag{58}$$

$$\sum_{j=1}^{3} C'_{1j}\gamma_{3j}\psi_j - C_0(\alpha_0 a)^{-1} - C'_0(\alpha_0 a)^{-2} = 0, \tag{59}$$

where $\psi_j = \psi_j(\alpha_j a)$ and

$$\beta_{1j} = (c_{44} + d_{44}\gamma_{2j})(\partial\psi_j/\partial a - \psi_j/a), \tag{60}$$

$$\beta_{2j} = (d_{44} + b_{44}\gamma_{2j})(\partial\psi_j/\partial a - \psi_j/a) + b_{77}\gamma_{2j}(\partial\psi_j/\partial a + \psi_j/a), \tag{61}$$

$$\beta_{3j} = \gamma_{3j}\alpha_0^{-1}\,\partial\psi_j/\partial a. \tag{62}$$

The solution of the eight equations (54)–(59) is

$$C_{1j} = (GA_{3j} + A_{1j})T/|\beta_{ij}|, \qquad C'_{1j} = G'A_{3j}T/|\beta_{ij}|, \tag{63}$$

$$C_0 = F_2\{[2(\alpha_0 a)^{-3} - (\alpha_0 a)^{-1}]F_1 + (\alpha_0 a)^{-2}\}T/D, \tag{64}$$

$$C'_0 = -F_2[2(\alpha_0 a)^{-2}F_1 + (\alpha_0 a)^{-1}]T/D, \tag{65}$$

where $|\beta_{ij}|$ is the determinant with elements β_{ij}, the A_{ij} are the cofactors of the β_{ij} in $|\beta_{ij}|$ and

$$D = \{[(\alpha_0 a)^{-1} - 2(\alpha_0 a)^{-3}]F_1 - (\alpha_0 a)^{-2}\}^2 + [2(\alpha_0 a)^{-2}F_1 + (\alpha_0 a)^{-1}]^2 \tag{66}$$

$$|\beta_{1j}|F_1 = \sum_{j=1}^{3}\gamma_{3j}\psi_j A_{3j}, \qquad |\beta_{1j}|F_2 = \sum_{j=1}^{3}\gamma_{3j}\psi_j A_{1j}, \tag{67}$$

$$G = -F_2\{[4 + (\alpha_0 a)^4]F_1 + (\alpha_0 a)^3 + 2(\alpha_0 a)\}/(\alpha_0 a)^6 D, \qquad G' = -F_2/(\alpha_0 a)^2 D. \tag{68}$$

APPLICATION

The only values of the constants b and d that are known, at this point in time, are those for alkali halides determined by Askar *et al.* [4]. We shall use, here, their isotropized values of b_{44}, b_{77}, d_{44} and c_{44} for sodium chloride [5] as corrected subsequently by Lee:

$$\begin{aligned} b_{44} &= 0{\cdot}222 \times 10^4 \text{ dyn cm}^4/\text{C}^2, \\ b_{77} &= 0{\cdot}218 \times 10^4 \text{ dyn cm}^4/\text{C}^2 \\ d_{44} &= -1{\cdot}61 \times 10^6 \text{ dyn cm/C}, \\ c_{44} &= 1{\cdot}49 \times 10^{11} \text{ dyn/cm}^2. \end{aligned} \tag{69}$$

In addition, we require numerical values of the dielectric constant, K, and mass density, ρ, of sodium chloride, and the permittivity, ε_0, and velocity of electromagnetic waves, c, in a vacuum:

$$\begin{aligned} K &= 5{\cdot}6 \text{ [6, p. 69]}, \\ \rho &= 2{\cdot}214 \text{ gm/cm}^3 \text{ [7, p. 88]}, \\ \varepsilon_0 &= 8{\cdot}854 \times 10^{-21} \text{ C}^2/\text{dyn cm}^2 \text{[6, p. 68]}, \\ c &= 2.998 \times 10^{10} \text{ cm/sec [2, p. 11]}. \end{aligned} \tag{70}$$

We shall calculate the radiation associated with the fundamental mode of rotatory vibration of the sphere. Since the radiation is extremely small, the frequency of that mode is very nearly that of the corresponding mode of the purely elastic sphere. From (54), this frequency is determined by the lowest root of $\beta_{11} = 0$ with $d_{44} = 0$, i.e. [3]

$$\tan\alpha_1 a = 3\alpha_1 a/(3 - \alpha_1^2 a^2) \text{ or } \alpha_1 a = 5{\cdot}763. \tag{71}$$

Since, also, the frequency of the fundamental mode is low in comparison with those of the higher modes, the wave numbers, α_i, are determined very closely by (24), (25) and (27):

$$\alpha_1^2 = \rho\omega^2/c_{44}, \qquad \alpha_2^2 = K\omega^2/c^2, \qquad \alpha_3^2 = a_{11}c_{44}/[c_{44}(b_{44}+b_{77}) - d_{44}^2], \tag{72}$$

noting, again, that the sign of α_3^2 has been reversed after (27). Thus, we have, from (69), (70) and (72),

$$\alpha_2 a = \alpha_1 a(Kc_{44}/\rho)^{1/2}/c = 1{\cdot}180 \times 10^{-4}, \qquad \alpha_3 = 7{\cdot}485 \times 10^7\ \text{cm}^{-1}, \tag{73}$$

and, finally,

$$\alpha_0 a = a\omega/c = \alpha_1 a(c_{44}/\rho)^{1/2}/c = 4{\cdot}987 \times 10^{-5}. \tag{74}$$

The numerical values in (69)–(74) are all that are required to calculate the ψ_j from (31) and (32), the γ_{ij} from (33) and (34), the β_{ij} from (60)–(62) and, finally, the C_{1j}, C'_{1j}, C_0 and C'_0 from (63), (64) and (65) for insertion in the formulas (35)–(38) for u_φ, P_φ, A_φ and A_φ^0 and, thus, to complete the solution.

The rate of energy radiation, per unit area, from a point on the surface of the sphere, is equal to the normal component of the Pointing vector and this must be equal to the rate of working of the surface traction:

$$\begin{aligned}\mathbf{t}\cdot\dot{\mathbf{u}}]_{r=a} &= T\dot{\mathbf{u}}_\varphi]_{r=a}\sin\theta\cos\omega t\\ &= T\omega\sin^2\theta\cos\omega t\sum_{j=1}^{3}(-C_{1j}\sin\omega t + C'_{1j}\cos\omega t)\psi_j.\end{aligned} \tag{75}$$

The average rate over a period $2\pi/\omega$ is

$$(\omega/2\pi)\int_0^{2\pi/\omega}\mathbf{t}\cdot\dot{\mathbf{u}}]_{r=a}\,\mathrm{d}t = \tfrac{1}{2}T\omega\sin^2\theta\sum_{j=1}^{3}C'_{1j}\psi_j, \tag{76}$$

so that the radiation rate from the entire sphere is

$$(\omega/2\pi)\int_0^{2\pi/\omega}\mathrm{d}t\int_{\theta=0}^{\pi}\int_{\varphi=0}^{2\pi}\mathbf{t}\cdot\dot{\mathbf{u}}]_{r=a}a^2\sin\theta\,\mathrm{d}\theta\,\mathrm{d}\varphi = \tfrac{4}{3}\pi\omega a^2T\sum_{j=1}^{3}C'_{1j}\psi_j. \tag{77}$$

Now,

$$\omega a = c\alpha_0 a = 1{\cdot}495 \times 10^6\ \text{cm/sec}, \tag{78}$$

and

$$\sum_{j=1}^{3}aC'_{1j}\psi_j = 2{\cdot}394 \times 10^{-47}T\ \text{cm}^2, \tag{79}$$

with T in dynes/cm^2. Hence the rate of radiation from the sphere is

$$1{\cdot}50 \times 10^{-40}T^2\ \text{ergs/sec}. \tag{80}$$

The traction amplitude, T, may be expressed in terms of the maximum shear strain, $(2e_{r\varphi})_{\text{mix}}$, which is located at $\theta = \pi/2$ and $r = a_m$ where a_m is determined by

$$\partial e_{r\varphi}/\partial r = \tfrac{1}{2}\,\partial(\partial u_\varphi/\partial r - u_\varphi/r)/\partial r = 0, \tag{81}$$

in which u_φ for the purely elastic case may be used in view of its preponderant contribution to the strain. From (81) and (28), with $j = 1$, a_m is determined by the lowest root of

$$\tan \alpha_1 a_m = \alpha_1 a_m (9 - \alpha_1^2 a_m^2)/(9 - 4\alpha_1^2 a_m^2) \qquad \text{or} \qquad \alpha_1 a_m = 3{\cdot}342. \tag{82}$$

The amplitude of $(2e_{r\varphi})_{\max}$, at $\theta = \pi/2$, is then,

$$\begin{aligned}(2e_{r\varphi})_{\max} &= a_m{}^{-1} C_{11}\{[3(\alpha_1 a_m)^{-2} - 1]\sin \alpha_1 a_m - 3(\alpha_1 a_m)^{-1} \cos \alpha_1 a_m\} \\ &= 2{\cdot}708 \times 10^{-8} T.\end{aligned} \tag{83}$$

Hence, for $(2e_{r\varphi})_{\max} = 10^{-3}$,

$$T = 3{\cdot}69 \times 10^4 \text{ dyn/cm}^2. \tag{84}$$

Upon substituting this value of T in (80), we find that, if the maximum shear strain is 10^{-3}, the radiation rate from the sphere, when it is vibrating in its fundamental rotatory mode, is about 2×10^{-31} erg/sec; i.e. of the order of 10^{-38} watts.

REFERENCES

1. R. D. Mindlin, *Int. J. Solids Struct.* **4**, 637 (1968).
2. M. Born and E. Wolf, *Principles of Optics*. Pergamon Press (1959).
3. A. E. H. Love, *Theory of Elasticity*, 4th edition. Cambridge University Press (1927).
4. A. Askar, P. C. Y. Lee and A. S. Cakmak, *Phys. Rev.* **1B**, 3525 (1970).
5. A. Askar, P. C. Y. Lee and A. S. Cakmak, *Int. J. Solids Struct.* **7**, 523 (1971).
6. J. F. Nye, *Physical Properties of Crystals*. Oxford University Press (1957).
7. R. F. S. Hearmon, *Applied Anisotropic Elasticity*. Oxford University Press (1961).

Абстракт — Когда учитывается фактор градиента поляризации в аккумулированной энергии, в теории упругих диэлектриков, тогда оказывается малое сопряжение между энергиями электромагнитной и упругой, даже, в изотропных материалах. Находится, что электромагнитное излучение из изотропной, упругой, диэлектрической сферы, которая колебается по своей основной форме вращения с максимальной деформацией сдвига порядка 10^{-3}, является порядка 10^{-38} ватт; оно не зависит от радиуса сферы и пропорционально к квадрату деформации.

Int. J. Solids Structures, Vol. 11, pp. 321–328. Pergamon Press 1975. Printed in Great Britain

SOLUTION OF ST. VENANT'S TORSION PROBLEM BY POWER SERIES

R. D. MINDLIN

Department of Civil Engineering, Columbia University, New York, N.Y. 10027, U.S.A.

(*Received* 3 *June* 1973)

Abstract—It is shown how St. Venant's problem of torsion of elastic bars can be formulated so as to fit readily into the procedure for deriving approximate equations of motion of bars by expansion of the displacements in a double series of powers of the transverse coordinates. Applications are exhibited for bars with elliptic, equilateral triangular and rectangular cross-sections.

INTRODUCTION

Approximate equations of motion of elastic bars, obtained by expansion of displacements in a double power series of the transverse coordinates and subsequent truncation, are not as fully developed as those for plates. This is especially true for motions involving torsion, in which the warping of the cross-section introduces a complication which is present no matter how low the frequency and, in fact, persists even in the equilibrium state. In uniform extension or flexure of cylindrical or prismatic bars of isotropic materials, plane sections remain plane regardless of their contour, so that only linear terms in the power series expansion are necessary. In uniform torsion, on the other hand, plane sections remain plane only for the circular section. For all other sections axial warping occurs—and the character of the warping is different for different sections. There are exact solutions of the St. Venant torsion problem for a great many cross-sectional shapes; but they are not as easily incorporated in equations of motion obtained by power series expansion as they would be if they themselves had been similarly obtained.

In a previous paper[1] it was shown how the power series formulation of the problem of torsion of isotropic bars, by Bleustein and Stanley[2], can be extended to accommodate appropriate warping and produce suitable torsional rigidity without the introduction of correction factors: applied to moments of area of the section or otherwise. In the present paper, the direct connection of the series treatment with St. Venant's method of solution is demonstrated. The torsion function, instead of being governed by Laplace's equation, is expressed as a double power series, the coefficients of which are determined by simultaneous linear algebraic equations, deduced from a variational principle and equal in number to that of the terms of the series retained to represent the axial displacement. These terms are not necessarily all the early terms of the series up to a certain order, but are only a few isolated terms, judiciously chosen. For example, if the St. Venant torsion function is a polynomial (the simplest are for the elliptic and equilateral triangular sections) those are the only terms retained and the resulting torsional rigidity from the power series solution is exact. Otherwise, as in the case of the rectangular section for which the torsion function is an infinite series of transcendental functions, a limited selection of terms in the power series is retained as an approximation. Thus, for the square, a two-term approximation gives the torsional rigidity with an error of about 0·1 per cent. For other rectangular sections, with side-ratio k, say, a four-term approximation gives torsional rigidities with errors increasing from the 0·1 per cent for $k = 1$ to about 3 per cent around $k = 10$ and then diminishing to zero as k approaches infinity.

The criterion of acceptability of the choice of the terms to be retained in the power series is the torsional rigidity, as that determines the torque which balances the inertia in the equation of torsional motion and thus controls the long-wave-speed which, along with the length of the bar and the end-conditions, fixes the lower natural frequencies of torsional vibration.

The procedure described here can readily be applied to vibrations of anisotropic bars in which the torsional motion is coupled with extensional and flexural motions[3].

DERIVATION OF EQUATIONS

The St. Venant problem of torsion of cylindrical or prismatic bars of isotropic material may be stated as follows (Ref. [4], p. 311). With the generators of the surface parallel to the axis x_3 of a rectangular coordinate system x_i, $i = 1, 2, 3$, the displacement components are taken to be

$$u_1 = -\tau x_2 x_3, \quad u_2 = \tau x_3 x_1, \quad u_3 = \tau\varphi, \tag{1}$$

where τ is the twist and φ is a function of x_1 and x_2 satisfying Laplace's equation

$$\frac{\partial^2\varphi}{\partial x_1^2} + \frac{\partial^2\varphi}{\partial x_2^2} = 0 \tag{2}$$

in the interior of the bar and

$$\frac{\partial\varphi}{\partial\nu} = x_2\cos(x_1, \nu) - x_1\cos(x_2, \nu) \tag{3}$$

on the cylindrical or prismatic surface, to which $d\nu$ is the element of the outward drawn normal. The torsional rigidity is the ratio of the torque M to the twist and is given by

$$C = \frac{M}{\tau} = \mu\int\!\!\int_A \left(x_1^2 + x_2^2 + x_1\frac{\partial\varphi}{\partial x_2} - x_2\frac{\partial\varphi}{\partial x_1}\right) dx_1\, dx_2, \tag{4}$$

where μ is the modulus of rigidity and the integration is over the area of the section.

Instead of seeking solutions of (2) satisfying (3), we take φ as a double series of powers of the coordinates x_1 and x_2:

$$\varphi = \sum_{m=0}^{\infty}\sum_{n=0}^{\infty} x_1^m x_2^n u_3^{m,n}, \tag{5}$$

in which the $u_3^{m,n}$ are constants. The displacements (1) may then be written as

$$u_1 = \tau x_2 u_1^{0,1}, \quad u_2 = \tau x_1 u_2^{1,0}, \quad u_3 = \tau\sum_{m=0}^{\infty}\sum_{n=0}^{\infty} x_1^m x_2^n u_3^{m,n}, \tag{6}$$

where $u_1^{0,1} = -x_3$, $u_2^{1,0} = x_3$; so that the components of strain:

$$S_{ij} = \tfrac{1}{2}(u_{j,i} + u_{i,j}),$$

become $S_{11} = S_{22} = S_{33} = S_{12} = 0$ and

$$S_{3\alpha} = \sum_{m=0}^{\infty} \sum_{n=0}^{\infty} x_1{}^m x_2{}^n S_{3\alpha}^{m,n}, \ \alpha = 1, 2, \tag{7}$$

where

$$S_{31}^{0,1} = \tfrac{1}{2}\tau(u_3{}^{1,1} - 1), \ S_{32}^{1,0} = \tfrac{1}{2}\tau(u_3{}^{1,1} + 1) \tag{8}$$

and, otherwise,

$$S_{31}^{m,n} = \tfrac{1}{2}\tau(m+1)u_3{}^{m+1,n}, \ S_{32}^{m,n} = \tfrac{1}{2}\tau(n+1)u_3{}^{m,n+1}. \tag{9}$$

The components of stress are $T_{11} = T_{22} = T_{33} = T_{12} = 0$ and

$$T_{3\alpha} = 2\mu S_{3\alpha} = 2\mu \sum_{m=0}^{\infty} \sum_{n=0}^{\infty} x_1{}^m x_2{}^n S_{3\alpha}, \quad \alpha = 1, 2. \tag{10}$$

These components are substituted in the variational equation

$$\int_A \int_{-l}^{l} T_{ij,i} \delta u_j \, dx_3 \, dA + \int_S (t_j - \nu_i T_{ij}) \delta u_j \, dS = 0, \tag{11}$$

where $\int_A (\,)\, dA$ is the integral over the area of the cross-section, $2l$ is the length of the bar, $\int_S (\,)\, dS$ is the integral over the surface of the bar, including the end faces, and t_j is the surface traction. On substituting (6)–(10) in (11), and performing the integrations where possible, we find

$$j = 1: \int_{-l}^{l} F_1{}^{0,1} \delta u_1{}^{0,1} \, dx_3 + [t_1{}^{0,1} - T_{31}^{0,1}]_{-l}^{l} \delta u_1{}^{0,1} = 0, \tag{12}$$

$$j = 2: \int_{-l}^{l} F_2{}^{1,0} \delta u_2{}^{1,0} \, dx_3 + [t_2{}^{1,0} - T_{32}^{1,0}]_{-l}^{l} \delta u_2{}^{1,0} = 0, \tag{13}$$

$$j = 3: \sum_{m=0}^{\infty} \sum_{n=0}^{\infty} \left\{ \int_{-l}^{l} (m T_{31}^{m-1,n} + n T_{32}^{m,n-1} - F_3{}^{m,n}) \, dx_3 - [t_3{}^{m,n}]_{-l}^{l} \right\} \delta u_3{}^{m,n} = 0, \tag{14}$$

where

$$T_{3\alpha}^{m,n} = \int_A T_{3\alpha} x_1{}^m x_2{}^n \, dA, \ F_j^{m,n} = \oint_C t_j x_1{}^m x_2{}^n \, ds, \ t_j^{m,n} = \int_A t_j x_1{}^m x_2{}^n \, dA \tag{15}$$

and $\oint_C (\,)\, dS$ is the line integral around the bounding curve of the section.

The condition that the cylindrical surface be free of traction requires $F_j^{m,n} = 0$. Then, from (14), we find the equilibrium equations, one for each $u_3{}^{m,n}$,

$$m T_{31}^{m-1,n} + n T_{32}^{m,n-1} = 0 \tag{16}$$

and the requirement $[t_3{}^{m,n}]_{-l}^{l} = 0$, i.e. there can be no tension applied to the bar. The torque,

$t_2^{1,0} - t_1^{0,1}$, applied to the end faces is, from (12) and (13),

$$M = T_{32}^{1,0} - T_{31}^{0,1} = C\tau. \tag{17}$$

From the first of (15) and (7)–(10), we find

$$T_{31}^{m,n} = \mu\tau\left[-I^{m,n+1} + \sum_{p=0}^{\infty}\sum_{q=0}^{\infty}(p+1)I^{m+p,n+q}u_3^{p+1,q}\right], \tag{18}$$

$$T_{32}^{m,n} = \mu\tau\left[I^{m+1,n} + \sum_{p=0}^{\infty}\sum_{q=0}^{\infty}(q+1)I^{m+p,n+q}u_3^{p,q+1}\right], \tag{19}$$

where

$$I^{m+p,n+q} = \int_A x_1^{m+p}x_2^{n+q}\,\mathrm{d}A. \tag{20}$$

Upon substituting (18) and (19) in (16) and (17), we find the equilibrium equations

$$\sum_{p=0}^{\infty}\sum_{q=0}^{\infty}[m(p+1)I^{m+p-1,n+q}u_3^{p+1,q} + n(q+1)I^{m+p,n+q-1}u_3^{p,q+1}] = mI^{m-1,n+1} - nI^{m+1,n-1}$$

and the torsional rigidity

$$C = \mu\left\{I^{0,2} + I^{2,0} - \sum_{p=0}^{\infty}\sum_{q=0}^{\infty}[(p+1)I^{p,q+1}u_3^{p+1,q} - (q+1)I^{p+1,q}u_3^{p,q+1}]\right\}. \tag{21}$$

ELLIPSE

The St. Venant torsion function for the bar of elliptic cross-section with semi-principal axes a and b along x_1 and x_2, respectively, is

$$\varphi = a^2b^2x_1x_2/(a^4 - b^4).$$

Hence, in the power series for u_3, we retain only $u_3^{1,1}$. The equilibrium equations (16) then reduce to the single equation for $m = 1$, $n = 1$:

$$T_{31}^{0,1} + T_{32}^{1,0} = 0, \tag{22}$$

where, from (18) and (19),

$$T_{31}^{0,1} = \mu\tau I^{0,2}(u_3^{1,1} - 1), \quad T_{32}^{1,0} = \mu\tau I^{2,0}(u_3^{1,1} + 1).$$

Upon substituting these values in (22) and solving for $u_3^{1,1}$, we find

$$u_3^{1,1} = (I^{0,2} - I^{2,0})/(I^{0,2} + I^{2,0})$$

and this, inserted in (21), yields the torsional rigidity

$$C = 4\mu I^{0,2}I^{2,0}/(I^{0,2} + I^{2,0}).$$

From (20) we have, for the ellipse,

$$I^{0,2} = \pi ab^3/4, \; I^{2,0} = \pi a^3 b/4.$$

Hence,

$$C = \pi\mu a^3 b^3/(a^2 + b^2),$$

which is the St. Venant torsional rigidity (Ref. [4], p. 316).

EQUILATERAL TRIANGLE

If the cross-section is an equilateral triangle with sides given by the equation (Ref. [4], p. 319)

$$(x_1 - a)(x_1 - x_2\sqrt{3} + 2a)(x_1 + x_2\sqrt{3} + 2a) = 0,$$

the torsion function φ is

$$\varphi = (3x_1{}^2 x_2 - x_2{}^3)/6a. \tag{23}$$

Hence, in the power series we retain only $u_3{}^{2,1}$ and $u_3{}^{0,3}$; so that the equilibrium equations (16) become, with $m = 2$, $n = 1$ and $m = 0$, $n = 3$:

$$2T_{31}^{1,1} + T_{32}^{2,0} = 0, \; T_{31}^{0,2} = 0.$$

From (18) and (19),

$$T_{31}^{0,1} = \mu\tau(-I^{0,2} + 2I^{1,2}u_3{}^{2,1}), \; T_{32}^{1,0} = \mu\tau(I^{2,0} + I^{3,0}u_3{}^{2,1} + 3I^{1,2}u_3{}^{0,3}),$$

$$T_{31}^{1,1} = \mu\tau(-I^{0,2} + 2I^{2,2}u_3{}^{2,1}), \; T_{32}^{2,0} = \mu\tau(I^{3,0} + I^{4,0}u_3{}^{2,1} + 3I^{2,2}u_3{}^{0,3}),$$

$$T_{32}^{0,2} = \mu\tau(I^{1,2} + I^{2,2}u_3{}^{2,1} + 3I^{0,4}u_3{}^{0,3}).$$

The equilibrium equations then become a pair of simultaneous equations for $u_3{}^{2,1}$ and $u_3{}^{0,3}$:

$$(4I^{2,2} + I^{4,0})u_3{}^{2,1} + 3I^{2,2}u_3{}^{0,3} = 2I^{1,2} - I^{3,0},$$

$$I^{2,2}u_3{}^{2,1} + 3I^{0,4}u_3{}^{0,3} = -I^{1,2},$$

from which

$$\begin{aligned} u_3{}^{2,1} &= 3[I^{0,4}(2I^{1,2} - I^{3,0}) + I^{1,2}I^{2,2}]/\Delta, \\ u_3{}^{0,3} &= -[I^{1,2}(4I^{2,2} + I^{4,0}) + I^{2,2}(2I^{1,2} - I^{3,0})]/\Delta, \end{aligned} \tag{24}$$

where

$$\Delta = 3[I^{0,4}(4I^{2,2} + I^{4,0}) - I^{2,2}I^{2,2}].$$

The torsional rigidity, from (21) with only $u_3{}^{2,1}$ and $u_3{}^{0,3}$, is

$$C = \mu[I^{0,2} + I^{2,0} + (I^{3,0} - 2I^{1,2})u_3{}^{2,1} + 3I^{1,2}u_3{}^{0,3}].$$

Then, employing the values (24) of $u_3^{2,1}$ and $u_3^{0,3}$ and noting that, from (20),

$$I^{2,0}=I^{0,2}=3a^4\sqrt{3}/2,\; I^{1,2}=-I^{3,0}=3a^5\sqrt{3}/5,\; I^{4,0}=I^{0,4}=3I^{2,2}=9a^6\sqrt{3}/5,$$

we find

$$C=9\mu a^4\sqrt{3}/5. \tag{25}$$

If we insert the St. Venant torsion function (23) in the integrand in (4) and perform the integration over the triangular area, we find the same result as (25) for the torsional rigidity (Ref. [5], p. 266).

RECTANGLE

St. Venant's torsion function for the bar of rectangular section $x_1=\pm a$, $x_2=\pm b$ is (Ref. [4], p. 318)

$$\varphi=-x_1x_2+32b^2\sum_{n=1,3\ldots}^{\infty}\frac{(-)^{(n-1)/2}\sinh\dfrac{n\pi x_1}{2b}\sin\dfrac{n\pi x_2}{2b}}{n^3\pi^3\cosh\dfrac{n\pi a}{2b}}.$$

As an approximation, we retain only $u_3^{1,1}$, $u_3^{1,3}$, $u_3^{3,1}$, $u_3^{3,3}$ in the power series expression (5) for φ. Then the equilibrium equations (16) are

$$\begin{aligned}T_{31}^{0,1}+T_{32}^{0,1}=0,\; T_{31}^{0,3}+3T_{32}^{1,2}=0,\\ 3T_{31}^{2,1}+T_{32}^{3,0}=0,\; T_{31}^{2,3}+T_{32}^{3,2}=0,\end{aligned} \tag{26}$$

and we find components of stress, from (18) and (19),

$$\begin{aligned}
T_{31}^{0,1}&=\mu\tau(-I^{0,2}+I^{0,2}u_3^{1,1}+I^{0,4}u_3^{1,3}+3I^{2,2}u_3^{3,1}+3I^{2,4}u_3^{3,3}),\\
T_{32}^{1,0}&=\mu\tau(I^{2,0}+I^{2,0}u_3^{1,1}+3I^{2,2}u_3^{1,3}+I^{4,0}u_3^{3,1}+3I^{4,2}u_3^{3,3}),\\
T_{31}^{0,3}&=\mu\tau(-I^{0,4}+I^{0,4}u_3^{1,1}+I^{0,6}u_3^{1,3}+3I^{2,4}u_3^{3,1}+3I^{2,6}u_3^{3,3}),\\
T_{32}^{1,2}&=\mu\tau(I^{2,2}+I^{2,2}u_3^{1,1}+3I^{2,4}u_3^{1,3}+I^{4,2}u_3^{3,1}+3I^{4,4}u_3^{3,3}),\\
T_{31}^{2,1}&=\mu\tau(-I^{2,2}+I^{2,2}u_3^{1,1}+I^{2,4}u_3^{1,3}+3I^{4,2}u_3^{3,1}+3I^{4,4}u_3^{3,3}),\\
T_{32}^{3,0}&=\mu\tau(I^{4,0}+I^{4,0}u_3^{1,1}+3I^{4,2}u_3^{1,3}+I^{6,0}u_3^{3,1}+3I^{6,2}u_3^{3,3}),\\
T_{31}^{2,3}&=\mu\tau(-I^{2,4}+I^{2,4}u_3^{1,1}+I^{2,6}u_3^{1,3}+3I^{4,4}u_3^{3,1}+3I^{4,6}u_3^{3,3}),\\
T_{32}^{3,2}&=\mu\tau(I^{4,2}+I^{4,2}u_3^{1,1}+3I^{4,4}u_3^{1,3}+I^{6,2}u_3^{3,1}+3I^{6,4}u_3^{3,3}).
\end{aligned} \tag{27}$$

Upon substituting (27) in (26), we obtain a set of four simultaneous equations for the four $u_3^{m,n}$:

$$\begin{aligned}
a_{11}u_3^{1,1}+a_{21}u_3^{1,3}+a_{31}u_3^{3,1}+3a_{41}u_3^{3,3}&=c_1,\\
a_{21}u_3^{1,1}+a_{22}u_3^{1,3}+a_{32}u_3^{3,1}+3a_{42}u_3^{3,3}&=c_2,\\
a_{31}u_3^{1,1}+a_{32}u_3^{1,3}+a_{33}u_3^{3,1}+3a_{43}u_3^{3,3}&=c_3,\\
a_{41}u_3^{1,1}+a_{42}u_3^{1,3}+a_{43}u_3^{3,1}+3a_{44}u_3^{3,3}&=c_4,
\end{aligned} \tag{28}$$

where, with $I^{m,n} = 4a^{m+1}b^{n+1}/(m+1)(n+1)$ and $k = b/a$,

$$a_{11} = I^{0,2} + I^{2,0} = 4a^3b(1+k^2)/3, \qquad a_{32} = 3(I^{2,4} + I^{4,2}) = 4a^5b^3(1+k^2)/5,$$
$$a_{21} = I^{0,4} + 3I^{2,2} = 4a^3b^3(5+3k^2)/15, \qquad a_{42} = 3I^{4,4} + I^{2,6} = 4a^5b^5(63+25k^2)/525,$$
$$a_{31} = I^{4,0} + 3I^{2,2} = 4a^5b(3+5k^2)/15, \qquad a_{33} = 9I^{4,2} + I^{6,0} = 4a^7b(5+21k^2)/35,$$
$$a_{41} = I^{2,4} + I^{4,2} = 4a^5b^3(1+k^2)/15, \qquad a_{43} = 3I^{4,4} + I^{6,2} = 4a^7b^3(25+63k^2)/525,$$
$$a_{22} = 9I^{2,4} + I^{0,6} = 4a^3b^5(21+5k^2)/35, \qquad a_{44} = I^{4,6} + I^{6,4} = 4a^7b^5(1+k^2)/35,$$
$$c_1 = I^{0,2} - I^{2,0} = -4a^3b(1-k^2)/3, \qquad c_3 = 3I^{2,2} - I^{4,0} = -4a^5b(3-5k^2)/15,$$
$$c_2 = I^{0,4} - 3I^{2,2} = -4a^3b^3(5-3k^2)/15, \qquad c_4 = I^{2,4} - I^{4,2} = -4a^5b^3(1-k^2)/15.$$

The solution of the four equations (28) is

$$\begin{aligned} u_3^{1,1} &= (k^4-1)(14+375k^2+14k^4)/\Delta, \\ u_3^{1,3} &= 35k^2(1-6k^2-7k^4)/b^2\Delta, \\ u_3^{3,1} &= 35k^2(7+6k^2-k^4)/a^2\Delta, \\ 3u_3^{3,3} &= 245k^2(k^4-1)/a^2b^2\Delta \end{aligned} \tag{29}$$

where

$$\Delta = 2(1+k^2)^2(7+121k^2+7k^4).$$

The torsional rigidity, from (21), is

$$C = \mu(I^{0,2} + I^{2,0} - c_1u_3^{1,1} - c_2u_3^{1,3} - c_3u_3^{3,1} - 3c_4u_3^{3,3}),$$

or

$$C = k_1\mu(2a)^3(2b), \tag{30}$$

where

$$12k_1 = 1 + k^2 - \frac{(1-k^2)^2(7+212k^2+7k^4)}{(1+k^2)(7+121k^2+7k^4)} + \frac{7k^2(13-25k^2-25k^4+13k^6)}{(1+k^2)^2(7+121k^2+7k^4)}.$$

Values of k_1 for various values of k are listed in Table 1 along with the corresponding values (Ref.

Table 1. Torsional rigidity coefficient in equation (30): k_1 (four terms of power series approximation); k_1' (St. Venant solution)

$k = b/a$	k_1	k_1'	b/a	k_1	k_1'
1·0	0·1407	0·1406	4	0·286	0·281
1·2	0·166	0·166	5	0·298	0·291
1·5	0·196	0·196	10	0·322	0·312
2·0	0·230	0·229	20	0·330	0·323
2·5	0·251	0·249	100	0·333	0·331
3·0	0·266	0·263	∞	1/3	1/3

[5], p. 277) calculated from St. Venant's solution. It may be seen, from (29), that for the square section ($k = 1$), $u_3^{1,1}$ and $u_3^{3,3}$ are zero; the approximation reduces to a two-term one with

$$u_3^{3,1} = -u_3^{1,3} = 7/18a^2, \quad k_1 = 19/135.$$

This simple approximation yields a torsional rigidity in error by only about 0·1 per cent. As k increases from unity, the error in k_1 increases to about 3 per cent around $k = 10$ and then diminishes to zero as k approaches infinity, at which limit $u_3^{1,3}$, $u_3^{3,1}$ and $u_3^{3,3}$, in (29), are zero and $u_3^{1,1}$ is unity, yielding the correct form for φ:

$$\varphi = \tau x_1 x_2$$

at that limit. The errors may be reduced by retaining more terms in the series, the next group being $u_3^{1,5}$, $u_3^{3,5}$, $u_3^{5,5}$, $u_3^{5,3}$, $u_3^{5,1}$.

REFERENCES

1. R. D. Mindlin, Note on torsional rigidity of elastic bars. *Mech. Res. Communications* **1, No. 2**, 57–59 (1974).
2. J. L. Bleustein and R. M. Stanley, A dynamical theory of torsion, *Int. J. Solids Struct.* **6**, 569–586 (1970).
3. R. D. Mindlin, Low frequency vibrations of elastic bars, to be published.
4. A. E. H. Love, *A Treatise on the Mathematical Theory of Elasticity*, 4th edition. Cambridge University Press (1927); also Dover (1944).
5. S. Timoshenko and J. N. Goodier, *Theory of Elasticity*, 2nd edition. McGraw-Hill (1951).

Reprinted from *The Photoelastic Effect and its Applications*
J. Kestens (ed.),Springer-Verlag, Berlin, (1975) 376-388

Rotation-Dependence of the Photoelastic Effect

R.D. Mindlin

Department of Civil Engineering
Columbia University, New York, N.Y.

ABSTRACT

The discovery, by Nelson and Lax, of the rotation-dependence of the photoelastic effect is reviewed in the light of Toupin's formula for the reciprocal dielectric susceptibility. The influences of the polarization gradient and a non-primitive crystal structure are also examined.

INTRODUCTION

In a recent paper [1], Nelson and Lax showed that, although the photoelastic effect is independent of the rotation in isotropic media, it depends on the rotation, as well as on the strain (or stress) in all crystal systems except cubic. Previously, in a paper on the continuum theory of elastic dielectrics, Toupin [2] had given a formula, in a brief treatment of photoelasticity, for the deformation-dependent reciprocal dielectric susceptibility for general anisotropy. In the present paper, it is shown that Toupin's formula also leads to the rotation-dependence of the photoelastic effect in all crystal systems except cubic. In addition, it is shown that, if the contribution of the polarization gradient to the stored energy is taken into account, the rotation-dependence is extended to include cubic symmetry. Alternatively, if a non-primitive cubic crystal structure is considered a limited rotation-dependence is found. The treatment is fully linear.

LINEAR VERSION OF TOUPIN'S EQUATIONS OF PHOTOELASTICITY

Maxwell's equations of the electromagnetic field and Cauchy's stress equations of motion are employed in the forms

$$\varepsilon_{ijk}E_{k,j} + \dot{B}_i = 0, \tag{1}$$

$$\mu_0^{-1}\,\varepsilon_{ijk}B_{k,j} = \varepsilon_0\dot{E}_i + \dot{P}_i, \tag{2}$$

$$B_{i,i} = 0, \tag{3}$$

$$\varepsilon_0 E_{i,i} + P_{i,i} = 0, \tag{4}$$

$$T_{ij,i} = \rho\,\ddot{u}_j, \tag{5}$$

where $E_i, B_i, P_i, \varepsilon_0, \mu_0, \varepsilon_{ijk}, T_{ij}, u_j$ and ρ are respectively, the Maxwell electric self-field, magnetic flux density, electronic polarization density, dielectric permittivity and magnetic permeability of a vacuum, unit alternating tensor, stress, mechanical displacement and mass density.

To the Maxwell and Cauchy equations are adjoined Toupin's equation of equilibrium of electrical forces (omitting the external field, here) :

$$E_i^L + E_i = 0 \tag{6}$$

and the constitutive equations

$$\begin{aligned} -E_j^L &= \hat{a}_{jk}P_k + f_{jk\ell}S_{k\ell}\,, \\ T_{ij} &= f_{kij}P_k + c^P_{ijk\ell}S_{k\ell}\,, \end{aligned} \tag{7}$$

where E_i^L is the effective local electric field and S_{ij} is the strain:

$$S_{ij} = u_{(j,i)} = \frac{1}{2}(u_{j,i} + u_{i,j}). \tag{8}$$

In the constitutive equations, f_{ijk} is a piezoelectric stress-constant, $c^{P}_{ijk\ell}$ is the elastic stiffness at constant polarization and $\varepsilon_0\hat{a}_{ij}$ is the deformation-dependent reciprocal dielectric susceptibility. When fully linearized, Toupin's formula [2, (6.20)] for $\hat{a}_{ij}$ (his χ^{-1}_{ij}) is

$$\hat{a}_{ij}=a_{k\ell}(\delta_{ik}-\hat{u}_{k,i})(\delta_{j\ell}-\hat{u}_{\ell,j})\simeq a_{ij}-a_{ik}\hat{u}_{k,j}-a_{jk}\hat{u}_{k,i} \tag{9}$$

where $\varepsilon_0 a_{ij}$ is the ordinary reciprocal dielectric susceptibility $\hat{u}_{j,i}$ is the initial (small) displacement gradient and δ_{ij} is the Kronecker delta.

The formula (9) may be written as

$$\hat{a}_{ij} = a_{ij} - a_{ijk\ell}\,\hat{u}_{\ell,k}\,, \tag{10}$$

where

$$a_{ijk\ell} = a_{i\ell}\,\delta_{jk} + a_{j\ell}\,\delta_{ik}. \tag{11}$$

It will be observed that $a_{ijk\ell}$ is symmetric only in its first two indices, so that an alternative form for (10) is

$$\hat{a}_{ij}=a_{ij}-a_{ij(k\ell)}\hat{S}_{k\ell}-a_{ij[k\ell]}\hat{\omega}_{k\ell}\,, \tag{12}$$

where $\hat{S}_{k\ell}$ and $\hat{\omega}_{k\ell}$ are the initial strain and rotation, respectively :

$$\hat{S}_{k\ell}=\hat{u}_{(\ell,k)}=\frac{1}{2}(\hat{u}_{\ell,k}+\hat{u}_{k,\ell}),\quad \hat{\omega}_{k\ell}=\hat{u}_{[k\ell]}=\frac{1}{2}(\hat{u}_{\ell,k}-\hat{u}_{k,\ell}) \tag{13}$$

and $a_{ij(k\ell)}$ and $a_{ij[k\ell]}$ are the parts of $a_{ijk\ell}$ symmetric and antisymmetric, respectively, in its last two indices :

$$a_{ij(k\ell)}=\frac{1}{2}(a_{i\ell}\delta_{jk}+a_{j\ell}\delta_{ik}+a_{ik}\delta_{j\ell}+a_{jk}\delta_{i\ell})\,, \tag{14}$$

$$a_{ij[k\ell]}=\frac{1}{2}(a_{i\ell}\delta_{jk}+a_{j\ell}\delta_{ik}-a_{ik}\delta_{j\ell}-a_{jk}\delta_{i\ell}). \tag{15}$$

The antisymmetric tensor $\varepsilon_0 a_{ij[k\ell]}$, expressed in terms of the reciprocal susceptibility $\varepsilon_0 a_{ij}$, controls the rotation-dependence of the photoelastic effect and has the same form as Nelson and Lax's expression [1, (4.20)] for their "antisymmetric susceptibility" in terms of the dielectric tensor. From (14) and (15), the matrices of the $a_{ij(k\ell)}$ and $a_{ij[k\ell]}$ are as given in Tables I and II. The matrices of the a_{ij} for the seven crystal systems and isotropy are given in Table III. Tables II and III produce the same results as Nelson and Lax's Table II. Note that $a_{ij[k\ell]}$ vanishes for cubic and isotropic symmetries, but for none of the others ; so that, as found by Nelson and Lax, cubic and isotropic material symmetries are the only ones for which the photoelastic effect is independent of the rotation- at least within the framework of equations (1) - (9).

EQUATIONS OF MOTION AND DISPERSION RELATION

To obtain the equation of motion and dispersion relation in a classical form, first eliminate B_i Maxwell's equations by subtracting the curl of (1) from the time derivative of (2), with the result :

$$E_{j,ii} - E_{i,ij} = \varepsilon_0 \mu_0 \ddot{E}_j + \mu_0 \ddot{P}_j \; . \tag{16}$$

Then solve (6) and the first of (7) for P_i in terms of E_i and S_{ij} :

$$P_i = e_{ijk} S_{jk} + \varepsilon_0 \hat{\eta}_{ij} E_j \tag{17}$$

where

$$e_{ijk} = -\varepsilon_0 \hat{\eta}_{i\ell} f_{\ell jk} \; , \quad \varepsilon_0 \hat{\eta}_{ij} \hat{a}_{jk} = \delta_{ik} \; , \tag{18}$$

i.e. e_{ijk} is the usual piezoelectric stress-constant and $\hat{\eta}_{ij}$ is the deformation-dependent dielectric susceptibility. For initial displacement gradients small in comparison with unity,

$$\hat{\eta}_{ij} = \eta_{ij} + \eta_{ik} u_{j,k} + \eta_{jk} u_{i,k} = \eta_{ij} + \eta_{ij(k\ell)} \hat{S}_{k\ell} + \eta_{ij[k\ell]} \hat{\omega}_{k\ell} \; , \tag{19}$$

where η_{ij} is the ordinary dielectric susceptibility and

$$\eta_{ij(k\ell)} = \frac{1}{2}(\eta_{i\ell}\delta_{jk}+\eta_{j\ell}\delta_{ik}+\eta_{ik}\delta_{j\ell}+\eta_{jk}\delta_{i\ell}) \,, \tag{20}$$

$$-\eta_{ij[k\ell]} = \frac{1}{2}(\eta_{i\ell}\delta_{jk}+\eta_{j\ell}\delta_{ik}-\eta_{ik}\delta_{j\ell}-\eta_{jk}\delta_{i\ell}) \,; \tag{21}$$

i.e. $\eta_{ij(k\ell)}$ has the same form as $a_{ij(k\ell)}$ and $\eta_{ij[k\ell]}$ has the same form as the negative of $a_{ij[k\ell]}$, so that the matrices displayed in Tables I and II hold for η as well as a.

The dielectric constant is $\mathcal{K}_{ij} = \delta_{ij} + \eta_{ij}$ and we define a deformation-dependent dielectric constant, from (19) :

$$\hat{\mathcal{K}}_{ij}=\mathcal{K}_{ij}+\eta_{ij(k\ell)}\hat{S}_{k\ell}+\eta_{ij[k\ell]}\hat{\omega}_{k\ell} \,. \tag{22}$$

Then, upon substituting (17) in (16) and employing (8) and (22), we find

$$E_{j,ii}-E_{i,ij}=c_0^{-2}\hat{\mathcal{K}}_{jk}\ddot{E}_k+\mu_0 e_{jk\ell}\ddot{u}_{\ell,k} \tag{23}$$

where c_0 , $= (\varepsilon_0\mu_0)^{-1/2}$, is the velocity of light in a vacuum.

For the stress-equations of motion, we first write T_{ij} in the alternative form

$$T_{ij} = c^E_{ijk\ell}S_{k\ell}-e_{kij}E_k \,, \tag{24}$$

where

$$c^E_{ijk\ell}=c^P_{ijk\ell}-\varepsilon_0\eta_{mn}f_{mij}f_{nk\ell}=c^P_{ijk\ell}-a_{mn}e_{mij}e_{nk\ell} \tag{25}$$

and then find, from (24), (8) and (5), the displacement-equation of motion

$$c^E_{ijk\ell}u_{\ell,ki} - e_{kij}E_{k,i} = \rho\,\ddot{u}_j \,. \tag{26}$$

Except for the replacement of $\mathcal{K}_{ij}$ by $\hat{\mathcal{K}}_{ij}$, (23) and (26) are the classical equations of motion in terms of the Maxwell electric self-field, E_i , and the mechanical displacement u_i - coupled by the piezoelectric effect through the constant e_{ijk} .

Inserting

$$E_i = A_i e^{i\zeta(n_j x_j - vt)} , \quad u_i = B_i e^{i\zeta(n_j x_j - vt)} \tag{27}$$

in (23) and (26) and eliminating A_i and B_i, we find the determinantal dispersion relation

$$| \delta_{jk} - n_j n_k - c_0^{-2} v^2 \hat{\mathcal{K}}_{jk} - \psi_{jk} | = 0 , \tag{28}$$

where

$$\psi_{jk} = \mu_0 \rho^{-1} n_p n_i e_{jp\ell} e_{kiq} (\delta_{\ell q} - n_m n_n c^E_{m\ell nq} / \rho v^2)^{-1} . \tag{29}$$

The complete dispersion relation (28) includes the electromagnetic and acoustic branches. It reduces to a simple form for the electromagnetic branches alone if e_{ijk} is zero, so that ψ_{jk} is zero. Then, if we let $\hat{\mathcal{K}}_1$, $\hat{\mathcal{K}}_2$, $\hat{\mathcal{K}}_3$ be the principal values of $\hat{\mathcal{K}}_{ij}$ and ℓ, m, n be the direction cosines of the wave normal referred to the principal directions of $\hat{\mathcal{K}}_{ij}$ (not necessarily the principal direction of $\mathcal{K}_{ij}$) and define principal velocities a, b, c according to

$$a^2 = c_0^2/\hat{\mathcal{K}}_1 , \quad b^2 = c_0^2/\hat{\mathcal{K}}_2 , \quad c^2 = c_0^2/\hat{\mathcal{K}}_3 \tag{30}$$

we find, after discarding a root $v^2 = 0$,

$$v^4 - v^2[\ell^2(b^2+c^2) + m^2(c^2+a^2) + n^2(a^2+b^2)] + \ell^2 b^2 c^2 + m^2 c^2 a^2 + n^2 a^2 b^2 = 0 \tag{31}$$

This is the same equation as that given by Coker and Filon [3, p.16] for optical wave velocities in a crystal except that, here, the strain and rotation-dependence of the coefficients is specified by (22) and (30).

POLARIZATION GRADIENT

If the contribution of the polarization gradient to the energy of deformation and polarization is taken into account [4], Toupin's equation of equilibrium of electrical forces (6) becomes

$$E_j^L + E_{ij,i} + E_j = 0 \tag{32}$$

and E_j^L, E_{ij} and T_{ij} are given by the constitutive equations [4]

$$\begin{aligned} -E_j^L &= \hat{a}_{jk} P_k + j_{jk\ell} P_{\ell,k} + f_{jk\ell} S_{k\ell} \,, \\ E_{ij} &= j_{kij} P_k + \hat{b}_{ijk\ell} P_{\ell,k} + d_{ijk\ell} S_{k\ell} \,, \\ T_{ij} &= f_{kij} P_k + d_{k\ell ij} P_{\ell,k} + c^P_{ijk\ell} S_{k\ell} \,. \end{aligned} \tag{33}$$

Of interest, here, are the two deformation-dependent coefficients $\hat{a}_{ij}$ and $\hat{b}_{ijk\ell}$. The former is again given by (9) or (12) whereas the form of the latter analogous to (9) is

$$\begin{aligned} \hat{b}_{ijk\ell} &= b_{mnpq}(\delta_{im} - \hat{u}_{m,i})(\delta_{jn} - \hat{u}_{n,j})(\delta_{kp} - \hat{u}_{p,k})(\delta_{\ell q} - \hat{u}_{q,\ell}) \\ &\simeq b_{ijk\ell} - b_{ijkm}\hat{u}_{m,\ell} - b_{ijm\ell}\hat{u}_{m,k} - b_{imk\ell}\hat{u}_{m,j} - b_{mjk\ell}\hat{u}_{m,i} \end{aligned} \tag{34}$$

where $b_{ijk\ell}$ is independent of the deformation ; and analogous to (I2) :

$$\hat{b}_{ijk\ell} = b_{ijk\ell} - b_{ijk\ell(mn)}\hat{S}_{mn} - b_{ijk\ell[mn]}\hat{\omega}_{mn} \tag{35}$$

where

$$\begin{aligned} b_{ijk\ell(mn)} = \tfrac{1}{2}(&b_{ijkn}\delta_{\ell m} + b_{ijn\ell}\delta_{km} + b_{ink\ell}\delta_{jm} + b_{njk\ell}\delta_{im} \\ &+ b_{ijkm}\delta_{\ell n} + b_{ijm\ell}\delta_{kn} + b_{imk\ell}\delta_{jn} + b_{mjk\ell}\delta_{in}), \end{aligned} \tag{36}$$

$$\begin{aligned} b_{ijk\ell[mn]} = \tfrac{1}{2}(&b_{ijkn}\delta_{\ell m} + b_{ijn\ell}\delta_{km} + b_{ink\ell}\delta_{jm} + b_{njk\ell}\delta_{im} \\ &- b_{ijkm}\delta_{\ell n} - b_{ijm\ell}\delta_{kn} - b_{imk\ell}\delta_{jn} - b_{mjk\ell}\delta_{in}). \end{aligned} \tag{37}$$

Of particular interest is the non-vanishing of the rotation-dependence of $\hat{b}_{ijk\ell}$ in the case of cubic symmetric, for which

$$b_{ijk\ell}=\left(b_{11}-b_{12}-2b_{44}\right)\delta_{ijk\ell}+b_{12}\delta_{ij}\delta_{k\ell}+\left(b_{44}+b_{77}\right)\delta_{ik}\delta_{j\ell} + \left(b_{44}-b_{77}\right)\delta_{i\ell}\delta_{jk} \tag{38}$$

where $b_{11}, b_{12}, b_{44}, b_{77}$ are the two-index notations for b_{1111}, b_{1122}, b_{2323} and b_{3232}, respectively, and $\delta_{ijk\ell}$ is unity if all its indices are alike and zero otherwise. Upon substituting (38) in (37), we find

$$b_{ijk\ell[mn]} = (b_{11}-b_{12}-2b_{44})\ \hat{\omega}_{ijk\ell}\ , \tag{39}$$

where the components of $\hat{\omega}_{ijk\ell}$ are components of rotation $\hat{\omega}_{ij}$ as given in Table IV. Thus, rotation-dependence of the photoelastic effect is extended to cubic symmetry if the contribution of the polarization gradient to the stored energy of deformation and polarization is taken into account.

In the case of isotropy, $b_{11} = b_{12} + 2b_{44}$, so that there is no rotation-dependence of the photoelastic effect and all that remains of $\hat{b}_{ijk\ell}$ is

$$\begin{aligned}\hat{b}_{ijk\ell}&=b_{12}\delta_{ij}\delta_{k\ell}+\left(b_{44}+b_{77}\right)\delta_{ik}\delta_{j\ell}+\left(b_{44}-b_{77}\right)\delta_{i\ell}\delta_{jk}\\ &-2b_{12}\left(\delta_{ij}\hat{S}_{k\ell}+\delta_{k\ell}\hat{S}_{ij}\right)-2\left(b_{44}+b_{77}\right)\left(\delta_{ik}\hat{S}_{j\ell}+\delta_{j\ell}\hat{S}_{ik}\right)\\ &-2\left(b_{44}-b_{77}\right)\left(\delta_{i\ell}\hat{S}_{jk}+\delta_{jk}\hat{S}_{i\ell}\right).\end{aligned} \tag{40}$$

NaCl Structure

Consider, now, the case of a cubic crystal with a non-primitive structure of the NaCl - type, comprising atoms at the lattice points of two, interpenetrating face-centered cubic lattices. In the continuum representation, the displacements and polarizations of the two corresponding continua are identified by $u_i^{(\mathcal{K})}$ and $P_i^{(\mathcal{K})}$, $\mathcal{K} = 1,2$. We need consider, here, only the constitutive equations for the effective local electric fields in the two continua [5] :

$$-E_i^{L(\mathcal{K})} = \sum_{\lambda} \hat{a}_{ij}^{\mathcal{K}\lambda} P_j^{(\lambda)} + a^{\cdot\mathcal{K}} (u_i^{(2)} - u_i^{(1)}) ; \quad \mathcal{K}=1,2; \quad \lambda=1,2 \tag{41}$$

where, for cubic symmetry, $\hat{a}_{ij}^{\mathcal{K}\lambda} = \hat{a}_{11}^{\mathcal{K}\lambda} \delta_{ij}$. The $\varepsilon_0 \hat{a}_{11}^{11}$ and $\varepsilon_0 \hat{a}_{11}^{22}$ are the reciprocal dielectric susceptibilities of the two continua and $\hat{a}_{11}^{12}$, $\hat{a}_{11}^{21}$, is a coefficient of interaction between the two continua. Corresponding to (9), we have

$$\hat{a}_{ij}^{\mathcal{K}\lambda} = a_{pk}^{\mathcal{K}\lambda} \left(\delta_{pi} - \hat{u}_{p,i}^{(\mathcal{K})}\right)\left(\delta_{qj} - \hat{u}_{q,j}^{(\lambda)}\right) \simeq a_{11}^{\mathcal{K}\lambda} \left(\delta_{ij} - \hat{u}_{j,i}^{(\mathcal{K})} - \hat{u}_{i,j}^{(\lambda)}\right) \tag{42}$$

where $\hat{u}_{j,i}^{(\mathcal{K})}$ is the initial displacement gradient in the $\mathcal{K}^{th}$ continuum. Alternatively, (42) may be written as

$$\begin{aligned} \hat{a}_{ij}^{\mathcal{K}\lambda} = a_{11}^{\mathcal{K}\lambda} \delta_{ij} &- \frac{1}{2} a_{11}^{\mathcal{K}\lambda} \left(\delta_{ik}\delta_{j\ell} + \delta_{i\ell}\delta_{jk}\right)\left(\hat{S}_{k\ell}^{(\mathcal{K})} + \hat{S}_{k\ell}^{(\lambda)}\right) \\ &- \frac{1}{2} \left(\delta_{ik}\delta_{j\ell} - \delta_{i\ell}\delta_{jk}\right)\left(\hat{\omega}_{k\ell}^{(\mathcal{K})} - \hat{\omega}_{k\ell}^{(\mathcal{K})}\right) , \end{aligned} \tag{43}$$

where

$$\hat{S}_{ij}^{(\mathcal{K})} = \frac{1}{2}\left(\hat{u}_{j,i}^{(\mathcal{K})} + \hat{u}_{i,j}^{(\mathcal{K})}\right), \quad \hat{\omega}_{ij}^{(\mathcal{K})} = \frac{1}{2}\left(\hat{u}_{j,i}^{(\mathcal{K})} - \hat{u}_{i,j}^{(\mathcal{K})}\right) . \tag{44}$$

It is apparent, from (43), that the reciprocal dielectric susceptibility, $\varepsilon_0 \hat{a}_{ij}^{\mathcal{K}\lambda}$, and hence the photoelastic effect, is independent of the initial rotation within each continuum ($\mathcal{K} = \lambda$) but depends on the relative rotation of the two continua ($\mathcal{K} \neq \lambda$).

Table I. $a_{ij(k\ell)} = a_{ji(k\ell)} = a_{ij(\ell k)}$

$k\ell$ =	11	22	33	23,32	31,13	12,21
ij=11	$2a_{11}$	0	0	0	a_{13}	a_{12}
22	0	$2a_{22}$	0	a_{23}	0	a_{12}
33	0	0	$2a_{33}$	a_{23}	a_{13}	0
23,32	0	a_{23}	a_{23}	$\frac{1}{2}(a_{22}+a_{33})$	$\frac{1}{2}a_{12}$	$\frac{1}{2}a_{13}$
31,13	a_{13}	0	a_{13}	$\frac{1}{2}a_{12}$	$\frac{1}{2}(a_{33}+a_{11})$	$\frac{1}{2}a_{13}$
12,21	a_{12}	a_{12}	0	$\frac{1}{2}a_{13}$	$\frac{1}{2}a_{23}$	$\frac{1}{2}(a_{11}+a_{22})$

Table II. $a_{ij[k\ell]} = a_{ji[k\ell]} = -a_{ij[\ell k]}$

$k\ell$ =	11,22,33	23,-32	31,-13	12,-21
ij=11	0	0	$-a_{13}$	a_{12}
22	0	a_{23}	0	$-a_{12}$
33	0	$-a_{23}$	a_{13}	0
23,32	0	$\frac{1}{2}(a_{33}-a_{22})$	$\frac{1}{2}a_{12}$	$\frac{1}{2}a_{13}$
31,13	0	$-\frac{1}{2}a_{12}$	$\frac{1}{2}(a_{11}-a_{33})$	$\frac{1}{2}a_{13}$
12,21	0	$\frac{1}{2}a_{13}$	$-\frac{1}{2}a_{23}$	$\frac{1}{2}(a_{22}-a_{11})$

Table III. $a_{ij} = a_{ji}$

Triclinic			Monoclinic			Orthorhombic		
a_{11}	a_{12}	a_{13}	a_{11}	0	a_{13}	a_{11}	0	0
a_{12}	a_{22}	a_{23}	0	a_{22}	0	0	a_{22}	0
a_{13}	a_{23}	a_{33}	a_{13}	0	a_{33}	0	0	a_{33}

Tetragonal, Trigonal, Hexagonal			Cubic, Isotropic		
a_{11}	0	0	a_{11}	0	0
	a_{11}			a_{11}	0
0	0	a_{33}	0	0	a_{11}

Table IV $\hat{\omega}_{ijk\ell} = \hat{\omega}_{jik\ell} = \hat{\omega}_{k\ell ij}$

$k\ell$ =	11	22	33	23,32	31,13	12,21
ij = 11	0	0	0	0	$\hat{\omega}_{31}$	$\hat{\omega}_{21}$
22	0	0	0	$\hat{\omega}_{32}$	0	$\hat{\omega}_{12}$
33	0	0	0	$\hat{\omega}_{23}$	$\hat{\omega}_{13}$	0
23,32	0	$\hat{\omega}_{32}$	$\hat{\omega}_{23}$	0	0	0
31,13	$\hat{\omega}_{31}$	0	$\hat{\omega}_{13}$	0	0	0
12,21	$\hat{\omega}_{21}$	$\hat{\omega}_{12}$	0	0	0	0

Addendum

Professor Dill has supplied a concise and elegant proof that the rotation-dependence of the dielectric properties does not affect the birefringence, so that no new physical effect is implied. His result may be illustrated by an example in which the dispersion relation (28) is examined in detail.
Consider the case of a crystal of class hexagonal 6/m or 6/mmm or trigonal $\bar{3}$ or $\bar{3}$ m or tetragonal 4 mm of 4 mmm for which the piezoelectric constants e_{ijk} are zero and the matrix of susceptibilities is

$$\begin{pmatrix} \tau_{11} & 0 & 0 \\ 0 & \tau_{11} & 0 \\ 0 & 0 & \tau_{33} \end{pmatrix} \qquad (45)$$

From (22), for no superposed strain or rotation, $\hat{\mathcal{K}}_{ij} = \mathcal{K}_{ij}$ so that the dispersion relation (28), for waves in the direction of x_1, is

$$\begin{vmatrix} \mathcal{K}_{11} & 0 & 0 \\ 0 & c_o^2 - v^2\mathcal{K}_{11} & 0 \\ 0 & 0 & c_o^2 - v^2\mathcal{K}_{33} \end{vmatrix} = 0 \qquad (46)$$

Now, superpose a rotation $\hat{\omega}_{23} = \theta$ about the axis of x_1.
From (22), the rotation-dependent dielectric constants $\hat{\mathcal{K}}_{ij}$ become

$$\hat{\mathcal{K}}_{11} = \hat{\mathcal{K}}_{22} = \mathcal{K}_{11} \ ; \qquad \hat{\mathcal{K}}_{33} = \mathcal{K}_{33}$$

$$\hat{\mathcal{K}}_{23} = (\tau_{11}-\tau_{33})\ \theta = (\mathcal{K}_{11}-\mathcal{K}_{33})\ \theta\ , \quad \hat{\mathcal{K}}_{31} = \hat{\mathcal{K}}_{12} = 0\ ,$$

so that the dispersion relation (28)

$$\begin{vmatrix} \mathcal{K}_{11} & 0 & 0 \\ 0 & c_o^2 - v^2\mathcal{K}_{11} & v^2(\mathcal{K}_{33}-\mathcal{K}_{11})\theta \\ 0 & v_o^2(\mathcal{K}_{33}-\mathcal{K}_{11})\theta & c_o^2-v^2\mathcal{K}_{33} \end{vmatrix} = 0 \qquad (47)$$

It appears, upon superficial comparison of (46) and (47), that the velocities v are altered by the rotation. However, (47) is simply (46) in terms of dielectric constants, say $\mathcal{K}'_{ij}$ referred to rotated axes :

$$\mathcal{K}'_{ij} = \mathcal{K}_{rs} \ell_{ri} \ell_{sj}$$

For a rotation θ about x_1 :

$$\mathcal{K}'_{11}=\mathcal{K}_{11},\ \mathcal{K}'_{22}=\mathcal{K}_{11}\cos^2\theta+\mathcal{K}_{33}\sin^2\theta,\ \mathcal{K}'_{33}=K_{11}\sin^2\theta+\mathcal{K}_{33}\cos^2\theta,$$

$$\mathcal{K}'_{23}=(\mathcal{K}_{33}-\mathcal{K}_{11})\sin\theta\cos\theta,\ \mathcal{K}'_{31}=\mathcal{K}'_{12}=0$$

To the first degree in θ,

$$\mathcal{K}'_{22}=\mathcal{K}_{11}\ ,\ \mathcal{K}'_{33} = \mathcal{K}_{33}\ ,\ \mathcal{K}'_{23} = (\mathcal{K}_{33}-\mathcal{K}_{11})\ \theta$$

so that (28) becomes (47) when the components of $\mathcal{K}_{ij}$ are referred to the rotated axes. Hence, (46) and (47) yield the same velocities and, accordingly, the rotation-dependence of the dielectric constant produces no change in the birefringence.

REFERENCES

[1] D.F. Nelson and M. Lax,
Theory of the photoelastic interaction
Phy. Rev. B3 2778-2794 (1971)

[2] R.A. Toupin,
A dynamical theory of elastic dielectrics
Int. J. Engng. Sci. 1 101-126 (1963)

[3] E.G.Coker and L.N.G.Filon,
A Treatrise on Photo-Elasticity
Cambridge University Press (1931)

[4] R.D.Mindlin,
Polarization gradient in elastic dielectrics
Int. J. Solids Struct. 4 637-642 (1968)

[5] R.D.Mindlin,
A continuum theory of a diatomic, elastic dielectric
Int. J. Solids Struct. 8 369-383 401-408 (1972)

Int. J. Solids Structures, 1976, Vol. 12, pp. 27–49. Pergamon Press. Printed in Great Britain

LOW FREQUENCY VIBRATIONS OF ELASTIC BARS

R. D. MINDLIN
Department of Civil Engineering, Columbia University New York, NY 10027, U.S.A.

(Received 27 May 1975)

Abstract—The method of expansion of three-dimensional displacements in a double power series of the transverse coordinates is employed to find one-dimensional equations applicable to low frequency vibrations of uniform, elastic, isotropic and anisotropic bars. The axial displacements accompanying torsion are chosen specially for each cross-sectional shape of bar—resulting in the correct, or nearly correct, torsional rigidity. Applications are to bars of elliptic, triangular and rectangular sections, illustrating various independent and coupled extensional, flexural and torsional modes of motion.

1. INTRODUCTION

Poisson's[1] reduction of the three-dimensional equations of linear elasticity to one-dimensional equations of low frequency extensional, flexural and torsional motions of bars, by means of series expansion followed by truncation, has been the subject of many revisions, modernizations, amplifications and extensions to higher frequencies—most recently, for example, by Volterra[2], Medick[3, 4], Bleustein and Stanley[5], Dökmeci[6] and A. E. Green, Naghdi and Wenner[7]. (References to earlier papers may be found in the review articles by W. A. Green[8], Abramson, Plass and Ripperger[9], Miklowitz[10] and Redwood[11]). Before proceeding with developments at high frequencies, there remains a modification that can be made at the low frequency end of the torsional part of the motion and also affects high frequencies—as will be shown subsequently.

In the linear theory, the usual process of truncation of a power series expansion of the displacements, u_i, $i = 1, 2, 3$, involves the retention of all or part of a sequence of early terms; e.g. if the axis of the bar is along x_1,

$$u_i = u_i^{0,0} + x_2 u_i^{1,0} + x_3 u_i^{0,1} + x_2 x_3 u_i^{1,1}, \tag{1.1}$$

where the $u_i^{m,n}$ are functions of x_1 and time, t, only. These terms readily accommodate low frequency extensional and flexural motions since, in those cases, the three-dimensional displacements approach the approximate forms, as the frequency approaches zero and the wave length approaches infinity, so that, at the limit, the extensional and flexural rigidities are exact. However, in the case of torsion, only the bar of elliptic (including circular) section has displacements that can be represented by (1.1), or part of it, at the zero frequency, long wave limit. For all other sections, the axial displacement is more complicated.

Since each section has its peculiar axial displacement in torsion, it is necessary to eschew a restriction to one or a few terms as a universal approximation for all sections if an *ad hoc* correction factor is to be avoided. The series of transverse components of displacement, $u_2^{m,n}$, $u_3^{m,n}$, may be truncated as usual, but an infinite series of axial components can be retained in the derivation of a set of approximate equations applicable to all sections. Then, for each section to which the equations are applied, a select few terms, not necessarily early ones, may be chosen judiciously. As has been shown[12], the selection can be made so that the exact, St. Venant torsional rigidity is obtained for the infinity of sections for which the St. Venant torsion function is a polynomial in the transverse coordinates[13]. For other sections, a close approximation to the St. Venant result may be found; for example, the two term selection $u_1^{1,3}$, $u_1^{3,1}$ produces an error of only 0·1% for the square section.

As usual, the order of the system of one-dimensional differential equations governing the motion is twice the number of components of displacement retained. However, the additional axial components do not contribute to the order of the final system as they are accompanied by an equal number of linear, algebraic equations relating them to the lower order terms. The solution of this system of simultaneous equations serves to eliminate the additional displacements—in effect allows their free development—and leaves, finally, only a single, second order differential equation governing the torsional motion. The scheme is analogous to the elimination of the contour strains in the usual low frequency approximation for extensional vibrations—which is also adopted here: the contour stresses (those in the normal plane) are set equal to zero and the resulting equations are used to eliminate the contour strains from the stress-strain relations; leading, ultimately, to a single, second order differential equation governing the axial displacement.

In addition to the second order differential equations for low frequency torsion and extension, there are two pairs of second order equations for flexure—also with vanishing contour stresses in the low frequency approximation. However, the Bernoulli-Euler equations (the most elementary form) are not good approximations to nearly as high frequencies as are the elementary equations for torsion and extension. Accordingly, the Timoshenko shear deformation terms [14], which do not raise the order of the differential equations, are retained to the end that, in the case of anisotropic bars in which flexure may be coupled with torsion and extension, the various equations are valid up to comparable frequencies.

The rotatory inertia terms which, together with the Timoshenko shear deformation, produce the fundamental, high frequency axial shear modes, are, in general, omitted from the equations of flexure as they introduce complications that are better left to a treatment of high frequency vibrations. In the first place, anisotropy may produce coupling between flexure and extension or torsion and, except possibly for torsion of circular or near circular sections, the approximations for extension and torsion are not valid to frequencies as high as those of the fundamental axial shear modes. Secondly, the shape of the cross-section of the bar may be such that a contour mode (one in which the displacement is in the plane of the normal section) may have a low frequency cut-off lower than that of either or both of the fundamental axial shear modes. An extreme example is that of a strip or blade for which one or even more than one of the contour flexure modes may have a lower frequency than that of even the lower of the fundamental axial shear modes. Since the contour stresses are neglected, in the low frequency approximation, there can be no contour modes and their absence is one of the factors which must be taken into account in deciding whether or not to include one or both of the rotatory inertia terms. Except for certain types of approximation valid only at very long wave lengths [15–17], it is not permissible to omit intermediate branches; and, of course, it is foolish to employ approximate equations which take into account branches higher than those of influence in the frequency range of interest. Medick, in his studies of high frequency extensional vibrations of isotropic bars of rectangular section [3, 4], appears to have been the first properly to take into account such complications.

In Section 2, which follows, there is summarized a now familiar process of obtaining a doubly infinite series of one-dimensional equations of motion, end conditions and stress-strain-displacement relations from the three dimensional equations of linear elasticity. Next, in Section 3, the truncation of the series for the low frequency approximation is effected. The main features of the truncation are: the elimination of the contour modes; the retention of the fundamental axial shear deformations but not, in general, the accompanying rotatory inertias; the retention of an unlimited number of axial displacement terms and the establishment of an equal number of linear algebraic equations to relate them to the lower order displacements. In Section 4, the six stress-equations of motion and boundary conditions are stated for the six surviving dependent variables: three components of displacement and three of rotation. The next three sections contain applications to extensional, torsional and flexural motions of isotropic bars of elliptic, equilateral triangular and rectangular cross-sections as examples, increasing in complexity, of the selection of terms from the infinite series of axial displacements. The calculations of the torsional rigidities have been described before [12], but are summarized here for completeness and ready reference. Section 8 contains a treatment of the bar of elliptic section and the most general crystal anisotropy—resulting in the coupling of all four modes of motion: extension, torsion and two flexures. This is followed, in Section 9, by a detailed application to a free-ended quartz bar, of elliptic section, with the trigonal axis and a digonal axis of elastic symmetry parallel to the principal axes of the ellipse—in which case one of the flexures is coupled with extension and the other with torsion, but the two pairs are not coupled. In Section 10 an analysis by Voigt [18, p. 641] is reviewed to explain why, except for the elliptic section, the absence of the contour modes restricts the allowable anisotropies to those for which the normal section of the bar is a plane of elastic symmetry. Such a bar, of rectangular section, is treated in the final section.

2. EXPANSION IN POWER SERIES

From the linear theory of elasticity we need, for the present purposes, only the stress-strain-displacement relations and a variational equation of motion. The former are

$$T_{ij} = c_{ijkl}S_{kl} \quad \text{or} \quad T_r = c_{rs}S_s, \tag{2.1}$$

$$S_{ij} = \tfrac{1}{2}(u_{j,i} + u_{i,j}), \tag{2.2}$$

in which the T_{ij}, $i,j = 1, 2, 3$ (or T_r, $r = 1 \ldots 6$) are the components of stress; the S_{ij} (or S_s, $s = 1 \ldots 6$) are the components of strain, where

$$(S_1, S_2, S_3) = (S_{11}, S_{22}, S_{33}), \quad (S_4, S_5, S_6) = 2(S_{23}, S_{31}, S_{12});$$

the u_i are the components of displacement; and the c_{ijkl} (or c_{rs}) are the elastic stiffnesses.

The variational equation of motion, for a body occupying a volume V, bounded by a surface S with outward normal $\mathbf{n}$ on which act surface tractions t_j, is [19, p. 115]

$$\int_{t_0}^{t_1} \mathrm{d}t \int_V (T_{ij,i} - \rho\ddot{u}_j)\delta u_j \,\mathrm{d}V + \int_{t_0}^{t_1} \mathrm{d}t \int_S (t_j - n_i T_{ij})\delta u_j \,\mathrm{d}S = 0, \tag{2.3}$$

from which follow the stress-equations of motion,

$$T_{ij,i} = \rho\ddot{u}_j, \tag{2.4}$$

and boundary conditions sufficient, in the absence of singularities and discontinuities, to assure a unique solution of the fifteen equations (2.1), (2.2) and (2.4): specification, at each point P, on S, of one member of each of the three products

$$T_{nn}u_n, \quad T_{ns}u_s, \quad T_{nt}u_t, \tag{2.5}$$

where n, s, t are orthogonal directions at P.

The cross-sections of the uniform bar are positioned with their centroids on the axis of x_1 and their principal axes along x_2 and x_3. The components of displacement are expanded in a double series of powers of the transverse coordinates:

$$u_i = \sum_{m=0}^{\infty}\sum_{n=0}^{\infty} x_2{}^m x_3{}^n u_i{}^{m,n}(x_1, t) \equiv \sum_{m,n=0}^{\infty} x_2{}^m x_3{}^n u_i{}^{m,n}, \tag{2.6}$$

where $x_2{}^m$ and $x_3{}^n$ are the mth and nth powers of x_2 and x_3 while $u_i{}^{m,n}(x_1, t)$ is the amplitude (x_1 and t dependent) of the mnth term of the series. The single summation sign, for the double summation, is employed, in the sequel, to save space.

From (2.2) and (2.6), the components of strain may be expressed as

$$S_{ij} = \sum_{m,n=0}^{\infty} x_2{}^m x_3{}^n S_{ij}^{m,n} \quad \text{or} \quad S_r = \sum_{m,n=0}^{\infty} x_2{}^m x_3{}^n S_r{}^{m,n}, \tag{2.7}$$

where

$$S_{ij}^{m,n} = \tfrac{1}{2}[u_{j,i}^{m,n} + u_{i,j}^{m,n} + (m+1)(\delta_{2i}u_j{}^{m+1,n} + \delta_{2j}u_i{}^{m+1,n}) + (n+1)(\delta_{3i}u_j{}^{m,n+1} + \delta_{3j}u_i{}^{m,n+1})], \tag{2.8}$$

in which δ_{ij} is the Kronecker delta and the relation of the $S_r{}^{m,n}$ to the $S_{ij}^{m,n}$ is the same as that between the S_r and the S_{ij}. Note that, since the $u_i{}^{m,n}$ are independent of x_2 and x_3, the differential quotients $u_{j,i}^{m,n}$ are zero except if $i = 1$.

Upon substituting the expansion (2.6) in the variational equation (2.3), we find, for a bar with ends at $x_1 = \pm l$ and cross-sections with area A and contour C,

$$\sum_{m,n=0}^{\infty}\int_{t_0}^{t_1}\mathrm{d}t\int_{-l}^{l}(T_{1j}^{m,n}-mT_{2j}^{m-1,n}-nT_{3j}^{m,n-1}+F_j^{m,n}-\rho\ddot{U}^{m,n})\delta u_j^{m,n}\,\mathrm{d}x_1$$

$$+\sum_{m,n=0}^{\infty}\int_{t_0}^{t_1}\mathrm{d}t[t_j^{m,n}-T_{1j}^{m,n}]_{-l}^{l}\delta u_j^{m,n}=0, \tag{2.9}$$

where

$$T_{ij}^{m,n}=\int_A T_{ij}x_2{}^m x_3{}^n\,\mathrm{d}A, \tag{2.10}$$

$$F_j^{m,n}=\oint_C t_j x_2{}^m x_3{}^n\,\mathrm{d}s, \tag{2.11}$$

$$t_j^{m,n}=\int_A t_j x_2{}^m x_3{}^n\,\mathrm{d}A, \tag{2.12}$$

$$\ddot{U}_j^{m,n}=\sum_{p,q=0}^{\infty}I^{m+p,n+q}\ddot{u}_j^{p,q}, \tag{2.13}$$

$$I^{m+p,n+q}=\int_A x_2{}^{m+p}x_3{}^{n+q}\,\mathrm{d}A. \tag{2.14}$$

Note that, since x_2 and x_3 are principal axes of the section of the bar,

$$I^{1,0}=I^{0,1}=I^{1,1}=0. \tag{2.15}$$

In the case of a rectangular section $x_2=\pm b$, $x_3=\pm c$,

$$F_j^{m,n}=\int_{-c}^{c}[t_j x_2{}^m x_3{}^n]_{-b}^{b}\,\mathrm{d}x_3+\int_{-b}^{b}[t_j x_2{}^m x_3{}^n]_{-c}^{c}\,\mathrm{d}x_2 \tag{2.16}$$

and, for $m>1$ or $n>1$,

$$I^{m,n}=\begin{cases}4b^{m+1}c^{n+1}/(m+1)(n+1), & m \text{ and } n \text{ even}\\ 0, & m \text{ or } n \text{ odd.}\end{cases} \tag{2.17}$$

From (2.9) follow the stress-equations of motion of order m, n:

$$T_{1j,1}^{m,n}-mT_{2j}^{m-1,n}-nT_{3j}^{m,n-1}+F_j^{m,n}=\rho\ddot{U}^{m,n} \tag{2.18}$$

and the end conditions: one member of each of the products

$$T_{11}^{m,n}u_1{}^{m,n},\quad T_{12}^{m,n}u_2{}^{m,n},\quad T_{13}^{m,n}u_3{}^{m,n} \tag{2.19}$$

on $x_1=\pm l$.

Finally, the constitutive equations,

$$T_r^{m,n}=\sum_{p,q=0}^{\infty}I^{m+p,n+q}c_{rs}S_s^{p,q}, \tag{2.20}$$

are obtained by substituting (2.7) in (2.1) and the result in (2.10).

3. TRUNCATION OF SERIES FOR LOW FREQUENCIES

We begin the truncation by discarding, in the stress-strain relation (2.20), all $S_r^{m,n}$ for $m+n>1$ except those parts of $S_r^{m,n}$ which contain $u_1{}^{m,n}$ (but not $u_{1,1}^{m,n}$); i.e. we retain, at least temporarily, all orders of axial displacement and such displacements and displacement gradients as appear in the zero-order and first order strains: $S_r^{0,0}$, $S_r^{1,0}$, $S_r^{0,1}$. Thus, (2.20) becomes

$$T_r^{m,n}=\sum_{p,q=0}^{p+q<2}I^{m+p,n+q}c_{rs}S_s^{p,q}+\sum_{p+q>1}^{\infty}I^{m+p,n+q}[c_{r6}(p+1)u_1{}^{p+1,q}+c_{r5}(q+1)u_1{}^{p,q+1}]. \tag{3.1}$$

Next, in the expression (3.1) for the case $m = 0$, $n = 0$, i.e.

$$T_r^{0,0} = \sum_{p,q=0}^{p+q<2} I^{p,q} c_{rs} S_s^{p,q} + \sum_{p+q>1}^{\infty} I^{p,q}[c_{r6}(p+1)u_1^{p+1,q} + c_{r5}(q+1)u_1^{p,q+1}], \tag{3.2}$$

we set the zero-order contour stresses equal to zero:

$$T_2^{0,0} = T_3^{0,0} = T_4^{0,0} = 0 \tag{3.3}$$

and use the resulting equations to eliminate $S_2^{0,0}$, $S_3^{0,0}$ and $S_4^{0,0}$ from the zero-order stress-strain relations. By this process, we permit the free development of the zero-order contour strains. To obtain the solution of (3.3) for $T_1^{0,0}$, $T_5^{0,0}$ and $T_6^{0,0}$, we first multiply (3.2) by the compliances s_{rs}, where the matrix of the s_{rs} is the reciprocal of the stiffness matrix:

$$s_{rt} c_{rs} = \delta_{ts}, \quad \mathrm{r, s, t} = 1 \ldots 6, \tag{3.4}$$

to obtain

$$s_{rt} T_r^{0,0} = I^{0,0} S_t^{0,0} + \sum_{p+q>1}^{\infty} I^{p,q}[\delta_{t6}(p+1)u_1^{p+1,q} + \delta_{t5}(q+1)u_1^{p,q+1}]. \tag{3.5}$$

With (3.3), (3.5) become, for $t = 1, 5, 6$,

$$\begin{aligned}
&s_{11} T_1^{0,0} + s_{51} T_5^{0,0} + s_{61} T_6^{0,0} = I^{0,0} S_1^{0,0},\\
&s_{15} T_1^{0,0} + s_{55} T_5^{0,0} + s_{65} T_6^{0,0} = I^{0,0} S_5^{0,0} + \sum_{p+q>1}^{\infty} I^{p,q}(q+1)u_1^{p,q+1},\\
&s_{16} T_1^{0,0} + s_{56} T_5^{0,0} + s_{66} T_6^{0,0} = I^{0,0} S_6^{0,0} + \sum_{p+q>1}^{\infty} I^{p,q}(p+1)u_1^{p+1,q}.
\end{aligned} \tag{3.6}$$

The solution of (3.6) for $T_1^{0,0}$, $T_5^{0,0}$ and $T_6^{0,0}$ is

$$T_\alpha^{0,0} = I^{0,0} c_{\alpha\beta}^{0,0} S_\beta^{0,0} + \sum_{p+q>1}^{\infty} I^{p,q}[c_{\alpha 5}^{0,0}(q+1)u_1^{p,q+1} + c_{\alpha 6}^{0,0}(p+1)u_1^{p+1,q}], \quad \alpha,\beta = 1, 5, 6, \tag{3.7}$$

where

$$c_{\alpha\beta}^{0,0} = A_{\alpha\beta} / |s_{\alpha\beta}| \tag{3.8}$$

in which $|s_{\alpha\beta}|$ is the determinant

$$|s_{\alpha\beta}| = \begin{vmatrix} s_{11} & s_{15} & s_{16} \\ s_{51} & s_{55} & s_{56} \\ s_{61} & s_{65} & s_{66} \end{vmatrix} \tag{3.9}$$

and $A_{\alpha\beta}$ is the cofactor of element $\alpha\beta$ in $|s_{\alpha\beta}|$:

$$\begin{aligned}
&A_{11} = s_{55}s_{66} - s_{56}^2, && A_{15} = s_{56}s_{61} - s_{51}s_{66}, && A_{16} = s_{51}s_{65} - s_{55}s_{61},\\
&A_{51} = A_{15}, && A_{55} = s_{11}s_{66} - s_{16}^2, && A_{56} = s_{15}s_{61} - s_{11}s_{65},\\
&A_{61} = A_{16}, && A_{65} = A_{65}, && A_{66} = s_{11}s_{55} - s_{15}^2.
\end{aligned} \tag{3.10}$$

Note that the final, zero-order stress-strain relations (3.7) contain the axial tensile stress and strain:

$$T_1^{0,0}, \quad S_1^{0,0} = u_{1,1}^{0,0}, \tag{3.11}$$

for extensional motions, and the transverse shear stresses and the accompanying shear strains:

$$T_5^{0,0},\quad S_5^{0,0} = u_{3,1}^{0,0} + u_1^{0,1}, \tag{3.12}$$

$$T_6^{0,0},\quad S_6^{0,0} = u_{2,1}^{0,0} + u_1^{1,0}, \tag{3.13}$$

which participate in flexural motions. The strains $S_5^{0,0}$ and $S_6^{0,0}$ are the ones that are set equal to zero in the Bernoulli–Euler theory but are retained in Timoshenko's theory.

For later use, we record, here, the result, from (3.5), for the zero-order contour-shear strain:

$$S_4^{0,0} = 2S_{23}^{0,0} = (s_{14}T_1^{0,0} + s_{54}T_5^{0,0} + s_{64}T_6^{0,0})/I^{0,0}. \tag{3.14}$$

Proceeding, now, to the first order stress-strain relations, we retain only the flexural stresses and strains:

$$\begin{aligned} &T_1^{1,0},\quad S_1^{1,0} = u_{1,1}^{1,0}, \\ &T_1^{0,1},\quad S_1^{0,1} = u_{1,1}^{0,1}, \end{aligned} \tag{3.15}$$

and the torsional stresses and strains:

$$\begin{aligned} &T_5^{1,0},\quad S_5^{1,0} = u_{3,1}^{1,0} + u_1^{1,1}, \\ &T_6^{0,1},\quad S_6^{0,1} = u_{2,1}^{0,1} + u_1^{1,1}; \end{aligned} \tag{3.16}$$

i.e. we set

$$\begin{aligned} &T_2^{1,0} = T_3^{1,0} = T_4^{1,0} = T_6^{1,0} = 0, \\ &T_2^{0,1} = T_3^{0,1} = T_4^{0,1} = T_5^{0,1} = 0, \end{aligned} \tag{3.17}$$

in (3.1), and use these eight relations to eliminate (i.e. allow the free development of) the corresponding strains. Thus, we have, from (3.1), recalling that $I^{1,0} = I^{0,1} = I^{1,1} = 0$,

$$\begin{aligned} T_r^{1,0} &= I^{2,0}c_{rs}S_s^{1,0} + \sum_{p+q>1}^{\infty} I^{p+1,q}[c_{r6}(p+1)u_1^{p+1,q} + c_{r5}(q+1)u_1^{p,q+1}], \\ T_r^{0,1} &= I^{0,2}c_{rs}S_s^{0,1} + \sum_{p+q>1}^{\infty} I^{p,q+1}[c_{r6}(p+1)u_1^{p+1,q} + c_{r5}(q+1)u_1^{p,q+1}]; \end{aligned}$$

whence, upon multiplying by s_{rt}, we have

$$\begin{aligned} s_{rt}T_r^{1,0} &= I^{2,0}S_t^{1,0} + \sum_{p+q>1}^{\infty} I^{p+1,q}[\delta_{t6}(p+1)u_1^{p+1,q} + \delta_{t5}(q+1)u_1^{p,q+1}], \\ s_{rt}T_r^{0,1} &= I^{0,2}S_t^{0,1} + \sum_{p+q>1}^{\infty} I^{p,q+1}[\delta_{t6}(p+1)u_1^{p+1,q} + \delta_{t5}(q+1)u_1^{p,q+1}]. \end{aligned} \tag{3.18}$$

After expanding the left hand sides of (3.18) and employing (3.17), we can solve the first of (3.18) for $T_1^{1,0}$ and $T_5^{1,0}$ and the second of (3.18) for $T_1^{0,1}$ and $T_6^{0,1}$:

$$\begin{aligned} T_a^{1,0} &= I^{2,0}c_{ab}^{1,0}S_b^{1,0} + \sum_{p+q>1}^{\infty} I^{p+1,q}c_{a5}^{1,0}(q+1)u_1^{p,q+1}, \quad a, b = 1, 5, \\ T_c^{0,1} &= I^{0,2}c_{cd}^{0,1}S_d^{0,1} + \sum_{p+q>1}^{\infty} I^{p,q+1}c_{c6}^{0,1}(p+1)u_1^{p+1,q}, \quad c, d = 1, 6, \end{aligned} \tag{3.19}$$

where $c_{ab}^{1,0} = c_{ba}^{1,0}$, $c_{cd}^{0,1} = c_{dc}^{0,1}$ and

$$(c_{11}^{1,0}, c_{55}^{1,0}, c_{15}^{1,0}) = (s_{55}, s_{11}, -s_{15})/(s_{11}s_{55} - s_{15}^2), \tag{3.20}$$

$$(c_{11}^{0,1}, c_{66}^{0,1}, c_{16}^{0,1}) = (s_{66}, s_{11}, -s_{16})/(s_{11}s_{66} - s_{16}^2). \tag{3.21}$$

Employing the strain-displacement relations in (3.11)–(3.13) and (3.15) and (3.16) and adopting the notations

$$u_5^{m,n} = u_3^{m,n}, \quad u_6^{m,n} = u_2^{m,n}, \tag{3.22}$$

we may write the zero-order and first-order stress-displacement relations, for the low frequency motions, as follows:

$$T_\alpha^{0,0} = I^{0,0} c_{\alpha\beta}^{0,0} u_{\beta,1}^{0,0} + \sum_{p,q=0}^{\infty} I^{p,q}[c_{\alpha 5}^{0,0}(q+1)u_1^{p,q+1} + c_{\alpha 6}^{0,0}(p+1)u_1^{p+1,q}], \quad \alpha,\beta = 1,5,6 \tag{3.23}$$

$$T_a^{1,0} = I^{2,0} c_{ab}^{1,0} u_{b,1}^{1,0} + \sum_{p,q=0}^{\infty} I^{p+1,q} c_{a5}^{1,0}(q+1)u_1^{p,q+1}, a, b = 1,5, \tag{3.24}$$

$$T_c^{0,1} = I^{0,2} c_{cd}^{0,1} u_{d,1}^{0,1} + \sum_{p,q=0}^{\infty} I^{p,q+1} c_{c6}^{0,1}(p+1)u_1^{p+1,q}, \quad c,d = 1,6. \tag{3.25}$$

As the final step in the truncation process, we allow the free development of the higher order axial warping of the sections of the bar, at the long wave, low frequency limit, by setting the force associated with each of the $u_1^{m,n}$, $m+n>1$, equal to zero. From the equations of motion (2.18), with $j = 1$, we omit the axial stress gradients $T_{11,1}^{m,n}$ and the accelerations $\ddot{U}_1^{m,n}$, which vanish at infinite wave length and zero frequency, and we also omit the surface forces $F_1^{m,n}$, with the result:

$$mT_6^{m-1,n} + nT_5^{m,n-1} = 0, \ m+n>1. \tag{3.26}$$

These are the equations which are to be used to eliminate the higher order $u_1^{m,n}$.

The stress-strain-displacement relations, for the stresses in (3.26), are obtained by setting

$$T_1^{m,n} = T_2^{m,n} = T_3^{m,n} = T_4^{m,n} = 0, \ m+n>1, \tag{3.27}$$

in (3.1) and solving for $T_5^{m,n}$ and $T_6^{m,n}$, with the result:

$$T_\xi^{m,n} = \sum_{p,q=0}^{p+q<2} I^{m+p,n+q}\bar{c}_{\xi\eta}S_\eta^{p,q} + \sum_{p+q>1}^{\infty} I^{m+p,n+q}[\bar{c}_{\xi 6}(p+1)u_1^{p+1,q} + \bar{c}_{\xi 5}(q+1)u_1^{p,q+1}], \tag{3.28}$$

where $\xi = 5, 6$; $\eta = 5, 6$ and

$$(\bar{c}_{55}, \bar{c}_{66}, \bar{c}_{56}) = (s_{66}, s_{55}, -s_{56})/(s_{55}s_{66} - s_{56}^2). \tag{3.29}$$

In (3.28), $S_5^{0,0}$, $S_6^{0,0}$, $S_5^{1,0}$ and $S_6^{0,1}$ are given in terms of displacements by (3.12), (3.13) and (3.16), but $S_6^{1,0}$ and $S_5^{0,1}$ have to be calculated from (3.18):

$$\begin{aligned} I^{2,0}S_6^{1,0} &= s_{a6}T_a^{1,0} - \sum_{p+q>1}^{\infty} I^{p+1,q}(p+1)u_1^{p+1,q}, \ a = 1,5, \\ I^{0,2}S_5^{0,1} &= s_{c5}T_c^{0,1} - \sum_{p+q>1}^{\infty} I^{p,q+1}(q+1)u_1^{p,q+1}, \ c = 1,6, \end{aligned} \tag{3.30}$$

where $T_a^{1,0}$ and $T_c^{0,1}$ are given by (3.19).

In the isotropic case,

$$s_{15} = s_{16} = s_{56} = 0, \quad s_{55} = s_{66} = 1/\mu, \quad s_{11} = 1/E, \tag{3.31}$$

where μ is the shear modulus and E is Young's modulus. Then (3.23)–(3.25) become

$$T_1^{0,0} = I^{0,0}Eu_{1,1}^{0,0},$$

$$T_5^{0,0} = I^{0,0}\mu(u_{3,1}^{0,0} + u_1^{0,1}) + \mu \sum_{p+q>1}^{\infty} I^{p,q}(q+1)u_1^{p,q+1}, \quad (3.32)$$

$$T_6^{0,0} = I^{0,0}\mu(u_{2,1}^{0,0} + u_1^{1,0}) + \mu \sum_{p+q>1}^{\infty} I^{p,q}(p+1)u_1^{p+1,q},$$

$$T_1^{1,0} = I^{2,0}Eu_{1,1}^{1,0}, \quad (3.33)$$

$$T_5^{1,0} = I^{2,0}\mu u_{3,1}^{1,0} + \mu \sum_{p,q=0}^{\infty} I^{p+1,q}(q+1)u_1^{p,q+1},$$

$$T_1^{0,1} = I^{0,2}Eu_{1,1}^{0,1}, \quad (3.34)$$

$$T_6^{0,1} = I^{0,2}\mu u_{2,1}^{0,1} + \mu \sum_{p,q=0}^{\infty} I^{p,q+1}(p+1)u_1^{p+1,q},$$

and, from (3.14), $S_4^{0,0} = 0$. Also, (3.28) and (3.30) become

$$T_5^{m,n} = \mu \sum_{p,q=0}^{p+q<2} I^{m+p,n+q}S_5^{p,q} + \mu \sum_{p+q>1}^{\infty} I^{m+p,n+q}(q+1)u_1^{p,q+1},\ m+n>1, \quad (3.35)$$

$$T_6^{m,n} = \mu \sum_{p,q=0}^{p+q<2} I^{m+p,n+q}S_6^{p,q} + \mu \sum_{p+q>1}^{\infty} I^{m+p,n+q}(p+1)u_1^{p+1,q},\ m+n>1$$

and

$$I^{2,0}S_6^{1,0} = -\sum_{p+q>1}^{\infty} I^{p+1,q}(p+1)u_1^{p+1,q} \quad (3.36)$$

$$I^{0,2}S_5^{0,1} = -\sum_{p+q>1}^{\infty} I^{p,q+1}(q+1)u_1^{p,q+1}$$

respectively.

4. STRESS-EQUATIONS OF MOTION AND BOUNDARY CONDITIONS

There is a stress-equation of motion (or equilibrium) for each of the surviving components of displacement and rotation.

For the zero-order displacements $u_1^{0,0}$, $u_2^{0,0}$, $u_3^{0,0}$, we have, from (2.18) with $m=0$, $n=0$, the stress-equations of motion

$$T_{1,1}^{0,0} + F_1^{0,0} = \rho I^{0,0}\ddot{u}_1^{0,0},$$
$$T_{6,1}^{0,0} + F_2^{0,0} = \rho I^{0,0}\ddot{u}_2^{0,0}, \quad (4.1)$$
$$T_{5,1}^{0,0} + F_3^{0,0} = \rho I^{0,0}\ddot{u}_3^{0,0},$$

in which the higher order accelerations have been omitted.

For the rotations $u_1^{1,0}$ and $u_1^{0,1}$ about x_3 and x_2, respectively, we have, from (2.18) with $m=1$, $n=0$ and $m=0$, $n=1$ and $j=1$, the stress-equations of equilibrium

$$T_{1,1}^{1,0} - T_6^{0,0} + F_1^{1,0} = 0, \quad (4.2)$$

$$T_{1,1}^{0,1} - T_5^{0,0} + F_1^{0,1} = 0,$$

in which the rotatory inertias $\rho I^{2,0}\ddot{u}_1^{1,0}$ and $\rho I^{0,2}\ddot{u}_1^{0,1}$ and all the higher order inertias have been omitted.

The rotation about x_1 (the torsional rotation) is defined as

$$\theta = \tfrac{1}{2}(u_3{}^{1,0} - u_2{}^{0,1}) \tag{4.3}$$

(i.e. positive clockwise advancing in the positive direction along x_1) and we note that, since $S_{23}^{0,0} = \tfrac{1}{2}(u_3{}^{1,0} + u_2{}^{0,1})$,

$$u_3{}^{1,0} = S_{23}^{0,0} + \theta, \quad u_2{}^{0,1} = S_{23}^{0,0} - \theta. \tag{4.4}$$

The torsional stress-equation of motion is obtained by subtracting (2.18), with $m = 0$, $n = 1$, $j = 2$, from (2.18), with $m = 1$, $n = 0$, $j = 3$, and neglecting the contour shear acceleration $\ddot{S}_{2,3}^{0,0}$ and all the higher order accelerations. The result is

$$T_{5,1}^{1,0} - T_{6,1}^{0,1} + F_\theta = \rho(I^{2,0} + I^{0,2})\ddot{\theta}, \tag{4.5}$$

in which $F_\theta = F_3{}^{1,0} - F_2{}^{0,1}$.

The torque about the line of centroids of the bar is

$$M_1 = T_5{}^{1,0} - T_6{}^{0,1}, \tag{4.6}$$

positive M_1 producing positive θ. The couples $T_1{}^{1,0}$ and $T_1{}^{0,1}$ are the bending moments sometimes designated by

$$M_3 = -T_1{}^{1,0}, \quad M_2 = T_1{}^{0,1}, \tag{4.7}$$

i.e. positive counter-clockwise about the axes of x_3 and x_2, respectively, as viewed by an observer looking toward the origin.

The equilibrium equations (3.26), one for each $u_1{}^{m,n}$, $m + n > 1$, that is retained, constitute a set of linear algebraic equations which are to be solved for the $u_1{}^{m,n}$ in terms of the six dependent variables: the three displacements $u_1{}^{0,0}$, $u_2{}^{0,0}$, $u_3{}^{0,0}$ and the three rotations $u_1{}^{1,0}$, $u_1{}^{0,1}$, θ. The resulting expressions are then to be substituted in the constitutive equations (3.28), for the anisotropic case, or (3.32)–(3.34) for the isotropic case, to obtain the stresses exclusively in terms of the six dependent variables. This will permit the six stress-equations of motion, (4.1), (4.2) and (4.5), to be expressed solely in terms of the six variables—resulting in a twelfth order system of partial differential equations for the general, anisotropic case.

As for end conditions, all that remains of (2.19) is the prescription, at $x_1 = \pm l$, of one member of each of the six products

$$T_1{}^{0,0}u_1{}^{0,0}, \quad T_6{}^{0,0}u_2{}^{0,0}, \quad T_5{}^{0,0}u_3{}^{0,0}, \quad T_1{}^{1,0}u_1{}^{1,0}, \quad T_1{}^{0,1}u_1{}^{0,1}, \quad (T_5{}^{1,0} - T_6{}^{0,1})\theta. \tag{4.8}$$

There are also six surface-tractions:

$$F_1{}^{0,0}, \quad F_2{}^{0,0}, \quad F_3{}^{0,0}, \quad F_1{}^{1,0}, \quad F_1{}^{0,1}, \quad F_\theta, \tag{4.9}$$

to be prescribed on the cylindrical or prismatic surface of the bar. Alternatively, one or more of the six dependent variables may be specified—as a consequence of which the order of the system of differential equations is reduced, as each of the corresponding equations becomes simply a formula for a surface traction.

5. ISOTROPIC BAR, ELLIPTIC SECTION

In this and the next two articles, the equations of motion are derived for isotropic bars of elliptic, triangular and rectangular sections to illustrate successively more complicated contributions of the terms of the series of axial displacements.

For the ellipse, with semi-principal axes b and c, along x_2 and x_3, respectively, the axial displacement found from the St. Venant theory is[19, p. 317]

$$u_1 = -(b^2 - c^2)\theta_{,1}x_2x_3/(b^2 + c^2). \tag{5.1}$$

Accordingly, from the series of $u_1^{m,n}$, $m+n>1$, we retain only $u_1^{1,1}$. Then (3.26) become

$$T_6^{0,1}+T_5^{1,0}=0 \tag{5.2}$$

and (3.32)–(3.34) reduce to

$$\begin{aligned}
T_1^{0,0} &= I^{0,0}Eu_{1,1}^{0,0}\\
T_6^{0,0} &= I^{0,0}\mu(u_{2,1}^{0,0}+u_1^{1,0}),\\
T_5^{0,0} &= I^{0,0}\mu(u_{3,1}^{0,0}+u_1^{0,1}),\\
T_1^{1,0} &= I^{2,0}Eu_{1,1}^{1,0},\\
T_5^{1,0} &= I^{2,0}\mu(u_{3,1}^{1,0}+u_1^{1,1}) = I^{2,0}\mu(u_1^{1,1}+\theta_{,1}),\\
T_1^{0,1} &= I^{0,2}Eu_{1,1}^{0,1},\\
T_6^{0,1} &= I^{0,2}\mu(u_{2,1}^{0,1}+u_1^{1,1}) = I^{0,2}\mu(u_1^{1,1}-\theta_{,1}).
\end{aligned} \tag{5.3}$$

The second forms for $T_5^{1,0}$ and $T_6^{1,0}$ are obtained from (4.4), in which $S_{23}^{0,0}$, according to (3.14), is zero for the isotropic case. Upon substituting these two expressions in (5.2) and solving for $u_1^{1,1}$, we find

$$u_1^{1,1}=(I^{0,2}-I^{2,0})\theta_{,1}/(I^{0,2}+I^{2,0}), \tag{5.4}$$

from which the torque-twist relation is

$$T_5^{1,0}-T_6^{0,1}=4\mu I^{0,2}I^{2,0}\theta_{,1}/(I^{0,2}+I^{2,0}). \tag{5.5}$$

From (2.14),

$$I^{0,2}=\pi bc^3/4, \quad I^{2,0}=\pi b^3c/4. \tag{5.6}$$

Hence, the torsional rigidity is

$$C=(T_5^{1,0}-T_6^{0,1})/\theta_{,1}=\pi\mu b^3c^3/(b^2+c^2), \tag{5.7}$$

which is the St. Venant value ([19], p. 323).

The equation of torsional motion is, from (5.5) and (4.5),

$$4\mu I^{0,2}I^{2,0}\theta_{,11}+(I^{0,2}+I^{2,0})F_\theta=\rho(I^{0,2}+I^{2,0})^2\ddot{\theta}. \tag{5.8}$$

The equations of extensional and flexural motion are, from (5.3), (4.1) and (4.2):

Extensional,

$$I^{0,0}Eu_{1,11}^{0,0}+F_1^{0,0}=\rho I^{0,0}\ddot{u}_1^{0,0}, \tag{5.9}$$

Flexural (x_1-x_2 plane),

$$\begin{aligned}
I^{0,0}\mu(u_{2,11}^{0,0}+u_{1,1}^{1,0})+F_2^{0,0} &= \rho I^{0,0}\ddot{u}_2^{0,0},\\
I^{2,0}Eu_{1,11}^{1,0}-I^{0,0}\mu(u_{2,1}^{0,0}+u_1^{1,0})+F_1^{1,0} &= 0,
\end{aligned} \tag{5.10}$$

Flexural (x_1-x_3 plane),

$$\begin{aligned}
I^{0,0}\mu(u_{3,11}^{0,0}+u_{1,1}^{0,1})+F_3^{0,0} &= \rho I^{0,0}\ddot{u}_3^{0,0},\\
I^{0,2}Eu_{1,11}^{0,1}-I^{0,0}\mu(u_{3,1}^{0,0}+u_1^{0,1})+F_1^{0,1} &= 0.
\end{aligned} \tag{5.11}$$

There is no coupling between extensional, flexural and torsional motions and the torsional warping of the sections does not affect the extensional and flexural stiffnesses. Equations (5.10) and (5.11) are Timoshenko's beam equations[14] with the rotatory inertia terms and the shear-correction factors omitted. If the eccentricity of the elliptic cross-section is not large enough for the lowest contour-flexure mode to approach one of the fundamental axial shear modes, the rotatory inertia terms $\rho I^{2,0}\ddot{u}_1{}^{1,0}$ and $\rho I^{0,2}\ddot{u}_1{}^{0,1}$ may be added to the right hand sides of the rotational equations in (5.10) and (5.11). Concurrently, the Timoshenko shear-correction factors should be introduced by replacing μ by $\kappa_2{}^2\mu$ in (5.10) and by $\kappa_3{}^2\mu$ in (5.11) where κ_2 and κ_3 are chosen to make the frequencies of the axial shear modes match the corresponding ones of the three-dimensional theory[20]. The ranges of validity of the equations of flexure are thereby raised to frequencies about 20% above those of the axial shear modes. With the rotatory inertia terms omitted, "20% above" is reduced to about "50% below" and a slight improvement can be effected by retaining the shear-correction factor but adjusting it to obtain a match between the surface wave frequencies from the approximate and exact equations[21].

The lower the frequency, the less is the influence of the shear-deformation terms and, hence, the shear-correction factor. At low enough frequencies, say less than 10% of the fundamental axial shear frequency, the shear-deformation terms may be omitted entirely: thereby reducing (5.10) and (5.11) to Bernoulli-Euler equations. This is done by eliminating the shear terms

$$u_{2,1}^{0,0} + u_1{}^{1,0}, \quad u_{3,1}^{0,0} + u_1{}^{0,1}$$

between the deflection and rotation equations, following which the rotations $u_1{}^{1,0}$ and $u_1{}^{0,1}$ are replaced by $-u_{2,1}^{0,0}$ and $-u_{3,1}^{0,0}$, respectively, with the results:

$$I^{2,0}Eu_{2,1111}^{0,0} + \rho I^{0,0}\ddot{u}_2{}^{0,0} = F_2{}^{0,0} + F_{1,1}^{1,0}, \qquad (5.10)'$$

$$I^{0,2}Eu_{3,1111}^{0,0} + \rho I^{0,0}\ddot{u}_3{}^{0,0} = F_3{}^{0,0} + F_{1,1}^{0,1}, \qquad (5.11)'$$

which are the Bernoulli–Euler equations.

6. ISOTROPIC BAR, TRIANGULAR SECTION

If the cross-section of the isotropic bar is an equilateral triangle with sides defined by[19, p. 319]

$$(x_2 - a)(x_2 - x_3\sqrt{3} + 2a)(x_2 + x_3\sqrt{3} + 2a) = 0, \qquad (6.1)$$

the axial displacement in torsion, according to St. Venant, is[19, p. 319]

$$u_1 = (3x_2{}^2x_3 - x_3{}^3)\theta_{,1}/6a. \qquad (6.2)$$

Therefore, of the $u_1{}^{m,n}$, $m + n > 1$, we retain only $u_1{}^{2,1}$ and $u_1{}^{0,3}$; and the equations for calculating them are, from (3.26) with $m = 2$, $n = 1$ and $m = 0$, $n = 3$:

$$2T_6{}^{1,1} + T_5{}^{2,0} = 0, \quad T_5{}^{0,2} = 0. \qquad (6.3)$$

For the stresses, we have, from (3.32)–(3.34),

$$\begin{aligned}
T_1{}^{0,0} &= I^{0,0}Eu_{1,1}^{0,0},\\
T_5{}^{0,0} &= \mu[I^{0,0}(u_{3,1}^{0,0} + u_1{}^{0,1}) + I^{2,0}u_1{}^{2,1} + 3I^{0,2}u_1{}^{0,3}],\\
T_6{}^{0,0} &= \mu I^{0,0}(u_{2,1}^{0,0} + u_1{}^{1,0}),\\
T_1{}^{1,0} &= I^{2,0}Eu_{1,1}^{1,0}, \qquad (6.4)\\
T_5{}^{1,0} &= \mu(I^{2,0}u_{3,1}^{1,0} + I^{3,0}u_1{}^{2,1} + 3I^{1,2}u_1{}^{0,3}),\\
T_1{}^{0,1} &= I^{0,2}Eu_{1,1}^{0,1},\\
T_6{}^{0,1} &= \mu(I^{0,2}u_{2,1}^{0,1} + 2I^{1,2}u_1{}^{2,1}),
\end{aligned}$$

and, from (3.35), noting that, in the present case, $I^{m,n}=0$ if n is odd,

$$\begin{aligned} T_5^{2,0} &= \mu[I^{2,0}(u_{3,1}^{0,0}+u_1^{0,1})+I^{3,0}u_{3,1}^{1,0}+I^{4,0}u_1^{2,1}+3I^{2,2}u_1^{0,3}],\\ T_5^{0,2} &= \mu[I^{0,2}(u_{3,1}^{0,0}+u_1^{0,1})+I^{1,2}u_{3,1}^{1,0}+I^{2,2}u_1^{2,1}+3I^{0,4}u_1^{0,3}],\\ T_6^{1,1} &= \mu(I^{1,2}u_{2,1}^{0,1}+2I^{2,2}u_1^{2,1}), \end{aligned} \tag{6.5}$$

in which, from (6.1) and (2.14),

$$\begin{aligned} I^{0,0} &= 3\sqrt{3}a^2,\\ I^{0,2} &= I^{2,0} = 3\sqrt{3}a^4/2,\\ I^{1,2} &= -I^{3,0} = 3\sqrt{3}a^5/5,\\ I^{0,4} &= I^{4,0} = 3I^{2,2} = 9\sqrt{3}a^6/5. \end{aligned} \tag{6.6}$$

Upon substituting (6.6) in (6.5) and the result in (6.3), we may solve the latter for $u_1^{2,1}$ and $u_1^{0,3}$:

$$\begin{aligned} u_1^{2,1} &= \theta_{,1}/2a-(u_{3,1}^{0,0}+u_1^{0,1})/4a^2,\\ u_1^{0,3} &= -\theta_{,1}/6a-(u_{3,1}^{0,0}+u_1^{0,1})/4a^2, \end{aligned} \tag{6.7}$$

after noting, from (4.4) and (3.14), that $u_3^{1,0}=-u_2^{0,1}=\theta$. Then, inserting (6.7) in the second, fifth and seventh of (6.4) we find

$$\begin{aligned} T_5^{0,0} &= \frac{1}{2}I^{0,0}\mu(u_{3,1}^{0,0}+u_1^{0,1}),\\ T_5^{1,0}-T_6^{0,1} &= I^{0,4}\mu\theta_{,1}/a^2. \end{aligned} \tag{6.8}$$

The second of (6.8), with (6.6), yields the torsional rigidity:

$$C=(T_5^{1,0}-T_6^{0,1})/\theta_{,1}=9\sqrt{3}\mu a^4/5, \tag{6.9}$$

which is the result from the St. Venant theory [22, p. 266]; and the equation of torsional motion is, from (6.8), (6.6) and (4.5),

$$\mu I^{0,4}\theta_{,11}+a^2F_\theta=2a^2I^{0,2}\rho\ddot{\theta}. \tag{6.10}$$

The equations of extensional and flexural motions are, from (6.4), (6.7), (4.1) and (4.2),

Extensional,

$$I^{0,0}Eu_{1,11}^{0,0}+F_1^{0,0}=\rho I^{0,0}\ddot{u}_1^{0,0}, \tag{6.11}$$

Flexural (x_1-x_2 *plane*),

$$\begin{aligned} I^{0,0}\mu(u_{2,11}^{0,0}+u_{1,1}^{1,0})+F_2^{0,0} &= \rho I^{0,0}\ddot{u}_2^{0,0},\\ I^{2,0}Eu_{1,11}^{1,0}-I^{0,0}\mu(u_{2,1}^{0,0}+u_1^{1,0})+F_1^{1,0} &= 0, \end{aligned} \tag{6.12}$$

Flexural (x_1-x_3 *plane*),

$$\begin{aligned} \frac{1}{2}I^{0,0}\mu(u_{3,11}^{0,0}+u_{1,1}^{0,1})+F_{3,}^{0,0} &= \rho I^{0,0}\ddot{u}_3^{0,0},\\ I^{0,2}Eu_{1,11}^{0,1}-\frac{1}{2}I^{0,0}\mu(u_{3,1}^{0,0}+u_1^{0,1})+F_1^{0,1} &= 0. \end{aligned} \tag{6.13}$$

Inclusion of the axial displacements $u_1^{2,1}$ and $u_1^{0,3}$ does not affect the flexural motion in the $x_1 - x_2$ plane—which is a plane of geometrical symmetry—but the flexural motion in the $x_1 - x_3$ plane (parallel to a face of the triangular prism) *is* affected, as can be seen by the factor 1/2 in the expression for $T_5^{0,0}$ in terms of displacements, in (6.8), and, consequently, in (6.13). This effect diminishes as the frequency approaches zero, so that the two dispersion relations become the same.

If the rotatory inertia terms are included, the two shear-correction factors for flexure may be calculated from the circular frequency

$$\omega = (2\pi/3a)(\mu/3\rho)^{\frac{1}{2}} \tag{6.14}$$

(the same for both modes) obtained from the three-dimensional theory[23]. As a result, the two dispersion relations again become the same.

7. ISOTROPIC BAR, RECTANGULAR SECTION

For the rectangular section, the axial displacement in St. Venant torsion is not expressible as a finite polynomial. As an approximation we retain, of the $u_1^{m,n}$, $m+n>1$, only the four terms

$$u_1^{1,1}, \quad u_1^{1,3}, \quad u_1^{3,1}, \quad u_1^{3,3} \tag{7.1}$$

so that (3.26) become

$$\begin{aligned} T_6^{0,1} + T_5^{1,0} &= 0,\\ T_6^{0,3} + 3T_5^{1,2} &= 0,\\ 3T_6^{2,1} + T_5^{3,0} &= 0,\\ T_6^{2,3} + T_5^{3,2} &= 0; \end{aligned} \tag{7.2}$$

and the stresses, from (3.32)–(3.34) and (3.35) are

$$\begin{aligned} T_1^{0,0} &= I^{0,0}Eu_{1,1}^{0,0},\\ T_5^{0,0} &= I^{0,0}\mu(u_{3,1}^{0,0} + u_1^{0,1}),\\ T_6^{0,0} &= I^{0,0}\mu(u_{2,1}^{0,0} + u_1^{1,0}),\\ T_1^{1,0} &= I^{2,0}Eu_{1,1}^{1,0},\\ T_1^{0,1} &= I^{0,2}Eu_{1,1}^{0,1}, \end{aligned} \tag{7.3}$$

and

$$\begin{aligned} T_6^{0,1} &= \mu(-I^{0,2}\theta_{,1} + I^{0,2}u_1^{1,1} + I^{0,4}u_1^{1,3} + 3I^{2,2}u_1^{3,1} + 3I^{2,4}u_1^{3,3}),\\ T_5^{1,0} &= \mu(I^{2,0}\theta_{,1} + I^{2,0}u_1^{1,1} + 3I^{2,2}u_1^{1,3} + I^{4,0}u_1^{3,1} + 3I^{4,2}u_1^{3,3}),\\ T_6^{0,3} &= \mu(-I^{0,4}\theta_{,1} + I^{0,4}u_1^{1,1} + I^{0,6}u_1^{1,3} + 3I^{2,4}u_1^{3,1} + 3I^{2,6}u_1^{3,3}),\\ T_5^{1,2} &= \mu(I^{2,2}\theta_{,1} + I^{2,2}u_1^{1,1} + 3I^{2,4}u_1^{1,3} + I^{4,2}u_1^{3,1} + 3I^{4,4}u_1^{3,3}),\\ T_6^{2,1} &= \mu(-I^{2,2}\theta_{,1} + I^{2,2}u_1^{1,1} + I^{2,4}u_1^{1,3} + 3I^{4,2}u_1^{3,1} + 3I^{4,4}u_1^{3,3}),\\ T_5^{3,0} &= \mu(I^{4,0}\theta_{,1} + I^{4,0}u_1^{1,1} + 3I^{4,2}u_1^{1,3} + I^{6,0}u_1^{3,1} + 3I^{6,2}u_1^{3,3}),\\ T_6^{2,3} &= \mu(-I^{2,4}\theta_{,1} + I^{2,4}u_1^{1,1} + I^{2,6}u_1^{1,3} + 3I^{4,4}u_1^{3,1} + 3I^{4,6}u_1^{3,3}),\\ T_5^{3,2} &= \mu(I^{4,2}\theta_{,1} + I^{4,2}u_1^{1,1} + 3I^{4,4}u_1^{1,3} + I^{6,2}u_1^{3,1} + 3I^{6,4}u_1^{3,3}), \end{aligned} \tag{7.4}$$

in which the $I^{m,n}$ are calculated from (2.14) or (2.17) for the section $x_2 = \pm b$, $x_3 = \pm c$.

Upon substituting (7.4) in (7.2) and solving for the axial displacements, we find

$$\begin{aligned} u_1^{1,1} &= (k^4-1)(14+375k^2+14k^4)\theta_{,1}/\Delta, \\ u_1^{1,3} &= 35k^2(1-6k^2-7k^4)\theta_{,1}/c^2\Delta, \\ u_1^{3,1} &= 35k^2(7+6k^2-k^4)\theta_{,1}/b^2\Delta, \\ u_1^{3,3} &= 245k^2(k^4-1)\theta_{,1}/3b^2c^2\Delta, \end{aligned} \tag{7.5}$$

where

$$k = c/b, \quad \Delta = 2(k^2+1)^2(7+121k^2+7k^4). \tag{7.6}$$

The torsional rigidity is

$$C = (T_5^{1,0} - T_6^{0,1})/\theta_{,1} = k_1\mu(2b)^3(2c), \tag{7.7}$$

where

$$12k_1 = k^2+1-\frac{(k^2-1)^2(7+212k^2+7k^4)}{(k^2+1)(7+121k^2+7k^4)}+\frac{7k^2(13-25k^2-25k^4+13k^6)}{(k^2+1)^2(7+121k^2+7k^4)}. \tag{7.8}$$

Values of k_1 for various values of k are listed in Table 1 along with the corresponding

Table 1. Torsional rigidity coefficient for rectangular section; k_1: four terms of power series, k_1': St. Venant solution

k = c/b	k_1	k_1'	k = c/b	k_1	k_1'
1.0	0.1407	0.1406	4	0.286	0.281
1.2	0.166	0.166	5	0.298	0.291
1.5	0.196	0.196	10	0.322	0.312
2.0	0.230	0.229	20	0.330	0.323
2.5	0.251	0.249	100	0.333	0.331
3.0	0.266	0.263	∞	1/3	1/3

values [22, p. 277] calculated from St. Venant's solution. It may be seen, from (7.5), that, for the square section ($k = 1$), $u_1^{1,1}$ and $u_1^{3,3}$ are zero and the approximation reduces to

$$u_1^{3,1} = -u_1^{1,3} = 7/18b^2, \quad k_1 = 19/135. \tag{7.9}$$

This simple approximation yields a torsional rigidity differing from St. Venant's result by only about 0·1%. The fact that $u_1^{1,1}$ is zero for the square section accounts for the failure of the Bleustein-Stanley [5] equations for that section, as they used only the term $u_1^{1,1}$ in their approximation for the axial displacement. As k increases from unity, the error in torsional rigidity increases to about 3% around $k = 10$ and then diminishes to zero as k approaches infinity; at which limit $u_1^{1,3}$, $u_1^{3,1}$ and $u_1^{3,3}$, in (7.5), are zero and $u_1^{1,1} = \theta_{,1}$, corresponding to the correct form: $u_1 = x_2x_3\theta_{,1}$ of the St. Venant solution at that limit.

The equation of torsional motion is, from (7.7) and (4.5),

$$16k_1\mu b^3c\theta_{,11} + F_\theta = 2bc(b^2+c^2)\rho\ddot{\theta}. \tag{7.10}$$

The equations of extensional and flexural motion are the same as (5.9), (5.10) and (5.11) for the elliptic section—except that $I^{0,0}$, $I^{0,2}$ and $I^{2,0}$ are given by (2.14).

8. ANISOTROPIC BAR, ELLIPTIC SECTION

The bar treated in this article has the same section as in Section 5, but the material is now anisotropic. According to the St. Venant torsion theory, the axial displacement is

$$\begin{aligned} u_1 = &\left(\frac{s_{15}M_1}{2I^{2,0}}-\frac{s_{11}M_3}{I^{2,0}}\right)x_1x_2-\left(\frac{s_{16}M_1}{2I^{0,2}}-\frac{s_{11}M_2}{I^{0,2}}\right)x_1x_3+\left(\frac{s_{56}M_1}{4I^{2,0}}+\frac{s_{16}M_3}{2I^{2,0}}\right)x_2^2 \\ &-\left(\frac{s_{56}M_1}{4I^{0,2}}-\frac{s_{15}M_2}{2I^{0,2}}\right)x_3^2+\left[\frac{M_1}{4}\left(\frac{s_{55}}{I^{2,0}}-\frac{s_{66}}{I^{0,2}}\right)+\frac{s_{16}M_2}{2I^{0,2}}-\frac{s_{15}M_3}{2I^{2,0}}\right]x_2x_3 \end{aligned} \tag{8.1}$$

and the twist is[18, p. 638]

$$\theta_{,1} = \left(\frac{s_{55}}{I^{2,0}} + \frac{s_{66}}{I^{0,2}}\right)\frac{M_1}{4} - \frac{s_{16}M_2}{2I^{0,2}} - \frac{s_{15}M_3}{2I^{2,0}}, \tag{8.2}$$

where M_1 is the torque and M_2 and M_3 are the bending moments as defined in (4.6) and (4.7). From (8.1), we conclude that we have to retain $u_1^{1,1}$, $u_1^{0,2}$ and $u_1^{2,0}$ in the power series for $u_1^{m,n}$, $m+n>1$, and, hence, the equilibrium equations

$$T_6^{0,1} + T_5^{1,0} = 0, \quad T_6^{1,0} = 0, \quad T_5^{0,1} = 0 \tag{8.3}$$

from (3.26). Now, as may be seen in (3.17) *et seq.*, the last two of (8.3), which provide for the free development of $u_1^{0,2}$ and $u_1^{2,0}$, have already been incorporated in the truncated equations. Hence, we need apply only the first of (8.3) and include only $u_1^{1,1}$ in the series of $u_1^{m,n}$, $m+n>1$, in the expressions (3.23), (3.24) and (3.25), for the stresses, which become:

$$\begin{aligned} T_1^{0,0} &= I^{0,0}(c_{11}^{0,0}u_{1,1}^{0,0} + c_{15}^{0,0}u_{3,1}^{0,0} + c_{16}^{0,0}u_{2,1}^{0,0} + c_{15}^{0,0}u_1^{0,1} + c_{16}^{0,0}u_1^{1,0}),\\ T_5^{0,0} &= I^{0,0}(c_{51}^{0,0}u_{1,1}^{0,0} + c_{55}^{0,0}u_{3,1}^{0,0} + c_{56}^{0,0}u_{2,1}^{0,0} + c_{55}^{0,0}u_1^{0,1} + c_{56}^{0,0}u_1^{1,0}),\\ T_6^{0,0} &= I^{0,0}(c_{61}^{0,0}u_{1,1}^{0,0} + c_{65}^{0,0}u_{3,1}^{0,0} + c_{66}^{0,0}u_{2,1}^{0,0} + c_{65}^{0,0}u_1^{0,0} + c_{66}^{0,0}u_1^{1,0}), \end{aligned} \tag{8.4}$$

$$\begin{aligned} T_1^{1,0} &= I^{2,0}[c_{11}^{1,0}u_{1,1}^{1,0} + c_{15}^{1,0}(u_{3,1}^{1,0} + u_1^{1,1})] = I^{2,0}[c_{11}^{1,0}u_{1,1}^{1,0} + c_{15}^{1,0}(\hat{u}_1^{1,1} + \theta_{,1})],\\ T_5^{1,0} &= I^{2,0}[c_{51}^{1,0}u_{1,1}^{1,0} + c_{55}^{1,0}(u_{3,1}^{1,0} + u_1^{1,1})] = I^{2,0}[c_{51}^{1,0}u_{1,1}^{1,0} + c_{55}^{1,0}(\hat{u}_1^{1,1} + \theta_{,1})], \end{aligned} \tag{8.5}$$

$$\begin{aligned} T_1^{0,1} &= I^{0,2}[c_{11}^{0,1}u_{1,1}^{0,1} + c_{16}^{0,1}(u_{2,1}^{0,1} + u_1^{1,1})] = I^{0,2}[c_{11}^{0,1}u_{1,1}^{0,1} + c_{16}^{0,1}(\hat{u}_1^{1,1} - \theta_{,1})],\\ T_6^{0,1} &= I^{0,2}[c_{61}^{0,1}u_{1,1}^{0,1} + c_{66}^{0,1}(u_{2,1}^{0,1} + u_1^{1,1})] = I^{0,2}[c_{61}^{0,1}u_{1,1}^{0,1} + c_{66}^{0,1}(\hat{u}_1^{1,1} - \theta_{,1})] \end{aligned} \tag{8.6}$$

where $\hat{u}_1^{1,1} = u_1^{1,1} + S_{23}^{0,0}$. Then, substituting the second of (8.5) and (8.6) into the first of (8.3) and solving for $\hat{u}_1^{1,1}$, we find

$$\hat{u}_1^{1,1} = [(I^{0,2}c_{66}^{0,1} - I^{2,0}c_{55}^{1,0})\theta_{,1} - I^{0,2}c_{16}^{0,1}u_{1,1}^{0,1} - I^{2,0}c_{15}^{1,0}u_{1,1}^{1,0}]/(I^{0,2}c_{66}^{0,1} + I^{2,0}c_{55}^{1,0}). \tag{8.7}$$

With this value of $\hat{u}_1^{1,1}$, we have, from (8.5) and (8.6),

$$T_1^{1,0} = \frac{I^{2,0}}{I^{0,2}c_{66}^{0,1} + I^{2,0}c_{55}^{1,0}}\left[2I^{0,2}c_{66}^{0,1}c_{15}^{1,0}\theta_{,1} + \left(\frac{I^{2,0}}{s_{11}s_{55} - s_{15}^2} + I^{0,2}c_{11}^{1,0}c_{66}^{0,1}\right)u_{1,1}^{1,0} - I^{0,2}c_{15}^{1,0}c_{16}^{0,1}u_{1,1}^{0,1}\right], \tag{8.8}$$

$$T_1^{0,1} = \frac{I^{0,2}}{I^{0,2}c_{66}^{0,1} + I^{2,0}c_{55}^{1,0}}\left[-2I^{2,0}c_{55}^{1,0}c_{16}^{0,1}\theta_{,1} + \left(\frac{I^{0,2}}{s_{11}s_{66} - s_{16}^2} + I^{2,0}c_{11}^{0,1}c_{55}^{1,0}\right)u_{1,1}^{0,1} - I^{2,0}c_{15}^{1,0}c_{16}^{0,1}u_{1,1}^{1,0}\right], \tag{8.9}$$

$$T_5^{1,0} - T_6^{0,1} = \frac{2I^{0,2}I^{2,0}(2c_{55}^{1,0}c_{66}^{0,1}\theta_{,1} + c_{15}^{1,0}c_{66}^{0,1}u_{1,1}^{1,0} - c_{16}^{0,1}c_{55}^{1,0}u_{1,1}^{0,1})}{I^{0,2}c_{66}^{0,1} + I^{2,0}c_{55}^{1,0}}, \tag{8.10}$$

where we have used, from (3.20) and (3.21),

$$\begin{aligned} c_{11}^{1,0}c_{55}^{1,0} - (c_{15}^{1,0})^2 &= (s_{11}s_{55} - s_{15}^2)^{-1},\\ c_{11}^{0,1}c_{66}^{0,1} - (c_{16}^{0,1})^2 &= (s_{11}s_{66} - s_{16}^2)^{-1}. \end{aligned}$$

Now, (8.10) is the torque M_1; but, to get the torque-twist relation in a form comparable to (8.2), we must first solve (8.8) and (8.9) for $u_{1,1}^{0,1}$ and $u_{1,1}^{1,0}$ in terms of $\theta_{,1}$, $T_1^{0,1}$ $(=M_2)$ and $T_1^{1,0}$ $(=-M_3)$:

$$u_{1,1}^{0,1} = \left[\frac{2I^{2,0}c_{16}^{0,1}\theta_{,1}}{s_{11}s_{55} - s_{15}^2} + \left(\frac{I^{2,0}}{I^{0,2}(s_{11}s_{55} - s_{15}^2)} + c_{11}^{1,0}c_{66}^{0,1}\right)M_2 - c_{15}^{1,0}c_{16}^{0,1}M_3\right]/\Delta,$$

$$u_{1,1}^{1,0}=\left[-\frac{2I^{0,2}c_{15}^{1,0}\theta_{,1}}{s_{11}s_{66}-s_{16}^{2}}-\left(\frac{I^{0,2}}{I^{2,0}(s_{11}s_{66}-s_{16}^{2})}+c_{11}^{0,1}c_{55}^{1,0}\right)M_3+c_{15}^{1,0}c_{16}^{0,1}M_2\right]\Big/\Delta,$$

$$\Delta=\frac{I^{0,2}c_{11}^{1,0}}{s_{11}s_{66}-s_{16}^{2}}+\frac{I^{2,0}c_{11}^{0,1}}{s_{11}s_{55}-s_{15}^{2}}.$$

Upon substituting these values in (8.10) and solving for $\theta_{,1}$, we find exactly the Voigt result (8.2).

The equations of motion are obtained by substituting (8.4), (8.8)–(8.10) in (4.1), (4.2) and (4.5), with the result

$$a_{11}u_1^{0,0}+a_{12}u_2^{0,0}+a_{13}u_3^{0,0}+a_{14}u_1^{1,0}+a_{15}u_1^{0,1}+a_{16}\theta+F_1^{0,0}=\rho I^{0,0}\ddot{u}_1^{0,0}, \tag{8.11}$$

$$a_{21}u_1^{0,0}+a_{22}u_2^{0,0}+a_{23}u_3^{0,0}+a_{24}u_1^{1,0}+a_{25}u_1^{0,1}+a_{26}\theta+F_2^{0,0}=\rho I^{0,0}\ddot{u}_2^{0,0}, \tag{8.12}$$

$$a_{31}u_1^{0,0}+a_{32}u_2^{0,0}+a_{33}u_3^{0,0}+a_{34}u_1^{1,0}+a_{35}u_1^{0,1}+a_{36}\theta+F_3^{0,0}=\rho I^{0,0}\ddot{u}_3^{0,0}, \tag{8.13}$$

$$a_{41}u_1^{0,0}+a_{42}u_2^{0,0}+a_{43}u_3^{0,0}+a_{44}u_1^{1,0}+a_{45}u_1^{0,1}+a_{46}\theta-F_1^{1,0}=0, \tag{8.14}$$

$$a_{51}u_1^{0,0}+a_{52}u_2^{0,0}+a_{53}u_3^{0,0}+a_{54}u_1^{1,0}+a_{55}u_1^{0,1}+a_{56}\theta-F_1^{0,1}=0, \tag{8.15}$$

$$a_{61}u_1^{0,0}+a_{62}u_2^{0,0}+a_{63}u_3^{0,0}+a_{64}u_1^{1,0}+a_{65}u_1^{0,1}+a_{66}\theta-F_\theta=-\rho(I^{2,0}+I^{0,2})\ddot{\theta}, \tag{8.16}$$

where $a_{pq}=a_{qp}$ and

$$\begin{aligned}
&a_{11}=I^{0,0}c_{11}^{0,0}\partial^2\\
&a_{12}=I^{0,0}c_{16}^{0,0}\partial^2 \quad a_{22}=I^{0,0}c_{66}^{0,0}\partial^2\\
&a_{13}=I^{0,0}c_{15}^{0,0}\partial^2 \quad a_{23}=I^{0,0}c_{56}^{0,0}\partial^2 \quad a_{33}=I^{0,0}c_{55}^{0,0}\partial^2\\
&a_{14}=I^{0,0}c_{16}^{0,0}\partial \quad a_{24}=I^{0,0}c_{66}^{0,0}\partial \quad a_{34}=I^{0,0}c_{56}^{0,0}\partial\\
&a_{15}=I^{0,0}c_{15}^{0,0}\partial \quad a_{25}=I^{0,0}c_{56}^{0,0}\partial \quad a_{35}=I^{0,0}c_{55}^{0,0}\partial\\
&a_{16}=0 \quad a_{26}=0 \quad a_{36}=0\\
&a_{44}=-\alpha I^{2,0}\{I^{2,0}[c_{11}^{1,0}c_{55}^{1,0}-(c_{15}^{1,0})^2]+I^{0,2}c_{11}^{1,0}c_{66}^{0,1}\}\partial^2+I^{0,0}c_{66}^{0,0}\\
&a_{45}=\alpha I^{2,0}I^{0,2}c_{15}^{1,0}c_{16}^{0,1}\partial^2+I^{0,0}c_{56}^{0,0}\\
&a_{46}=-2\alpha I^{2,0}I^{0,2}c_{15}^{1,0}c_{66}^{0,1}\partial^2\\
&a_{55}=-\alpha I^{0,2}\{I^{0,2}[c_{11}^{0,1}c_{66}^{0,1}-(c_{16}^{0,1})^2]+I^{2,0}c_{11}^{0,1}c_{55}^{1,0}\}\partial^2+I^{0,0}c_{55}^{0,0}\\
&a_{56}=2\alpha I^{2,0}I^{0,2}c_{16}^{0,1}c_{55}^{1,0}\partial^2\\
&a_{66}=-4\alpha I^{2,0}I^{0,2}c_{55}^{1,0}c_{66}^{0,1}\partial^2\\
&\alpha=(I^{0,2}c_{66}^{0,1}+I^{2,0}c_{55}^{1,0})^{-1}
\end{aligned}$$

Recalling the definitions of $c_{\alpha\beta}^{0,0}$, $c_{ab}^{1,0}$ and $c_{cd}^{0,1}$ in (3.9), (3.20) and (3.21), it may be seen that all the coupling between modes is solely through the compliances s_{15}, s_{16} and s_{56}. The connections are depicted in Table 2. In general, extension, both flexures and torsion are coupled; although, since $a_{16}=0$, the coupling of torsion with extension is not direct, but through flexure. Torsion (θ) is coupled with the rotational component $u_1^{1,0}$ of flexure in the x_1-x_2 plane through a_{46}, i.e. through s_{15}, and with the rotational component $u_1^{0,1}$ of flexure in the x_1-x_3 plane through a_{56}, i.e. through s_{16}. Extension is coupled with both components, $u_2^{0,0}$ and $u_1^{1,0}$, of flexure in the x_1-x_2 plane through a_{12} and a_{14}, respectively, i.e. through s_{16}; and with both components, $u_3^{0,0}$ and $u_1^{0,1}$, of flexure in the x_1-x_3 plane through a_{13} and a_{15}, respectively, i.e. through s_{15}. The two flexures are coupled through s_{56}.

If s_{56} is not zero, but both s_{15} and s_{16} are zero—as, for example, in the case of monoclinic symmetry with x_1 the digonal axis, for which[24]

$$s_{15}=s_{16}=s_{25}=s_{26}=s_{35}=s_{36}=s_{45}=s_{46}=0 \tag{8.17}$$

Table 2. Coupling coefficients: anisotropic bar, elliptic section

		Extension	Flexure $x_1 - x_2$		Flexure $x_1 - x_3$		Torsion
		$u_1^{0,0}$	$u_2^{0,0}$	$u_1^{1,0}$	$u_3^{0,0}$	$u_1^{0,1}$	θ
Extension	$u_1^{0,0}$	(8.11)	a_{12} s_{16}	a_{14} s_{16}	a_{13} s_{15}	a_{15} s_{15}	
Flexure $x_1 - x_2$	$u_2^{0,0}$	a_{12} s_{16}	(8.12)	a_{24} $c_{66}^{0,0}$	a_{23} s_{56}	a_{25} s_{56}	
	$u_1^{1,0}$	a_{14} s_{16}	a_{24} $c_{66}^{0,0}$	(8.14)	a_{34} s_{56}	a_{45} s_{56}	a_{46} s_{15}
Flexure $x_1 - x_3$	$u_3^{0,0}$	a_{13} s_{15}	a_{23} s_{56}	a_{34} s_{56}	(8.13)	a_{35} $c_{55}^{0,0}$	
	$u_1^{0,1}$	a_{15} s_{15}	a_{25} s_{56}	a_{45} s_{56}	a_{35} $c_{55}^{0,0}$	(8.15)	a_{56} s_{16}
Torsion	θ			a_{46} s_{15}		a_{56} s_{16}	(8.16)

the two flexures are coupled, but torsion and extension are independent. If s_{56} is not zero and either s_{15} or s_{16} is not zero, all four modes of motion are coupled. For example, suppose that s_{56} and s_{15} are not zero, but s_{16} is zero, as is the case of classes 3 and $\bar{3}$ of the trigonal system with x_3 the trigonal axis, in which case[24]

$$\begin{gathered} s_{16} = s_{26} = s_{34} = s_{35} = s_{36} = s_{45} = 0, \\ s_{11} = s_{22}, \quad s_{13} = s_{23}, \quad s_{44} = s_{55}, \quad s_{66} = 2(s_{11} - s_{12}), \\ s_{25} = -s_{15} = \frac{1}{2} s_{46}, \quad s_{14} = -s_{24} = \frac{1}{2} s_{56}. \end{gathered} \tag{8.18}$$

Then, with x_1 along the axis of the bar, extension is coupled, through a_{13} and a_{15}, with $x_1 - x_3$ flexure which is coupled, through a_{23}, a_{25}, a_{43} and a_{45}, with $x_1 - x_2$ flexure which is coupled, through a_{46}, with torsion.

If s_{56} is zero, crystal symmetry conditions require that either or both of s_{15} and s_{16} must be zero and thus a_{23}, a_{25}, a_{43} and a_{45}, which depend on s_{56} and the product of s_{15} and s_{16}, must be zero. Consequently, if $s_{56} = 0$, the coupling between the two flexures is absent. If s_{56} and only one of s_{15} and s_{16} are zero, there is coupling between extension and one of the flexures and between torsion and the other flexure, but no coupling between the two pairs. For example, if it is s_{15} that is zero, as in the case of classes, 4, $\bar{4}$ and $4/m$ of the tetragonal system, for which[24]

$$\begin{gathered} s_{14} = s_{15} = s_{24} = s_{25} = s_{34} = s_{35} = s_{36} = s_{45} = s_{46} = s_{56} = 0 \\ s_{11} = s_{22}, \quad s_{23} = s_{13}, \quad s_{26} = -s_{16}, \quad s_{44} = s_{55} \end{gathered} \tag{8.19}$$

if x_3 is the tetragonal axis, then extension is coupled with flexure in the $x_1 - x_2$ plane and torsion is coupled with flexure in the $x_1 - x_3$ plane.

In all of the foregoing, the axis of the bar is that of x_1. If the axis of the bar is that of x_2 or x_3, subscripts 1, 2, 3 are permuted cyclically and, independently, subscripts 4, 5, 6 are permuted cyclically.

Reduction of the ellipse to a circle does not affect the coupling.

9. QUARTZ BAR, ELLIPTIC SECTION

As an illustration of cyclical permutation, when the axis of the bar is not along x_1, and as an instance of coupling between torsion and one flexure and between extension and the other flexure, but no coupling between the two pairs, consider the case of an alpha-quartz bar of elliptic section with x_3 the trigonal axis, x_1 a digonal axis and x_2 the axis of centroids of the sections. The restrictions on the compliances, for this case of crystallographic class 32 (and also $3m$ and $\bar{3}m$ in the absence of piezoelectric properties) of the trigonal system, are[24]

$$s_{15} = s_{16} = s_{25} = s_{26} = s_{34} = s_{35} = s_{36} = s_{45} = s_{46} = 0,$$
$$s_{11} = s_{22}, \quad s_{23} = s_{31}, \quad s_{24} = -s_{14}, \tag{9.1}$$
$$s_{44} = s_{55}, \quad s_{56} = 2s_{14}, \quad s_{66} = 2(s_{11} - s_{12}).$$

As x_2 is along the axis of the bar, s_{26}, s_{24} and s_{64} replace s_{15}, s_{16} and s_{56}, respectively, in Table 2 and, from (9.1), s_{26} and s_{64} are zero but s_{24} is not. Furthermore, in Table 2, $u_1{}^{0,0}$, $u_2{}^{0,0}$, $u_3{}^{0,0}$, $u_1{}^{1,0}$, $u_1{}^{0,1}$ become $u_2{}^{0,0}$, $u_3{}^{0,0}$, $u_1{}^{0,0}$, $u_2{}^{1,0}$, $u_2{}^{0,1}$, respectively. Thus, the coupling is between extension ($u_2{}^{0,0}$) and flexure ($u_3{}^{0,0}$, $u_2{}^{1,0}$) in the $x_2 - x_3$ plane and between torsion (θ) and flexure ($u_1{}^{0,0}$, $u_2{}^{0,1}$) in the $x_2 - x_1$ plane.

The equations of motion of the "extensional" and "torsional" groups become, respectively,

$$c_{22}^{0,0}\partial^2 u_2{}^{0,0} + c_{24}^{0,0}\partial^2 u_3{}^{0,0} + c_{24}^{0,0}\partial u_2{}^{1,0} + F_2{}^{0,0}/I^{0,0} = \rho\ddot{u}_2{}^{0,0},$$
$$c_{24}^{0,0}\partial^2 u_2{}^{0,0} + c_{44}^{0,0}\partial^2 u_3{}^{0,0} + c_{44}^{0,0}\partial u_2{}^{1,0} + F_3{}^{0,0}/I^{0,0} = \rho\ddot{u}_3{}^{0,0}, \tag{9.2}$$
$$c_{24}^{0,0}\partial u_2{}^{0,0} + c_{44}^{0,0}\partial u_3{}^{0,0} + [c_{44}^{0,0} - c_{22}^{1,0}(I^{2,0}/I^{0,0})\partial^2]u_2{}^{1,0} + F_1{}^{1,0}/I^{0,0} = 0,$$

and

$$c_{66}^{0,0}\partial^2 u_1{}^{0,0} + c_{66}^{0,0}\partial u_2{}^{0,1} + F_1{}^{0,0} = \rho\ddot{u}_1{}^{0,0}$$
$$c_{66}^{0,0}\partial u_1{}^{0,0} - \alpha(I^{0,2}/I^{0,0})\{I^{0,2}[c_{22}^{0,1}c_{44}^{0,1} - (c_{24}^{0,1})^2] + I^{2,0}c_{22}^{0,1}c_{66}^{1,0}\}\partial^2 u_2{}^{0,1},$$
$$+ c_{66}^{0,0}u_2{}^{0,1} + 2\alpha(I^{2,0}I^{0,2}/I^{0,0})c_{24}^{0,1}c_{66}^{1,0}\partial^2\theta - F_2{}^{0,1}/I^{0,0} = 0, \tag{9.3}$$
$$2\alpha I^{2,0}I^{0,2}c_{24}^{0,1}c_{66}^{1,0}\partial^2 u_2{}^{0,1} - 4\alpha I^{2,0}I^{0,2}c_{66}^{1,0}c_{44}^{0,1}\partial^2\theta - F_\theta = -\rho(I^{2,0} + I^{0,2})\ddot{\theta},$$

where $\alpha = (I^{0,2}c_{44}^{0,1} + I^{2,0}c_{66}^{1,0})^{-1}$.

Considering the extensional group (9.2), we set $F_1{}^{0,0}$, $F_2{}^{0,0}$, $F_1{}^{1,0}$ equal to zero and

$$(u_2{}^{0,0}, u_3{}^{0,0}, u_2{}^{1,0}) = (A, B, C)e^{i(\eta x_2 - \omega t)} \tag{9.4}$$

and find

$$(\rho\omega^2 - \eta^2 c_{22}^{0,0})A - \eta^2 c_{24}^{0,0}B + i\eta c_{24}^{0,0}C = 0,$$
$$-\eta^2 c_{24}^{0,0}A + (\rho\omega^2 - \eta^2 c_{44}^{0,0})B + i\eta c_{44}^{0,0}C = 0, \tag{9.5}$$
$$i\eta c_{24}^{0,0}A + i\eta c_{44}^{0,0}B + (c_{44}^{0,0} + c_{22}^{1,0}\eta^2 I^{2,0}/I^{0,0})C = 0,$$

The dispersion relation is obtained by setting the determinant of the coefficients of A, B, C equal to zero, resulting in a biquadratic equation on the frequency:

$$[1 + (s_{22}s_{44} - s_{24}^2)\bar{\eta}^2/s_{22}^2]\Omega^4 - [1 + \bar{\eta}^2(1 + s_{44}/s_{22})]\bar{\eta}^2\Omega^2 + \bar{\eta}^6 = 0 \tag{9.6}$$

or, alternatively, a bicubic on the wave number:

$$\bar{\eta}^6 + (1 + s_{44}/s_{22})\Omega^2\bar{\eta}^4 - [1 - (s_{22}s_{44} - s_{24}^2)\Omega^2/s_{22}^2]\Omega^2\bar{\eta}^2 + \Omega^4 = 0, \tag{9.7}$$

where

$$\Omega^2 = \rho\omega^2 h^2 s_{22}, \quad \bar{\eta} = \eta h, \quad h^2 = I^{2,0}/I^{0,0} = b^2/4$$

and we note that, in the present case of trigonal symmetry,

$$(c_{22}^{0,0}, c_{44}^{0,0}, c_{24}^{0,0}) = (s_{44}, s_{22}, s_{24})/(s_{22}s_{44} - s_{24}^2),$$
$$c_{22}^{1,0} = [c_{22}^{0,0}c_{44}^{0,0} - (c_{24}^{0,0})^2]/c_{44}^{0,0} = 1/s_{22}.$$

The coupling between extension and flexure is through s_{24}. If s_{24} is set equal to zero, (9.6) and (9.7) reduce to

$$(\Omega^2 - \bar{\eta}^2)\,[\Omega^2 - \bar{\eta}^4/(1 + s_{44}\bar{\eta}^2/s_{22})] = 0. \tag{9.8}$$

The first factor in (9.8) gives Bernoulli's result for dispersionless longitudinal waves; the second factor gives the Bernoulli–Euler dispersion relation for low frequency flexural waves with Timoshenko's correction for shear deformation: $s_{44}\bar{\eta}^2/s_{22}$.

In the case of a bar of finite length, the boundary conditions most convenient for reproduction in the laboratory (but, unfortunately, not the simplest mathematically) are those for free ends:

$$T_2^{0,0} = T_4^{0,0} = T_2^{1,0} = 0 \quad \text{on} \quad x_2 = \pm l \tag{9.9}$$

or, again on $x_2 = \pm l$, from (8.4) and (8.5) with one cyclical permutation of subscripts:

$$\begin{aligned} c_{22}^{0,0} u_{2,2}^{0,0} + c_{24}^{0,0} u_{3,2}^{0,0} + c_{24}^{0,0} u_2^{1,0} &= 0,\\ c_{24}^{0,0} u_{2,2}^{0,0} + c_{44}^{0,0} u_{3,2}^{0,0} + c_{44}^{0,0} u_2^{1,0} &= 0,\\ u_{2,2}^{1,0} &= 0. \end{aligned} \tag{9.10}$$

The three conditions (9.10) are such that all three roots $\bar{\eta}^2$, of (9.7), are required for each Ω. Two of these roots are positive and one negative, so that two of the η are real and one is imaginary. For every Ω and each η_i, $i = 1, 2, 3$, the simultaneous equations (9.5) define a set of amplitude ratios A:B:C. For each η_i, let

$$\alpha_{2i} = B/A, \quad \alpha_{3i} = C/A. \tag{9.11}$$

Then (9.4) may be written as

$$\begin{aligned} u_2^{0,0} &= h(A_1 \sin \eta_1 x_2 + A_2 \sin \eta_2 x_2 + A_3 \sinh \eta_3 x_2),\\ u_3^{0,0} &= h(A_1\alpha_{21} \sin \eta_1 x_2 + A_2\alpha_{22} \sin \eta_2 x_2 + A_3\alpha_{23} \sinh \eta_3 x_2),\\ u_2^{1,0} &= A_1\alpha_{31} \cos \eta_1 x_2 + A_2\alpha_{32} \cos \eta_2 x_2 + A_3\alpha_{33} \cosh \eta_3 x_2, \end{aligned} \tag{9.12}$$

with

$$\begin{aligned} \alpha_{2i} &= \frac{s_{22}}{s_{24}}\left(1 \mp \frac{\bar{\eta}_i^2}{\Omega^2}\right),\\ \alpha_{3i} &= \pm\frac{(s_{22}s_{44} - s_{24}^2)\Omega^2}{s_{22}s_{24}\bar{\eta}_i} - \frac{(s_{22} + s_{44})\bar{\eta}_i}{s_{24}} \pm \frac{s_{22}\bar{\eta}_i^3}{s_{24}\Omega^2}, \end{aligned} \tag{9.13}$$

where the upper signs are for $i = 1, 2$ and the lower signs are for $i = 3$.

In (9.12) the extensional motion is symmetric and the flexural motion is antisymmetric with respect to the center of the bar. The converse set would be obtained by interchanging sine and cosine (trigonometric and hyperbolic).

Upon substituting (9.12) in the boundary conditions (9.10), we find

$$\begin{aligned} A_1\beta_{11} \cos \eta_1 l + A_2\beta_{12} \cos \eta_2 l + A_3\beta_{13} \cosh \eta_3 l &= 0,\\ A_1\beta_{21} \cos \eta_1 l + A_2\beta_{22} \cos \eta_2 l + A_3\beta_{23} \cosh \eta_3 l &= 0,\\ A_1\beta_{31} \sin \eta_1 l + A_2\beta_{32} \sin \eta_2 l + A_3\beta_{33} \sinh \eta_3 l &= 0 \end{aligned} \tag{9.14}$$

where

$$\begin{aligned} \beta_{1i} &= [s_{44}\bar{\eta}_i + s_{24}(\bar{\eta}_i\alpha_{2i} + \alpha_{3i})]/(s_{22}s_{44} - s_{24}^2),\\ \beta_{2i} &= [s_{24}\bar{\eta}_i + s_{22}(\bar{\eta}_i\alpha_{2i} + \alpha_{3i})]/(s_{22}s_{44} - s_{24}^2),\\ \beta_{3i} &= \bar{\eta}_i\alpha_{3i}. \end{aligned} \tag{9.15}$$

Finally, the frequency equation is obtained by setting the determinant of the coefficients of the A_i, in (9.14), equal to zero:

$$B_1 \tan \eta_1 l + B_2 \tan \eta_2 l + B_3 \tanh \eta_3 l = 0, \tag{9.16}$$

where

$$\begin{aligned} B_1 &= \beta_{31}(\beta_{12}\beta_{23} - \beta_{22}\beta_{13}), \\ B_2 &= \beta_{32}(\beta_{13}\beta_{21} - \beta_{23}\beta_{11}), \\ B_3 &= \beta_{33}(\beta_{11}\beta_{22} - \beta_{21}\beta_{12}). \end{aligned} \tag{9.17}$$

The solution for the torsional group (9.3) is similar as it involves only coupling between a dispersionless mode (in this case torsion instead of extension) and a flexural mode (in this case in the $x_2 - x_1$ plane instead of the $x_2 - x_3$ plane) including the Timoshenko shear correction. Since, in (9.3),

$$c_{24}^{0,1} = -s_{24}/(s_{22}s_{44} - s_{24}^2),$$

the coupling coefficient again depends on s_{24}.

10. ANISOTROPIC BARS: RESTRICTIONS

There are certain situations in which the contour stresses (T_{22}, T_{33}, T_{23}, in a bar with axis along x_1) do not vanish as the frequency approaches zero; in which cases the equations deduced in Sections 3 and 4 are not valid. The survival of contour stresses in torsional equilibrium may be seen by inspecting Voigt's compatibility equations [18, p. 641] which must be satisfied by a St. Venant torsion function (φ) and the Airy plane strain function (χ) for St. Venant-type torsion of anisotropic bars.

If x_1 is the axis of centroids of the cross-sections of the bar and if the strains are independent of x_1, the two stress functions are defined by

$$\begin{gathered} T_{31} = -\frac{\partial \varphi}{\partial x_2}, \quad T_{12} = \frac{\partial \varphi}{\partial x_3} \\ T_{22} = \frac{\partial^2 \chi}{\partial x_3^2}, \quad T_{33} = \frac{\partial^2 \chi}{\partial x_2^2}, \quad T_{23} = -\frac{\partial^2 x}{\partial x_2 \partial x_3} \end{gathered} \tag{10.1}$$

in consequence of which the equilibrium equations

$$T_{ij,i} = 0 \tag{10.2}$$

are satisfied. The strains, related to the stresses according to

$$S_{ij} = s_{ijkl} T_{kl}, \tag{10.3}$$

must satisfy the compatibility equations

$$\epsilon_{ikm}\epsilon_{jln} S_{ij,kl} = 0 \tag{10.4}$$

where ϵ_{ijk} is the unit alternating tensor. Upon substituting (10.1) in (10.3) and the result in (10.4), we find Voigt's compatibility equations on φ and χ [18, p. 643]:

$$\begin{aligned} D_{11}\varphi + D_{12}\chi &= -2\theta_{,1} - (s_{15}\kappa_2 - s_{16}\kappa_3)/s_{11}, \\ D_{12}\varphi + D_{22}\chi &= 0, \end{aligned} \tag{10.5}$$

where κ_2 and κ_3 are the curvatures in the $x_1 - x_2$ and $x_2 - x_3$ planes and

$$D_{11} = \bar{s}_{55}\frac{\partial^2}{\partial x_2^2} - 2\bar{s}_{56}\frac{\partial^2}{\partial x_2 \partial x_3} + \bar{s}_{66}\frac{\partial^2}{\partial x_3^2},$$

$$D_{22} = \bar{s}_{33}\frac{\partial^4}{\partial x_2^4} - 2\bar{s}_{34}\frac{\partial^4}{\partial x_2^3 \partial x_3} + (2\bar{s}_{23} + \bar{s}_{44})\frac{\partial^4}{\partial x_2^2 \partial x_3^2} - 2\bar{s}_{24}\frac{\partial^4}{\partial x_2 \partial x_3^3} + \bar{s}_{22}\frac{\partial^3}{\partial x_3^4}, \tag{10.6}$$

$$D_{12} = -\bar{s}_{35}\frac{\partial^3}{\partial x_2^3} + (\bar{s}_{45} + \bar{s}_{36})\frac{\partial^3}{\partial x_2^2 \partial x_3} - (\bar{s}_{25} + \bar{s}_{46})\frac{\partial^3}{\partial x_2 \partial x_3^2} + \bar{s}_{26}\frac{\partial^3}{\partial x_3^3},$$

in which

$$\bar{s}_{rs} = s_{rs} - s_{r1}s_{1s}/s_{11}; \quad r,s = 2,3,4,5,6. \tag{10.7}$$

It may be seen that, if $D_{12}\varphi = 0$, χ may be taken to be zero so that there need be no contour stresses accompanying torsion and flexure. This is the case if φ is quadratic in x_2 and x_3, as it is for the elliptic section, or if

$$\bar{s}_{25} + \bar{s}_{46} = \bar{s}_{45} + \bar{s}_{36} = \bar{s}_{35} = \bar{s}_{26} = 0. \tag{10.8}$$

It is unlikely that (10.8) would be satisfied unless at least the line of centroids of the cross-sections is an axis of two-fold symmetry, in which case

$$s_{15} = s_{16} = s_{25} = s_{26} = s_{35} = s_{36} = s_{45} = s_{46} = 0. \tag{10.9}$$

Thus, for the truncated equations to hold for anisotropic bars, the normal section must be either elliptic or a plane of elastic symmetry.

11. ANISOTROPIC BAR, RECTANGULAR SECTION

The rectangular section is bounded by $x_2 = \pm b$, $x_3 = \pm c$, as in Article 7, but now the material is anisotropic with x_1 a digonal axis of symmetry so that (10.9) hold and (3.8), (3.20), (3.21) and (3.29) reduce to

$$\begin{gathered} c_{15}^{0,0} = c_{16}^{0,0} = c_{15}^{1,0} = c_{16}^{0,1} = 0, \\ c_{11}^{0,0} = c_{11}^{1,0} = c_{11}^{0,1} = 1/s_{11}, \quad c_{55}^{1,0} = 1/s_{55}, \quad c_{66}^{0,1} = 1/s_{66}, \\ (c_{55}^{0,0}, c_{66}^{0,0}, c_{56}^{0,0}) = (\bar{c}_{55}, \bar{c}_{66}, \bar{c}_{56}) = (s_{66}, s_{55}, -s_{56})/\Delta_{56}, \\ \Delta_{56} = s_{55}s_{66} - s_{56}^2. \end{gathered} \tag{11.1}$$

Again, from the $u_1^{m,n}$, $m+n>1$, we retain only $u_1^{1,1}$, $u_1^{1,3}$, $u_1^{3,1}$ and $u_1^{3,3}$ along with the corresponding equilibrium equations (7.2). Then, employing (11.1), we find, from (3.7), (3.11), (3.12) and (3.13), the zero order stresses

$$\begin{aligned} T_1^{0,0} &= I^{0,0}u_{1,1}^{0,0}/s_{11}, \\ T_5^{0,0} &= I^{0,0}[s_{66}(u_{3,1}^{0,0} + u_1^{0,1}) - s_{56}(u_{2,1}^{0,0} + u_1^{1,0})]/\Delta_{56}, \\ T_6^{0,0} &= I^{0,0}[s_{55}(u_{2,1}^{0,0} + u_1^{1,0}) - s_{56}(u_{3,1}^{0,0} + u_1^{0,1})]/\Delta_{56}, \end{aligned}$$

and, from (3.19), (3.15) and (3.16), the first order stresses

$$T_1^{0,1} = I^{0,2}u_{1,1}^{0,1}/s_{11}, \quad T_1^{1,0} = I^{2,0}u_{1,1}^{1,0}/s_{11}, \tag{11.3}$$

and

$$\begin{aligned} s_{66}T_6^{0,1} &= -I^{0,2}\theta_{,1} + I^{0,2}\hat{u}_1^{1,1} + I^{0,4}u_1^{1,3} + 3I^{2,2}u_1^{3,1} + 3I^{2,4}u_1^{3,3}, \\ s_{55}T_5^{1,0} &= I^{2,0}\theta_{,1} + I^{2,0}\hat{u}_1^{1,1} + 3I^{2,2}u_1^{1,3} + I^{4,0}u_1^{3,1} + 3I^{4,2}u_1^{3,3}, \end{aligned} \tag{11.4}$$

in which

$$\hat{u}_1^{1,1} = u_1^{1,1} + s_{14}u_{1,1}^{0,0}/s_{11} \tag{11.5}$$

and we have employed (4.4), (3.14) and (3.7).

For the higher order stresses, we find, from (3.28),

$$\begin{aligned}
s_{66}T_6^{0,3} &= I^{0,4}\theta_{,1} + I^{0,4}\hat{u}_1^{1,1} + (s_{55}s_{56}/\Delta_{56})(I^{0,6}u_1^{1,3} + 3I^{2,4}u_1^{3,1} + 3I^{2,6}u_1^{3,3}) - (s_{56}^2 I^{0,4}/\Delta_{56}I^{0,2}) \\
&\quad \times (I^{0,4}u_1^{1,3} + 3I^{2,2}u_1^{3,1} + 3I^{2,4}u_1^{3,3}), \\
s_{55}T_5^{1,2} &= I^{2,2}\theta_{,1} + I^{2,2}\hat{u}_1^{1,1} + (s_{55}s_{56}/\Delta_{56})(3I^{2,4}u_1^{1,3} + I^{4,2}u_1^{3,1} + 3I^{4,4}u_1^{3,3}) - (s_{56}^2 I^{2,2}/\Delta_{56}I^{2,0}) \\
&\quad \times (3I^{2,2}u_1^{1,3} + I^{4,0}u_1^{3,1} + 3I^{4,2}u_1^{3,3}), \\
s_{66}T_6^{2,1} &= -I^{2,2}\theta_{,1} + I^{2,2}\hat{u}_1^{1,1} + (s_{55}s_{66}/\Delta_{56})(I^{2,4}u_1^{1,3} + 3I^{4,2}u_1^{3,1} + 3I^{4,4}u_1^{3,3}) - (s_{56}^2 I^{2,2}/\Delta_{56}I^{0,2}) \\
&\quad \times (I^{0,4}u_1^{1,3} + 3I^{2,2}u_1^{3,1} + 3I^{2,4}u_1^{3,3}), \\
s_{55}T_5^{3,0} &= I^{4,0}\theta_{,1} + I^{4,0}\hat{u}_1^{1,1} + (s_{55}s_{66}/\Delta_{56})(3I^{4,2}u_1^{1,3} + I^{6,0}u_1^{3,1} + 3I^{6,2}u_1^{3,3}) - (s_{56}^2 I^{4,0}/\Delta_{56}I^{2,0}) \\
&\quad \times (3I^{2,2}u_1^{1,3} + I^{4,0}u_1^{3,1} + 3I^{4,2}u_1^{3,3}), \\
s_{66}T_6^{2,3} &= -I^{2,4}\theta_{,1} + I^{2,4}\hat{u}_1^{1,1} + (s_{55}s_{66}/\Delta_{56})(I^{2,6}u_1^{1,3} + 3I^{4,4}u_1^{3,1} + 3I^{4,6}u_1^{3,3}) - (s_{56}^2 I^{2,4}/\Delta_{56}I^{0,2}) \\
&\quad \times (I^{0,4}u_1^{1,3} + 3I^{2,2}u_1^{3,1} + 3I^{2,4}u_1^{3,3}), \\
s_{55}T_5^{3,2} &= I^{4,2}\theta_{,1} + I^{4,2}\hat{u}_1^{1,1} + (s_{55}s_{56}/\Delta_{56})(3I^{4,4}u_1^{1,3} + I^{6,2}u_1^{3,1} + 3I^{6,4}u_1^{3,3}) - (s_{56}^2 I^{4,2}/\Delta_{56}I^{2,0}) \\
&\quad \times (3I^{2,2}u_1^{1,3} + I^{4,0}u_1^{3,1} + 3I^{4,2}u_1^{3,3}).
\end{aligned} \tag{11.6}$$

The equations of extensional and flexural motions are obtained by substituting (11.2) and (11.3) in (4.1) and (4.2) with the results:

Extensional:

$$(I^{0,0}/s_{11})u_{1,11}^{0,0} + F_1^{0,0} = \rho I^{0,0}\ddot{u}_1^{0,0} \tag{11.7}$$

Flexural ($x_1 - x_2$ *plane*):

$$\begin{aligned}
&(I^{0,0}/\Delta_{56})[s_{55}(u_{2,11}^{0,0} + u_{1,1}^{1,0}) - s_{56}(u_{3,11}^{0,0} + u_{1,1}^{0,1})] + F_2^{0,0} = \rho I^{0,0}\ddot{u}_2^{0,0}, \\
&(I^{2,0}/s_{11})u_{1,11}^{1,0} - (I^{0,0}/\Delta_{56})[s_{55}(u_{2,1}^{0,0} + u_1^{1,0}) - s_{56}(u_{3,1}^{0,0} + u_1^{0,1})] + F_1^{1,0} = 0
\end{aligned} \tag{11.8}$$

Flexural ($x_1 - x_3$ *plane*):

$$\begin{aligned}
&(I^{0,0}/\Delta_{56})[s_{66}(u_{3,11}^{0,0} + u_{1,1}^{0,1}) - s_{56}(u_{2,11}^{0,0} + u_{1,1}^{1,0})] + F_3^{0,0} = \rho I^{0,0}\ddot{u}_3^{0,0} \\
&(I^{0,2}/s_{11})u_{1,11}^{0,1} - (I^{0,0}/\Delta_{56})[s_{66}(u_{3,1}^{0,0} + u_1^{0,1}) - s_{56}(u_{2,1}^{0,0} + u_1^{1,0})] + F_1^{0,1} = 0.
\end{aligned} \tag{11.9}$$

Thus, the extensional mode is independent and the two flexural modes are coupled through s_{56}.

To find the appropriate form of the equation of torsional motion, we must first express the torque in terms of the twist. This may be done, as in Section 7, by substituting the stresses (11.4) and (11.6), corresponding to (7.4), in (7.2) and solving for $\hat{u}_1^{1,1}$, $u_1^{1,3}$, $u_1^{3,1}$ and $u_1^{3,3}$ in terms of $\theta_{,1}$, corresponding to (7.5). These solutions may then be inserted in (11.4) from which the torque, $T_5^{1,0} - T_6^{0,1}$, may be expressed in terms of the twist $\theta_{,1}$. As in the case of the St. Venant theory of torsional equilibrium ([19], p. 324), the algebra is simplified in the case $s_{56} = 0$; for then the solutions for $\hat{u}_1^{1,1}$, $u_1^{1,3}$, $u_1^{3,1}$ and $u_1^{3,3}$, in terms of $\theta_{,1}$ become the same as (7.5) except that $u_1^{1,1}$ is replaced by $\hat{u}_1^{1,1}$ and k^2 is replaced by

$$\hat{k} = s_{55}c^2/s_{66}b^2. \tag{11.10}$$

As a result, the expression for the torsional rigidity, analogous to (7.7) is

$$\hat{C} = (T_5^{1,0} - T_6^{0,1})/\theta_{,1} = \hat{k}_1 s_{55}^{-1}(2b)^3(2c), \tag{11.11}$$

where $\hat{k}_1$ is given by (7.8) with k replaced by $\hat{k}$. Thus Table 1 holds for the case $s_{56} = 0$ if k and k_1 are replaced by $\hat{k}$ and $\hat{k}_1$.

Finally, the equation of torsional motion (for the case $s_{56} = 0$) is, from (4.5) and (11.11),

$$16\hat{k}_1 s_{55}^{-1} b^3 c\theta_{,11} + F_\theta = 2bc(b^2 + c^2)\rho\ddot{\theta}. \tag{11.12}$$

It may be observed that the torsional mode is independent of extension (as well as flexure) despite the appearance of $u_1^{0,0}$ in (11.5), as the whole of $\hat{u}_1^{1,1}$ is eliminated from $T_5^{1,0}$ and $T_6^{0,1}$ through the equilibrium equations (7.2). This property holds whether or not $s_{56} = 0$.

REFERENCES

1. S. D. Poisson, Mémoire sur l'équilibre et le mouvement des corps élastiques. *Mém. de l'Acad. Sci., Paris, Ser.* 2, **8**, 357 (1829).
2. E. Volterra, Second approximation of method of internal constraints and its application, *Int. J. Mech. Scis.* **3**, 47 (1961).
3. M. A. Medick, One-dimensional theories of wave propagation and vibrations in elastic bars of rectangular cross section. *J. Appl. Mech.* **33**, *Trans. ASME* **88** Ser. E, 489 (1966).
4. M. A. Medick, On dispersion of longitudinal waves in rectangular bars. *J. Appl. Mech.* **34**, *Trans. ASME* **89**, Ser. E, 714 (1967).
5. J. L. Bleustein and R. M. Stanley, A dynamical theory of torsion. *Int. J. Solids Structures* **6**, 569 (1970).
6. M. C. Dökmeci, A general theory of elastic beams. *Int. J. Solids Structures* **8**, 1205 (1972).
7. A. E. Green, P. M. Naghdi and M. L. Wenner, On the theory of rods—I. Derivation from the three-dimensional equations. *Proc. Roy. Soc. London, Ser. A.* **337**, 451 (1974).
8. W. A. Green, In *Progress in Solid Mechanics* (Edited by R. A. Hill and I. N. Sneddon), Vol. 1, p. 225. North Holland, Amsterdam (1960).
9. H. N. Abramson, H. J. Plass and E. A. Ripperger, In *Advances in Applied Mechanics*, Vol. 5, pp. 111–195. Academic Press, New York (1958).
10. J. Miklowitz, In *Applied Mechanics Surveys.* (Edited by H. N. Abramson, H. Liebowitz, J. M. Crowley and S. Juhasz). pp. 809–839. Spartan Books, Washington, D.C. (1966).
11. M. R. Redwood, *Mechanical Waveguides.* Pergamon Press, New York (1960).
12. R. D. Mindlin, Solution of St. Venant's torsion problem by power series. *Int. J. Solids Structures.* **11**, 321 (1975).
13. B. de St. Venant, Mémoire sur la torsion des prismes, *Mémoires des Savants étrangers* **14**, 233 (1855).
14. S. P. Timoshenko, On the correction for shear of the differential equation for transverse vibrations of prismatic bars, *Phil. Mag. Ser.* 6, **41**, 744 (1921).
15. R. D. Mindlin and M. Forray, Thickness-shear and flexural vibrations of contoured crystal plates. *J. Appl. Phys.* **25**, 12 (1954).
16. R. D. Mindlin, Bechmann's number for harmonic overtones of thickness-twist vibrations of rotated-Y-cut quartz plates. *J. Acoust. Soc. Am.* **41**, 969 (1967).
17. D. W. Haines, Bechmann's number for harmonic overtones of thickness-shear vibrations of rotated-Y-cut quartz plates, *Int. J. Solids Structures* **5**, 1 (1969).
18. W. Voigt, *Lehrbuch der Kristallphysik.* Teubner, Leipzig (1928).
19. A. E. H. Love, *A Treatise on the Mathematical Theory of Elasticity*, 4th Edn. Cambridge University Press (1927).
20. R. D. Mindlin and H. Deresiewicz, Timoshenko's shear coefficient for flexural vibrations of beams. *Proc. 2nd U.S. Nat. Cong. Appl. Mech.* Vol. 1, pp. 175–178 (1954).
21. R. D. Mindlin, Influence of rotatory inertia and shear on flexural motions of isotropic, elastic plates. *J. Appl. Mech.* **18**, 31 (1951).
22. S. P. Timoshenko and J. N. Goodier, *Theory of Elasticity*, 2nd Edn. McGraw-Hill, New York (1951).
23. R. D. Mindlin, Note on axial-shear and contour vibrations prisms (forthcoming).
24. J. F. Nye, *Physical Properties of Crystals.* Oxford University Press (1959).

Int. J. Solids Structures, 1976, Vol. 12, pp. 99–105. Pergamon Press. Printed in Great Britain

NOTE ON AXIAL-SHEAR AND CONTOUR VIBRATIONS OF PRISMS

R. D. MINDLIN
Department of Civil Engineering, Columbia University, New York, NY 10027, U.S.A.

(*Received* 10 *June* 1975)

Abstract—Exact solutions of the equations of the linear theory of elasticity are given for axial-shear modes of vibration of an isotropic, prismatic bar whose normal section is an equilateral triangle or has the equilateral triangle as a module. A family of contour modes is also described for bars with a rhombic section formed of two equilateral triangles and with sections having the rhombus as a module.

INTRODUCTION

Axial-shear and contour modes of vibration of a cylindrical or prismatic bar are those in which the displacements are parallel and perpendicular, respectively, to the generators of the bar and unvarying along its length. The associated frequencies are the long wave limits of certain of the upper branches of the dispersion relation for waves along the bar. Accordingly, exact solutions of the equations of elasticity for such modes are of interest as they supply data for adjusting or testing various approximate equations or solutions. Thus, the high frequency range of Timoshenko's one-dimensional equations of flexural vibrations of bars is improved by employing the exact frequencies of the fundamental axial-shear modes to calculate shear-correction factors[1]. A higher axial-shear mode and a contour mode play similar roles in the Bleustein-Stanley one-dimensional equations of torsional vibrations[2]. Approximate solutions of the three-dimensional equations, for example the finite element solutions by Talbot and Przemieniecki[3], can also profit by comparison of the computed limiting frequencies of the upper extensional, flexural and torsional modes with the exact results for the corresponding axial-shear and contour modes.

Recent applications[3, 4] of approximate methods to the problem of vibrations of an isotropic, elastic prism with an equilateral triangular cross-section motivated a quest for the exact solutions for the axial-shear modes. Those modes are governed by the same differential equation (the two-dimensional wave equation) and boundary conditions (vanishing normal derivative) as are the motions of water waves in a shallow basin or box[5] and the two-dimensional acoustic vibrations of a gas in a cylindrical or prismatic chamber[6]. Both of these phenomena have been subjects of investigation for almost one hundred and fifty years; so it seems unlikely that simple solutions like those for the equilateral triangle should not have been discovered long ago. However, no published solutions appear to be available.

For the special case of a prism with Poisson's ration 1/4 and a rhombic normal section formed of two equilateral triangles, a solution is given for a family of contour modes—modes with displacements normal to the generators of the prism and uniform along its length. Whereas the axial-shear modes are equivoluminal, the contour modes in the rhombic prism are composed of coupled equivoluminal and dilatational waves. The period of the lowest mode of the family was given in a previous paper[7].

The solutions for both the triangular and rhombic sections satisfy traction-free conditions on all planes parallel to the faces of the prisms and regularly spaced at intervals of the altitude of the triangles. Accordingly, a great variety of vibrating prisms with traction-free faces can be formed by juxtaposition of prisms with the triangular or rhombic section. Examples are exhibited.

EQUILATERAL TRIANGULAR SECTION

The simplicity of the solutions for axial-shear vibrations of a prism with an equilateral triangular section stems from the fact that a straight-crested, sinusoidal, shear wave, with its displacement parallel to the generators of the prism and its wave normal parallel or perpendicular to a side of the triangle, reflects from the other sides in such a way that the wave returns upon itself: in the original direction in the case of the "parallel" wave and in the opposite direction in the case of the "perpendicular" wave, as indicated in Fig. 1. The paths of parallel rays, in each case, have the same length for a full circuit and a pair of waves, traveling in opposite directions, forms a steady vibration.

Thus, in a prism with its generators parallel to the axis of z, we are led to consider waves with displacements only axial and of the forms

$$w = A \sin \gamma (y \sin \alpha \pm x \cos \alpha \pm ct \pm \epsilon), \tag{1}$$

where c is the velocity of shear waves, $\epsilon = 0$ or $\pi/2$, and α is the angle the wave normal makes with the axis of x. As the components of displacement, u and v, in the plane of the normal section are zero and the third component is independent of z, the components of stress are

$$T_{xx} = T_{yy} = T_{zz} = T_{xy} = 0, \quad T_{yz} = \mu \partial w/\partial y, \quad T_{zx} = \mu \partial w/\partial x, \tag{2}$$

where μ is the shear modulus. Then the displacement equations of motion, in linear, isotropic elasticity, reduce to

$$c^2(\partial^2 w/\partial x^2 + \partial^2 w/\partial y^2) = \partial^2 w/\partial t^2. \tag{3}$$

The boundary of the normal section is given by the equation

$$x(x - \sqrt{3}y + h)(x + \sqrt{3}y + h) = 0 \tag{4}$$

i.e. an equilateral triangle with altitude h and sides $2h/\sqrt{3}$. From (2) and (4), the conditions for traction-free lateral surfaces of the prism are

$$\begin{aligned} \partial w/\partial x &= 0 \quad \text{on} \quad x = 0, \\ \partial w/\partial x - \sqrt{3}\,\partial w/\partial y &= 0 \quad \text{on} \quad x = \sqrt{3}y - h, \\ \partial w/\partial x + \sqrt{3}\,\partial w/\partial y &= 0 \quad \text{on} \quad x = -\sqrt{3}y - h. \end{aligned} \tag{5}$$

Upon expanding (1) and combining members of the resulting sum for various combinations of the plus and minus signs and values of ϵ and α, we find three independent forms descriptive of waves with wave normals parallel or perpendicular to the sides of the triangle.

For the waves parallel to the sides, $\alpha = \pi/2, \pm\pi/6$ and we have

$$w = (A_1 \cos \gamma y + A_2 \cos \tfrac{1}{2}\sqrt{3}\gamma x \cos \tfrac{1}{2}\gamma y)e^{i\omega t}, \tag{6}$$

$$w = (B_1 \sin \gamma y + B_2 \cos \tfrac{1}{2}\sqrt{3}\gamma x \sin \tfrac{1}{2}\gamma y)e^{i\omega t}. \tag{7}$$

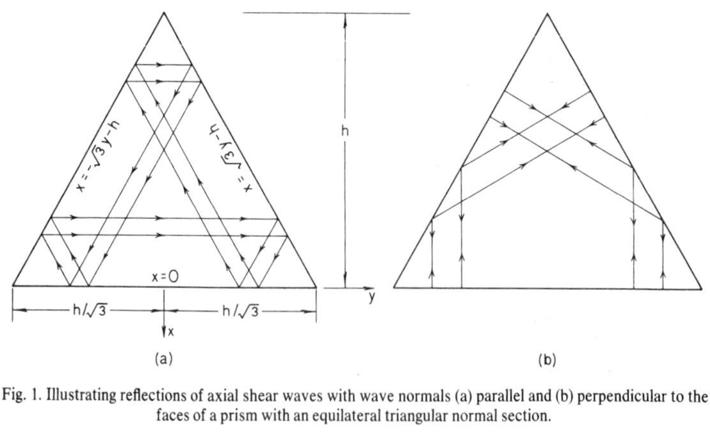

Fig. 1. Illustrating reflections of axial shear waves with wave normals (a) parallel and (b) perpendicular to the faces of a prism with an equilateral triangular normal section.

For the waves perpendicular to the sides, $\alpha = 0, \pm\pi/3$ and we have only

$$w = (C_1 \cos \gamma y + C_2 \cos \tfrac{1}{2}\gamma x \cos \tfrac{1}{2}\sqrt{3}\gamma y)e^{i\omega t} \tag{8}$$

as there can be no companion solution odd in y with straight-crested waves.

Substitution of (6), (7) and (8) in the differential eqn (3) and the boundary conditions (5) yields

Solution A (waves parallel to sides, displacement even in y):

$$\begin{gathered} w = A[\cos \gamma y + 2(-1)^n \cos \tfrac{1}{2}\sqrt{3}\gamma x \cos \tfrac{1}{2}\gamma y]e^{i\omega t}, \\ \omega = \gamma c = 2n\pi c/\sqrt{3}h, \quad n = 1, 2, 3 \ldots \end{gathered} \tag{9}$$

Solution B (waves parallel to sides, displacement odd in y):

$$\begin{gathered} w = B[\sin \gamma y - 2(-1)^n \cos \tfrac{1}{2}\sqrt{3}\gamma x \sin \tfrac{1}{2}\gamma y]e^{i\omega t}, \\ \omega = \gamma c = 2n\pi c/\sqrt{3}h, \quad n = 1, 2, 3 \ldots \end{gathered} \tag{10}$$

Solution C (waves perpendicular to sides):

$$\begin{gathered} w = C[\cos \gamma x + 2(-1)^n \cos \tfrac{1}{2}\gamma x \cos \tfrac{1}{2}\sqrt{3}\gamma y]e^{i\omega t}, \\ \omega = \gamma c = 2n\pi c/h, \quad n = 1, 2, 3 \ldots \end{gathered} \tag{11}$$

The shapes of the first two modes of solutions A and C are illustrated in Figs. 2 and 4. In Fig. 3, for solution B, the third mode is included as it is the lowest axial-shear mode associated with torsional vibrations of the prism; the frequency of the mode $n = 3$ is the exact "warping cut-off frequency" required for the Bleustein-Stanley equations [2].

The frequencies found here do not conform with limiting frequencies of branches shown in Fig. 13 of [3], but the authors agree that the scale of the figure should be reduced by a factor of 2·54/2 (one inch to two centimeters).

SECTIONS WITH EQUILATERAL TRIANGLE AS MODULE

In addition to satisfying the boundary conditions (5), the solutions (9)–(11) satisfy the conditions of zero traction on planes parallel to the lateral faces of the triangular prism and regularly spaced at intervals of the altitude, h, of the triangle, i.e.

$$\begin{aligned} \partial w/\partial x &= 0 \quad \text{on} \quad x = \pm(p-1)h, & p &= 1, 2, 3, \ldots \\ \partial w/\partial x - \sqrt{3}\,\partial w/\partial y &= 0 \quad \text{on} \quad x = \sqrt{3}y \pm (2q-1)h, & q &= 1, 2, 3, \ldots \\ \partial w/\partial x + \sqrt{3}\,\partial w/\partial y &= 0 \quad \text{on} \quad x = -\sqrt{3}y \pm (2r-1)h, & r &= 1, 2, 3, \ldots \end{aligned} \tag{12}$$

Examples of sections for which these conditions are satisfied, for solutions (9)–(11), are illustrated in Figs. 5–7.

RHOMBIC SECTION

The simplest example of the axial-shear solutions described in the preceding section, beyond the single triangle, is the case of the rhombic section given by the equation

$$(x + \sqrt{3}y + h)(x + \sqrt{3}y - h)(x - \sqrt{3}y + h)(x - \sqrt{3}y - h) = 0, \tag{13}$$

i.e. a pair of juxtaposed equilateral triangles. The paths of the rays are illustrated in Figs. 8(a) and (b). In addition, there is a simple family of contour modes if Poisson's ratio of the material is 1/4. A contour mode is one in which the displacements are perpendicular to the generators of the prism and unvarying along its length. Whereas axial-shear modes involve only equivoluminal waves, the contour modes in the rhombic prism comprise coupled equivoluminal and dilatational

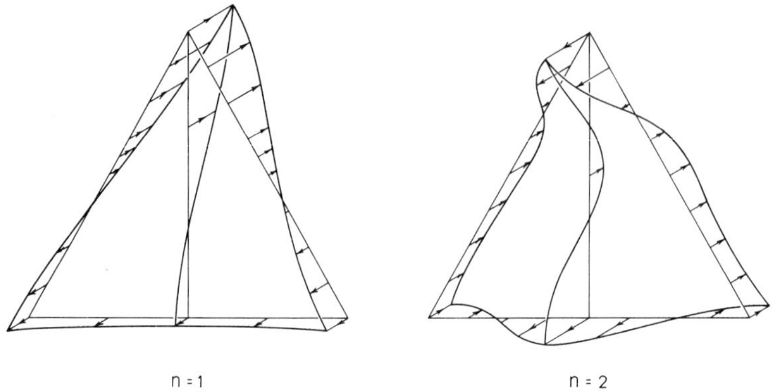

Fig. 2. Axial displacements: Solution A, $n = 1, 2$.

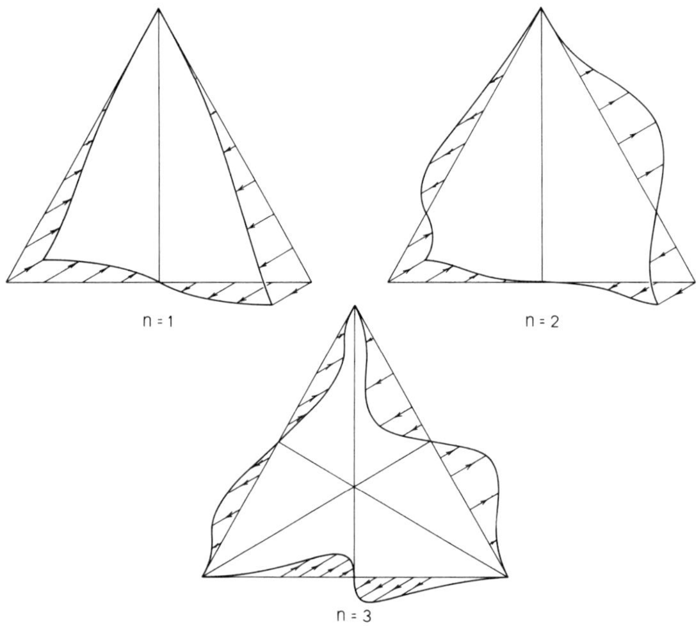

Fig. 3. Axial displacements: Solution B, $n = 1, 2, 3$.

waves. The only known equivoluminal contour modes are the Lamé modes [8] in isotropic prisms with square-module sections and analogous modes [9] in anisotropic prisms with rectangular-module sections.

If Poisson's ratio is 1/4, a dilatational wave, with its normal (and displacement) parallel to the long diagonal of the rhombic section, reflects, on incidence at a boundary, as an equivoluminal wave with its normal parallel (and displacement perpendicular) to the short diagonal. Further, an equivoluminal wave traveling parallel to the short diagonal reflects as a dilatational wave parallel to the long diagonal. Parallel rays form closed rectangular paths of equal length as shown in Fig. 8(c). Pairs of such waves, traveling in opposite directions, form steady vibrations with all displacements parallel to the long diagonal. Accordingly, the components of displacement v and

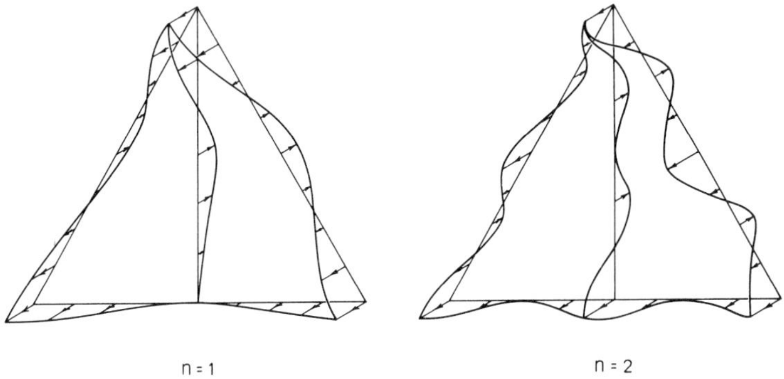

Fig. 4. Axial displacements: Solution C, $n = 1, 2$.

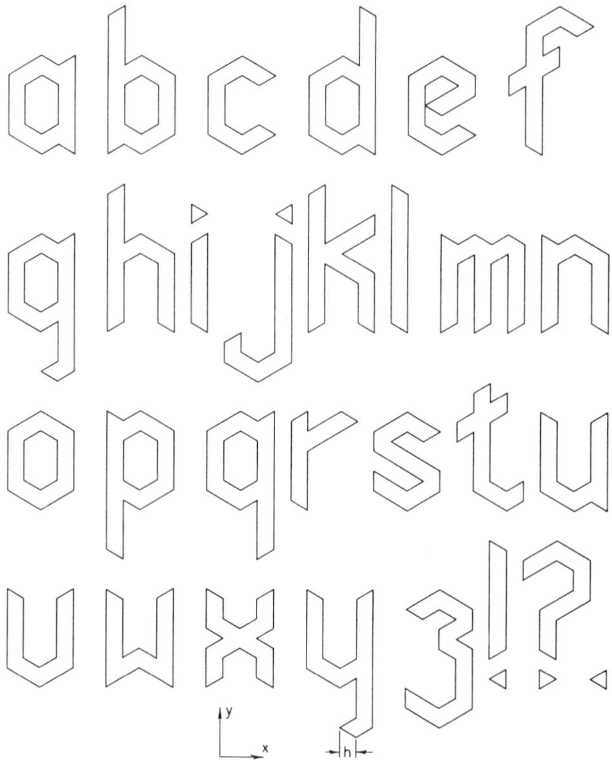

Fig. 5. Examples of normal sections of prisms for which the traction-free boundary conditions (12) are satisfied by the solutions (9)–(11) for axial-shear modes of vibrations.

w, parallel to y and z, are zero and the remaining component has the form

$$u = (D_1 \cos \alpha x + D_2 \cos \beta y)e^{i\omega t}. \tag{14}$$

The components of stress are then

$$\tfrac{1}{3}T_{xx} = T_{yy} = T_{zz} = \mu \partial u/\partial x, \quad T_{yz} = T_{zx} = 0, \quad T_{xy} = \mu \partial u/\partial y \tag{15}$$

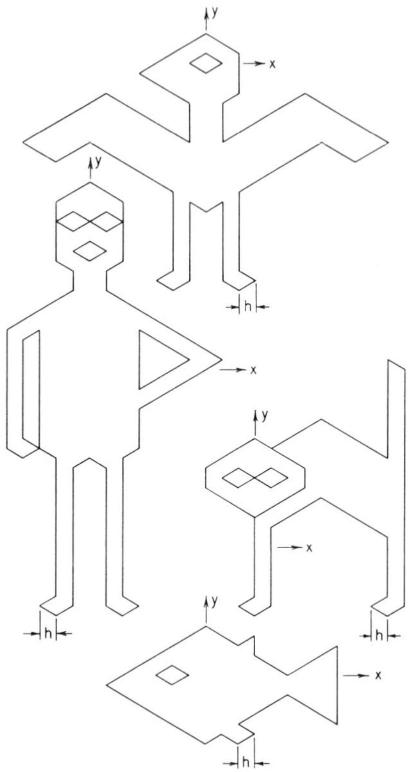

Fig. 6. Examples of normal sections of prisms for which the traction-free boundary conditions (12) are satisfied by the solutions (9)–(11) for axial-shear modes of vibrations.

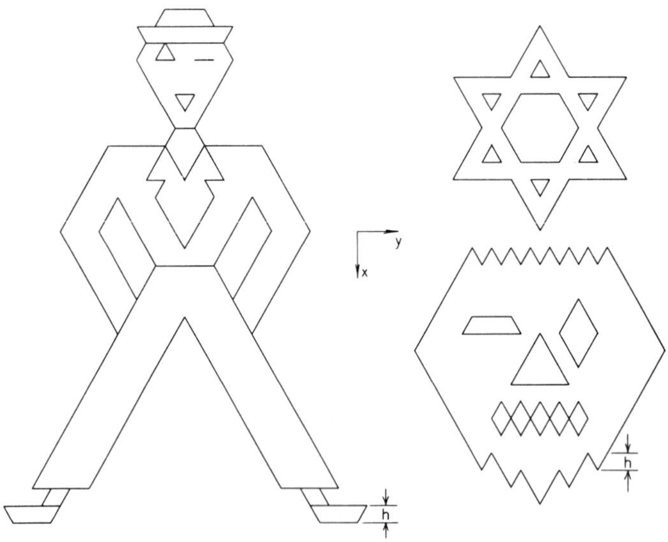

Fig. 7. Examples of normal sections of prisms for which the traction-free boundary conditions (12) are satisfied by the solutions (9)–(11) for axial-shear modes of vibrations.

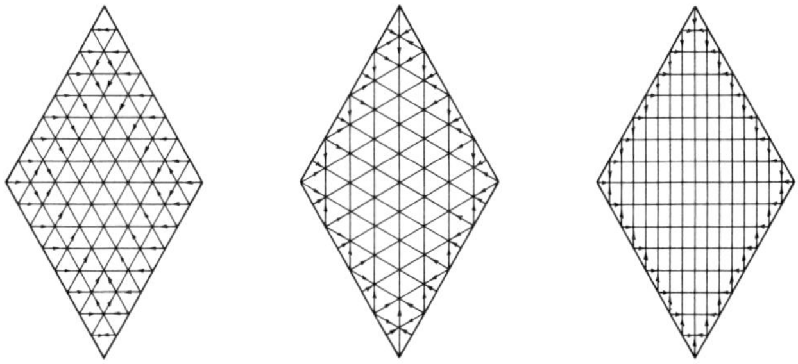

(a) (b) (c)

Fig. 8. Illustrating paths of rays in the rhombic section of a prism: in axial-shear modes comprising equivoluminal waves with normals (a) parallel and (b) perpendicular to the faces of the prism; (c) in contour modes comprising coupled equivoluminal and dilatational waves with normals parallel to the short and long diagonals, respectively.

and the displacement equations of motion reduce to

$$c^2(3\partial^2 u/\partial x^2 + \partial^2 u/\partial y^2) = \delta^2 u/\delta t^2. \tag{16}$$

Although both tensile and shear stresses contribute, the conditions for traction-free faces of the prism are simply

$$\begin{aligned} \sqrt{3}\,\partial u/\partial x + \partial u/\partial y &= 0 \quad \text{on} \quad x + \sqrt{3}y = \pm h, \\ \sqrt{3}\,\partial u/\partial x - \partial u/\partial y &= 0 \quad \text{on} \quad x - \sqrt{3}y = \pm h, \end{aligned} \tag{17}$$

as the conditions for vanishing normal and tangential traction are the same on each face.

Upon substituting (14) in (16) and (17), we find

$$u = D[\cos \alpha x + (-1)^n \cos \beta y]e^{i\omega t},$$
$$\alpha = \beta/\sqrt{3} = n\pi/h, \quad \omega = \beta c = n\pi c\sqrt{3}/h, \quad n = 1, 2, 3, \ldots \tag{18}$$

The period of the fundamental mode, $n = 1$, in terms of the length of a side, $L = 2h/\sqrt{3}$, is

$$T_1 = \omega_1/2\pi = L/c, \tag{19}$$

a result which was obtained previously [7] simply as the transit time of an equivoluminal wave across the short diagonal or a dilatational wave across the long diagonal.

As in the case of the triangular section, the solution (18), as well as the solutions (9)–(11), produces traction-free planes parallel to the faces of the rhombic prism at intervals h. Thus, a great variety of cross-sectional shapes is also accommodated by the solution (18), although more limited than in the case of the axial shear modes, as there are only two, instead of three, families of traction-free planes.

REFERENCES

1. R. D. Mindlin and H. Deresiewicz, Timoshenko's shear coefficient for flexural vibrations of beams. *Proc. 2nd U.S. Natl. Cong. Appl. Mech.* pp. 175–178. Am. Soc. Mech. Engrs. (1954).
2. J. L. Bleustein and R. M. Stanley, A dynamical theory of torsion. *Int. J. Solids Struct.* **6**, 569 (1970).
3. R. J. Talbot and J. S. Przemieniecki, Finite element analysis of frequency spectra for elastic waveguides. *Int. J. Solids Struct.* **11**, 115 (1975).
4. R. D. Mindlin, Low frequency vibrations of elastic bars. *Int. J. Solids Struct.* **12**, 27 (1976).
5. H. Lamb, *Hydrodynamics*, p. 282. Cambridge University Press (1932).
6. Lord Rayleigh, *Theory of Sound*, Vol. 2, p. 297. Macmillan, London (1929).
7. R. D. Mindlin, Note on vibrations of prisms and polygonal plates. *Problems of Hydrodynamics and Continuum Mechanics*, pp. 485–488. Society for Industrial and Applied Mathematics, Philadelphia (1969).
8. G. Lamé, *Lečons sur la Théorie Mathématique de l'Élasticité des Corps Solides*, 2nd Edn, p. 170. Gauthier-Villars, Paris (1866).
9. R. D. Mindlin, Simple modes of vibrations of crystals. *J. Appl. Phys.* **27**, 1462 (1956).

Torsion and Flexure of Piezoelectric Crystal Bars

R. D. Mindlin
Columbia University
New York, N. Y. 10027

Summary

In the absence of exact solutions of the three-dimensional equations of the mathematical theory of piezoelectricity for vibrations of piezoelectric crystal bars, approximate solutions are used. There are many methods of approximation for bar vibration problems: approximate solutions of the three-dimensional equations (by energy methods, finite element methods, etc.) and either exact or approximate solutions of approximate two-dimensional or one-dimensional equations deduced from the three-dimensional equations by expansion in series of algebraic or transcendental functions of transverse coordinates, followed by truncation.

No matter what method of approximation is employed, it is desirable to have available a few exact solutions of the three-dimensional equations with which to test the validity and accuracy of the method—at least in special, usually limiting cases. One of these is the low frequency limit, i.e., static deformation. This not only has interest as a check on an approximation but also has direct use in very low frequency applications.

In uniform extension and flexure of a uniform isotropic bar of any cross section, plane sections remain plane and, as a consequence, traction-free conditions on any cylindrical or prismatic boundary are automatically satisfied by a very simple, exact solution: the same for all sections. In the case of torsion, however, each cross-section has its peculiar axial displacement and solutions are not as simple. Nevertheless, there have been a great many problems solved for isotropic bars of various cross-sectional shapes. In the case of anisotropic bars, there are possible complications of coupling of torsion with flexure and transverse, plane strain—depending on the particular anisotropy; but a few solutions are known. Solutions are rare, though, for bars that are both anisotropic and piezoelectric.

Described in this talk are: the formulation of a system of exact equations for uniform torsion and flexure of piezoelectric, crystal bars, in terms of three stress-functions; the solution of the equations for the case of coupled torsion and flexure of an elliptic cylinder of the most general anisotropy; and the solution for the torsion of bars of rectangular cross-section with certain restricted anisotropies.

Programs for the HP-65 pocket calculator have been written for the computation of the torsional rigidities.

REFERENCES

1. W. Voigt, *Lehrbuch der Kristallphysik*, B. G. Teubner, Leipzig, (1928). Edwards Brothers, Inc., reprint, Ann Arbor, p. 641 et seq. (1946).
2. R. D. Mindlin, "Low Frequency Vibrations of Elastic Bars," Int. J. Solids and Structures, *12*, pp. 27–49, (1976).

The development of approximate equations of motion of piezoelectric bars lags far behind corresponding equations for plates. I think the reason for this is that, if you are developing two-dimensional approximate equations of plates, you don't have to take into account the contour of the plate in order to get the equations of motion. The shape of the plate comes into the boundary conditions later on. In the analogous one-dimensional schemes for rods or bars or beams, however, you have to take into account the cross-sectional shape of the bar in developing the equations of motion.

At very low frequencies and where flexural motions and extensional motions occur without torsion, as in the case of isotropic materials, plane cross-sections remain essentially plane and there is no difficulty. In the case of torsion, however, except for the circular section, even at low frequencies cross sections warp and the axial displacement of cross-sectional planes depends very sensitively on the shape of the cross section of the bar.

Now, in crystals, the torsional motions may be coupled with the flexural and extensional motions and, in developing equations of motion of bars that apply to any cross-sectional shape and any anisotropy, including piezoelectric properties, you have to take into account what goes on in torsion.

Just as in the theory of plates, you can't get general exact solutions of the three-dimensional equations. Consequently, you reduce the system to approximate two-dimensional equations for plates and one-dimensional equations for rods. At least, that's one expedient; but, in any kind of approximation that you may make, whether it's finding approximate equations or using a finite element method with the exact equations or any other method, it's always very comforting to have an exact solution, for some limiting frequency, against which you can test your approximation. The obvious thing to do, when you're starting, is to treat the low frequency limit and that is the static case. Therefore, today I will talk about exact, static solutions of the three-dimensional equations of piezoelectricity in the case of torsion, which is automatically coupled with flexure in anisotropic, piezoelectric bars.

Slide I contains the equations of static, linear piezoelectricity written in a form suitable for treating static problems by means of stress functions. Slide 2 contains the stress equations of equilibrium and the charge equation of electrostatics, from which the "stress" functions ψ_1, ψ_2 and ψ_3 are obtained. It should be noted that the axis of the bar is along x_1 and the solution is independent of this coordinate. Slide 3 contains the relevant compatibility equations including the electric ones. Slide 4 contains the differential equations on the "stress" functions ψ_i. Slide 5 contains the definitions of the differential operators appearing in Slide 4.

Slide 6 contains the form of the solution in the case of the elliptic cross-section. Slide 7 gives the twist and curvatures in terms of the twisting couple and two bending couples for the elliptic cross-section. These relations are just what you need for the low frequency limit of coupled torsional and flexural vibrations.

Slide 8 gives the form of the solution for the rectangular cross-section. Slide 9 gives the torsional rigidity of the rectangular cross-section. In the case of the rectangular cross-section, the solution is exact only for two crystal symmetries, both of which are in the hexagonal system; but the solution is nearly exact, except for a couple of the g_{ip}, for almost any crystal class for which the x_1-axis is a two-fold axis of symmetry. It may be noted that, for the torsional rigidity of an *isotropic* bar of rectangular cross-section, $a_1 = \pi/2$, $b_1 = 192/\pi^5$, $a_2 = 0$, $b_2 = 0$ and all the other terms exist on account of the anisotropy and the piezoelectric coupling.

With results like the foregoing you're in a position now to test, at the low frequency limit, approximate equations of motion that you might concoct for the coupled extensional, flexural and torsional vibrations of piezoelectric crystal bars.

Equations of Piezoelectricity

	Elastic		*Electric*
Divergence	$T_{ij,i} = 0$		$D_{i,i} = 0$
Constitutive	$S_{ij} = s_{ijkl}T_{kl} + g_{kij}D_k$	$\leftrightarrow$	$E_i = -g_{ijk}T_{jk} + \beta_{ij}D_j$
Gradient	$S_{ij} = \frac{1}{2}(u_{j,i} + u_{i,j})$		$E_i = -\varphi_{,i}$
Compatibility	$\nabla \times \mathbf{S} \times \nabla = 0$		$\nabla \times \mathbf{E} = 0$

Figure 1

Divergence (Equilibrium) Equations

Elastic *Electric*

$$\frac{\partial T_{21}}{\partial x_2} + \frac{\partial T_{31}}{\partial x_3} = 0 \quad \begin{cases} T_{21} = \dfrac{\partial \psi_1}{\partial x_3} \\[2ex] T_{31} = -\dfrac{\partial \psi_1}{\partial x_2} \end{cases} \qquad \frac{\partial D_2}{\partial x_2} + \frac{\partial D_3}{\partial x_3} = 0 \quad \begin{cases} D_2 = \dfrac{\partial \psi_3}{\partial x_3} \\[2ex] D_3 = -\dfrac{\partial \psi_3}{\partial x_2} \end{cases}$$

$$\begin{aligned} \frac{\partial T_{22}}{\partial x_2} + \frac{\partial T_{32}}{\partial x_3} &= 0 \\ \frac{\partial T_{23}}{\partial x_2} + \frac{\partial T_{33}}{\partial x_3} &= 0 \end{aligned} \quad \begin{cases} T_{22} = \dfrac{\partial^2 \psi_2}{\partial x_3^2} \\[2ex] T_{33} = \dfrac{\partial^2 \psi_2}{\partial x_2^2} \\[2ex] T_{23} = -\dfrac{\partial^2 \psi_2}{\partial x_2 \partial x_3} \end{cases} \qquad \psi_i = \text{“Stress” Functions}$$

Figure 2

Compatibility Equations

Elastic | *Electric*

$$S_{11} = a_1 + a_2 x_2 + a_3 x_3 \qquad E_1 = b_1$$

$$\frac{\partial S_{31}}{\partial x_2} - \frac{\partial S_{21}}{\partial x_3} = a_0 \qquad \frac{\partial E_3}{\partial x_2} - \frac{\partial E_2}{\partial x_3} = 0$$

$$\frac{\partial^2 S_{33}}{\partial x_2^2} + \frac{\partial^2 S_{22}}{\partial x_3^2} = 2\frac{\partial^2 S_{23}}{\partial x_2 \partial x_3}$$

Figure 3

Equations on "Stress" Functions

$$\mathscr{D}_{11}^2\psi_1 + \mathscr{D}_{12}^3\psi_2 + \mathscr{D}_{13}^2\psi_3 = \text{const.}$$

$$\mathscr{D}_{21}^3\psi_1 + \mathscr{D}_{22}^4\psi_2 + \mathscr{D}_{23}^3\psi_3 = 0$$

$$\mathscr{D}_{31}^2\psi_1 + \mathscr{D}_{32}^3\psi_2 + \mathscr{D}_{33}^2\psi_3 = \text{const.}$$

Figure 4

$$\mathscr{D}_{11}^2 = \bar{s}_{55}\partial_2^2 - 2\bar{s}_{56}\partial_2\partial_3 + \bar{s}_{66}\partial_3^2$$

$$\mathscr{D}_{22}^4 = \bar{s}_{33}\partial_2^4 - 2\bar{s}_{34}\partial_2^3\partial_3 + (2\bar{s}_{23} + \bar{s}_{44})\partial_2^2\partial_3^2 - 2\bar{s}_{24}\partial_2\partial_3^2 + \bar{s}_{22}\partial_3^4$$

$$\mathscr{D}_{33}^2 = -\bar{\beta}_{33}\partial_2^2 + 2\bar{\beta}_{23}\partial_2\partial_3 - \bar{\beta}_{22}\partial_3^2$$

$$\mathscr{D}_{23}^3 = \mathscr{D}_{32}^3 = -\bar{g}_{33}\partial_2^3 + (\bar{g}_{34} + \bar{g}_{23})\partial_2^2\partial_3 - (\bar{g}_{32} + \bar{g}_{24})\partial_2\partial_3^2 + \bar{g}_{22}\partial_3^3$$

$$\mathscr{D}_{31} = \mathscr{D}_{13} = \bar{g}_{35}\partial_2^2 - (\bar{g}_{25} + \bar{g}_{36})\partial_2\partial_3 + \bar{g}_{26}\partial_3^2$$

$$\mathscr{D}_{12}^3 = \mathscr{D}_{21}^3 = -\bar{s}_{53}\partial_2^3 + (\bar{s}_{54} + \bar{s}_{63})\partial_2^2\partial_3 - (\bar{s}_{52} + \bar{s}_{64})\partial_2\partial_3^2 + \bar{s}_{62}\partial_3^3$$

$$\bar{s}_{pq} = s_{pq} - [s_{1p}(\beta_{11}s_{1q} + g_{11}g_{1q}) - g_{1p}(s_{11}g_{1q} - s_{1q}g_{11})]/(\beta_{11}s_{11} + g_{11}^2)$$

$$\bar{\beta}_{ij} = \beta_{ij} - [g_{i1}(\beta_{1j}g_{11} - \beta_{11}g_{j1}) + \beta_{i1}(\beta_{j1}s_{11} + g_{j1}g_{11})]/(\beta_{11}s_{11} + g_{11}^2)$$

$$\bar{g}_{ip} = g_{ip} - [g_{i1}(\beta_{11}s_{1p} + g_{11}g_{1p}) + \beta_{i1}(s_{11}g_{1p} - s_{1p}g_{11})]/(\beta_{11}s_{11} + g_{11}^2)$$

$$i, j = 1, 2, 3; \qquad p, q = 1, 2, 3, 4, 5, 6.$$

Figure 5

$$\text{Elliptic Section:} \quad \frac{x_2^2}{b^2} + \frac{x_3^2}{c^2} = 1$$

$$\psi_1 = C_1\left(\frac{x_2^2}{b^2} + \frac{x_3^2}{c^2} - 1\right)$$

$$\psi_2 = 0$$

$$\psi_3 = C_3\left(\frac{x_2^2}{b^2} + \frac{x_3^2}{c^2} - 1\right)$$

Figure 6

Elliptic Section: Twist (τ) and Curvatures (κ_2, κ_3)

$$\tau = \frac{M_1}{4}\left(\frac{s_{66} + g_{16}^2/\beta_{11}}{I_2} + \frac{s_{55} + g_{15}^2/\beta_{11}}{I_3}\right) - \frac{M_2(s_{16} + g_{11}g_{16}/\beta_{11})}{2I_2} - \frac{M_3(s_{15} + g_{11}g_{15}/\beta_{11})}{2I_3}$$

$$\kappa_2 = \frac{M_3(s_{11} + g_{11}^2/\beta_{11})}{I_3} - \frac{M_1(s_{15} + g_{11}g_{15}/\beta_{11})}{2I_3} + \frac{2C_3(g_{31} - \beta_{31}g_{11}/\beta_{11})}{b^2}$$

$$\kappa_3 = -\frac{M_2(s_{11} + g_{11}^2/\beta_{11})}{I_2} + \frac{M_1(s_{16} + g_{11}g_{16}/\beta_{11})}{2I_2} + \frac{2C_3(g_{21} - \beta_{21}g_{11}/\beta_{11})}{c^2}$$

Figure 7

Rectangular Section: $x_2 = \pm b, x_3 = \pm c$

$$\psi_1 = \sum_{n=1,3,5\ldots}\left(A_n' \cosh\frac{n\alpha_1\pi x_3}{2b} + A_n'' \cosh\frac{n\alpha_2\pi x_3}{2b} + A_n^0\right)\cosh\frac{n\pi x_2}{2b}$$

$$\psi_2 = 0$$

$$\psi_3 = \sum_{n=1,3,5\ldots}\left(C_n' \cosh\frac{n\alpha_1\pi x_3}{2b} + C_n'' \cosh\frac{n\alpha_2\pi x_3}{2b} + C_n^0\right)\cosh\frac{n\pi x_2}{2b}$$

$$(\bar{\beta}_{22}\bar{s}_{66} + \bar{g}_{26}^2)\alpha^4 - (\bar{\beta}_{22}\bar{s}_{55} + \bar{\beta}_{33}\bar{s}_{66} + 2\bar{g}_{35}\bar{g}_{26})\alpha^2 + (\bar{\beta}_{33}\bar{s}_{55} + \bar{g}_{35}^2) = 0$$

Figure 8

Rectangular Section: Torsional Rigidity

$$\frac{M_1}{\tau} = \frac{16\bar{\beta}_{33}b^3c}{\bar{s}_{55}\bar{\beta}_{33} + \bar{g}_{35}^2}\left[\frac{1}{3} - \sum_{n=1,3,5\ldots}(b_1 \tanh na_1r - b_2 \tanh na_2r)/rn^5\right]$$

$$\left.\begin{aligned} a_m &= \pi|\alpha_m|/2 \\ b_m &= 2^6(\gamma_m - \bar{g}_{35}^2/\bar{\beta}_{33})/\pi^5|\alpha_m|(\gamma_1 - \gamma_2) \\ \gamma_m &= (\alpha_m^2\bar{g}_{26} - \bar{g}_{35})/(\alpha_m^2\bar{\beta}_{22} - \bar{\beta}_{33}) \end{aligned}\right\} m = 1, 2$$

$$r = c/b$$

Figure 9

Discussion

L. Wilson: You mentioned that in certain crystal classes the operators multiplying the Airy stress function vanish. Can you say what classes.

Reply: Two in the hexagonal system. I forget just which two.

F. R. Norwood: Is the work you are presenting here in Ref. 2?

Reply: No, I never wrote this up. Reference 2 would be this without the piezoelectric coupling; just the anisotropic case.

H. F. Tiersten: Have you started work on the expansion for the dynamic part?

Reply: Yes, in fact I have it all done. It is a double power series expansion, but it looks so messy and I haven't had time to make it look simple yet.

H. F. Tiersten: Where did you truncate it?

Reply: It you look at the paper I have in the References you will see how I truncated it. It is truncated differently depending on the shape of the cross-section. I keep the full double series expansion for the final equations and then judiciously select terms, and they don't start at the lowest order necessarily. I judiciously select terms of the double series expansion which will give the appropriate warping of the section. I show how to do that for the non-piezoelectric case in Ref. 2.

H. F. Tiersten: And the electrical variable goes along with it too.

Reply: Yes.

J. Miklowitz: Underlying all of this there will be some complex arms, I assume, that will pierce the $\omega = 0$ plane, which is effectively what you are exploiting. So in a sense you could exploit those, could you not, in some sort of an expansion just a little higher than $\omega = 0$ to get some correlation with these.

Reply: As a matter of fact I tried that for plates and I never got very far with it. I don't really know how to do it.

J. Miklowitz: Can't you do it really in the two-dimensional problem? Wouldn't it simplify in the sense that you usually get for instance in a rod, for the isotropic rod, you get $(\sin z) + z = 0$.

Reply: Yes but the same as in a plate.

J. Miklowitz: So what probably will happen is you will only need the first couple of terms, it would seem, because the imaginary parts of the modes get large and hence won't be needed in an edge wave sense.

Reply: Oh yes, but they don't get so large. You know they are fairly closely spaced.

J. Miklowitz: Well yes, but they are getting large.

Reply: Well as I say, I tried it and I never got anywhere with it. You try it.

J. Miklowitz: I'll send you the write-up.

L. Wilson: I can mention some progress in work on the dynamic case for a circular rod that is piezoelectric in any of the hexagonal classes that I can get exact solutions for. Also for circular rods in crystal classes such as that of which sapphire is a member, and lithium niobate is in a piezoelectric class of. I can get approximate solutions for all the modes. And I have also, along with Bob Thurston, done some work on low frequency expansions for sapphire for the longitudinal and torsional modes.

Reply: For circular sections.

L. Wilson: Yes.

Reply: Yes, there is no trouble with rods of circular cross-section. All this fancy business is for noncircular rods. Actually it was aimed at for rectangular bars, but the elliptic comes out so nice and simple that this works fine.

M. Wnuk: I am not sure that I understood the beginning of the statement you made about anisotropy but I think it's a general anisotropy and anything you want you can read into it.

Reply: Yes.

M. Wnuk: Now, wouldn't this solution simplify a little bit if you were a little more specific and chose an anisotropy or a certain kind of anisotropy and then zero in on this?

Reply: For the bar of elliptic cross-section that isn't necessary. You can get it completely and then put in the particular crystal classes later. As I mentioned, for the bar of rectangular cross-section I couldn't do the general one because it was coupled in with the Airy stress function, and there I did take particular crystal classes.

M. Wnuk: Well I think perhaps you could get rid of some of these operators by assuming orthotropic, or another kind of anisotropy, to match the shape of the boundary.

Reply: As I said, the only case in which I could get rid of all those coefficients was in the two hexagonal classes.

A. A. Comparini: In the dynamic problem when you look at the elliptic cross-section those operators do not drop out do they, even for the hexagonal classes, you don't get this simplification with the elliptic cross-section when you look at the dynamic problem, do you?

Reply: Yes, you still do, even in the dynamical case. You get some additional coupling, you see this does not include the longitudinal, but in the dynamical case the flexural, of course, is coupled with the torsional motion. The extensional does not couple.

W. J. Spencer: Probably the largest dollar volume business, in response to a question earlier in piezoelectric devices, at least that I am aware of, is in the quartz crystal business. It has been around a long time; it has been steady dollar business in the 50–75 million dollar range. Although crystal oscillators have been replaced in highly precise time keeping applications by atomic clocks, they find use in many systems where timing accuracy of a part in 9 million are required. As you are aware, there is a blooming new industry in the crystal watch business, and nobody really knows yet how large that is going to be or how low the crystal price is going to be. It has the potential of being a multihundred million dollar business and it will certainly make use of the kind of resonators using $\omega > 0$ above the solutions that Prof. Mindlin talked about.

Reset from
Application of Elastic Waves in Electrical Devices, Non-Destructive Testing and Seismology
J.D. Achenbach, Y.-H. Pao and H.F. Tiersten (eds.)
Northwestern University, Evanston, Illinois
(1976) 83-99

Computers & Structures, Vol. 7, pp. 639–650. Pergamon Press 1977. Printed in Great Britain

FREQUENCIES OF VIBRATION OF BEAMS BY SCHELKUNOFF ITERATION WITH THE HP-65 CALCULATOR

R. D. MINDLIN
Columbia University, NY 10027, U.S.A

(Received 6 June 1976)

Abstract—Three programs are given for computing frequencies of free vibrations of uniform beams with the HP-65 programmable, pocket calculator: Bernoulli beams with all 10 combinations of homogeneous end-conditions, two-span Bernoulli beams on hinged supports and Timoshenko beams with 9 of the 10 combinations of homogeneous end-conditions. The programs illustrate the capacity and versatility of the calculator and various strengths and weaknesses of Schelkunoff's iterative method for solving trigonometric equations.

1. SCHELKUNOFF ITERATION

The problem of finding the roots of an equation in which at least one term is a trigonometric function:

$$\tan x = f(x) \quad \text{or} \quad \sin x = f(x) \quad \text{or} \quad \cos x = f(x),$$

occurs frequently in vibration problems. Schelkunoff's iterative solution of such equations proceeds as follows:†

For $\tan x = f(x)$, let

$$x = q\pi + \delta, \quad -\pi/2 < \delta < \pi/2, \quad q = 1, 2, 3, \ldots.$$

Then $\tan \delta = f(q\pi + \delta)$ or

$$\delta = \tan^{-1}[f(q\pi + \delta)].$$

As a first approximation, take $\delta_1 = \tan^{-1}[f(q\pi)]$ and, as successive approximations,

$$\delta_{i+1} = \tan^{-1}[f(q\pi + \delta_i)], \quad i = 1, 2, 3, \ldots$$

The computation ends when $i = j$ such that

$$|\delta_{j+1} - \delta_j| < \epsilon$$

where ϵ is chosen for the desired accuracy. Then

$$x = q\pi + \delta_{j+1}.$$

The process is repeated for as many values of q as are required.

For $\sin x = f(x)$, the equation is written as

$$\tan x = f(x)/\cos x$$

and the solution proceeds as before, with $f(x)$ replaced by $f(x)/\cos x$.

For $\cos x = f(x)$, let $x = x' - \pi/2$. The equation becomes

$$\tan x' = f(x' - \pi/2)/\cos x'.$$

Let $x' = q\pi + \delta, \quad -\pi/2 < \delta < \pi/2, \quad q = 1, 2, 3 \ldots$. The procedure is again the same as before but, this time, to find

$$x' = q\pi + \delta_{j+1} \text{ so that}$$
$$x = q\pi + \delta_{j+1} - \pi/2.$$

The method is particularly suited to the HP-65 programmable, pocket calculator because the number of steps per cycle, to be added to those for $f(x)$, is small and no trial roots are required. As a result of the latter property, time consuming guessing and manual keystrokes are avoided. Once the function $f(x)$ has been defined, the only input is the integer q.

Depending on the form of $f(x)$, difficulties may or may not occur. Several examples are encountered in the three problems of vibrations of beams treated in the sequel:

(A) Frequencies of free vibration of uniform, Bernoulli beams with homogeneous end-conditions. Here $f(x)$ is non-oscillatory and real for all real x and, consequently, there are no difficulties.

(B) Frequencies of free vibration of two-span, uniform, Bernoulli beams on hinged supports. In this case a family of roots is skipped in a straightforward application of the procedure; but an interchange of the roles of two tangent functions results in the generation of the missing roots. In addition, some cases of very slow convergence are encountered and provision is made for distinguishing between these and non-convergent iterations.

(C) Frequencies of free vibration of uniform, Timoshenko beams with homogeneous end-conditions. This is an example of the use of more than one 100-step program card for the complete program. The phenomenon of isolated, skipped roots is exhibited and a part of the program is used to illustrate an oscillatory iteration.

The three beam-problems reveal the surprisingly large mathematical capacity of a 100-step magnetic card—a consequence of the small number of steps required for the iterative procedure (a minimum of 19) and the capability and versatility of the calculator. In the case of example A, all 10 admissible combinations of homogeneous end-conditions are accommodated on one card and 9 of the 10 on two cards for the far more complicated example C.

The differential equations, boundary values and many of the solutions are to be found on a few pages of

†S. A. Schelkunoff showed me this method in 1950. Experts in numerical analysis say it may have been in use much earlier, but I have found no references to it in textbooks.

Timoshenko's "*Vibration Problems in Engineering*"[1]; and his notation is used, for the most part.

2. BERNOULLI BEAMS

Bernoulli's equation of motion for flexural vibrations of uniform beams is

$$r^2 v^2 \frac{\partial^4 y}{\partial x^4} = -\frac{\partial^2 y}{\partial t^2},$$

where $y(x)$ is the deflection, r is the radius of gyration of the cross-section and v is the bar velocity:

$$v = \sqrt{(Eg/\gamma)},$$

in which E is Young's modulus and γ/g is the mass density. There are four admissible, homogeneous end-conditions:

at a hinged end (H):

$$y = 0, \quad \frac{\partial^2 y}{\partial x^2} = 0,$$

at a free end (F):

$$\frac{\partial^2 y}{\partial x^2} = 0, \quad \frac{\partial^3 y}{\partial x^3} = 0,$$

at a clamped end (C):

$$y = 0, \quad \frac{\partial y}{\partial x} = 0,$$

at a guided end (G):

$$\frac{\partial y}{\partial x} = 0, \quad \frac{\partial^3 y}{\partial x^3} = 0;$$

totalling $4+3+2+1 = 10$ combinations: one condition at each end. The frequencies of free vibration for all ten cases are given in cycles per second (Hz), by

$$f_n = \alpha_n^2 rv/2\pi l^2, \quad n = 1, 2, 3 \ldots.$$

where l is the length of the beam and the α_n are the roots of transcendental equations, as follows:

Case	*End-conditions*	*Frequency equation*
I and II	CH or FH	$\tan\alpha = \tanh\alpha$
III and IV	CG or FG	$\tan\alpha = -\tanh\alpha$
V	CF	$\cos\alpha = -\operatorname{sech}\alpha$
VI and VII	FF or CC	$\cos\alpha = \operatorname{sech}\alpha$
VIII and IX	HH or GG	$\sin\alpha = 0$
X	HG	$\cos\alpha = 0.$

There are only six transcendental equations as there are four duplications. Furthermore, the last two yield explicit solutions, leaving only four equations requiring iteration.

The 100-step program on Form 1 and the instructions given on Form 2 yield the frequencies for all ten cases. The results for the beam

$$l = 100\,\text{cm}, \quad r = 0.25\,\text{cm}, \quad v = 4\times 10^5\,\text{cm/sec},$$

with $\epsilon = 10^{-6}$, are given in Table 1. The execution times (sec) are those for computations made in the order given, i.e. $n = 1, 2, 3, 4, 5$ with the initial value of δ_i that of the δ_{j+1} of the preceding n. Execution times depend on the initial values of δ_i as well as on ϵ.

3. TWO-SPAN BERNOULLI BEAMS ON HINGED SUPPORTS

The span-lengths of the two-span beams are l_1 and l_2 and the frequencies may be expressed as

$$f_n = \alpha_n^2 rv/2\pi l_1^2,$$

where the α_n are the roots of [1]

$$\cot\alpha - \coth\alpha + \cot\beta - \coth\beta = 0, \quad \beta = \rho\alpha, \quad \rho = l_2/l_1.$$

The iterative procedure, in this case, distinguishes between two families of roots: those found by solving for α and those found by solving for β. For the former, let

$$\alpha = m\pi + \delta, \quad -\pi/2 < \delta < \pi/2, \quad m = 1, 2, 3 \ldots$$

so that the iteration formula is

$$\delta_{j+1} = \tan^{-1}[\coth\rho(m\pi+\delta_i) - \cot\rho(m\pi+\delta_i) + \coth(m\pi+\delta_i)]^{-1}$$

and the computation ends with

$$\alpha_m = m\pi + \delta_{j+1}.$$

For the other family,

$$\beta = n\pi + \delta, \quad -\pi/2 < \delta < \pi/2, \quad n = 1, 2, 3 \ldots$$

with the iteration formula

$$\delta_{j+1} = \tan^{-1}[\coth\rho^{-1}(n\pi+\delta_i) - \coth\rho^{-1}(n\pi+\delta_i) + \coth(n\pi+\delta_i)]^{-1}$$

and the result

$$\beta_n = n\pi + \delta_{j+1}$$

so that

$$\alpha_n = \rho^{-1}(n\pi + \delta_{j+1}).$$

The two series of values, α_m and α_n, may finally be arranged in ascending order: α_p, $p = 1, 2, 3 \ldots.$

The 100 step program on Form 3 and the instructions given on Form 4 were used to find the results, in Table 2, for the case

$$l_1 = 100\,\text{cm}, \quad \rho = 0.75, \quad r = 0.25\,\text{cm}, \quad v = 4\times 10^5\,\text{cm/sec}.$$

The selection of roots α_m or α_n is effected by setting Flag 1 off (f^{-1} SF1) for α_m or on (f SF1) for α_n. The execution times in Table 2 (to the nearest 5 sec, with $\epsilon = 10^{-6}$) were obtained in the order $m = 1, 2, 3\ldots$ and $n = 1, 2, 3\ldots$ with the initial value of δ_i (in Register R_6) the final value of δ_{j+1} of the preceding root. Two exceptionally long execution times are encountered: 215 sec for $m = 4$ and 380 sec for $n = 3$. If the root for $m = 4$ had been the first to be computed after the calculator was switched on, the execution time would have been less than 5 sec inasmuch as the initial value of δ_i would have been zero and that happens to be the exact, final value—so that only 1 cycle

Table 1. Bernoulli beams with homogeneous end-conditions

Cases I & II: $\tan\alpha = \tanh\alpha$			
n	f_n (Hz)	α_n	sec
1	24.539	3.9266	11
2	79.523	7.0686	6
3	165.915	10.2102	6
4	283.725	13.3518	6
5	432.951	16.4934	6
⋮	⋮	⋮	
k	$5\alpha_k^2/\pi$	$(k-5)\pi+\alpha_5$	

Cases III & IV: $\tan\alpha = -\tanh\alpha$			
n	f_n (Hz)	α_n	sec
1	8.902	2.3650	13
2	48.106	5.4978	9
3	118.791	8.6394	6
4	220.893	11.7810	6
5	354.411	14.9226	6
⋮	⋮	⋮	
k	$5\alpha_k^2/\pi$	$(k-5)\pi+\alpha_5$	

Case V: $\cos\alpha = -\operatorname{sech}\alpha$			
n	f_n (Hz)	α_n	sec
1	5.596	1.8751	37
2	35.069	4.6940	20
3	98.194	7.8548	12
4	192.421	10.9955	12
5	318.086	14.1372	11
⋮	⋮	⋮	
k	$5\alpha_k^2/\pi$	$(2k-1)\pi/2$	

Cases VI & VII: $\cos\alpha = \operatorname{sech}\alpha$			
n	f_n (Hz)	α_n	sec
1	Note 1		
2	35.608	4.7300	19
3	98.155	7.8532	12
4	192.424	10.9956	12
5	318.086	14.1372	11
⋮	⋮	⋮	
k	$5\alpha_k^2/\pi$	$(2k-1)\pi/2$	

Cases VIII & IX: $\sin\alpha = 0$			
n	f_n (Hz)	α_n	sec
1	15.708	π	1
2	62.832	2π	1
3	141.372	3π	1
4	251.327	4π	1
5	392.699	5π	1
⋮	⋮	⋮	
k	$5\alpha_k^2/\pi$	$k\pi$	

Case X: $\cos\alpha = 0$			
n	f_n (Hz)	α_n	sec
1	3.927	$\pi/2$	1
2	35.343	$3\pi/2$	1
3	98.175	$5\pi/2$	1
4	192.423	$7\pi/2$	1
5	318.086	$9\pi/2$	1
⋮	⋮	⋮	
k	$5\alpha_k^2/\pi$	$(2k-1)\pi/2$	

Note 1: 0, by inspection, for VI; absent for VII.

Table 2. Two-span Bernoulli beams on hinged supports

m	f_m, f_p (Hz)	p	α_m, α_p	sec
1	18.579	1	3.4167	40
2	71.249	3	6.6908	30
3	158.510	5	9.9797	45
4	251.327	6	12.5664	215
5	406.617	8	15.9839	40
6	590.210	10	19.2572	30
7	809.027	12	22.5461	45
⋮	⋮		⋮	
k	$5\alpha_m^2/\pi$		$\alpha_{m-4}+4\pi$	

n	f_n, f_p (Hz)	p	α_n, α_p	sec
1	36.474	2	4.7872	50
2	122.402	4	8.7697	55
3	289.140	7	13.4786	380
4	479.306	9	17.3539	50
5	724.518	11	21.3361	55
⋮	⋮		⋮	
k	$5\alpha_n^2/\pi$		$\alpha_{n-3}+4\pi$	

of the iteration would have been required. This is an extreme case of the effect of the initial value of δ_i on the execution time.

To determine, at an intermediate stage, whether or not an exceptionally long iteration is converging, provision is made, in the program, for switching to manual iteration. The change is made by stopping the execution (R/S on the keyboard) and setting Flag 2 on (f SF2) thereby bypassing the instruction (program steps 71, 72) to return to the start (LBL, E, program steps 7, 8) of the iteration loop for another cycle if $|\delta_{i+1} - \delta_i| > \epsilon$. Each loop of the iteration can then be actuated independently by key E to display the current f_m or f_n, following which RCL 6, RCL 7, RCL 8 display the current δ_{i+1}, δ_i and α_m or α_n, respectively. The convergence, though slow for $m = 4, 8, 12\ldots$ and $n = 3, 6, 9\ldots$ can be confirmed accordingly. Automatic iteration is reinstated by keying in f^{-1} SF2, E. The display does not have to be watched continuously to note the end of the iteration. If the operating calculator is placed near an AM radio, switched on and tuned between stations, silence indicates the end of the run.

4. TIMOSHENKO BEAM

Timoshenko's beam equations [1] may be written in the form

$$\frac{\bar{\omega}^2 r^2}{\tau}\frac{\partial^2\psi}{\partial x^2}+\bar{\omega}^2\left(\frac{\partial y}{\partial x}-\psi\right)=\frac{\partial^2\psi}{\partial t^2},$$

$$\bar{\omega}^2 r^2\left(\frac{\partial^2 y}{\partial x^2}-\frac{\partial\psi}{\partial x}\right)=\frac{\partial^2 y}{\partial t^2},$$

in which the additional symbols are the rotation, ψ, the circular frequency, $\bar{\omega}$, of the fundamental axial shear mode and the material and shape factor τ:

$$\bar{\omega}=\frac{1}{r}\left(\frac{k'Gg}{\gamma}\right)^{1/2},\quad \tau=\frac{k'G}{E},$$

where G is the shear modulus and k' is Timoshenko's shear correction factor. The equations take into account the shear deformation, $\partial y/\partial x-\psi$, and the rotatory acceleration, $\partial^2\psi/\partial t^2$, not accommodated by Bernoulli's equation. The latter may be recovered by first eliminating the shear deformation term between the two Timoshenko equations and then setting

$$\frac{\partial y}{\partial x}-\psi=0,\quad \frac{\partial^2\psi}{\partial t^2}=0.$$

The admissible, homogeneous end-conditions are of the same four types as in the Bernoulli equation, but, now, at a hinged end (H):

$$y=0,\quad \frac{\partial\psi}{\partial x}=0,$$

at a free end (F):

$$\frac{\partial y}{\partial x}-\psi=0,\quad \frac{\partial\psi}{\partial x}=0,$$

at a clamped end (C):

$$y=0,\quad \psi=0,$$

at a guided end (G):

$$\frac{\partial y}{\partial x}-\psi=0,\quad \psi=0.$$

Again, there are ten pairs of the four conditions—one condition at each end of the beam; but, in this case, the transcendental equations are more complicated and the program for frequencies of free vibration, even though twice as long as the one for Bernoulli's equation, accommodates only nine combinations. The cantilever (clamped-free) is not included.

The functions

$$y=(A_{11}\sin k_1x+A_{12}\sin k_2x+A_{21}\cos k_1x+A_{22}\cos k_2x)\,e^{i\omega t},$$

$$\psi=(A'_{11}\cos k_1x+A'_{12}\cos k_2x-A'_{21}\sin k_1x-A'_{22}\sin k_2x)\,e^{i\omega t}$$

are solutions of the two differential equations if

$$A'_{ij}=A_{ij}(\kappa_j^2-\Omega^2)/\kappa_j r,\quad i=1,2,;\quad j=1,2,$$

where the $\kappa_i(=k_ir)$ are solutions of the biquadratic equation

$$\kappa^4-(1+\tau)\Omega^2\kappa^2+\tau\Omega^2(\Omega^2-1)=0$$

and $\Omega=\omega/\bar{\omega}$. The larger root, κ_1^2, of the biquadratic equation is always positive and the smaller root is positive for $\Omega>1$, zero for $\Omega=1$ and negative for $\Omega<1$. Hence, the wave-numbers k_2 are real for $\Omega>1$ and imaginary for $\Omega<1$. The larger wave-number k_1 (shorter wave-length) corresponds to predominantly flexural motion ($y>r\psi$) and the shorter wave number, k_2, corresponds to predominantly axial shear motion ($r\psi>y$).

Upon substitution of the functions y and ψ in the homogeneous end-conditions for a beam of length l, the results which follow are obtained.

(i) *HH*: *Beam hinged at both ends* (x = 0 *at the center*)

In this case, the (predominantly) flexural modes ($y>r\psi$) are not coupled with the (predominantly) axial shear modes ($r\psi>y$). The results are expressed as the ratios, l_m/r or l_n/r, of the lengths, l_m or l_n, of beams to the radius of gyration r (or other representative length in the cross section). l_m and l_n are the lengths of beams which will vibrate with m half-wave-lengths of predominantly flexural deformation, or n half-wave-lengths of predominantly axial shear-deformation, at a circular frequency ω expressed as the ratio $\Omega(=\omega/\bar{\omega})$ to the fundamental axial shear frequency:

Flexure:

$$l_m/r=m\pi/\kappa_1,\quad m=1,2,3,\ldots,\Omega>0,$$

$$y(x)=-(-1)^m y(-x),\quad \psi(x)=(-1)^m\psi(-x),$$

Axial shear:

$$l_m/r=n\pi/\kappa_2,\quad n=1,2,3,\ldots,\Omega>1,$$

$$y(x)=-(-1)^n y(-x),\quad \psi(x)=(-1)^n\psi(-x).$$

Given Ω and τ,κ_1 and κ_2 are determined by the biquadratic equation and then the l_m/r and l_n/r are fixed by the numbers of half-wave-lengths m and n. In addition, there is the solution, $\Omega=1$ (for which $\kappa_2=0$) and $n=0$, for which l_n/r is indeterminate. This is the fundamental, pure ($y=0$) axial shear mode at circular frequency $\bar{\omega}$: possible in a hinged-hinged beam of any length.

(ii) *GG*: *Beam guided at both ends* (x = 0 *at the center*)

The results are the same as for the HH beam except that odd and even are interchanged and the solution $\Omega=1$, $n=0$ is excluded.

(iii) *HG*: *Beam hinged at one end and guided at the other* (x = 0 *at one end*)

The l_m/r and l_n/r are half of the l_m/r and l_n/r, respectively, for the HH beam with m and n odd, only, and the solution $\Omega=1$, $n=0$ excluded.

(iv) *FF*: *Beam free at both ends* (x = 0 *at the center*)

In this case, reflections of waves at the ends of the beam cause coupling of the predominantly flexural modes with the predominantly axial shear modes; so that there are only two families of roots instead of the four of the HH beam in which there is no such coupling. The two families are identified by m even (y odd in x coupled with ψ even in x) and m odd (y even in x coupled with ψ odd in x). For a given τ and Ω, the ratios of the corresponding lengths of beam, l_m, to the radius of gyration, r, are

$$l_m/r = 2\alpha_m/k_1$$

where the α_m are the roots of

(a) for m even:

$$\tan\alpha = \varphi_{FF}\tan\rho\alpha, \quad \Omega \geq 1,$$
$$\tan\alpha = -\varphi_{FF}\tanh\rho\alpha, \quad \Omega \leq 1,$$

(b) for m odd:

$$\tan\alpha = \varphi_{FF}^{-1}\tan\rho\alpha, \quad \Omega > 1,$$
$$\tan\alpha = -\tau\alpha, \quad \Omega = 1,$$
$$\tan\alpha = \varphi_{FF}^{-1}\tanh\rho\alpha, \quad \Omega < 1,$$

where

$$\rho = \hat{\kappa}_2/\kappa_1, \quad \hat{\kappa}_2 = |\kappa_2^2|^{1/2}, \quad \varphi_{FF} = \hat{\kappa}_2(\kappa_2^2 - \Omega^2)/\kappa_1(\kappa_1^2 - \Omega^2).$$

(v) *FG*: *Beam free at one end and guided at the other* (x = 0 *at one end*)

The l_m/r are half the l_m/r for FF, m odd.

(vi) *FH*: *Beam free at one end and hinged at the other* (x = 0 *at one end*)

The l_m/r are half the l_m/r FF, m even.

(vii) *CC*: *Beam clamped at both ends* (x = 0 *at the center*)

Again, for a given τ and Ω, the l_m/r are equal to 2α; but, now, the α are the roots of

(a) for m even:

$$\tan\alpha = \varphi_{CC}\tan\rho\alpha, \quad \Omega \geq 1,$$
$$\tan\alpha = -\varphi_{CC}\tanh\rho\alpha, \quad \Omega \leq 1,$$

(b) for m odd:

$$\tan\alpha = \varphi_{CC}^{-1}\tan\rho\alpha, \quad \Omega > 1,$$
$$\tan\alpha = -\tau^{-1}\alpha, \quad \Omega = 1,$$
$$\tan\alpha = \varphi_{CC}^{-1}\tanh\rho\alpha, \quad \Omega < 1,$$

where

$$\varphi_{CC} = \rho^2/\varphi_{FF}.$$

(viii) *CG*: *Beam clamped at one end, guided at the other* (x = 0 *at one end*)

The l_m/r are half the l_m/r for CC, m odd.

(ix) *CH*: *Beam clamped at one end, hinged at the other* (x = 0 *at one end*)

The l_m/r are half the l_m/r for CC, m even.

The computations of the roots of the biquadratic equation and the ten transcendental equations are provided for in the two programs on Forms 5 and 6 and the instructions on Form 7. The first program contains the steps for calculating the roots of the biquadratic equation, for given τ and Ω, and the successive values of l_m/r and l_n/r for the HH beam. The branches $m = 1, 2, 3, \ldots$, for flexure are hyperbola-like with $\Omega \to \infty$ as $l_m/r \to 0$ and $l_m/r \to \infty$ as $\Omega \to 0$. The branches $n = 1, 2, 3, \ldots$, for shear, are also hyperbola-like with $\Omega \to \infty$ as $l_n/r \to 0$, but $l_n/r \to \infty$ as $\Omega \to 1$ and there is one branch $\Omega = 1$, $n = 0$ for all l_n/r. The program produces a blink display for $\Omega \leq 1$ for HH shear.

Part of the frequency spectrum for the HH beam is illustrated in Fig. 1 (taken from a previous paper [2]) for $\tau = 0.283$ and the ranges $1 \leq \Omega \leq 1.2$ and $0 \leq l/h \leq 32$, where

$$h = 2r\sqrt{3},$$

i.e. h is the depth of a beam of rectangular section with radius of gyration r. τ is the same as g of Ref. [2]. The change of notation is made to avoid possible confusion with the downshift prefix g on the HP-65.

Also on the first program card are preliminary calculations (for storage in the calculator and later recall by program card No. 2) to be used in the computations for the FF and CC beams. In addition to κ_1 and Ω^2 already calculated and stored for the HH beam, the stored results are φ_{FF} and $-\tau$ for the FF beam or φ_{CC} and $-\tau^{-1}$ for the CC beam, and, finally, ρ and the set for radian angular mode for both beams.

The program on the second card includes various conditional devices which, in conjunction with the stored quantities for the first card, select one of the ten transcendental equations which is then solved by Schelkunoff's method. The selection of the proper equation for Ω less than, equal to or greater than unity is automatic. The choice of m odd or even is made with the setting of Flag 1 off or on (Form 7 Instructions 10 or 11). The choice between the free-free beam and the clamped-clamped beam is made with the first program card (Form 7 Instructions 7 and 8). The results of calculations for the FF beam are given in Fig. 2 for $\tau = 0.283$ and the ranges $0.8 \leq \Omega \leq 1.2$ and $0 < l/h \leq 32$. Again $h = 2r\sqrt{3}$ and Fig. 2 is taken from a previous paper [3]. The full lines are the branches m even and the dashed lines m odd.

Schelkunoff's scheme again uses

$$\alpha = q\pi + \delta, \quad q = 1, 2, 3\ldots$$

but, in this case, the relation between q and the order m of a branch of one of the transcendental equations is more complicated. For $\Omega \leq 1$,

$$m \text{ odd} = 2q - 1, \quad m \text{ even} = 2q.$$

As Ω increases above $\Omega = 1$, each branch m even crosses and recrosses the branch $m - 1$. At each crossing, q for each of the two branches diminishes by unity. For the computations to follow a given branch, the input integer, q, must be diminished by unity after the frequency rises above an intersection of branches. This is illustrated in Table 3 which gives the results of computations for branches $m = 23$ to $m = 26$ in the range $0.99 \leq \Omega \leq 1.01$ at intervals of $\Omega = 0.0025$. Branches 23 and 24 cross between $\Omega = 1.0050$ and $\Omega = 1.0075$ and q decreases from 12 to 11. Branches 25 and 26 cross between $\Omega = 1.0025$ and $\Omega = 1.0050$ and q decreases from 13 to 12. As a result, at

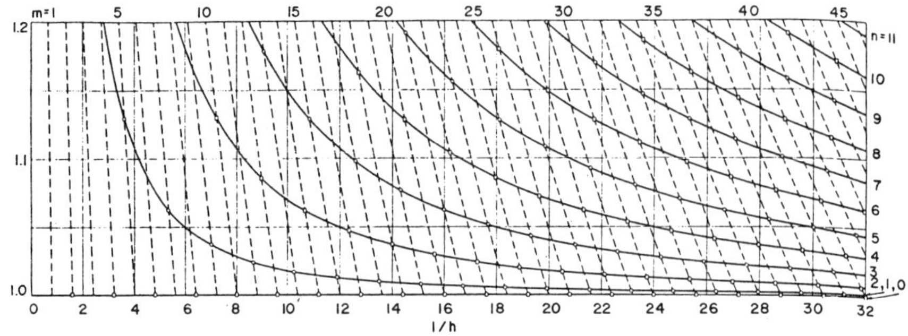

Fig. 1. Frequency spectrum: hinged-hinged Timoshenko beams.

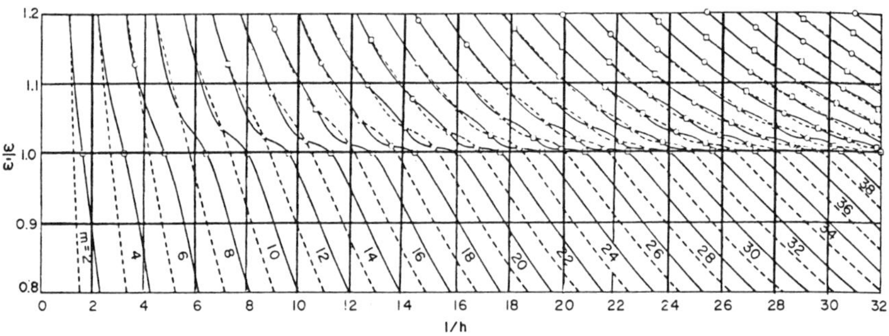

Fig. 2. Frequency spectrum: free-free Timoshenko beams.

Table 3. Free-free and clamped-clamped Timoshenko beams

		Free-Free				Clamped-Clamped			
Ω	m =	23	24	25	26	23	24	25	26
1.0100	q	11	11	12	12	11	11	12	12
	ℓ_m/h	18.20	17.67	19.76	19.19	18.26	17.49	19.84	19.07
	ℓ_m/r	63.05	61.20	68.46	66.49	63.25	60.58	68.74	66.06
1.0075	q	11	11	12	12	11	11	12	12
	ℓ_m/h	18.28	17.86	19.85	19.32	18.30	17.54	19.89	19.12
	ℓ_m/r	63.32	61.86	68.17	66.94	63.39	60.78	68.90	66.62
1.0050	q	12	12	12	12	12	12	12	12
	ℓ_m/h	18.34	skip	skip	19.79	18.34	18.93	skip	skip
	ℓ_m/r	63.55			68.56	63.53	65.59		
1.0025	q	12	12	13	13	12	12	13	13
	ℓ_m/h	18.41	19.07	20.00	20.65	18.38	19.17	19.98	20.76
	ℓ_m/r	63.76	66.07	69.28	71.52	63.67	66.40	69.20	71.93
1.0000	q	12	12	13	13	12	12	13	13
	ℓ_m/h	18.46	19.22	20.06	20.82	18.42	19.22	20.02	20.82
	ℓ_m/r	63.96	66.57	69.50	72.11	63.81	66.57	69.35	72.11
0.9950	q	12	12	13	13	12	12	13	13
	ℓ_m/h	18.57	19.36	20.18	20.97	18.50	19.30	20.11	20.91
	ℓ_m/r	64.34	67.08	69.90	72.65	64.08	66.86	69.65	72.43
0.9900	q	12	12	13	13	12	12	13	13
	ℓ_m/h	18.68	19.48	20.29	21.09	18.58	19.38	20.10	21.00
	ℓ_m/r	64.70	67.47	70.29	73.07	64.36	67.15	69.95	72.74

$\Omega = 1.0050$, q is 12 for both branch 23 and branch 25 of the family m odd and for both branch 24 and branch 26 of the family m even. Since, for each family, a single input of $q = 12$ can produce only one root, the other root is skipped. Which of the two roots in each family is the one skipped depends on the initial value of δ_i. (In the computations for Table 3, the initial value of δ_i for $q = 12$ was the final value for $q = 11$). The skipped root can sometimes be recovered by proceeding to the next root and then returning. In the present case, advance to $q = 13$ and return to $q = 12$ produces

$$l_{24}/h = 18.54, \quad l_{24}/r = 64.21;$$
$$l_{25}/h = 19.93, \quad l_{25}/r = 69.04$$

to fill the blank spaces for the FF beam in Table 3; but that procedure does not always work; it depends on a fortuitous initial value of δ_i. In any event, the most important information in the neighborhood of the intersection would be the intersection point itself, and the coordinates of all those points are given, more readily, by[2].

$$\Omega = \tau^{1/2}(m^2 + n^2)[(\tau m^2 - n^2)(m^2 - \tau n^2)]^{-1/2},$$
$$l/h = (\pi/2)[(\tau m^2 - n^2)(m^2 - \tau n^2)/3\tau(1+\tau)(m^2+n^2)]^{1/2},$$

where the integers m and n must be both even or both odd. The coordinates of the intersections of the branches m and n for the hinged-hinged beam, in Fig. 1, are given by the same formulas. The intersections covered in Table 3 are for $m = 23, 25$, $n = 1$, yielding

$$\Omega = 1.00552, \quad l/r = 63.50, \quad l/h = 18.33;$$
$$\Omega = 1.00467, \quad l/r = 69.07, \quad l/h = 19.94,$$

as calculated with the program on Forms 9 and 10.

The results for the clamped-clamped beam are much like those for the free-free beam. In fact, the intersections of the branches m and $m-1$, in $\Omega > 1$, are at the same points for the two cases and, as $\Omega \to 0$, the transcendental equations for the region $\Omega < 1$ approach the same limit:

$$\cos\alpha = \operatorname{sech}\alpha,$$

i.e. the equation for Cases VI and VII of Bernoulli's low frequency approximation. As a consequence of these coincidences both above and below $\Omega = 1$, the two sets of numerical results do not differ very much between, as may be observed in Table 3. Note that, for the clamped-clamped beam, the skipped root at $\Omega = 1.0050$ for $q = 12$, m even, following $q = 11$, m even, is on branch 26 instead of on branch 24—as it was for the free-free beam. The skipped root for branch 25 can be recovered, as in the case of the free-free beam, by advancing to $q = 13$ and returning to $q = 12$, with the result

$$l_{25}/h = 19.03, \quad l_{25}/r = 69.05.$$

However, the same procedure does not work for branch 26; the result for branch 24 is duplicated; instead. One alternative for obtaining a root in the neighborhood is to advance to a nearby frequency and find

$$\Omega = 1.0051, \quad l_{26}/h = 19.37, \quad l_{26}/r = 67.09.$$

The other alternative is to use the algebraic formulas for the coordinates of the intersection points.

5. OSCILLATORY ITERATION

In Schelkunoff's method, the iteration becomes oscillatory if, in

$$\delta_{i+1} = \tan^{-1}[f(q\pi + \delta_i)]$$

any δ_{i+1} duplicates, within a difference ϵ, a former value before $i = j$ for which $|\delta_{j+1} - \delta_j| < \epsilon$; i.e. if $\delta_1 = \tan^{-1}[f(q\pi + \delta_{i+k})]$, $k = 1, 2, 3, \ldots$; $i + k < j + 1$.

An example, for $k = 1$, is found in the equation

$$\tan\alpha = -20\tan 1.75\alpha$$

which can be produced with fictitious inputs on card No. 2 of the program for the Timoshenko beam, as specified in the instructions on Form 8. After the program has run at least 30 sec (10 iterations) the current value of δ_i, following each successive iteration (3 sec) is either ± 1.547678132, so that

$$|\delta_{i+1} - \delta_i| = 3.095356264$$

and never satisfies $|\delta_{j+1} - \delta_j| < 10^{-6}$.

REFERENCES

1. S. Timoshenko, *Vibration Problems in Engineering* 2nd Edn, pp. 331–347. D. Van Nostrand, New York (1937).
2. R. D. Mindlin, Thickness-shear and flexural vibrations of crystal plates. *J. Appl. Phys.* **22**, 316–323 (1951).
3. R. D. Mindlin, Forced thickness-shear and flexural vibrations of piezoelectric crystal plates. *J. Appl. Phys.* **23**, 83–88 (1952).

HP-65 Program Form

SWITCH TO W/PRGM. PRESS [f] [PRGM] TO CLEAR MEMORY.

Bernoulli Beam

KEY ENTRY	CODE SHOWN	COMMENTS
g	35	
RAD	42	Set for radian mode
g	35	
π	02	π
×	71	$n\pi = \alpha_n$ for Cases VIII & IX and
STO 4	33 04	$\alpha_n + \pi/2$ for Case X: both in R_4
R/S	84	
LBL	23	Start of iteration
E	15	
10 RCL 5	34 05	δ_{i+1} (= δ_i for next cycle)
STO 6	33 06	δ_i
RCL 4	34 04	$n\pi$
+	61	$n\pi + \delta_i = \alpha_i$
STO 7	33 07	α_i in R_7
f^{-1}	32	To
TF 1	61	skip to
GTO	22	Step 41 for
1	01	Cases I, II, III, IV.
g	35	Start of α'_i (for Cases V, VI, VII)
20 π	02	π
2	02	2
÷	81	$\pi/2$
STO 8	33 08	$\pi/2$ in R_8
−	51	$\alpha_i - \pi/2 = \alpha'_i$
f^{-1}	32	Start of sech α'_i
LN	07	$e^{\alpha'_i}$
E↑	41	To store $e^{\alpha'_i}$ in the Stack
g	35	
1/x	04	$e^{-\alpha'_i}$
30 +	61	$e^{\alpha'_i} + e^{-\alpha'_i}$
2	02	2
÷	81	$\frac{1}{2}(e^{\alpha'_i} + e^{-\alpha'_i}) = \cosh\alpha'_i$
g	35	
1/x	04	sech α'_i (for Cases V, VI, VII)
RCL 7	34 07	α_i
f	31	
COS	05	$\cos\alpha_i$
÷	81	sech $\alpha'_i/\cos\alpha_i$
GTO	22	Skip to Step 55
40 2	02	for Cases V, VI, VII
LBL	23	Start of tanh α_i
1	01	for Cases I, II, III, IV
RCL 7	34 07	α_i
2	02	2
×	71	$2\alpha_i$
f^{-1}	32	
LN	07	$e^{2\alpha_i}$
1	01	1
−	51	$e^{2\alpha_i} - 1$
50 E↑	41	To store $e^{2\alpha_i} - 1$ in the Stack

KEY ENTRY	CODE SHOWN	COMMENTS
E↑	41	To store $e^{2\alpha_i} - 1$ in the Stack
2	02	2
+	61	$e^{2\alpha_i} + 1$
÷	81	tanh α_i (for Cases I & II)
LBL	23	Resume from Step 40
2	02	
f	31	−tanh α for
TF 2	81	Cases III & IV and
g NOP	35 01	−sech $\alpha'_i/\cos\alpha_i$ for
60 CHS	42	Case V
f^{-1}	32	
TAN	06	$\tan^{-1}[\pm\tanh\alpha_i$ or sech $\alpha'_i/\cos\alpha_i] = \delta_{i+1}$
STO 5	33 05	δ_{i+1} in R_5
RCL 6	34 06	δ_i
−	51	$\delta_{i+1} - \delta_i$
g	35	
ABS	06	$\lvert\delta_{i+1} - \delta_i\rvert$
EEX	43	
CHS	42	
70 6	06	$\epsilon = 10^{-6}$ (User's choice)
g x<y	35 22	If $\lvert\delta_{i+1} - \delta_i\rvert < \epsilon$, skips to Step 74
GTO	22	otherwise returns to Step 8
E	15	for another cycle
RCL 4	34 04	$n\pi$
RCL 5	34 05	δ_{j+1} (j = last i)
+	61	$n\pi + \delta_{j+1} = \alpha_n$ or α'_n
STO 5	33 05	α_n for I, II, III, IV or α'_n for V, VI, VII in R_5
f^{-1}	32	To separate
TF 1	61	α_n for V, VI, VII
80 GTO	22	from
B	12	α_n for I, II, III, IV
RCL 8	34 08	$\pi/2$
−	51	$n\pi + \delta_{j+1} - \pi/2 = \alpha_n$ for V, VI, VII
STO 6	33 06	α_n for V, VI, VII in R_6
LBL	23	Start of computation of f_n
B	12	
RCL 1	34 01	ℓ
÷	81	α_n/ℓ
f^{-1}	32	
90 $\sqrt{x}$	09	α_n^2/ℓ^2
RCL 2	34 02	r
×	71	$\alpha_n^2 r/\ell^2$
RCL 3	34 03	v
×	71	$\alpha_n^2 rv/\ell^2$
g	35	
π	02	π
2	02	2
×	71	2π
÷	81	$\alpha_n^2 rv/2\pi\ell^2 = f_n$
100 RTN	24	

REGISTERS	
R_1	ℓ
R_2	r
R_3	v
R_4	$n\pi$ $= \alpha_n$ for VIII, IX $= \alpha_n + \frac{\pi}{2}$ for X
R_5	δ_{i+1} and α_n for I, II, III, IV
R_6	δ_i and α_n for V, VI, VII
R_7	α_i
R_8	$\pi/2$
R_9	used

LABELS	
A	GO
B	used
C	
D	
E	used
0	
1	used
2	used
3	
4	
5	
6	
7	
8	
9	

FLAGS	
1	used
2	used

TO RECORD PROGRAM INSERT MAGNETIC CARD WITH SWITCH SET AT W/PRGM.

Form 1. Program for Bernoulli beams with homogeneous end-conditions.

HP-65 User Instructions

Bernoulli Beam
GO | VIII, IX, X

STEP	INSTRUCTIONS	INPUT DATA/UNITS	KEYS		OUTPUT DATA/UNITS
1	Enter Program				
2	Input: length of beam	ℓ (cm)	STO	1	
3	Input: radius of gyration	r (cm)	STO	2	
4	Input: bar velocity	v (cm/sec)	STO	3	
5	Set for Cases I & II: Fixed-Hinged & Free-Hinged		f^{-1}	SF1	
	(Flags 1 and 2 OFF)		f^{-1}	SF2	
6	Compute frequency of nth mode	n	A	R/S	f_n (Hz)
7	Display n^{th} root of $\tan\alpha = \tanh\alpha$ (optional)		RCL	5	α_n
	For new n, go to Step 6				
8	Set for Cases III & IV: Fixed-Guided & Free-Guided		f^{-1}	SF1	
	(Flag 1 OFF, Flag 2 ON)		f	SF2	
9	Compute frequency of nth mode	n	A	R/S	f_n (Hz)
10	Display n^{th} root of $\tan\alpha = -\tanh\alpha$ (optional)		RCL	5	α_n
	For new n, go to Step 9				
11	Set for Case V: Fixed-Free		f	SF1	
	(Flags 1 and 2 ON)		f	SF2	
12	Compute frequency of nth mode	n	A	R/S	f_n (Hz)
13	Display n^{th} root of $\cos\alpha = -$sechα (optional)		RCL	6	α_n
	For new n, go to Step 12				
14	Set for Cases VI & VII: Free-Free & Fixed-Fixed		f	SF1	
	(Flag 1 ON, Flag 2 OFF)		f^{-1}	SF2	
15	Compute frequency of nth mode	n	A	R/S	f_n (Hz)
16	Display n^{th} root of $\cos\alpha =$ sechα (optional)		RCL	6	α_n
	For new n, go to Step 15				
	Cases VIII & IX: Hinged-Hinged & Guided-Guided				
17	Compute n^{th} root of $\sin\alpha = 0$ (required)	n	A		α_n
18	Compute frequency of nth mode		B		f_n (Hz)
	For new n, go to Step 17				
	Case X: Hinged-Guided				
19	Compute n^{th} root of $\cos\alpha = 0$ (required)	$n - \frac{1}{2}$	A		α_n
20	Compute frequency of nth mode		B		f_n (Hz)
	For new n, go to Step 19				
	Any order of Cases and integers n is permissible.				
	For new beam, go to one or more of Steps 2, 3, 4.				

Form 2. Instructions for program for Bernoulli beams with homogeneous end-conditions.

HP-65 Program Form

SWITCH TO W/PRGM. PRESS [f] [PRGM] TO CLEAR MEMORY.

Two-span Bernoulli Beam

KEY ENTRY	CODE SHOWN	COMMENTS
g	35	
RAD	42	Set for radian mode
g	35	
π	02	π
×	71	mπ for f^{-1} SF1, nπ for f SF1
STO 5	33 05	mπ or nπ in R_5
LBL	23	
E	15	Start of iteration
RCL 6	34 06	δ_{i+1} (= δ_i for next cycle)
10 STO 7	33 07	δ_i
RCL 5	34 05	mπ or nπ
+	61	$m\pi+\delta_i=\alpha_i$ or $n\pi+\delta_i=\beta_i$
RCL 2	34 02	ρ
f	31	From here on, "... or ..."
TF 1	61	signifies f^{-1} SF1 or f SF1 on
g	35	the keyboard
1/x	04	ρ^{-1}
×	71	$\alpha_i\rho=\beta_i$ or $\beta_i\rho^{-1}=\alpha_i$
STO 8	33 08	β_i or α_i in R_8
20 2	02	2
×	71	$2\beta_i$ or $2\alpha_i$
f^{-1}	32	
LN	07	$e^{2\beta_i}$ or $e^{2\alpha_i}$
1	01	1
+	61	$e^{2\beta_i}+1$ or $e^{2\alpha_i}+1$
E↑	41	To store $e^{2\beta_i}+1$ or $e^{2\alpha_i}+1$
E↑	41	in the stack
2	02	2
−	51	$e^{2\beta_i}-1$ or $e^{2\alpha_i}-1$
30 ÷	81	$\coth\beta_i$ or $\coth\alpha_i$
RCL 8	34 08	β_i or α_i
f	31	
TAN	06	$\tan\beta_i$ or $\tan\alpha_i$
g	35	
1/x	04	$\cot\beta_i$ or $\cot\alpha_i$
−	51	$\coth\beta_i-\cot\beta_i$ or $\beta_i\to\alpha_i$
STO 8	33 08	$\coth\beta_i-\cot\beta_i$ or $\beta_i\to\alpha_i$ in R_8
RCL 5	34 05	mπ or nπ
RCL 7	34 07	δ_i
40 +	61	α_i or β_i
2	02	2
×	71	$2\alpha_i$ or $2\beta_i$
f^{-1}	32	
LN	07	$e^{2\alpha_i}$ or $e^{2\beta_i}$
1	01	1
+	61	$e^{2\alpha_i}+1$ or $e^{2\beta_i}+1$
E↑	41	To store $e^{2\alpha_i}+1$ or $e^{2\beta_i}+1$
E↑	41	in the stack
2	02	2
50 −	51	$e^{2\alpha_i}-1$ or $e^{2\beta_i}-1$

KEY ENTRY	CODE SHOWN	COMMENTS
÷	81	$\coth\alpha_i$ or $\coth\beta_i$
RCL 8	34 08	$\coth\beta_i-\cot\beta_i$ or $\beta_i\to\alpha_i$
+	61	$\coth\beta_i-\cot\beta_i+\coth\alpha_i$ or $\beta_i\rightleftarrows\alpha_i$
g	35	
1/x	04	$(\coth\beta_i-\cot\beta_i+\coth\alpha_i)^{-1}$ or $\beta_i\rightleftarrows\alpha_i$
f^{-1}	32	$\tan^{-1}(\coth\beta_i-\cot\beta_i+\coth\alpha_i)^{-1}$
TAN	06	$=\delta_{i+1}$ (or $\beta_i\rightleftarrows\alpha_i$)
STO 6	33 06	δ_{i+1} in R_6
RCL 7	34 07	δ_i
60 f	31	f SF2 on the keyboard sets for
TF 2	81	manual iteration
GTO	22	f^{-1} SF2 on the keyboard sets for
2	02	automatic iteration
−	51	$\delta_{i+1}-\delta_i$
g	35	
ABS	06	$\lvert\delta_{i+1}-\delta_i\rvert$
EEX	43	
CHS	42	
6	06	$\epsilon=10^{-6}$ (User's choice)
70 g x≤y	35 22	If $\lvert\delta_{i+1}-\delta_i\rvert<\epsilon$, skips to Step 73
GTO	22	otherwise returns to Step 7
E	15	for another cycle
LBL	23	Skips from Step 59 to here
2	02	for manual iteration (f SF2)
RCL 5	34 05	mπ or nπ
RCL 6	34 06	δ_{j+1} (j = last i)
+	61	$m\pi+\delta_{j+1}=\alpha_m$ or $n\pi+\delta_{j+1}=\beta_n$
f^{-1}	32	Goes to Step 84 for Family I
TF1	61	Goes to Step 82 for Family II
80 GTO	22	
1	01	
RCL 2	34 02	ρ
÷	81	$\beta_n/\rho=\alpha_n$ (Family II, f SF1)
LBL	23	Skips from Step 77 to here
1	01	for Family I, f^{-1} SF1
STO 8	33 08	α_m or α_n in R_8
RCL 1	34 01	ℓ_1
÷	81	$(\alpha_m$ or $\alpha_n)/\ell_1$
f^{-1}	32	
90 $\sqrt{x}$	09	$(\alpha_m^2$ or $\alpha_n^2)/\ell_1^2$
RCL 3	34 03	r
×	71	$(\alpha_m^2$ or $\alpha_n^2)r/\ell_1^2$
RCL 4	34 04	v
×	71	$(\alpha_m^2$ or $\alpha_n^2)rv/\ell_1^2$
g	35	
π	02	π
2	02	2
×	71	2π
÷	81	$(\alpha_m^2$ or $\alpha_n^2)rv/2\pi\ell_1^2=f_m$ or f_n
100 RTN	24	

REGISTERS	
R_1	ℓ_1
R_2	ρ
R_3	r
R_4	v
R_5	mπ or nπ
R_6	δ_{i+1}
R_7	δ_i
R_8	used, α_m or α_n
R_9	used

LABELS	
A	GO
B	
C	
D	
E	used
0	
1	used
2	used
3	
4	
5	
6	
7	
8	
9	

FLAGS	
1	used
2	used

TO RECORD PROGRAM INSERT MAGNETIC CARD WITH SWITCH SET AT W/PRGM

Form 3. Program for two-span Bernoulli beams on hinged supports.

HP-65 User Instructions

Two-span Bernoulli Beam

GO | | | | Manual

STEP	INSTRUCTIONS	INPUT DATA/UNITS	KEYS		OUTPUT DATA/UNITS
1	Enter Program				
2	Input: length of Span 1	ℓ_1 (cm)	STO	1	
3	Input: ratio of span-lengths (ℓ_2/ℓ_1)	ρ	STO	2	
4	Input: radius of gyration	r (cm)	STO	3	
5	Input: bar velocity	v (cm/sec)	STO	4	
6	Set for Family I		f^{-1}	SF1	
7	Compute m^{th} frequency of Family I	m	A		f_m (Hz)
8	Display m^{th} root of Family I (optional)		RCL	8	α_m
	For another m for Family I, go to Step 7				
9	Set for Family II		f	SF1	
10	Compute n^{th} frequency of Family II	n	A		f_n (Hz)
11	Display n^{th} root of Family II (optional)		RCL	8	α_n
	for another n for Family II, go to Step 10				
	For manual iteration:				
	Key in f SF2 and press E repeatedly.				
	To return to automatic iteration:				
	Key in f^{-1} SF2 and press E.				

Form 4. Instructions for program for two-span Bernoulli beams on hinged supports.

HP-65 Program Form

SWITCH TO W/PRGM. PRESS [f] [PRGM] TO CLEAR MEMORY.

Timoshenko Beam — Page 1 of 2

KEY ENTRY	CODE SHOWN	COMMENTS
1	01	1
+	61	$1+\tau$
RCL 1	34 01	Ω
f^{-1}	32	
$\sqrt{x}$	09	Ω^2
STO 3	33 03	Ω^2 in R_3
×	71	$(1+\tau)\Omega^2$
2	02	2
÷	81	$\frac{1}{2}(1+\tau)\Omega^2 = C_1$
10 STO 4	33 04	C_1 in R_4
1	01	1
RCL 3	34 03	Ω^2
−	51	$1-\Omega^2$
RCL 3	34 03	Ω^2
×	71	$\Omega^2(1-\Omega^2)$
RCL 2	34 02	τ
×	71	$\tau\Omega^2(1-\Omega^2)$
RCL 4	34 04	C_1
f^{-1}	32	
20 $\sqrt{x}$	09	C_1^2
÷	81	$\tau\Omega^2(1-\Omega^2)/C_1^2 = C_2$
1	01	1
+	61	$1+C_2$
f	31	
$\sqrt{x}$	09	$\sqrt{1+C_2}$
RCL 4	34 04	C_1
×	71	$C_1\sqrt{1+C_2}$
STO 5	33 05	$C_1\sqrt{1+C_2}$ in R_5
RCL 4	34 04	C_1
30 +	61	$C_1(1+\sqrt{1+C_2}) = \kappa_1^2$
STO 6	33 06	κ_1^2 in R_6
f	31	
$\sqrt{x}$	09	κ_1
STO 1	33 01	κ_1 in R_1
RCL 4	34 04	C_1
RCL 5	34 05	$C_1\sqrt{1+C_2}$
−	51	$C_1(1-\sqrt{1+C_2}) = \kappa_2^2$
STO 7	33 07	κ_2^2 in R_7
g	35	
40 ABS	06	$\lvert\kappa_2^2\rvert$
f	31	
$\sqrt{x}$	09	$\lvert\kappa_2^2\rvert^{1/2} = \hat{\kappa}_2$
STO 5	33 05	$\hat{\kappa}_2$ in R_5
RCL 3	34 03	Ω^2
RCL 7	34 07	κ_2^2
−	51	$\Omega^2-\kappa_2^2$
RCL 3	34 03	Ω^2
RCL 6	34 06	κ_1^2
−	51	$\Omega^2-\kappa_1^2$
50 ÷	81	$(\Omega^2-\kappa_2^2)/(\Omega^2-\kappa_1^2)$
RCL 5	34 05	$\hat{\kappa}_2$
RCL 1	34 01	κ_1
÷	81	$\hat{\kappa}_2/\kappa_1 = \rho$
STO 5	33 05	ρ in R_5
×	71	$\rho(\Omega^2-\kappa_2^2)/(\Omega^2-\kappa_1^2) = \varphi_{FF}$
STO 4	33 04	φ_{FF} in R_4
RTN	24	
LBL	23	Start of computation of l_m/r
B	12	for HH flexure
60 g	35	
π	02	π
×	71	$m\pi$
RCL 1	34 01	κ_1
÷	81	$m\pi/\kappa_1 = l_m/r$ for HH flexure
RTN	24	
LBL	23	Start of computation of l_n/r
C	13	for HH axial shear
g	35	
π	02	π
70 ×	71	$n\pi$
RCL 7	34 07	κ_2^2
f	31	
$\sqrt{x}$	09	κ_2 (or Blink if $\Omega<1$)
÷	81	$n\pi/\kappa_2 = l_n/r$ for HH axial
RTN	24	shear (or Blink if $\Omega=0$)
LBL	23	Start of preparation for
D	14	Card #2
g	35	
RAD	42	Set for radian mode
80 RCL 2	34 02	τ
CHS	42	$-\tau$
STO 2	33 02	$-\tau$ in R_2
R/S	84	Preparation complete for FF
g	35	Start of additional steps for CC
1/x	04	$-1/\tau$
STO 2	33 02	$-1/\tau$ in R_2
0	00	0
RCL 4	34 04	φ_{FF}
g x=y	35 23	
90 STO 4	33 04	$\varphi_{FF} = \varphi_{CC} = 0$ in R_4
RTN	24	Stops here if $\varphi_{CC} = 0$
g	35	
1/x	04	φ_{FF}^{-1}
RCL 5	34 05	ρ
f^{-1}	32	
$\sqrt{x}$	09	ρ^2
×	71	$\rho^2\varphi_{FF}^{-1} = \varphi_{CC}$
STO 4	33 04	φ_{CC} in R_4
RTN	24	
100 g NOP	35 01	

REGISTERS	
R_1	Ω
	κ_1
R_2	τ or
	$-\tau$ for FF
	$-1/\tau$ for CC
R_3	Ω^2
R_4	C_1
	φ_{FF} or
	φ_{CC}
R_5	$C_1\sqrt{1+C_2}$
	$\hat{\kappa}_2$
	ρ
R_6	κ_1^2
R_7	κ_2^2
R_8	not used
R_9	not used

LABELS	
A	GO
B	HH FLEX
C	HH SHEAR
D	FF or CC
E	
0	
1	
2	
3	
4	
5	
6	
7	
8	
9	

FLAGS	
1	
2	

TO RECORD PROGRAM INSERT MAGNETIC CARD WITH SWITCH SET AT W/PRGM.

Form 5. Part 1 of program for Timoshenko beams with homogeneous end-conditions.

HP-65 Program Form

SWITCH TO W/PRGM. PRESS [f] [PRGM] TO CLEAR MEMORY.

Timoshenko Beam — Page 2 of 2

KEY ENTRY	CODE SHOWN	COMMENTS
g	35	
π	02	π
×	71	$q\pi$
STO 6	33 06	$q\pi$ in R_6
LBL	23	Start of iteration
E	15	
RCL 8	34 08	δ_{i+1} (= δ_i for next cycle)
STO 7	33 07	δ_i
RCL 6	34 06	$q\pi$
10 +	61	$q\pi+\delta_i$
1	01	1
RCL 3	34 03	Ω^2
g x=y	35 23	If $\Omega=1$, goes to Step 23 for
f	31	m even (f SF1) or to Step 18
TF1	61	for m odd (f^{-1} SF1)
GTO	22	If $\Omega\neq1$, goes to Step 23 for
2	02	m odd or even
RCL 2	34 02	$-\tau^{\pm1}$ (upper sign on $\tau^{\pm1}$ is for
g R↑	35 09	FF beam, lower for CC beam)
20 ×	71	$-\tau^{\pm1}(q\pi+\delta_i)$
GTO	22	$-\tau^{\pm1}(q\pi+\delta_i)$ goes to Step 72
3	03	for $\Omega=1$, m odd (f^{-1}SF1)
LBL	23	Continues computation for
2	02	$\Omega=1$, m even, and $\Omega\neq1$
RCL 5	34 05	ρ
g R↑	35 09	
×	71	$\rho(q\pi+\delta_i)$
STO 2	33 02	$\rho(q\pi+\delta_i)$ in R_2
1	01	1
30 RCL 3	34 03	Ω^2
g x>y	35 24	If $\Omega<1$, goes to Step 34
GTO	22	If $\Omega>1$, goes to Step 61
4	04	
RCL 2	34 02	$\rho(q\pi+\delta_i)$
2	02	2
×	71	$2\rho(q\pi+\delta_i)$
f^{-1}	32	
LN	07	$e^{2\rho(q\pi+\delta_i)}$
1	01	1
40 −	51	$e^{2\rho(q\pi+\delta_i)}-1$
E↑	41	To retain $e^{2\rho(q\pi+\delta_i)}-1$
E↑	41	in the stack
2	02	2
+	61	$e^{2\rho(q\pi+\delta_i)}+1$
÷	81	$\tanh\rho(q\pi+\delta_i)$
RCL 4	34 04	φ (= φ_{FF} or φ_{CC})
f^{-1}	32	If Flag 1 is off (m odd),
TF1	61	$\varphi\to\varphi^{-1}$. If Flag 1 is on (m even),
g	35	no change. (Both for $\Omega<1$).
50 1/x	04	φ^{-1} (for m odd).
×	71	$\varphi^{\pm1}\tanh\rho(q\pi+\delta_i)$
f	31	In the sequel, upper signs of $\mp\varphi^{\pm1}$
TF1	61	are for m even, lower for m odd.
CHS	42	Change from $+\varphi^{\pm1}$ to $-\varphi^{\pm1}$
g NOP	35 01	for m even (f SF1)
f^{-1}	32	$\tan^{-1}[\mp\varphi^{\pm1}\tanh\rho(q\pi+\delta_i)] = \delta_{i+1}$
TAN	06	for $\Omega=1$, m even, or $\Omega<1$.
STO 8	33 08	δ_{i+1} for $\Omega=1$, m even or $\Omega<1$ in R_8
GTO	22	Computation for preceding δ_{i+1}
60 5	05	resumes at Step 80.
LBL	23	Computation of δ_{i+1} for $\Omega>1$
4	04	resumes here from Step 33.
RCL 2	34 02	$\rho(q\pi+\delta_i)$
f	31	
TAN	06	$\tan\rho(q\pi+\delta_i)$
RCL 4	34 04	φ
f^{-1}	32	Steps 67–70 are the same as
TF1	61	Steps 47–50, but for $\Omega>1$.
g	35	
70 1/x	04	φ^{-1} (for m odd)
×	71	$\varphi^{\pm1}\tan\rho(q\pi+\delta_i)$
LBL	23	$-\tau^{\pm1}(q\pi+\delta_i)$ from Step 22
3	03	(for $\Omega=1$, m odd).
f^{-1}	32	$\tan^{-1}[-\tau^{\pm1}(q\pi+\delta_i)]$ for $\Omega=1$, m odd, or
TAN	06	$\tan^{-1}[\varphi^{\pm1}\tan\rho(q\pi+\delta_i)]$ for $\Omega>1 = \delta_{i+1}$
STO 8	33 08	δ_{i+1} for $\Omega=1$, m odd or for $\Omega>1$ in R_8
RCL 3	34 03	Ω^2
1	01	1
g x>y	35 24	If $\Omega>1$, goes to Step 82
80 LBL	23	If $\Omega<1$, continues from Step 60.
5	05	
RCL 8	34 08	δ_{i+1} from Step 58 or Step 76.
RCL 7	34 07	δ_i
−	51	$\delta_{i+1}-\delta_i$
g	35	
ABS	06	$\lvert\delta_{i+1}-\delta_i\rvert$
EEX	43	
CHS	42	
6	06	$10^{-6}=\epsilon$ (User's choice)
90 g x≤y	35 22	If $\lvert\delta_{i+1}-\delta_i\rvert>\epsilon$, returns to
GTO	22	Step 5 for another cycle;
E	15	Otherwise proceeds to end.
RCL 6	34 06	$q\pi$
RCL 8	34 08	δ_{j+1} (j = last i)
+	61	$q\pi+\delta_{j+1}$
RCL 1	34 01	κ_1
÷	81	$(q\pi+\delta_{j+1})/\kappa_1$
2	02	2
×	71	$2(q\pi+\delta_{j+1})/\kappa_1 = l_m/r$
100 RTN	24	

REGISTERS	
R_1	κ_1
R_2	$-\tau^{\pm1}$
	$\rho(q\pi+\delta_i)$
R_3	Ω^2
R_4	φ_{FF} or φ_{CC}
R_5	ρ
R_6	$q\pi$
R_7	δ_i
R_8	δ_{i+1}
R_9	used

LABELS	
A	Loop
B	
C	
D	
E	GO
0	
1	
2	used
3	used
4	used
5	used
6	
7	
8	
9	

FLAGS	
1	used
2	

TO RECORD PROGRAM INSERT MAGNETIC CARD WITH SWITCH SET AT W/PRGM.

Form 6. Part 2 of program for Timoshenko beams with homogeneous end-conditions.

HP-65 User Instructions

Timoshenko Beam #1: GO | HH FLEX | HH SHEAR | PREP FF, CC |

Timoshenko Beam #2: | | | | FF, CC

STEP	INSTRUCTIONS	INPUT DATA/UNITS	KEYS		OUTPUT DATA/UNITS
1	Enter Card #1				
2	Input: Frequency Ratio	Ω	STO	1	
3	Input: Material and Shape Factor	τ	STO	2	
4	Compute Constants		A		
5	ℓ_m/r for HH Flex $[y(x)=-(-1)^m y(-x),\ \psi(x)=(-1)^m \psi(-x)]$	m	B		ℓ_m/r
	For new m go to Step 5. For new Ω go to Step 2.				
6	ℓ_n/r for HH Shear $[y(x)=-(-1)^n y(-x),\ \psi(x)=(-1)^n \psi(-x)]$	n	C		ℓ_n/r
	For new n go to Step 6. For new Ω go to Step 2.				
	Note: Blink display for $\Omega \leq 1$, HH Shear.				
	Note: Order of m, n, Steps 5, 6 immaterial.				
7	To make ready for FF Beam, after Step 4, 5 or 6.		D		
8	To make ready for CC Beam, after Step 7.		R/S		
9	Enter Card #2 to calculate ℓ_m/r:				
	For FF Beam, after Step 7 or				
	For CC Beam, after Steps 7 and 8.				
10	For branches m odd, set Flag 1 off.		f^{-1}	SF1	
11	For branches m even, set Flag 1 on.		f	SF1	
12	Input integer q and compute ℓ_m/r	q	A		ℓ_m/r
	For new q go to Step 12 for same flag setting,				
	otherwise go to appropriate Step 10 or 11.				
	For new frequency, return to Card #1, Steps 1,2,3,4				
	and Step 7 for FF Beam or Steps 7 and 8 for CC Beam.				
	Notes for FF and CC Beams:				
	To identify order m of branch:				
	for $\Omega \leq 1$, m odd $=2q-1$, $y(x)=y(-x)$, $\psi(x)=-\psi(-x)$.				
	m even $=2q$, $y(x)=-y(-x)$, $\psi(x)=\psi(-x)$.				
	Branches are continuous into and in $\Omega > 1$.				
	Integers q need not be entered in sequence.				
	Order of Flag setting is immaterial except:				
	for $\Omega = 1$, all required m odd must precede any m even				

Form 7. Instructions for program for Timoshenko beams with homogeneous end-conditions.

HP-65 User Instructions

Oscillatory Iteration #2: GO | | | | Manual

STEP	INSTRUCTIONS	INPUT DATA/UNITS	KEYS		OUTPUT DATA/UNITS
1	Enter Card #2 of Timoshenko Beam program				
2	Set radian mode		g	RAD	
3	Set Flag 1 ON		f	SF1	
4	Input: $\Omega^2 > 1$	1.1	STO	3	
5	Input: φ	−2.0	STO	4	
6	Input: ρ	1.75	STO	5	
7	Input: q and execute	2	A		
8	Wait about 30 seconds (10 iterations)				
9	Stop iteration		R/S		
10	Display $\delta_i = \pm 1.547678132$		RCL	7	δ_i
11	Resume iteration		E		
12	Wait about 3 seconds (1 cycle)				
13	Stop iteration		R/S		
14	Display $\delta_i = \mp 1.547678132$		RCL	7	δ_i
	Execute Steps 11-14 repeatedly				

Form 8. Instruction for use of Part 2 of Timoshenko beam program to illustrate oscillatory iteration.

HP-65 Program Form

SWITCH TO W/PRGM. PRESS [f] [PRGM] TO CLEAR MEMORY.

Intersection Points

KEY ENTRY	CODE SHOWN	COMMENTS
LBL	23	
A	11	
RCL 2	34 02	m
f^{-1}	32	
$\sqrt{x}$	09	m^2
STO 2	33 02	m^2 in R_2
RCL 3	34 03	n
f^{-1}	32	
$\sqrt{x}$	09	n^2
STO 3	33 03	n^2 in R_3
+	61	m^2+n^2
STO 4	33 04	m^2+n^2 in R_4
RCL 1	34 01	τ
RCL 2	34 02	m^2
×	71	τm^2
RCL 3	34 03	n^2
−	51	τm^2-n^2
RCL 2	34 02	m^2
RCL 3	34 03	n^2
RCL 1	34 01	τ
×	71	τn^2
−	51	$m^2-\tau n^2$
×	71	$(\tau m^2-n^2)(m^2-\tau n^2)$
RCL 1	34 01	τ
÷	81	$(\tau m^2-n^2)(m^2-\tau n^2)/\tau$
STO 6	33 06	$(\tau m^2-n^2)(m^2-\tau n^2)/\tau$ in R_6
RCL 4	34 04	m^2+n^2
RCL 6	34 06	$(\tau m^2-n^2)(m^2-\tau n^2)/\tau$
f	31	
$\sqrt{x}$	09	$[(\tau m^2-n^2)(m^2-\tau n^2)/\tau]^{1/2}$
STO 6	33 06	$[(\tau m^2-n^2)(m^2-\tau n^2)/\tau]^{\frac{1}{2}}$ in R_6
÷	81	$(m^2+n^2)[\tau/(\tau m^2-n^2)(m^2-\tau n^2)]^{\frac{1}{2}}=\Omega$
STO 7	33 07	Ω in R_7
1	01	1
RCL 1	34 01	τ
+	61	$1+\tau$
RCL 4	34 04	m^2+n^2
×	71	$(1+\tau)(m^2+n^2)$
f	31	
$\sqrt{x}$	09	$[(1+\tau)(m^2+n^2)]^{\frac{1}{2}}$
RCL 6	34 06	$[(\tau m^2-n^2)(m^2-\tau n^2)/\tau]^{\frac{1}{2}}$
÷	81	$[\tau(1+\tau)(m^2+n^2)/(\tau m^2-n^2)(m^2-\tau n^2)]^{\frac{1}{2}}$
g	35	
1/x	04	$[(\tau m^2-n^2)(m^2-\tau n^2)/\tau(1+\tau)(m^2+n^2)]^{\frac{1}{2}}$
g	35	
π	02	π
×	71	ℓ/r
STO 8	33 08	ℓ/r in R_8
RCL 7	34 07	Ω
RTN	24	
LBL	23	
B	12	
RCL 8	34 08	ℓ/r
RTN	24	
LBL	23	
C	13	
RCL 8	34 08	ℓ/r
1	01	
2	02	12
f	31	
$\sqrt{x}$	09	$\sqrt{12}$
÷	81	$\ell/r\sqrt{12}=\ell/h$
RTN	24	

REGISTERS
R_1
R_2
R_3
R_4
R_5
R_6
R_7
R_8
R_9

LABELS	
A	Ω
B	ℓ/r
C	ℓ/h
D	
E	
0	
1	
2	
3	
4	
5	
6	
7	
8	
9	

FLAGS
1
2

TO RECORD PROGRAM INSERT MAGNETIC CARD WITH SWITCH SET AT W/PRGM.

Form 9. Program for coordinates of intersection points of branches in Figs. 1 and 2.

HP-65 User Instructions

Coordinates of Intersection Points

[Ω] [ℓ/r] [ℓ/h]

STEP	INSTRUCTIONS	INPUT DATA/UNITS	KEYS	OUTPUT DATA/UNITS
1	Enter Program			
2	Input: τ	τ	[STO] [1]	
3	Input: m	m	[STO] [2]	
4	Input: n	n	[STO] [3]	
5	Compute Ω		[A]	Ω
6	Compute ℓ/r		[B]	ℓ/r
7	Compute ℓ/h		[C]	ℓ/h
	For new m and/or n, go to Step 3.			
	Note: Either or both of Steps 6 and 7 may be chosen and their order is immaterial.			

Form 10. Instructions for program for coordinates of intersection points of branches in Figs. 1 and 2.

On Thickness Twist Vibrations of a Quartz Strip

R. D. Mindlin
COLUMBIA UNIVERSITY, NEW YORK, U.S.A.

In a series of papers, beginning in 1951 [1], approximate, two-dimensional equations of motion of crystal plates were developed which extended the range of applications of such equations to frequencies as high as those of the fundamental thickness-shear modes of vibration. Despite a reduction from three to two dimensions, formidable difficulties remain in attempts to find useful solutions of the equations, satisfying prescribed boundary conditions, owing to the fact that practical applications call for free edges of the plates. The obstacles are not unexpected as similar difficulties arise with the simpler equations of Cauchy [2] even though those are applicable only to much lower frequencies. However, simple solutions can be found by restricting them to one-dimensional forms (i. e. cylindrical modes) applicable, strictly speaking, to infinitely long strips, rather than to fully bounded plates; and, further, by neglecting certain terms deemed to be unimportant. To be useful, such solutions must be tested in one way or another: by comparison with experiment or, in the unusual case, with an exact solution of the three-dimensional equations.

In the case of AT-cut, quartz plates, which are the most widely used in high frequency applications, the important cylindrical modes of motion are those excited by an electric field normal to the middle plane of the plate [3]. There are five such modes: one type of each of thickness-shear, flexure and thickness-twist and two types of face-shear. The face-shear modes are coupled to the others through an elastic constant, c_{56} which is small in comparison with the other constants in the equations governing the cylindrical modes. If c_{56} is set equal to zero, the face-shear modes are uncoupled, leaving a pair of coupled thickness-shear and flexure equations and a single thickness-twist equation. As noted previously [4], the resulting thickness-shear and flexure equations predict experimental results for those modes [1] about as well as do the equations that include the coupling with face-shear [3]; but it was also noted [4] that the solution of the thickness-twist equation had not been tested. Subsequently, however, an exact solution of the three-dimentional equations was found for thickness-twist modes [5]. It is the purpose of this paper to show the relation between the two solutions.

The approximate equation governing the thickness-twist modes is [4, Eq. (11)]

$$\gamma_{55}\frac{\partial^2\psi_1}{\partial x_3} - \frac{\pi^2 c_{66}}{4b^2}\psi_1 = \rho\,\frac{\partial^2\psi_1}{\partial t^2} \tag{1}$$

where γ_{55} and c_{66} are elastic constants referred to rotated coordinates, ρ is the mass density, $x_2 = \pm b$ defines the faces of the plate and ψ_1 is the rotation, about x_3, of the normal to the faces; i.e. $b\psi_1$ is the approximation to the displacement parallel to x_1. The function

$$\psi_1 = (A\cos\zeta x_3 + B\sin\zeta x_3)\, e^{i\omega t} \tag{2}$$

yields, on its substitution in (1), the circular frequency

$$\omega = \left[\frac{\pi^2 c_{66}}{4\rho b^2} + \frac{\gamma_{55}\zeta^2}{\rho}\right]^{\frac{1}{2}}. \tag{3}$$

The free boundary conditions on the strip are

$$\frac{\partial\psi_1}{\partial x_3} = 0 \text{ on } x_3 = \pm c \tag{4}$$

from which

$$\zeta = n\pi/2b,\ n = 0, 1, 2, 3\ldots$$

so that

$$\omega = \frac{\pi}{2b}\left[\frac{c_{66}}{\rho}\right]^{\frac{1}{2}}\left[1 + \frac{\gamma_{55} n^2 b^2}{c_{66} c^2}\right]^{\frac{1}{2}}. \tag{5}$$

The solution of the three-dimensional equations, for the case of thickness-twist modes in a strip bounded by free surfaces at $x_2 = \pm b$ and $x_3 = \pm c$, cannot be expressed in simple form. However, if the faces at $x_3 = \pm c$ are replaced by faces, again perpendicular to the plane of x_2 and x_3, but at a dihedral angle

$$\alpha = \arctan\,(c_{56}/c_{66}) \tag{6}$$

with the plane of x_1 and x_2, an exact solution of simple form can be obtained [5] with circular frequencies

$$\omega = \frac{m\pi}{2b}\left[\frac{c_{66}}{\rho}\right]^{\frac{1}{2}}\left[1 + \frac{\gamma_{55} n^2 b^2}{c_{66} m^2 c^2}\right]^{\frac{1}{2}} \tag{7}$$

where m is the number of nodes between $x_2 = \pm b$. Comparing (5) and (7), with $m = 1$, which pertains in the present case, we see that neglect of the c_{56} coupling term, in the approximate two-dimensional equations, results in precisely the same frequencies as those obtained from the exact solution of the three-dimensional equations for the case of edges bevelled at an angle α as given in (6). For the AT-cut of quarts, α is approximately 5°.

References

1. R. D. Mindlin, Thickness-shear and flexural vibrations of crystal plates, J. Appl. Phys., Vol. 22, pp. 316-323, 1951.
2. A. L. Cauchy, Sur l'équilibre et le mouvement d' une plaque élastique dont l' élasticité n'est pas la même dans tous les sens, Exercices de Mathématique, Vol. 4, pp. 1-14, 1829.
3. R. D. Mindlin and D. C. Gazis, Strong resonances of rectangular AT-cut quartz plates, Proc. Fourth U. S. National Congress of Applied Mechanics, pp. 305-310, 1962.
4. R. D. Mindlin and W. J. Spencer, Anharmonic, thickness-twist overtones of thickness-shear and flexural vibrations of rectangular, AT-cut quartz plates, J. Acoustical Soc. of America, Vol. 42, pp. 1268-1277, 1967.
5. R. D. Mindlin, Thickness-twist vibrations of a quartz strip, Int. J. Solids and Structures, Vol. 7, pp. 1-4, 1971.

Reprinted from
Some Aspects of Mechanics of Continua
B. Sen Memorial Volume
Department of Mathematics, Jadavpur University
Calcutta, Part I (1977) 29-31

R. D. MINDLIN
(*New York, USA*)

A VARIATIONAL PRINCIPLE FOR THE EQUATIONS OF PIEZOELECTROMAGNETISM IN A COMPOUND CONTINUUM

Just as a diatomic crystal lattice is composed of a pair of interpenetrating Bravais lattices, so its continuum representation is a compound contiinuum comprising two simple continua occupying the same space. At each point of the compound continuum there are two displacements, say $u_i^{(1)}$ and $u_i^{(2)}$, one for each of the component continua. If the material is a dielectric, two electronic polarizations, $P_i^{(k)}$, $k = 1, 2$ are also associated with each point. If, in addition, the material is ionic, there is a single ionic polarization, at each point, proportional, in a linear theory, to the relative displacement of the two continua at that point.

In the equations to be considered, the stored energy density of deformation and polarization, W^L, is assumed to be a quadratic function of the small strains, $S_{ij}^{(k)}$, the electronic polarizations, $P_i^{(k)}$, and polarization gradients $P_{i,j}^{(k)}$, of the individual continua, and of the relative displacement, u_i^*, and relative rotation, ω_{ij}^*, of the two continua at each point:

$$W^L = W^L(S_{ij}^{(1)}, S_{ij}^{(2)}, P_i^{(1)}, P_i^{(2)}, P_{i,j}^{(1)}, P_{i,j}^{(2)}, u_i^*, \omega_{ij}^*), \tag{1}$$

where

$$S_{ij}^{(k)} = {}^1/_2\,(u_{j,i}^{(k)} + u_{i,j}^{(k)}), \qquad k = 1, 2, \tag{2}$$

$$u_i^* = u_i^{(2)} - u_i^{(1)}, \tag{3}$$

$$\omega_{ij}^* = {}^1/_2\,(u_{j,i}^* - u_{i,j}^*). \tag{4}$$

The total potential energy density, W, is assumed to be the sum of W^L and the energy density of the electromagnetic field:

$$W = W^L + {}^1/_2\varepsilon_0 E_i^{ms} E_i^{ms} + {}^1/_2\mu_0^{-1} B_i B_i, \tag{5}$$

where E_i^{ms} is the Maxwell electric self-field, B_i is the magnetic flux density and ε_0 and μ_0 are the dielectric permittivity and magnetic permeability, respectively, of a vacuum.

We also define an electric enthalpy density

$$W^E = W - E_i^{ms} D_i, \tag{6}$$

where

$$D_i = \varepsilon_0 E_i^{ms} + P_i^{(1)} + P_i^{(2)} + q_* u_i^*, \tag{7}$$

in which $q_* u_i^*$ is the ionic polarization and q_* is the ionic charge per unit volume. (The inclusion of $q_* u_i^*$ at this point, to account for ionicity, was suggested by P.C.Y. Lee.) Finally, the kinetic energy density is taken to be

$$T = \sum_{\kappa} {}^1/_2 \rho^{(k)} \dot{u}_i^{(k)} \dot{u}_i^{(k)}, \qquad k = 1, 2, \tag{8}$$

where the dot designates differentiation with respect to time and the $\rho^{(k)}$ are the mass densities of the two continua. Then, as shown previously [1, 2], the field equations corresponding to these energies are

$$T^{(k)}_{ij,i} + (-1)^k [T^*_{ij,i} - T_j{}^* + q_* (E^{ms}_j + E_j{}^0)] + f^{(k)}_j = \rho^{(k)} \ddot{u}^{(k)}_j, \quad k = 1, 2, \tag{9}$$

$$E^{(k)}_{ij,i} + E^{(k)}_j + E^{ms}_j + E_j{}^0 = 0, \qquad k = 1, 2, \tag{10}$$

$$\varepsilon_0 E^{ms}_{i,i} + P^{(1)}_{i,i} + P^{(2)}_{i,i} + q_* u^*_{i,i} = 0, \tag{11}$$

$$\varepsilon_{ijk} E_{k,j} + \dot{B}_i = 0, \tag{12}$$

$$\mu_0^{-1} \varepsilon_{ijk} B_{k,j} - \varepsilon_0 \dot{E}^{ms}_i - \dot{P}^{(1)}_i - \dot{P}^{(2)}_i - q_* \dot{u}_i{}^* = 0, \tag{13}$$

$$B_{i,i} = 0, \tag{14}$$

where $f_j{}^{(k)}$ and $E_j{}^0$ are the body forces and external electric field ε_{ijk} is the unit atlernating tensor and

$$T^{(k)}_{ij} = \frac{\partial W^L}{\partial S^{(k)}_{ij}}, \quad E^{(k)}_i = -\frac{\partial W^L}{\partial P^{(k)}_i}, \quad E^{(k)}_{ij} = \frac{\partial W^L}{\partial P^{(k)}_{j,i}}, \quad T_i{}^* = \frac{\partial W^L}{\partial u_i{}^*},$$
$$T_{ij}{}^* = \frac{\partial W^L}{\partial \omega_{ij}{}^*}. \tag{15}$$

The electromagnetic variables may be expressed as usual, in terms of a scalar and a vector potential:

$$E_i{}^{ms} = -\varphi_{,i} - \dot{A}_i, \tag{16}$$

$$B_i = \varepsilon_{ijk} A_{k,j}, \tag{17}$$

related through the Lorentz condition

$$\alpha \equiv \varepsilon_0 \mu_0 \dot{\varphi} + A_{i,i} = 0. \tag{18}$$

Then (9) — (14) become

$$T^{(k)}_{ij,i} + (-1)^k [T^*_{ij,i} - T_j{}^* - q_* (\varphi_{,j} + \dot{A}_j - E_j{}^0)] + f^{(k)}_j = \rho^{(k)} \ddot{u}^{(k)}_j, \quad k = 1, 2, \tag{19}$$

$$E^{(k)}_{ij,i} + E^{(k)}_j - \varphi_{,j} - \dot{A}_j + E_j{}^0 = 0, \qquad k = 1, 2, \tag{20}$$

$$-\varepsilon_0 \varphi_{,ii} + P^{(1)}_{i,i} + P^{(2)}_{i,i} + q_* (u^{(2)}_{i,i} - u^{(1)}_{i,i}) = -\varepsilon_0{}^2 \mu_0 \ddot{\varphi}, \tag{21}$$

$$\mu_0^{-1} A_{i,jj} + \dot{P}^{(1)}_i + \dot{P}^{(2)}_i + q_* (\dot{u}^{(2)}_i - \dot{u}^{(1)}_i) = \varepsilon_0 \ddot{A}_i. \tag{22}$$

In a vacuum, (19)—(22) reduce to

$$\varphi_{,ii} = \varepsilon_0 \mu_0 \ddot{\varphi}, \qquad A_{i,jj} = \varepsilon_0 \mu_0 \ddot{A}_i. \tag{23}$$

It is the purpose of this paper to exhibit a variational principle for which (19) — (23) are the Euler equations.

With (16) and (17), the electric enthalpy density becomes

$$W^E = W^L - {}^1\!/_2 \varepsilon_0 (\varphi_{,i} \varphi_{,i} + \dot{A}_i \dot{A}_i) - \varepsilon_0 \varphi_{,i} \dot{A}_i + {}^1\!/_2 (A_{j,i} A_{j,i} - A_{j,i} A_{i,j}) +$$
$$+ \varphi_{,i} (P^{(1)}_i + P^{(2)}_i) + \dot{A}_i (P^{(1)}_i + P^{(2)}_i) + q_* (\varphi_{,i} u_i{}^* + A_i u_i{}^*), \tag{24}$$

in which the identity

$$\varepsilon_{ijk} \varepsilon_{ilm} A_{k,j} A_{m,l} = A_{j,i} A_{j,i} - A_{j,i} A_{i,j} \tag{25}$$

has been employed.

The variational principle for equations (19) — (23) then follows: In a region V bounded by a surface S, with outward normal n, separating V from an outer vacuum V',

$$\delta \int_{t_0}^{t_1} dt \int_{V+V'} \left(T - W^E - \frac{1}{2}\mu_0^{-1}\alpha^2\right) dV +$$

$$+ \int_{t_0}^{t_1} dt \int_V \sum_k (f_i^{(k)}\delta u_i^{(k)} + E_i{}^0 \delta P_i^{(k)} + q_* E_i{}^0 \delta u_i{}^*)\, dV = 0 \tag{26}$$

for independent variations $\delta u_i{}^{(1)}$, $\delta u_i{}^{(2)}$, $\delta P_i{}^{(1)}$, $\delta P_i{}^{(2)}$, $\delta\varphi$ and δA_i between fixed limits at times t_0 and t_1. The use of the electric enthalpy density W^E, in the Lagrangian, follows Toupin's [3] variational principle for his nonlinear equations of the classical elastic dielectric. For the use of α^2 in the Lagrangian, see [4].

Upon substituting (8) for T, (24) for W^E and (18) for α, in (26), performing the variation operation, employing (15) and then eliminating time and space derivatives of $\delta u_i{}^{(k)}$, $\delta P_i{}^{(k)}$, $\delta\varphi$ and δA_i by means of the chain rule of differentiation, integration by parts (noting that the variations vanish at times t_0 and t_1) and the divergence theorem, we find

$$\sum_k \int_{t_0}^{t_1} dt \int_V \{T_{ij,i}^{(k)} + (-1)^k [T_{ij,i}^* - T_j{}^* - q_* (\varphi,_j + \dot{A}_j - E_j{}^0] +$$

$$+ f_j^{(k)} - \rho^{(k)} \ddot{u}_j^{(k)}\} \delta u_j^{(k)} dV +$$

$$+ \sum_k \int_{t_0}^{t_1} dt \int_V (E_{ij,i}^{(k)} + E_j^{(k)} - \varphi,_j - \dot{A}_j + E_j{}^0)\, \delta P_j^{(k)} dV +$$

$$+ \int_{t_0}^{t_1} dt \int_V [-\varepsilon_0 \varphi,_{ii} + P_{i,i}^{(1)} + P_{i,i}^{(2)} + q_* (u_{i,i}^{(2)} - u_{i,i}^{(1)}) + \varepsilon_0{}^2 \mu_0 \ddot{\varphi}]\, \delta\varphi dV +$$

$$+ \int_{t_0}^{t_1} dt \int_V [\mu_0^{-1} A_{i,jj} + \dot{P}_i^{(1)} + \dot{P}_i^{(2)} + q_* (\dot{u}_i^{(2)} - \dot{u}_i^{(1)}) - \varepsilon_0 \ddot{A}_i]\, \delta A_i dV +$$

$$+ \int_{t_0}^{t_1} dt \int_{V'} [-\varepsilon_0 (\varphi,_{ii} - \varepsilon_0 \mu_0 \ddot{\varphi})\, \delta\varphi + \mu_0^{-1} (A_{i,jj} - \varepsilon_0 \mu_0 \ddot{A}_i)\, \delta A_i]\, dV -$$

$$- \sum_k \int_{t_0}^{t_1} dt \int_S \{n_i [T_{ij}^{(k)} + (-1)^k T_{ij}{}^*]\, \delta u_j^{(k)} + n_i E_{ij}^{(k)} \delta P_j^{(k)}\}\, dS -$$

$$- \int_{t_0}^{t_1} dt \int_S n_i \{-\varepsilon_0 [\varphi,_i + \dot{A}_i] + P_i^{(1)} + P_i^{(2)} + q_* (u_i^{(2)} - u_i^{(1)})\}\, \delta\varphi dS -$$

$$- \int_{t_0}^{t_1} dt \int_S n_j \mu_0^{-1} [A_{i,j} - A_{j,i}]\, \delta A_i dS = 0, \tag{27}$$

where the bold-face brackets [ ] designate the discontinuity of the enclosed quantity across S.

From the volume integrals in (27), the Euler equations of the variational principle are seen to be the field equations (19) — (23), as required.

From the surface integrals in (27), admissible sets of boundary conditions are

$$u_j^{(k)} \quad \text{or} \quad n_i [T_{ij}^{(k)} + (-1)^k T_{ij}{}^*], \qquad k = 1, 2; \tag{28}$$

and $$P_j^{(k)} \quad \text{or} \quad n_i E_{ij}^{(k)}, \qquad k = 1, 2; \tag{29}$$

and $$\varphi \quad \text{or} \quad n_i \{-\varepsilon_0 [\varphi_{,i} + \dot{\underset{\sim}{A}}_i] + P_i^{(1)} + P_i^{(2)} + q_* u_j{}^*\}; \tag{30}$$

and $$A_i \quad \text{or} \quad n_i [\underset{\sim}{A}_{j,i} - \underset{\approx}{A}_{ij}]; \tag{31}$$

or, in the case of each of the three vector boundary conditions (28), (29) and (31), a component of the left hand alternative and the resultant of the right hand alternative in the plane at right angles, or vice versa. Such sets of boundary conditions are sufficient to assure a unique solution of the field equations provided that, across all surfaces (including the boundary) the discontinuities, if any, of $n_i B_i$ and $\varepsilon_{ijk} E_{k,j}^{ms}$ are specified [5] as well as the discontinuities, if any, of the right hand alternatives in (28) — (31).

REFERENCES

1. *Mindlin R. D.* A continuum theory of a diatomic, elastic dielectric.— Intern. J. Solids and Struct., 1972, 8, 369—383.
2. *Mindlin R. D.* Coupled elastic and electromagnetic fields in a diatomic, dielectric continuum.— Intern. J. Solids and Struct., 1972, 8, 401—408.
3. *Toupin R. A.* The Elastic Dielectric.— J. Ration Mech. and Anal., 1956, 5, 849—915.
4. *Yourgrau W., Mandelstam S.* Variational Principles in Dynamics and Quantum Theory. 2nd ed. New York, Pitman Publ. Corp., 1960.
5. *Born M., Wolf E.* Principles of Optics. New York. Pergamon Press, 1959, p. 3.

Reprinted from
Kompleksnii Analiz i ego Prilozhenia
I.N. Vekua 70th Birthday Volume
Academy of Sciences U.S.S.R.
Nauka, Moscow (1978) 397-400

Computers & Structures Vol. 10, pp. 751–759
Pergamon Press Ltd., 1979. Printed in Great Britain

VIBRATIONS OF QUARTZ PLATES WITH THE HP-67 POCKET CALCULATOR

R. D. MINDLIN
89 Deer Hill Drive, Ridgefield, CT 06877, U.S.A.

(*Received* 20 *August* 1978; *received for publication* 5 *October* 1978)

Abstract—Two mathematical solutions of a problem of coupled thickness-shear and flexural vibrations of the AT-cut quartz plate, with a pair of parallel, free edges, are presented. In one, some 27 yr old, the dispersion relation for waves in an infinite plate is approximated by a biquadratic equation, following which the edge conditions lead to a transcendental equation. In the other, the exact transcendental equation is employed for the infinite plate and no approximation is introduced until the setting of the edge conditions. Calculations of frequency vs dimensions for both methods are programmed for the HP-67 pocket calculator.

1. INTRODUCTION

Cauchy's[1] approximate equations of motion of anisotropic plates which, in the case of flexure, are limited in validity to frequencies no higher than about one-tenth that of the lowest thickness-shear mode, were extended, in 1951[2], to accommodate frequencies up to at least 25% higher than the frequency of that mode. An example was given illustrating the effects of free edges and the resulting coupling of flexure with thickness-shear and also exhibiting a close correspondence with measurements of frequency vs length-to-thickness ratio made by Sykes[3] with AT-cut quartz plates.

The higher order approximation was obtained, essentially, by retaining only the initial terms of an expansion of displacements in power series of the thickness-coordinate and integrating weighted stress–displacement relations and stress-equations of motion over the plate-thickness: thus reducing all functions and equations to dependence on only the two coordinates in the plane of the plate and, of course, the time. In the illustrative example, the roots of a biquadratic equation yielded an approximation to the dispersion relation for straight-crested waves in the infinite plate while the introduction of the conditions for free edges led to a transcendental equation the computation of the roots of which was a major chore with the calculators available at that time. Now, however, the program for the entire solution fits easily on a single magnetic card of a programmable, pocket calculator.

The biquadratic equation, referred to in the preceding paragraph, yields an approximation to the flexure and thickness-shear branches of the exact transcendental equation, for vibrations of the infinite rotated-Y-cut quartz plate, first established by Ekstein[4] and subsequently explored in detail[5, 6]. For plates with free edges, there is the possibility of using the branches from the exact solution for the infinite plate, expanding in terms of the associated, exact, characteristic functions and postponing any approximations to the final step of application of the free-edge conditions which produce another transcendental equation. A result, in addition to improved accuracy, would be survival of the exact variation of displacement, stress, strain and curvature, through the thickness, contributed by the exact characteristic functions—as many of them as are retained in the series expansion. Although this procedure was proposed as early as 1956[7], recommended again in 1960[8] and restated in Tiersten's text book in 1969[9], an example seems not to have been computed and it is only recently that the calculation has been feasible with a pocket calculator, at least for the simplest case analogous to that considered in 1951. Instead of one biquadratic equation and one transcendental equation to be solved, as before, the new scheme involves the solution of two biquadratic equations, a transcendental equation for each of the first two branches of Ekstein's dispersion relation and another transcendental equation from the boundary conditions—all with far more complicated coefficients. However, the programs for computations of frequency vs length-to-thickness ratio, for both schemes, fit on three magnetic cards of the HP-67 pocket calculator.

In what follows, the 1951 solution is identified as Case A and the new solution, by the method proposed in 1956, is identified as Case B.

2. CASE A: POWER SERIES EXPANSION

This section contains a brief review, with some changes of notation, of the equations and solution published in 1951[2].

We consider a rotated-Y-cut quartz plate, referred to rectangular coordinates x_i, $i = 1, 2, 3$, with free faces at $x_2 = \pm h$. The plate has free edges, at $x_1 = \pm l$, perpendicular to the two-fold axis of elastic symmetry of the plate[10]. As observed experimentally by Sykes[3], the frequencies of vibration of the plate are almost independent of the x_3-dimension and, hence, of the x_3 coordinate in the equations. In the case of complete absence of dependence on x_3 and the displacement u_3 in the x_3 direction, the general equations derived in[2] reduce to

$$\left(c_{11} - \frac{c_{12}^2}{c_{22}}\right)\frac{\partial^2\psi}{\partial x_1^2} - \frac{\pi^2 c_{66}}{4h^2}\left(\psi + \frac{\partial\eta}{\partial x_1}\right) = \rho\frac{\partial^2\psi}{\partial t^2},$$
$$\frac{\pi^2 c_{66}}{12}\left(\frac{\partial\psi}{\partial x_1} + \frac{\partial^2\eta}{\partial x_1^2}\right) = \rho\frac{\partial^2\eta}{\partial t^2}, \qquad (1)$$

which are eqns (23) of[2]. In these equations the c_{pq} are elasticity constants referred to the rotated-Y coordinates, ρ is the mass density and ψ and η derive from the early terms of a power series expansion of the displacement components:

$$u_1 = x_2\psi(x_1, t), \qquad u_2 = \eta(x_1, t). \tag{2}$$

The pertinent solution of (1) is

$$\begin{aligned}\eta &= (A_1 \sin \xi_1 x_1 + A_2 \sin \xi_2 x_1)\, e^{i\omega t},\\ \psi &= (B_1 \cos \xi_1 x_1 + B_2 \cos \xi_2 x_1)\, e^{i\omega t},\end{aligned} \tag{3}$$

where the amplitudes A_j and B_j are related according to

$$B_j/A_j = \sqrt{(3)}(\kappa_j^2 - \Omega^2)/\kappa_j h, \qquad j = 1, 2, \tag{4}$$

in which

$$\Omega^2 = \omega^2/\bar{\omega}^2, \qquad \bar{\omega}^2 = \pi^2 c_{66}/4\rho h^2; \tag{5}$$

i.e. $\bar{\omega}$ is the circular frequency of the fundamental thickness-shear mode. Also, the $\kappa_j(=\xi_j h/\sqrt{(3)})$ are the roots of the biquadratic equation

$$\kappa^4 - (1+g)\Omega^2\kappa^2 + g\Omega^2(\Omega^2 - 1) = 0, \tag{6}$$

with

$$g = \pi^2 c_{66}/12(c_{11} - c_{12}^2/c_{22}). \tag{7}$$

The roots of (6) yield the approximation to the thickness-shear and flexure branches of Ekstein's exact dispersion relation for vibrations of the infinite plate, as mentioned in the introduction.

Application of the conditions for free edges:

$$\frac{\partial \eta}{\partial x_1} + \psi = 0, \qquad \frac{\partial \psi}{\partial x_1} = 0, \qquad \text{on } x_1 = \pm l, \tag{8}$$

leads to the transcendental frequency equation

$$\tan \alpha = \begin{cases} \varphi \tan \gamma\alpha, & \Omega \geqslant 1 \\ -\varphi \tanh \gamma\alpha, & \Omega \leqslant 1 \end{cases} \tag{9}$$

where

$$\alpha = \kappa_1\sqrt{(3)}l/h, \qquad \gamma = \hat{\kappa}_2/\kappa_1, \qquad \hat{\kappa}_2 = |\kappa_2{}^2|^{1/2},$$
$$\varphi = \gamma(\kappa_2{}^2 - \Omega^2)/(\kappa_1{}^2 - \Omega^2). \tag{10}$$

Choice of Ω fixes $\kappa_1{}^2$ and $\kappa_2{}^2$ in (6) and φ in (10); following which (9) can be solved for a sequence of values of α and, hence, for l/h, for the chosen Ω. The process is then repeated for a sequence of values of Ω.

As mentioned in[2], (1) have the same form as the equations of a Timoshenko beam, so that the solution (3)–(10) is also that for vibrations of such a beam with free ends. This result has already been programmed for the HP-65 pocket calculator[11] employing Schelkunoff iteration for the solution of (9). The program given here on Card No. 1 for the HP-67 pocket calculator, with Instructions identified as Case A, has the added features of choice of ranges of l/h and Ω and increment of Ω, so that, after these quantities and the material constants have been entered on the keyboard or by a data card, only a single key need be pressed to initiate and complete the computation of the entire spectrum of Ω vs l/h within the chosen ranges without further keyboard entries except, possibly a change of the increment of Ω.

The results, with constants for the AT cut of quartz:

$$c_{11} = 86.05, \qquad c_{22} = 129.9, \qquad c_{66} = 29.34,$$
$$c_{12} = -10.49 \times 10^{10}\,\text{dyn cm}^2,$$

calculated from Mason's data, as used in[2] along with $\rho = 2.65\,\text{gm/cm}^3$, are, of course, the same as in[2]. An example, the family of curves designated as A along with Sykes's data, is shown in Fig. 1 which is taken from Fig. 3 of[2].

3. EXACT SOLUTION FOR THE INFINITE PLATE

The three-dimensional displacement-equations of motion and stress-displacement relations are, in a usual notation,

$$c_{ijkl}u_{k,li} = \rho\ddot{u}_j, \tag{11}$$

$$T_{ij} = c_{ijkl}u_{k,l}, \tag{12}$$

with $c_{ijkl} = c_{klij} = c_{jikl}$. For the rotated-Y-cuts[10] of quartz, with x_1 a two-fold axis of crystal symmetry and x_2 normal to the middle plane of the plate, the following eight constants vanish:

$$c_{ij12} = c_{ij13} = 0, \qquad ij = 11, 22, 33, 23. \tag{13}$$

To restrict wave solutions of (11) to straight-crested

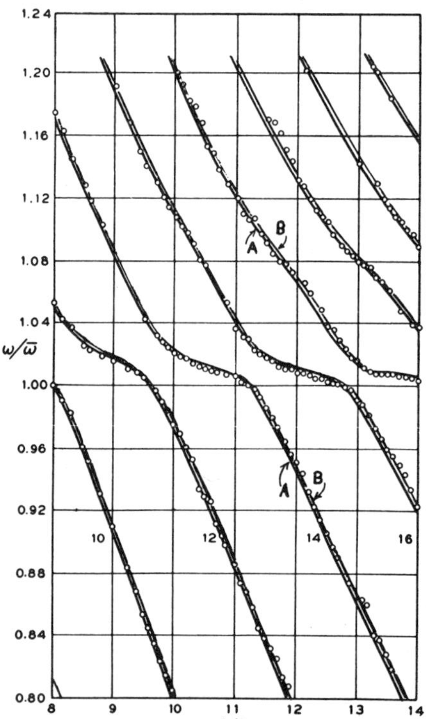

Fig. 1. Examples of curves of frequency ratio, $\Omega(=\omega/\bar{\omega})$, vs width-to-thickness ratio, l/h, computed for Cases A and B, compared with experimental data obtained by Sykes.

waves in the x_1-direction, the displacements are made independent of x_3. Finally, to eliminate coupling of thickness-shear and flexure with face-shear, we set

$$c_{1123} = c_{2223} = c_{1312} = u_3 = 0. \qquad (14)$$

The assumptions (14) are made to conform with the solution in[2]. They were not made by Ekstein and, although they do not simplify the solution greatly, they do shorten the computation.

The portions of the equations of motion (11) that remain after (13) and (14) are

$$\begin{aligned} c_{11}u_{1,11} + c_{66}u_{1,22} + (c_{12} + c_{66})u_{2,12} &= \rho\ddot{u}_1, \\ (c_{12} + c_{66})u_{1,12} + c_{66}u_{2,11} + c_{22}u_{2,22} &= \rho\ddot{u}_2, \end{aligned} \qquad (15)$$

where $c_{11} = c_{1111}$, $c_{22} = c_{2222}$, $c_{12} = c_{1122}$, $c_{66} = c_{1212}$. These equations are what (1) approximate. The accompanying stress–displacement relations of interest are

$$\begin{aligned} T_{11} &= c_{11}u_{1,1} + c_{12}u_{2,2}, \\ T_{22} &= c_{12}u_{1,1} + c_{22}u_{2,2}, \\ T_{12} &= c_{66}(u_{2,1} + u_{1,2}). \end{aligned} \qquad (16)$$

As solutions of (15), we start with

$$\begin{aligned} u_1 &= C_1 \cos \xi x_1 \sin \alpha x_2\, e^{i\omega t}, \\ u_2 &= C_2 \sin \xi x_1 \cos \alpha x_2\, e^{i\omega t}. \end{aligned} \qquad (17)$$

Upon substituting (17) in (15) we find

$$\frac{C_2}{C_1} = \frac{\rho\omega^2 - c_{11}\xi^2 - c_{66}\alpha^2}{(c_{12} + c_{66})\xi\alpha} = \frac{(c_{12} + c_{66})\xi\alpha}{\rho\omega^2 - c_{66}\xi^2 - c_{22}\alpha^2} = \gamma(say). \qquad (18)$$

The second of (18) may be regarded as a quadratic in α^2 for given ω and ξ. Call α_1^2, α_2^2 the two roots. Then (17) become, omitting $e^{i\omega t}$ here and in the sequel,

$$\begin{aligned} u_1 &= (B_1 \sin \alpha_1 x_2 + B_2 \sin \alpha_2 x_2) \cos \xi x_1, \\ u_2 &= (B_1\gamma_1 \cos \alpha_1 x_2 + B_2\gamma_2 \cos \alpha_2 x_2) \sin \xi x_1, \end{aligned} \qquad (19)$$

where

$$\gamma_i = \frac{\rho\omega^2 - c_{11}\xi^2 - \alpha_i^2}{(c_{12} + c_{66})\xi\alpha_i} = \frac{(c_{12} + c_{66})\xi\alpha_i}{\rho\omega^2 - c_{66}\xi^2 - c_{22}\alpha_i^2}, \qquad i = 1, 2. \qquad (20)$$

To clear the traction from the faces of the plate, we require

$$T_{12} = T_{22} = 0 \text{ on } x_2 = \pm h. \qquad (21)$$

Upon substituting (19) in (16) and the result in (21), we find

$$\frac{B_2}{B_1} = -\frac{(c_{12}\xi + c_{22}\gamma_1\alpha_1) \sin \alpha_1 h}{(c_{12}\xi + c_{22}\gamma_2\alpha_2) \sin \alpha_2 h} = -\frac{(\alpha_1 + \gamma_1\xi) \cos \alpha_1 h}{(\alpha_2 + \gamma_2\xi) \cos \alpha_2 h} = \zeta\,(say). \qquad (22)$$

The second of (22) is the transcendental equation

$$\frac{\tan \alpha_1 h}{\tan \alpha_2 h} = \frac{(c_{12}\xi + c_{22}\gamma_2\alpha_2)(\alpha_1 + \gamma_1\xi)}{(c_{12}\xi + c_{22}\gamma_1\alpha_1)(\alpha_2 + \gamma_2\xi)} \qquad (23)$$

which, along with the quadratic (18) in α^2, is the dispersion relation, ω vs ξ, for waves in the infinite plate; i.e. straight-crested waves travelling in the direction of the two-fold axis x_1, decoupled from face-shear. Up to this point, aside from the decoupling (which would be intrinsic for orthotropic, hexagonal or isotropic symmetry) there is no approximation within the framework of the linear theory of elasticity. If the coupling with face-shear were included for the quartz plate, the biquadratic equation (18) would be a bicubic and the transcendental equation (23) would have three transcendental functions instead of two.

The dispersion relation (23) has an infinity of branches, the first two of which are approximated by the biquadratic (6). Let ξ_n, $n = 1, 2, 3 \ldots$, be the wave number associated with the nth branch of (23). Then (20) and (22) become

$$\gamma_{in} = \frac{\rho\omega^2 - c_{11}\xi_n^2 - c_{66}\alpha_{in}^2}{(c_{12} + c_{66})\xi_n\alpha_{in}} = \frac{(c_{12} + c_{66})\xi_n\alpha_{in}}{\rho\omega^2 - c_{66}\xi_n^2 - c_{22}\alpha_{in}^2}, \qquad (24)$$

$$\zeta_n = -\frac{(c_{12}\xi_n + c_{22}\gamma_{1n}\alpha_{in}) \sin \alpha_{1n} h}{(c_{12}\xi_n + c_{22}\gamma_{2n}\alpha_{2n}) \sin \alpha_{2n} h} = -\frac{(\alpha_{1n} + \gamma_{1n}\xi_n) \cos \alpha_{1n} h}{(\alpha_{2n} + \gamma_{2n}\xi_n) \cos \alpha_{2n} h}, \qquad (25)$$

where the first subscript, $i(=1, 2)$, refers to the two roots, α_{1n}^2 and α_{2n}^2, for each ξ_n of

$$c_{22}c_{66}\alpha_{in}^4 + \{[c_{11}c_{22} + c_{66}^2 - (c_{12} + c_{66})^2]\xi_n^2 - (c_{22} + c_{66})\rho\omega^2\}\,\alpha_{in}^2 + c_{11}c_{66}\xi_n^4 - (c_{12} + c_{66})\xi_n^2\rho\omega^2 + \rho^2\omega^4 = 0 \qquad (26)$$

and $n(=1, 2, 3\ldots)$ refers to the roots ξ_n of

$$\frac{\tan \alpha_{1n} h}{\tan \alpha_{2n} h} = \frac{(c_{12}\xi_n + c_{22}\gamma_{2n}\alpha_{2n})(\alpha_{1n} + \gamma_{1n}\xi_n)}{(c_{12}\xi_n + c_{22}\gamma_{1n}\alpha_{1n})(\alpha_{2n} + \gamma_{2n}\xi_n)} \qquad (27)$$

all for each ω. Then (19) become

$$\begin{aligned} u_1 &= \sum_{n=1}^{\infty} A_n(\sin \alpha_{1n}x_2 + \zeta_n \sin \alpha_{2n}x_2) \cos \xi_n x_1, \\ u_2 &= \sum_{n=1}^{\infty} A_n(\gamma_{1n} \cos \alpha_{1n}x_2 + \zeta_n\gamma_{2n} \cos \alpha_{2n}x_2) \sin \xi_n x_1. \end{aligned} \qquad (28)$$

These are the characteristic functions for vibrations of the infinite plate with free faces at $x_2 = \pm h$. We have now to set the conditions for free edges at $x_1 = \pm l$.

4. EDGE CONDITIONS

From Hamilton's variational principle, we have[12], for independent variations δu_j in a volume V,

$$\int_V (T_{ij,i} - \rho\ddot{u}_j)\delta u_j \, dV + \int_S (t_j - \nu_i T_{ij})\delta u_j \, dS = 0, \qquad (29)$$

where t_j is the traction across the boundary S with outward normal ν_i. The integrand of the volume integral vanishes according to (11) and (12), t_j is zero by prescription and $\nu_i T_{ij}$ is zero on $x_2 = \pm h$ by (21). Thus, (29) reduces to

$$\int_{-h}^{h} T_{11}\delta u_1\,dx_2 = 0, \qquad \int_{-h}^{h} T_{12}\delta u_2\,dx_2 = 0, \qquad \text{on } x_1 = \pm l. \tag{30}$$

Now, from (28),

$$\delta u_1 = \sum_{n=1}^{\infty} f_1^{(n)}\delta A_n, \qquad \delta u_2 = \sum_{n=1}^{\infty} f_2^{(n)}\delta A_n, \tag{31}$$

where

$$\begin{aligned} f_1^{(n)} &= (\sin\alpha_{1n}x_2 + \zeta_n \sin\alpha_{2n}x_2)\cos\xi_n x_1,\\ f_2^{(n)} &= (\gamma_{1n}\cos\alpha_{1n}x_2 + \zeta_n\gamma_{2n}\cos\alpha_{2n}x_2)\sin\xi_n x_1. \end{aligned} \tag{32}$$

Also, from (28) and (16),

$$\begin{aligned} T_{11} &= \mp\sum_{n=1}^{\infty} A_n g_{11}^{(n)}\sin\xi_n l,\\ T_{12} &= \sum_{n=1}^{\infty} A_n g_{12}^{(n)}\cos\xi_n l, \qquad \text{on } x_1 = \pm l, \end{aligned} \tag{33}$$

where

$$\begin{aligned} g_{11}^{(n)} &= \sum_{i=1}^{2} \zeta_n^{i-1}(c_{11}\xi_n + c_{12}\gamma_{in}\alpha_{in})\sin\alpha_{in}x_2,\\ g_{12}^{(n)} &= \sum_{i=1}^{2} \zeta_n^{i-1}c_{66}(\alpha_{in} + \xi_n\gamma_{in})\cos\alpha_{in}x_2. \end{aligned} \tag{34}$$

Now, write

$$\begin{aligned} h_1^{m,n} &= \int_{-h}^{h} g_{11}^{(m)} f_1^{(n)}\sin\xi_m l\,dx_2,\\ h_2^{m,n} &= \int_{-h}^{h} g_{12}^{(m)} f_2^{(n)}\cos\xi_m l\,dx_2. \end{aligned} \tag{35}$$

Then, for independent variations δu_j, (30) become

$$\begin{aligned} &A_1h_1^{1,1} + A_2h_1^{2,1} + A_3h_1^{3,1} + A_4h_1^{4,1} + \cdots = 0,\\ &A_1h_1^{1,2} + A_2h_1^{2,2} + A_3h_1^{3,2} + A_rh_1^{4,2} + \cdots = 0,\\ &\quad\vdots \end{aligned} \tag{36}$$

and

$$\begin{aligned} &A_1h_2^{1,1} + A_2h_2^{2,1} + A_3h_2^{3,1} + A_4h_2^{4,1} + \cdots = 0,\\ &A_1h_2^{1,2} + A_2h_2^{2,2} + A_3h_2^{3,2} + A_4h_2^{4,2} + \cdots = 0,\\ &\quad\vdots \end{aligned} \tag{37}$$

subject to the requirements of balance of momentum and moment of momentum which, in the present case, reduce to

$$\int_{-h}^{h} T_{12}\,dx_2 = 0, \qquad \int_{-h}^{h} x_2T_{11}\,dx_2 = 0, \qquad \text{on } x_1 = \pm l. \tag{38}$$

The vanishing of the determinant of the coefficients of the A_n, in (36) and (37), for as many of the branches as are retained, and subject to the conditions (38), yield $\omega/\bar{\omega}$ vs l/h.

For comparison with Case A, only the first two branches of (27) are required, so that (33) become

$$\begin{aligned} T_{11} &= \mp\sum_{i=1}^{2}\sum_{n=1}^{2} A_n\zeta_n^{i-1}(c_{11}\xi_n + c_{12}\gamma_{in}\alpha_{in})\sin\alpha_{in}x_2\sin\xi_n l,\\ T_{12} &= \sum_{i=1}^{2}\sum_{i=1}^{2} A_n\zeta_n^{i-1}(\alpha_{in} + \xi_n\gamma_{in})\cos\alpha_{in}x_2\cos\xi_n l, \end{aligned} \tag{39}$$

on $x_1 = \pm l$.

As there are now only two of the A_n, (36) and (37) are superfluous in view of the requirement that (38) be satisfied. Upon substituting (39) in (38), we find

$$\begin{aligned} &\sum_{i=1}^{2}\sum_{n=1}^{2} A_n\beta_{1n}\sin\xi_n l = 0,\\ &\sum_{i=1}^{2}\sum_{n=1}^{2} A_n\beta_{2n}\cos\xi_n l = 0, \end{aligned} \tag{40}$$

where

$$\begin{aligned} \beta_{1n} &= \zeta_n^{i-1}\alpha_{in}^{-2}(c_{11}\xi_n + c_{12}\gamma_{in}\alpha_{in})(\sin\alpha_{in}h - \alpha_{in}h\cos\alpha_{in}h),\\ \beta_{2n} &= \zeta_n^{i-1}\alpha_{in}^{-1}(\alpha_{in} + \xi_n\gamma_{in})\sin\alpha_{in}h. \end{aligned} \tag{41}$$

Elimination of the A_n from (40) produces the equation

$$\beta_{11}\beta_{22}\tan\xi_1 l = \beta_{12}\beta_{21}\tan\xi_2 l, \tag{42}$$

the solution of which, with (26) and (27), determines the frequency spectrum $\omega/\bar{\omega}$ vs l/h for Case B.

5. COMPUTATION OF CASE B

The equations in the preceding section were prepared for computation with the HP-67 pocket calculator as follows. With Card No. 1, the roots of (6) are calculated as

$$\bar{\xi}_n^2 = 12\,\kappa_n^2/\pi^2 = 4\xi_n^2h^2/\pi^2, \qquad n = 1, 2; \qquad \bar{\xi}_1^2 > \bar{\xi}_2^2,$$

instead of κ_1^2 and κ_2^2, simply by pressing Key B instead of Key A, as indicated on the Instruction form; and these results are used, subsequently, as trial roots of (27).

Equation (26) is expressed as

$$\hat{c}_{22}\hat{\alpha}_{in}^4 + [\hat{c}_{11}\hat{c}_{22} - 2\hat{c}_{12} - \hat{c}_{12}^2 - (1 + \hat{c}_{22})\hat{\Omega}_n^2]\,\hat{\alpha}_{in}^2 + \hat{c}_{11} - (1 + \hat{c}_{11})\hat{\Omega}_n^2 - \hat{\Omega}_n^4 = 0, \tag{43}$$

where

$$\hat{c}_{pq} = c_{pq}/c_{66}, \qquad \hat{\Omega}_n^2 = \Omega^2/\bar{\xi}_n^2, \qquad \hat{\alpha}_{in}^2 = \alpha_{in}^2/\xi_n^2,$$
$$i = 1, 2; \qquad j = 1, 2.$$

Thus, $\hat{\alpha}_{1n}^2$ and $\hat{\alpha}_{2n}^2$ are the two roots of (43) for each of the two n. Note that $\bar{\xi}_1$ is real but $\bar{\xi}_2$ is real above $\Omega = 1$ and imaginary below $\Omega = 1$. This applies to both the approximate values from (6) and the exact values from (27). Also, the α_{i2} are real, but the α_{i1} are imaginary. However, what appear in (27) are

$$\begin{aligned} \gamma_{in}\hat{\alpha}_{in} &= (\hat{\Omega}_n^2 - \hat{c}_{11} - \hat{\alpha}_{in}^2)/(1 + \hat{c}_{12}),\\ \gamma_{in}/\hat{\alpha}_{in} &= (1 + \hat{c}_{12})/(\hat{\Omega}_n^2 - 1 - \hat{c}_{22}\hat{\alpha}_{in}^2) \end{aligned} \tag{44}$$

and these are both real. We define

$$\chi_{in} = (\hat{c}_{12} + \hat{c}_{22}\gamma_{in}\hat{\alpha}_{in})/(1 + \gamma_{in}/\hat{\alpha}_{in}), \tag{45}$$

$$\bar{\alpha}_{in} = \pi|\hat{\alpha}_{in}^2\bar{\xi}_n^2|^{1/2}/2. \tag{46}$$

Then (27) becomes

$$\chi_{11}\bar{\alpha}_{21}\tanh\bar{\alpha}_{11}=\chi_{21}\bar{\alpha}_{11}\tanh\bar{\alpha}_{21},\qquad n=1,\qquad (47)$$

$$\chi_{12}\bar{\alpha}_{22}\tan\bar{\alpha}_{12}=\chi_{22}\bar{\alpha}_{12}\tan\bar{\alpha}_{22},\qquad n=2.\qquad (48)$$

These are the exact dispersion relations for the first two branches for waves in the infinite plate.

As already remarked, for a given choice of Ω, Card No. 1 is employed to calculate and store trial values of $\bar{\xi}_1^2$ and $\bar{\xi}_2^2$ from the approximate equation (6). These are used in Card No. 2 to form the $\hat{\Omega}_n^2$ and then to calculate the $\hat{\alpha}_{in}^2$ from (43) and the χ_{in} from (44) and (45). If these values do not satisfy (47) and (48), new values of $\bar{\xi}_n^2$ are chosen automatically and the process is automatically repeated and corrected by the secant iterative method until (47) and (48) are satisfied within 10^{-6}. All this by simply pressing Key B after inserting Card No. 2. The final values of $\bar{\xi}_n^2$, $\hat{\alpha}_{in}^2$ and the quantities used in (47) and (48) remain stored for use in formulating (42) and solving it for l/h with Card No. 3.

For (42), we define

$$\psi_{in}=(\hat{c}_{11}+\hat{c}_{12}\gamma_{in}\hat{\alpha}_{in})/\hat{\alpha}_{in}^2(1+\gamma_{in}/\hat{\alpha}_{in}),\qquad (49)$$

$$T_1=(\bar{\alpha}_{21}^{-1}\tanh\bar{\alpha}_{21}-1)/(\bar{\alpha}_{11}^{-1}\tanh\bar{\alpha}_{11}-1),\qquad (50)$$

$$T_2=(\bar{\alpha}_{22}^{-1}\tan\bar{\alpha}_{22}-1)/(\bar{\alpha}_{12}^{-1}\tan\bar{\alpha}_{12}-1),\qquad (51)$$

$$r=|\bar{\xi}_2^2/\bar{\xi}_1^2|^{1/2},\qquad \Delta=\xi_1 l=\frac{1}{2}\pi\bar{\xi}_1 l/h.\qquad (52)$$

These quantities are computed with Card No. 3 and employed to calculate

$$\varphi=\frac{\psi_{12}(1-T_2)^{-1}+\psi_{22}(1-T_2^{-1})^{-1}}{r[\psi_{11}(1-T_1)^{-1}+\psi_{21}(1-T_1^{-1})^{-1}]}\qquad (53)$$

by means of which (42) is expressed as

$$\tan\Delta=\begin{cases}\varphi\tan r\Delta, & \Omega>1,\\ \varphi\tanh r\Delta, & \Omega<1,\end{cases}\qquad (54)$$

which is solved automatically by Schelkunoff iteration[11] for all the values of l/h in the chosen interval for the Ω selected. Then a new Ω is entered on the keyboard (the $\hat{c}_{pq}$ and the limits of l/h remain stored), Card No. 1 is re-inserted and the process is repeated to obtain the next set of values of l/h for the new Ω.

The results of calculations, again with Mason's values of the c_{pq}, are shown in Fig. 1 as the family of curves marked B. The values of l/h, for a given Ω, are everywhere slightly larger than those from the 1951 calculations (marked A). This is primarily due to the difference in the ξ_n, for the infinite plate, between the approximate (6) and the exact (47) and (48).

The program printouts, shown on the following pages, were made with the HP-97 calculator so that the Key Codes are for the HP-97. The corresponding Key Codes for the HP-67 are given in Appendix E of the HP-67 Owners Handbook and Programming Guide.

The readouts of the HP-67, or printouts of the HP-97, with Card No. 1, for Case A, begin with Ω_{min} followed by pairs of l/h and m (mode order) in the range $l/h_{min}\leq l/h\leq l/h_{max}$; then $\Omega+\Delta\Omega$ followed by pairs of l/h and m; then $\Omega+2\Delta\Omega$, etc. up to Ω_{max}.

For Case B, the readout or printout of Card No. 1 is only the Ω entered in Instruction Step 7′. Card No. 2 has no readout or printout. The readouts or printouts of Card No. 3 are the pairs of l/h and m, in the range $l/h_{min}\leq l/h\leq l/h_{max}$ for the chosen Ω.

REFERENCES

1. A. L. Cauchy, Sur l'équilibre et le mouvement d'une plaque élastique dont l'élasticité n'est pas la même dans tous les sens. *Exercices de Mathématique* 4, 1–14 (1829).
2. R. D. Mindlin, Thickness-shear and flexural vibrations of crystal plates. *J. Appl. Phys.* **22**, 316–323 (1951).
3. R. A. Sykes, Modes of motion in quartz crystals, the effects of coupling and methods of design. *Quartz Crystals for Electrical Circuits*, (Edited by Raymond A. Heising), Chap. VI, pp. 205–208. D. Van Nostrand, New York (1946).
4. H. Ekstein, High frequency vibrations of thin crystal plates. *Physical Review* **68**, 11–23 (1945).
5. E. G. Newman and R. D. Mindlin, Vibrations of a monoclinic crystal plate. *J. Acoustical Soc. Am.* **29**, 1206–1218 (1957).
6. R. K. Kaul and R. D. Mindlin, Frequency spectrum of a monoclinic crystal plate, *J. Acoustical Soc. Am.* **34**, 1902–1910 (1962).
7. R. D. Mindlin, Mathematical theory of vibrations of crystals: expansion in series of characteristic functions of infinite plates. *Proc. 10th Ann. Symp. Frequency Control*, p. 16. Signal Corps Engineering Laboratories, Fort Monmouth, New Jersey (1956).
8. R. D. Mindlin, Waves and vibrations in isotropic, elastic plates. *Structural Mechanics*, (Edited by J. N. Goodier and N. J. Hoff), pp. 199–232. New York (1960).
9. H. F. Tiersten, *Linear Piezoelectric Plate Vibrations*, pp. 137–138. Plenum Press, London (1969).
10. W. P. Mason, *Piezoelectric Crystals and Their Application to Ultrasonics*. Van Nostrand, New York (1950).
11. R. D. Mindlin, Frequencies of vibration of beams by Schelkunoff iteration on the HP-65 calculator. *Comput. Structures* **7**, 639–650 (1977).
12. A. E. H. Love, *Theory of Elasticity*, 4th Edn, p. 166. Cambridge University Press (1927).

STEP	INSTRUCTIONS	INPUT DATA/UNITS	KEYS		OUTPUT DATA/UNITS
	To compute Ω vs. ℓ/h for Case A:				
1	Insert Card No. 1				
2	Enter $\hat{c}_{11}$ $(= c_{11}/c_{66})$	$\hat{c}_{11}$	STO	A	
3	Enter $\hat{c}_{22}$ $(= c_{22}/c_{66})$	$\hat{c}_{22}$	STO	B	
4	Enter $\hat{c}_{12}$ $(= c_{12}/c_{66})$	$\hat{c}_{12}$	STO	C	
5	Enter ℓ/h_{min}	ℓ/h_{min}	STO	7	
6	Enter ℓ/h_{max}	ℓ/h_{max}	STO	E	
7	Enter Ω_{min}	Ω_{min}	STO	D	
8	Enter $\Delta\Omega$	$\Delta\Omega$	STO	6	
9	Enter Ω_{max}	Ω_{max}	STO	5	Ω
10	Compute Ω vs ℓ/h		A		ℓ/h
					m
	To compute Ω vs. ℓ/h for Case B:				
1′-6′	Steps 1′-6′ same as 1-6 for Case A (skip if after Case A)				
7′	Enter Ω	Ω	STO	D	
8′	Compute (6 sec.)		B		Ω
9′	Insert Card No. 2				
10′	Compute (3 min.)		B		
11′	Insert Card No. 3				
12′	Compute ℓ/h for Ω in Step 7′ (1 min. to first ℓ/h)		B		ℓ/h
	For another Ω, insert Card No. 1 and go to 7′.				m
	(Omit $\Omega = 1$. Use $\Omega = 1.0001$).				
	Note: times are approximate.				

Form 1. Instructions for use of program cards.

```
001 *LBLA       21 11
002 CF0      16 22 00
003 GTOE        22 15
004 *LBLB       21 12
005 SF0      16 21 00
006 *LBLE       21 15
007 RCLC        36 13
008 X²             53
009 RCLB        36 12
010 ÷             -24
011 RCLA        36 11
012 -             -45
013 CHS           -22
014 Pi          16-24
015 X²             53
016 ÷             -24
017 1              01
018 2              02
019 ×             -35
020 1/X            52
021 STOI        35 46
022 1              01
023 +             -55
024 RCLD        36 14
025 DSP4       -63 04
026 PRTX          -14
027 SPC         16-11
028 X²             53
029 STOD        35 14
030 ×             -35
031 2              02
032 ÷             -24
033 STO1        35 01
034 1              01
035 RCLD        36 14
036 -             -45
037 RCLD        36 14
038 ×             -35
039 RCLI        36 46
040 ×             -35
041 RCL1        36 01
042 X²             53
043 ÷             -24
044 1              01
045 +             -55
046 √X             54
047 RCL1        36 01
048 ×             -35
049 STO0        35 00
050 +             -55
051 F0?      16 23 00
052 GSBC        23 13
053 STO4        35 04
054 RCL1        36 01
055 RCL0        36 00
056 -             -45
057 F0?      16 23 00
058 GSBC        23 13
059 P⇄S         16-51
060 STO4        35 04
061 P⇄S         16-51
062 RCLI        36 46
063 1              01
064 +             -55
065 3              03
066 ×             -35
067 √X             54
068 Pi          16-24
069 ÷             -24
070 RCL7        36 07
071 ×             -35
072 DSP0       -63 00
073 RND         16 24
074 STO9        35 09
075 DSP4       -63 04
076 F0?      16 23 00
077 RTN            24
078 P⇄S         16-51
079 RCL4        36 04
080 ABS         16 31
081 √X             54
082 STO5        35 05
083 RCLD        36 14
084 RCL4        36 04
085 -             -45
086 RCLD        36 14
087 P⇄S         16-51
088 RCL4        36 04
089 -             -45
090 ÷             -24
091 RCL4        36 04
092 √X             54
093 P⇄S         16-51
094 RCL5        36 05
095 X⇄Y           -41
096 ÷             -24
097 STO0        35 00
098 ×             -35
099 STO1        35 01
100 P⇄S         16-51
101 *LBLa    21 16 11
102 RCL9        36 09
103 GSBD        23 14
104 RCL7        36 07
105 X⇄Y           -41
106 X≤Y?        16-35
107 GTOb     22 16 12
108 1              01
109 ST-9     35-45 09
110 GTOa     22 16 11
111 *LBLb    21 16 12
112 1              01
113 ST+9     35-55 09
114 RCL9        36 09
115 GSBD        23 14
116 RCL7        36 07
117 X⇄Y           -41
118 X≤Y?        16-35
119 GTOb     22 16 12
120 *LBLc    21 16 13
121 PRTX          -14
122 *LBLd    21 16 14
123 RCL9        36 09
124 2              02
125 ×             -35
126 DSP0       -63 00
127 PRTX          -14
128 DSP4       -63 04
129 1              01
130 ST+9     35-55 09
131 RCL9        36 09
132 GSBD        23 14
133 RCLE        36 15
134 X⇄Y           -41
135 X≤Y?        16-35
136 GTOc     22 16 13
137 RCLD        36 14
138 √X             54
139 RCL6        36 06
140 +             -55
141 STOD        35 14
142 SPC         16-11
143 RCL5        36 05
144 X>Y?        16-34
145 GTOA        22 11
146 RTN            24
147 *LBLC       21 13
148 Pi          16-24
149 X²             53
150 ÷             -24
151 1              01
152 2              02
153 ×             -35
154 RTN            24
155 *LBLD       21 14
156 P⇄S         16-51
157 RAD         16-22
158 Pi          16-24
159 ×             -35
160 STO6        35 06
161 3              03
162 9              09
163 STOI        35 46
164 *LBL0       21 00
165 DSZI     16 25 46
166 GTOe     22 16 15
167 GTOd     22 16 14
168 SPC         16-11
169 *LBLe    21 16 15
170 RCL8        36 08
171 STO9        35 09
172 RCL6        36 06
173 +             -55
174 RCL0        36 00
175 ×             -35
176 STO3        35 03
177 1              01
178 RCLD        36 14
179 X>Y?        16-34
180 GTO1        22 01
181 RCL3        36 03
182 2              02
183 ×             -35
184 e^x            33
185 1              01
186 -             -45
187 ENT↑          -21
188 ENT↑          -21
189 2              02
190 +             -55
191 ÷             -24
192 RCL1        36 01
193 ×             -35
194 CHS           -22
195 TAN⁻¹       16 43
196 STO8        35 08
197 GTO2        22 02
198 *LBL1       21 01
199 RCL3        36 03
200 TAN            43
201 RCL1        36 01
202 ×             -35
203 TAN⁻¹       16 43
204 STO8        35 08
205 *LBL2       21 02
206 RCL9        36 09
207 -             -45
208 ABS         16 31
209 EEX           -23
210 CHS           -22
211 6              06
212 X≤Y?        16-35
213 GTO0        22 00
214 RCL6        36 06
215 RCL8        36 08
216 +             -55
217 P⇄S         16-51
218 RCL4        36 04
219 √X             54
220 ÷             -24
221 3              03
222 √X             54
223 ÷             -24
224 RTN            24
```

Program for Card No. 1. To calculate approximate dispersion relation for infinite plate and Ω vs *l*/*h* for plate of finite width (Case A). Also, to calculate two roots of approximate dispersion relation, for a chosen Ω, to use as trial roots with Card No. 2 (Case B).

001	*LBLB	21 12	057	SF2	16 21 02	113	RCL6	36 06	169	STO5	35 05
002	CF0	16 22 00	058	CF3	16 22 03	114	+	-55	170	*LBL5	21 05
003	CF1	16 22 01	059	GTOD	22 14	115	STO2	35 02	171	RCL3	36 03
004	CF2	16 22 02	060	*LBL9	21 09	116	RCL6	36 06	172	F1?	16 23 01
005	SF3	16 21 03	061	P⇄S	16-51	117	RCL5	36 05	173	RCL2	36 02
006	RCL9	36 09	062	SPC	16-11	118	-	-45	174	RCLB	36 12
007	P⇄S	16-51	063	RTN	24	119	STO3	35 03	175	×	-35
008	STO7	35 07	064	*LBLA	21 11	120	*LBL1	21 01	176	CHS	-22
009	P⇄S	16-51	065	SF2	16 21 02	121	RCL3	36 03	177	1	01
010	*LBLD	21 14	066	RCLD	36 14	122	F2?	16 23 02	178	-	-45
011	RAD	16-22	067	RCL4	36 04	123	RCL2	36 02	179	RCL1	36 01
012	GSBA	23 11	068	X=0?	16-43	124	RCL4	36 04	180	+	-55
013	STOI	35 46	069	GTO0	22 00	125	×	-35	181	RCLC	36 13
014	ABS	16 31	070	÷	-24	126	ABS	16 31	182	1	01
015	EEX	-23	071	STO1	35 01	127	√X	54	183	+	-55
016	CHS	-22	072	RCLB	36 12	128	Pi	16-24	184	÷	-24
017	6	06	073	1	01	129	×	-35	185	1/X	52
018	X>Y?	16-34	074	+	-55	130	2	02	186	1	01
019	GTO7	22 07	075	×	-35	131	÷	-24	187	-	-55
020	RCL4	36 04	076	RCLC	36 13	132	STO5	35 05	188	STO9	35 09
021	EEX	-23	077	X²	53	133	F0?	16 23 00	189	RCL3	36 03
022	CHS	-22	078	+	-55	134	GTO2	22 02	190	F1?	16 23 01
023	2	02	079	RCLC	36 13	135	2	02	191	RCL2	36 02
024	+	-55	080	2	02	136	×	-35	192	CHS	-22
025	STO4	35 04	081	×	-35	137	e^x	33	193	RCLA	36 11
026	GSBA	23 11	082	+	-55	138	1	01	194	-	-45
027	RCLI	36 46	083	RCLA	36 11	139	-	-45	195	RCL1	36 01
028	-	-45	084	RCLB	36 12	140	ENT↑	-21	196	+	-55
029	EEX	-23	085	×	-35	141	ENT↑	-21	197	RCLC	36 13
030	CHS	-22	086	-	-45	142	2	02	198	1	01
031	2	02	087	RCLB	36 12	143	+	-55	199	+	-55
032	÷	-24	088	÷	-24	144	÷	-24	200	÷	-24
033	RCLI	36 46	089	2	02	145	RCL5	36 05	201	RCLB	36 12
034	X⇄Y	-41	090	÷	-24	146	÷	-24	202	×	-35
035	÷	-24	091	STO6	35 06	147	F1?	16 23 01	203	RCLC	36 13
036	CHS	-22	092	RCLA	36 11	148	GTO3	22 03	204	+	-55
037	RCL4	36 04	093	1	01	149	STO6	35 06	205	RCL9	36 09
038	+	-55	094	+	-55	150	SF1	16 21 01	206	÷	-24
039	EEX	-23	095	RCL1	36 01	151	GTO1	22 01	207	F2?	16 23 02
040	CHS	-22	096	×	-35	152	*LBL3	21 03	208	GTO6	22 06
041	2	02	097	LSTX	16-63	153	RCL6	36 06	209	STO0	35 00
042	-	-45	098	X²	53	154	÷	-24	210	RCL9	36 09
043	STO4	35 04	099	-	-45	155	STO5	35 05	211	STO8	35 08
044	GTOD	22 14	100	RCLA	36 11	156	GTO5	22 05	212	CF1	16 22 01
045	*LBL7	21 07	101	-	-45	157	*LBL2	21 02	213	SF2	16 21 02
046	RCL4	36 04	102	RCLB	36 12	158	TAN	43	214	GTO5	22 05
047	ABS	16 31	103	÷	-24	159	LSTX	16-63	215	*LBL6	21 06
048	√X	54	104	RCL6	36 06	160	÷	-24	216	STO1	35 01
049	*LBL0	21 00	105	X²	53	161	F1?	16 23 01	217	RCL0	36 00
050	F3?	16 23 03	106	÷	-24	162	GTO4	22 04	218	÷	-24
051	GTOC	22 13	107	1	01	163	STO6	35 06	219	1/X	52
052	GTO9	22 09	108	+	-55	164	SF1	16 21 01	220	RCL5	36 05
053	*LBLC	21 13	109	√X	54	165	GTO1	22 01	221	-	-45
054	P⇄S	16-51	110	RCL6	36 06	166	*LBL4	21 04	222	RTN	24
055	SF0	16 21 00	111	×	-35	167	RCL6	36 06	223	R/S	51
056	CF1	16 22 01	112	STO5	35 05	168	÷	-24			

Program for Card No. 2. To calculate coefficients and solutions of equations for a point on each of the first two branches of the exact dispersion relation, for the infinite plate, for the Ω chosen previously (Case B).

```
001 *LBLB     21 12
002  DSP4    -63 04
003   RAD     16-22
004   SF0  16 21 00
005   CF1  16 22 01
006   SF2  16 21 02
007 *LBL1     21 01
008  RCL5     36 05
009  RCL6     36 06
010     x       -35
011     1        01
012     -       -45
013  RCL6     36 06
014     1        01
015     -       -45
016     ÷       -24
017  STO5     35 05
018 *LBL2     21 02
019  RCL1     36 01
020   F0?  16 23 00
021  RCL0     36 00
022  STO8     35 08
023  RCL9     36 09
024   F0?  16 23 00
025  RCL8     36 08
026  STO8     35 08
027  RCL0     36 00
028     x       -35
029  RCLC     36 13
030     -       -45
031  RCLC     36 13
032     x       -35
033  RCLB     36 12
034     ÷       -24
035  RCLA     36 11
036     +       -55
037  RCL8     36 08
038     ÷       -24
039  STO8     35 08
040  RCL3     36 03
041   F0?  16 23 00
042  RCL2     36 02
043  RCL8     36 08
044     ÷       -24
045   1/X        52
046     1        01
047  RCL5     36 05
048   F1?  16 23 01
049   1/X        52
050     -       -45
051     ÷       -24
052   F1?  16 23 01
053  GTO3     22 03
054  STO6     35 06
055   CF0  16 22 00
056   SF1  16 21 01
057  GTO2     22 02
058 *LBL3     21 03
059  RCL6     36 06
060     +       -55
061   F2?  16 23, 02
062  GTO4     22 04
063  GTO5     22 05
064 *LBL4     21 04
065  STO5     35 05
066   SF0  16 21 00
067   CF1  16 22 01
068   P⇄S     16-51
069  GTO1     22 01
070 *LBL5     21 05
071   P⇄S     16-51
072  RCL5     36 05
073     ÷       -24
074  STO5     35 05
075   P⇄S     16-51
076  RCL4     36 04
077   P⇄S     16-51
078  RCL4     36 04
079     ÷       -24
080   ABS     16 31
081    √X        54
082  STO6     35 06
083  RCL5     36 05
084     ÷       -24
085   1/X        52
086  STO5     35 05
087  RCLD     36 14
088   P⇄S     16-51
089  RCL7     36 07
090   P⇄S     16-51
091  STO9     35 09
092     0        00
093  STO8     35 08
094 *LBLa  21 16 11
095  RCL9     36 09
096  GSBA     23 11
097  RCL7     36 07
098   X⇄Y       -41
099  X≤Y?     16-35
100  GTOb  22 16 12
101     1        01
102  ST-9  35-45 09
103  GTOa  22 16 11
104 *LBLb  21 16 12
105     1        01
106  ST+9  35-55 09
107  RCL9     36 09
108  GSBA     23 11
109  RCL7     36 07
110   X⇄Y       -41
111  X≤Y?     16-35
112  GTOb  22 16 12
113 *LBLc  21 16 13
114  PRTX       -14
115   P⇄S     16-51
116  STO1     35 01
117   P⇄S     16-51
118 *LBLd  21 16 14
119  RCL9     36 09
120     2        02
121     x       -35
122  DSP0    -63 00
123  PRTX       -14
124  DSP4    -63 04
125     1        01
126  ST+9  35-55 09
127  RCL9     36 09
128  GSBA     23 11
129  RCLE     36 15
130   X⇄Y       -41
131  X≤Y?     16-35
132  GTOc  22 16 13
133   SPC     16-11
134   P⇄S     16-51
135  RCL1     36 01
136   P⇄S     16-51
137   RTN        24
138 *LBLA     21 11
139    Pi     16-24
140     x       -35
141  STO0     35 00
142     3        03
143     9        09
144  STOI     35 46
145 *LBLC     21 13
146  DSZI  16 25 46
147  GTOe  22 16 15
148   SPC     16-11
149  GTOd  22 16 14
150 *LBLe  21 16 15
151  RCL8     36 08
152  STO2     35 02
153  RCL0     36 00
154     +       -55
155  RCL6     36 06
156     x       -35
157  STO1     35 01
158     1        01
159  RCLD     36 14
160  X>Y?     16-34
161  GTOD     22 14
162  RCL1     36 01
163     2        02
164     x       -35
165    e^x       33
166     1        01
167     -       -45
168  ENT↑       -21
169  ENT↑       -21
170     2        02
171     +       -55
172     ÷       -24
173  RCL5     36 05
174     x       -35
175 TAN⁻¹     16 43
176  STO8     35 08
177  GTOE     22 15
178 *LBLD     21 14
179  RCL1     36 01
180   TAN        43
181  RCL5     36 05
182     x       -35
183 TAN⁻¹     16 43
184  STO8     35 08
185 *LBLE     21 15
186  RCL2     36 02
187     -       -45
188   ABS     16 31
189   EEX       -23
190   CHS       -22
191     6        06
192  X≤Y?     16-35
193  GTOC     22 13
194  RCL0     36 00
195  RCL8     36 08
196     +       -55
197  RCL4     36 04
198    √X        54
199     ÷       -24
200     2        02
201     x       -35
202    Pi     16-24
203     ÷       -24
204   RTN        24
205   R/S        51
```

Program for Card No. 3. To calculate coefficients and solution of equations for a sequence of roots l/h for the chosen Ω (Case B).

Reprinted from *Mechanics Today*, **5**, S. Nemat-Nasser (ed.)
E. Reissner 65th Birthday Volume
Pergamon Press, Oxford, (1980) 315-328

XX

On Reissner's Equations for Sandwich Plates

R. D. Mindlin†

Columbia University, New York, New York

Summary. The domain of applicability of Reissner's sandwich plate equations is examined by means of a reduction from a higher-order approximation.

1 INTRODUCTION

In one of a series of pioneering papers on the equilibrium of elastic plates, Eric Reissner showed that his equations for flexure of homogeneous plates, which include the effect of transverse shear deformation, apply to sandwich plates when only the transverse shear stiffness of the core and the in-plane stiffness of the face-plates are taken into account. Excluded were the transverse shear stiffness of the face-plates, the flexural stiffnesses of both the core and the face-plates and the effects of the offsets of the middle planes of the face-plates from the interfacial planes. He prefaced the derivation with a precautionary statement: "While the assumptions made in what follows should give an accurate picture (within the linear theory of bending) for combinations such as a foamy core substance and aluminum face-plates they will not be sufficiently accurate for plates composed for instance of two different kinds of wood." His warning is here quantified by exhibiting what has to be neglected in passing to Reissner's equations from a higher-order approximation. The more inclusive equations are derived by joining separate equations for core and face-plates through continuity conditions across the interfaces. Equations of motion are employed as they are more revealing than equations of equilibrium.

†James Kip Finch Emeritus Professor of Applied Science.

What follows is taken from lecture notes written some 25 years ago in response to students' questions about Reissner's sandwich-plate equations.

2 NOTATION

The middle plane of the sandwich plate is taken as the plane of rectangular coordinates x_1 and x_3. The axis of x_2 is oriented vertically and positive upward. The interfaces between the face-plates and the core of the sandwich are at $x_2 = \pm c$ and each face-plate has thickness $2f$. Quantities associated with the upper and lower face-plates are identified by prime and double prime, respectively. The tractions $\tau_{ij}(i,j = 1, 2, 3)$ acting on the horizontal surfaces of the core are designated by

$$[\tau_{2j}]_{\pm c}(= \tau_{ij} \text{ on } x_2 = \pm c);$$

those acting on the horizontal surfaces of the upper and lower face-plates by

$$[\tau'_{2j}]_{\pm f}(= \tau_{ij} \text{ on } x_2 = \pm f + c + f),\ [\tau''_{2j}]_{\pm f}(= \tau_{ij} \text{ on } x_2 = \pm f - c - f),$$

respectively. The displacements in the core, upper face-plate, and lower face-plate are u_i, u'_i, u''_i, respectively. In the approximate equations to be employed, it is assumed that

$$\begin{aligned} u_1 &= x_2\psi_1, & u_2 &= v, & u_3 &= x_3\psi_3, \\ u'_1 &= U' - v'_{,1}, & u'_2 &= v', & u'_3 &= U'_3 - v'_{,3}, \\ u''_1 &= U''_1 - v''_{,1}, & u''_2 &= v'', & u''_3 &= U''_3 - v''_{,3}, \end{aligned}$$

where the comma notation denotes the derivative and the right-hand sides are functions of x_1, x_3 and time (t) only; i.e. classical displacements, both flexural (v', v'') and extensional (U'_1, U'_3, U''_1, U''_3) are assumed for the face-plates [2, Chap. 22] and Timoshenko-type [3, p. 337] flexural displacements (ψ_1, ψ_3, v) are assumed for the core.

The Young's modulus, shear modulus, Poisson's ratio, and density of the core are designated by E_c, μ_c, v_c, ρ_c, respectively, and, of the face-plates, by E_f, μ_f, v_f, and ρ_f.

3 EQUATIONS OF CORE AND FACE-PLATES

With indices α, β, γ ranging over 1 and 3 only and the summation convention employed for repeated indices, the stress-equations of motion for the core are

[2, pp. 488, 496]

$$M_{\alpha\beta,\alpha} - Q_\beta + c[\tau_{2\beta}]_c + c[\tau_{2\beta}]_{-c} = \tfrac{2}{3}c^3\rho_c\ddot{\psi}_\beta, \tag{3.1}$$

$$Q_{\alpha,\alpha} + [\tau_{22}]_c - [\tau_{22}]_{-c} = 2c\rho_c\ddot{v}, \tag{3.2}$$

where $M_{\alpha\beta}$ are bending and twisting moments, Q_α are transverse shear forces—all per unit length. The right-hand sides of (3.1) are the rotatory inertias [2, p. 430]. The associated stress-displacement equations are [4]

$$M_{\alpha\beta} = D_c[\nu_c\delta_{\alpha\beta}\psi_{\gamma,\gamma} + (1-\nu_c)\psi_{(\alpha,\beta)}], \tag{3.3}$$

$$Q_\alpha = 2c\kappa^2\mu_c(v_{,\alpha} + \psi_\alpha), \tag{3.4}$$

where $\psi_{(\alpha,\beta)} = \tfrac{1}{2}(\psi_{\beta,\alpha} + \psi_{\alpha,\beta})$; $v_{,\alpha} + \psi_\alpha$ are the transverse shear deformations (ψ, here, is the negative of the ψ employed by Timoshenko [3, p. 337]); $\delta_{\alpha\beta}$ is the Kronecker delta, κ is a correction factor to be determined later, and D_c is the flexural rigidity of the core:

$$D_c = 2c^3E_c/3(1-\nu_c^2). \tag{3.5}$$

Regarding the face-plates, we suppose that the circular frequency, ω_f, of the fundamental thickness-shear mode of a free face-plate is much greater than that, ω_s, of the sandwich:

$$\omega_f = \pi(\mu_f/\rho_f)^{1/2}/2f \gg \omega_s, \tag{3.6}$$

so that the transverse shear deformation of the face-plates and their rotatory inertias about their middle planes can be neglected. The equations of flexural motion of the face-plates are, then, the classical ones [2, pp. 488, 496, 465]:

$$M^i_{\alpha\beta,\alpha} - Q^i_\beta + f[\tau^i_{2\beta}]_f + f[\tau^i_{2\beta}]_{-f} = 0, \tag{3.7}$$

$$Q^i_{\alpha,\alpha} + [\tau^i_{22}]_f - [\tau^i_{22}]_{-f} = 2f\rho_f\ddot{v}^i, \tag{3.8}$$

$$M^i_{\alpha\beta} = -D_f[\nu_f\delta_{\alpha\beta}v^i_{,\gamma\gamma} + (1-\nu_f)v^i_{,\alpha\beta}], \tag{3.9}$$

where the superscript i stands for prime or double prime and D_f is the flexural rigidity of a face-plate; i.e. (3.5) with c replaced by f.

The extensional motions of the face-plates are governed by Poisson's equations [2, p. 497]:

$$N^i_{\alpha\beta,\alpha} + [\tau^i_{2\beta}]_f - [\tau^i_{2\beta}]_{-f} = 2f\rho_f\ddot{U}^i_\beta, \tag{3.10}$$

$$N^i_{\alpha\beta} = (3D_f/f^2)[\nu_f\delta_{\alpha\beta}U^i_{\gamma,\gamma} + (1-\nu_f)U^i_{(\alpha,\beta)}], \tag{3.11}$$

where the $N^i_{\alpha\beta}$ are the forces, per unit length, in the planes of the face-plates.

4 EQUATIONS OF THE SANDWICH

The five sets of equations (one set of flexure equations (3.1)–(3.4) including the contributions of rotatory inertia and shear, one set of classical flexure equations (3.7)–(3.9), and one set of classical extension equations (3.10)–(3.11) for each of the face-plates) are joined through conditions of continuity of displacements and tractions at the two interfaces $x_2 = \pm c$.

As the changes of thicknesses of the core and face-plates are neglected, the conditions of continuity of vertical displacements are

$$v = v' = v''. \tag{4.1}$$

For continuity of normal components of tractions across the interfaces, we have

$$[\tau_{22}]_c = [\tau'_{22}]_{-f}, \qquad [\tau_{22}]_{-c} = [\tau''_{22}]_f.$$

Subtracting and employing (3.8) and (4.1), we find

$$[\tau_{22}]_c - [\tau_{22}]_{-c} = Q'_{\beta,\beta} + Q''_{\beta,\beta} + p - 4f\rho_f \ddot{v} \tag{4.2}$$

where $p\,(= [\tau'_{22}]_f - [\tau''_{22}]_{-f})$ is the transverse loading on the sandwich. From (3.7),

$$Q'_\beta + Q''_\beta = M'_{\alpha\beta,\alpha} + M''_{\alpha\beta,\alpha} + f[\tau''_{2\beta}]_f + f[\tau''_{2\beta}]_{-f} \tag{4.3}$$

in which $f[\tau'_{2\beta}]_f$ and $f[\tau''_{2\beta}]_{-f}$, the external tangential loads on the sandwich, have been omitted. The last two terms in (4.3) are called "offset" terms; they represent the contribution of the offset of the interfacial tangential tractions from the middle planes of the face-plates. From (3.9) and (4.1),

$$M'_{\alpha\beta,\alpha\beta} + M''_{\alpha\beta,\alpha\beta} = -2D_f v_{,\alpha\alpha\beta\beta} = -2D_f \nabla^4 v, \tag{4.4}$$

where $\nabla^4 = \nabla^2\nabla^2$ and ∇^2 is the two dimensional Laplacian in the plane of x_1 and x_3.

Substituting (4.3) and (4.4) in (4.2), we have

$$[\tau_{22}]_c - [\tau_{22}]_{-c} = -2D_f\nabla^4 v + f\{[\tau''_{2\alpha}]_f + [\tau'_{2\alpha}]_{-f}\}_{,\alpha} + p - 4f\rho_f \ddot{v}. \tag{4.5}$$

Next, from (4.5) and (3.2),

$$Q_{\alpha,\alpha} - 2D_f\nabla^4 v + f\{[\tau''_{2\alpha}]_f + [\tau'_{2\alpha}]_{-f}\}_{,\alpha} + p = 2c\rho_c(1 + R)\ddot{v}, \tag{4.6}$$

where $R = 2f\rho_f/c\rho_c$, i.e. the ratio of the mass of both face-plates to the mass of the core. To express the offset terms in (4.6) in terms of displacements, we note, from (3.10),

$$\{[\tau''_{2\alpha}]_f + [\tau''_{2\alpha}]_{-f}\}_{,\alpha} = N'_{\alpha\beta,\alpha\beta} - N''_{\alpha\beta,\alpha\beta} - 2f\rho_f(\ddot{U}'_\alpha - \ddot{U}''_\alpha)_{,\alpha} \tag{4.7}$$

(taking into account that the external tangential tractions are neglected). The terms on the right-hand side of (4.7) are the static and dynamic offset terms. Next, we set the conditions of continuity of tangential displacements across the interfaces:

$$U'_\alpha + fv'_{,\alpha} = c\psi_\alpha, \qquad U''_\alpha - fv''_{,\alpha} = -c\psi_\alpha, \tag{4.8}$$

in which the $fv^i_{,\alpha}$ are additional offset terms. Then, from (4.8) and (4.1),

$$U'_{\alpha,\alpha} - U''_{\alpha,\alpha} = 2c\Delta - 2f\nabla^2 v, \tag{4.9}$$

where $\Delta = \psi_{\alpha,\alpha}$. As for the remaining terms on the right-hand side of (4.7), we have, from (3.11) and (4.9),

$$N'_{\alpha\beta,\alpha\beta} - N''_{\alpha\beta,\alpha\beta} = (6D_f/f^2)(c\nabla^2\Delta - f\nabla^4 v). \tag{4.10}$$

Finally, inserting (4.9) and (4.10) in (4.7) and the result in (4.6), we find, after using (3.4),

$$\begin{aligned} 2c\kappa^2\mu_c(\Delta + \nabla^2 v) - 2D_f\nabla^4 v + p + (6D_f/f)(c\nabla^2\Delta - f\nabla^4 v) \\ = 2c\rho_c(1 + R)\ddot{v} + 4f^2\rho_f(c\ddot{\Delta} - f\nabla^2\ddot{v}). \end{aligned} \tag{4.11}$$

This is the equation of transverse motion of the sandwich. The terms with coefficient μ_c account for the horizontal (Δ) and vertical ($\nabla^2 v$) shear stiffnesses of the core; the next term derives from the flexural rigidities of the face-plates; p is the transverse loading per unit area; and the remaining terms on the left-hand side of (4.11) are static offset terms. On the right-hand side, $2c\rho_c\ddot{v}$ is the transverse inertia of the core, while that times R accounts for the transverse inertia of the face-plates; the remaining terms are dynamic offset terms.

Turning, now, to the rotational motion of the sandwich, we apply, first, the conditions of continuity of tangential tractions across the interfaces:

$$[\tau'_{2\alpha}]_{-f} = [\tau_{2\alpha}]_c, \; [\tau''_{2\alpha}]_f = [\tau_{2\alpha}]_{-c}. \tag{4.12}$$

Hence,

$$[\tau_{2\alpha}]_c + [\tau_{2\alpha}]_{-c} = [\tau'_{2\alpha}]_{-f} + [\tau''_{2\alpha}]_f,$$

so that, analogous to (4.7), (4.9), and (4.10),

$$[\tau_{2\alpha}]_c + [\tau_{2\alpha}]_{-c} = N'_{\alpha\beta,\beta} - N''_{\alpha\beta,\beta} - 2f\rho_f(\dot{u}'_\alpha - \dot{u}''_\alpha), \tag{4.13}$$

$$\ddot{u}'_\alpha - \ddot{u}''_\alpha = 2c\ddot{\psi}_\alpha - 2f\ddot{u}_{2,\alpha}, \tag{4.14}$$

$$N'_{\alpha\beta,\alpha} - N''_{\alpha\beta,\alpha} = 3D_f f^{-2}[(1 + \nu_f)c\Delta_{,\beta} + (1 - \nu_f)c\nabla^2\psi_\beta - 2f\nabla^2 u_{2,\beta}]. \tag{4.15}$$

Upon inserting (4.14) and (4.15) in (4.13) and the result in (3.1) and then applying (3.3) and (3.4), we find the equations of flexural motion of the sandwich:

$$\tfrac{1}{2}D_c[(1-\nu_c)\nabla^2\psi_\alpha + (1+\nu_c)\Delta_{,\alpha}] + 3c^2 f^{-2} D_f[(1-\nu_f)\nabla^2\psi_\alpha + (1+\nu_f)\Delta_{,\alpha}]$$
$$- 2c\kappa^2\mu_c(v_{,\alpha} + \psi_\alpha) - 6cf^{-1}D_f\nabla^2 v_{,\alpha}$$
$$= \tfrac{2}{3}c^3\rho_c(1+3R)\ddot{\psi}_\alpha - 4cf^2\rho_f\ddot{v}_{,\alpha}. \qquad (4.16)$$

In (4.16), the two sets of terms on the first line are the contributions, to the flexural rigidity of the sandwich, of the flexural rigidity of the core and the extensional stiffness of the face-plates, respectively. The first term of the second line accounts for the shear stiffness of the core and the remaining term on the left-hand side of (4.16) is a static offset term. On the right-hand side, $\frac{2}{3}c^3\rho_c\ddot{\psi}_\alpha$ is the rotatory inertia of the core and that times $3R$ is the rotatory inertia of the face-plates about the middle plane of the sandwich. The remaining term is a dynamic offset term.

Equations (4.11) and (4.16) are the displacement equations of motion of the sandwich. To complete them it is only necessary to specify the correction factor κ.

5 CORRECTION FACTOR

A formula for the correction factor κ may be derived [4] by equating the frequency of the thickness-shear mode obtained from the sandwich equations with that of the lowest thickness-shear mode calculated from the three-dimensional equations of elasticity. For the sandwich, set

$$\psi_1 = A\exp(i\omega_s t), \quad \psi_3 = v = p = 0.$$

Then, from (4.11) and (4.16),

$$\omega_s^2 = 3\kappa^2\mu_c/c^2\rho_c(1+3R). \qquad (5.1)$$

For the three-dimensional theory, we have the equations of motion

$$\mu_c u^c_{j,ii} + \mu_c(1-2\nu_c)^{-1}u^c_{i,ij} = \rho_c\ddot{u}^c_j \qquad (5.2)$$

for the core and similar ones for the face-plates. For the core, we set

$$u^c_1 = A\sin\beta_c x_2\, e^{i\omega t}, \qquad u^c_2 = u^c_3 = 0$$

and find

$$\omega^2 = \beta_c^2\mu_c/\rho_c. \qquad (5.3)$$

As for the face-plates, we need consider only the upper one, owing to symmetry. We set

$$u_1^f = (B \sin \beta_f x_2 + C \cos \beta_f x_2)\, e^{i\omega t}, \qquad u_2^f = u_3^f = 0$$

in the equations of motion analogous to (5.2) and find

$$\omega^2 = \beta_f^2 \mu_f / \rho_f \tag{5.4}$$

so that

$$\beta_c^2 \mu_c \rho_f = \beta_f^2 \mu_f \rho_c. \tag{5.5}$$

For the boundary conditions, we require continuity of displacement and traction across the interface:

$$u_1^c = u_1^f, \qquad \mu_c u_{1,2}^c = \mu_f u_{1,2}^f \text{ on } x_2 = c$$

and vanishing traction on the upper surface:

$$u_{1,2}^f = 0 \text{ on } x_2 = c + 2f.$$

Upon substituting the expressions for the displacements of u_1^c and u_1^f in the three boundary conditions and eliminating the constants A, B, and C from the resulting equations we find the frequency equation

$$\mu_c \beta_c [\tan \beta_f c \tan \beta_f (c + 2f) + 1] = \mu_f \beta_f [\tan \beta_f (c + 2f) - \tan \beta_f c\,] \tan \beta_c c. \tag{5.6}$$

From (5.4) and (3.6), $2f\beta_f = \pi\omega/\omega_f$, i.e. π times the ratio of the thickness-shear frequency of the sandwich to that of the free face-plate. If $\omega/\omega_f \ll 1$,

$$\tan \beta_f (c + 2f) \approx (\tan \beta_f c + 2\beta_f f)/(1 - 2\beta_f f \tan \beta_f c)$$

and the frequency equation (5.6) reduces to

$$R \beta_c c \tan \beta_c c = 1. \tag{5.7}$$

Upon equating (5.1) and (5.3), we have

$$\kappa^2 = \beta_c^2 c^2 (1 + 3R)/3, \tag{5.8}$$

where $\beta_c c$ is the lowest root of (5.6)—or (5.7) if ω/ω_f is small. In the latter case, for $R = 0, 1, \infty$, $\kappa^2 = \pi^2/12$, 0.99, and 1.00, respectively. Thus, for small ω/ω_f and for an even moderately large mass ratio of face-plates to core ($R > 1$) κ may be taken equal to unity. The reason no correction factor is required is that the "exact" displacement, u_1^c, of the core is practically linear across the thickness of the core and that is just what is assumed in the derivation of the approximate equations.

6 STRAIGHT-CRESTED WAVES

The relative magnitudes of the contributions of the various terms in the equations of motion (4.11) and (4.16) may be determined by examining the solution for straight-crested waves. A sandwich of a 1-inch-thick cellular, cellulose acetate core and $\frac{1}{32}$-inch aluminum alloy face-plates is an example of the type for which Reissner designed his equations. Then

$$c/f = 32, \; \mu_f/\mu_c = 2000, \; \rho_f/\rho_c = 35 \tag{6.1}$$

and, for simplicity, we take $\nu_c = \nu_f$. First of all, $R = 2.19$ so that, from (5.7) and (5.8), $\kappa = 1$. Then, from (3.6) and (5.1), $\omega_f/\omega_s = 440$ so that the omission of transverse shear and rotatory inertia from the equations for the face-plates is amply justified. Hundreds of modes with resonance frequencies between ω_s and ω_f would take precedence.

For the straight-crested waves, we take

$$\psi_1 = A\cos\xi x_1\, e^{i\omega t}, \qquad v = Bc\sin\xi x_1\, e^{i\omega t},\; \psi_3 = p = 0. \tag{6.2}$$

Substitution of (6.2) in (4.11) and (4.16), followed by elimination of A and B, results in the dispersion relation:

$$\frac{1 + \dfrac{2}{3(1-\nu)}\left(1 + \dfrac{6\mu_f f}{\mu_c c}\right)\bar{\xi}^2 - \frac{1}{3}(1+3R)\Omega^2}{1 - \dfrac{4}{1-\nu}\dfrac{\mu_f}{\mu_c}\dfrac{f^2}{c^2}\bar{\xi}^2 + \dfrac{f}{c}R\Omega^2} = \frac{1 - \dfrac{4}{1-\nu}\dfrac{\mu_f}{\mu_c}\dfrac{f^2}{c^2}\bar{\xi}^2 + \dfrac{f}{c}R\Omega^2}{1 - (1+R)\dfrac{\Omega^2}{\bar{\xi}^2} + \dfrac{16}{3(1-\nu)}\dfrac{f^3}{c^3}\dfrac{\mu_f}{\mu_c}\bar{\xi}^2 - \dfrac{f^2}{c^2}R\Omega^2} \tag{6.3}$$

where $\bar{\xi} = \xi c$ and $\Omega = \omega c(\rho_c/\mu_c)^{1/2}$, i.e. respectively, π times the ratio of the core thickness to the wave length and $\pi/2$ times the ratio of the frequency to the frequency of the lowest thickness-shear mode of the free core. Note that the reference frequency is not that of the sandwich, which is lower and is given by (5.1).

In the numerator of the left-hand side of (6.3), the first term (unity) stems from the shear stiffness of the core. The two terms in the first parentheses are from the contributions of the flexural rigidity of the core (unity) and the extensional stiffness of the face-plates, respectively, to the flexural rigidity of the sandwich. The ratio of the latter to the former is 375, which justifies

Reissner's omission of the flexural rigidity of the core. In the second parentheses the first term, unity, comes from the rotatory inertia of the core and the second, $3R$, from the rotatory inertia of the face-plates about the middle plane of the sandwich. Thus, although the flexural rigidity of the core may be negligible, its rotatory inertia may not be small in comparison with that of the face-plates about the middle plane of the sandwich. However, the contributions of both rotatory inertias may be negligible in comparison with the shear stiffness of the core if $\Omega \ll 1$.

In the denominator of the left-hand side of (6.3) (and the numerator of the right-hand side) the first term, unity, stems from the shear stiffness of the core and the remaining two terms are static and dynamic offset terms. With the values in (6.1) and $\nu = 0.29$, the static term is $11\bar{\xi}^2$ so that, except for very long wave lengths, it is of the same order of magnitude as the shear term. The dynamic offset term is not as important, especially at low frequency.

In the denominator of the right-hand side of (6.3), the first term, unity, stems from the shear stiffness of the core. In the parentheses following, the terms unity and R are from the transverse inertias of the core and facings, respectively; both must usually be retained in dynamical problems. The next term comes from both the flexural rigidity of the face-plates and the static offset, introduced in (4.8), in the proportion of three of the latter to one of the former. The two together, in the present example, contribute $0.46\bar{\xi}^2$—a not inconsiderable amount in comparison with the unity of the shear contribution, except at very long wave lengths. The last term, a small one, is a dynamic offset term also stemming from the offset term in (4.8).

The dispersion relation (6.3) has two real branches—a low-frequency "flexural" branch and a high-frequency "thickness-shear" branch, the two connected by an imaginary segment of the latter, with a cut-off frequency given by (5.1). Even if the offset terms and the flexural rigidities of the core and face-plates are omitted (Reissner's approximation) there are still two real branches. The real part of the upper branch disappears if, in addition, the rotatory inertia terms are omitted.

7 INITIAL AND BOUNDARY CONDITIONS

To establish initial and boundary conditions sufficient to assure a unique solution of the equations of motion, it is convenient to have stress equations of motion and stress-displacement relations for the sandwich rather than (3.1)–(3.4) for the core and (3.7)–(3.11) for the face-plates separately.

We define stress-displacement relations:

$$Q_\alpha = 2c\kappa^2\mu_c(v_{,\alpha} + \psi_\alpha), \tag{7.1}$$

$$M_{\alpha\beta} = D_c[\nu_c\delta_{\alpha\beta}\psi_{\gamma,\gamma} + (1-\nu_c)\psi_{(\beta,\alpha)}], \tag{7.2}$$

$$\mathscr{M}_{\alpha\beta} = -2D_f[\nu_f\delta_{\alpha\beta}v_{,\gamma\gamma} + (1-\nu_f)v_{,\alpha\beta}], \tag{7.3}$$

$$\mathscr{N}_{\alpha\beta} = 6D_f f^{-2}[\nu_f\delta_{\alpha\beta}(c\psi_{\gamma,\gamma} - fv_{,\gamma\gamma}) + (1-\nu_f)(c\psi_{(\beta,\alpha)} - fv_{,\alpha\beta})]. \tag{7.4}$$

Then, the stress equations of motion of the sandwich are

$$M_{\alpha\beta,\alpha} + c\mathscr{N}_{\alpha\beta,\alpha} - Q_\beta = \tfrac{2}{3}c^3\rho_c(1+3R)\ddot{\psi}_\beta - 4cf^2\rho_f\ddot{v}_{,\beta}, \tag{7.5}$$

$$\begin{aligned} &Q_{\alpha,\alpha} + \mathscr{M}_{\alpha\beta,\alpha\beta} + f\mathscr{N}_{\alpha\beta,\alpha\beta} + p \\ &\quad = 2c\rho_c(1+R)\ddot{v} + 4f^2\rho_f(c\ddot{\psi}_{\alpha,\alpha} - f\ddot{v}_{,\alpha\alpha}). \end{aligned} \tag{7.6}$$

Substitution of (7.1)–(7.4) in (7.5) and (7.6) reproduces (4.11) and (4.16).

A potential energy density, W, is defined by

$$2W = Q_\alpha(v_{,\alpha} + \psi_\alpha) + M_{\alpha\beta}\psi_{\alpha,\beta} + \mathscr{M}_{\alpha\beta}v_{,\alpha\beta} + \mathscr{N}_{\alpha\beta}(c\psi_{\alpha,\beta} - fv_{,\alpha\beta}) \tag{7.7}$$

which, from (7.1)–(7.4), becomes

$$\begin{aligned} 2W &= 2c\kappa^2\mu_c(v_{,\alpha} + \psi_\alpha)(v_{,\alpha} + \psi_\alpha) + D_c[\nu_c\psi_{\alpha,\alpha}\psi_{\beta,\beta} + (1-\nu_c)\psi_{(\beta,\alpha)}\psi_{(\beta,\alpha)}] \\ &\quad + 2D_f[\nu_f v_{,\alpha\alpha}v_{,\beta\beta} + (1-v_f)v_{,\alpha\beta}v_{,\alpha\beta}] + 6f^{-2}D_f\nu_f(c\psi_{\alpha,\alpha} - fv_{,\alpha\alpha})^2 \\ &\quad + 6f^{-2}D_f(1-\nu_f)(c\psi_{(\beta,\alpha)} - fv_{,\alpha\beta})(c\psi_{(\beta,\alpha)} - fv_{,\alpha\beta}). \end{aligned} \tag{7.8}$$

Necessary and sufficient conditions for W to be positive definite are

$$E_c > 0, \qquad E_f > 0, \qquad -1 < \nu_c < 1, \qquad -1 < \nu_f < 1.$$

We also define a positive definite kinetic energy density, K, by

$$\begin{aligned} 2K &= \tfrac{2}{3}c^3\rho_c(1+3R)\dot{\psi}_\alpha\dot{\psi}_\alpha + 2c\rho_c(1+R)\dot{v}^2 + 4f^3\rho_f\dot{v}_{,\alpha}\dot{v}_{,\alpha} \\ &\quad - 8cf^2\rho_f\dot{v}_{,\alpha}\dot{\psi}_\alpha. \end{aligned} \tag{7.9}$$

Following a procedure similar to Neumann's for the three-dimensional equations of elasticity [2, p. 176], we consider two sets of quantities Q^a_α, $M^a_{\alpha\beta}$, $\mathscr{M}^a_{\alpha\beta}$, $\mathscr{N}^a_{\alpha\beta}$, p^a, ψ^a_α, v^a and Q^b_α, $M^b_{\alpha\beta}$, $\mathscr{M}^b_{\alpha\beta}$, $\mathscr{N}^b_{\alpha\beta}$, p^b, ψ^b_α, v^b, each of which satisfies (7.1)–(7.6). Then the difference set:

$$Q_\alpha = Q^a_\alpha - Q^b_\alpha, M_{\alpha\beta} = M^a_{\alpha\beta}, \ldots, v = v^a - v^b, \tag{7.10}$$

also satisfies the same equations since the equations are linear. We suppose that, at time t_0,

$$v = 0, \qquad \dot{v} = 0, \qquad \psi_\alpha = 0, \qquad \dot{\psi}_\alpha = 0, \qquad v_{,\alpha} = 0, \qquad \dot{v}_{,\alpha} = 0. \tag{7.11}$$

This makes the difference potential and kinetic energy densities, W and K, zero at $t = t_0$.

Consider the identity

$$\begin{aligned}
&\int_{t_0}^{t} dt \int_A [Q_{\alpha,\alpha} + \mathscr{M}_{\alpha\beta,\alpha\beta} + f\mathscr{N}_{\alpha\beta,\alpha\beta} + p - 2c\rho_c(1 + R)\ddot{v} \\
&\quad - 4f^2\rho_f(c\ddot{\psi}_{\alpha,\alpha} - f\ddot{v}_{,\alpha\alpha})]\dot{v}\, dA \\
&\quad + \int_{t_0}^{t} dt \int_A [M_{\alpha\beta,\alpha} + c\mathscr{N}_{\alpha\beta,\alpha} - Q_\beta - \tfrac{2}{3}c^3\rho_c(1 + 3R)\ddot{\psi}_\beta \\
&\quad + 4cf^2\rho_f\ddot{v}_{,\beta}]\dot{\psi}_\beta\, dA = 0,
\end{aligned} \tag{7.12}$$

where A is the area of the plate. Now,

$$\begin{aligned}
&\int_{t_0}^{t} dt \int_A [\tfrac{2}{3}c^3\rho_c(1 + 3R)\ddot{\psi}_\beta\dot{\psi}_\beta + 2c\rho_c(1 + R)\ddot{v}\dot{v}]\, dA \\
&\qquad = \int_A [\tfrac{1}{3}c^3\rho_c(1 + 3R)\dot{\psi}_\beta\dot{\psi}_\beta + c\rho_c(1 + R)\dot{v}^2]\, dA,
\end{aligned} \tag{7.13}$$

since $\dot{\psi}$ and $\dot{v}$ are zero at $t = t_0$. Also,

$$\begin{aligned}
&\int_{t_0}^{t} dt \int_A [c\ddot{v}_{,\alpha}\dot{\psi}_\alpha - c\ddot{\psi}_{\alpha,\alpha}\dot{v} + f\ddot{v}_{,\alpha\alpha}\dot{v}]\, dA \\
&= \int_{t_0}^{t} dt \int_A [c\ddot{v}_{,\alpha}\dot{\psi}_\alpha - (c\ddot{\psi}_\alpha\dot{v})_{,\alpha} + c\ddot{\psi}_\alpha\dot{v}_{,\alpha} + (f\ddot{v}_{,\alpha}\dot{v})_{,\alpha} - f\ddot{v}_\alpha v_{,\alpha}]\, dA \\
&= \int_A (c\dot{v}_{,\alpha}\dot{\psi}_\alpha - \tfrac{1}{2}f\dot{v}_{,\alpha}\dot{v}_{,\alpha})\, dA - \int_{t_0}^{t} dt \oint_C n_\alpha(c\ddot{\psi}_\alpha - f\ddot{v}_{,\alpha})\dot{v}\, ds,
\end{aligned} \tag{7.14}$$

where, in the line integral which results from the use of the divergence theorem, C is the bounding curve or polygon of the plate, s is the coordinate along C and n is the unit, outward-drawn normal to C. Then, adding (7.13) and (7.14) and noting (7.9) and (7.11), we find that the inertia terms in (7.12) amount to

$$\int_A K\, dA + 4f^2\rho_f \int_{t_0}^{t} dt \oint_C n_\alpha(c\ddot{\psi}_\alpha - f\ddot{v}_{,\alpha})\dot{v}\, ds. \tag{7.15}$$

Now, return to (7.12) and observe that

$$\int_A [(M_{\alpha\beta,\alpha} + c\mathcal{N}_{\alpha\beta,\alpha} - Q_\beta)\dot{\psi}_\beta + (Q_{\alpha,\alpha} + \mathcal{M}_{\alpha\beta,\alpha\beta} + f\mathcal{N}_{\alpha\beta,\alpha\beta})\dot{v}]\,dA$$

$$= -\int_A [(M_{\alpha\beta} + c\mathcal{N}_{\alpha\beta})\dot{\psi}_{\beta,\alpha} + Q_\beta(\dot{\psi}_\alpha + \dot{v}_{,\alpha}) - (\mathcal{M}_{\alpha\beta} + f\mathcal{N}_{\alpha\beta})\dot{v}_{,\alpha\beta}]\,dA$$

$$+ \int_A [(M_{\alpha\beta} + c\mathcal{N}_{\alpha\beta})\dot{\psi}_\beta + (\mathcal{M}_{\alpha\beta,\beta} + f\mathcal{N}_{\alpha\beta,\beta} + Q_\alpha)\dot{v} - (\mathcal{M}_{\alpha\beta} + f\mathcal{N}_{\alpha\beta})\dot{v}_{,\beta}]_{,\alpha}\,dA$$

$$= -\int_A \dot{W}\,dA - \oint_C n_\alpha(\mathcal{M}_{\alpha\beta} + f\mathcal{N}_{\alpha\beta})\dot{v}_{,\beta})\,ds$$

$$+ \oint_C [n_\alpha(M_{\alpha\beta} + c\mathcal{N}_{\alpha\beta})\dot{\psi}_\beta + n_\alpha(Q_\alpha + \mathcal{M}_{\alpha\beta,\beta} + f\mathcal{N}_{\alpha\beta,\beta})\dot{v}]\,ds; \tag{7.16}$$

but

$$\oint_C n_\alpha(\mathcal{M}_{\alpha\beta} + f\mathcal{N}_{\alpha\beta})\dot{v}_{,\beta}\,ds = \oint_C \left[(\mathcal{M}_{nn} + f\mathcal{N}_{nn})\frac{d\dot{v}}{dn} + (\mathcal{M}_{ns} + f\mathcal{N}_{ns})\frac{d\dot{v}}{ds}\right]ds \tag{7.17}$$

and

$$\oint_C (\mathcal{M}_{ns} + f\mathcal{N}_{ns})\frac{d\dot{v}}{ds}\,ds$$

$$= [(\mathcal{M}_{ns} + f\mathcal{N}_{ns})\dot{v}]_0^0 - \oint_C \left(\frac{d\mathcal{M}_{ns}}{ds} + f\frac{d\mathcal{N}_{ns}}{ds}\right)\dot{v}\,ds \tag{7.18}$$

where $\mathcal{M}_{nn}$ and $\mathcal{M}_{ns}$ are bending and twisting couples, respectively, $\mathcal{N}_{nn}$ and $\mathcal{N}_{ns}$ are normal and tangential components, respectively, and $[\ ]_0^0$ indicates the jump around a corner.

Upon assembling the results in (7.15)–(7.18), we find that (7.12) becomes

$$\int_A (W + K)\,dA = \int_{t_0}^t dt \int_A \rho\dot{v}\,dA + \int_{t_0}^t dt \oint_C [n_\alpha(M_{\alpha\beta} + c\mathcal{N}_{\alpha\beta})\dot{\psi}_\beta + V\dot{v}]\,ds$$

$$- \int_{t_0}^t dt \oint_C (\mathcal{M}_{nn} + f\mathcal{N}_{nn})\frac{d\dot{v}}{dn}\,ds - \int_{t_0}^t [(\mathcal{M}_{ns} + f\mathcal{N}_{ns})\dot{v}]_0^0\,dt, \tag{7.19}$$

where

$$V = Q_n + \frac{d\mathscr{M}_{ns}}{ds} + f\frac{d\mathscr{N}_{ns}}{ds} + n_\alpha[\mathscr{M}_{\alpha\beta,\beta} + f\mathscr{N}_{\alpha\beta,\beta} - 4f^2\rho_f(c\ddot{\psi}_\alpha - f\ddot{v}_{,\alpha})].$$

By the usual argument [2, p. 176], conditions sufficient for a unique solution of the equations of motion, (4.11) and (4.16), are: specification of p and the initial values of v, ψ_α, $v_{,\alpha}$ and $\dot{v}$, $\dot{\psi}_\alpha$, $\dot{v}_{,\alpha}$ throughout and, on the boundary, four quantities at each point and one at each corner. Specifically, at each point, either

1. a component of ψ_α or of $n_\alpha(M_{\alpha\beta} + c\mathscr{N}_{\alpha\beta})$ and
2. the orthogonal component of ψ_α or of $n_\alpha(M_{\alpha\beta} + c\mathscr{N}_{\alpha\beta})$ and
3. v or V and
4. dv/dn or $\mathscr{M}_{nn} + f\mathscr{N}_{nn}$

and, at each corner, v or $\mathscr{M}_{ns} + f\mathscr{N}_{ns}$.

8 LOW FREQUENCIES AND EQUILIBRIUM

The equations (4.11), (4.16) and (7.1)–(7.9) are valid up to frequencies almost as high as the second root of (5.6) or (5.7). However, at low frequencies, say $\Omega < 0.1$, or at equilibrium, Reissner's approximation holds. Then the offset terms, the flexural rigidities of the core and face-plates and the rotatory inertia may be omitted and (4.11) and (4.16) reduce to

$$2c\kappa^2\mu_c(v_{,\alpha\alpha} + \psi_{\alpha,\alpha}) + p = 2c\rho_c(1 + R)\ddot{v}, \tag{8.1}$$

$$3c^2f^{-2}D_f[(1 - \nu_f)\psi_{\beta,\alpha\alpha} + (1 + \nu_f)\psi_{\alpha,\alpha\beta}] - 2c\kappa^2\mu_c(v_{,\alpha} + \psi_\alpha) = 0, \tag{8.2}$$

and (7.1)–(7.9) reduce to

$$Q_\alpha = 2c\kappa^2\mu_c(v_{,\alpha} + \psi_\alpha), \tag{8.3}$$

$$\mathscr{N}_{\alpha\beta} = 6D_f f^{-2}c[\nu_f\delta_{\alpha\beta}\psi_{\gamma,\gamma} + (1 - \nu_f)\psi_{(\beta,\alpha)}], \tag{8.4}$$

$$c\mathscr{N}_{\alpha\beta,\alpha} - Q_\beta = 0, \tag{8.5}$$

$$Q_{\alpha,\alpha} + p = 2c\rho_c(1 + R)\ddot{v}, \tag{8.6}$$

$$2W = Q_\alpha(v_{,\alpha} + \psi_\alpha) + cN_{\alpha\beta}\psi_{\alpha,\beta}, \tag{8.7}$$

$$K = c\rho_c(1 + R)\dot{v}^2. \tag{8.8}$$

Finally, (7.19) reduces to

$$\int_A (W + K)\,dA = \int_{t_0}^{t} dt \int_A p\,dA + \int_{t_0}^{t} dt \oint_C (cn_\alpha \mathcal{N}_{\alpha\beta}\dot{\psi}_\beta + Q_n \dot{v})\,ds \tag{8.9}$$

so that the initial conditions reduce to specification of v, ψ_α, $\dot{v}$, $\dot{\psi}_\alpha$ and the boundary conditions reduce to Reissner's three.

9 REFERENCES

1. Reissner, E., "On Bending of Elastic Plates," *Q. Appl. Math.* **5** (1947) 55–68.
2. Love, A. E. H., *A Treatise on the Theory of Elasticity*, 4th ed., Cambridge (1927).
3. Timoshenko, S., *Vibration Problems in Engineering*, 2nd ed., D. van Nostrand (1937).
4. Mindlin, R. D., "Influence of Rotatory Inertia and Shear on Flexural Motions of Isotropic, Elastic Plates," *J. Appl. Mech.* **18** (1951) 31–38.

Int. J. Solids Structures Vol. 16, pp. 785–791
Pergamon Press Ltd., 1980. Printed in Great Britain

HIGH FREQUENCY VIBRATIONS OF QUARTZ PLATES BY EXPANSION IN SERIES OF EKSTEIN FUNCTIONS†

R. D. MINDLIN
89 Deer Hill Drive, Ridgefield, CT 06877, U.S.A.

(*Received* 2 *February* 1980)

Abstract—A solution is presented for high frequency vibrations of anisotropic, elastic plates applicable to rotated-Y-cuts of quartz with a pair of free edges. The solution is based on an expansion in a series of Ekstein's exact, normal functions for the infinite plate—retaining the first three terms: flexure, thickness-shear, face-shear.

1. INTRODUCTION

There is no difficulty, today, in obtaining exact solutions of the three-dimensional, linear equations of elasticity for problems of vibrations of infinite plates with a pair of free faces. For the isotropic plate, this was first accomplished by Rayleigh in 1889[1]; and Ekstein, in 1945[2], gave a solution for an anistropic plate applicable to the technologically important rotated-Y-cuts of quartz. Although many years passed before the resulting frequency equations were analyzed in detail, their properties are now well understood.

At a free edge of a plate with free faces, each of the infinity of modes of the infinite plate reflects, except for a few special cases, as an infinity of modes; and therein lies the difficulty with finding solutions for plates with free faces and even only a single free edge or pair of parallel free edges. The classical method of overcoming the difficulty has been the employment of approximate, two-dimensional equations of motion which already satisfy, at least approximately, the conditions of free faces so that the conditions for free edges are analogous to those for free faces in the three-dimensional case. A solution of the approximate equations for the infinite plate contains, usually, only a small number of modes approximating the corresponding ones of the infinity of modes from the three-dimensional equations. Each of the approximate modes reflects, at a free edge, at most as the total, small number; so that the frequency equation is finite. The approximate equations are generally obtained, especially in the case of the high frequency range, by expanding the three-dimensional displacements or strains or stresses in series of powers or polynomials or trigonometric functions of the thickness coordinate of the plate, followed by an averaging across the thickness which eliminates one of the three independent spatial variables from the equations of motion. Thus, all variations across the thickness, throughout the plate, are assumed to be of simple form and then are averaged. An alternative is to expand in a series of the exact functions, obtained from the exact solution of the three-dimensional equations for the infinite plate, and postpone all approximations to the boundary conditions at the edges. At that stage, only a small number of the exact functions may be retained. Then averaging over the thickness is performed only at the edges as the last step. This procedure was carried out in detail in a previous paper[3], retaining only the first two exact modes (flexure and thickness-shear) of the infinite plate. Although that application was to the AT-cut of quartz, the coupling with the face-shear mode was omitted in order to compare with an earlier approximation. The coupling with the face-shear is included in the present application.

2. EKSTEIN FUNCTIONS

We consider, first, vibrations of an infinite plate. In an x_i, $i = 1,2,3$ rectangular coordinate system, the plate is bounded by free surfaces $x_2 = \pm h$ and vibrates in modes of motion with

†This investigation was supported by the U.S. Army Research Office under a contract, with Princeton University, supervised by Professor P.C.Y. Lee.

straight crests along x_3. Anticipating that there will be n modes of propagation ($n = 1, 2, \ldots, \infty$), along the plate, each of which contains three wave lengths across the thickness, we may write, for the components of displacement essentially antisymmetric with respect to the middle of the plate:

$$
\begin{aligned}
u_1 &= \sum_{n=1}^{\infty} \sum_{i=1}^{3} A_{in} \cos \xi_n x_1 \sin \eta_{in} x_2, \\
u_2 &= \sum_{n=1}^{\infty} \sum_{i=1}^{3} B_{in} \sin \xi_n x_1 \cos \eta_{in} x_2, \\
u_3 &= \sum_{n=1}^{\infty} \sum_{i=1}^{3} C_{in} \sin \xi_n x_1 \cos \eta_{in} x_2.
\end{aligned}
\tag{1}
$$

With displacements of the form (1), the equations of motion for rotated-Y-cuts of quartz [4] reduce to

$$
\begin{aligned}
&c_{11}u_{1,11} + c_{66}u_{1,22} + \rho\omega^2 u_1 + (c_{12} + c_{66})u_{2,12} + (c_{14} + c_{56})u_{3,12} = 0, \\
&(c_{12} + c_{66})u_{1,12} + c_{66}u_{2,11} + c_{22}u_{2,22} + \rho\omega^2 u_2 + c_{56}u_{3,11} + c_{24}u_{3,22} = 0, \\
&(c_{14} + c_{56})u_{1,12} + c_{56}u_{2,11} + c_{24}u_{2,22} + c_{55}u_{3,11} + c_{44}u_{3,22} + \rho\omega^2 u_3 = 0.
\end{aligned}
\tag{2}
$$

Upon substituting (1) in (2), we find

$$
\begin{aligned}
&(\bar{c}_{11}\bar{\xi}_n^2 + \bar{\eta}_{in}^2 - \Omega^1)A_{in} + (1 + \bar{c}_{12})\bar{\xi}_n\bar{\eta}_{in}B_{in} + (\bar{c}_{14} + \bar{c}_{56})\bar{\xi}_n\bar{\eta}_{in}C_{in} = 0, \\
&(1 + \bar{c}_{12})\bar{\xi}_n\bar{\eta}_{in}A_{in} + (\bar{\xi}_n^2 + \bar{c}_{22}\bar{\eta}_{in}^2 - \Omega^2)B_{in} + (\bar{c}_{56}\bar{\xi}_n^2 + \bar{c}_{24}\bar{\eta}_{in}^2)C_{in} = 0, \\
&(\bar{c}_{14} + \bar{c}_{56})\bar{\xi}_n\bar{\eta}_{in}A_{in} + (\bar{c}_{56}\bar{\xi}_n^2 + \bar{c}_{24}\bar{\eta}_{in}^2)B_{in} + (\bar{c}_{55}\bar{\xi}_n^2 + \bar{c}_{44}\bar{\eta}_{in}^2 - \Omega^2)C_{in} = 0,
\end{aligned}
\tag{3}
$$

where

$$
\bar{\xi}_n = 2\xi_n h/\pi, \quad \bar{\eta}_{in} = 2\eta_{in}h/\pi, \quad \bar{c}_{pq} = c_{pq}/c_{66}, \quad \Omega^2 = 4h^2\rho\omega^2/\pi c_{66}. \tag{4}
$$

$\bar{\xi}_n$ and $\bar{\eta}_{in}$ are the ratios of the thickness of the plate to the half wave lengths along and through the plate, respectively; Ω is the ratio of the frequency to the cut-off frequency of the fundamental thickness-shear mode. From (3),

$$
\begin{vmatrix}
\bar{c}_{11}\bar{\xi}_n^2 + \bar{\eta}_{in}^2 - \Omega^2 & 1 + \bar{c}_{12} & (\bar{c}_{14} + \bar{c}_{56})\bar{\xi}_n\bar{\eta}_{in} \\
1 + \bar{c}_{12} & \bar{\xi}_n^2 + \bar{c}_{22}\bar{\eta}_{in}^2 - \Omega^2 & \bar{c}_{56}\bar{\xi}_n^2 + \bar{c}_{24}\bar{\eta}_{in}^2 \\
(\bar{c}_{14} + \bar{c}_{56})\bar{\xi}_n\bar{\eta}_{in} & \bar{c}_{56}\bar{\xi}_n^2 + \bar{c}_{24}\bar{\eta}_{in}^2 & \bar{c}_{55}\bar{\xi}_n^2 + \bar{c}_{44}\bar{\eta}_{in}^2 - \Omega^2
\end{vmatrix} = 0 \tag{5}
$$

and $B_{in}/A_{in} = \beta_{in}$, say, and $C_{in}/A_{in} = \gamma_{in}$, say, so that

$$
\begin{aligned}
\beta_{in} &= \frac{(\Omega^2 - \bar{c}_{11}\bar{\xi}_n^2 - \bar{\eta}_{in}^2)(\bar{c}_{56}\bar{\xi}_n^2 + \bar{c}_{24}\bar{\eta}_{in}^2) + (1 + \bar{c}_{12})(\bar{c}_{14} + \bar{c}_{56})\bar{\xi}_n^2\bar{\eta}_{in}^2}{\bar{\xi}_n\bar{\eta}_{in}[(1 + \bar{c}_{12})(\bar{c}_{56}\bar{\xi}_n^2 + \bar{c}_{24}\bar{\eta}_{in}^2) - (\bar{c}_{14} + \bar{c}_{56})(\bar{\xi}_n^2 + \bar{c}_{22}\bar{\eta}_{in}^2 - \Omega^2)]}, \\
\gamma_{in} &= \frac{(\Omega^2 - \bar{c}_{11}\bar{\xi}_n^2 - \bar{\eta}_{in}^2)(\Omega^2 - \bar{\xi}_n^2 - \bar{c}_{22}\bar{\eta}_{in}^2) - (1 + \bar{c}_{12})^2\bar{\xi}_n^2\bar{\eta}_{in}^2}{\bar{\xi}_n\bar{\eta}_{in}[(1 + \bar{c}_{12})(\bar{c}_{56}\bar{\xi}_n^2 + \bar{c}_{24}\bar{\eta}_{in}^2) - (\bar{c}_{14} + \bar{c}_{56})(\bar{\xi}_n^2 + \bar{c}_{22}\bar{\eta}_{in}^2 - \Omega^2)]}
\end{aligned}
\tag{6}
$$

The components of traction on $x_2 = \pm h$ are given by

$$T_{21} = c_{56}u_{3,1} + c_{66}(u_{2,1} + u_{1,2}),$$
$$T_{22} = c_{12}u_{1,1} + c_{22}u_{2,2} + c_{24}u_{3,2}, \tag{7}$$
$$T_{23} = c_{14}u_{1,1} + c_{24}u_{2,2} + c_{44}u_{3,2},$$

whence, on $x_2 = \pm h$,

$$T_{21} = \sum_{n=1}^{\infty}\sum_{i=1}^{3} A_{in}\lambda_{in} \cos \xi_n x_1 \cos \eta_{in} h,$$
$$T_{22} = \mp \sum_{n=1}^{\infty}\sum_{i=1}^{3} A_{in}\mu_{in} \sin \xi_n x_1 \sin \eta_{in} h, \tag{8}$$
$$T_{23} = \mp \sum_{n=1}^{\infty}\sum_{i=1}^{3} A_{in}\nu_{in} \sin \xi_n x_1 \sin \eta_{in} h,$$

where

$$\lambda_{in} = c_{66}\eta_{in} + c_{66}\xi_n\beta_{in} + c_{56}\xi_n\gamma_{in}$$
$$\mu_{in} = c_{21}\xi_n + c_{22}\eta_{in}\beta_{in} + c_{24}\xi_{in}\gamma_{in}, \tag{9}$$
$$\nu_{in} = c_{14}\xi_n + c_{24}\eta_{in}\beta_{in} + c_{44}\eta_{in}\gamma_{in}.$$

Hence, for each n, the surfaces $x_2 = \pm h$ are free of traction if

$$A_{1n}\lambda_{1n} \cos \eta_{1n}h + A_{2n}\lambda_{2n} \cos \eta_{2n}h + A_{3n}\lambda_{3n} \cos \eta_{3n}h = 0,$$
$$A_{1n}\mu_{1n} \sin \eta_{1n}h + A_{2n}\mu_{2n} \sin \eta_{2n}h + A_{3n}\mu_{3n} \sin \eta_{3n}h = 0, \tag{10}$$
$$A_{1n}\nu_{1n} \sin \eta_{1n}h + A_{2n}\nu_{2n} \sin \eta_{2n}h + A_{3n}\nu_{3n} \sin \eta_{3n}h = 0,$$

from which the Ekstein dispersion relation is

$$\begin{aligned} E = {} & \lambda_{1n}(\mu_{2n}\nu_{3n} - \mu_{3n}\nu_{2n}) \cos \eta_{1n}h \sin \eta_{2n}h \sin \eta_{3n}h \\ & + \lambda_{2n}(\mu_{3n}\nu_{1n} - \mu_{1n}\nu_{3n}) \sin \eta_{1n}h \cos \eta_{2n}h \sin \eta_{3n}h \\ & + \lambda_{3n}(\mu_{1n}\nu_{2n} - \mu_{2n}\nu_{1n}) \sin \eta_{1n}h \sin \eta_{2n}h \cos \eta_{3n}h = 0 \end{aligned} \tag{11}$$

and the Ekstein displacement components are

$$u_1 = \sum_{n=1}^{\infty}\sum_{i=1}^{3} D_n\zeta_{in} \cos \xi_n x_1 \sin \eta_{in} x_2,$$
$$u_2 = \sum_{n=1}^{\infty}\sum_{i=1}^{3} D_n\zeta_{in}\beta_{in} \sin \xi_n x_1 \cos \eta_{in} x_2, \tag{12}$$
$$u_3 = \sum_{n=1}^{\infty}\sum_{i=1}^{3} D_n\zeta_{in}\gamma_{in} \sin \xi_n x_1 \cos \eta_{in} x_2$$

in which $\zeta_{1n} = 1$ and $\zeta_{2n} = A_{2n}/A_{1n}$, $\zeta_{3n} = A_{3n}/A_{1n}$, i.e.

$$\zeta_{2n} = (\mu_{3n}\nu_{1n} - \mu_{1n}\nu_{3n}) \sin \eta_{1n}h/(\mu_{2n}\nu_{3n} - \mu_{3n}\nu_{2n}) \sin \eta_{2n}h,$$

$$\zeta_{3n} = (\mu_{1n}\nu_{2n} - \mu_{2n}\nu_{1n}) \sin \eta_{1n}h/(\mu_{2n}\nu_{3n} - \mu_{3n}\nu_{2n}) \sin \eta_{3n}h. \quad (13)$$

The Ekstein dispersion relation (11) has been studied in some detail in [5, 6].

3. FREQUENCIES OF A PLATE OF FINITE LENGTH

The components of traction on planes x_1 = constant are

$$T_{11} = c_{11}u_{1,1} + c_{12}u_{2,2} + c_{14}u_{3,2},$$

$$T_{12} = c_{56}u_{3,1} + c_{66}(u_{2,1} + u_{1,2}), \quad (14)$$

$$T_{13} = c_{55}u_{3,1} + c_{56}(u_{2,1} + u_{1,2}).$$

From (12) and (14),

$$T_{11} = -\sum_{n=1}^{\infty}\sum_{i=1}^{3} D_n\zeta_{in}a_{in} \sin \xi_n x_1 \sin \eta_{in}x_2,$$

$$T_{12} = \sum_{n=1}^{\infty}\sum_{i=1}^{3} D_n\zeta_{in}\lambda_{in} \cos \xi_n x_1 \cos \eta_{in}x_2, \quad (15)$$

$$T_{13} = \sum_{n=1}^{\infty}\sum_{i=1}^{3} D_n\zeta_{in}b_{in} \cos \xi_n x_1 \cos \eta_{in}x_2,$$

where

$$a_{in} = c_{11}\xi_n + c_{12}\eta_{in}\beta_{in} + c_{14}\eta_{in}\gamma_{in},$$

$$b_{in} = c_{56}\eta_{in} + c_{56}\xi_n\beta_{in} + c_{55}\xi_n\gamma_{in}. \quad (16)$$

We now restrict n to the first three branches of Ekstein's dispersion relation (11), i.e. the flexure, thickness-shear and face-shear branches, and set the resultant force and couple on $x_1 = \pm l$, per unit length along x_3, equal to zero. Thus, on $x_1 = \pm l$,

$$\int_{-h}^{h} x_2T_{11}\, dx_2 = 0, \quad \int_{-h}^{h} T_{12}\, dx_2 = 0, \quad \int_{-h}^{h} T_{13}\, dx_2 = 0, \quad (17)$$

from which

$$\sum_{n=1}^{3} D_nL_{1n} \sin \xi_n l = 0,$$

$$\sum_{i=1}^{3} D_nL_{2n} \cos \xi_n l = 0, \quad (18)$$

$$\sum_{i=1}^{3} D_nL_{3n} \cos \xi_n l = 0,$$

where

$$L_{1n} = \sum_{i=1}^{3} \zeta_{in} a_{in} \eta_{in}^{-2} (\sin \eta_{in} h - \eta_{in} h \cos \eta_{in} h),$$

$$L_{2n} = \sum_{i=1}^{3} \zeta_{in} \lambda_{in} \eta_{in}^{-1} \sin \eta_{in} h, \tag{19}$$

$$L_{3n} = \sum_{i=1}^{3} \zeta_{in} b_{in} \eta_{in}^{-1} \sin \eta_{in} h.$$

Upon eliminating the three D_n from (18), we obtain, finally, the equation governing the frequency spectrum of the plate of length $2l$ in the x_1-direction, thickness $2h$ in the x_2-direction and of infinite width in the x_3-direction:

$$\begin{aligned} F = {} & L_{11}(L_{22}L_{33} - L_{32}L_{23}) \sin \xi_1 l \cos \xi_2 l \cos \xi_3 l \\ & + L_{12}(L_{23}L_{31} - L_{33}L_{21}) \cos \xi_1 l \sin \xi_2 l \cos \xi_3 l \\ & + L_{13}(L_{21}L_{32} - L_{31}L_{22}) \cos \xi_1 l \cos \xi_2 l \sin \xi_3 l = 0. \end{aligned} \tag{20}$$

4. SOLUTION OF EQUATIONS

The equations to be solved are (5) and (11), for the Ekstein branches, and then (20) for the frequency spectrum. As both ξ_n and η_{in} appear in both (5) and (11) and, since (11) is transcendental, the two equations have to be solved simultaneously by successive approximations; and this must be done separately for each of the three branches for a given frequency ratio Ω. Trial values of the ξ_n, for the flexure and thickness-shear branches were found from an approximate equation obtained previously[7]:

$$\kappa^4 \bar{\xi}_n^4 + (1+g)\Omega^2 \kappa^2 \bar{\xi}_n^2 + g\Omega^2(\Omega^2 - 1) = 0, \quad n = 1,2, \tag{21}$$

where

$$\kappa^2 = \pi^2/12, \quad g = \kappa^2 c_{66}/(c_{11} - c_{12}^2/c_{22}). \tag{22}$$

For the trial values of $\bar{\xi}_3$, for the face-shear branch, the approximation

$$\bar{\xi}_3^2 = \Omega^2/\bar{c}_{55} \tag{23}$$

was employed.

With each of the three trial values of $\bar{\xi}_n$ and a chosen frequency ratio Ω, the three roots $\bar{\eta}_{in}^2$ of (5) were obtained and, along with their $\bar{\xi}_n$, employed to calculate a trial value of E from (11). Then the next approximation to $\bar{\xi}_n$ was computed from (5) and (11) by the secant method and the process, beginning with (5), was repeated until either E was less than a certain small quantity or passed through zero. In the latter case, the final value of $\bar{\xi}_n$ was computed by linear interpolation and the associated values of the $\bar{\eta}_{in}$ were finally obtained from (5). Thus, three $\bar{\xi}_n$ and three $\bar{\eta}_{in}$ for each of them were computed for the given frequency. The twelve quantities were then employed to compute the coefficients in (20) and to supply the $\xi_n h$ parts of the arguments $\xi_n l = (\xi_n h)(l/h)$. The remaining part, l/h, of the arguments was then increased, from a starting value, in small, equal increments until F passed through zero. Linear interpolation then gave an abscissa l/h on a branch of the frequency spectrum for the chosen frequency. l/h was then increased in successive, small increments until the next branch was encountered, as indicated by F again passing through zero. After covering the range of l/h desired, the next frequency was chosen and the entire procedure was repeated, beginning with trial values of the $\bar{\xi}_n$ from (21) and (23).

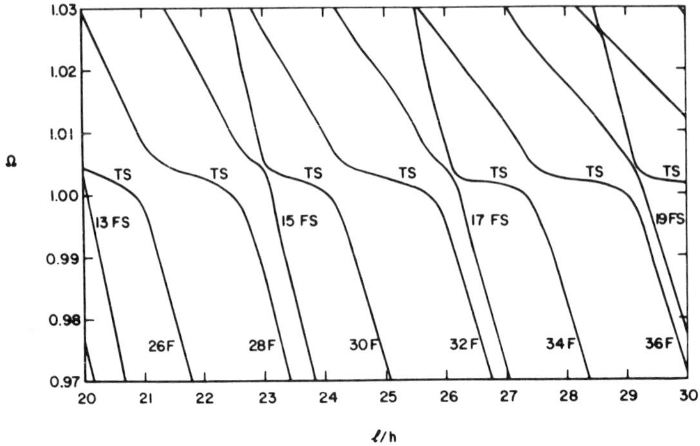

Fig. 1. Frequency spectrum of Flexure (F), Thickness-Shear (TS) and Face-Shear (FS) Modes of Vibration of AT-Cut Quartz Plates.

It should be noted that, from (5), in the range of the present application, the three $\bar{\eta}_{i1}$ for the flexure branch are imaginary; the three $\bar{\eta}_{i2}$ for the thickness-shear branch are real; while, for the face-shear branch, $\bar{\eta}_{13}$ and $\bar{\eta}_{33}$ are imaginary while $\bar{\eta}_{23}$ is real. Thus, the real form of (11) is different for each of the first three branches of Ekstein's equation.

In the function F, in (20), $\bar{\xi}_1$ and $\bar{\xi}_3$ are real but $\bar{\xi}_2$ is imaginary for $\Omega < 1$, zero for $\Omega = 1$ and real for $\Omega > 1$; so that F also takes three different real forms depending, in this case, on the frequency. It should be noted that, although $\sin \xi_2 l$ is zero when $\Omega = 1$, the entire second term in (20) does not then disappear. This is because of the presence of ξ_2 in the denominator of L_{12}.

The results of a typical computation are shown in Fig. 1 for the AT-cut of quartz for which, as calculated from Bechmann's constants [8]:

$$\bar{c}_{11} = 2.98969 \qquad \bar{c}_{12} = -0.284719$$

$$\bar{c}_{22} = 4.47269 \qquad \bar{c}_{14} = -0.125973$$

$$\bar{c}_{44} = 1.33083 \qquad \bar{c}_{24} = -0.196478$$

$$\bar{c}_{55} = 2.37159 \qquad \bar{c}_{56} = 0.0873254.$$

The ranges and intervals for the computation were

$$l/h\text{: } 20(0.1)30$$
$$\Omega\text{: } 0.97(0.0025)1(0.001)1.03.$$

About 150 min of computing time were required on a TRS-80 microcomputer to solve for the approximately 400 points on which Fig. 1 is based. The figure illustrates how the coupling of the thickness-shear modes with the face-shear modes can disturb the flat, thickness-shear frequency terraces as, for example, occurs in $22.5 < l/h < 24.1$ and in $25.7 < l/h < 27.3$ but not in $24.1 < l/h < 25.7$.

REFERENCES

1. Lord Rayleigh, On the free vibrations of an infinite plate of homogeneous isotropic elastic matter. *Proc. London Math. Soc.* **20**, 225–234 (1889).
2. H. Ekstein, High frequency vibrations of thin crystal plates. *Phys. Review* **68**, 11–23 (1945).

3. R. D. Mindlin, Vibrations of quartz plates with the HP-67 pocket calculator. *Comput. and Structures* **10**, 751–759 (1979).
4. H. F. Tiersten, *Linear Piezoelectric Plate Vibrations*. Plenum Press, p. 56 (1969).
5. E. G. Newman and R. D. Mindlin, Vibrations of a monoclinic crystal plate. *J. Acoustical Soc. Am.* **29**, 1206–1218 (1957).
6. R. K. Kaul and R. D. Mindlin, Frequency spectrum of a monoclinic crystal plate. *J. Acoustical Soc. Am.* **34**, 1902–1910 (1962).
7. R. D. Mindlin, Thickness-shear and flexural vibrations of crystal plates. *J. Appl. Phys.* **22**, 316–323 (1951).
8. R. Bechmann, Elastic and piezoelectric constants of alpha quartz. *Phys. Rev.* **110**, 1060–1061 (1958).

Int. J. Solids Structures Vol. 18, No. 9, pp. 809–817, 1982
Printed in Great Britain.
0020–7683/82/090809–09$03.00/0
Pergamon Press Ltd.

THIRD OVERTONE QUARTZ RESONATOR

R. D. MINDLIN
89 Deer Hill Drive, Ridgefield, CT 06877, U.S.A.

(*Received* 30 *October* 1981)

Abstract—The Lee–Nikodem equations of motion of elastic plates are solved for the case of vibrations of an AT-cut quartz strip, with free faces and edges, at frequencies up to and including the third harmonic thickness-shear overtone.

1. INTRODUCTION

About 30 years ago, A. W. Warner[1] developed a high precision crystal-plate resonator utilizing the third harmonic overtone of thickness-shear vibration, i.e. a mode involving a thickness-shear motion with three nodes across the thickness of the plate rather than the one node of the fundamental thickness-shear mode. At about the same time, equations were developed which extended the classical (Lagrange–Germain–Cauchy) range of frequencies to include that of the fundamental thickness-shear mode; but it was not until much later Lee and Nikodem[2, 3] formulated equations suitable for studying vibrations at frequencies of the harmonic overtone modes of thickness-shear.

In the present paper, the Lee–Nikodem third-order equations are solved for a case of rotated-Y-cut quartz plates with free faces and a pair of parallel, free edges. The results of computations for the AT-cut plate are presented for vibrations in the neighborhood of the frequency of the fundamental thickness-shear mode and in the neighborhood of the third harmonic overtone. The differences between the two exhibit some of the reasons for the higher stability of the latter.

2. LEE–NIKODEM EQUATIONS

To obtain two-dimensional equations of motion of plates from the three-dimensional equations of linear elasticity, Lee and Nikodem start with an expansion of the three-dimensional, rectangular components of displacement, u_j, $j = 1, 2, 3$, in series of trigonometric functions of the thickness-coordinate, x_2, of the plate:

$$u_j = \sum_{n=0}^{\infty} u_j^{(n)} \cos n\beta, \tag{1}$$

where the $u_j^{(n)}$ are independent of x_2 and

$$\beta = \pi(1 - x_2/b)/2, \tag{2}$$

in which b is the half-thickness of the plate. The functions $\cos n\beta$ give the shapes of the simple thickness-modes of an infinite, isotropic plate with free faces at $x_2 = \pm b$.

The expression (1), for the u_j, is substituted in the variational equation of motion[4]:

$$\int_V (T_{ij,i} - \rho\ddot{u}_j)\,\delta u_j \,\mathrm{d}V = 0, \tag{3}$$

where the T_{ij} are the components of stress, ρ is the mass density and V is the volume. The integration is performed over the thickness of the plate and leads to stress-equations of motion of order n; which are, omitting the terms accounting for surface tractions,

$$T_{ij}^{(n)} - (n\pi/2b)\,\bar{T}_{2j}^{(n)} = e_n \rho \ddot{u}_j^{(n)}, \tag{4}$$

where

$$T_{ij}^{(n)} = b^{-1} \int_{-b}^{b} T_{ij} \cos n\beta \, dx_2, \qquad \bar{T}_{ij}^{(n)} = b^{-1} \int_{-b}^{b} T_{ij} \sin n\beta \, dx_2, \tag{5}$$

and $e_n = 2$ for $n = 0$ and $e_n = 1$ for $n > 0$. (Corrections of [3] by a factor of 2, for $n = 0$, were kindly supplied by Prof. Lee).

The three-dimensional strain-displacement relations,

$$S_{ij} = (u_{j,i} + u_{i,j})/2, \tag{6}$$

become, with (1),

$$S_{ij} = \sum_{n=0}^{\infty} (S_{ij}^{(n)} \cos n\beta + \bar{S}_{ij}^{(n)} \sin n\beta), \tag{7}$$

where

$$S_{ij}^{(n)} = (u_{j,i}^{(n)} + u_{i,j}^{(n)})/2, \qquad \bar{S}_{ij}^{(n)} = n\pi(\delta_{2i}u_j^{(n)} + \delta_{2j}u_i^{(n)})/4b \tag{8}$$

and δ_{ij} is the Kronecker delta.

The three-dimensional stress–strain relations,

$$T_{ij} = c_{ijkl}S_{kl}, \quad i, j, k, l = 1, 2, 3 \quad \text{or} \quad T_p = c_{pq}S_q, \quad p, q = 1, \ldots, 6, \tag{9}$$

become, from (5) and (6),

$$T_{ij}^{(n)} = c_{ijkl}\left(e_n S_{kl}^{(n)} + \sum_{m=1}^{\infty} A_{mn}\bar{S}_{kl}^{(m)}\right), \qquad \bar{T}_{ij}^{(n)} = c_{ijkl}\left(\bar{S}_{kl}^{(n)} + \sum_{m=0}^{\infty} A_{nm}S_{kl}^{(m)}\right), \tag{10}$$

where

$$A_{mn} = 0, m + n \text{ even}; \quad 4m/(m^2 - n^2)\pi, m + n \text{ odd}. \tag{11}$$

The components of stress (10) are derivable from a strain energy density, U, according to

$$T_{ij}^{(n)} = \partial U/\partial S_{ij}^{(n)}, \qquad \bar{T}_{ij}^{(n)} = \partial U/\partial \bar{S}_{ij}^{(n)}, \tag{12}$$

where

$$2U = c_{ijkl} \sum_{n=0}^{\infty} \left[e_n S_{ij}^{(n)} S_{kl}^{(n)} + \bar{S}_{ij}^{(n)} \bar{S}_{kl}^{(n)} + \sum_{m=0}^{\infty} (A_{mn} S_{ij}^{(n)} \bar{S}_{kl}^{(m)} + A_{nm} \bar{S}_{ij}^{(n)} S_{kl}^{(m)}) \right]. \tag{13}$$

3. REDUCTION TO A SPECIAL CASE

The example to be studied is one of steady vibrations at frequencies high enough to include the third harmonic overtone of the thickness-shear family of modes of an AT-cut quartz plate bounded by free faces at $x_2 = \pm b$ and free edges at $x_1 = \pm a$. The modes are to be straight-crested along x_3 and antisymmetric with respect to both x_1 and x_2. Thus, we take, of (1), only

$$\begin{aligned} u_1 &= (u_1^{(1)} \cos \beta + u_1^{(3)} \cos 3\beta)\, e^{i\omega t}, \\ u_2 &= (u_2^{(0)} + u_2^{(2)} \cos 2\beta)\, e^{i\omega t}, \\ u_3 &= (u_3^{(0)} + u_3^{(2)} \cos 2\beta)\, e^{i\omega t}, \end{aligned} \tag{14}$$

in which the $u_j^{(n)}$ depend only on x_1. The second term in u_1 accommodates the third harmonic overtone thickness-shear mode.

What remain of the stress-equations of motion (4) are

$$
\begin{aligned}
T^{(0)}_{12,1} + 2\rho\omega^2 u_2^{(0)} &= 0, & T^{(0)}_{13,1} + 2\rho\omega^2 u_3^{(0)} &= 0,\\
T^{(1)}_{11,1} - (\pi/2b)\bar{T}^{(1)}_{21} + \rho\omega^2 u_1^{(1)} &= 0, & T^{(2)}_{12,1} - (\pi/b)\bar{T}^{(2)}_{22} + \rho\omega^2 u_2^{(2)} &= 0,\\
T^{(2)}_{13} - (\pi/b)\bar{T}^{(2)}_{23} + \rho\omega^2 u_3^{(2)} &= 0, & T^{(3)}_{11,1} - (3\pi/2b)\bar{T}^{(3)}_{21} + \rho\omega^2 u_1^{(3)} &= 0
\end{aligned}
\tag{15}
$$

and the only non-zero components of strain are, from (8) and (14),

$$
\begin{aligned}
S_5^{(0)} &= u^{(0)}_{3,1}, & S_6^{(0)} &= u^{(0)}_{2,1},\\
S_1^{(1)} &= u^{(1)}_{1,1}, & \bar{S}_6^{(1)} &= (\pi/2b)u_1^{(1)},\\
S_5^{(2)} &= u^{(2)}_{3,1}, & S_6^{(2)} &= u^{(2)}_{2,1},\\
\bar{S}_2^{(2)} &= (\pi/b)u_2^{(2)}, & \bar{S}_4^{(2)} &= (\pi/b)u_3^{(2)},\\
S_1^{(3)} &= u^{(3)}_{1,1}, & \bar{S}_6^{(3)} &= (3\pi/2b)u_1^{(3)}.
\end{aligned}
\tag{16}
$$

Nine constants of elasticity, referred to axes in and normal to the plane of the plate (with x_1 an axis of two-fold symmetry of the elastic properties of quartz) enter into the present example. As computed by Ballato[5] from Bechmann's[6] principal constants, they are (in $N/m^2 \times 10^{-9}$):

$$
\begin{aligned}
c_{11} &= 86.74 & c_{12} &= -8.260543013 & c_{55} &= 68.80698505\\
c_{22} &= 129.7663387 & c_{24} &= 5.700423178 & c_{66} &= 29.01301496\\
c_{44} &= 38.61152627 & c_{14} &= -3.654869573 & c_{56} &= 2.533571817.
\end{aligned}
$$

Of the remaining twelve constants, four (c_{13}, c_{23}, c_{33}, c_{43}) do not enter into the present example, as the modes are independent of x_3; and the others ($c_{15}, c_{25}, c_{35}, c_{45}, c_{16}, c_{26}, c_{36}, c_{46}$) are zero for the rotated-Y-cuts of quartz.

Lee and Nikodem introduce a low frequency correction factor k_1 and a high frequency correction factor k_2. The former appears as a factor of A_{10} in the strain energy density appropriate to the present example:

$$
\begin{aligned}
2U = {} & 2(c_{55}S_5^{(0)}S_5^{(0)} + c_{66}S_6^{(0)}S_6^{(0)} + 2c_{56}S_5^{(0)}S_6^{(0)}) + c_{11}S_1^{(1)}S_1^{(1)}\\
& + c_{55}S_5^{(2)}S_5^{(2)} + c_{66}S_6^{(2)}S_6^{(2)} + 2c_{56}S_5^{(2)}S_6^{(2)} + c_{11}S_1^{(3)}S_1^{(3)}\\
& + c_{22}\bar{S}_2^{(2)}\bar{S}_2^{(2)} + c_{44}\bar{S}_4^{(2)}\bar{S}_4^{(2)} + 2c_{24}\bar{S}_2^{(2)}\bar{S}_4^{(2)} + c_{66}\bar{S}_6^{(1)}\bar{S}_6^{(1)} + c_{66}\bar{S}_6^{(3)}\bar{S}_6^{(3)}\\
& + 2k_1A_{10}(c_{66}S_6^{(0)} + c_{56}S_5^{(0)})\bar{S}_6^{(1)} + 2A_{12}(c_{66}S_6^{(2)} + c_{56}S_5^{(2)})\bar{S}_6^{(1)}\\
& + 2A_{21}(c_{12}\bar{S}_2^{(2)} + c_{14}\bar{S}_4^{(2)})S_1^{(1)} + 2A_{23}(c_{12}\bar{S}_2^{(2)} + c_{14}\bar{S}_4^{(2)})S_1^{(3)}\\
& + 2A_{30}(c_{66}S_6^{(0)} + c_{56}S_5^{(0)})\bar{S}_6^{(3)} + 2A_{32}(c_{66}S_6^{(2)} + c_{56}S_5^{(2)})\bar{S}_6^{(3)},
\end{aligned}
\tag{17}
$$

where

$$
\begin{aligned}
A_{10} &= 4/\pi, & A_{12} &= -4/3\pi & A_{21} &= 8/3\pi,\\
A_{23} &= -8/5\pi, & A_{30} &= 4/3\pi, & A_{32} &= 12/5\pi.
\end{aligned}
\tag{18}
$$

The correction factor k_2, in the present example, is inserted as a divisor of the term $2\rho\omega^2 u_2^{(0)}$ in the first of (15).

Adjusted values of k_1 and k_2, as supplied by Professor Lee, are

$$
k_1^2 = \pi^2/8, \qquad k_2^{1/2} = 0.901. \tag{19}
$$

From (12), (17) and (16), the surviving stress-displacement relations are

$$
\begin{aligned}
T_{13}^{(0)} &= 2[c_{55}u_{3,1}^{(0)} + c_{56}(u_{2,1}^{(0)} + k_1 b^{-1}u_1^{(1)} + b^{-1}u_1^{(3)})],\\
T_{12}^{(0)} &= 2[c_{56}u_{3,1}^{(0)} + c_{66}(u_{2,1}^{(0)} + k_1 b^{-1}u_1^{(1)} + b^{-1}u_1^{(3)})],\\
T_{11}^{(1)} &= c_{11}u_{1,1}^{(1)} + (8/3b)(c_{12}u_2^{(2)} + c_{14}u_3^{(2)}),\\
T_{13}^{(2)} &= c_{55}u_{3,1}^{(2)} + c_{56}[u_{2,1}^{(2)} - (2/3b)u_1^{(1)} + (18/5b)u_1^{(3)}],\\
T_{12}^{(2)} &= c_{56}u_{3,1}^{(2)} + c_{66}[u_{2,1}^{(2)} - (2/3b)u_1^{(1)} + (18/5b)u_1^{(3)}],\\
T_{11}^{(3)} &= c_{11}u_{1,1}^{(3)} - (8/5b)(c_{12}u_2^{(2)} + c_{14}u_3^{(2)}),\\
\bar{T}_{12}^{(1)} &= (4k_1/\pi)(c_{56}u_{3,1}^{(0)} + c_{66}u_{2,1}^{(0)}) + (\pi/2b)c_{66}u_1^{(1)} - (4/3\pi)(c_{56}u_{3,1}^{(2)} + c_{66}u_{2,1}^{(2)}),\\
\bar{T}_{22}^{(2)} &= (8/\pi)c_{12}(u_{1,1}^{(1)}/3 - u_{1,1}^{(3)}/5) + (\pi/b)(c_{22}u_2^{(2)} + c_{24}u_3^{(2)}),\\
T_{23}^{(1)} &= (8/\pi)c_{14}(u_{1,1}^{(1)}/3 - u_{1,1}^{(3)}/5) + (\pi/b)(c_{24}u_2^{(2)} + c_{44}u_3^{(2)}),\\
\bar{T}_{12}^{(3)} &= (4/3\pi)(c_{56}u_{3,1}^{(0)} + c_{66}u_{2,1}^{(0)}) + (12/5\pi)(c_{56}u_{3,1}^{(2)} + c_{66}u_{2,1}^{(2)}) + (3\pi/2b)c_{66}u_1^{(3)}.
\end{aligned}
\tag{20}
$$

The displacement equations of motion, to be solved, are obtained by substituting the stress-displacement relations (20) into the stress-equations of motion (15)—with k_2 inserted in the first of (15), as mentioned previously.

Finally, the edge conditions are

$$
T_{13}^{(0)} = T_{12}^{(0)} = T_{11}^{(1)} = T_{13}^{(2)} = T_{12}^{(2)} = T_{11}^{(3)} = 0 \quad \text{on} \quad x_1 = \pm a. \tag{21}
$$

4. DISPERSION RELATION

In (14) we take, omitting the factor $e^{i\omega t}$,

$$
\begin{aligned}
u_2^{(0)} &= A_2^{(0)} \sin \xi x_1, & u_2^{(2)} &= A_2^{(2)} \sin \xi x_1,\\
u_3^{(0)} &= A_3^{(0)} \sin \xi x_1, & u_3^{(2)} &= A_3^{(2)} \sin \xi x_1,\\
u_1^{(1)} &= A_1^{(1)} \cos \xi x_1, & u_1^{(3)} &= A_1^{(3)} \cos \xi x_1
\end{aligned}
\tag{22}
$$

and substitute first in (20) and the result in (15) to produce a set of six simultaneous, homogeneous, linear algebraic equations in the six amplitudes $A_j^{(n)}$ of (22):

$$
\begin{aligned}
a_{11}A_2^{(0)} + a_{12}A_3^{(0)} + a_{13}A_1^{(1)} + 0 + 0 + a_{16}A_1^{(3)} &= 0\\
a_{12}A_2^{(0)} + a_{22}A_3^{(0)} + a_{23}A_1^{(1)} + 0 + 0 + a_{26}A_1^{(3)} &= 0\\
a_{13}A_2^{(0)} + a_{23}A_3^{(0)} + a_{33}A_1^{(1)} + a_{34}A_2^{(2)} + a_{35}A_3^{(2)} + 0 &= 0\\
0 + 0 + a_{34}A_1^{(1)} + a_{44}A_2^{(2)} + a_{45}A_3^{(2)} + a_{46}A_1^{(3)} &= 0\\
0 + 0 + a_{35}A_1^{(1)} + a_{45}A_2^{(2)} + a_{55}A_3^{(2)} + a_{56}A_1^{(3)} &= 0\\
a_{16}A_2^{(0)} + a_{26}A_3^{(0)} + 0 + a_{46}A_2^{(2)} + a_{56}A_3^{(2)} + a_{66}A_1^{(3)} &= 0.
\end{aligned}
\tag{23}
$$

The coefficients a_{pq}, made dimensionless and real by some manipulations of the equations, are

$$
\begin{aligned}
&a_{11} = 2(z^2 - \Omega^2/k_2), \quad a_{12} = 2\bar{c}_{56}z^2, \quad a_{13} = 4k_1z^2/\pi, \quad a_{16} = 4z^2/\pi,\\
&a_{22} = 2(\bar{c}_{55}z^2 - \Omega^2), \quad a_{23} = 4k_1\bar{c}_{56}z^2/\pi, a_{26} = 4\bar{c}_{56}z^2/\pi,\\
&a_{33} = (\bar{c}_{11}z^2 + 1 - \Omega^2)z^2, \quad a_{34} = -4(1 + 4\bar{c}_{12})z^2/3\pi, \quad a_{35} = 4(4\bar{c}_{14} - \bar{c}_{56})z^2/3\pi,\\
&a_{44} = z^2 + 4\bar{c}_{22} - \Omega^2, \quad a_{45} = 4\bar{c}_{24} + \bar{c}_{56}z^2, \quad a_{46} = 4(9 + 4\bar{c}_{12})z^2/5\pi,\\
&a_{55} = \bar{c}_{55}z^2 + 4\bar{c}_{44} - \Omega^2, \quad a_{56} = 4(4\bar{c}_{14} + 9\bar{c}_{56})z^2/5\pi,\\
&a_{66} = (\bar{c}_{11}z^2 + 9 - \Omega^2)z^2,
\end{aligned}
\tag{24}
$$

where

$$z = 2\xi b/\pi,\quad \Omega = \omega/\bar{\omega},\quad \bar{\omega}^2 = \pi^2 c_{66}/4\rho b^2,\quad \bar{c}_{pq} = c_{pq}/c_{66};$$

i.e. z is the ratio of the thickness of the plate to the half-wave-length along the plate and Ω is the ratio of the frequency to that of the fundamental thickness-shear mode of the infinite plate.

The determinant of the coefficients of the $A_j^{(n)}$, set equal to zero:

$$|a_{pq}| = 0, \tag{25}$$

in which $a_{14} = a_{15} = a_{24} = a_{25} = a_{36} = 0$, $a_{pq} = a_{qp}$, produces the dispersion relation Ω vs z: a sextic, algebraic equation in z^2. The equation is the same as (43) of [3] except for the factors 2 in a_{11}, a_{12}, a_{22} as already noted above in connection with (4). Also, here, all the elements a_{pq} are real as a result of multiplication of the third and sixth rows and columns by z.

The six branches of the dispersion relation, computed on the HP-85 micro-computer, are illustrated in Fig. 1. The characters of the branches are indicated by their identifying symbols:

F = Flexure
FS = Face-shear
1_1 = 1st Thickness-shear (in x_1-direction)
2_3 = 2nd Thickness-shear (in x_3-direction)
3_1 = 3rd (Harmonic) Thickness-shear (in x_1-direction)
2_2 = 2nd Thickness-stretch (in x_2-direction).

The subscripts in the symbols 1_1, 2_3, 3_1, 2_2, designate the direction of displacement (or predominant displacement) at $z = 0$, whereas the numbers themselves give the number of nodes between $x_2 = \pm b$. Thus: in 2_3 the displacement at $z = 0$ is predominantly in the direction of x_3 with two nodes across the thickness of the plate. Note that the roots z for branches F and FS are real for all Ω, but the roots for the remaining four branches may be real or imaginary, depending on the frequency. If imaginary, the variation of displacement along x_1 is exponential or hyperbolic rather than trigonometric.

The zigzags in the curves in Fig. 1 result from the spacing of dots on the cathode ray tube display of the HP-85. The figure is the HP-85's hard copy of the CRT display. The roots z were actually computed to an accuracy of 10^{-9}—a precision required for their subsequent use in solving (34) and (37). Intervals of 0.02 in Ω were employed for Fig. 1, resulting in a computation time, for the range $0 < \Omega < 4$, of about 6 hr or about 18 sec per root. The secant iterative method

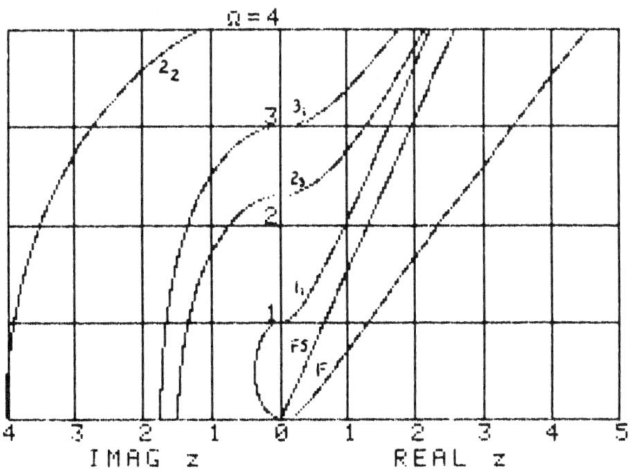

Fig. 1. Dispersion curves for waves in an infinite AT-cut quartz plate.

was used, with starting values given by the following approximate formulas, followed by increments of 10^{-6} in z_n^2:

$$\left.\begin{matrix} F: z_1^2 \\ 1_1: z_3^2 \end{matrix}\right\} = 6.42258(1+G)\Omega^2[1 \pm (1+K)^{1/2}]/\pi^2, \tag{26}$$

$$G = \pi^2/12(\bar{c}_{11} - \bar{c}_{12}^2/\bar{c}_{22}), \quad K = 4G(\Omega^{-2} - 1)/(1+G)^2, \tag{27}$$

$$FS: z_2^2 = 0.44119\Omega^2, \tag{28}$$

$$2_3: z_4^2 = \begin{cases} 2.229(\Omega^2/\Omega_4^2 - 1), & \Omega < \Omega_4, \\ 0.42395(\Omega^2/\Omega_4^2 - 1), & \Omega > \Omega_4, \end{cases} \tag{29}$$

$$2_2: z_6^2 = 16(\Omega^2/\Omega_6^2 - 1), \quad \Omega < \Omega_6, \tag{30}$$

$$\Omega_4^2, \Omega_6^2 = 2\{\bar{c}_{22} + \bar{c}_{44} \mp [(\bar{c}_{22} - \bar{c}_{44})^2 + 4\bar{c}_{24}^2]^{1/2}\}, \tag{31}$$

$$3_1: z_5^2 = \begin{cases} 0.33799(\Omega^2 - 9), & \Omega < 3, \\ 0.40651(\Omega^2 - 9), & \Omega > 3. \end{cases} \tag{32}$$

These trial roots match closely or exactly the roots of the sextic at $z = 0$ and at $\Omega = 0, 3$ (except z_6 at $\Omega = 3$) resulting in trial values adequate for convergence of the iteration for all $0 < \Omega < 4$.

5. FREQUENCY SPECTRUM

For each of the roots z_n^2 of (25), five amplitude ratios, say

$$A_2^{(0)}/A_1^{(1)} = \alpha_{1n}, \quad A_3^{(0)}/A_1^{(1)} = \alpha_{2n}, \quad A_2^{(2)}/A_1^{(1)} = \alpha_{3n}, \quad A_3^{(2)}/A_1^{(1)} = \alpha_{4n}, \quad A_1^{(3)}/A_1^{(1)} = \alpha_{5n}, \tag{33}$$

may be found from five of the six equations (23). Thus, with the third of (23) omitted, we may write

$$\begin{aligned} a_{11}(z_n\alpha_{1n}) + a_{12}(z_n\alpha_{2n}) + 0 + 0 + a_{16}\alpha_{5n} &= -a_{13} \\ a_{12}(z_n\alpha_{1n}) + a_{22}(z_n\alpha_{2n}) + 0 + 0 + a_{26}\alpha_{5n} &= -a_{23} \\ 0 + 0 + a_{44}(z_n\alpha_{3n}) + a_{45}(z_n\alpha_{4n}) + a_{46}\alpha_{5n} &= -a_{34}, \\ 0 + 0 + a_{45}(z_n\alpha_{3n}) + a_{55}(z_n\alpha_{4n}) + a_{56}\alpha_{5n} &= -a_{35} \\ a_{16}(z_n\alpha_{1n}) + a_{26}(z_n\alpha_{2n}) + a_{46}(z_n\alpha_{3n}) + a_{56}(z_n\alpha_{4n}) + a_{66}\alpha_{5n} &= 0. \end{aligned} \tag{34}$$

This form is chosen because the $z_n\alpha_{1n}$, $z_n\alpha_{2n}$, $z_n\alpha_{3n}$, $z_n\alpha_{4n}$ and α_{5n} are real for all Ω, as are also the a_{pq}—as arranged previously.

With the six z_n from (25) and the thirty α_{pn} determined from (34), we may now write, in place of (22):

$$\begin{aligned} u_2^{(0)} &= \sum_{n=1}^{6} A_n\alpha_{1n} \sin \xi_n x_1, & u_3^{(0)} &= \sum_{n=1}^{6} A_n\alpha_{2n} \sin \xi_n x_1, \\ u_1^{(1)} &= \sum_{n=1}^{6} A_n \cos \xi_n x_1, & u_2^{(2)} &= \sum_{n=1}^{6} A_n\alpha_{3n} \sin \xi_n x_1, \\ u_3^{(2)} &= \sum_{n=1}^{6} A_n\alpha_{4n} \sin \xi_n x_1, & u_1^{(3)} &= \sum_{n=1}^{6} A_n\alpha_{5n} \cos \xi_n x_1. \end{aligned} \tag{35}$$

Upon substituting the displacements (35) in the formulas (20) for the stresses and the results in the edge conditions (21), we have the six equations:

$$\sum_{n=1}^{6} A_n b_{mn} = 0, \qquad m = 1, \ldots, 6, \tag{36}$$

where

$$
\begin{aligned}
b_{1n} &= (\bar{c}_{56} z_n \alpha_{1n} + \bar{c}_{55} z_n \alpha_{2n} + 2k_1 \bar{c}_{56}/\pi + 2\bar{c}_{56}\alpha_{5n}/\pi) \cos \hat{z}_n l, \\
b_{2n} &= (z_n \alpha_{1n} + \bar{c}_{56} z_n \alpha_{2n} + 2k_1/\pi + 2\alpha_{5n}/\pi) \cos \hat{z}_n l, \\
b_{3n} &= (-\bar{c}_{11} z_n^2 + 16 z_n \alpha_{3n}/3 + 16\bar{c}_{14} z_n \alpha_{4n}/3)\hat{z}_n^{-1} \sin \hat{z}_n l, \\
b_{4n} &= (-4\bar{c}_{56}/3\pi + \bar{c}_{56} z_n \alpha_{3n} + \bar{c}_{55} z_n \alpha_{4n} + 36\bar{c}_{56}\alpha_{5n}/5\pi) \cos \hat{z}_n l, \\
b_{5n} &= (-4/3\pi + z_n \alpha_{3n} + \bar{c}_{56} z_n \alpha_{4n} + 36\alpha_{5n}/5\pi) \cos \hat{z}_n l, \\
b_{6n} &= (16\bar{c}_{12} z_n \alpha_{3n}/5\pi + 16\bar{c}_{14} z_n \alpha_{4n}/5\pi + \bar{c}_{11} z_n^2 \alpha_{5n})\hat{z}_h^{-1} \sin \hat{z}_n l,
\end{aligned}
$$

in which $\hat{z}_n = \pi z_n/2 = \xi_n b$, $l = a/b$ and the b_{mn} are real for all Ω.

The roots l of the equation obtained by setting the 6×6 determinant of the coefficients of the A_n, in (36), equal to zero:

$$|b_{mn}| = 0, \tag{37}$$

produce the data for plotting a frequency spectrum Ω vs a/b.

The results of computations in the two ranges

$$0.99 < \Omega < 1.01, \qquad 16 < a/b < 24$$

and

$$2.995 < \Omega < 3.005, \qquad 18 < a/b < 22$$

are illustrated in Figs. 2 and 3. To construct these figures, the six roots z_n of the sextic (25) were first computed for a given Ω. Then the five linear equations (34) were solved for the α_{pn} for each of the six z_n and the resulting combinations of α_{pn} and z_n substituted in the transcendental equation (37), after which the range of l (= a/b) was traversed in steps of 0.1 for Fig. 2 and 0.025 for Fig. 3 and the values of $|b_{mn}|$ computed at each step. A change of sign of $|b_{mn}|$ indicated a straddled root l which was then determined to 10^{-3} by successive linear interpolations. The process was then repeated at intervals of Ω of 5×10^{-5}. Figures 2 and 3 required about 58 and 49 hr of computation, respectively, on the HP-85.

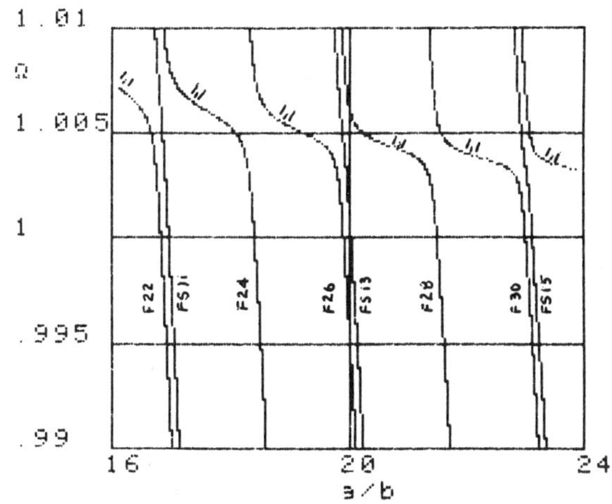

Fig. 2. Frequency spectrum—AT-cut quartz strip with free edges—in the neighborhood of the fundamental thickness-shear mode.

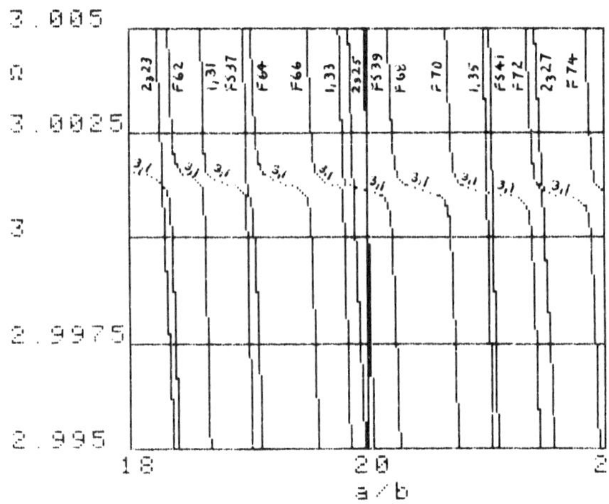

Fig. 3. Frequency spectrum—AT-cut quartz strip with free edges—in the neighborhood of the third harmonic thickness-shear overtone mode.

In Fig. 2:

F 22, . . . , 30 are overtones of flexure
FS 11, . . . , 15 are overtones of face-shear
$1_1 1$ is the 1st thickness-shear (fundamental).

In Fig. 3:

F 62, . . . , 74 are overtones of flexure
FS 37, ... , 41 are overtones of face-shear
$1_1 33$, . . . , 35 are anharmonic overtones of the 1st (fundamental) thickness-shear
$2_3 23$, . . . , 27 are anharmonic overtones of the 2nd transverse thickness-shear
$3_1 1$ is the 3rd harmonic thickness-shear overtone.

The numbers following the symbols F, FS, 1_1, 2_3 and 3_1 designate both the order of the overtone and the approximate number of half-wave-lengths between $x_1 = \pm a$.

Figure 2 illustrates the well known phenomenon of strong coupling of the 1st thickness-shear fundamental with flexure overtones and weak coupling with face-shear overtones. Figure 3 shows that the 3rd harmonic thickness-shear mode has moderately strong coupling with flexure overtones and weak coupling with face-shear overtones and, in addition, weak coupling with transverse thickness-shear overtones. As for the interaction of the 3rd harmonic thickness-shear overtone with the anharmonic overtones of the fundamental thickness shear, the coupling is moderately strong at small a/b (thick plates and low order anharmonic overtones) and diminishes as a/b increases (thin plates and increasing order of anharmonic overtones).

Finally, the minimum absolute values of the slopes of the segments $1_1 1$ are much larger than those of $3_1 1$. For large a/b, the ratio of those slopes is approximated by the ratio of the curvatures of branches 3_1 and 1_1 at $z = 0$ in Fig. 1. The exact values of those curvatures, in the three-dimensional theory, were given by Ekstein[7, eqn (56)]:

$$\kappa_n = [\mathrm{d}^2\Omega/\mathrm{d}z^2]_{z=0} = k + C \cot(n\pi/2c_2^{1/2}) + D \cot(n\pi/2c_3^{1/2}),$$

where

$$k = (\bar{c}_{11} + A + B), \qquad n = 1, 3, 5, \ldots,$$
$$A = [(1 + \bar{c}_{12}) \cos\theta + (\bar{c}_{14} + \bar{c}_{56}) \sin\theta]^2/(1 - c_2),$$
$$B = [(\bar{c}_{14} + \bar{c}_{56}) \cos\theta - (1 + \bar{c}_{12}) \sin\theta]^2/(1 - c_3),$$
$$C = 4[(c_2 + \bar{c}_{12}) \cos\theta + (c_2\bar{c}_{56} + \bar{c}_{14})]^2/n^2\pi c_2^{1/2}(1 - c_2)^2,$$
$$D = 4[(c_3 + \bar{c}_{12}) \sin\theta - (c_3\bar{c}_{56} + \bar{c}_{14})]^2/n^2\pi c_2^{1/2}(1 - c_3)^2,$$
$$c_2, c_3 = \{\bar{c}_{22} + \bar{c}_{44} \pm [(\bar{c}_{22} - \bar{c}_{44})^2 + 4\bar{c}_{24}^2]^{1/2}\}/2,$$
$$\tan\theta = \bar{c}_{24}/(c_2 - \bar{c}_{44}).$$

For the present case, the curvature ratio κ_1/κ_3 is 4.7 and that is the ratio of the slopes.

The large ratio of slopes and the absence, at large a/b, of strong coupling with all overtones except those of flexure (which, at such high overtones, have very small amplitudes) are important contributors to the high stability of third harmonic overtone resonators.

Acknowledgement—This work was sponsored by the Department of the Navy, Office of Naval Research, Structural Mechanics Program, Contract N00014-80-C-0559.

REFERENCES

1. A. W. Warner, High frequency crystal units for primary frequency standards. *Proc. I.R.E.* **40**, 1030 (1952).
2. P. C. Y. Lee and Z. Nikodem, An approximate theory for high-frequency vibrations of elastic plates. *Int. J. Solids Structures* **8**, 581 (1972).
3. Z. Nikodem and P. C. Y. Lee, Approximate theory of vibrations of crystal plates at high frequencies. *Int. J. Solids Structures* **10**, 177 (1974).
4. A. E. H. Love, *A Treatise on the Mathematical Theory of Elasticity*, 4th Edn, p. 166. Cambridge, University Press, London (1927).
5. A. Ballato, private communication.
6. R. Bechmann, Elastic and piezoelectric constants of alpha-quartz. *Phys. Rev.* **110**, 1060 (1950).
7. H. Ekstein, High frequency vibrations of thin crystal plates. *Phys. Rev.* **68**, 11 (1945).

Int. J. Solids Structures Vol. 20, No. 2, pp. 141–157, 1984
Printed in Great Britain.

0020–7683/84 $3.00 + .00
Pergamon Press Ltd.

FREQUENCIES OF PIEZOELECTRICALLY FORCED VIBRATIONS OF ELECTRODED, DOUBLY ROTATED, QUARTZ PLATES†

R. D. MINDLIN
P. O. Box 385, Grantham, NH 03753, U.S.A.

(Received 16 May 1983)

Abstract—Two-dimensional equations of motion of piezoelectric crystal plates, obtained from the three-dimensional equations of linear piezoelectricity by expansion in power series of the thickness coordinate of the plate, are solved for forced vibrations of electroded SC-cut quartz plates. Results of computations are given for frequencies of simple thickness modes of vibration, for the dispersion of straight-crested waves and for frequencies of vibration of a strip, along with its dimensional ratios for minimal coupling of the fundamental thickness-shear mode with overtones of flexure, face-shear and thickness-twist.

1. INTRODUCTION

A review of vibrational properties of doubly rotated quartz plates may be found in a recent article by Kusters[1]. Solutions of the three-dimensional equations of piezoelectricity for the simple thickness modes of vibration of doubly rotated plates appear in Tiersten's book[2] and in a comprehensive article by Ballato[3]. Simple thickness modes are those in which the oscillatory displacements are functions of only the thickness coordinate of the plate. Solutions of two dimensional plate-equations for doubly rotated quartz strips, in which the displacements depend on a coordinate in the plane of the stip, were presented by Lee and Wu[4] for the purely elastic case. Some of their results are extended, in the present paper, to account for the effects of piezoelectric coupling and the mass of electrode coatings.

In the following Section the three-dimensional, linear equations of piezoelectricity are exhibited. The equations are solved, in Section 3, for the simple thickness modes and the results of computations of the first three frequencies are given for the electroded SC-cut. A brief review of the derivation of two dimensional plate-equations by expansion in power series is given in Section 4 and restricted to the first order in Section 5. The solution of the two-dimensional equations for the simple thickness modes and the solution of the equations governing the correction factors for the simple thickness frequencies are described in Section 6. Section 7 contains the derivation of the dispersion relation for forced, straight-crested waves in the electroded plate and the graphical presentation of the results of computations of the branches for the SC-cut. The final Section includes the solution for the forced vibrations of an electroded strip with mixed edge-conditions and the graphs of portions of the frequency spectrum. Also, in Section 8 is a sketch which transforms a frequency spectrum for mixed edge-conditions approximately to one for free edges. The results exhibit discrete ranges of dimensional ratios favorable to the avoidance of coupling of the fundamental thickness-shear mode with flexure, extension and face-shear overtones and the associated activity dips. An appendix contains formulas for computing the material constants of quartz referred to doubly rotated axes along with the results for the SC-cut.

2. THREE-DIMENSIONAL EQUATIONS OF PIEZOELECTRICITY

We begin with a brief review of the three-dimensional, linear equations of piezoelectricity[2] from which are to be deduced two-dimensional plate-equations and frequencies of simple thickness modes which are required for the computation of correction factors appearing in the two-dimensional equations.

The three-dimensional field equations are the stress equations of motion and the charge equation of electrostatics:

$$T_{ij,i} = \rho \ddot{u}_j, \quad D_{i,i} = 0, \tag{1}$$

†Investigation supported by the Office of Naval Research.

where ρ and the T_{ij}, u_j and D_i are, respectively, the mass density and the components of stress, mechanical displacement and electric displacement.

The constitutive equations are

$$T_{ij} = c_{ijkl}S_{kl} - e_{kij}E_k, \quad D_i = e_{ijk}S_{jk} + \epsilon_{ij}E_j, \tag{2}$$

where the c_{ijkl}, e_{kij} and ϵ_{ij} are the components of elastic stiffness, piezoelectric strain constant and dielectric permittivity, respectively. S_{ij} and E_i are the components of strain and electric field which are expressed in terms of the u_i and the electric potential ϕ by

$$S_{ij} = \frac{1}{2}(u_{j,i} + u_{i,j}), \quad E_i = -\phi_{,i}. \tag{3}$$

Combining (1)–(3), we have the equations of motion,

$$\begin{aligned} c_{ijkl}u_{k,li} + e_{kij}\phi_{,ki} &= \rho\ddot{u}_j, \\ e_{kij}u_{k,ij} - \epsilon_{ij}\phi_{,ij} &= 0, \end{aligned} \tag{4}$$

which can be derived from a variational principle, for a region V bounded by a surface S with outward normal $\mathbf{n}$:

$$\delta\int_{t_0}^{t_1} \mathrm{d}t \int_V (K-H)\,\mathrm{d}V + \int_{t_0}^{t_1} \mathrm{d}t \int_S (t_j\delta u_j + \sigma\delta\phi)\,\mathrm{d}S = 0 \tag{5}$$

where the t_j and σ are the surface traction and surface charge. K and H are the kinetic energy and electric enthalpy densities:

$$K = \frac{1}{2}\rho\dot{u}_i\dot{u}_i, \quad H = \frac{1}{2}c_{ijkl}S_{ij}S_{kl} - \frac{1}{2}\epsilon_{ij}E_iE_j - e_{ijk}E_iS_{jk}, \tag{6}$$

from which

$$T_{ij} = \partial H/\partial S_{ij}, \quad D_i = -\partial H/\partial E_i. \tag{7}$$

Upon substitution of (6) and (7) in (5), the latter can be converted to the variational equation of motion:

$$\int_{t_0}^{t_1} \mathrm{d}t \int_V [(T_{ij,i} - \rho\ddot{u}_j)\delta u_j + D_{i,i}\delta\phi]\,\mathrm{d}V + \int_{t_0}^{t_1} \mathrm{d}t \int_S [(t_j - n_iT_{ij})\delta u_j + (\sigma - n_iD_i)\delta\phi]\,\mathrm{d}S = 0, \tag{8}$$

which produces the field equations (1) and the boundary conditions

$$n_iT_{ij} = t_j \quad \text{or} \quad u_i = \bar{u}_i \quad \text{on} \quad S, \tag{9}$$

where $\bar{u}_i$ is the surface displacement, and

$$n_iD_i = \sigma \quad \text{or} \quad \phi = \bar{\phi} \quad \text{on} \quad S, \tag{10}$$

where $\bar{\phi}$ is the surface potential. As an alternative to (9), a component of n_iT_{ij} and the resultant of u_i in the plane at right angles, or *vice versa*, may be specified.

3. SIMPLE THICKNESS-MODES

Simple thickness-modes of vibration of a plate are those in which the three components of displacement are independent of the coordinates parallel to the middle plane of the plate.

We consider an infinite plate bounded by surfaces at $x_2 = \pm b$ which are coated with electrodes each of thickness $2b'$ and mass density ρ'. A uniform alternating voltage $V\,e^{i\omega t}$ is applied to the electrodes so that the voltage drop across the thickness of the plate is $2V$. The response of the plate is independent of x_1 and x_3 whence (4), (in the reduced notation whereby pairs of indices 11, 22, 33, 23 or 32, 31 or 13, 12 or 21 become 1, 2, 3, 4, 5, 6, respectively) reduce to

$$\begin{aligned}
c_{66}u_{1,22} + c_{26}u_{2,22} + c_{46}u_{3,22} + e_{26}\phi_{,22} &= \rho\ddot{u}_1,\\
c_{26}u_{1,22} + c_{22}u_{2,22} + c_{24}u_{3,22} + e_{22}\phi_{,22} &= \rho\ddot{u}_2,\\
c_{46}u_{1,22} + c_{24}u_{2,22} + c_{44}u_{3,22} + e_{24}\phi_{,22} &= \rho\ddot{u}_3,\\
e_{26}u_{1,22} + e_{22}u_{2,22} + e_{24}u_{3,22} - \epsilon_{22}\phi_{,22} &= 0.
\end{aligned} \tag{11}$$

The boundary conditions (9) and (10), on $x_2 = \pm b$, become

$$(T_6, T_2, T_4) = \mp 2\rho' b'(\ddot{u}_1, \ddot{u}_2, \ddot{u}_3), \quad \phi = \pm V\,e^{i\omega t} \tag{12}$$

or

$$\begin{aligned}
c_{66}u_{1,2} + c_{26}u_{2,2} + c_{46}u_{3,2} + e_{26}\phi_{,2} &= \mp 2\rho' b'\ddot{u}_1,\\
c_{26}u_{1,2} + c_{22}u_{2,2} + c_{24}u_{3,2} + e_{22}\phi_{,2} &= \mp 2\rho' b'\ddot{u}_2,\\
c_{46}u_{1,2} + c_{24}u_{2,2} + c_{44}u_{3,2} + e_{24}\phi_{,2} &= \mp 2\rho' b'\ddot{u}_3,\\
\phi &= \pm V\,e^{i\omega t}.
\end{aligned} \tag{13}$$

From the fourth of (11),

$$\phi = (e_{26}u_1 + e_{22}u_2 + e_{24}u_3)/\epsilon_{22} + Ax_2 + B; \tag{14}$$

but the constant B may be omitted as a constant ϕ produces no electric field.

Now, substitute (14) in the first three of (11) and get the same form of equations except with c_{pq} replaced by $\bar{c}_{pq}$, where

$$\bar{c}_{pq} = c_{pq} + e_{2p}e_{2q}/\epsilon_{22}. \tag{15}$$

Thus:

$$\begin{aligned}
\bar{c}_{66}u_{1,22} + \bar{c}_{26}u_{2,22} + \bar{c}_{46}u_{3,22} &= \rho\ddot{u}_1,\\
\bar{c}_{26}u_{1,22} + \bar{c}_{22}u_{2,22} + \bar{c}_{24}u_{3,22} &= \rho\ddot{u}_2,\\
\bar{c}_{46}u_{1,22} + \bar{c}_{24}u_{2,22} + \bar{c}_{44}u_{3,22} &= \rho\ddot{u}_3.
\end{aligned} \tag{16}$$

Now, take

$$u_i = A_i \sin\eta x_2\,e^{i\omega t}, \quad \phi = \phi_0 \sin\eta x_2\,e^{i\omega t} \tag{17}$$

and substitute in (16) to get

$$\begin{aligned}
(\bar{\eta}^2 - \bar{\Omega}^2)A_1 + \check{c}_{26}\bar{\eta}^2 A_2 + \check{c}_{46}\bar{\eta}^2 A_3 &= 0,\\
\check{c}_{26}\bar{\eta}^2 A_1 + (\check{c}_{22}\bar{\eta}^2 - \bar{\Omega}^2)A_2 + \check{c}_{24}\bar{\eta}^2 A_3 &= 0\\
\check{c}_{46}\bar{\eta}^2 A_1 + \check{c}_{24}\bar{\eta}^2 A_2 + (\check{c}_{44}\bar{\eta}^2 - \bar{\Omega}^2)A_3 &= 0
\end{aligned} \tag{18}$$

where

$$\bar{\eta} = \eta b, \quad \bar{\Omega}^2 = \rho\omega^2 b^2/\bar{c}_{66}, \quad \check{c}_{pq} = \bar{c}_{pq}/c_{66}. \tag{19}$$

The determinant of the coefficients of the A_i in (18), set equal to zero, is a bicubic equation in η with three roots η_j^2, $j = 1, 2, 3$. Then (17) can be extended to

$$u_i = \sum_{j=1}^{3} A_{ij} \sin \eta_j x_2 \tag{20}$$

and (14) to

$$\phi = \sum_{j=1}^{3} \epsilon_{22}^{-1}(e_{26}A_{1j} + e_{22}A_{2j} + e_{24}A_{3j}) \sin \eta_j x_2 + Ax_2, \tag{21}$$

with $e^{i\omega t}$ omitted here and in the sequel.

For each η_j, define

$$\alpha_{2j} = A_{2j}/A_{1j}, \quad \alpha_{4j} = A_{3j}/A_{1j}, \quad \alpha_{6j} = 1, \quad B_j = A_{1j}/b. \tag{22}$$

Then

$$\begin{aligned} u_1 &= \sum_j B_j b\alpha_{6j} \sin \eta_j x_2, \\ u_2 &= \sum_j B_j b\alpha_{2j} \sin \eta_j x_2, \\ u_3 &= \sum_j B_j b\alpha_{4j} \sin \eta_j x_2, \\ \phi &= \sum_j B_j b\epsilon_{22}^{-1}(e_{26}\alpha_{6j} + e_{22}\alpha_{2j} + e_{24}\alpha_{4j}) \sin \eta_j x_2 + Ax_2. \end{aligned} \tag{23}$$

We may find α_{2j} and α_{4j} from the second and third of (18) with η replaced by η_j:

$$\begin{aligned} (\bar{c}_{22}\bar{\eta}_j^2 - \bar{\Omega}^2)\alpha_{2j} + \bar{c}_{24}\bar{\eta}_j^2\alpha_{4j} &= -\bar{c}_{26}\bar{\eta}_j^2, \\ \bar{c}_{24}\bar{\eta}_j^2\alpha_{2j} + (\bar{c}_{44}\bar{\eta}_j^2 - \bar{\Omega}^2)\alpha_{4j} &= -\bar{c}_{46}\bar{\eta}_j^2, \end{aligned} \tag{24}$$

from which

$$\begin{aligned} \alpha_{2j} &= [(\bar{c}_{24}\bar{c}_{46} - \bar{c}_{26}\bar{c}_{44})\bar{\eta}_j^4 + \bar{c}_{26}\bar{\eta}_j^2\bar{\Omega}^2]/\alpha, \\ \alpha_{4j} &= [(\bar{c}_{24}\bar{c}_{26} - \bar{c}_{46}\bar{c}_{22})\bar{\eta}_j^4 + \bar{c}_{46}\bar{\eta}_j^2\bar{\Omega}^2]/\alpha, \\ \alpha &= (\bar{c}_{22}\bar{\eta}_j^2 - \bar{\Omega}^2)(\bar{c}_{44}\bar{\eta}_j^2 - \bar{\Omega}^2) - \bar{c}_{24}^2\bar{\eta}_j^4. \end{aligned} \tag{25}$$

Upon substituting (25) in (23) and the result in the boundary conditions (13), we find

$$\begin{aligned} \beta_{61}B_1 + \beta_{62}B_2 + \beta_{63}B_3 &= -e_{26}V/b\bar{c}_{66}, \\ \beta_{21}B_1 + \beta_{22}B_2 + \beta_{23}B_3 &= -e_{22}V/b\bar{c}_{66}, \\ \beta_{41}B_1 + \beta_{42}B_2 + \beta_{43}B_3 &= -e_{24}V/b\bar{c}_{66}, \end{aligned} \tag{26}$$

where

$$\beta_{ai} = \sum_b \alpha_{bi}[\bar{c}_{ab}\bar{\eta}_i \cos \bar{\eta}_i - (\delta_{ab}R\bar{\Omega}^2 + \bar{e}_{ab}) \sin \bar{\eta}_i] \tag{27}$$

in which a and b range over 2, 4, 6 and i over 1, 2, 3; δ_{ab} is the Krönecker delta and

$$\bar{e}_{ab} = e_{2a}e_{2b}/\epsilon_{22}\bar{c}_{66}, \quad R = 2\rho' b'/\rho b, \tag{28}$$

i.e. R is the ratio of the mass of both electrodes to the mass of the crystal.

The determinant of the coefficients of the B_i in (26), set equal to zero:

$$|\beta_{ai}| = 0, \tag{29}$$

is a transcendental equation whose roots $\bar{\Omega}$ give the frequencies of the simple thickness-modes. Except for notation, division by $\bar{c}_{pq}$ to make each element dimensionless and the additional term $\delta_{ab}R\bar{\Omega}^2$, which accounts for the inertia of the electrodes, (29) is the same as Tiersten's equation (9.69), ([2], p.92).

For the SC-cut (see the Appendix) and $R = 0.01$, the first three roots of (29), converted from the $\bar{\Omega}$ of (19) to

$$\Omega = (2\omega b/\pi)(\rho/c_{66})^{1/2} = (2\bar{\Omega}/\pi)(\bar{c}_{66}/c_{66})^{1/2}, \tag{30}$$

are

$$\begin{aligned} \Omega_2 &= 1.7512 \\ \Omega_4 &= 1.0235 \\ \Omega_6 &= 0.9304 \end{aligned} \tag{31}$$

where Ω_2, Ω_4, Ω_6 give the fundamental frequencies of the essentially thickness-stretch, – twist, – shear modes, respectively; "essentially" because each of the three modes has contributions from all three components of displacement, u_1, u_2, u_3, one of which predominates in each mode.

4. EXPANSION IN POWER SERIES

The two-dimensional equations of motion of piezoelectric crystal plates, to be used in the sequel, are deduced from the three-dimensional equations by a procedure based on expansions in series of powers of the thickness coordinate of the plate. The process was developed in stages: beginning in 1952[5], extended in 1962[6] and revised in 1972[7]. As revised, the start is with expansions of mechanical displacement and electric potential in series of powers of the thickness coordinate x_2:

$$u_i = \sum_n x_2{}^n u_i^{(n)}, \quad \phi = \sum_n x_2{}^n \phi^{(n)}, \tag{32}$$

where $u_i^{(n)}$ and $\phi^{(n)}$ are independent of x_2. The three-dimensional strain and electric field are, from (3) and (32),

$$S_{ij} = \frac{1}{2}(u_{j,i} + u_{i,j}) = \sum_n x_2{}^n S_{ij}^{(n)}, \quad E_i = -\phi_{,i} = \sum_n x_2{}^n E_i^{(n)}, \tag{33}$$

in which the two-dimensional strain and electric field of order n are

$$\begin{aligned} S_{ij}^{(n)} &= \frac{1}{2}[u_{j,i}^{(n)} + u_{i,j}^{(n)} + (n+1)(\delta_{i2}u_j^{(n+1)} + \delta_{2j}u_i^{(n+1)})], \\ E_i^{(n)} &= -\phi_{,i}^{(n)} - (n+1)\delta_{i2}\phi^{(n+1)}. \end{aligned} \tag{34}$$

The kinetic energy density and electric enthalpy density of the plate are defined as

$$\bar{K} = \int_{-b}^{b} K \, dx_2, \quad \bar{H} = \int_{-b}^{b} H \, dx_2, \tag{35}$$

whence, from (6), (32) and (34),

$$\bar{K} = \frac{1}{2}\rho \sum_m \sum_n B_{mn} \dot{u}_j^{(m)} \dot{u}_j^{(n)}$$

$$\bar{H} = \frac{1}{2}\sum_m \sum_n B_{mn}(c_{ijkl} S_{ij}^{(m)} S_{kl}^{(n)} - \epsilon_{ij} E_i^{(m)} E_j^{(n)} - 2e_{ijk} E_i^{(m)} S_{jk}^{(n)}) \tag{36}$$

where

$$B_{mn} = 2b^{m+n+1}/(m+n+1),\ m+n \quad \text{even;} \quad 0, m+n \quad \text{odd.} \tag{37}$$

The constitutive equations of order n are

$$T_{ij}^{(n)} = \int_{-b}^{b} x_2{}^n T_{ij}\,\mathrm{d}x_2 = \sum_m B_{mn}(c_{ijkl} S_{kl}^{(m)} - e_{kij} E_k^{(m)}) = \partial \bar{H}/\partial S_{ij}^{(n)},$$

$$D_i^{(n)} = \int_{-b}^{b} x_2{}^n D_i\,\mathrm{d}x_2 = \sum_m B_{mn}(e_{ijk} S_{jk}^{(m)} + \epsilon_{ij} E_j^{(m)}) = -\,\partial \bar{H}/\partial E_i^{(n)}. \tag{38}$$

The variational equation of motion becomes

$$\begin{aligned}&\sum_n \int_{t_0}^{t_1} \mathrm{d}t \int_A (T_{ij,i}^{(n)} - nT_{2j}^{(n-1)} + B_n T_j^{(n)} - \rho \sum_m B_{mn} \ddot{u}_j^{(m)})\delta u_j^{(n)}\,\mathrm{d}A \\ &+ \sum_n \int_{t_0}^{t_1} \mathrm{d}t \int_A (D_{i,i}^{(n)} - nD_2^{(n-1)} + B_n D^{(n)})\delta\phi^{(n)}\,\mathrm{d}A \\ &+ \sum_n \int_{t_0}^{t_1} \mathrm{d}t \oint_C [(B_n t_j^{(n)} - n_c T_{cj}^{(n)})\delta u_j^{(n)} + (B_n d^{(n)} - n_c D_c^{(n)})\delta\phi^{(n)}]\,\mathrm{d}s = 0,\end{aligned} \tag{39}$$

where the index c ranges over 1 and 3 only, s is the coordinate along the edge curve C,

$$B_n = 2b^{2n+1}/(2n+1) \tag{40}$$

and the face-tractions $T_j^{(n)}$, face charges $D^{(n)}$, edge tractions $t^{(n)}$ and edge charges $d^{(n)}$ are defined by

$$T_j^{(n)} = B_n^{-1}[x_2{}^n T_{2j}]_{-b}^{b}, \quad D^{(n)} = B_n^{-1}[x_2{}^n D_2]_{-b}^{b}$$

$$t_j^{(n)} = B_n^{-1} \int_{-b}^{b} x_2{}^n (n_c T_{cj})_C\,\mathrm{d}x_2, \quad d^{(n)} = B_n^{-1} \int_{-b}^{b} x_2{}^n (n_c D_c)_C\,\mathrm{d}x_2. \tag{41}$$

Finally, the field equations of order n are

$$T_{ij,i}^{(n)} - nT_{2j}^{(n-1)} + B_n T_j^{(n)} = \rho \sum_m B_{mn} \ddot{u}_j^{(m)},$$

$$D_{i,i}^{(n)} - nD_2^{(n-1)} + B_n D^{(n)} = 0, \tag{42}$$

with edge conditions, on C,

$$n_c T_{cj}^{(n)} = B_n t_j^{(n)} \quad \text{or} \quad u_j^{(n)} = \bar{u}_j^{(n)} \tag{43}$$

where $\bar{u}_j^{(n)}$ is the edge displacement of order n, and

$$n_c D_c^{(n)} = B_n d^{(n)} \quad \text{or} \quad \phi^{(n)} = \bar{\phi}^{(n)} \tag{44}$$

where $\bar{\phi}^{(n)}$ is the edge potential of order n. As an alternative to (43), a component of $n_i T_{in}^{(n)}$ and the resultant of $u_j^{(n)}$ in the plane at right angles, or *vice versa*, may be specified.

5. FIRST ORDER APPROXIMATION

The equations of the preceding section are now to be restricted so as to exclude simple thickness modes higher than the fundamental stretch, twist and shear modes—which involve only terms of orders $n = 0, 1$. The components of strain and electric field, of these orders, are, from (34),

$$\begin{aligned} S_{11}^{(0)} &= S_1^{(0)} = u_{1,1}^{(0)} & 2S_{23}^{(0)} &= S_4^{(0)} = u_{2,3}^{(0)} + u_3^{(1)} & E_1^{(0)} &= -\phi_{,1}^{(0)} \\ S_{22}^{(0)} &= S_2^{(0)} = u_2^{(1)} & 2S_{31}^{(0)} &= S_5^{(0)} = u_{1,3}^{(0)} + u_{3,1}^{(0)} & E_2^{(0)} &= -\phi^{(1)} \\ S_{33}^{(0)} &= S_3^{(0)} = u_{3,3}^{(0)} & 2S_{12}^{(0)} &= S_6^{(0)} = u_{2,1}^{(0)} + u_1^{(1)} & E_3^{(0)} &= -\phi_{,3}^{(0)} \end{aligned} \tag{45}$$

$$\begin{aligned} S_{11}^{(1)} &= S_1^{(1)} = u_{1,1}^{(1)} & 2S_{23}^{(1)} &= S_4^{(1)} = u_{2,3}^{(1)} + 2u_3^{(2)} & E_1^{(1)} &= -\phi_{,1}^{(1)} \\ S_{22}^{(1)} &= S_2^{(1)} = 2u_2^{(2)} & 2S_{31}^{(1)} &= S_5^{(1)} = u_{1,3}^{(1)} + u_{3,1}^{(1)} & E_2^{(1)} &= -2\phi^{(2)} \\ S_{33}^{(1)} &= S_3^{(1)} = u_{3,3}^{(1)} & 2S_{12}^{(1)} &= S_6^{(1)} = u_{2,1}^{(1)} + 2u_1^{(2)} & E_3^{(1)} &= -\phi_{,3}^{(1)}. \end{aligned} \tag{46}$$

The strain $S_{22}^{(1)}$ is associated with the second thickness-stretch mode—which is not to be included. We set $T_{22}^{(1)}$ and $\ddot{u}_2^{(2)}$ equal to zero so as to permit free development of $S_{22}^{(1)}$ without contribution to the kinetic or potential energy. From (38),

$$T_{ij}^{(1)} = (2b^3/3)(c_{ijkl}S_{kl}^{(1)} - e_{kij}E_k^{(1)}) \tag{47}$$

so that we set

$$T_{22}^{(1)} = (2b^3/3)(c_{22kl}S_{kl}^{(1)} - e_{k22}E_k^{(1)}) = 0. \tag{48}$$

Write these two equations as

$$3T_{ij}^{(1)}/2b^3 = (c_{ijkl}S_{kl}^{(1)} - c_{ij22}S_{22}^{(1)}) + c_{ij22}S_{22}^{(1)} - e_{kij}E_k^{(1)} \tag{49}$$

and

$$S_{22}^{(1)} = -c_{22kl}S_{kl}^{(1)}/c_{2222} + S_{22}^{(1)} + e_{k22}E_k^{(1)}/c_{2222}, \tag{50}$$

respectively. Then substitute the expression for $S_{22}^{(1)}$, in (50), for the $S_{22}^{(1)}$ outside the parentheses in (49) and collect terms to obtain, in place of (47),

$$T_{ij}^{(1)} = (2b^3/3)(c_{ijkl}^{(1)}S_{kl}^{(1)} - e_{kij}^{(1)}E_k^{(1)}) \tag{51}$$

where

$$c_{ijkl}^{(1)} = c_{ijkl} - c_{ij22}c_{22kl}/c_{2222}, \qquad e_{kij}^{(1)} = e_{kij} - e_{k22}c_{ij22}/c_{2222}. \tag{52}$$

It may be verified that (51) satisfies (48).

The remaining terms in (46) of order $n = 2$ are $u_1^{(2)}$, $u_3^{(2)}$ and $\phi^{(2)}$. These are to be omitted and $e_{ijk}^{(1)}$ substituted for e_{ijk} in $D_i^{(1)}$.

At this stage, the constitutive equations of the first order approximation are

$$\begin{aligned} T_{ij}^{(0)} &= 2b(c_{ijkl}S_{kl}^{(0)} - e_{kij}E_k^{(0)}), & D_i^{(0)} &= 2b(e_{ijk}S_{jk}^{(0)} + \epsilon_{ij}E_j^{(0)}), \\ T_{ij}^{(1)} &= (2b^3/3)(c_{ijkl}^{(1)}S_{kl}^{(1)} - e_{kij}^{(1)}E_k^{(1)}), & D_i^{(1)} &= (2b^3/3)(e_{ijk}^{(1)}S_{jk}^{(1)} + \epsilon_{ij}E_j^{(1)}), \end{aligned} \tag{53}$$

where

$$\begin{aligned} S_{ij}^{(0)} &= \frac{1}{2}(u_{j,i}^{(0)} + u_{i,j}^{(0)} + \delta_{i2}u_j^{(1)} + \delta_{2j}u_i^{(1)}), & E_i^{(0)} &= -\phi_{,i}^{(0)} - \delta_{i2}\phi^{(1)} \\ S_{ij}^{(1)} &= \frac{1}{2}(u_{j,i}^{(1)} + u_{i,j}^{(1)}), & E_i^{(1)} &= -\phi_{,i}^{(1)}. \end{aligned} \tag{54}$$

In the reduced notation (53) are

$$T_p^{(0)} = 2b(c_{pq}S_q^{(0)} - e_{kp}E_k^{(0)}), \quad D_i^{(0)} = 2b(e_{ip}S_p^{(0)} + \epsilon_{ij}E_j^{(0)}),$$

$$T_p^{(1)} = (2b^3/3)(c_{pq}^{(1)}S_q^{(1)} - e_{kp}^{(1)}E_k^{(1)}), \quad D_i^{(1)} = (2b^3/3)(e_{ip}^{(1)}S_p^{(1)} + \epsilon_{ij}E_j^{(1)}) \tag{55}$$

and the corresponding electric enthalpy density is

$$\bar{H} = b(c_{pq}S_p^{(0)}S_q^{(0)} - \epsilon_{ij}E_i^{(0)}E_j^{(0)} - 2e_{kp}E_k^{(0)}S_p^{(0)})$$
$$+ (b^3/3)(c_{pq}^{(1)}S_p^{(1)}S_q^{(1)} - \epsilon_{ij}E_i^{(1)}E_j^{(1)} - 2\, e_{kp}^{(1)}E_k^{(1)}S_p^{(1)}). \tag{56}$$

It is important that the simple thickness frequencies from the approximate equations match exactly those from the three-dimensional equations. As the approximate equations now stand, the match is not exact owing to the difference between the exact trigonometric distributions of displacements, from (23), and the approximate linear distributions from the early terms of (32). To compensate, correction coefficients are inserted as multipliers of the thickness-stretch and -shear strains $S_2^{(0)}$, $S_4^{(0)}$, $S_6^{(0)}$. Thus

$$S_p^{(0)} \to \kappa_p S_p^{(0)} \quad \text{(not summed)} \tag{57}$$

where

$$\kappa_p = \begin{cases} \kappa_p, & p = 2, 4, 6 \\ 1, & p = 1, 3, 5 \end{cases} \tag{58}$$

and the three κ_p are to be determined so that the three simple thickness frequencies from the approximate equations match those from the three-dimensional equations.

The revised electric enthalpy density is

$$\bar{H} = b(c_{pq}^{(0)}S_p^{(0)}S_q^{(0)} - \epsilon_{ij}E_i^{(0)}E_j^{(0)} - 2e_{kp}^{(0)}E_k^{(0)}S_p^{(0)})$$
$$+ (b^3/3)(c_{pq}^{(1)}S_p^{(1)}S_q^{(1)} - \epsilon_{ij}E_i^{(1)}E_j^{(1)} - 2e_{kp}^{(1)}E_k^{(1)}S_p^{(1)}), \tag{59}$$

where

$$c_{pq}^{(0)} = \kappa_p\kappa_q c_{pq}, \quad e_{kp}^{(0)} = \kappa_p e_{kp} \quad \text{(not summed)} \tag{60}$$

and the $S_p^{(n)}$ and $E_i^{(n)}$ are

$$\begin{array}{lll} S_1^{(0)} = u_{1,1}^{(0)} & S_4^{(0)} = u_{2,3}^{(0)} + u_3^{(1)} & E_1^{(0)} = -\phi_{,1}^{(0)} \\ S_2^{(0)} = u_2^{(1)} & S_5^{(0)} = u_{1,3}^{(0)} + u_{3,1}^{(0)} & E_2^{(0)} = -\phi^{(1)} \\ S_3^{(0)} = u_{3,3}^{(0)} & S_6^{(0)} = u_{2,1}^{(0)} + u_1^{(1)} & E_3^{(0)} = -\phi_{,3}^{(0)} \end{array} \tag{61}$$

$$\begin{array}{lll} S_1^{(1)} = u_{1,1}^{(1)} & S_4^{(1)} = u_{2,3}^{(1)} & E_1^{(1)} = -\phi_{,1}^{(1)} \\ & S_5^{(1)} = u_{1,3}^{(1)} + u_{3,1}^{(1)} & \\ S_3^{(1)} = u_{3,3}^{(1)} & S_6^{(1)} = u_{2,1}^{(1)} & E_3^{(1)} = -\phi_{,3}^{(1)}. \end{array}$$

The constitutive equations are

$$T_p^{(0)} = \partial\bar{H}/\partial S_p^{(0)} = 2b(c_{pq}^{(0)}S_q^{(0)} - e_{kp}^{(0)}E_k^{(0)}),$$

$$T_p^{(1)} = \partial\bar{H}/\partial S_p^{(0)} = (2b/3)(c_{pq}^{(1)}S_q^{(1)} - e_{kp}^{(1)}E_k^{(1)}),$$

$$D_i^{(0)} = -\partial\bar{H}/\partial E_i^{(0)} = 2b(e_{ip}^{(0)}S_p^{(0)} + \epsilon_{ij}E_j^{(0)}), \tag{62}$$

$$D_i^{(1)} = -\partial\bar{H}/\partial E_i^{(1)} = (2b^3/3)(e_{ip}^{(1)}S_p^{(1)} + \epsilon_{ij}E_j^{(1)});$$

and, finally, the field equations are

$$
\begin{gathered}
T_{ij,i}^{(0)} + 2bT_j^{(0)} = 2b\rho\ddot{u}_j^{(0)}, \\
T_{ij,i}^{(1)} - T_{2j}^{(0)} + (2b^3/3)T_j^{(1)} = (2b^3/3)\rho\ddot{u}_j^{(1)}, \\
D_{i,i}^{(0)} + 2bD^{(0)} = 0, \\
D_{i,i}^{(1)} - D_2^{(0)} + (2b^3/3)D^{(1)} = 0.
\end{gathered} \tag{63}
$$

The inclusion of the thickness-stretch mode, in (59)–(63), is what distinguishes them from the equations in [7].

6. CORRECTION FACTORS

In this section the procedure is established for computing the values of the correction factors κ_a, $a = 2, 4, 6$, so as to make the simple thickness-frequencies, from the first order equations, the same as the corresponding ones obtained in Section 3 from the three-dimensional equations.

To find the simple thickness-frequencies from the first-order equations, we set

$$
u_j^{(0)} = 0, \quad u_j^{(1)} = A_j^0\, e^{i\omega t}, \quad \phi^{(0)} = 0, \quad \phi^{(1)} = b^{-1}V\, e^{i\omega t} \tag{64}
$$

in the equations of motion.

In general, from (41),

$$
T_j^{(n)} = B_n^{-1}[x_2{}^nT_{2j}]_{-b}^{b}
$$

and, for electrode coatings of density ρ' and half-thickness b',

$$
T_{2j}]_{\pm b} = \mp 2\rho' b' \ddot{u}_j]_{\pm b}. \tag{65}
$$

Then

$$
\begin{gathered}
T_j^{(0)} = (2b)^{-1}[\mp 2\rho' b' \ddot{u}_j]_{\pm b} = -R\rho\ddot{u}_j^{(0)}, \\
T_j^{(1)} = (3/2b^3)[\mp x_2 2\rho' b' \ddot{u}_j]_{\pm b} = -3R\rho\ddot{u}_j^{(1)}
\end{gathered} \tag{66}
$$

where, again, $R = 2\rho' b'/\rho b$.

With (64) and (66), the stress equations of motion (63) reduce to

$$
\begin{gathered}
(T_6^{(0)}, T_2^{(0)}, T_4^{(0)}) = -(2b^3/3)(1+3R)\rho(\ddot{u}_1^{(1)}, \ddot{u}_2^{(1)}, \ddot{u}_3^{(1)}) \\
D_2^{(0)} = (2b^3/3)D^{(1)},
\end{gathered} \tag{67}
$$

in which

$$
\begin{aligned}
T_6^{(0)} &= 2b(c_{66}^{(0)}u_1^{(1)} + c_{26}^{(0)}u_2^{(1)} + c_{46}^{(0)}u_3^{(1)} + e_{26}^{(0)}\phi^{(1)}), \\
T_2^{(0)} &= 2b(c_{26}^{(0)}u_1^{(1)} + c_{22}^{(0)}u_2^{(1)} + c_{24}^{(0)}u_3^{(1)} + e_{22}^{(0)}\phi^{(1)}), \\
T_4^{(0)} &= 2b(c_{46}^{(0)}u_1^{(1)} + c_{24}^{(0)}u_2^{(1)} + c_{44}^{(0)}u_3^{(1)} + e_{24}^{(0)}\phi^{(1)}), \\
D_2^{(0)} &= 2b(e_{26}^{(0)}u_1^{(1)} + e_{22}^{(0)}u_2^{(1)} + e_{24}^{(0)}u_3^{(1)} - \epsilon_{22}\phi^{(1)}).
\end{aligned} \tag{68}
$$

The last of (67) and (68) contribute to a formula for the current, per unit area, through the crystal:

$$
I = (2b^3/3)\, \mathrm{d}D^{(1)}/\mathrm{d}t. \tag{69}
$$

From (64) and the first three of (67) and (68), we find

$$\begin{aligned}(1-\bar{\Omega}^2/\bar{\kappa}_6^2)\kappa_6 A_1^0+\bar{c}_{26}\kappa_2 A_2^0+\bar{c}_{46}\kappa_4 A_3^0&=-e_{26}V/bc_{66},\\ \bar{c}_{26}\kappa_6 A_1^0+(\bar{c}_{22}-\bar{\Omega}^2/\bar{\kappa}_2^2)\kappa_2 A_2^0+\bar{c}_{24}\kappa_4 A_3^0&=-e_{22}V/bc_{66},\\ \bar{c}_{46}\kappa_6 A_1^0+\bar{c}_{24}\kappa_2 A_2^0+(\bar{c}_{44}-\bar{\Omega}^2/\bar{\kappa}_4^2)\kappa_4 A_3^0&=-e_{24}V/bc_{66},\end{aligned} \tag{70}$$

where

$$\bar{c}_{pq}=c_{pq}/c_{66},\quad \bar{\kappa}_a^2=12\bar{\kappa}_a^2/\pi^2,\quad \bar{\Omega}^2=(1+3R)\omega^2/(\pi^2 c_{66}/4\rho b^2). \tag{71}$$

The determinant of the coefficients of the A_j^0, in (70), set equal to zero, yields

$$A\bar{\Omega}^6+B\bar{\Omega}^4+C\bar{\Omega}^2+D=0, \tag{72}$$

where

$$\begin{aligned}A&=1/\bar{\kappa}_2^2\bar{\kappa}_4^2\bar{\kappa}_6^2,\\ B&=-(\bar{c}_{22}\bar{\kappa}_2^2+\bar{c}_{44}\bar{\kappa}_4^2+\bar{\kappa}_6^2)/\bar{\kappa}_2^2\bar{\kappa}_4^2\bar{\kappa}_6^2,\\ C&=(\bar{c}_{44}-\bar{c}_{46}^2)/\bar{\kappa}_2^2+(\bar{c}_{22}-\bar{c}_{26}^2)/\bar{\kappa}_4^2+(\bar{c}_{22}\bar{c}_{44}-c_{24}^2)/\bar{\kappa}_6^2,\\ D&=-(\bar{c}_{22}\bar{c}_{44}+2\bar{c}_{26}\bar{c}_{24}\bar{c}_{46}-\bar{c}_{22}c_{46}^2-\bar{c}_{44}\bar{c}_{26}^2-\bar{c}_{24}^2).\end{aligned} \tag{73}$$

The bicubic (72) must be satisfied by each of

$$\bar{\Omega}_a^2=\Omega_a^2(1+3R),\quad a=2,4,6 \tag{74}$$

where the Ω_a are the exact roots from (29). Thus, we have the three simultaneous, nonlinear equations on the κ_a:

$$\begin{aligned}A\bar{\Omega}_2^6+B\bar{\Omega}_2^4+C\bar{\Omega}_2^2+D&=0,\\ A\bar{\Omega}_4^6+B\bar{\Omega}_4^4+C\bar{\Omega}_4^2+D&=0,\\ A\bar{\Omega}_6^6+B\bar{\Omega}_6^4+C\bar{\Omega}_6^2+D&=0.\end{aligned} \tag{75}$$

We find, from (75),

$$\begin{aligned}A&=-D/\bar{\Omega}_2^2\bar{\Omega}_4^2\bar{\Omega}_6^2,\\ B&=-A(\bar{\Omega}_2^2+\bar{\Omega}_4^2+\bar{\Omega}_6^2),\\ C&=A(\bar{\Omega}_2^2\bar{\Omega}_4^2+\bar{\Omega}_4^2\bar{\Omega}_6^2+\bar{\Omega}_6^2\bar{\Omega}_2^2)\end{aligned} \tag{76}$$

and note that the right hand sides of (76) are independent of the κ_a. From the third of (73) and of (76), $\bar{\kappa}_2$ may be expressed in terms of $\bar{\kappa}_4$ and $\bar{\kappa}_6$ (and the Ω_a and $\bar{c}_{pq}$); and this expression may be used to eliminate $\bar{\kappa}_2$ from the second of (76), leaving that equation as a quadratic in $1/\bar{\kappa}_4^2$ with coefficients functions of $\bar{\kappa}_6$. Then $1/\bar{\kappa}_4^2$ can be obtained as the algebraically larger root of the quadratic—a function of $\bar{\kappa}_6$, say

$$1/\bar{\kappa}_4^2=f(\bar{\kappa}_6). \tag{77}$$

With this and the third of (76), we may express $\bar{\kappa}_2$ in terms of $\bar{\kappa}_6$ alone, say

$$1/\bar{\kappa}_2^2=g(\bar{\kappa}_6). \tag{78}$$

Upon substituting (77) and (78) in the first of (76), we have

$$f(\bar{\kappa}_6)g(\bar{\kappa}_6)/\bar{\kappa}_6^2+D/\bar{\Omega}_2^2\bar{\Omega}_4^2\bar{\Omega}_6^2=0, \tag{79}$$

an equation on $\bar{\kappa}_6$, alone, which may be solved, iteratively, for $\bar{\kappa}_6$; following which $\bar{\kappa}_2$ and $\bar{\kappa}_4$ may be obtained from (77) and (78). For the SC-cut and $R = 0.01$, we find

$$\begin{aligned} \bar{\kappa}_2 &= 1.004933, \\ \bar{\kappa}_4 &= 1.005039, \\ \bar{\kappa}_6 &= 1.005083. \end{aligned} \tag{80}$$

7. DISPERSION RELATION

We consider, again, an electroded, infinite plate under a uniform voltage drop $V\,e^{i\omega t}/b$ across the thickness; but, in addition to the forced, simple thickness-modes, of the preceding section, we now permit variation of displacements along x_1:

$$\begin{aligned} u_1^{(0)} &= A_1 b \sin \xi x_1\, e^{i\omega t}, & u_1^{(1)} &= A_4 \cos \xi x_1\, e^{i\omega t}, \\ u_2^{(0)} &= A_2 b \sin \xi x_1\, e^{i\omega t}, & u_2^{(1)} &= A_5 \cos \xi x_1\, e^{i\omega t}, \\ u_3^{(0)} &= A_3 b \sin \xi x_1\, e^{i\omega t}, & u_3^{(1)} &= A_6 \cos \xi x_1\, e^{i\omega t}, \\ \phi^{(0)} &= A_7 b \sin \xi x_1\, e^{i\omega t}, & \phi^{(1)} &= (V/b)\, e^{i\omega t}. \end{aligned} \tag{81}$$

When independent of x_3, the stress-displacement equations become

$$\begin{aligned} T_p^{(0)} &= 2b[c_{p1}^{(0)} u_{1,1}^{(0)} + c_{p2}^{(0)} u_2^{(1)} + c_{p4}^{(0)} u_3^{(1)} + c_{p5}^{(0)} u_{3,1}^{(0)} + c_{p6}^{(0)}(u_{2,1}^{(0)} + u_1^{(1)}) + e_{1p}^{(0)} \phi_{,1}^{(0)} + e_{2p}^{(0)} \phi^{(1)}], \\ T_p^{(1)} &= (2b^3/3)(c_{p1}^{(1)} u_{1,1}^{(1)} + c_{p5}^{(1)} u_{3,1}^{(1)} + c_{p6}^{(1)} u_{2,1}^{(1)} + e_{1p}^{(1)} \phi_{,1}^{(1)}), \\ D_i^{(0)} &= 2b[e_{i1}^{(0)} u_{1,1}^{(0)} + e_{i2}^{(0)} u_2^{(1)} + e_{i4}^{(0)} u_3^{(1)} + e_{i5}^{(0)} u_{3,1}^{(0)} + e_{i6}^{(0)}(u_{2,1}^{(0)} + u_1^{(1)}) - \epsilon_{i1} \phi_{,1}^{(0)} - \epsilon_{i2} \phi^{(1)}], \\ D_i^{(1)} &= (2b^3/3)(e_{i1}^{(1)} u_{1,1}^{(1)} + e_{i5}^{(1)} u_{3,1}^{(1)} + e_{i6}^{(1)} u_{2,1}^{(1)} - \epsilon_{i1} \phi_{,1}^{(1)}). \end{aligned} \tag{82}$$

When these expressions and those for $T_j^{(0)}$ and $T_j^{(1)}$, from (66), are inserted in the equations of motion (63), the displacement equations of motion become:

$$\begin{aligned} & c_{11}^{(0)} u_{1,11}^{(0)} + c_{12}^{(0)} u_{2,1}^{(1)} + c_{14}^{(0)} u_{3,1}^{(1)} + c_{15}^{(0)} u_{3,11}^{(0)} + c_{16}^{(0)}(u_{2,11}^{(0)} + u_{1,1}^{(1)}) + e_{11}^{(0)} \phi_{,11}^{(0)} + e_{21}^{(0)} \phi_{,1}^{(1)} = (1+R)\rho \ddot{u}_1^{(0)} \\ & c_{61}^{(0)} u_{1,11}^{(0)} + c_{62}^{(0)} u_{2,1}^{(1)} + c_{64}^{(0)} u_{3,1}^{(1)} + c_{65}^{(0)} u_{3,11}^{(0)} + c_{66}^{(0)}(u_{2,11}^{(0)} + u_{1,1}^{(1)}) + e_{16}^{(0)} \phi_{,11}^{(0)} + e_{26}^{(0)} \phi_{,1}^{(1)} = (1+R)\rho \ddot{u}_2^{(0)} \\ & c_{51}^{(0)} u_{1,11}^{(0)} + c_{52}^{(0)} u_{2,1}^{(1)} + c_{54}^{(0)} u_{3,1}^{(1)} + c_{55}^{(0)} u_{3,11}^{(0)} + c_{56}^{(0)}(u_{2,11}^{(0)} + u_{1,1}^{(1)}) + e_{15}^{(0)} \phi_{,11}^{(0)} + e_{25}^{(0)} \phi_{,1}^{(1)} = (1+R)\rho \ddot{u}_3^{(0)} \\ & c_{11}^{(1)} u_{1,11}^{(1)} + c_{15}^{(1)} u_{3,11}^{(1)} + c_{16}^{(1)} u_{2,11}^{(1)} + e_{11}^{(1)} \phi_{,11}^{(1)} \\ & - 3b^{-2}[c_{61}^{(0)} u_{1,1}^{(0)} + c_{62}^{(0)} u_2^{(1)} + c_{64}^{(0)} u_3^{(1)} + c_{65}^{(0)} u_{3,1}^{(0)} + c_{66}^{(0)}(u_{2,1}^{(0)} + u_1^{(1)}) + e_{16}^{(0)} \phi_{,1}^{(0)} + e_{26}^{(0)} \phi^{(1)}] = (1+3R)\rho \ddot{u}_1^{(1)} \\ & c_{61}^{(1)} u_{1,11}^{(1)} + c_{65}^{(1)} u_{3,11}^{(1)} + c_{66}^{(1)} u_{2,11}^{(1)} + e_{16}^{(1)} \phi_{,11}^{(1)} - 3b^{-2}[c_{21}^{(0)} u_{1,1}^{(0)} + c_{22}^{(0)} u_2^{(1)} + c_{24}^{(0)} u_3^{(1)} + c_{25}^{(0)} u_{3,1}^{(0)} + c_{26}^{(0)}(u_{2,1}^{(0)} \\ & + u_1^{(1)}) + e_{12}^{(0)} \phi_{,1}^{(0)} + e_{22}^{(0)} \phi^{(1)}] = (1+3R)\rho \ddot{u}_2^{(1)} \\ & -3b^{-2}[c_{41}^{(0)} u_{1,1}^{(0)} + c_{42}^{(0)} u_2^{(1)} + c_{44}^{(0)} u_3^{(1)} + c_{45}^{(0)} u_{3,1}^{(0)} + c_{46}^{(0)}(u_{2,1}^{(0)} + u_1^{(1)}) + e_{14}^{(0)} \phi_{,1}^{(0)} + e_{24}^{(0)} \phi^{(1)}] = (1+3R)\rho \ddot{u}_3^{(1)} \\ & e_{11}^{(0)} u_{1,11}^{(0)} + e_{12}^{(0)} u_{2,1}^{(1)} + e_{14}^{(0)} u_{3,1}^{(1)} + e_{15}^{(0)} u_{3,11}^{(0)} + e_{16}^{(0)}(u_{2,11}^{(0)} + u_{1,1}^{(1)}) - \epsilon_{11} \phi_{,11}^{(0)} - \epsilon_{12} \phi_{,1}^{(1)} = D^{(0)} \\ & e_{11}^{(1)} u_{1,11}^{(1)} + e_{15}^{(1)} u_{3,11}^{(1)} + e_{16}^{(1)} u_{2,11}^{(1)} - \varepsilon_{11} \phi_{,1}^{(1)} - 3b^{-2}[e_{21}^{(0)} u_{1,1}^{(0)} + e_{22}^{(0)} u_2^{(1)} + e_{24}^{(0)} u_3^{(1)} \\ & + e_{25}^{(0)} u_{3,1}^{(0)} + e_{26}^{(0)}(u_{2,1}^{(0)} + u_1^{(1)}) - \epsilon_{21} \phi_{,1}^{(0)} - \epsilon_{22} \phi^{(1)} = D^{(1)}. \end{aligned} \tag{83}$$

We first find a particular solution of (83) for the constant forcing term $\phi^{(1)}$. This is the same as the solution for simple thickness modes, of the preceding section, except that the frequency ω need not be a resonance frequency. Then the amplitudes $A_1{}^0, A_2{}^0, A_3{}^0$ may be obtained from (70).

The complementary solution is found by substituting (81) into (83) with $\phi^{(1)}$ zero. The eighth of (83) may be set aside as it serves only at a later stage to yield the surface charge and the current through the crystal. In the seventh of (83), $D^{(0)}$ is zero from symmetry; and the equation may be solved for A_7:

$$A_7 = (e_{11}^{(0)}\bar{\xi}A_1 + e_{16}^{(0)}\bar{\xi}A_2 + e_{15}^{(0)}\bar{\xi}A_3 + e_{16}^{(0)}A_4 + e_{12}^{(0)}A_5 + e_{14}^{(0)}A_6)/\epsilon_{11}\bar{\xi} \tag{84}$$

and used to replace A_7 in the remaining six equations. The result is

$$a_{ij}A_j = 0, \quad a_{ij} = a_{ji}; \quad i, j = 1 \ldots 6 \tag{85}$$

where

$$\begin{aligned}
&a_{11} = \bar{c}_{11}\bar{\xi}^2 - R'\hat{\Omega}^2 && a_{12} = \bar{c}_{16}\bar{\xi}^2\kappa_6 && a_{23} = \bar{c}_{56}\bar{\xi}^2\\
&a_{22} = (\bar{c}_{66}\bar{\xi}^2 - R'\hat{\Omega}/\kappa_6^2)\kappa_6 && a_{13} = \bar{c}_{15}\bar{\xi}^2 && a_{24} = \bar{c}_{66}\bar{\xi}\kappa_6\\
&a_{33} = \bar{c}_{55}\bar{\xi}^2 - R'\hat{\Omega}^2 && a_{14} = \bar{c}_{16}\bar{\xi}\kappa_6 && a_{25} = \bar{c}_{26}\bar{\xi}\kappa_2\\
&a_{44} = (\bar{c}_{66} + \hat{c}_{11}\bar{\xi}^2 - \hat{\Omega}^2/\kappa_6^2)\kappa_6 && a_{15} = \bar{c}_{12}\bar{\xi}\kappa_2 && a_{26} = \bar{c}_{46}\bar{\xi}\kappa_4\\
&a_{55} = (\bar{c}_{22} + \hat{c}_{66}\bar{\xi}^2 - \hat{\Omega}^2/\kappa_2^2)\kappa_2 && a_{16} = \bar{c}_{14}\bar{\xi}\kappa_4 &&\\
&a_{66} = (\bar{c}_{44} + \hat{c}_{55}\bar{\xi}^2 - \hat{\Omega}^2/\kappa_4^2)\kappa_4 && &&\\
&a_{34} = \bar{c}_{56}\bar{\xi}\kappa_6 && a_{35} = \bar{c}_{25}\bar{\xi}\kappa_2 && a_{36} = \bar{c}_{45}\bar{\xi}\kappa_4\\
&a_{45} = (\bar{c}_{26} + \hat{c}_{16}\bar{\xi}^2)\kappa_2 && a_{46} = (\bar{c}_{46} + \hat{c}_{15}\bar{\xi}^2)\kappa_4 && a_{56} = (\bar{c}_{24} + \hat{c}_{56}\bar{\xi}^2)\kappa_4
\end{aligned} \tag{86}$$

in which

$$\hat{\Omega}^2 = (\pi^2/12)(1+3R)\omega^2/\omega_0^2, \quad \omega_0^2 = \pi^2 c_{66}/4\rho b^2, \quad R' = 3(1+R)(1+3R),$$
$$\bar{c}_{pq} = (c_{pq} - e_{1p}e_{1q}/\epsilon_{11})/c_{66}, \quad \hat{c}_{pq} = c_{pq}^{(1)}/3\kappa_r\kappa_s c_{66}, \quad \bar{\xi} = \xi b. \tag{87}$$

In $\hat{c}_{pq}$, when p (or q) = 1, 5, 6 then r (or s) = 2, 4, 6 respectively.

Resonance occurs when the determinant of the coefficients of the A_j, in (85), is equal to zero:

$$|a_{ij}| = 0. \tag{88}$$

This is a sextic equation in $\bar{\xi}^2$ yielding a six-branched dispersion relation as illustrated in Fig. 1 for the SC-out and $R = 0.01$.

In Fig. 1, the ordinate is Ω as given in (30) and the abscissa is $X = 2\xi b/\pi$. The six branches are identified according to the behavior of the corresponding modes at long wave-lengths ($\xi \to 0$):

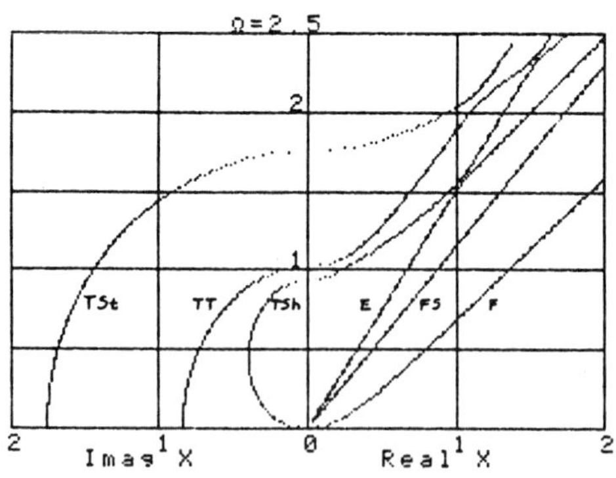

Fig. 1. Plot of dispersion relation (88).

F = flexure
FS = face-shear
E = extension
TSh = thickness-shear
TT = thickness-twist
TSt = thickness-stretch

The computation of the roots of the sextic (88), for Fig. 1, required about 90 hr on the HP-85 microcomputer. In order to reduce the computation time for subsequent use in the interval $0.91 < \Omega < 1.07$, an approximation was introduced. In that interval, each of the six branches was divided into three segments and a cubic, quadratic or linear curve, as appropriate, was fitted to each of the eighteen segments so that the roots along each curve could be expressed explicitly. For each frequency, the time for computation and storage of the six roots was thereby reduced by a factor of about 1250. The result is illustrated in Fig. 2.

The equations of motion are satisfied by (81) and (64) for each of the ξ_n, $n = 1 \dots 6$. Hence, we can write

$$
\begin{aligned}
u_1^{(0)} &= \sum_{n=1}^{6} A_{1n} b \sin(\xi_n x_1 + \epsilon)\, e^{i\omega t}, & u_1^{(1)} &= A_1{}^0 + \sum_{n=1}^{6} A_{4n} \cos(\xi_n x_1 + \epsilon)\, e^{i\omega t} \\
u_2^{(0)} &= \sum_{n=1}^{6} A_{2n} b \sin(\xi_n x_1 + \epsilon)\, e^{i\omega t}, & u_2^{(1)} &= A_2{}^0 + \sum_{n=1}^{6} A_{5n} \cos(\xi_n x_1 + \epsilon)\, e^{i\omega t} \\
u_3^{(0)} &= \sum_{n=1}^{6} A_{3n} b \sin(\xi_n x_1 + \epsilon)\, e^{i\omega t}, & u_3^{(1)} &= A_3{}^0 + \sum_{n=1}^{6} A_{6n} \cos(\xi_n x_1 + \epsilon)\, e^{i\omega t} \\
\phi^{(0)} &= \sum_{n=1}^{6} A_{7n} b \sin(\xi_n x_1 + \epsilon)\, e^{i\omega t}, & \phi^{(1)} &= (V/b)\, e^{i\omega t},
\end{aligned}
\tag{89}
$$

where $\epsilon = 0$ or $\pi/2$.

8. VIBRATIONS OF A STRIP

A strip, bounded by edges at, say, $x_1 = \pm a$, may be subject to a variety of edge conditions. The only one that can be attained physically without difficulty is the traction-free condition. However, the simplest condition mathematically is one for which the six branches of the dispersion relation are not coupled. In the present case, this can be satisfied by

$$u_1^{(0)} = u_2^{(0)} = u_3^{(0)} = \phi^{(0)} = T_1^{(1)} = T_6^{(1)} = T_5^{(1)} = D_1^{(1)} = 0 \quad \text{on} \quad x_1 = \pm a; \tag{90}$$

a combination which is admissible according to (43) and (44). The conditions (90) are the

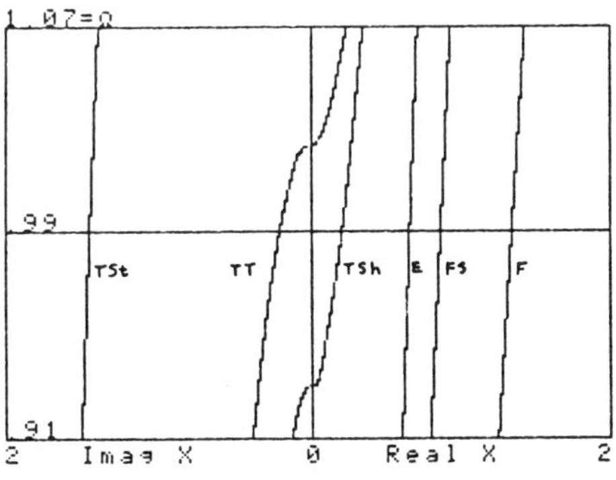

Fig. 2. Portion of dispersion curves of Fig. 1 with expanded frequency scale.

analogue of those for the simply supported beam in the Bernoulli–Euler or Timoshenko beam theory, in which only the bending moment $T_1^{(1)}$ and the deflection $u_2^{(0)}$, of the functions in (90), are present and required to be zero at the ends of the beam.

With (89), (90) can be satisfied by

$$\sin \xi_n a = 0 \tag{91}$$

for $\epsilon = 0$ and

$$\cos \xi_n a = 0 \tag{92}$$

for $\epsilon = \pi/2$ for each branch separately—for real ξ_n. Thus, the boundary conditions (90) can be satisfied by (91) or (92), for real ξ_n and each n separately if

$$X = 2\xi_n b/\pi = mb/a, \quad m = 1, 2, 3\ldots \tag{93}$$

This equation and (88) can be represented graphically by replotting the real part of Fig. 1 or Fig. 2 with the reciprocals of the abscissae of each of the six curves (i.e. a plot of Ω vs a/mb) and then multiplying the resulting abscissae by the integers m to produce a set of equally spaced Ω vs a/b curves (overtone branches) for each of the six replotted dispersion curves. The result is shown in Fig. 3 for the range $0 < a/b < 24$ and the frequency range of Fig. 2. The range $17 < a/b < 23$ is shown in Fig. 4 with the abscissa expanded by a factor of 4. The numbers following *F*, *FS*, *E*, *TSh* and *TT*, in Figs. 3 and 4, are the values of m, in (93), identifying the order of the overtone. No thickness-stretch branches appear, as their ξ is imaginary in the frequency range displayed.

The computation for the strip with free edges is more complicated; but the main effect on the frequency spectrum, of the coupling of the modes and overtones at the free edges, is the elimination of intersections of overtone branches. The general features of the frequency spectrum of the coupled modes can readily be sketched, without computation, over the grid of the spectrum of uncoupled modes and overtones. Figure 5 is such a sketch on the background of Fig. 4. If m odd or even has been chosen correctly for each set of overtones, the only error in a sketch—such as in Fig. 5—is in the strength of the coupling; i.e. whether the coupled branches lie close to the intersections (weak coupling) or far from the intersections (strong coupling).

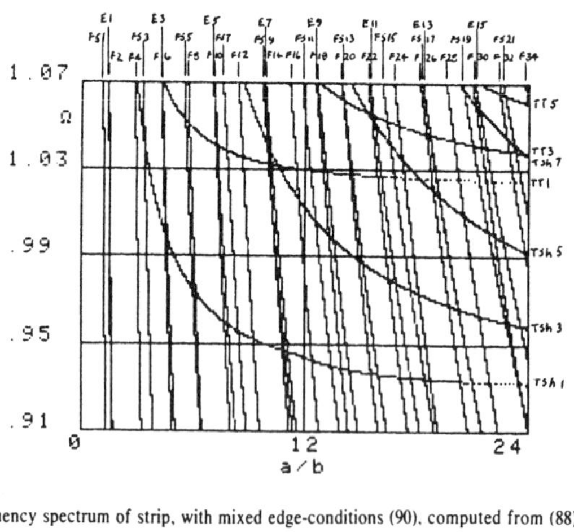

Fig. 3. Frequency spectrum of strip, with mixed edge-conditions (90), computed from (88) and (93).

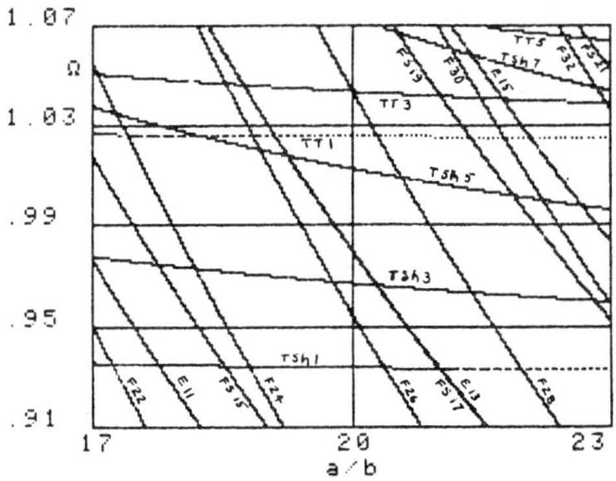

Fig. 4. Portion of frequency spectrum of Fig. 3 with expanded *a*/*b*.

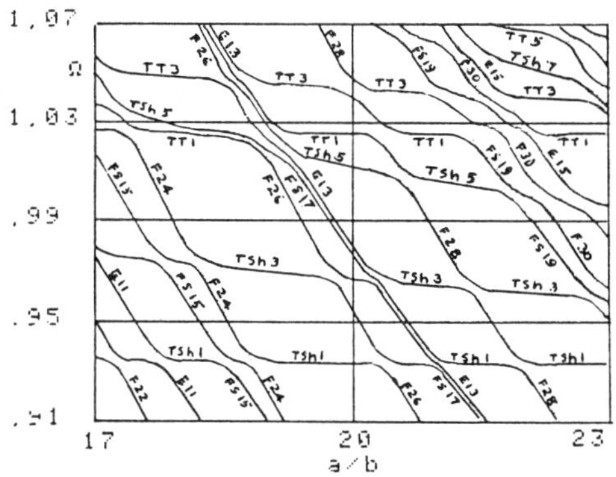

Fig. 5. Frequency spectrum of strip with free edges, sketched, without computation, on grid of branches illustrated in Fig. 4.

In both Figs. 4 and 5, it may be seen that there is an interval of a/b, along the branch *TSh*-1 between *F*-24 and *F*-26, where there are no *E* and *FS* overtone branches. Halfway into that interval, at about $a/b = 19.6$, would be a favorable dimensional ratio for a minimum of activity dip resulting from coupling of thickness-shear with flexure, extension and face-shear. Similar ratios occur at $a/b = 32.2, 44.8, 62.1, 71.5\ldots$; again for $R = 0.01$ and the SC-cut.

REFERENCES

1. J. A. Kusters, The SC-cut crystal—an overview. 1981 *Ultrasonics Symp. Proc.* **1**, 402–409. IEEE, New York (1981).
2. H. F. Tiersten, *Linear Piezoelectric Plate Vibrations*. Plenum Press, New York (1969).
3. A. Ballato, *Doubly Rotated Thickness Mode Plate Vibrators*, *Physical Acoustics* (Edited by W. P. Mason and R. N. Thurston), Vol. XIII, Chap. 5, pp. 115–183. Academic Press, New York (1977).
4. P. C. Y. Lee and K.-M. Wu, Coupled resonances in doubly rotated crystal plates. *Presented at* 33*rd Annual Frequency Control Symp.* U.S. Army Electronics Research and Development Command, Fort Monmouth, New Jersey (1979).
5. R. D. Mindlin, Forced thickness-shear and flexural vibrations of piezoelectric crystal plates. *J. Appl. Phys.* **23**, 83–88 (1952).
6. H. F. Tiersten and R. D. Mindlin, Forced vibrations of piezoelectric crystal plates. *Quart. Appl. Math.* **20**, 107–119 (1962).
7. R. D. Mindlin, High frequency vibrations of piezoelectric crystal plates. *Int. J. Solids Structures* **8**, 895–906 (1972).
8. R. Bechmann, Elastic and piezoelectric constants of alpha-quartz. *Physical Rev.* **110**, 1060–1061 (1958).

APPENDIX

Formulas for material constants of doubly rotated quartz plates

α-Quartz has an axis of three-fold symmetry, say X_3, and three axes of two-fold symmetry, one of which is designated as X_1 in a right-handed, rectangular coordinate system X_i; $i = 1, 2, 3$. Rotate the X_i system a positive angle ϕ about X_3 and a positive angle θ about X_1 to a new orientation x_i. The direction cosines, l_{ij}, of the x_i axes with respect to the X_i axes:

$$\begin{array}{cccc} & X_1 & X_2 & X_3 \\ x_1 & l_{11} & l_{12} & l_{13} \\ x_2 & l_{21} & l_{22} & l_{23} \\ x_3 & l_{31} & l_{32} & l_{33} \end{array} \qquad \text{(a)}$$

are

$$\begin{array}{lll} l_{11} = \cos\phi & l_{12} = \sin\phi & l_{13} = 0 \\ l_{21} = -\sin\phi\cos\theta & l_{22} = \cos\phi\cos\theta & l_{23} = \sin\theta \\ l_{31} = \sin\phi\sin\theta & l_{32} = -\cos\phi\sin\theta & l_{33} = \cos\theta. \end{array} \qquad \text{(b)}$$

The quartz plate is cut with its thickness parallel to x_2 and a pair of edges parallel to x_1.

The elastic, c_{rstu}, piezoelectric, e_{rst}, and dielectric, ϵ_{rs}, constants, referred to the rotated axes x_i, expressed in terms of the constants c^0_{ijk}, e^0_{ijk}, ϵ^0_{ij}, referred to the axes X_i, are

$$\begin{aligned} c_{rstu} &= c^0_{ijkl} l_{ri} l_{sj} l_{tk} l_{ul}, \\ e_{rst} &= e^0_{ijk} l_{ri} l_{sj} l_{tk}, \\ \mathrm{e}_{rs} &= \epsilon^0_{ij} \mathrm{l}_{ri} \mathrm{l}_{sj}, \end{aligned} \qquad \text{(c)}$$

summed over repeated indices i, j, k, l. For α-quartz,

$$\begin{aligned} &c^0_{22} = c^0_{11}, \quad c^0_{23} = c^0_{13}, \quad c^0_{24} = -c^0_{14} = -c^0_{56}, \quad c^0_{55} = c^0_{44}, \quad c^0_{66} = (c^0_{11} - c^0_{12})/2, \\ &c^0_{15} = c^0_{16} = c^0_{25} = c^0_{26} = c^0_{34} = c^0_{35} = c^0_{36} = c^0_{45} = c^0_{46} = 0, \\ &e^0_{11} = -e^0_{12} = -e^0_{26}, \quad e^0_{14} = -e^0_{25}, \\ &e^0_{13} = e^0_{15} = e^0_{16} = e^0_{21} = e^0_{22} = e^0_{23} = e^0_{24} = e^0_{31} = e^0_{32} = e^0_{33} = e^0_{34} = e^0_{35} = e^0_{36} = 0, \\ &\epsilon^0_{11} = \epsilon^0_{22}, \quad \epsilon^0_{12} = \epsilon^0_{23} = \epsilon^0_{31} = 0. \end{aligned} \qquad \text{(d)}$$

Thus, we have, from (c) and (d),

$$\begin{aligned} c_{rstu} = {} & c^0_{11}\left[l_{r1}l_{s1}l_{t1}l_{u1} + l_{r2}l_{s2}l_{t2}l_{u2} + \frac{1}{2}(l_{r1}l_{s2} + l_{r2}l_{s1})(l_{t1}l_{u2} + l_{t2}l_{u1})\right] \\ & + c^0_{12}\left[l_{r1}l_{s1}l_{t2}l_{u2} + l_{r2}l_{s2}l_{t1}l_{u1} - \frac{1}{2}(l_{r1}l_{s2} + l_{r2}l_{s1})(l_{t1}l_{u2} + l_{t2}l_{u1}\right] \\ & + c^0_{13}(l_{r1}l_{s1}l_{t3}l_{u3} + l_{r3}l_{s3}l_{t1}l_{u1} + l_{r2}l_{s2}l_{t3}l_{u3} + l_{r3}l_{s3}l_{t2}l_{u2}) \\ & + c^0_{14}[(l_{r1}l_{s1} - l_{r2}l_{s2})(l_{t2}l_{u3} + l_{t3}l_{u2}) + (l_{t1}l_{u1} - l_{t2}l_{u2})(l_{r2}l_{s3} + l_{r3}l_{s2}) \\ & \qquad + (l_{r3}l_{s1} + l_{r1}l_{s3})(l_{t1}l_{u2} + l_{t2}l_{u1}) + (l_{r1}l_{s2} + l_{r2}l_{s1})(l_{t3}l_{u1} + l_{t1}l_{u3})] \\ & + c^0_{33}(l_r l_{s3} + l_{t3}l_{u3}) \\ & + c^0_{44}[(l_{r2}l_{s3} + l_{r3}l_{s2})(l_{t2}l_{u3} + l_{t3}l_{u2}) + (l_{r3}l_{s1} + l_{r1}l_{s3})(l_{t3}l_{u1} + l_{t1}l_{u3})], \end{aligned}$$

$$\begin{aligned} e_{rst} = {} & e^0_{11}(l_{r1}l_{s1}l_{t1} - l_{r1}l_{s2}l_{t2} - l_{r2}l_{s1}l_{t2} - l_{r2}l_{s2}l_{t1}) \\ & + e^0_{14}(l_{r1}l_{s2}l_{t3} + l_{r1}l_{s3}l_{t2} - l_{r2}l_{s3}l_{t1} - l_{r2}l_{s1}l_{t3}), \\ \epsilon_{rs} = {} & \epsilon^0_{11}(l_{r1}l_{s1} + l_{r2}l_{s2}) + \epsilon^0_{33}l_{43}l_{s3}. \end{aligned}$$

Bechmann's values for α-quartz are [8]

$$\begin{array}{llll} c^0_{11} = 86.74 & c^0_{14} = -17.91 & e^0_{11} = 0.171 & e^0_{14} = -0.0406 \\ c^0_{12} = 6.98 & c^0_{33} = 107.2 & \epsilon^0_{11} = 39.21 & \epsilon^0_{33} = 41.03 \\ c^0_{13} = 11.91 & c^0_{44} = 57.94 & & \end{array}$$

in units of $10^9 N/m^2$, C/m^2 and $10^{-12} F/m$ for the c^0_{pq}, e^0_{ip} and ϵ^0_{ij}, respectively.

For the SC-cut quartz plate ($\phi = 21.93°$, $\theta = 33.93°$), the constants are listed in Table 1.

Table 1. Doubly-rotated quartz plate

```
        Φ= 21.93°   θ= 33.93°

      CONSTANT                    UNITS
Cpq=ELASTIC STIFFNESS:  10^9 N/m²
Eip=PIEZOELECTRIC CONSTANT:  C/m²
Kij=DIELECTRIC CONST:10^-12 F/m


C11= 86.7400000001
C12= 1.71298689814
C13= 17.177013102
C14=-.484711423
C15=-13.5533039091
C16=-9.1177467817
C22= 115.702789532
C23=-3.8751539075
C24= 8.8720683789
C25= .8851190534
C26= 18.8309418361
C33= 109.807518283
C34= 3.3715867534
C35= 12.6681848557
C36=-9.71319505441
C44= 42.1548460928
C45=-9.71319505431
C46= .88511905334
C55= 59.1161712855
C56= 5.59622323258
C66= 38.7038287147

E11= 7.01240600627E-2
E12=-8.58822822078E-2
E13= 1.57582221448E-2
E14= .017175794557
E15= 8.70538637456E-2
E16=-.129403404157
E21=-.129403404157
E22= 8.90858979159E-2
E23= 4.03175062407E-2
E24=-5.99309706679E-2
E25= 6.04272390907E-2
E26=-2.94726979707E-2
E31= 8.70538637458E-2
E32=-5.99309706681E-2
E33=-2.71228930779E-2
E34= 4.03175062408E-2
E35=-4.06513620922E-2
E36= 1.98272390908E-2

K11= 39.21
K22= 39.7770473805
K23= .8429018334
K33= 40.4629526195
K12=K13=0
```

Int. J. Solids Structures Vol. 21, No. 6, pp. 597–607, 1985
Printed in the U.S.A.

0020-7683/85 \$3.00+.00
1985 Pergamon Press Ltd.

VIBRATIONS OF DOUBLY-ROTATED-CUT QUARTZ PLATES WITH MONOCLINIC SYMMETRY†

R. D. MINDLIN
P.O. Box 385, Grantham, NH 03753, U.S.A.

(*Received* 15 *December* 1983)

Abstract—For certain doubly rotated cuts of quartz, the elastic stiffness constants have the same symmetry and absolute values as those for certain rotated-Y-cuts; but 4 of the 13 constants have signs reversed. Mathematical solutions of corresponding problems for the two types of cut have the same form but the numerical results may be the same or different according as the constants with changed signs enter the solution as even or odd powers or products. Examples of both are exhibited.

1. EQUATIONS FOR DOUBLY-ROTATED-CUT QUARTZ PLATES

Alpha-quartz has an axis of three-fold symmetry, say X_3, and three axes of two-fold symmetry one of which is designated as X_1 in a right-handed, rectangular coordinate system X_i, $i = 1, 2, 3$. A doubly rotated set of axes is obtained by rotating the X_i system a positive angle θ about X_1 and a positive angle ϕ about X_3 to a new orientation x_i. The direction cosines l_{ij}, of the x_i axes with respect to the X_i axes are

$$\begin{array}{cccc} & X_1 & X_2 & X_3 \\ x_1 & l_{11} = \cos\phi & l_{12} = \sin\phi & l_{13} = 0 \\ x_2 & l_{21} = -\sin\phi\cos\theta & l_{22} = \cos\phi\cos\theta & l_{23} = \sin\theta \\ x_3 & l_{31} = \sin\phi\sin\theta & l_{32} = -\cos\phi\sin\theta & l_{33} = \cos\theta. \end{array} \tag{1}$$

Rotated-Y-cut and doubly-rotated-cut plates are cut with faces perpendicular to x_2, as shown in Fig. 1.

The elastic stiffness constants c_{rstu}, $r, s, t, u = 1, 2, 3$ (or, in the reduced indicial notation: c_{pq}, $p, q = 1 \ldots 6$), referred to the rotated axes x_i, are expressed in terms of the constants c^0_{ijkl}, referred to the X_i, by

$$c_{rstu} = c^0_{ijkl} l_{ri} l_{sj} l_{tk} l_{ul} \tag{2}$$

summed over $i, j, k, l = 1, 2, 3$.

Of the 21 possible constants $c^0_{pq}(= c^0_{qp})$, referred to the X_i coordinates, only six are independent inasmuch as, for α-quartz[1],

$$c^0_{22} = c^0_{11},\ c^0_{55} = c^0_{44},\ c^0_{23} = c^0_{13},\ c^0_{14} = c^0_{56} = -c^0_{24},\ c^0_{66} = \tfrac{1}{2}(c^0_{11} - c^0_{12}),$$
$$c^0_{15} = c^0_{25} = c^0_{35} = c^0_{45} = c^0_{16} = c^0_{26} = c^0_{36} = c^0_{46} = c^0_{34} = 0. \tag{3}$$

The values of the remaining c^0_{pq}, as given by Bechmann[2], are

$$\begin{array}{ll} c^0_{11} = 86.74 & c^0_{12} = 6.98 \\ c^0_{33} = 107.2 & c^0_{13} = 11.91 \\ c^0_{44} = 57.94 & c^0_{14} = -17.91 \end{array} \tag{4}$$

in units of 10^{10} dyn/cm^2 or 10^9 N/m^2.

† Investigation supported by the Office of Naval Research.

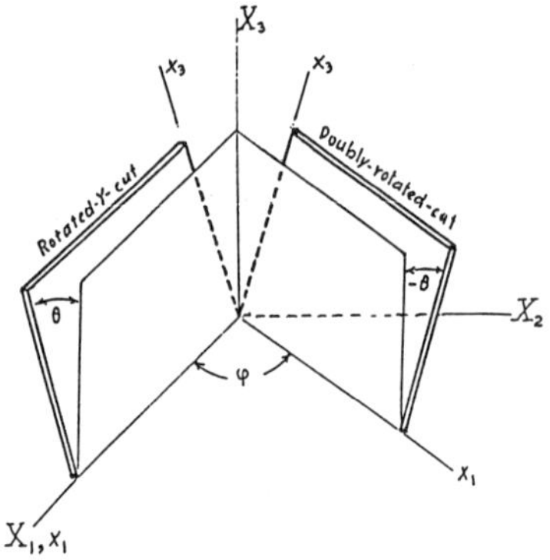

Fig. 1. Rotated-Y-cut and doubly-rotated-cut quartz plates. X_1 and X_3 are digonal and trigonal axes of symmetry, respectively.

From (2) and (3) we have

$$
\begin{aligned}
c_{rstu} = {} & c_{11}^0\,[l_{r1}l_{s1}l_{t1}l_{u1} + l_{r2}l_{s2}l_{t2}l_{u2} + \tfrac{1}{2}(l_{r1}l_{s2} + l_{r2}l_{s1})(l_{t1}l_{u2} + l_{t2}l_{u1})] \\
& + c_{33}^0 l_{r3}l_{s3}l_{t3}l_{u3} \\
& + c_{44}^0[(l_{r2}l_{s3} + l_{r3}l_{s2})(l_{t2}l_{u3} + l_{t3}l_{u2}) + (l_{r3}l_{s1} + l_{r1}l_{s3})(l_{t3}l_{u1} + l_{t1}l_{u3})] \\
& + c_{12}^0\,[l_{r1}l_{s1}l_{t2}l_{u2} + l_{r2}l_{s2}l_{t1}l_{u1} - \tfrac{1}{2}(l_{r1}l_{s2} + l_{r2}l_{s1})(l_{t1}l_{u2} + l_{t2}l_{u1})] \\
& + c_{13}^0[l_{r3}l_{s3}(l_{t1}l_{u1} + l_{t2}l_{u2}) + l_{t3}l_{u3}(l_{r1}l_{s1} + l_{r2}l_{s2})] \\
& + c_{14}^0[(l_{r2}l_{s3} + l_{r3}l_{2})(l_{t1}l_{u1} - l_{t2}l_{u2}) + (l_{t2}l_{u3} + l_{t3}l_{u2})(l_{r1}l_{s1} - l_{r2}l_{s2}) \\
& + (l_{r1}l_{s2} + l_{r2}l_{s1})(l_{t3}l_{u1} + l_{t1}l_{u3}) + (l_{t1}l_{u2} + l_{t2}l_{u1})(l_{r3}l_{s1} + l_{r1}l_{s3})].
\end{aligned}
\tag{5}
$$

Finally, upon substituting (1) in (5), we find

$$c_{11} = c_{11}^0$$

$$c_{12} = c_{12}^0 \cos^2\theta + c_{13}^0 \sin^2\theta + c_{14}^0 \sin 2\theta \cos 3\phi$$

$$c_{13} = c_{12}^0 \sin^2\theta + c_{13}^0 \cos^2\theta - c_{14}^0 \sin 2\theta \cos 3\phi$$

$$c_{14} = (c_{13}^0 - c_{12}^0) \sin\theta \cos\theta + c_{14}^0 \cos 2\theta \cos 3\phi$$

$$c_{15} = c_{14}^0 \cos\theta \sin 3\phi$$

$$c_{16} = c_{14}^0 \sin\theta \sin 3\phi$$

$$c_{22} = c_{11}^0 \cos^4\theta + c_{33}^0 \sin^4\theta + (c_{44}^0 + \tfrac{1}{2}c_{13}^0) \sin^2 2\theta + 4\, c_{14}^0 \sin\theta \cos^3\theta \cos 3\phi$$

$$c_{23} = \tfrac{1}{4}(c_{11}^0 + c_{33}^0 - 2\, c_{13}^0 - 4\, c_{44}^0) \sin^2 2\theta + c_{13}^0 + \tfrac{1}{2}c_{14}^0 \sin 4\theta \cos 3\phi$$

$$
\begin{aligned}
c_{24} = {} & - c_{11}^0 \sin\theta \cos^3\theta + c_{33}^0 \sin^3\theta \cos\theta + \tfrac{1}{2}(c_{44}^0 + \tfrac{1}{2}c_{13}^0) \sin 4\theta \\
& - c_{14}^0 \cos\theta \cos 3\theta \cos 3\phi
\end{aligned}
$$

$$c_{25} = c_{14}^0 (3 \sin^2\theta - 1) \cos\theta \sin 3\phi$$

$$c_{26} = 3\, c_{14}^0 \sin\theta \cos^2\theta \sin 3\phi$$

$$c_{33} = c_{11}^0 \sin^4\theta + c_{33}^0 \cos^4\theta + (c_{44}^0 + \tfrac{1}{2}c_{13}^0) \sin^2 2\theta + 4\, c_{14}^0 \sin^3\theta \cos\theta \cos 3\phi$$

$$
\begin{aligned}
c_{34} = {} & - c_{11}^0 \sin^3\theta \cos\theta + c_{33}^0 \sin\theta \cos^3\theta - \tfrac{1}{2}(c_{44}^0 + \tfrac{1}{2}c_{13}^0) \sin 4\theta \\
& - c_{14}^0 \sin\theta \sin 3\theta \cos 3\phi
\end{aligned}
$$

$$
\begin{aligned}
c_{35} &= -3\,c_{14}^0 \sin^2\theta \cos\theta \sin 3\phi \\
c_{36} &= c_{14}^0\,(3\cos^2\theta - 1)\sin\theta \sin 3\phi \\
c_{44} &= \tfrac{1}{4}(c_{11}^0 + c_{33}^0 - 2\,c_{13}^0 - 4\,c_{44}^0)\sin^2 2\theta + c_{44}^0 + \tfrac{1}{2}c_{14}^0 \sin 4\theta \cos 3\phi \\
c_{45} &= c_{14}^0\,(3\cos^2\theta - 1)\sin\theta \sin 3\phi \\
c_{46} &= c_{14}^0\,(3\sin^2\theta - 1)\cos\theta \sin 3\phi \\
c_{55} &= \tfrac{1}{2}(c_{11}^0 - c_{12}^0)\sin^2\theta + c_{44}^0\cos^2\theta - c_{14}^0 \sin 2\theta \cos 3\phi \\
c_{56} &= -\tfrac{1}{4}(c_{11}^0 - c_{12}^0 - 2\,c_{44}^0)\sin 2\theta + c_{14}^0\cos 2\theta \cos 3\phi \\
c_{66} &= \tfrac{1}{2}(c_{11}^0 - c_{12}^0)\cos^2\theta + c_{44}^0\sin^2\theta + c_{14}^0 \sin 2\theta \cos 3\phi
\end{aligned}
\qquad (6)
$$

These are the c_{pq} which appear in the stress–strain relations referred to the x_i:

$$
T_{ij} = c_{ijkl}S_{kl} \quad \text{or} \quad T_p = c_{pq}S_q \qquad (7)
$$

in which the strains, S_{ij} or S_p, in terms of displacements, u_i, are

$$
\begin{aligned}
S_{11} &= S_1 = u_{1,1} \qquad & 2\,S_{23} &= S_4 = u_{3,2} + u_{2,3} \\
S_{22} &= S_2 = u_{2,2} \qquad & 2\,S_{31} &= S_5 = u_{1,3} + u_{3,1} \\
S_{33} &= S_3 = u_{3,3} \qquad & 2\,S_{12} &= S_6 = u_{2,1} + u_{1,2}.
\end{aligned}
\qquad (8)
$$

Upon substituting (8) in (7) and the result in the stress-equations of motion:

$$
T_{ij,i} = \rho\ddot{u}_j, \qquad (9)
$$

we find the displacement-equations of motion:

$$
D_{ij}u_j = \rho\ddot{u}_j \qquad (10)
$$

or

$$
\begin{aligned}
D_{11}u_1 + D_{12}u_2 + D_{13}u_3 &= \rho\ddot{u}_1, \\
D_{21}u_1 + D_{22}u_2 + D_{23}u_3 &= \rho\ddot{u}_2, \\
D_{31}u_1 + D_{32}u_2 + D_{33}u_3 &= \rho\ddot{u}_3,
\end{aligned}
\qquad (11)
$$

in which the $D_{ij}(= D_{ji})$ are the differential operators

$$
\begin{aligned}
D_{11} &= c_{11}\partial_1^2 + c_{66}\partial_2^2 + c_{55}\partial_3^2 + 2\,c_{56}\partial_2\partial_3 + 2\,c_{15}\partial_3\partial_1 + 2\,c_{16}\partial_1\partial_2, \\
D_{22} &= c_{22}\partial_2^2 + c_{44}\partial_3^2 + c_{66}\partial_1 + 2\,c_{64}\partial_3\partial_1 + 2\,c_{26}\partial_1\partial_2 + 2\,c_{24}\partial_2\partial_3, \\
D_{33} &= c_{33}\partial_3^2 + c_{55}\partial_1^2 + c_{44}\partial_2^2 + 2\,c_{45}\partial_1\partial_2 + 2\,c_{34}\partial_2\partial_3 + 2\,c_{35}\partial_3\partial_1, \\
D_{23} &= c_{56}\partial_1^2 + c_{24}\partial_2^2 + c_{34}\partial_3^2 + (c_{23} + c_{44})\partial_2\partial_3 + (c_{36} + c_{45})\partial_3\partial_1 + (c_{25} + c_{64})\partial_1\partial_2, \\
D_{31} &= c_{64}\partial_2^2 + c_{35}\partial_3^2 + c_{15}\partial_1^2 + (c_{31} + c_{55})\partial_3\partial_1 + (c_{14} + c_{56})\partial_1\partial_2 + (c_{36} + c_{45})\partial_2\partial_3, \\
D_{12} &= c_{45}\partial_3^2 + c_{16}\partial_1^2 + c_{26}\partial_2^2 + (c_{12} + c_{66})\partial_1\partial_2 + (c_{25} + c_{64})\partial_2\partial_3 + (c_{14} + c_{56})\partial_3\partial_1,
\end{aligned}
\qquad (12)
$$

where $\partial_i\partial_j = \partial^2/\partial x_i\partial x_j$, $\partial_i^2 = \partial^2/\partial x_i^2$.

For traction-free planes parallel to the coordinate planes, it is required that on x_1 = constant:

$$\begin{aligned}
T_{11} &= T_1 = c_{11}S_1 + c_{12}S_2 + c_{13}S_3 + c_{14}S_4 + c_{15}S_5 + c_{16}S_6 = 0,\\
T_{12} &= T_6 = c_{61}S_1 + c_{62}S_2 + c_{63}S_3 + c_{64}S_4 + c_{65}S_5 + c_{66}S_6 = 0,\\
T_{13} &= T_5 = c_{51}S_1 + c_{52}S_2 + c_{53}S_3 + c_{54}S_4 + c_{55}S_5 + c_{56}S_6 = 0;
\end{aligned} \tag{13}$$

on x_2 = constant:

$$\begin{aligned}
T_{21} &= T_6 = c_{61}S_1 + c_{62}S_2 + c_{63}S_3 + c_{64}S_4 + c_{65}S_5 + c_{66}S_6 = 0,\\
T_{22} &= T_2 = c_{21}S_1 + c_{22}S_2 + c_{23}S_3 + c_{24}S_4 + c_{25}S_5 + c_{26}S_6 = 0,\\
T_{23} &= T_4 = c_{41}S_1 + c_{42}S_2 + c_{43}S_3 + c_{44}S_4 + c_{45}S_5 + c_{46}S_6 = 0;
\end{aligned} \tag{14}$$

on x_3 = constant:

$$\begin{aligned}
T_{31} &= T_5 = c_{51}S_1 + c_{52}S_2 + c_{53}S_3 + c_{54}S_5 + c_{55}S_5 + c_{56}S_6 = 0,\\
T_{32} &= T_4 = c_{41}S_1 + c_{42}S_2 + c_{43}S_3 + c_{44}S_4 + c_{45}S_5 + c_{46}S_6 = 0,\\
T_{33} &= T_3 = c_{31}S_1 + c_{32}S_2 + c_{33}S_3 + c_{34}S_4 + c_{35}S_5 + c_{36}S_6 = 0.
\end{aligned} \tag{15}$$

2. ROTATED-Y-CUTS VS 60° DOUBLY-ROTATED-CUTS

Case A. If $\phi = 0$, $\theta \neq 0$ (the rotated-Y-cuts) then $\sin 3\phi = 0$, $\cos 3\phi = 1$ and, from (6),

$$c_{15} = c_{16} = c_{25} = c_{26} = c_{35} = c_{36} = c_{45} = c_{46} = 0. \tag{16}$$

The remaining 13 constants are those for monoclinic symmetry with x_1 the digonal axis.

Case B. If $\phi = 60°$, $\theta \neq 0$ (doubly-rotated cuts), then $\sin 3\phi = 0$, $\cos 3\phi = -1$ and (16) again holds so that the symmetry is the same as for rotated-Y-cuts. Even if θ is the same in A and B, all the surviving constants in B (except for c_{11} which remains fixed) are different from the corresponding ones in A as the last term in each c_{pq} has its sign reversed. However, if θ in B is the negative of θ in A, nine of the constants are the same for the two cuts and the remaining four have the same absolute values in A and B but are of opposite sign.

To summarize the properties of the two sets of constants $c_{pq}(\phi, \theta)$:

$$\begin{aligned}
c_{pq}(0, \theta) &= c_{pq}(60°, -\theta) \text{ for } pq = 11, 12, 13, 22, 23, 33, 44, 55, 66,\\
c_{pq}(0, \theta) &= -c_{pq}(60°, -\theta) \text{ for } pq = 14, 24, 34, 56.
\end{aligned} \tag{17}$$

The displacement equations of motion reduce to

$$\begin{aligned}
&(c_{11}\partial_1^2 + c_{66}\partial_2^2 + c_{55}\partial_3^2 + 2\, c_{56}\partial_2\partial_3)u_1 + [(c_{12} + c_{66})\partial_1\partial_2 + (c_{14} + c_{56})\partial_3\partial_1]u_2\\
&\qquad + [(c_{13} + c_{55})\partial_3\partial_1 + (c_{14} + c_{56})\partial_1\partial_2]u_3 = \rho\ddot{u}_1,\\
&[(c_{12} + c_{66})\partial_1\partial_2 + (c_{14} + c_{56})\partial_3\partial_1]u_1 + (c_{22}\partial_2^2 + c_{44}\partial_3^2 + c_{66}\partial_1^2 + 2\, c_{24}\partial_2\partial_3)u_2\\
&\qquad + [c_{56}\partial_1^2 + c_{24}\partial_2^2 + c_{34}\partial_3^2 + (c_{23} + c_{44})\partial_2\partial_3]u_3 = \rho\ddot{u}_2,\\
&[(c_{12} + c_{66})\partial_1\partial_2 + (c_{14} + c_{56})\partial_3\partial_1]u_1 + [c_{56}\partial_1^2 + c_{24}\partial_2^2 + c_{34}\partial_3^2 + (c_{23} + c_{44})\partial_2\partial_3]u_2\\
&\qquad + (c_{33}\partial_3^2 + c_{55}\partial_1^2 + c_{44}\partial_2^2 + 2\, c_{34}\partial_2\partial_3)u_3 = \rho\ddot{u}_3,
\end{aligned} \tag{18}$$

and the conditions (13), (14), (15) for traction-free boundaries reduce to:

on x_1 = constant:

$$
\begin{aligned}
T_{11} &= T_1 = c_{11}u_{1,1} + c_{12}u_{2,2} + c_{13}u_{3,3} + c_{14}(u_{3,2} + u_{2,3}) = 0,\\
T_{12} &= T_6 = c_{56}(u_{1,3} + u_{3,1}) + c_{66}(u_{2,1} + u_{1,2}) = 0, \qquad (19)\\
T_{13} &= T_5 = c_{55}(u_{1,3} + u_{3,1}) + c_{56}(u_{2,1} + u_{1,2}) = 0;
\end{aligned}
$$

on x_2 = constant:

$$
\begin{aligned}
T_{21} &= T_6 = c_{56}(u_{1,3} + u_{3,1}) + c_{66}(u_{2,1} + u_{1,2}) = 0,\\
T_{22} &= T_2 = c_{12}u_{1,1} + c_{22}u_{2,2} + c_{23}u_{3,3} + c_{24}(u_{3,2} + u_{2,3}) = 0, \qquad (20)\\
T_{23} &= T_4 = c_{14}u_{1,1} + c_{24}u_{2,2} + c_{34}u_{3,3} + c_{44}(u_{3,2} + u_{2,3}) = 0;
\end{aligned}
$$

on x_3 = constant:

$$
\begin{aligned}
T_{31} &= T_5 = c_{55}(u_{1,3} + u_{3,1}) + c_{56}(u_{2,1} + u_{1,2}) = 0,\\
T_{32} &= T_4 = c_{14}u_{1,1} + c_{24}u_{2,2} + c_{34}u_{3,3} + c_{44}(u_{3,2} + u_{2,3}) = 0, \qquad (21)\\
T_{33} &= T_3 = c_{13}u_{1,1} + c_{23}u_{2,2} + c_{33}u_{3,3} + c_{34}(u_{3,2} + u_{2,3}) = 0.
\end{aligned}
$$

Any solution of the equations of motion (and boundary conditions, if any) referred to rotated axes with $\phi = 0$, $\theta = \theta'$ is the same solution, at least in form, referred to axes with $\phi = 60°$, $\theta = -\theta'$. Whether or not the solutions are the same numerically depends on the occurrence of c_{14}, c_{24}, c_{34}, c_{56} as even or odd powers or products in the resulting formulas.

In the following sections, we review solutions obtained previously for $c_{pq}(0, \theta)$ and determine if the transition to $c_{pq}(60°, -\theta)$ changes the numerical results.

3. PLANE WAVES IN A PLATE

In a plate with faces at $x_2 = \pm b$, we consider waves propagating in the direction of the two-fold axis of symmetry x_1:

$$
\begin{aligned}
u_1 &= A_1 \sin(\xi_2 x_2 + \xi_3 x_3) \sin(\xi_1 x_1 - \omega t),\\
u_2 &= -A_2 \cos(\xi_2 x_2 + \xi_3 x_3) \cos(\xi_1 x_1 - \omega t), \qquad (22)\\
u_3 &= -A_3 \cos(\xi_2 x_2 + \xi_3 x_3) \cos(\xi_1 x_1 - \omega t).
\end{aligned}
$$

Upon substituting (22) in (19) and setting the determinant of the coefficients of the A_j equal to zero, we find the equation

$$
| \lambda_{ij} - \delta_{ij} V^2 | = 0, \qquad \lambda_{ij} = \lambda_{ji}, \qquad (23)
$$

in which δ_{ij} is the Kronecker delta,

$$
\begin{aligned}
\lambda_{11} &= \bar{c}_{11} + \beta^2 + \bar{c}_{55}\Gamma^2 + 2\,\bar{c}_{56}\beta\Gamma, & \lambda_{23} &= \bar{c}_{56} + \bar{c}_{24}\beta^2 + \bar{c}_{34}\Gamma^2 + (\bar{c}_{23} + \bar{c}_{44})\beta\Gamma,\\
\lambda_{22} &= 1 + \bar{c}_{22}\beta^2 + \bar{c}_{44}\Gamma^2 + 2\,\bar{c}_{24}\beta\Gamma, & \lambda_{31} &= (\bar{c}_{14} + \bar{c}_{56})\beta + (\bar{c}_{13} + \bar{c}_{55})\Gamma, \qquad (24)\\
\lambda_{33} &= \bar{c}_{55} + \bar{c}_{44}\beta^2 + \bar{c}_{33}\Gamma^2 + 2\,\bar{c}_{34}\beta\Gamma, & \lambda_{12} &= (1 + \bar{c}_{12})\beta + (\bar{c}_{14} + \bar{c}_{56})\Gamma.
\end{aligned}
$$

$$
\bar{c}_{pq} = c_{pq}/c_{66} \qquad \beta = \xi_2/\xi_1, \qquad \Gamma = \xi_3/\xi_1, \qquad V^2 = \Omega^2/\bar{\xi}_1^2 = \rho\omega^2/c_{66}\xi_1^2. \qquad (25)
$$

In (25), β and Γ are the ratios of the wave length along x_1 to the wave lengths along x_2 and x_3, respectively; V is the ratio of the velocity to the velocity $v = (c_{66}/\rho)^{1/2}$; $\bar{\xi}_1(= 2\xi_1 b/\pi)$ is the ratio of the thickness, $2b$, of the plate to the half-wave-length along x_1; and Ω is the ratio of the circular frequency ω to the frequency $\pi v/2b$.

For given β and Γ, (23) is a bicubic in the velocity ratio V:

$$V^6 + BV^4 + CV^2 + D = 0, \tag{26}$$

in which

$$\begin{aligned} B &= -(\lambda_{11} + \lambda_{22} + \lambda_{33}), \\ C &= \lambda_{22}\lambda_{33} + \lambda_{33}\lambda_{11} + \lambda_{11}\lambda_{22} - \lambda_{23}^2 - \lambda_{31}^2 - \lambda_{12}^2, \\ D &= \lambda_{11}\lambda_{23}^2 + \lambda_{22}\lambda_{31}^2 + \lambda_{33}\lambda_{12}^2 - \lambda_{11}\lambda_{22}\lambda_{33} - 2\,\lambda_{23}\lambda_{31}\lambda_{12}. \end{aligned} \tag{27}$$

The coefficients of the bicubic are different for $c_{14}, c_{24}, c_{34}, c_{56}$ positive and negative. Hence, for given β and Γ, the roots of (26) yield different sets of velocity ratios V_1, V_2, V_3 for the rotated-Y-cut with $c_{pq}(0, \theta)$ and the doubly-rotated-cut with $c_{pq}(60°, -\theta)$. An example is illustrated in Fig. 2 in which either β is the abscissa and $\Gamma = 10$ or vice versa. In either case, the lowest velocity ratios V_3 exhibit little difference for the two cuts—and this is the branch which would contribute predominantly to the fundamental thickness-shear mode of the plate. However, the differences for the upper velocities are larger, at least for large β and Γ—as much as 13% for $\beta = \Gamma = 10$. These differences survive any boundary conditions that may be applied.

4. EKSTEIN'S SOLUTION

It will be observed, in Fig. 2, that the velocity ratios are the same for Case A and Case B if Γ (or β) is zero. This is the situation for modes with straight crests along x_3 (or x_2). In the case $\Gamma = 0$, λ_{23} and λ_{31} change sign, in the passage from Case A to Case B, but they enter the coefficients of the bicubic (26) only as their product and as squares—resulting in no change in roots. To examine whether this persists after the introduction of free faces of the plate, we consider Ekstein's solution[3] for modes with straight crests along x_3 in a plate with free faces on $x_2 = \pm b$.

With $\Gamma = 0$, and fixed ξ_1 and V, (23) yields three roots β_n^2, $n = 1, 2, 3$. Thus, for steady state vibrations, (22) may be written as

$$\begin{aligned} u_1 &= \sum_{n=1}^{3} A_{1n} \sin \xi_1\beta_n x_2 \sin \xi_1 x_1\, e^{i\omega t}, \\ u_2 &= -\sum_{n=1}^{3} A_{2n} \cos \xi_1\beta_n x_2 \cos \xi_1 x_1\, e^{i\omega t}, \\ u_3 &= -\sum_{n=1}^{3} A_{3n} \cos \xi_1\beta_n x_2 \cos \xi_1 x_1\, e^{i\omega t}. \end{aligned} \tag{28}$$

Then the boundary conditions (20):

$$T_{2j} = 0, \quad j = 1, 2, 3, \quad \text{on} \quad x_2 = \pm b, \tag{29}$$

result in Ekstein's frequency equation; which may be written in the form[4]

$$|\mu_{in}| = 0, \qquad i, n = 1, 2, 3, \tag{30}$$

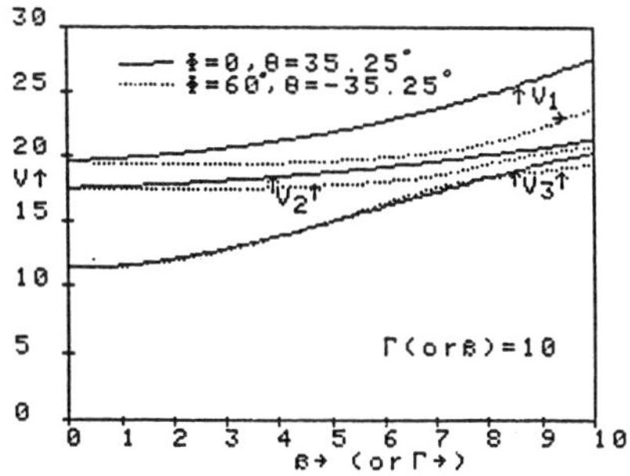

Fig. 2. Comparison of wave velocities for $c_{pq}(0, \theta)$ and $c_{pq}(60°, -\theta)$ as functions of the ratios, β and Γ, of wave lengths in the x_2 and x_3 directions to the wave length in the direction x_1 of the wave normal.

where

$$\begin{aligned} \mu_{1n} &= (\beta_n L_{1n} + L_{2n} + \tilde{c}_{56} L_{3n}) \cot \beta_n \xi_1 b, \\ \mu_{2n} &= \tilde{c}_{12} L_{1n} + \beta_n(\tilde{c}_{22} L_{2n} + \tilde{c}_{24} L_{3n}), \\ \mu_{3n} &= \tilde{c}_{14} L_{1n} + \beta_n(\tilde{c}_{24} L_{2n} + \tilde{c}_{44} L_{3n}), \end{aligned} \tag{31}$$

$$L_{in} = \operatorname{cof}(\lambda_{ni} - \delta_{ni} V_n^2)/\operatorname{cof}(\lambda_{nn} - V_n^2). \tag{32}$$

In the passage from $c_{pq}(0, \theta)$ to $c_{pq}(60°, -\theta)$, the L_{in} and, hence, the μ_{in} (which depend on c_{14}, c_{24}, c_{56}) change sign for subscripts 13, 31, 23, 32 while the remaining terms in (31) and (32) do not change. But those μ_{in} which do change appear only as product pairs in (30) and, hence, the roots of (30) do not change. These roots are usually depicted graphically as a many branched dispersion relation between Ω (as ordinate) and $\bar{\xi}_1$ (as abscissa):

$$\Omega = \Omega(\bar{\xi}_1) \tag{33}$$

as illustrated in [4]. Alternatively, the abscissa could be $1/\bar{\xi}_1$:

$$\hat{\Omega} = \hat{\Omega}(1/\bar{\xi}_1). \tag{34}$$

Suppose the plate has additional bounding planes $x_1 = \pm a$ at which the conditions are uniformly point-mixed, e.g. vanishing u_2, T_{11}, T_{13} corresponding to "simply supported" in the elementary theory of flexural vibrations of plates. For real roots of (30), these conditions are satisfied by $\xi_1 = m\pi/2a$, where m is an even integer; so that, for real roots, the dispersion relation converts to

$$\hat{\Omega} = \hat{\Omega}(a/mb). \tag{35}$$

Elimination of m from the abscissa requires only that each branch of the dispersion relation (35) be replaced by a sequence of branches obtained by multiplication of its abscissa by a sequence of integers. In this way, the branches of the dispersion relation for the infinite plate are converted to the branches of the frequency spectrum, Ω vs a/b, of the "simply supported" plate. As the process does not involve c_{14}, c_{24} and c_{56} anew, the frequency spectrum is not altered by a change of $c_{pq}(0, \theta)$ to $c_{pq}(60°, -\theta)$.

There is no closed solution of the three-dimensional equations for the case of *free* boundaries at $x_1 = \pm a$ and the situation there is not obvious inasmuch as c_{14} and c_{56}

enter into the traction-free conditions

$$T_{11} = T_{12} = T_{13} = 0 \quad \text{on} \quad x_1 = \pm a \tag{36}$$

as may be seen in (19).

5. EFFECT OF FREE EDGES

As a substitute for the unavailable extension of Ekstein's solution of the three-dimensional equations to accommodate a pair of parallel, free edges, there exists a solution of two-dimensional approximate equations[5]. For the case of straight crested flexural waves travelling in the direction of x_1 in a plate with free faces at $x_2 = \pm b$, the three dimensional displacements are approximated by

$$u_1 = x_2\psi_1(x_1)e^{i\omega t}, \qquad u_2 = U_2(x_1)e^{i\omega t}, \qquad u_3 = U_3(x_1)e^{i\omega t} \tag{37}$$

and the differential equations governing them are

$$\begin{aligned} \kappa\, c_{56}U_{3,11} + \kappa^2\, c_{66}(U_{2,11} + \psi_{1,1}) &= -\rho\omega^2 U_2, \\ c_{55}U_{3,11} + \kappa\, c_{56}(U_{2,11} + \psi_{1,1}) &= -\rho\omega^2 U_3, \\ \gamma_{11}\psi_{1,11} - 3b^{-2}[\kappa\, c_{56}U_{3,1} + \kappa^2\, c_{66}(U_{2,1} + \psi_1)] &= -\rho\omega^2\psi_1, \end{aligned} \tag{38}$$

where

$$\kappa^2 = \pi^2/12, \qquad \gamma_{11} = c_{11} - c_{12}^2/c_{22} - (c_{14} - c_{12}c_{24})^2/(c_{44} - c_{24}^2/c_{22}). \tag{39}$$

There is no change of sign of γ_{11} with change of sign of c_{14} and c_{24}; so only c_{56}, in (38), changes sign with the passage from $c_{pq}(0, \theta)$ to $c_{pq}(60°, -\theta)$.

The displacements are taken as

$$U_2 = A_2 b \sin \xi x_1, \qquad U_3 = A_3 b \sin \xi x_1, \qquad \psi_1 = A_4 \cos \xi x_1. \tag{40}$$

Then, from (38),

$$\begin{aligned} (\hat{\xi}^2 - 3\Omega^2)A_2 + \hat{c}_{56}\hat{\xi}^2 A_3 + \hat{\xi}A_4 &= 0, \\ \hat{c}_{56}\hat{\xi}^2 A_2 + (\hat{c}_{55}\hat{\xi}^2 - 3\Omega^2)A_3 + \hat{c}_{56}\hat{\xi}A_4 &= 0, \\ \hat{\xi}A_2 + \hat{c}_{56}\hat{\xi}A_3 + (\hat{\gamma}_{11}\hat{\xi}^2 + 1 - \Omega^2)A_4 &= 0, \end{aligned} \tag{41}$$

where

$$\hat{\xi} = \xi b, \qquad \hat{c}_{55} = c_{55}/\kappa^2\, c_{66}, \qquad \hat{c}_{56} = c_{56}/\kappa\, c_{66}, \qquad \hat{\gamma}_{11} = \gamma_{11}/3\kappa^2\, c_{66}. \tag{42}$$

The determinant of the coefficients of the A_i in (41), set equal to zero, is the equation

$$\begin{aligned} &\hat{\gamma}_{11}(\hat{c}_{55} - \hat{c}_{56}^2)\hat{\xi}^6 - \Omega^2[3\hat{\gamma}_{11}(1 + \hat{c}_{55}) + \hat{c}_{55} - \hat{c}_{56}^2]\hat{\xi}^4 \\ &\quad + 3\Omega^2[\Omega^2 - \hat{c}_{55}(1 - \Omega^2) + 3\hat{\gamma}_{11}\Omega^2 + \hat{c}_{56}^2]\hat{\xi}^2 + 9\Omega^4(1 - \Omega^2) = 0, \end{aligned} \tag{43}$$

which, for a fixed frequency ratio Ω, is a bicubic in $\hat{\xi}^2$ whose roots are independent of change of sign of c_{14}, c_{24}, c_{56}. Thus, as in the three-dimensional case, the dispersion relation does not change with passage from $c_{pq}(0, \theta)$ to $c_{pq}(60°, -\theta)$.

For each Ω, (43) has three roots $\hat{\xi}_n^2$, $n = 1, 2, 3$, and (41) has three sets of amplitude ratios $A_2:A_3:A_4$. Let $\overline{A}_n$, $n = 1, 2, 3$, be the value of A_4 for the nth root $\hat{\xi}_n^2$; and let

$$\alpha_{2n} = \frac{A_2}{A_4} = \hat{\xi}_n(\hat{c}_{56}^2\hat{\xi}_n^2 + 3\,\Omega^2 - \hat{c}_{55}\hat{\xi}_n^2)/\Delta_n,$$

$$\alpha_{3n} = \frac{A_3}{A_4} = 3\,\hat{c}_{56}\hat{\xi}_n\,\Omega^2/\Delta_n, \tag{44}$$

$$\Delta_n = (\hat{\xi}_n^2 - 3\,\Omega^2)(\hat{c}_{55}\hat{\xi}_n^2 - 3\,\Omega^2) - \hat{c}_{56}^2\hat{\xi}_n^4$$

for each root $\hat{\xi}_n^2$. Then (40) may be written as

$$U_2 = b\sum_{n=1}^{3}\overline{A}_n\alpha_{2n}\sin\xi_n x_1,$$

$$U_3 = b\sum_{n=1}^{3}\overline{A}_n\alpha_{3n}\sin\xi_n x_1, \tag{45}$$

$$\psi_1 = \sum_{n=1}^{3}\overline{A}_n\cos\xi_n x_1.$$

The conditions for free edges at $x_1 = \pm a$ are: the horizontal and vertical shears, N_5 and Q_1, and the bending moment, M_1, vanish. Thus, on $x_1 = \pm a$,

$$N_5 = 2b[c_{55}U_{3,1} + \kappa\,c_{56}(U_{2,1} + \psi_1)] = 0,$$

$$Q_1 = 2b\kappa[c_{56}U_{3,1} + \kappa\,c_{66}(U_{2,1} + \psi_1)] = 0, \tag{46}$$

$$M_1 = (2b^3/3)\gamma_{11}\psi_{1,1} = 0.$$

Upon substituting (45) into (46), we obtain

$$\sum_{n=1}^{3}\overline{A}_n\overline{\alpha}_{1n}\cos\xi_n a = 0,$$

$$\sum_{n=1}^{3}\overline{A}_n\overline{\alpha}_{2n}\cos\xi_n a = 0, \tag{47}$$

$$\sum_{n=1}^{3}\overline{A}_n\xi_n\sin\xi_n a = 0,$$

where

$$\overline{\alpha}_{1n} = c_{55}\alpha_{3n}\hat{\xi}_n + \kappa\,c_{56}(\alpha_{2n}\hat{\xi}_n + 1),$$
$$\overline{\alpha}_{2n} = c_{56}\alpha_{3n}\hat{\xi}_n + \kappa\,c_{66}(\alpha_{2n}\hat{\xi}_n + 1). \tag{48}$$

The frequency equation is obtained by setting the determinant of the coefficients of the $\overline{A}_n$ in (47) equal to zero:

$$\tilde{A}_1\tan\xi_1 a + \tilde{A}_2\tan\xi_2 a + \tilde{A}_3\tan\xi_3 a = 0 \tag{49}$$

where

$$\tilde{A}_1 = \hat{\xi}_1(\overline{\alpha}_{12}\overline{\alpha}_{23} - \overline{\alpha}_{22}\overline{\alpha}_{13}),$$

$$\tilde{A}_2 = \hat{\xi}_2(\overline{\alpha}_{13}\overline{\alpha}_{21} - \overline{\alpha}_{23}\overline{\alpha}_{11}), \tag{50}$$

$$\tilde{A}_3 = \hat{\xi}_3(\overline{\alpha}_{11}\overline{\alpha}_{22} - \overline{\alpha}_{21}\overline{\alpha}_{12}).$$

Upon substituting (44) in (48) and the result in (50), the frequency equation (49) becomes

$$\hat{\xi}_1 \Delta_1(\hat{\xi}_2^2 - \hat{\xi}_3^2) \tan \xi_1 a + \hat{\xi}_2 \Delta_2(\hat{\xi}_3^2 - \hat{\xi}_1^2) \tan \xi_2 a + \hat{\xi}_3 \Delta_3(\hat{\xi}_1^2 - \hat{\xi}_2^2) \tan \xi_3 a = 0: \quad (51)$$

an equation which does not change when $c_{pq}(0, \theta)$ is replaced by $c_{pq}(60°, -\theta)$.

6. VIBRATIONS OF A STRIP

An exact solution of the three-dimensional equations exists for coupled thickness-twist and face-shear modes of vibration in a rotated-Y-cut strip with a parallelogrammic cross-section and all four faces free of traction[6]. The displacements are $u_2 = u_3 = 0$ and, omitting a factor $e^{i\omega t}$,

$$\begin{aligned} u_1 = {} & A \sin \xi_2 x_2 \cos \xi_3(\tilde{c}_{56} x_2 - x_3) + B \sin \xi_2 x_2 \sin \xi_3(\tilde{c}_{56} x_2 - x_3) \\ & + C \cos \xi_2 x_2 \cos \xi_3(\tilde{c}_{56} x_2 - x_3) + D \cos \xi_2 x_2 \sin \xi_3(\tilde{c}_{56} x_2 - x_3), \end{aligned} \quad (52)$$

where $\tilde{c}_{56} = c_{56}/c_{66}$, as before in (25).

The equations of motion (18) are satisfied if

$$\rho\omega^2 = c_{66}\xi_2^2 + \gamma_{55}\xi_3^2, \quad \gamma_{55} = c_{55} - c_{56}^2/c_{66}; \quad (53)$$

and the faces at $x_2 = \pm b$ satisfy the traction-free conditions (20) if

$$2\xi_2 b = m\pi \quad (54)$$

where m is an odd integer for solutions A and B and an even integer for solutions C and D.

A pair of planes parallel to the x_1-axis, making dihedral angles α with the $x_1 - x_2$ plane and distant $2c \cos \alpha$ part, as illustrated in Fig. 3, are free of traction if

$$\alpha = \arctan \tilde{c}_{56} \quad (55)$$

and

$$2\xi_3 c = n\pi \quad (56)$$

where n is an even integer for solutions A and C and an odd integer for solutions B and D.

The frequencies are

$$\omega = \frac{m\pi}{2b}\left(\frac{c_{66}}{\rho}\right)^{1/2}\left(1 + \frac{n^2\gamma_{55}b^2}{m^2 c_{66} c^2}\right)^{1/2}. \quad (57)$$

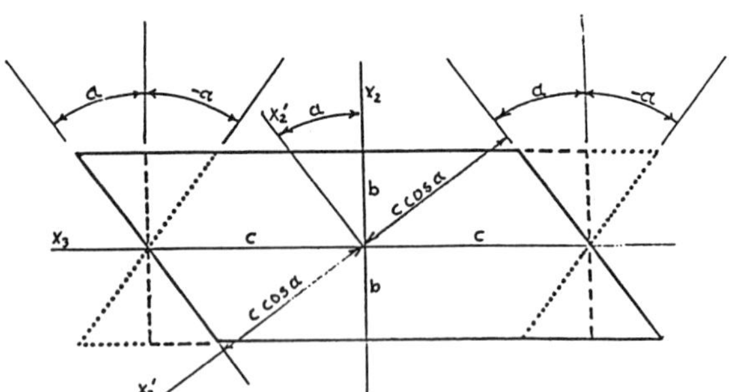

Fig. 3. Cross sections of strips.

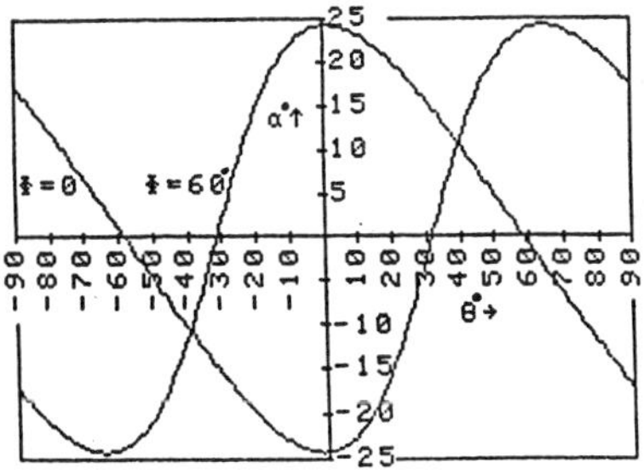

Fig. 4. Variation of dihedral angles, α, between face and edge planes of strip for $c_{pq}(0, \theta)$ and $c_{pq}(60°, -\theta)$ as functions of θ.

When $c_{pq}(0, \theta)$ changes to $c_{pq}(60°, -\theta)$, the frequencies do not change as c_{56} enters as c_{56}^2; but the mode-shape (52) changes and α, in (55), is reversed in sign so that the cross section changes, as illustrated in Fig. 3. The values of $\pm\alpha$ for the full range of values of θ are illustrated in Fig. 4.

REFERENCES

1. W. P. Mason, *Piezoelectric Crystals and their Application to Ultrasonics*, D. Van Nostrand Company, New York (1950).
2. R. Bechmann, Elastic and piezoelectric constants of alpha quartz. *Phys. Rev.* **110,** 1060–1061 (1958).
3. H. Ekstein, High frequency vibrations of thin crystal plates. *Phys. Rev.* **68,** 11–23 (1945).
4. R. K. Kaul and R. D. Mindlin, Frequency spectrum of a monoclinic crystal plate. *J. Acoust. Soc. Am.* **34,** 1902–1910 (1962).
5. R. D. Mindlin and D. C. Gazis, Strong resonances of rectangular AT-cut quartz plates, *Proc. 4th U.S. National Congress of Applied Mechanics* 305–310 (1962).
6. R. D. Mindlin, Thickness-twist vibrations of a quartz strip. *Int. J. Solids Structures* **7,** 1–4 (1971).

Int. J. Solids Structures Vol. 22, No. 12, pp. 1423–1430, 1986
Printed in Great Britain.

0020–7683/86 $3.00+.00
Pergamon Journals Ltd.

ON VIBRATIONS OF RECTANGULAR PARALLELEPIPEDS

R. D. MINDLIN
P.O. Box 385, Grantham, New Hampshire, U.S.A.

(*Received* 25 *September* 1985)

Abstract—An exact, closed-form solution of the three-dimensional equations of elasticity is given for vibrations of rectangular parallelepipeds with traction-free faces.

LAMÉ POTENTIAL FUNCTIONS

The problem of finding exact, closed-form solutions of the three-dimensional equations of linear elasticity, for steady vibrations of homogeneous, isotropic, rectangular parallelepipeds with all faces free of traction, was introduced by Lamé in 1852[1]. He exhibited a solution of that type for a family of modes with locally equivoluminal deformation but, for the case of coupled equivoluminal and dilatational deformations, he proposed a solution in which normal components of traction remain on all six faces. Some progress was made toward a solution with all faces traction free with the discovery, in 1960[2], of an exact, closed-form solution for a family of coupled modes in bars of rectangular cross-section with the four side-faces of the bars traction free. In this paper, that solution is extended to accommodate a family of modes in rectangular parallelepipeds with all six faces free of traction. Whereas the solution for the bars could be expressed in terms of three Lamé potential functions, 18 are required for the parallelepipeds, as follows:

$$
\begin{aligned}
\varphi &= A_1 \sin\xi_2 x \sin\eta_3 y \sin\zeta_3 z + A_2 \sin\xi_3 x \sin\eta_2 y \sin\zeta_3 z \\
&\quad + A_3 \sin\xi_3 x \sin\eta_3 y \sin\zeta_2 z \\
H_x &= B_2 \sin\xi_3 x \cos\eta_1 y \cos\zeta_3 z + C_3 \sin\xi_3 x \cos\eta_3 y \cos\zeta_1 z \\
&\quad + D_1 \sin\xi_1 x \cos\eta_3 y \cos\zeta_3 z + E_1 \sin\xi_2 x \cos\eta_3 y \cos\zeta_2 z \\
&\quad + F_1 \sin\xi_2 x \cos\eta_2 y \cos\zeta_3 z \\
H_y &= B_3 \cos\xi_3 x \sin\eta_3 y \cos\zeta_1 z + C_1 \cos\xi_1 x \sin\eta_3 y \cos\zeta_3 z \\
&\quad + D_2 \cos\xi_3 x \sin\eta_1 y \cos\zeta_3 z + E_2 \cos\xi_2 x \sin\eta_2 y \cos\zeta_3 z \\
&\quad + F_2 \cos\xi_3 x \sin\eta_2 y \cos\zeta_2 z \\
H_z &= B_1 \cos\xi_1 x \cos\eta_3 y \sin\zeta_3 z + C_2 \cos\xi_3 x \cos\eta_1 y \sin\zeta_3 z \\
&\quad + D_3 \cos\xi_3 x \cos\eta_3 y \sin\zeta_1 z + E_3 \cos\xi_3 x \cos\eta_2 y \sin\zeta_2 z \\
&\quad + F_3 \cos\xi_2 x \cos\eta_3 y \sin\zeta_2 z
\end{aligned}
\tag{1}
$$

in which a time-dependent factor $e^{i\omega t}$ is omitted.

The Lamé functions are governed by the equations

$$
\begin{aligned}
v_1^2\nabla^2\varphi + \omega^2\varphi &= 0 \\
v_2^2\nabla^2(H_x, H_y, H_z) + \omega^2(H_x, H_y, H_z) &= 0
\end{aligned}
\tag{2}
$$

where v_1 and v_2 are the velocities of dilatational and equivoluminal waves, respectively, in an infinite medium

$$v_1^2 = (\lambda + 2\mu)/\rho, \qquad v_2^2 = \mu/\rho \tag{3}$$

and λ and μ are Lamé's constants of elasticity, related to Poisson's ratio, ν, in a form to be employed subsequently

$$1 - \mu/(\lambda + 2\mu) = 1/2(1 - \nu) = \kappa^2. \tag{4}$$

Upon substituting eqn (1) in eqn (2) we find the requirements

$$\begin{aligned}
&\xi_2^2 + \eta_3^2 + \zeta_3^2 = \xi_3^2 + \eta_2^2 + \zeta_3^2 = \xi_3^2 + \eta_3^2 + \zeta_2^2 = \omega^2/v_1^2\\
&\xi_3^2 + \eta_1^2 + \zeta_3^2 = \xi_3^2 + \eta_3^2 + \zeta_1^2 = \xi_1^2 + \eta_3^2 + \zeta_3^2\\
&= \xi_2^2 + \eta_3^2 + \zeta_2^2 = \xi_2^2 + \eta_2^2 + \zeta_3^2 = \xi_3^2 + \eta_2^2 + \zeta_2^2 = \omega^2/v_2^2.
\end{aligned}$$

DISPLACEMENTS AND STRESSES

The components of displacement are related to the Lamé functions according to

$$\begin{aligned}
u &= \partial\varphi/\partial x + \partial H_z/\partial y - \partial H_y/\partial z\\
v &= \partial\varphi/\partial y + \partial H_x/\partial z - \partial H_z/\partial x\\
w &= \partial\varphi/\partial z + \partial H_y/\partial x - \partial H_x/\partial y
\end{aligned} \tag{6}$$

whence

$$\begin{aligned}
u = {} & A_1\xi_2 \cos\xi_2 x \sin\eta_3 y \sin\zeta_3 z + (C_1\zeta_3 - B_1\eta_3)\cos\xi_1 x \sin\eta_3 y \sin\zeta_3 z\\
& + A_2\xi_3 \cos\xi_3 x \sin\eta_2 y \sin\zeta_3 z + (D_2\zeta_3 - C_2\eta_1)\cos\xi_3 x \sin\eta_1 y \sin\zeta_3 z\\
& + A_3\xi_3 \cos\xi_3 x \sin\eta_3 y \sin\zeta_2 z + (B_3\zeta_1 - D_3\eta_3)\cos\xi_3 x \sin\eta_3 y \sin\zeta_1 z\\
& + E_2\zeta_3 \cos\xi_2 x \sin\eta_2 y \sin\zeta_3 z + F_2\zeta_2 \cos\xi_3 x \sin\eta_2 y \sin\zeta_2 z\\
& - E_3\eta_2 \cos\xi_3 x \sin\eta_2 y \sin\zeta_2 z - F_3\eta_3 \cos\xi_2 x \sin\eta_3 y \sin\zeta_2 z
\end{aligned}$$

$$\begin{aligned}
v = {} & A_2\eta_2 \sin\xi_3 x \cos\eta_2 y \sin\zeta_3 z + (C_2\xi_3 - B_2\zeta_3)\sin\xi_3 x \cos\eta_1 y \sin\zeta_3 z\\
& + A_3\eta_3 \sin\xi_3 x \cos\eta_3 y \sin\zeta_3 z + (D_3\xi_3 - C_3\zeta_1)\sin\xi_3 x \cos\eta_3 y \sin\zeta_1 z\\
& + A_1\eta_3 \sin\xi_2 x \cos\eta_3 y \sin\zeta_3 z + (B_1\xi_1 - D_1\zeta_3)\sin\xi_1 x \cos\eta_3 y \sin\zeta_3 z\\
& + E_3\xi_3 \sin\xi_3 x \cos\eta_2 y \sin\zeta_2 z + F_3\xi_2 \sin\xi_2 x \cos\eta_3 y \sin\zeta_2 z\\
& - E_1\zeta_2 \sin\xi_2 x \cos\eta_3 y \sin\zeta_2 z - F_1\zeta_3 \sin\xi_2 x \cos\eta_2 y \sin\zeta_3 z
\end{aligned}$$

$$\begin{aligned}
w = {} & A_3\zeta_2 \sin\xi_3 x \sin\eta_3 y \cos\zeta_2 z + (C_3\eta_3 - B_3\xi_3)\sin\xi_3 x \sin\eta_3 y \cos\zeta_1 z\\
& + A_1\zeta_3 \sin\xi_2 x \sin\eta_3 y \cos\zeta_3 z + (D_1\eta_3 - C_1\xi_1)\sin\xi_1 x \sin\eta_3 y \cos\zeta_3 z\\
& + A_2\zeta_3 \sin\xi_3 x \sin\eta_2 y \cos\zeta_3 z + (B_2\eta_1 - D_2\xi_3)\sin\xi_3 x \sin\eta_1 y \cos\zeta_3 z\\
& + E_1\eta_3 \sin\xi_2 x \sin\eta_3 y \cos\zeta_2 z + F_1\eta_2 \sin\xi_2 x \sin\eta_2 y \cos\zeta_3 z\\
& - E_2\xi_2 \sin\xi_2 x \sin\eta_2 y \cos\zeta_3 z - F_2\xi_3 \sin\xi_3 x \sin\eta_2 y \cos\zeta_2 z.
\end{aligned} \tag{7}$$

The components of stress are calculated from the displacements by

$$\begin{aligned}
\sigma_{xx} &= \lambda\Delta + 2\mu\partial u/\partial x, \qquad & \sigma_{yz} &= \mu(\partial w/\partial y + \partial v/\partial z)\\
\sigma_{yy} &= \lambda\Delta + 2\mu\partial v/\partial y, \qquad & \sigma_{xz} &= \mu(\partial u/\partial z + \partial w/\partial x)
\end{aligned}$$

$$\sigma_{zz} = \lambda\Delta + 2\mu\partial w/\partial z, \qquad \sigma_{xy} = \mu(\partial v/\partial x + \partial u/\partial y) \tag{8}$$

where $\Delta = \partial u/\partial x + \partial v/\partial y + \partial w/\partial z$.

Then, the normal components of stress are given by

$$\begin{aligned}
\mu^{-1}\sigma_{xx} = &-[A_1(\xi_1^2 - \eta_3^2 - \zeta_3^2)\sin\xi_2 x + 2(C_1\zeta_3 - B_1\eta_3)\xi_1\sin\xi_1 x]\sin\eta_3 y\sin\zeta_3 z\\
&-[A_2(\eta_1^2 - 2\eta_2^2 + \xi_3^2 - \zeta_3^2)\sin\xi_3 x + 2E_2\zeta_3\xi_2\sin\xi_2 x]\sin\eta_2 y\sin\zeta_3 z\\
&-[A_3(\zeta_1^2 - 2\zeta_2^2 - \eta_3^2 + \xi_3^2)\sin\xi_3 x - 2F_3\eta_3\xi_2\sin\xi_2 x]\sin\eta_3 y\sin\zeta_2 z\\
&-2[D_2\zeta_3 - C_2\eta_1]\xi_3\sin\xi_3 x\sin\eta_1 y\sin\zeta_3 z\\
&-2[B_3\zeta_1 - D_3\eta_3]\xi_3\sin\xi_3 x\sin\eta_3 y\sin\zeta_1 z\\
&-2[F_2\zeta_2 - E_3\eta_2]\xi_3\sin\xi_3 x\sin\eta_2 y\sin\zeta_2 z
\end{aligned}$$

$$\begin{aligned}
\mu^{-1}\sigma_{yy} = &-[A_1(\xi_1^2 - 2\xi_2^2 - \zeta_3^2 + \eta_3^2)\sin\eta_3 y - 2F_1\zeta_3\eta_2\sin\eta_2 y]\sin\xi_2 x\sin\zeta_3 z\\
&-[A_2(\eta_1^2 - \zeta_3^2 - \xi_3^2)\sin\eta_2 y + 2(C_2\xi_3 - B_2\zeta_3)\eta_1\sin\eta_1 y]\sin\xi_3 x\sin\zeta_3 z\\
&-[A_3(\zeta_1^2 - 2\zeta_2^2 + \eta_3^2 - \xi_3^2)\sin\eta_3 y + 2E_3\xi_3\eta_2\sin\eta_2 y]\sin\xi_3 x\sin\zeta_2 z\\
&-2[D_3\xi_3 - C_3\zeta_1]\eta_3\sin\xi_3 x\sin\eta_3 y\sin\zeta_1 z\\
&-2[B_1\xi_1 - D_1\zeta_3]\eta_3\sin\xi_1 x\sin\eta_3 y\sin\zeta_3 z\\
&-2[F_3\xi_2 - E_1\zeta_2]\eta_3\sin\xi_2 x\sin\eta_3 y\sin\zeta_2 z
\end{aligned}$$

$$\begin{aligned}
\mu^{-1}\sigma_{zz} = &-[A_1(\xi_1^2 - 2\xi_2^2 + \zeta_3^2 - \eta_3^2)\sin\zeta_3 z + 2E_1\eta_3\zeta_2\sin\zeta_2 z]\sin\xi_2 x\sin\eta_3 y\\
&-[A_2(\eta_1^2 - 2\eta_2^2 - \xi_3^2 + \zeta_3^2)\sin\zeta_3 z - 2F_2\xi_3\zeta_2\sin\zeta_2 z]\sin\xi_3 x\sin\eta_2 y\\
&-[A_3(\zeta_1^2 - \eta_3^2 - \xi_3^2)\sin\zeta_2 z + 2(C_3\eta_3 - B_3\xi_3)\zeta_1\sin\zeta_1 z]\sin\xi_3 x\sin\eta_3 y\\
&-2[D_1\eta_3 - C_1\xi_1]\zeta_3\sin\xi_1 x\sin\eta_3 y\sin\zeta_3 z\\
&-2[B_2\eta_1 - D_2\xi_3]\zeta_3\sin\xi_3 x\sin\eta_1 y\sin\zeta_3 z\\
&-2[F_1\eta_2 - E_2\xi_2]\zeta_3\sin\xi_2 x\sin\eta_2 y\sin\zeta_3 z
\end{aligned} \tag{9}$$

and the tangential components of stress are given by

$$\begin{aligned}
\mu^{-1}\sigma_{yz} = &\{2A_1\zeta_3\eta_3\sin\xi_2 x + [D_1(\eta_3^2 - \zeta_3^2) + B_1\xi_1\zeta_3 - C_1\xi_1\eta_3]\sin\xi_1 x\}\cos\eta_3 y\cos\zeta_3 z\\
&+\{2A_2\zeta_3\eta_2\sin\xi_3 x - [E_2\xi_2\eta_2 - F_1(\eta_2^2 - \zeta_3^2)]\sin\xi_2 x\}\cos\eta_2 y\cos\zeta_3 z\\
&+\{2A_3\zeta_2\eta_3\sin\xi_3 x + [F_3\xi_2\zeta_2 + E_1(\eta_3^2 - \zeta_2^2)]\sin\xi_2 x\}\cos\eta_3 y\cos\zeta_2 z\\
&+[B_2(\eta_1^2 - \zeta_3^2) + C_2\xi_3\zeta_3 - D_2\xi_3\eta_1]\sin\xi_3 x\cos\eta_1 y\cos\zeta_3 z\\
&+[C_3(\eta_3^2 - \zeta_1^2) + D_3\xi_3\zeta_1 - B_3\xi_3\eta_3]\sin\xi_3 x\cos\eta_3 y\cos\zeta_1 z\\
&+(E_3\xi_3\zeta_2 - F_2\xi_3\eta_2)\sin\xi_3 x\cos\eta_2 y\cos\zeta_2 z
\end{aligned}$$

$$\begin{aligned}
\mu^{-1}\sigma_{zx} = &\{2A_1\xi_2\zeta_3\sin\eta_3 y + [F_1\eta_2\xi_2 + E_2(\zeta_3^2 - \xi_2^2)]\sin\eta_2 y\}\cos\xi_2 x\cos\zeta_3 z\\
&+\{2A_2\xi_3\zeta_3\sin\eta_2 y + [D_2(\zeta_3^2 - \xi_3^2) + B_2\eta_1\xi_3 - C_2\eta_1\zeta_3]\sin\eta_1 y\}\cos\xi_3 x\cos\zeta_3 z\\
&+\{2A_3\xi_3\zeta_2\sin\eta_3 y - [E_3\eta_2\zeta_2 - F_2(\zeta_2^2 - \xi_3^2)]\sin\eta_2 y\}\cos\xi_3 x\cos\zeta_2 z\\
&+[B_3(\zeta_1^2 - \zeta_3^2) + C_3\eta_3\xi_3 - D_3\eta_3\zeta_1]\cos\xi_3 x\sin\eta_3 y\cos\zeta_1 z\\
&+[C_1(\zeta_3^2 - \xi_1^2) + D_1\eta_3\xi_1 - B_1\eta_3\zeta_3]\cos\xi_1 x\sin\eta_3 y\cos\zeta_3 z\\
&+(E_1\eta_3\xi_2 - F_3\eta_3\zeta_2)\cos\xi_2 x\sin\eta_3 y\cos\zeta_2 z
\end{aligned}$$

$$\begin{aligned}\mu^{-1}\sigma_{xy} = &\{2A_1\xi_2\eta_3\sin\zeta_3 z - [E_1\zeta_2\xi_2 - F_3(\xi_2^2-\eta_3^2)\sin\zeta_2 z\}\cos\xi_2 x\cos\eta_3 y\\ &+\{2A_2\xi_3\eta_2\sin\zeta_3 z + [F_2\zeta_2\eta_2 + E_3(\xi_3^2-\eta_2^2)]\sin\zeta_2 z\}\cos\xi_3 x\cos\eta_2 y\\ &+\{2A_3\xi_3\eta_3\sin\zeta_2 z + [D_3(\xi_3^2-\eta_3^2) + B_3\zeta_1\eta_3 - C_3\zeta_1\xi_3]\sin\zeta_1 z\}\cos\xi_3 x\cos\eta_3 y\\ &+[B_1(\xi_1^2-\eta_3^2) + C_1\zeta_3\eta_3 - D_1\zeta_3\xi_1]\cos\xi_1 x\cos\eta_3 y\sin\zeta_3 z\\ &+[C_2(\xi_3^2-\eta_1^2) + D_2\zeta_3\eta_1 - B_2\zeta_3\xi_3]\cos\xi_3 x\cos\eta_1 y\sin\zeta_3 z\\ &+(E_2\zeta_3\eta_2 - F_1\zeta_3\xi_2)\cos\xi_2 x\cos\eta_2 y\sin\zeta_3 z. \end{aligned} \tag{10}$$

BOUNDARY CONDITIONS

The conditions for the boundaries, $x=\pm a$, $y=\pm b$, $z=\pm c$ of a rectangular parallelepiped, to be traction free are

$$\begin{aligned}\sigma_{xx}=\sigma_{xy}=\sigma_{xz}=0 &\quad\text{on}\quad x=\pm a\\ \sigma_{yx}=\sigma_{yy}=\sigma_{yz}=0 &\quad\text{on}\quad y=\pm b\\ \sigma_{zx}=\sigma_{zy}=\sigma_{zz}=0 &\quad\text{on}\quad z=\pm c.\end{aligned} \tag{11}$$

The tangential components of traction vanish on the boundaries if

$$\xi_i = l_i\pi/2a, \qquad \eta_i = m_i\pi/2b, \qquad \zeta_i = n_i\pi/2c \tag{12}$$

for $i = 1, 2, 3$ and l_i, m_i, n_i odd integers. These integers give the integral numbers of half wavelengths of displacement between opposing faces of the body. From eqns (3)–(5) and (12), we find a frequency ratio Ω given by

$$\begin{aligned}\kappa^2\Omega^2 &= l_1^2 - l_2^2 = a^2(m_1^2-m_2^2)/b^2 = a^2(n_1^2-n_2^2)/c^2\\ &= l_2^2 - l_3^2 = a^2(m_2^2-m_3^2)/b^2 = a^2(n_2^2-n_3^2)/c^2\end{aligned} \tag{13}$$

where

$$\Omega^2 = \omega^2/\omega_a^2, \qquad \omega_a^2 = \pi^2 v_2^2/4a^2. \tag{14}$$

We note, from eqn (13), the requirements

$$l_1^2 - l_2^2 = l_2^2 - l_3^2, \qquad m_1^2 - m_2^2 = m_2^2 - m_3^2, \qquad n_1^2 - n_2^2 = n_2^2 - n_3^2 \tag{15}$$

and the results

$$\frac{b^2}{a^2} = \frac{m_1^2-m_2^2}{l_1^2-l_2^2} = \frac{m_2^2-m_3^2}{l_2^2-l_3^2}, \qquad \frac{c^2}{a^2} = \frac{n_1^2-n_2^2}{l_1^2-l_2^2} = \frac{n_2^2-n_3^2}{l_2^2-l_3^2}. \tag{16}$$

Although the Lamé functions in eqn (1) produce displacements antisymmetric with respect to all three coordinate planes, symmetry with respect to any or all of the coordinate planes would result from interchange of appropriate sines and cosines. Then the corresponding odd integers, among l_i, m_i, n_i, would be replaced by even integers. Thus, both the frequency, in eqn (13), and the dimensional ratios, in eqn (16), depend on differences between the squares of two integers, say P_i and Q_i, $Q_i > P_i$, both odd or both even. Such differences are positive integral multiples of 4

$$Q_i^2 - P_i^2 = 4N, \qquad N = 1,2,3\ldots \tag{17}$$

It was shown in Ref. [2] that all such P_i and Q_i are given by

$$P_i = N/M_i - M_i, \qquad Q_i = N/M_i + M_i, \qquad M_i = 1,2,3\ldots \leqslant N^{1/2}. \tag{18}$$

These results gave the integers eligible for use in Ref. [2], but here there is the additional restriction, eqn (15). Thus, only those values of N are admissible that give at least two pairs of integers, say Q_i, P_i and Q_j, P_j, which satisfy

$$Q_i^2 - P_i^2 = Q_j^2 - P_j^2, \qquad P_i = Q_j. \tag{19}$$

In the range $0 < N < 1000$, only 34 values of N, as listed in Table 1, give Q and P, from eqn (18), which satisfy eqn (19). One of them ($N = 210$) gives two such combinations and another ($N = 840$) gives three.

For the boundary conditions on the normal components of traction to be satisfied, the six expressions in brackets in each of σ_{xx}, σ_{yy}, σ_{zz}, in eqn (9), must vanish on $x = \pm a$, $y = \pm b$, $z = \pm c$, respectively. These conditions constitute a system of 18 simultaneous, homogeneous equations on the 18 constants $A_1 \ldots F_3$, the solution of which is

$$\frac{A_2}{A_1} = \frac{a_{23}a_{31}e_{13}e_{21}f_{32}f_{13}}{a_{32}a_{13}e_{23}e_{31}f_{12}f_{23}}, \qquad \frac{A_3}{A_1} = -\frac{a_{31}e_{12}f_{13}}{a_{13}e_{31}f_{23}}$$

$$\begin{aligned}
B_1/A_1 &= -a_{11}c_{31}d_{21}/\Delta_1, & C_1/A_1 &= -a_{11}b_{21}d_{31}/\Delta_1\\
B_2/A_2 &= -a_{22}c_{12}d_{32}/\Delta_2, & C_2/A_2 &= -a_{22}b_{32}d_{12}/\Delta_2\\
B_3/A_3 &= -a_{33}c_{23}d_{13}/\Delta_3, & C_3/A_3 &= -a_{33}b_{13}d_{23}/\Delta_3
\end{aligned}$$

$$\begin{aligned}
D_1/A_1 &= a_{11}b_{21}c_{31}/\Delta_1, & \Delta_1 &= b_{11}c_{31}d_{21} + b_{21}c_{11}d_{31}\\
D_2/A_2 &= a_{22}b_{32}c_{12}/\Delta_2, & \Delta_2 &= b_{22}c_{12}d_{32} + b_{33}c_{22}d_{12}\\
D_3/A_3 &= a_{33}b_{13}c_{23}/\Delta_3, & \Delta_3 &= b_{33}c_{23}d_{13} + b_{13}c_{33}d_{23}
\end{aligned}$$

$$\begin{aligned}
E_1/A_1 &= -a_{31}/e_{31}, & F_1/A_1 &= -a_{21}/f_{21}\\
F_2/A_2 &= -a_{12}/e_{12}, & F_2/A_2 &= -a_{32}/f_{32}\\
E_3/A_3 &= -a_{23}/e_{23}, & F_3/A_3 &= -a_{13}/f_{13}
\end{aligned} \tag{20}$$

provided that

$$\begin{aligned}
\hat{l}_3^2(\hat{l}_3^2 - \hat{n}_3^2)(\hat{m}_3^2 - \hat{l}_3^2) + \hat{m}_3^2(\hat{m}_3^2 - \hat{l}_3^2)(\hat{n}_3^2 - \hat{m}_3^2)&\\
+ \hat{n}_3^2(\hat{l}_3^2 - \hat{n}_3^2)(\hat{n}_3^2 - \hat{m}_3^2) + \hat{l}_3^2\hat{m}_3^2\hat{n}_3^2 = 0&
\end{aligned} \tag{21}$$

where

$$\hat{l}_i = l_i/a, \qquad \hat{m}_i = m_i/b, \qquad \hat{n}_i = n_i/c \tag{22}$$

and

$$a_{11} = (\hat{l}_1^2 - \hat{m}_3^2 - \hat{n}_3^2)(-1)^{(l_2-l_3)/2},\ b_{11} = -2\hat{l}_1\hat{m}_3,\ c_{11} = 2\hat{n}_3\hat{l}_1$$

$$a_{12} = (\hat{m}_1^2 - 2\hat{m}_2^2 + \hat{l}_3^2 - \hat{n}_3^2)(-1)^{(l_3-l_2)/2},\ e_{12} = 2\hat{n}_3\hat{l}_2$$

$$a_{13} = (\hat{n}_1^2 - 2\hat{n}_2^2 - \hat{m}_3^2 + \hat{l}_3^2)(-1)^{(l_3-l_2)/2},\ f_{13} = -2\hat{l}_2\hat{m}_3$$
$$d_{12} = \hat{n}_3,\ c_{12} = -\hat{m}_1,\ b_{13} = \hat{n}_1,\ d_{13} = -\hat{m}_3,\ f_{12} = \hat{n}_2,\ e_{13} = -\hat{m}_2$$

$$a_{21} = (\hat{l}_1^2 - 2\hat{l}_2^2 - \hat{n}_3^2 + \hat{m}_3^2)(-1)^{(m_3-m_2)/2},\ f_{21} = -2\hat{m}_2\hat{n}_3$$
$$a_{22} = (\hat{m}_1^2 - \hat{n}_3^2 - \hat{l}_3^2)(-1)^{(m_2-m_1)/2},\ b_{22} = -2\hat{m}_1\hat{n}_3,\ c_{22} = 2\hat{l}_3\hat{m}_1$$
$$a_{23} = (\hat{n}_1^2 - 2\hat{n}_2^2 + \hat{m}_3^2 - \hat{l}_3^2)(-1)^{(m_3-m_2)/2},\ e_{23} = 2\hat{l}_3\hat{m}_2$$
$$d_{23} = \hat{l}_3,\ c_{23} = -\hat{n}_1,\ b_{21} = \hat{l}_1,\ d_{21} = -\hat{n}_3,\ f_{23} = \hat{l}_2,\ e_{21} = -\hat{n}_2$$

$$a_{31} = (\hat{l}_1^2 - 2\hat{l}_2^2 + \hat{n}_3^2 - \hat{m}_3^2)(-1)^{(n_3-n_2)/2},\ e_{31} = 2\hat{m}_3\hat{n}_2$$
$$a_{32} = (\hat{m}_1^2 - 2\hat{m}_2^2 - \hat{l}_3^2 + \hat{n}_3^2)(-1)^{(n_3-n_2)/2},\ f_{32} = -2\hat{n}_3\hat{l}_3$$
$$a_{33} = (\hat{n}_1^2 - \hat{m}_3^2 - \hat{l}_3^2)(-1)^{(n_2-n_1)/2},\ b_{33} = -2\hat{n}_1\hat{l}_3,\ c_{33} = 2\hat{m}_3\hat{n}_1$$
$$a_{31} = \hat{m}_3,\ c_{31} = -\hat{l}_1,\ b_{32} = \hat{m}_1,\ d_{32} = -\hat{l}_3,\ f_{31} = \hat{m}_2,\ e_{32} = -\hat{l}_2. \quad (23)$$

EXAMPLES

It remains only to select differences of squares of integers, from eqns (17) and (18), to produce the frequencies according to eqn (13) and dimensional ratios according to eqn (16)—subject to restrictions (15), as in Table 1, and, finally, eqn (21).

The density of modes and shapes of parallelepipeds, included in the solution, is low in comparison with that in the solution for the bar in Ref. [2]. There, with no restriction on length, the allowable ratio of width to thickness of a bar was the square root of the ratio of any two of the $Q_i^2 - P_i^2$ differences calculated from eqn (18). In the range $0 < N \leqslant 1000$, there are 3551 such differences. But, in the present solution, only 37 combinations of two of them, as listed in Table 1, are allowed as a result of eqn (15). Three such combinations are required for the l_i, m_i, n_i of any solution and the admissibility of such triplets is severely restricted by eqn (21). That equation may be simplified by taking any one of the differences, say $\hat{m}_3^2 - \hat{n}_3^2$, equal to zero. As a result, eqn (21) reduces to

$$b^2/a^2 = 2m_3^2/l_3^2 \quad (24)$$

and, since we have taken $b^2/c^2 = m_3^2/n_3^2$, we find

$$c^2/a^2 = 2n_3^2/l_3^2. \quad (25)$$

Then, with eqn (16), we can write $2(l_2^2/l_3^2 - 1) = m_2^2/m_3^2 - 1 = n_2^2/n_3^2 - 1$. These relations facilitate the selection of l_i, m_i, n_i from Table 1 or its extension to higher N.

A few examples of modes in square and rectangular plates and bars are given below.

Square, thick plate:

$N_1 = 30,\quad l_1 = 17,\quad l_2 = 13,\quad l_3 = 7$

$N_2 = 540,\quad m_1 = 69,\quad m_2 = 51,\quad m_3 = 21$

$N_3 = 540,\quad n_1 = 69,\quad n_2 = 51,\quad n_3 = 21$

$a^2{:}b^2{:}c^2 = 1{:}18{:}18$

$\kappa^2\Omega^2 = 120.$

Table 1. $4N = Q_i^2 - P_i^2 = Q_j^2 - P_j^2$, $P_i = Q_j$

N	Q_i	P_i	P_j
6	7	5	1
24	14	10	2
30	17	13	7
54	21	15	3
60	23	17	7
84	31	25	17
96	28	20	4
120	34	26	14
150	35	25	5
180	49	41	31
210	47	37	23
210	41	29	1
216	42	30	6
240	46	34	14
270	51	39	21
294	49	35	7
330	71	61	49
336	62	50	34
384	56	40	8
480	68	52	28
486	63	45	9
504	79	65	47
540	69	51	21
546	97	85	71
600	70	50	10
630	73	53	17
720	98	82	62
726	77	55	11
750	85	65	35
756	93	75	51
840	127	113	97
840	94	74	46
840	82	58	2
864	84	60	12
924	89	65	23
960	92	68	28
990	119	101	79

Square bar:

$$N_1 = 750, \quad l_1 = 85, \quad l_2 = 65, \quad l_3 = 35$$
$$N_2 = 60, \quad m_1 = 23, \quad m_2 = 17, \quad m_3 = 7$$
$$N_3 = 60, \quad n_1 = 23, \quad n_2 = 17, \quad n_3 = 7$$
$$a^2 : b^2 : c^2 = 25:2:2$$
$$\kappa^2\Omega^2(c^2/a^2) = 240.$$

Rectangular bar:

$$N_1 = 30, \quad l_1 = 17, \quad l_2 = 13, \quad l_3 = 7$$
$$N_2 = 60, \quad m_1 = 23, \quad m_2 = 17, \quad m_3 = 7$$
$$N_3 = 540, \quad n_1 = 69, \quad n_2 = 51, \quad n_3 = 21$$
$$a^2 : b^2 : c^2 = 1:2:18$$
$$\kappa^2\Omega^2 = 120.$$

Rectangular plate:

$$N_1 = 750, \quad l_1 = 85, \quad l_2 = 65, \quad l_3 = 35$$
$$N_2 = 540, \quad m_1 = 69, \quad m_2 = 51, \quad m_3 = 21$$

$$N_3 = 60, \quad n_1 = 23, \quad n_2 = 17, \quad n_3 = 7$$
$$a^2:b^2:c^2 = 25:18:2$$
$$\kappa^2\Omega^2(c^2/a^2) = 240.$$

As mentioned earlier, some or all of the sines and cosines in the Lamé functions may be interchanged with the result that some or all of the l_i, m_i, n_i would change from odd to even integers.

REFERENCES

1. G. Lamé, *Lecons sur la Théorie Mathématique de l'Élasticité des Corps Solides.* Gauthier Villars, Paris (1852), also 2nd Edn (1866).
2. R. D. Mindlin and E. A. Fox, Vibrations and waves in elastic bars of rectangular cross section. *J. Appl. Mech.* **27,** 513–520 (1960).

MECHANICS RESEARCH COMMUNICATIONS Vol.13(6), 349-357, 1986. Printed in the USA.
0093-6413/86 $3.00 + .00

FLEXURAL VIBRATIONS OF RECTANGULAR PLATES WITH FREE EDGES

R. D. Mindlin
Grantham, New Hampshire 03753, U.S.A.

(Received 23 October 1986; accepted for print 23 October 1986)

1. Introduction

In this paper, equations of flexural vibrations of isotropic, elastic plates, with rotatory inertia and shear deformation terms included [1], are solved for rectangular plates with all four edges free. The solutions are exact and in closed form. Two main restrictions apply: (1) the length/width ratio must be a ratio of integers; (2) for each value of Poisson's ratio, all the allowable modes must have the same frequency. In each case, that frequency lies between about 1.9 and 2.7 times the frequency, say ω_0, of the fundamental thickness-shear mode of the infinite plate. However, the equations, with the appropriate shear-correction factor, are intended for use at frequencies no higher than about $1.2\omega_0$. To improve the accuracy at the higher frequencies, two additional correction factors are introduced to make the equations produce the exact values of frequency and wave-number, obtained from Rayleigh's [2] exact dispersion relation, at the frequency and for the Poisson's ratio required by the solution for the rectangular plate.

2. Equations

The notation is that employed in [1] except for ω which is here the circular frequency.

Strains:

$$\Gamma_x = \psi_{x,x}, \quad \Gamma_y = \psi_{y,y}, \quad \Gamma_{yx} = \psi_{y,x} + \psi_{x,y}, \quad \Gamma_{xz} = w_{,x} + \psi_x, \quad \Gamma_{yz} = w_{,y} + \psi_y,$$

where ψ_x and ψ_y are the rectangular components of rotation about axes in the plane of the plate and w is the deflection. Γ_x and Γ_y are the thickness-curvatures, Γ_{yx} is the thickness-twist and Γ_{xz} and

Γ_{yz} are thickness-shears. The comma notation for derivative is employed.

Strain and kinetic energy-densities:

$$W = \tfrac{1}{2}[\kappa^2 Gh(\Gamma_{xz}^2 + \Gamma_{yz}^2) + \kappa_c \mu D(\Gamma_x + \Gamma_y)^2 + (1-\mu)D(\Gamma_x^2 + \Gamma_y^2 + \tfrac{1}{2}\Gamma_{yx}^2)],$$
$$T = \tfrac{1}{2}\kappa_p \rho h \dot{w}^2 + \tfrac{1}{24}\rho h^3(\dot{\psi}_x^2 + \dot{\psi}_y^2),$$

in which $D = Gh^3/6(1-\mu)$ is the plate modulus, G is the shear modulus, h is the thickness of the plate, μ is Poisson's ratio, ρ is the mass density and the over-dot designates time derivative. There are three correction factors: κ for thickness-shear, κ_c for thickness curvature and κ_p for tranverse inertia. κ_c and κ_p are the additional correction factors referred to in the Introduction; they do not appear in [1] and, in what follows, are taken as unity in Section 4.

Stresses:

$$\begin{aligned}
M_x &= \partial W/\partial \Gamma_x = D(\nu \psi_{x,x} + \kappa_c \mu \psi_{y,y}),\\
M_y &= \partial W/\partial \Gamma_y = D(\nu \psi_{y,y} + \kappa_c \mu \psi_{x,x}),\\
M_{yx} &= \partial W/\partial \Gamma_{yx} = \tfrac{1}{2}(1-\mu)D(\psi_{y,x} + \psi_{x,y}),\\
Q_x &= \partial W/\partial \Gamma_{xz} = \kappa^2 Gh(w_{,x} + \psi_x),\\
Q_y &= \partial W/\partial \Gamma_{yz} = \kappa^2 Gh(w_{,y} + \psi_y),
\end{aligned} \tag{1}$$

in which

$$\nu = 1 - \mu + \kappa_c \mu. \tag{2}$$

Stress-equations of motion:

$$\begin{aligned}
M_{x,x} + M_{yx,y} - Q_x + \tfrac{1}{12}\rho\omega^2 h^3 \psi_x &= 0,\\
M_{yx,x} + M_{y,y} - Q_y + \tfrac{1}{12}\rho\omega^2 h^3 \psi_y &= 0,\\
Q_{x,x} + Q_{y,y} + \kappa_p \rho \omega^2 h w &= 0.
\end{aligned} \tag{3}$$

Substitution of (1) in (3) produces the displacement-equations of motion:

$$\begin{aligned}
\tfrac{1}{2} D[(1-\mu)\nabla^2\psi_x + (\nu+\kappa_c\mu)\Phi_{,x}] - \kappa^2 Gh(w_{,x}+\psi_x) + \tfrac{1}{12}\rho\omega^2 h^3\psi_x &= 0,\\
\tfrac{1}{2} D[(1-\mu)\nabla^2\psi_y + (\nu+\kappa_c\mu)\Phi_{,y}] - \kappa^2 Gh(w_{,y}+\psi_y) + \tfrac{1}{12}\rho\omega^2 h^3\psi_y &= 0,\\
\kappa^2 Gh(\nabla^2 w + \Phi) + \kappa_\rho\rho\omega^2 h w &= 0,
\end{aligned} \qquad (4)$$

where $\Phi = \psi_{x,x} + \psi_{y,y}$ and ∇^2 is the two-dimensional Laplacian.

3. General solution

To find the flexure and thickness-shear dispersion relations, substitute (omitting a factor $e^{i\omega t}$ here and in what follows)

$$\psi_x = A\xi \sin\xi x \cos\eta y, \quad \psi_y = A\eta\cos\xi x \sin\eta y, \quad w = B\cos\xi x \cos\eta y$$

in (4) to obtain

$$\begin{aligned}
(\nu S\delta^2 + 1 - RS\delta_0^4)A - B &= 0,\\
\delta^2 A - (\delta^2 - \kappa_\rho S\delta_0^4)B &= 0,
\end{aligned} \qquad (5)$$

where

$$\delta^2 = \xi^2 + \eta^2, \quad R = h^2/12, \quad S = D/\kappa^2 Gh, \quad \delta_0^4 = \rho\omega^2 h/D.$$

R and S identify the contributions of the rotatory inertia and transverse shear deformation, respectively. The determinant of the coefficients of A and B in (5), set equal to zero, is a biquadratic equation in δ with roots:

$$2\nu(\delta_1^2, \delta_2^2) = \delta_0^4\{R + \nu\kappa_\rho S \pm [(R - \nu\kappa_\rho S)^2 + 4\nu\kappa_\rho\delta_0^{-4}]^{1/2}\}. \qquad (6a)$$

In terms of the ratio, Ω, of the frequency ω to the frequency ω_0 of the thickness-shear mode in an infinite plate:

$$\Omega = \omega/\omega_0, \quad \omega_0 = (\pi/h)(G/\rho)^{1/2},$$

we have

$$\delta_1^2, \delta_2^2 = \frac{\pi^2\Omega^2}{\nu h^2}\left\{\frac{1-\mu}{4} + \frac{\nu\kappa_\rho}{2\kappa^2} \pm \left[\left(\frac{1-\mu}{4} - \frac{\nu\kappa_\rho}{2\kappa^2}\right)^2 + \frac{6\nu\kappa_\rho(1-\mu)}{\pi^2\Omega^2}\right]^{1/2}\right\}. \qquad (6b)$$

These are the flexure branch (δ_1) and the thickness-shear branch (δ_2) of the dispersion relation for waves in an infinite plate according to the two-dimensional plate-equations. Also, from (5), $A/B = 1 - \sigma$, where $\sigma = \kappa_\rho\delta_0^4 S/\delta^2$. Upon employing the two values of δ from

(6), we find two values of σ:

$$\sigma_1, \sigma_2 = (\delta_2^2, \delta_1^2)\,\nu/(R\delta_0^4 - S^{-1}). \tag{7}$$

When κ_c and κ_p are unity, (6) and (7) reduce to the corresponding formulas (59) and (60) in [1].

Assembling these results, we have, for the flexure and thickness-shear modes of the solution,

$$\begin{aligned} \psi_x &= B_1(1-\sigma_1)\xi_1 \sin\xi_1 x \cos\eta_1 y + B_2(1-\sigma_2)\xi_2 \sin\xi_2 x \cos\eta_2 y, \\ \psi_y &= B_1(1-\sigma_1)\eta_1 \cos\xi_1 x \sin\eta_1 y + B_2(1-\sigma_2)\eta_2 \cos\xi_2 x \sin\eta_2 y, \\ w &= B_1 \cos\xi_1 x \cos\eta_1 y + B_2 \cos\xi_2 x \cos\eta_2 y, \end{aligned} \tag{8}$$

in which

$$\xi_1^2 + \eta_1^2 = \delta_1^2, \quad \xi_2^2 + \eta_2^2 = \delta_2^2. \tag{9}$$

To (8) we add two thickness-twist modes:

$$\begin{aligned} \psi_x &= B_3\eta_1 \sin\xi_2 x \cos\eta_1 y + B_4\eta_2 \sin\xi_1 x \cos\eta_2 y, \\ \psi_y &= -B_3\xi_2 \cos\xi_2 x \sin\eta_1 y - B_4\xi_1 \cos\xi_1 x \sin\eta_2 y, \end{aligned} \tag{10}$$

and $w=0$. Substituting (10) in (4), we find

$$\xi_2^2 + \eta_1^2 = \xi_1^2 + \eta_2^2 = 2(R\delta_0^4 - S^{-1})/(1-\mu) = (\pi^2\Omega^2 - 12\kappa^2)/h^2. \tag{11}$$

We note that, if $\kappa^2 = \pi^2/12$, (11) is exactly the dispersion relation for the first order thickness-twist branch of the three-dimensional equations (a special case of Love waves [3]). Also, the thickness-shear branch δ_2 of the dispersion relation (6) then has the correct cut-off frequency $\Omega=1$ at $\delta_2=0$ as required by the three dimensional equations [2]. Note, also, that κ_c and κ_p do not affect the thickness-twist branch.

The components of stress contributed by all four modes are found by substituting (8) and (10) in (1):

$$\begin{aligned} M_x &= D[B_1(1-\sigma_1)(\nu\xi_1^2 + \kappa_c\mu\eta_1^2)\cos\xi_1 x \cos\eta_1 y + B_2(1-\sigma_2)(\nu\xi_2^2 + \kappa_c\mu\eta_2^2)\cos\xi_2 x \cos\eta_2 y \\ &\quad + B_3(1-\mu)\xi_2\eta_1 \cos\xi_2 x \cos\eta_1 y + B_4(1-\mu)\xi_1\eta_2 \cos\xi_1 x \cos\eta_2 y], \\ M_y &= D[B_1(1-\sigma_1)(\nu\eta_1^2 + \kappa_c\mu\xi_1^2)\cos\xi_1 x \cos\eta_1 y + B_2(1-\sigma_2)(\nu\eta_2^2 + \kappa_c\mu\xi_2^2)\cos\xi_2 x \cos\eta_2 y \\ &\quad - B_3(1-\mu)\xi_2\eta_1 \cos\xi_2 x \cos\eta_1 y - B_4(1-\mu)\xi_1\eta_2 \cos\xi_1 x \cos\eta_2 y], \\ M_{yx} &= -(1-\mu)D[B_1(1-\sigma_1)\xi_1\eta_1 \sin\xi_1 x \sin\eta_1 y + B_2(1-\sigma_2)\xi_2\eta_2 \sin\xi_2 x \sin\eta_2 y \\ &\quad + \tfrac{1}{2}B_3(\eta_1^2 - \xi_2^2)\sin\xi_2 x \sin\eta_1 y + \tfrac{1}{2}(\eta_2^2 - \xi_1^2)\sin\xi_1 x \sin\eta_2 y], \end{aligned} \tag{12}$$

$$Q_x = -\kappa^2 Gh[B_1\sigma_1\xi_1 \sin\xi_1 x \cos\eta_1 y + B_2\sigma_2\xi_2 \sin\xi_2 x \cos\eta_2 y - B_3\eta_1 \sin\xi_2 x \cos\eta_1 y - B_4\eta_2 \sin\xi_1 x \cos\eta_2 y],$$
$$Q_y = -\kappa^2 Gh[B_1\sigma_1\eta_1 \cos\xi_1 x \sin\eta_1 y + B_2\sigma_2\eta_2 \cos\xi_2 x \sin\eta_2 y + B_3\xi_2 \cos\xi_2 x \sin\eta_1 y + B_4\xi_1 \cos\xi_1 x \sin\eta_2 y].$$

The free edge conditions to be satisfied, for a plate bounded by $x=\pm a$, $y=\pm b$, are

$$M_{yx} = Q_x = M_x = 0 \quad \text{on } x = \pm a, \tag{13}$$
$$M_{yx} = Q_y = M_y = 0 \quad \text{on } y = \pm b. \tag{14}$$

The first two of (13) and (14) are satisfied if

$$\sin\xi_i a = 0, \quad \sin\eta_i b = 0, \quad i = 1,2$$

i.e.

$$\xi_1 a = m\pi/2, \quad \eta_1 b = n\pi/2, \quad \xi_2 a = p\pi/2, \quad \eta_2 b = q\pi/2, \quad m,n,p,q \text{ even.} \tag{15}$$

The remaining conditions, those on M_x and M_y, result in

$$\begin{aligned}
B_1(1-\sigma_1)(\nu\xi_1^2 + \kappa_c\mu\eta_1^2)\cos\xi_1 a + B_3(1-\mu)\xi_2\eta_1 \cos\xi_2 a &= 0,\\
B_2(1-\sigma_2)(\nu\xi_2^2 + \kappa_c\mu\eta_2^2)\cos\xi_2 a + B_4(1-\mu)\xi_1\eta_2 \cos\xi_1 a &= 0,\\
B_1(1-\sigma_1)(\nu\eta_1^2 + \kappa_c\mu\xi_1^2)\cos\eta_1 b - B_4(1-\mu)\xi_1\eta_2 \cos\eta_2 b &= 0,\\
B_2(1-\sigma_2)(\nu\eta_2^2 + \kappa_c\mu\xi_2^2)\cos\eta_2 b - B_3(1-\mu)\xi_2\eta_1 \cos\eta_1 b &= 0.
\end{aligned} \tag{16}$$

The determinant of the coefficients of the B_i in (16), set equal to zero, reduces to

$$\xi_1^2\xi_2^2 - \eta_1^2\eta_2^2 = 0, \tag{17}$$

which, with (11), requires

$$\begin{aligned}
\xi_1^2 = \eta_1^2 \quad &\text{or} \quad m/a = n/b,\\
\xi_2^2 = \eta_2^2 \quad &\text{or} \quad p/a = q/b.
\end{aligned} \tag{18}$$

Hence, the length/width ratios of the plate and the mode shapes are restricted to

$$a/b = m/n = p/q. \tag{19}$$

If sines and cosines of like argument, in (8) and (10), are interchanged, four solutions are obtained with m and n or p and q even or odd. Accordingly, the length/width ratio, a/b, can be any ratio of integers. It may also be seen, from (8), and (10), that fixed edge conditions, $\psi_x = \psi_y = w = 0$ could be satisfied on either or both $x = \pm a$ and $y = \pm b$. Of course, mixed conditions of moments and deflection, such as $M_x = M_{yx} = w = 0$ on $x = \pm a$ or $M_y = M_{yx} = w = 0$ on $y = \pm b$ (i.e. "simple supports") can be satisfied without restrictions like (17).

From (6) and (9),

$$\delta_1^2 + \delta_2^2 = \xi_1^2 + \xi_2^2 + \eta_1^2 + \eta_2^2 = \pi^2 \Omega^2 [(1-\mu)/2 + \nu \kappa_p / \kappa^2] / \nu h^2, \tag{20}$$

but, from (11),

$$\xi_1^2 + \xi_2^2 + \eta_1^2 + \eta_2^2 = 2(\pi^2 \Omega^2 - 12\kappa^2)/h^2. \tag{21}$$

Hence, from (20) and (21), with $\kappa^2 = \pi^2/12$, the frequency is given by

$$\Omega = \{1 - [(1-\mu)/4 + 6\nu\kappa_p/\pi^2]/\nu\}^{-1/2}. \tag{22}$$

From (18) and (21), the length/thickness ratio is

$$2a/h = [(\Omega^2 - 1)/(m^2 + p^2)]^{1/2}. \tag{23}$$

Finally, the Poisson ratio corresponding to a ratio m/p may be found as follows. First, from (9), (18) and (15),

$$\delta_1^2 = 2\xi_1^2 = m^2\pi^2/2a^2, \quad \delta_2^2 = 2\xi_2^2 = p^2\pi^2/2a^2. \tag{24}$$

Then, from (6) and (24):

$$[(\delta_1^2 - \delta_2^2)/(\delta_1^2 + \delta_2^2)]^2 = [(m^2 - p^2)/(m^2 + p^2)]^2 = \{1 - [24(1-\mu)\kappa_p]\}/[\pi^2(1-\mu) + 24\kappa_p \nu],$$

from which

$$\mu = [R - 1 + 24\kappa_p(\nu R - \nu + 1)/\pi^2]/(R - 1 + 24\kappa_p/\pi^2), \tag{25}$$

where

$$R = [(m^2/p^2 - 1)/(m^2/p^2 + 1)]^2. \tag{26}$$

4. Solutions without additional correction factors

If $\kappa_c=\kappa_p=1$, (22) and (25) reduce to

$$\Omega=\{1-[(1-\mu)/4+6/\pi^2]\}^{-1/2}, \tag{27}$$

$$\mu=(\pi^2 R-\pi^2-24R)/(\pi^2 R-\pi^2-24), \tag{28}$$

and (23) remains unchanged.

A set of compatible values of the parameters may be obtained as follows. Choose m and p first as they must be integers. Then μ, Ω and 2a/h are fixed from (28), (27) and (23). Finally, choose n and q so as to satisfy p/q=m/n, according to (19), and this fixes the length/width ratio, a/b, also according to (19). Typical results over a range of Poisson's ratios are listed in Table I for m,n,p,q all even.

Table I

m,p	μ	Ω	2a/h	n,q	a/b	n/q	a/b	n,q	a/b
32,16	0.1314	2.3910	16.4732	4,2	8	8,4	4	12,6	8/3
34,16	0.2145	2.2605	18.5350	34,16	1	68,32	1/2	102,48	1/3
36,16	0.2876	2.1618	20.5550	18,8	2	36,16	1	54,24	2/3
38,16	0.3520	2.0848	22.5391	38,16	1	76,32	1/2	114,48	1/3
40,16	0.4087	2.0234	24.4919	10,4	4	20,8	2	30,12	4/3

The frequencies listed in Table I are all higher than the limit below which the dispersion relation (6), with $\kappa_c=\kappa_p=1$, gives close approximations to the corresponding branches of the Rayleigh dispersion relation [2], i.e. the first two branches of

$$\tan(\alpha h/2)/\tan(\beta h/2)=4\alpha\beta\zeta_i^2/(\zeta_i^2-\beta^2)^2, \tag{29}$$

where

$$\alpha^2=(\pi^2/h^2)\Omega^2/k^2-\zeta_i^2,\ \beta^2=(\pi^2/h^2)\Omega^2-\zeta_i^2,\ k^2=2(1-\mu)/(1-2\mu).$$

The ζ_i, i=1,2 correspond to the δ_i, i=1,2 of (6). An example with

$\mu=0.3$ is illustrated in Fig.1 in which the dashed lines marked

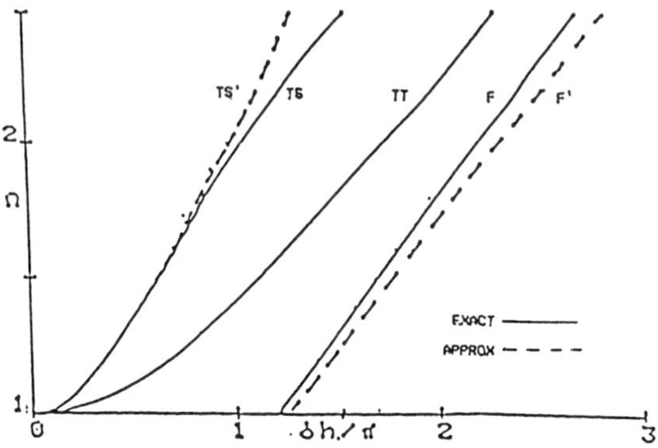

Fig. 1

F' and TS' are from δ_1 and δ_2, respectively, and the full lines marked F and TS (for which δ is to be replaced by ζ in the abscissa of the figure) are from (29). In particular, at $\Omega=2$,

$$\text{Exact: } h\zeta_1/\pi=2.19288, \qquad h\zeta_2/\pi=1.04718,$$
$$\text{Approx: } h\delta_1/\pi=2.30215, \qquad h\delta_2/\pi=0.98159,$$

so that the approximate δ_1 and δ_2 are in error by about 5% and 6%, respectively. The errors increase with increasing Ω and orders m, n, p, q. As already mentioned, the thickness-twist dispersion relation (11), with $\kappa^2=\pi^2/12$, is the same as that from the three-dimensional equations. That branch is the curve marked TT in Fig.1.

5. Solution with additional correction factors

To improve the accuracy of the preceding solution by computing appropriate values of κ_c and κ_p proceed as follows. Set

$$M=h^2(\zeta_1^2+\zeta_2^2)/\pi^2, \quad N=h^4\zeta_1^2\zeta_2^2/\pi^4 \tag{30}$$

and find from (6), with ζ_1 and ζ_2 replacing δ_1 and δ_2,

$$\kappa_p = \frac{\kappa^2}{2\Omega^2}\left[M + \left(M^2 - \frac{4N\Omega^2}{\Omega^2-1}\right)^{1/2}\right], \quad \kappa_c = \frac{1-\mu}{\mu}\left[\frac{\Omega^2(\Omega^2-1)\kappa_p}{2\kappa^2 N} - 1\right]. \tag{31}$$

The computation then proceeds:

Step 1. Choose m and p and find μ and Ω from (28) and (27).

Step 2. Find ζ_1 and ζ_2 from (29) and M and N from (30).

Step 3. Find κ_p and κ_c from (31).

Step 4. Find new Ω, 2a/h and μ from (22), (23) and (25).

Step 5. With the new Ω and μ , return to Step 2 and repeat the cycle as many times as are necessary to find $\kappa_p, \kappa_c, \Omega$, 2a/h and μ, to the desired accuracy.

Step 6. Finally, the choice of n and q (satisfying n/q=m/p) fixes a/b according to (19).

Table II

m, p	κ_c	κ_p	μ	Ω	2a/h
32, 16	0.5141	0.8981	0.1691	2.0957	19.4258
34, 16	0.5501	0.8779	0.2818	1.9585	22.3141
36, 16	0.4122	0.8623	0.4506	1.8604	25.1121

Table II contains the improved values, to $\pm 10^{-4}$, of the first three rows of Table I, along with the correction factors κ_c and κ_p. The last two rows from Table I are omitted as the corrected values of the Poisson ratios approach unity.

References

1. R.D.Mindlin, J. Appl. Mech. vol. 18, 23 (1951).
2. J.W.Strutt (Lord Rayleigh), Proc. Lond. Math. Soc. vol. 20, 225 (1888-1889).
3. A.E.H.Love, Some Problems of Geodynamics, Cambridge (1911).

Permissions

The editors and publisher wish to thank the original publishers of the papers of Raymond D. Mindlin for granting permission to reprint specific papers in this collection.

Académie des Sciences
(No. 3)

American Institute of Physics
(Nos. 1, 4, 12, 17, 26, 27, 32, 33, 35, 39, 43, 44, 45, 46, 50, 53, 54, 57, 69, 70, 71, 85, 86, 89)

American Mathematical Society
(No. 5)

American Society of Civil Engineers
(Nos. 2, 6, 11, 15)

American Society of Mechanical Engineers
(Nos. 9, 10, 18, 28, 29, 30, 34, 37, 38, 40, 41, 47, 49, 51, 52, 55, 56, 58, 59, 60, 61, 62, 63, 72, 73, 76, 92)

Brown University
(Nos. 22, 65, 66, 68)

Centro Internazionale Scienze Mecchaniche
(No. 97)

Jadavpur University
(Nos. 93, 119)

Kluwer Academic Publishers
(No. 105)

Macmillan Publishing Company
(No. 77)

Nauka
(Nos. 79, 120)

Northwestern University
(No. 117)

Optical Society of America
(No. 8)

Pergamon Journals, Ltd.
(Nos. 19, 80, 81, 82, 83, 84, 87, 88, 95, 96, 98, 99, 100, 101, 102, 103, 104, 106, 107, 108, 109, 110, 111, 112, 114, 115, 118, 122, 124, 125, 127, 128, 129)

Pergamon Press
(Nos. 64, 123)
Society for Experimental Mechanics
(Nos. 24, 25, 75)
Society for Industrial and Applied Mathematics
(Nos. 67, 94)
Springer-Verlag
(Nos. 74, 78, 91, 113)
Taylor & Francis Ltd.
(No. 13)
University of Illinois
(No. 42)
University of Toronto Press
(No. 90)
John Wiley & Sons
(No. 14)